HIGH PERFORMANCE CONTROL OF AC DRIVES WITH MATLAB®/ SIMULINK

HIGH PERFORMANCE CONTROL OF AC DRIVES WITH MATLAB®/ SIMULINK

Second Edition

Haitham Abu-Rub
Texas A&M University at Qatar, Doha, Qatar

Atif Iqbal
Qatar University, Doha, Qatar

Jaroslaw Guzinski
Gdansk University of Technology, Gdansk, Poland

WILEY

This edition first published 2021

Edition History
John Wiley & Sons, Ltd (1e 2012)

Registered Office(s)
John Wiley & Sons, Inc., 111 River Street, Hoboken, NJ 07030, USA
John Wiley & Sons Ltd, The Atrium, Southern Gate, Chichester, West Sussex, PO19 8SQ, UK

Editorial Office
The Atrium, Southern Gate, Chichester, West Sussex, PO19 8SQ, UK

For details of our global editorial offices, customer services, and more information about Wiley products visit us at www.wiley.com.

Wiley also publishes its books in a variety of electronic formats and by print-on-demand. Some content that appears in standard print versions of this book may not be available in other formats.

Library of Congress Cataloging-in-Publication Data applied for

ISBN: 9781119590781

Cover design by Wiley
Cover image: © Voyata/iStock/Getty Images Plus/Getty Images

Set in 10/12pt Times by SPi Global, Pondicherry, India

Printed and bound by CPI Group (UK) Ltd, Croydon, CR0 4YY

C9781119590781_300321

Dedicated to my parents, my wife Beata, and my children Fatima, Iman, Omar, and Muhammad.

—Haitham Abu-Rub

Dedicated to my parents, parents in-laws, my wife Shadma, and my kids Abuzar, Noorin, and Abu Baker who have inspired me to write this book.

—Atif Iqbal

Dedicated to my parents, my wife Anna, and my son Jurek.

—Jaroslaw Guzinski

Contents

Acknowledgment xiv

Biographies xvi

Preface to Second Edition xviii

Preface to First Edition xx

About the Companion Website xxii

1 Introduction to High-Performance Drives 1
1.1 Preliminary Remarks 1
1.2 General Overview of High-Performance Drives 6
1.3 Challenges and Requirements for Electric Drives for Industrial Applications 10
1.3.1 Power Quality and LC Resonance Suppression 11
1.3.2 Inverter Switching Frequency 12
1.3.3 Motor-Side Challenges 12
1.3.4 High dv/dt *and Wave Reflection* 12
1.3.5 Use of Inverter Output Filters 13
1.4 Wide Bandgap (WBG) Devices Applications in Electric Motor Drives 14
1.4.1 Industrial Prototype Using WBG 15
1.4.2 Major Challenges for WBG Devices for Electric Motor Drive Applications 15
1.5 Organization of the Book 16
References 19

2 Mathematical and Simulation Models of AC Machines 23
2.1 Preliminary Remarks 23
2.2 DC Motors 23
2.2.1 Separately Excited DC Motor Control 24
2.2.2 Series DC Motor Control 27
2.3 Squirrel Cage Induction Motor 28
2.3.1 Space Vector Representation 28
2.3.2 Clarke Transformation (ABC to $\alpha\beta$) 29
2.3.3 Park Transformation ($\alpha\beta$ to dq) 32
2.3.4 Per Unit Model of Induction Motor 33
2.3.5 Double Fed Induction Generator (DFIG) 36
2.4 Mathematical Model of Permanent Magnet Synchronous Motor 39
2.4.1 Motor Model in dq Rotating Frame 40
2.4.2 Example of Motor Parameters for Simulation 42

2.4.3 PMSM Model in Per Unit System 42
2.4.4 PMSM Model in $\alpha - \beta$ ($x - y$)-Axis 44
2.5 Problems 45
References 45

3 Pulse-Width Modulation of Power Electronic DC–AC Converter 47
Atif Iqbal, Arkadiusz Lewicki, and Marcin Morawiec
3.1 Preliminary Remarks 47
3.2 Classification of PWM Schemes for Voltage Source Inverters 48
3.3 Pulse-Width Modulated Inverters 49
3.3.1 Single-Phase Half-Bridge Inverters 49
3.3.2 Single-Phase Full-Bridge or H-Bridge Inverters 55
3.4 Three-Phase PWM Voltage Source Inverter 60
3.4.1 Carrier-Based Sinusoidal PWM 67
3.4.2 Third-Harmonic Injection Carrier-Based PWM 67
3.4.3 MATLAB/Simulink Model for Third-Harmonic Injection PWM 72
3.4.4 Carrier-Based PWM with Offset Addition 72
3.4.5 Space Vector PWM (SVPWM) 74
3.4.6 Discontinuous Space Vector PWM 79
3.4.7 MATLAB/Simulink Model for Space Vector PWM 84
3.4.8 Space Vector PWM in Overmodulation Region 93
3.4.9 MATLAB/Simulink Model to Implement Space Vector PWM in Overmodulation Regions 99
3.4.10 Harmonic Analysis 100
3.4.11 Artificial Neural Network-Based PWM 100
3.4.12 MATLAB/Simulink Model of Implementing ANN-Based SVPWM 103
3.5 Relationship Between Carrier-Based PWM and SVPWM 104
3.5.1 Modulating Signals and Space Vectors 105
3.5.2 Relationship Between Line-to-Line Voltages and Space Vectors 106
3.5.3 Modulating Signals and Space Vector Sectors 107
3.6 Low-Switching Frequency PWM 107
3.6.1 Types of Symmetries and Fourier Analysis 109
3.6.2 Selective Harmonics Elimination in a two-Level VSI 109
3.6.3 MATLAB Code 114
3.7 Multilevel Inverters 116
3.7.1 Neutral-Point-Clamped (Diode-Clamped) Multilevel Inverters 116
3.7.2 Flying Capacitor-Type Multilevel Inverter 120
3.7.3 Cascaded H-Bridge Multilevel Inverter 126
3.8 Space Vector Modulation and DC-Link Voltage Balancing in Three-Level Neutral-Point-Clamped Inverters 128
3.8.1 The Output Voltage of Three-Level NPC Inverter in the Case of the DC-Link Voltage Unbalance 128
3.8.2 The Space Vector PWM for NPC Inverters 134
3.8.3 MATLAB/Simulink of SVPWM 137
3.9 Space Vector PWM for Multilevel-Cascaded H-Bridge Converter with DC-Link Voltage Balancing 138
3.9.1 Control of a Multilevel CHB Converter 141
3.9.2 The Output Voltage of a Single H-Bridge 142

3.9.3 *Three-Level CHB Inverter* 143
3.9.4 *The Space Vector Modulation for Three-Level CHB Inverter* 145
3.9.5 *The Space Vector Modulation for Multilevel CHB Inverter* 149
3.9.6 *MATLAB/Simulink Simulation of SVPWM* 150
3.10 Impedance Source or Z-source Inverter 150
3.10.1 *Circuit Analysis* 154
3.10.2 *Carrier-Based Simple Boost PWM Control of a Z-source Inverter* 156
3.10.3 *Carrier-Based Maximum Boost PWM Control of a Z-source Inverter* 157
3.10.4 *MATLAB/Simulink Model of Z-source Inverter* 159
3.11 Quasi Impedance Source or qZSI Inverter 159
3.11.1 *MATLAB/Simulink Model of qZ-source Inverter* 164
3.12 Dead Time Effect in a Multiphase Inverter 164
3.13 Summary 169
Problems 169
References 170

4 Field-Oriented Control of AC Machines **177**
4.1 Introduction 177
4.2 Induction Machines Control 178
4.2.1 *Control of Induction Motor Using V/f Methods* 178
4.2.2 *Vector Control of Induction Motor* 182
4.2.3 *Direct and Indirect Field-Oriented Control* 188
4.2.4 *Rotor and Stator Flux Computation* 188
4.2.5 *Adaptive Flux Observers* 189
4.2.6 *Stator Flux Orientation* 190
4.2.7 *Field Weakening Control* 191
4.3 Vector Control of Double Fed Induction Generator (DFIG) 192
4.3.1 *Introduction* 192
4.3.2 *Vector Control of DFIG Connected with the Grid (αβ Model)* 194
4.3.3 *Variables Transformation* 194
4.3.4 *Simulation Results* 198
4.4 Control of Permanent Magnet Synchronous Machine 198
4.4.1 *Introduction* 198
4.4.2 *Vector Control of PMSM in dq Axis* 200
4.4.3 *Vector Control of PMSM in α−β Axis Using PI Controller* 203
4.4.4 *Scalar Control of PMSM* 207
Exercises 208
Additional Tasks 208
Possible Tasks for DFIG 208
Questions 208
References 209

5 Direct Torque Control of AC Machines **211**
Truc Phamdinh
5.1 Preliminary Remarks 211

5.2 Basic Concept and Principles of DTC 212
5.2.1 Basic Concept 212
5.2.2 Principle of DTC 214
5.3 DTC of Induction Motor with Ideal Constant Machine Model 220
5.3.1 Ideal Constant Parameter Model of Induction Motors 220
5.3.2 Direct Torque Control Scheme 222
5.3.3 Speed Control with DTC 225
5.3.4 MATLAB/Simulink Simulation of Torque Control and Speed Control with DTC 225
5.4 DTC of Induction Motor with Consideration of Iron Loss 240
5.4.1 Induction Machine Model with Iron Loss Consideration 240
5.4.2 MATLAB/SIMULINK Simulation of the Effects of Iron Losses in Torque Control and Speed Control 243
5.4.3 Modified Direct Torque Control Scheme for Iron Loss Compensation 254
5.5 DTC of Induction Motor with Consideration of Both Iron Losses and Magnetic Saturation 259
5.5.1 Induction Machine Model with Consideration of Iron Losses and Magnetic Saturation 259
5.5.2 MATLAB/Simulink Simulation of Effects of Both Iron Losses and Magnetic Saturation in Torque Control and Speed Control 260
5.6 Modified Direct Torque Control of Induction Machine with Constant Switching Frequency 275
5.7 Direct Torque Control of Sinusoidal Permanent Magnet Synchronous Motors (SPMSM) 276
5.7.1 Introduction 276
5.7.2 Mathematical Model of Sinusoidal PMSM 276
5.7.3 Direct Torque Control Scheme of PMSM 278
5.7.4 MATLAB/Simulink Simulation of SPMSM with DTC 278
References 296

6 Nonlinear Control of Electrical Machines Using Nonlinear Feedback 299
Zbigniew Krzeminski and Haitham Abu-Rub
6.1 Introduction 299
6.2 Dynamic System Linearization Using Nonlinear Feedback 300
6.3 Nonlinear Control of Separately Excited DC Motors 301
6.3.1 MATLAB/Simulink Nonlinear Control Model 303
6.3.2 Nonlinear Control Systems 303
6.3.3 Speed Controller 304
6.3.4 Controller for Variable m 304
6.3.5 Field Current Controller 306
6.3.6 Simulation Results 306
6.4 Multiscalar Model (MM) of Induction Motor 306
6.4.1 Multiscalar Variables 307

6.4.2 *Nonlinear Linearization of Induction Motor Fed by Voltage Controlled VSI* 308
6.4.3 *Design of System Control* 310
6.4.4 *Nonlinear Linearization of Induction Motor Fed by Current Controlled VSI* 311
6.4.5 *Stator-Oriented Nonlinear Control System (based on Ψ_s, i_s)* 314
6.4.6 *Rotor–Stator Fluxes-Based Model* 315
6.4.7 *Stator-Oriented Multiscalar Model* 316
6.4.8 *Multiscalar Control of Induction Motor* 318
6.4.9 *Induction Motor Model* 319
6.4.10 *State Transformations* 320
6.4.11 *Decoupled IM Model* 321
6.5 MM of Double-Fed Induction Machine (DFIM) 322
6.6 Nonlinear Control of Permanent Magnet Synchronous Machine 325
6.6.1 *Nonlinear Control of PMSM for a dq Motor Model* 327
6.6.2 *Nonlinear Vector Control of PMSM in $\alpha-\beta$ Axis* 329
6.6.3 *PMSM Model in $\alpha-\beta$ (x−y) Axis* 329
6.6.4 *Transformations* 329
6.6.5 *Control System* 333
6.6.6 *Simulation Results* 334
6.7 Problems 334
References 334

7 Five-Phase Induction Motor Drive System 337
7.1 Preliminary Remarks 337
7.2 Advantages and Applications of Multiphase Drives 338
7.3 Modeling and Simulation of a Five-Phase Induction Motor Drive 339
7.3.1 *Five-Phase Induction Motor Model* 339
7.3.2 *Five-Phase Two-Level Voltage Source Inverter Model* 345
7.3.3 *PWM Schemes of a Five-Phase VSI* 380
7.4 Direct Rotor Field-Oriented Control of Five-Phase Induction Motor 396
7.4.1 *MATLAB/Simulink Model of Field-Oriented Control of Five-Phase Induction Machine* 398
7.5 Field-Oriented Control of Five-Phase Induction Motor with Current Control in the Synchronous Reference Frame 402
7.6 Direct Torque Control of a Five-Phase Induction Motor 404
7.6.1 *Control of Inverter Switches Using DTC Technique* 404
7.6.2 *Virtual Vector for Five-Phase Two-Level Inverter* 405
7.7 Model Predictive Control (MPC) 420
7.7.1 *MPC Applied to a Five-Phase Two-Level VSI* 421
7.7.2 *MATLAB/Simulink of MPC for Five-Phase VSI* 422
7.7.3 *Using Eleven Vectors with $\gamma = 0$* 423
7.7.4 *Using Eleven Vectors with $\gamma = 1$* 425
7.8 Summary 426
7.9 Problems 426
References 427

8 Sensorless Speed Control of AC Machines **433**
8.1 Preliminary Remarks 433
8.2 Sensorless Control of Induction Motor 433
8.2.1 Speed Estimation Using Open-Loop Model and Slip Computation 434
8.2.2 Closed-Loop Observers 434
8.2.3 MRAS (Closed-Loop) Speed Estimator 443
8.2.4 The Use of Power Measurements 446
8.3 Sensorless Control of PMSM 448
8.3.1 Control System of PMSM 450
8.3.2 Adaptive Backstepping Observer 450
8.3.3 Model Reference Adaptive System for PMSM 452
8.3.4 Simulation Results 454
8.4 MRAS-Based Sensorless Control of Five-Phase Induction Motor Drive 454
8.4.1 MRAS-Based Speed Estimator 458
8.4.2 Simulation Results 460
References 464

9 Selected Problems of Induction Motor Drives with Voltage Inverter and Inverter Output Filters **469**
9.1 Drives and Filters – Overview 469
9.2 Three-Phase to Two-Phase Transformations 471
9.3 Voltage and Current Common Mode Component 473
9.3.1 MATLAB/Simulink Model of Induction Motor Drive with PWM Inverter and Common Mode Voltage 474
9.4 Induction Motor Common Mode Circuit 477
9.5 Bearing Current Types and Reduction Methods 478
9.5.1 Common Mode Choke 480
9.5.2 Common Mode Transformers 482
9.5.3 Common Mode Voltage Reduction by PWM Modifications 483
9.6 Inverter Output Filters 489
9.6.1 Selected Structures of Inverter Output Filters 489
9.6.2 Inverter Output Filters Design 494
9.6.3 Motor Choke 503
9.6.4 MATLAB/Simulink Model of Induction Motor Drive with PWM Inverter and Differential Mode LC Filter 506
9.7 Estimation Problems in the Drive with Filters 509
9.7.1 Introduction 509
9.7.2 Speed Observer with Disturbances Model 511
9.7.3 Simple Observer Based on Motor Stator Models 514
9.8 Motor Control Problems in the Drive with Filters 516
9.8.1 Introduction 516
9.8.2 Field-Oriented Control 518
9.8.3 Nonlinear Field-Oriented Control 522
9.8.4 Nonlinear Multiscalar Control 526

9.9 Predictive Current Control in the Drive System with Output Filter 530
9.9.1 Control System 530
9.9.2 Predictive Current Controller 534
9.9.3 EMF Estimation Technique 536
9.10 Problems 541
Questions 544
References 545

10 Medium Voltage Drives – Challenges and Trends **549**
Haitham Abu-Rub, Sertac Bayhan, Shaikh Moinoddin, Mariusz Malinowski, and Jaroslaw Guzinski
10.1 Introduction 549
10.2 Medium Voltage Drive Topologies 551
10.3 Challenges and Requirements of MV Drives 561
10.3.1 Power Quality and LC Resonance Suppression 561
10.3.2 Inverter Switching Frequency 561
10.3.3 Motor Side Challenges 562
10.4 Summary 569
References 569

11 Current Source Inverter Fed Drive **575**
Marcin Morawiec and Arkadiusz Lewicki
11.1 Introduction 575
11.2 Current Source Inverter Structure 576
11.3 Pulse Width Modulation of Current Source Inverter 578
11.4 Mathematical Model of the Current Source Inverter Fed Drive 582
11.5 Control System of an Induction Machine Supplied by a Current Source Inverter 583
11.5.1 Open-Loop Control 583
11.5.2 Direct Field Control of Induction Machine 584
11.6 Control System Model in Matlab/Simulink 587
References 591

Index **593**

Acknowledgment

We would like to take this opportunity to express our sincere appreciation to all the people who were directly or indirectly helpful in making this book a reality. Our thanks go to our colleague and students at Texas A&M University at Qatar, Qatar University, Aligarh Muslim University, and Gdansk University of Technology. Our special thanks go to Prof. Abullah Kouzou, Dr. M. Rizwan Khan, and Mr. M. Arif Khan for assisting us in this work. We are also grateful to Dr. Khalid Khan, for his assistance in preparing MATLAB®/Simulink models and converting C/C++ files into MATLAB®. We are also thankful to Dr. Shaikh Moinoddin for his help especially in Chapters 3, 7, and 10 of the book.

We are indebted to our family members for their continuous support, patience, and encouragement without which this project would not have been completed. We would also like to express our appreciation and sincere gratitude to the staff of Wiley for their help and cooperation.

Above all we are grateful to almighty, the most beneficent and merciful who provides us confidence and determination in accomplishing this work.

Haitham Abu-Rub, Atif Iqbal, and Jaroslaw Guzinski

Biographies

Haitham Abu-Rub is a full professor holding two PhDs from Gdansk University of Technology (1995) and from Gdansk University (2004). Dr. Abu Rub has long teaching and research experiences at many universities in many countries including Qatar, Poland, Palestine, the USA, and Germany.

Since 2006, Dr. Abu-Rub has been associated with Texas A&M University at Qatar, where he has served for five years as the chair of Electrical and Computer Engineering Program and has been serving as the Managing Director of the Smart Grid Center at the same university.

His main research interests are energy conversion systems, smart grid, renewable energy systems, electric drives, and power electronic converters.

Dr. Abu-Rub is the recipient of many prestigious international awards and recognitions, such as the American Fulbright Scholarship and the German Alexander von Humboldt Fellowship. He has co-authored around 400 journal and conference papers, five books, and five book chapters. Dr. Abu-Rub is an IEEE Fellow and Co-Editor in Chief of the *IEEE Transactions on Industrial Electronics*.

Atif Iqbal, *Fellow IET (UK), Fellow IE (India) and Senior Member IEEE,* DSc (Poland), PhD (UK)-Associate Editor Industrial Electronics and IEEE ACCESS, Editor-in-Chief, I-manager's *Journal of Electrical Engineering*, and Former Associate Editor *IEEE Trans. on Industry Application*, is a full professor at the Department of Electrical Engineering, Qatar University, and former full professor at Electrical Engineering, Aligarh Muslim University (AMU), Aligarh, India. He is a recipient of Outstanding Faculty Merit Award academic year 2014–2015 and Research excellence awards 2015 and 2019 at Qatar University, Doha, Qatar. He received his BSc (Gold Medal) and MSc Engineering (Power System & Drives) degrees in 1991 and 1996, respectively, from the Aligarh Muslim University (AMU), Aligarh, India, and PhD in 2006 from Liverpool John Moores University, Liverpool, UK. He obtained DSc (Habilitation) from Gdansk University of Technology in Control, Informatics and Electrical Engineering in 2019. He has been employed as a lecturer in the Department of Electrical Engineering, AMU, Aligarh, since 1991 where he served as full professor until August 2016. He is recipient of Maulana Tufail Ahmad Gold Medal for standing first at BSc Engg. (Electrical) Exams in 1991 from AMU. He has received several best research papers awards, e.g. at IEEE ICIT-2013, IET-SEISCON-2013, SIGMA 2018, IEEE CENCON 2019, and IEEE ICIOT 2020. He has published widely in International Journals and Conferences, his research findings related to power electronics, variable speed drives, and renewable energy sources. Dr. Iqbal has authored/co-authored more than 390 research papers and two books and three chapters in two other books. He has supervised several large R&D projects worth more than multimillion USD. He has supervised and co-supervised several PhD and Masters students. His principal area of research interest is smart grid, complex energy transition, active distribution network, electric vehicles drivetrain, sustainable development and energy security, distributed energy generation, and variable speed drives.

Jaroslaw Guzinski received MSc, PhD, and DSc degrees from the Electrical Engineering Department at Technical University of Gdansk, Poland, in 1994, 2000, and 2011, respectively. Since 2016, he has been associate professor at Gdansk University of Technology. Currently he is the head of the Department of Electric Drives and Energy Conversion at the same university. From 2006 to 2009, he has been involved in the European Commission Project PREMAID Marie Curie, "Predictive Maintenance and Diagnostics of Railway Power Trains," coordinated by Alstom Transport, France. From 2010 to 2014, he has been a consultant in the prestigious project of integration of renewable energy sources and smart grid for building unique laboratory LINTE^2. In 2012, he was awarded by Polish Academy of Sciences – Division IV: Engineering Sciences for his monograph "Electric drives with induction motors and inverters output filters – selected problems." He obtained scholarships from the Socrates/Erasmus program and was granted with three scientific projects supported by the Polish government in the area of sensorless control and diagnostic for drives with LC filters. He has authored and co-authored more than 150 journal and conference papers. He is an inventor of some solutions for speed sensorless drives with LC filters (six patents). His interests include sensorless control of electrical machines, multiphase drives (five-phase), inverter output filters, renewable energy, and electrical vehicles. Dr. Guzinski is a Senior Member of IEEE.

Preface to Second Edition

The first edition of the book was widely accepted by universities, students, and researchers. Second edition of the book endeavors to enrich the content of the book by updating the existing content, adding some new advancements on the motor drive technology, and responding to the suggestions of many readers. The second edition of the book enhances the content of almost all the chapters. The first chapter gives an overview of the electric motor drive technology using new switching devices. The new power switching devices fabricated using wide band gap technology are becoming popular in motor drives application. The new devices are silicon carbide (SiC) and gallium nitride (GaN) that offer low switching and conduction losses while offering operation at a higher temperature and reduced converters size. High power under medium voltage (MV) drives demand low switching frequency to reduce the losses in order to improve efficiency. Low switching frequency pulse-width modulation (PWM) is attractive for high-power drive systems. Selective harmonic elimination (SHE) is a low switching frequency PWM technique that selectively eliminates certain low-order harmonics by switching the devices at a prespecified switching angle or switching instant. SHE is further elaborated in this edition of the book. Multilevel inverters are used in high-power drives because of lower waveforms distortion and obtaining lower *dv/dt*. Phase-shifted and level-shifted PWM techniques are generally used for controlling multilevel inverters. Switching loss formula for a five-level cascaded H-bridge inverter is also presented in this edition of the book. Design of passive components of an impedance source network is added as well. Space vector PWM technique applied to three-level neutral point clamped inverter and cascaded H-bridge is added in this edition of the book.

Multiphase drives are also enhanced in this edition. Direct torque control is a high-performance drive control that controls torque and speed of AC motors. The key strategy lies in selecting proper inverter voltage according to the torque and stator flux errors. The inverter models are different in a multiphase inverter compared to three-phase inverter. Thus, the application of DTC in a multiphase machine is different from a three-phase machine. DTC applied to a five-phase induction motor is included in this edition of the book.

A new chapter is added on the state-of-art solutions and advances in MV drives technologies. The choice and deployment of MV drives in industries are associated with numerous requirements related to the front-end converter and inverter. Solutions are discussed that offer high efficiency, low price, size and weight, minimum harmonic distortion, reduction in *dv/dt*, mitigation of common mode voltage (CMV), avoiding torsional vibration, transformerless solutions, fault detection capability, and condition monitoring. The newly added Chapter 10 presents a comprehensive overview of the design and research trends of MV drives in addition to presenting the challenges and requirements associated with the use of this type of drives.

The new subject of the current-source-inverter-fed drive has been added in Chapter 11. In this new chapter, the structure of current source inverter based on insulated gate bipolar transistor (IGBT) transistors is presented. Moreover, the PWM strategy of the inverter output current is also given. The special attention is given to control system of induction machine supplied by current source inverter even to open loop control as well as for direct field control.

Sections 3.9 and 3.10 in Chapter 3 are added by Professor Arkadiusz Lewicki and Professor Marcin Morawiec, Gdansk University of Technology, Gdansk, Poland.

Many individuals have sent their suggestions and advice on the first edition of this book. We acknowledge with gratitude the generous and critical comments made by the individuals that helped us to improve the second edition of the book.

We are grateful to our PhD students and researchers who contributed in improving the second edition of the book. The whole book is proofread and typos and language error are ratified by Professor Abdallah Kouzou, Djelfa University, Algeria. Other researchers and PhD students involved in the preparation of this edition of the book are Dr. Khaliqurrahman, Aligarh Muslim University; Dr. Salman Ahmed, Islamic University of Science & Technology, India; Dr. Marif Duala, University of Malaya; Mr. Syed Rahman, Texas A&M University; Dr. Mohammad Meraj and Dr. Pandav Kiran Maroti, Qatar University; and Dr. Shady Khalil, Texas A&M University at Qatar.

Professor Haitham Abu-Rub is grateful to the emotional support and inspiration of his wife Beata and children: Fatima, Iman, Omar, and Muhammad. Professor Atif Iqbal is thankful to his wife Shadma, elder son Abuzar, daughter Noorin, and youngest son Abu Baker who are constant source of inspiration, their support helped to complete the second edition of the book. Professor Jaroslaw Guzinski appreciates the huge support of his whole family, friends from Qatar and India, as well as the wonderful collaborators and authorities of the Electrical and Control Engineering of the Gdansk University of Technology.

Preface to First Edition

The book describes the concept of advanced control strategies of AC machine drives along with their complete simulation models using MATLAB/Simulink. Electrical Motors consume the most energy of the electricity generated worldwide. Thus, there exists a huge scope of saving energy by devising efficient operation schemes of these electrical machines. One approach could be the special design of motors with high-energy efficiency. Other approach lies in the proper control of the machines. The electrical motors employed in various applications run at fixed speed. However, it is found that by varying the speed of motors depending upon the load requirements, the efficiency can be improved significantly; thus, the variable speed operation is extremely important in obtaining highly efficient operations of the machines. As a result, the speed control of a machine for industrial and household applications is most vital for limiting greenhouse gas emission and offering an environment-friendly solution. Controlling the operation of an electrical machine by varying its speed, in literature, is called "variable speed drives" or "adjustable speed drives."

This book discusses the advanced technology used to obtain variable speed AC drives. This book also describes the basic modeling procedures of power electronic converters and AC machines. The mathematical model thus obtained will be used to develop a simulation model using MATLAB/Simulink. The Pulse Width Modulation (PWM) techniques for voltage source inverters and their simulation models are described in one chapter. The AC machines that are taken up for discussion are the most popular squirrel cage induction machine, permanent magnet synchronous machine, and the double-fed induction machine. The book illustrates the advance control techniques of electric drives such as "field-oriented control," "direct torque control," "feedback linearization control," "sensorless operation," and advances in "multiphase (more than three-phase) drives." A separate chapter is dedicated to a five-phase motor drive system. The effect of using an output LC filter at the inverter side on the motor drive control is elaborated on in another chapter.

These control techniques are in general called "high-performance drives" as they offer extremely fast and precise dynamic and steady-state response of electric machines. Thus, this book describes the most important and industrially accepted advanced control technology of AC machines. The book encompasses these diverse topics in a single volume.

This book features exhaustive simulation models based on MATLAB/Simulink. MATLAB/Simulink is an integral part of taught courses at undergraduate and postgraduate programs and is also extensively used in industries. Thus, the simulation models will provide a handy tool to students, practicing engineers, and researchers to verify the algorithms, techniques, and models. Once familiar with the models presented in the book, students and practicing engineers can develop and verify their own algorithms and techniques.

The book is useful for students studying electric drives/motor control at UG/PG levels. The prerequisite will be the basic courses of electric machines, power electronics, and controls. Lecturers can find tutorial materials and solutions to the problems set out in the book on the companion website: http://www.wiley.com/go/aburub_control. The contents of the book will also be useful to researchers and practicing engineers, as well as specialists.

About the Companion Website

This book is accompanied by a companion website:

www.wiley.com/go/aburubcontrol2e

1

Introduction to High-Performance Drives

1.1 Preliminary Remarks

The function of an electric drives system is the controlled conversion of electrical energy to a mechanical form, and vice versa, via a magnetic field. Electric drive is a multidisciplinary field of study, requiring proper integration of knowledge of electrical machines, actuators, power electronic converters, sensors and instrumentation, control hardware and software, and communication links (Figure 1.1). There have been continued developments in the field of electric drives since the inception of the first principle of electrical motors by Michael Faraday in 1821 [1]. The world dramatically changed after the first induction machine was patented (US Patent 381968) by Nikola Tesla in 1888 [2]. Initial research focused on machine design with the aim of reducing the weight per unit power and increasing the efficiency of the motor. Constant efforts by researchers have led to the development of energy-efficient industrial motors with reduced volume machines. The market is saturated with motors reaching high efficiency of almost 95–96%, resulting in no more significant complaints from users [3]. AC motors are broadly classified into three groups: synchronous, asynchronous (induction), and electronically commutated motors. Asynchronous motors are induction motors with a field wound circuit or with squirrel cage rotors. Synchronous motors run at synchronous speeds decided by the supply frequency ($N_s = 120f/P$) and are classified into three major types: rotor excited, permanent magnets, and synchronous reluctance types. Electronic commutated machines use the principle of DC machines but replace the mechanical commutator with inverter-based commutations. There are two main types of motors that are classified under this category: brushless DC motors and switched reluctance motors. There are several other variations of these basic configurations

High Performance Control of AC Drives With MATLAB®/Simulink, Second Edition.
Haitham Abu-Rub, Atif Iqbal, and Jaroslaw Guzinski.

Companion website: www.wiley.com/go/aburubcontrol2e

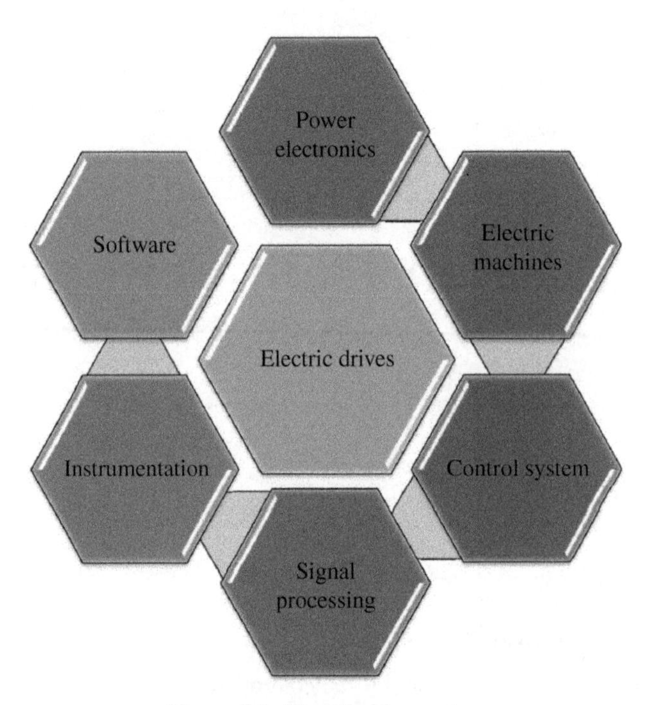

Figure 1.1 Electric drive system

of electric machines used for specific applications, such as stepper motors, hysteresis motors, permanent magnet-assisted synchronous reluctance motors, hysteresis-reluctance motors, universal motors, claw pole motors, frictionless active bearing-based motors, linear induction motors, etc. Active magnetic bearing systems work on the principle of magnetic levitation and, therefore, do not require working fluid, such as grease or lubricating oils. This feature is highly desirable in special applications, such as artificial heart or blood pumps, as well as in the oil and gas industry.

Induction motors are called the workhorse of industry due to their widespread use in industrial drives. They are the most rugged and cheap motors available off the shelf. However, their dominance is challenged by permanent magnet synchronous motors (PMSM), because of their high power density and high efficiency due to reduced rotor losses. Nevertheless, the use of PMSMs is still restricted to the high-performance application area, due to their moderate ratings and high cost. PMSMs were developed after the invention of Alnico, a permanent magnet material, in 1930. The desirable characteristics of permanent magnets are their large coercive force and high reminiscence. The former characteristics prevent demagnetization during start and short conditions of motors, and the latter maximizes the air gap flux density. The most used permanent magnet material is neodymium–boron–iron (NdBFe), which has almost 50 times higher B-H energy compared to Alnico. The major shortcomings of permanent magnet machines are the nonadjustable flux, irreversible demagnetization, and expensive rare-earth magnet resources. Variable flux permanent magnet (VFPM) machines have been developed to incorporate the adjustable flux feature. This variable flux feature offers flexibility by optimizing efficiency over the whole machine operation range, enhancing torque at low speed,

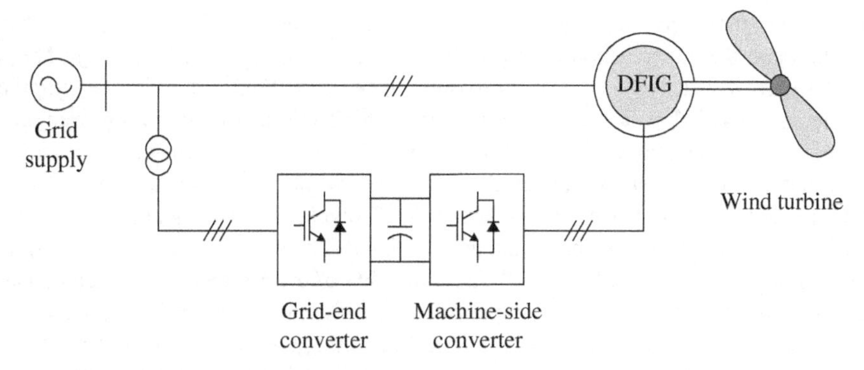

Figure 1.2 General view of a DFIG connected to wind system and utility grid

extending the high speed operating range, and reducing the likelihood of an excessively high back-electromotive force (EMF) being induced at high speed during inverter fault conditions. The VFPMs are broadly classified into hybrid-excited machines (they have the field coils and the permanent magnets) and mechanically adjusted permanent magnet machines. Detailed reviews on the variable flux machines are given in [4]. The detailed reviews on the advances on electric motors are presented in [5–16].

Another popular class of electrical machine is the double-fed induction machine (DFIM) with a wound rotor. The DFIM is frequently used as an induction generator in wind energy systems. The double-fed induction generator (DFIG) is a rotor-wound, three-phase induction machine that is connected to the AC supply from both stator and rotor terminals (Figure 1.2). The stator windings of the machine are connected to the utility grid without using power converters, and the rotor windings are fed by an active front-end converter. Alternatively, the machine can be fed by current or voltage source inverters with controlled voltage magnitude and frequency [17–22].

In the control schemes of DFIM, two output variables on the stator side are generally defined. These variables could be electromagnetic torque and reactive power, active and reactive power, or voltage and frequency, with each pair of variables being controlled by different structures.

The machine is popular and widely adopted for high-power wind generation systems and other types of generators with similar variable-speed high-power sources (e.g. hydro systems). The advantage of using this type of machine is that the required converter capacity is up to three times lower than those that connect the converter to the stator side. Hence, the costs and losses in the conversion system are drastically reduced [17].

A DFIG can be used either in an autonomous generation system (stand-alone) or, more commonly, in parallel with the grid. If the machine is working autonomously, the stator voltage and frequency are selected as the controlled signals. However, when the machine is connected to the infinite bus, the stator voltage and frequency are dictated by the grid system. In the grid-interactive system, the controlled variables are the active and reactive powers [23–25]. Indeed, there are different types of control strategies for this type of machine; however, the most widely used is vector control, which has different orientation frames similar to the squirrel cage induction motor; however, the most popular of these is the stator orientation scheme.

Power electronics converters are used as an interface between the stiff voltage and frequency grid system and the electric motors to provide adjustable voltage and frequency. This is the most vital part of a drive system that provides operational flexibility. The development in power electronic switches is steady, and nowadays high-frequency low-loss power semiconductor devices are available for manufacturing efficient power electronic converters. The power electronic converter can be used as DC-DC (buck, buck–boost, boost converters), AC-DC (rectifiers), DC-AC (inverters), and AC-AC (cycloconverters and matrix converters) modes. In AC drive systems, inverters are used with two-level output or multilevel output (particularly for higher-power applications). The input side of the inverter system can consist of a diode-based, uncontrolled rectifier or controlled rectifier for regeneration capability called back-to-back or active front-end converter. The conventional two-level inverter has the disadvantages of the poor source-side (grid-side) power factor and distorted source current. The situation is improved by using back-to-back converters or matrix converters in drive systems.

The output-side (AC) voltage/current waveforms are improved by employing the appropriate pulse width modulation (PWM) technique, in addition to using a multilevel inverter system. In modern motor drives, the transistor-based [insulated gate bipolar transistor (IGBT), integrated gate-commutated thyristor (IGCT), MOSFET] converters are most commonly used. The increase in transistors switching frequency and decrease in transistor switching times are a source of some serious problems. The high dv/dt and the common-mode (CM) voltage generated by the inverter PWM control result in undesirable bearing currents, shaft voltages, motor terminal overvoltages, reduced motor efficiency, acoustic noise, and electromagnetic interference (EMI) problems, which are aggravated by the long length of the cable between the converter and the motor. To alleviate such problems, generally, the passive LC filters are installed on the converter output. However, the use of an LC filter introduces unwanted voltage drops and causes a phase shift between the filter input and output voltages and currents. These can negatively influence the operation of the whole drive system, especially when sophisticated speed, sensorless control methods are employed, requiring some estimation and control modifications for an electric drive system with an LC filter at its output. With the LC filter, the principal problem is that the motor input voltages and currents are not precisely known; hence, additional voltage and current sensors are employed. Since the filter is an external element of the converter, the requirement of additional voltage and current sensors poses technical and economical problems in converter design. The more affordable solution is to develop proper motor control and use estimation techniques in conjunction with LC filter-based drive [26–30].

The simulation tool is a significant step for performing advanced control for industry. However, for practical implementation, the control platform for the electric drive system is provided with microcontrollers (mCs), digital signal processors (DSPs), and/or field-programmable gate arrays (FPGAs). These control platforms offer the flexibility of control and make possible the implementation of complex control algorithms, such as field-oriented control (FOC), direct torque control (DTC), nonlinear control, and artificial-intelligence-based control. The first microprocessor, the Intel 4004 (US Patent # 3821715), was invented by Intel engineers Federico Faggin, Ted Hoff, and Stan Mazor in November 1971 [31]. Since then, the development of faster and more capable microprocessors and μCs has grown tremendously. A microcontroller is a single IC containing the processor core, the memory, and the peripherals. Microprocessors are used for general-purpose applications, such as in PCs, laptops, and other electronic items, and are used in embedded applications for actions such as motor control. The first DSP was produced by Texas Instruments, TMS32010, in 1983 [32], followed by several DSPs being produced and used for several applications, ranging from motor control to multimedia applications to image processing. Texas Instruments has developed some specific DSPs for electric drive applications, such as the TMS320F2407, TMS320F2812, and TMS320F28335. These DSPs have dedicated pins for PWM signal generation that serves by controlling power converters. Nowadays, control algorithms implement more powerful programmable logic components called FPGAs, the first of which, XC2064, was invented by Xilinx co-founders Ross Freeman and Bernard Vonderschmitt in 1985. FPGAs are composed of logic blocks with memory elements that can be reconfigured to obtain different logic gates. These reconfigurable logic gates can be configured to perform complex combinational functions. The first FPGA XC2064 consisted of 64 configurable logic blocks, with two three-input lookup tables. In 2010, an extended processing platform was developed for FPGAs that combines the features of an advanced reduced instruction set machine (ARM) high-end microcontroller (32-bit processor, memory, and I/O) with an FPGA fabric for easier use in embedded applications. Such configurations make it possible to implement a combination of serial and parallel processing to address the challenges in designing today's embedded systems [33].

The primitive electric drive system uses a fixed-speed drive supplied from the grid, while mostly employing the DC motor. Adjustable-speed drive ASD systems offer more flexible control and increased drive efficiency when compared to the fixed speed drive. DC motors inherently offer decoupled flux and torque control, with fast dynamic response and simple control mechanism. However, the operating voltage of the DC machines is limited by the mechanical commutator's withstand voltage; in addition besides, the maintenance requirement is high due to its brush and commutator arrangement. DC drives are now increasingly replaced by AC drives due to the advent of the high-performance control of AC motors, such as vector control, direct torque control (DTC), and predictive control, offering precise position control and an extremely fast dynamic response [34]. The major advantages of AC drives over DC drives include their robustness, compactness, economy, and low maintenance requirements.

Biologically inspired artificial intelligence techniques are now being increasingly used for electric drive control and are based on artificial neural networks (ANN), fuzzy logic control (FLC), adaptive neuro-fuzzy inference system (ANFIS), and genetic algorithm (GA) [35, 36]. A new class of electric drive controls, called brain emotional learning-based intelligent controller (BELBIC), is reported in the literature [37]. The control relies on the emotion processing mechanisms in the brain, with the decisions made on the basis of an emotional search. This emotional intelligence controller offers a simple structure with a high autolearning feature that does not require any motor parameters for self-performance. The high-performance drive control requires some sort of parameter estimation of motors, in addition to the current, speed, and flux information for a feedback closed-loop control. Sensors are used to acquire the information and are subsequently used in the controller. The speed sensors are the most delicate part in the whole drive system, thus extensive research efforts are being made to eliminate the speed sensors from the drive system, with the resulting drive system becoming a 'sensorless' drive. In sensorless drive schemes, existing current and voltage sensors are used to compute the speed of the machine, and the computed speed is used for closed-loop control. The literature on sensorless drives is too vast to list; however, a comprehensive review is available in [38–40]. A sensorless drive offers the advantages of a compact drive with reduced maintenance, reduced cost, and its ability to withstand harsh environmental conditions. Despite impressive progress in drive automation, there are still a number of persistent challenges, including a very low speed near to zero, operation at zero speed with full load condition, and an overly high-speed operation.

Network-based control and remote control of the drive systems are still in progress. Plug-and-play types of electric drives are an important area that can serve the applications that have a direct impact on the quality of life, such as renewable energy, automotive applications, and biomedical applications. Integrated converter-motor drive systems for compact design, as well as reduced EMI due to cabling wave reflection, are also in progress. More diversity in machine design with rare-earth-free motors is the subject of research, and high air-gap flux density machines using superconductors are the direction of research in electric drive systems.

1.2 General Overview of High-Performance Drives

High-performance drive refers to the drive system's ability to offer precise control, in addition to rapid dynamic response and a good, steady-state response. High-performance drives are considered for safety-critical applications due to their precision of control [41]. Since the inception of AC machines, several techniques have evolved to control their speed, torque, and flux. The basic controlling parameters are the voltage and frequency of the applied voltage/current to the motor. The grid supplies fixed magnitude and frequency voltages/currents and are thus not suitable for obtaining controlled operation of machines. Hence, power electronic converters are used as an interface between the grid supply and the electric motors. These power electronic converters, in most cases, are AC-DC-AC converters for AC machine drives. Other alternatives are direct AC-AC converters, such as cycloconverters and matrix converters. However, these direct AC-AC converters suffer from some serious drawbacks, including the limited output frequency, as low as one-third in cycloconverters, and the limited output voltage magnitude, which is limited to 86% of the input voltage magnitude in matrix converters. Moreover, the control is extremely complex for direct AC-AC converters. Thus, invariably AC-DC-AC converters are more commonly called 'inverters,' and are used to feed the motors for adjustable-speed applications. This book will describe the modeling procedures of the inverters, followed by the illustration of their existing control techniques. The basic energy processing technique in an inverter is called 'Pulse Width Modulation' (PWM); hence, PWM will be discussed at length.

The control of AC machines can be broadly classified into 'scalar' and 'vector' controls (Figure 1.3). Scalar controls are easy to implement and offer a relatively steady-state response, even though the dynamics are sluggish. To obtain high precision and good dynamics, as well as a steady-steady response, 'vector' control approaches are to be employed with closed-loop feedback control. Thus, this book focuses on the 'vector' based approaches, namely 'Field Oriented Control,' 'Direct Torque Control,' 'Non-linear Control,' and 'Predictive Control.'

It is well known that the variable-speed drive offers significant energy savings a huge industrial setup. Thus, by employing variable-speed drives in the industry, there exists a huge scope for energy saving. The older installations relied on DC machines for variable-speed applications, because of their inherent decoupled torque and flux control with minimum electronics involved; however, in the early 1970s, the principle of decoupled torque and flux control, more commonly called 'field-oriented control' or 'vector control,' was achieved in more robust

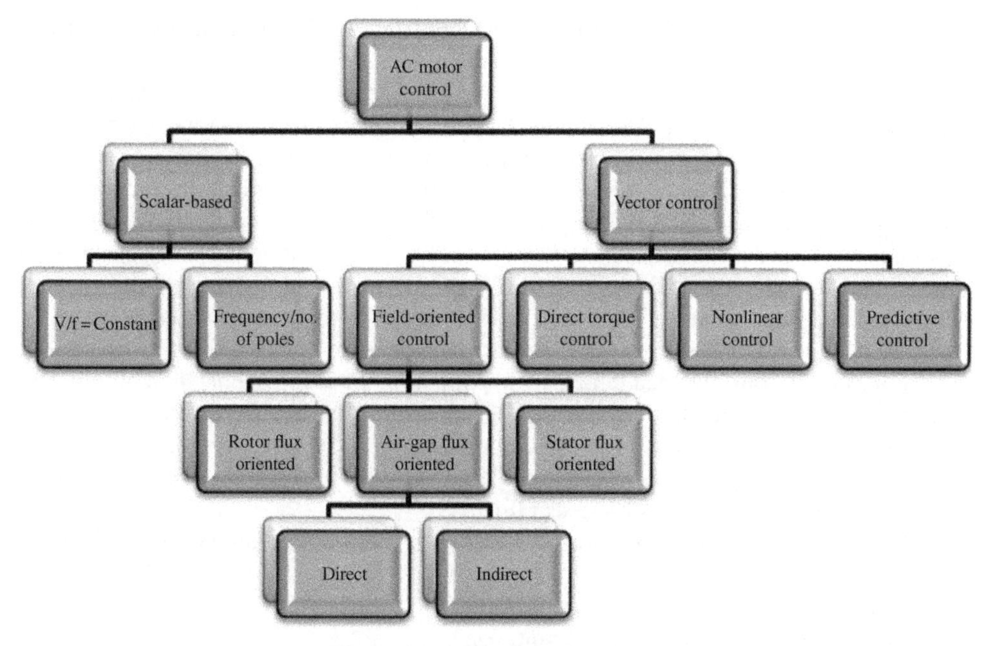

Figure 1.3 Motor control schemes

induction machines. Later, it was realized that such control was also possible in synchronous machines. However, the pace of development in variable-speed AC machine drives was slow and steady until the early 1980s, when the microprocessor era began and the realization of complex control algorithms became feasible [34, 35].

The FOC principle relies on the instantaneous control of stator current space vectors. The research on FOC is still active, with the idea of incorporating more advanced features for highly precise and accurate control, such as sensorless operation, and utilization of online parameter adaptations. The effect of parameter variations, magnetic saturation, and stray-load losses on the behaviour of field-oriented controlled drives is the subject of research in obtaining robust sensorless drives.

Theoretical principles of 'direct torque control' for high-performance drives were introduced in the mid-and second half of the 1980s. Compared with FOC, which had its origin at the beginning of the 1970s, DTC is a significantly newer concept. It took almost 20 years for the vector control to gain acceptance by the industry. In contrast, the concept of DTC has been received by the industry relatively quickly, in only 10 years. While FOC predominantly relies on the mathematical modelling of an induction machine, DTC makes direct use of physical interactions that take place within the integrated system of the machine and its supply. The DTC scheme requires simple signal processing methods, relying entirely on the nonideal nature of the power source that is used to supply an induction machine, within the variable-speed drive system (two-level or three-level voltage source inverters, matrix converters, etc.). It can, therefore, be applied to power electronic converter-fed machines only. The on–off control of converter switches is used for the decoupling of the nonlinear structure of the AC machines. The most frequently discussed and used power electronic converter in DTC drives is a voltage source inverter.

DTC takes a different look at the machine and the associated power electronic converter. First, it is recognized that regardless of how the inverter is controlled, it is by default a voltage source rather than a current source. Next, it dispenses with one of the main characteristics of the vector control, indirect flux, and torque control by means of two stator current components. In essence, DTC recognizes that if flux and torque can be controlled indirectly by these two current components, then there is no reason why it should not be possible to control flux and torque directly, without intermediate current control loops.

DTC is inherently sensorless. Information about actual rotor speed is not necessary in the torque mode of operation, because of the absence of coordinate transformation. However, correct estimations of stator flux and torque are important for the accurate operation of hysteresis controllers. An accurate mathematical model of an induction machine is, therefore, essential in DTC. The accuracy of DTC is also independent of the rotor's parameters variations. Only the variation of stator resistance, due to a change in thermal operating conditions, causes problems for high-performance DTC at low speeds [38].

In summary, the main features of DTC and its differences from the vector control are:

- direct control of flux and torque;
- indirect control of stator currents and voltages;
- absence of coordinate transformation;
- absence of separate voltage modulation block, usually required in vector drives;
- ability to know only the sector in which the stator flux linkage space vector is positioned, rather than the exact position of it (necessary in vector drives for coordinate transformation);
- absence of current controllers;
- inherently sensorless control since speed information is not required in the torque mode of operation;
- in its basic form, the DTC scheme is sensitive to only variation in stator resistance.

The research on the direct torque is still active, and the effects of nonlinearity in the machine models are being explored; the flexibility and simple implementation of the algorithms will be the focus of research in the near future. The use of artificial intelligence is another direction of research in this area. It is important to emphasize that many manufacturers offer variable-speed drives based on the 'field-oriented control' and 'DTC' principles and are readily available in the market.

The main disadvantage of vector control methods is the presence of nonlinearity in the mechanical part of the equation during the change of rotor flux linkage. Direct use of vector methods to control an induction machine fed by a current source inverter provides a machine model with high complexity, which is necessary to obtain precise control systems. Although positive results from field-oriented/vector control have been observed, attempts to obtain new, beneficial, and more precise control methods are continuously made. One such development is the 'non-linear control' of an induction machine. There are a few methods that are encompassed in this general term of 'non-linear control,' such as 'feedback linearization control' or 'input-output decoupling control,' and 'multi-scalar model-based non-linear control.' Multiscalar-based nonlinear control or MM control was presented for the first time in 1987 [38, 39] and is discussed in the book. The multiscalar model-based control relies on the choice of specific state variables and, thus, the model obtained completely decoupled mechanical and electromagnetic subsystems. It has been shown that it is possible to have nonlinear control and decoupling between electromagnetic torque and the square of the linear combination of a stator current vector and the vector of rotor linkage flux.

When a motor is fed by voltage source inverters, and when the rotor flux linkage magnitude is kept constant, the nonlinear control system control is equivalent to the vector control method. In many other situations, the nonlinear control leads to more system structure simplicity and good overall drive response [35, 38, 39].

The use of variables transformation to obtain nonlinear model variables makes the control strategy easy to perform because only four state variables have been obtained with a relatively simple nonlinearity form [38]. This makes it possible to use this method in the case of change flux vector, as well as to obtain simple system structures. In such systems, it is possible to change the rotor flux linkage with the operating point without affecting the dynamic of the system. The relations occurring between the new variables make it possible to obtain novel control structures that guarantee a good response of the drive system, which is convenient for the economical operation of drive systems in which this flux is reduced if the load is decreased. The use of variables transformation to obtain MM makes the control strategy easier than the vector control method, because four variables are obtained within simple nonlinearity form. This makes it possible to use this method in the field-weakening region (high-speed applications) more easily when compared to the vector control methods. Extensive research has been conducted on the nonlinear control theory of induction machines, leading to a number of suggested improvements. It is expected that more such control topology will evolve in time.

High-performance control of AC machines requires the information of several electromagnetic and mechanical variables, including currents, voltages, fluxes, speed, and position. Currents and voltage sensors are robust and provide sufficiently accurate measurements and so are adopted for the closed-loop control. The speed sensors are more delicate and often pose serious threats to control issues, so speed sensorless operation is sought in many applications that require closed-loop control. Several schemes have been developed recently to extract speed and position information without using speed sensors. Similarly, rotor flux information is typically obtained using 'observer' systems. Much research efforts occurred throughout the 1990s to develop robust and precise observer systems. Improvements have been offered by the development of the methods, including the 'model reference adaptive system,' the 'Kalman filters,' and the 'Luenberger observers,' [40–42].

Initially, observers were designed based on the assumption of a linear magnetic circuit and were later improved by taking into account different nonlinearities. The methods developed so far still suffer from stability problems around zero frequency. They fail to prove global stability for sensorless AC drives. This has led many researchers to conclude that globally asymptotically stable model-based systems for sensorless induction motor drives may not exist. Indeed, most investigations on sensorless induction motor drives today focus on providing sustained operation at a high dynamic performance at very low speed, particularly at zero speed or at zero stator frequency. Two advanced methodologies are competing to reach this goal. The first category includes the methods that model the induction motor by its state equations. A sinusoidal flux density distribution in the air gap is then assumed, neglecting space harmonics and other secondary effects. The approach defines the class of fundamental models. They are implemented either as open-loop structures, such as the stator model, or as closed-loop observers. The adaptive flux observer is a combination of the nonlinear observer with a speed adaptation process. This structure is now receiving considerable attention, and many new solutions follow a similar generic approach [41].

Three-phase electric power generation, transmission, distribution, and utilization have been well known for over a century. It was realized that generation and transmission of power with more than three phases do not offer significant advantages in terms of power density (generation) and right-of-way and initial cost (transmission). A five-phase induction motor drive system was first tested in 1969 [43]. The supply to a five-phase drive system was made possible by using a five-phase voltage source inverter, since simply adding an extra leg increases the output phases in an inverter. It was realized that the five-phase induction motor drive systems offered some distinct advantages over three-phase drive system counterparts, such as reduced torque pulsation and enhanced frequency of pulsation, reduced harmonic losses, reduced volume for the same power output, reduced DC link current harmonics, greater fault tolerance, and better noise characteristics.

In addition, there is a significant advantage on the power converter end, due to the reduced power per leg, the power semiconductor switch rating reduces, thus avoiding their series/parallel combination and eliminating the problem of static and dynamic voltage sharing. Furthermore, the stress on the power semiconductor switches reduces due to the reduced dv/dt. The attractive features of multiphase drive systems mean enormous research efforts have been made globally in the last decade to develop commercially feasible and economically viable solutions. Niche application areas are then identified for multiphase drive systems, such as ship propulsion, traction, 'more electric aircraft' fuel pumps, and other safety-critical applications. Due to their complex control structure, their widespread use in a general-purpose application is still not accepted. One of the commercial applications of a 15-phase induction motor drive system is in the British naval ship 'Destroyer II.' Similar drive systems are under preparation for US naval ships and will be commissioned soon. Nevertheless, there are many challenges still to be met before the widespread use of multiphase drive systems, especially in general-purpose electric drive systems [44].

1.3 Challenges and Requirements for Electric Drives for Industrial Applications

Industrial automation requires precisely controlled electric drive systems. The challenges and requirements for electric drive systems depend upon the specific applications being used. Among different classes of electric drives, medium voltage drives (0.2–40 MW at the voltage level of 2.3–13.8 kV) are more popular for use in the industry, such as in the oil and gas sector, rolling and milling operations, production and process plants, petrochemical industry, cement industry, metal industry, marine drive, and traction drive. However, only 3% of the existing medium voltage (MV) motors are variable-speed drives, with the rest of these running at a fixed speed [45]. The installation of properly speed-controlled MV drives will significantly reduce losses and total drive costs, as well as improve power quality in any industrial setup. There are several challenges associated with the controlled MV drives that are related to the line/source side (e.g. power quality, resonance, and power factor), motor side (e.g. dv/dt, torsional vibration the and travelling wave reflections), and power semiconductor switching devices (e.g. switching losses and voltage and current handling capability). The power rectifier at the source side produces distorted currents at the source, in addition to poor power factors, thus posing a challenge to the designer of the controlled electric drive system. The PWM of inverter generates a CM voltage on the motor side, which poses another challenge. The rating of the power semiconductor devices is also an important factor to be considered while designing an electric drive. High-quality voltage and current waveforms at the converter input and output are essential in all types of electric drive systems.

The power quality is a factor of the converter topology used and refers to the characteristic of the load, the size and type of the filter, the switching frequency, and the adopted control strategy. The switching losses of power converter devices contribute to the major portion of the drive losses; therefore, operation at a low switching frequency makes it possible to increase the maximum power of the inverter. However, an increase in the switching frequency of an inverter increases the harmonic distortion of the input and output waveforms. Another solution is to use multilevel inverters that deliver waveforms with better harmonic spectrum and lower dv/dt, which limits the insulation stress of the motor windings. However, increasing the number of switching devices in multilevel inverters tends to reduce the overall reliability and efficiency of the power converter. On the other hand, an inverter with a lower level of output voltage requires a large LC output filter to reduce the motor winding insulation stress. The challenge then is to reduce output voltage and current waveform distortions when the low switching frequency is used to ensure high power quality and high efficiency.

The maximum voltage blocking capability of modern power semiconductor switching devices is nearly 6.5 kV. This sets the maximum voltage limit of the inverter and the motor in an electric drive system. Referring to the two-level voltage source inverter and the maximum conducting current (600 A) of available voltage IGBT switches, the obtained maximum apparent power is less than 1 MVA [45]. To overcome the limits of inverter ratings, series and/or parallel combinations of power devices are suggested. In such instances, extra measurements are required to balance the current between devices during turning on and turning off. Due to inherent differing device characteristics, more losses are generated, requiring a reduction in the derating of the inverter. There is a requirement to find a solution to increase the power range of inverter, while avoiding the problems associated with series/parallel connections of switches. One possible solution is using machines and converters with a high phase number (more than three). Drive systems using motors and converters of high phase orders have gained popularity in recent times, and extensive research efforts are being put into developing such commercially feasible drives.

Essential requirements for general-purpose electric drives for industrial application include high efficiency, low cost, compact size, high reliability, fault protection features, easy installation, maintenance, and, in some applications, high dynamic performance, better power quality at the input side, precise position control, and regeneration capability.

1.3.1 Power Quality and LC Resonance Suppression

Unwanted harmonics are introduced in power grids due to the power electronic switching converters on the load side, which poses a serious problem that needs to be effectively solved.

Uncontrolled diode-based rectifiers at the source side of the inverter draw distorted currents from the grid and cause notches in voltage waveforms. This results in several problems, such as computer data loss, malfunction of communication equipment and protective equipment, and noise. Therefore, many standards define the limit of harmonics injected into the power grid, including IEEE 519-1999, IEC 1000-3-2 International Standard, 1995, and IEC 61000-3-2 International Standard, 2000. Current research and industrial applications tend to comply with these international standards.

The LC line-side filter is used for current harmonic reduction or power factor compensation. Such LC filters may exhibit resonance phenomena. The supply system at the MV level has very low impedance; therefore, lightly damped LC resonances may cause undesired oscillations or overvoltages. This may shorten the life of the switching devices or other components of the rectifier circuits. Effective solutions should guarantee low harmonics and low dv/dt using just a reactor instead of an LC filter or using a small filter to solve the problem of LC resonance.

1.3.2 Inverter Switching Frequency

The use of high switching frequency devices in power converters causes rapid voltages and current transitions. This leads to serious problems in the drive system, such as the generation of unwanted CM currents, EMI, shaft voltage, with consequent generation of bearing currents and deterioration of the winding insulation of motors and transformers.

Switching losses are a crucial issue that should be taken into account in designing electric drives because they pose a limit on the switching frequency value and the output power level of the power converters. The switching losses of semiconductor devices contribute to a major portion of total device losses. A reduction of switching frequencies increases the maximum output power of the power converter. However, the reduction of switching frequency may cause an increase in harmonic distortion of the line and motor side. Hence, a trade-off exists between these two conflicting requirements.

1.3.3 Motor-Side Challenges

Fast switching transients of the power semiconductors devices at high commutation voltages cause high switching losses and poor harmonic spectrum of the output voltages, generating additional losses in the machine. The problems are aggravated due to the long length of the cables between the converters and motors, as well as generating bearing currents due to switching transients.

1.3.4 High dv/dt *and Wave Reflection*

The high switching frequency of power devices causes high dv/dt at the rising and falling edges of the inverter output voltage waveform. This may cause the failure of the motor winding insulation due to partial discharges and high stress. High dv/dt also produces rotor shaft voltages, which creates current flow into the shaft bearing through the stray coupling capacitors, ultimately leading to motor bearings failure. This is a common problem in adjustable-speed drive systems in the industry.

The wave reflections are caused by the mismatch between the cable, the inverter, and the motor wave impedance. Wave reflections can double the voltage on the motor terminals at each switching transient, if the cable length exceeds a certain limit. The critical cable length for 500 V/μs is in the 100-m range, while for 10,000 V/μs it is in the 5-m range [45]. The d*v*/d*t* also causes electromagnetic emission in the cables between the inverter and the motor. Expensive shielded cables are used to avoid these effects; nevertheless, the electromagnetic emission may affect the operation of nearby sensitive electronic equipment, which is called electromagnetic interference.

The design of the filters should achieve international standards (e.g. IEEE 519-1999). Filters also reduce the switching frequency. The widely used passive filters are designed to ensure very low total harmonic distortion (THD) in both motor and line sides. A large inductor in the LC filter must be used in most high-power drive systems, resulting in undesirable voltage drops across the inductor. The increase in capacitor size reduces LC resonant frequency, which is then affected by the parallel connection of the filter capacitor and motor magnetizing inductance.

In an electric drive with an LC filter, instability could appear due to electric resonance between L and C parameters. Such phenomena are mostly observed in overmodulation regions when some inverter output voltage harmonics are close to LC filter resonant frequency. Active damping techniques could be then employed to resolve the problem of instability, while at the same time suppressing the LC resonance.

1.3.5 Use of Inverter Output Filters

The output voltage quality at the inverter side can be improved by using active and passive filters. Today, passive filtering is widely used at the output of the inverter to improve the voltage waveform. Such filters are hardware circuits that are installed on the output of the converter structure. The most common approach is the use of filters based on resistors, inductors, and capacitors (LC filters). In order to reduce the overvoltages that can occur, because of wave reflection at the motor terminals when long cables are used, differential-mode LC filters are used. The cable length is important in determining the output performance of the drive system; however, the cable layout on the user end is generally unknown to the inverter manufacturer. Moreover, such filter components are decided according to the switching frequency of the inverter. When an inverter output filter is installed in the electric drive, the voltage drops and the phase shifts between the filter's input while output voltages and currents appear. This complicates the control system design, particularly for low-speed conditions. The control systems are generally designed assuming the inverter's output voltages and currents are equal to the motor input values. In the case of a discrepancy between voltages and currents, the region of proper motor operation is limited. Therefore, in a control system of electric drives with an inverter output filter, it is essential to provide modification in the measurement circuits or in the control algorithms. A simple way to improve the performance of the electric drive with inverter output filter is to introduce additional sensors for motor voltages and current measurements. Such a solution is not practical because it requires changes in the inverter structure so, in this case, an accepted solution is to keep the inverter structure unchanged but to modify a control algorithm [46].

1.4 Wide Bandgap (WBG) Devices Applications in Electric Motor Drives

Due to the high efficiency of the wide bandgap (WBG) semiconductor device-based power converters, the WBG device-based power converters have been highly researched for their applications for the electric motor drive applications. The distinct merits of the WBG devices over the conventional Si-based devices are: the low losses, high switching operation with their ability to handle higher temperature operations. The two key materials for the WBG-based devices are silicon carbide (SiC) and gallium nitride (GaN). Presently, GaN devices are manufactured up to a voltage level of 650 V; however, the SiC-based systems have been developed with much higher levels with an operating temperature of 200°C [47, 48]. The converters based on WBG devices have found numerous applications in the motor drive especially for high-speed operation, high-ambient temperature operation, and operation of low inductance motors. The WBG devices enable high-power, high-frequency operation with the ability to handle the high-temperature environment. The specially designed very high-speed motor requires the output voltage of a higher fundamental frequency of kHz. Similarly, the low-inductance motors require the converter operation of 50–100 kHz for the lower current ripple [48]. For these two types of motor operations, the Si-based converters have a limited-frequency operation (>20 kHz) [49]. Furthermore, for the motor drives for high ambient temperature operation, the Si-based devise cannot sustain such high-temperature operations. The maximum junction temperature for Si is 200°C, whereas, for the SiC and GaN, it is theoretically up to 300°C and 600°C, respectively [50, 51]. Therefore, for the motor operation with high-speed and higher-temperature operations, the WBG-based devices provide a suitable alternative to Si-based devices.

1. **Application of WBG devices for low-inductance motors:** For the low-inductance motors, the converters are operated with high-frequency PWM with frequency varying from 50 to 100 kHz. The high-frequency PWM operation is necessary to minimize the ripple current, which is unwanted and will cause power loss as well as responsible for higher torque pulsation. Slotless motor is one of the typical examples of a low-inductance motor with inductance in the range of 10–100 μH [52]. These type of motors have ironless composite stator. These motors are very much suitable for the application that requires positioning accuracy, torque linearity, higher power-to-weight ratio, and small cogging toque. Some applications that require these features are electric vehicles, aerospace, portable industrial application, defense, and precision manufacturing [53]. Another type of motor with low inductance is a surface-mounted permanent magnet (SMPM), which has effectively longer airgap. High-speed PM brushless DC (BLDC) motors have a compact size, are efficient and reliable [54, 55].

   With few kW of power rating, the low-inductance motors can be operated with conventional Si-based MOSFET, which can operate up to 50 kHz; however, for higher power ratings, Si-based IGBTs are preferred due to their higher-power handling capabilities over MOSFET. However, the switching frequency of Si-based IGBT is limited to 20 kHz, which will not be able to reduce the current ripple of low-inductance motors. Therefore, the WBG-based devices make the possibility of the use of low-inductance motor at higher power rating.

2. **High-speed motors:** Due to the higher power density of the high-speed machines, they are gaining their popularity in the field of EVs, gas compressors, energy storage systems, industrial air compressors, and blowers, etc. High-speed motors (10,000–20,000 rpm) have been used for the petroleum industries. For the MW operations, with the standard 60 Hz supply, the motor is made to run at 1800–3600 rpm, whereas the speed of the compressor is about 15,000–20,000 rpm. To match both speeds, the gearbox is used to increase the speed of the motor. Now, with the use of WBG devices, the motor can be made to run at higher speeds with the increase of fundamental frequency in kHz. This also enhances the reliability of the overall system by the removal of the gearbox.
3. **High-temperature applications:** The maximum junction temperature of power switches could be 200°C for Si and more than 300°C for SiC and GaN [48]. Therefore, the WBG devices can operate at higher operating temperatures compared to the Si.

1.4.1 Industrial Prototype Using WBG

The products based on WBG devices are being developed by several companies such as Mitsubishi Electric, KSB Group, Beckhoff, etc. Mitsubishi Electric designed a drive system for EV with a SiC inverter for a PM motor. It was found out that the 60 kW system has less than 50% losses compared to the conventional Si-based system [56]. Another 22 kW synchronous reluctance motor-based high-efficiency system was designed using SiC-based devices [57]. The use of WBG devices reduces the overall system volume by more than 25% compared to the conventional system design. Similarly, a 22 kW, 1050 V system was designed by John Deere employing WBG devices for their inverter part [57]. The designed inverter has a power density, which is almost 20–25% more than that of conventional design. In addition to these drive systems, a 1.6 MW, 15,000 rpm motor has been driven using a SiC-based inverter for a gas compression, which yields the overall efficiency of 94% without using any gearbox [58]. In 2018, a 650 V, 100 A WBG device-based electric drive train system was developed by Tesla [59].

1.4.2 Major Challenges for WBG Devices for Electric Motor Drive Applications

As the cost of the WBG devices is high compared to the conventional Si-based devices, these switches must be operated at very high switching frequencies to achieve better utilization with the higher power density and improved efficiency. These benefits of the WBG-based drive system can be an important aspect to eliminate the higher cost. However, at the higher switching frequency, the additional challenges develop in the system in terms of EMI generation, high dv/dt, which will affect the performance of the motor insulation and bearings. The high value of dv/dt will create voltage stresses of higher magnitude in the insulation of the motor. This can cause premature failure due to partial discharge in the windings. Similarly, the high dv/dt can cause the flow of bearing current, which will reduce the life span of bearings.

Another important aspect of the WBG device is their packing for high-temperature and high-voltage operation. For the compact high-voltage WBG-based modules, the consideration regarding the parasitic components, terminal connections, and insulation must be taken care of with higher priorities as these factors affect the performance of the overall system. Similarly, the design of the overall power module must be optimized in order to reduce the stray inductance between different loops of the circuits. The gate drive design has also a crucial role in order to improve the performance of the overall system. As the chip area for the WBG devices is small compared to the Si-based devices, the chance of short-circuiting in WBG-based converter is more. For the optimal control of the WBG devices, the response of the gate driver circuit should be higher than that of the WBG devices in order to protect them from the short-circuit situations. The faster response from the gate driver circuit protects the WBG devices from the thermal breakdown. Other types of protection required for WBG devices are voltage clamp protection and soft turn-off after a short-circuit fault. With the higher switching speed and higher dv/dt, the presence of intrinsic capacitances increases the noise. Therefore, the design of the gate driver circuit should also consider noise immunity. Furthermore, the gate voltage required for the WBG MOSFET falls in the range of (−5 V to +50 V) compared to the gate voltage of Si-based MOSFET, which is 10–15 V [60–63].

1.5 Organization of the Book

This book consists of nine different chapters dealing with different issues of high-performance AC drives along with the MATLAB/Simulink models. Chapter 1 discusses the components of AC drive system and presents an overview of high-performance drives.

The classification of electrical machines and their state-of-the-art control strategies, including FOC and DTC, are elaborated on. The persisting challenges for the industrial application of AC drives are further illustrated in Section 1.3. Section 1.4 is devoted to the overview of applications of newly developed WBG devices in electro motor drives. This chapter gives an overall view of high-performance drives.

Chapter 2 discusses the basic modelling procedures of different types of electrical machines. For the sake of completeness, DC machine modeling is presented in Section 2.2, followed by the modeling of the squirrel cage induction machine, which is presented based on the space vector approach. The obtained dynamic model is then converted into a per-unit system for further simulation and generalization of this approach. This is followed by the modelling of DFIM and permanent magnet synchronous machine. The simulation using MATLAB/Simulink is also presented.

Chapter 3 describes the PWM control of DC-AC converter system. The basic modelling of a two-level inverter based on the space vector approach is discussed, followed by different PWM approaches such as carrier-based sinusoidal PWM, harmonic injection schemes, offset addition methods, selective harmonic elimination PWM, and space vector PWM techniques. This is followed by a discussion on multilevel inverter operation and control. Three most popular topologies (diode clamped, flying capacitor, and cascaded H-bridge) are illustrated. A comparison between phase-shifted and level-shifted PWM methods is given. A new class of inverter, most popular in renewable energy application, called the Z-source inverter and its modified form called the quasi Z-source are discussed in this chapter. The design procedure for the passive network of impedance source inverter is given. The chapter includes also knowledge of space vector modulation and DC-link voltage balancing in three-level neutral point clamped (NPC) inverters. The discussed techniques are further simulated using MATLAB/Simulink and the simulation models are presented.

Chapter 4 is dedicated to the FOC or vector control of AC machines, including the squirrel cage induction machine, the DFIM, and the permanent magnet synchronous machine. For consistency, scalar control (*v/f* = constant) is also presented. Different types of vector control are presented along with their simulation models. Wide speed control range from low to high (field weakening) is elaborated on. The field weakening region is discussed in detail, with the aim of producing high torque at high speed. Vector control of DFIG with grid interface is also described, as is the basic rotor flux estimation scheme using the Luenberger observer system.

Principles of DTC for high-performance drives are introduced in Chapter 5. DTC takes a different look at the induction machine and the associated power electronic converter. First, it recognizes that, regardless of how the inverter is controlled, it is by default a voltage source rather than a current source. Next, it dispenses with one of the main characteristics of vector control, indirect flux, and torque control, by means of two stator current components. In essence, DTC recognizes that if flux and torque can be controlled indirectly by these two current components, then it should not be possible to control flux and torque directly, without intermediate current control loops. This concept is discussed in Chapter 5. The main features, advantages, shortcomings, and implementation of DTC are elaborated on. A simulation model for implementing d DTC is presented.

High-performance drives are a solution intended to embed separately excited DC motor characteristics into AC machines. This goal has been almost achieved with the inception of vector control principle. The main disadvantage of vector control methods is the presence of nonlinearity in the mechanical part of the equation during the change of rotor flux linkage. Although good results from vector control have been observed, attempts to obtain new control methods are still being made. Nonlinear control of induction motors is another alternative to obtain decoupled dynamic control of torque and flux. This method of control to obtain high-performance drive is presented in Chapter 6. Such a control technique introduces a novel mathematical model for induction motors, which makes it possible to avoid using sin/cos transformation of state variables. The model consists of two completely decoupled subsystems, mechanical and electromagnetic. It has been shown that in such a situation it is possible to have nonlinear control and decoupling between electromagnetic torque and the rotor linkage flux. Nonlinear control of induction machine based on the multiscalar model is discussed. Nonlinear control of a separately excited DC motor is also presented. Nonlinear control of nonlinear induction machine and permanent magnet synchronous machine is illustrated. The discussed techniques are supported by their simulation model using MATLAB/Simulink.

Chapter 7 is devoted to a five-phase induction motor drive system. The advantages and applications of a multiphase (more than three phases) system are described. The chapter discusses the dynamic modelling of a five-phase induction machine, followed by space vector model of a five-phase voltage source inverter. The PWM control of a five-phase voltage source inverter is elaborated on. The vector control principle of a five-phase induction motor in conjunction with the current control in the stationary reference frame and the synchronously rotating reference frame is presented. Finite-state model predictive control applied to a five-phase voltage source inverter for current control is also presented. DTC of a five-phase induction motor is presented. The simulation models of a five-phase induction motor and five-phase voltage source inverter are illustrated using MATLAB/Simulink.

Chapter 8 describes the speed sensorless operation of high-performance drive systems. Speed sensors are the most delicate component of a drive system, which are susceptible to faults and malfunctioning. A more robust drive system is obtained by replacing the physical speed sensors with the observer system that computes the speed and uses the information for the closed-loop control. Several observer systems and their tuning are elaborated on in this chapter. A model reference adaptive-speed estimator system for a three-phase and a five-phase induction machine is described along with their simulation model. The sensorless control scheme of a permanent magnet synchronous machine is also discussed. Model reference adaptive-speed estimator system for a three-phase PMSM is also illustrated.

Nowadays, electric drives with induction motors and voltage source-type inverters are commonly used as adjustable-speed drives in industrial systems. The inverters are built with the insulated gate bipolar transistors, IGBT, whose dynamic parameters are high, i.e. the on- and off-switch times are extremely short. Fast switching of power devices causes high dv/dt at the rising and falling edges of the inverter output waveform. High dv/dt in modern inverters is the source of numerous disadvantageous effects in the drive systems. The main negative effects are faster motor bearings degradation, overvoltages on motor terminals, failure or degradation of the motor winding insulation due to partial discharges, an increase of motor losses, and a higher EMI level. The prevention or limiting of the negative effects of dv/dt is possible if proper passive or active filers are installed in the drive. Particularly passive filters are preferable for industrial applications. This issue is described in detail in Chapter 9. The problems due to the use of passive LC filters at the output of the inverter and their solutions are discussed.

Medium voltage drives have found extensive applications in several industries, such as in the oil and gas, petrochemical, mining, water/waste, pulp/paper, cement, chemical, power generation, metal production and processes, traction, and marine drive sectors. To improve power quality, system response and to reduce operation cost and energy loss, the installed MV drives should be adjustable-speed drives (ASD). The state of the art and the challenges posed by MVD are discussed in Chapter 10.

Chapter 11 describes the current source inverter-fed drives. The structure of the current source inverter based on IGBT transistors is presented in this chapter. The PWM strategy of the inverter output current is given. Special attention is paid to the control system of induction machine supplied by current source inverter with open loop as well as direct field control.

Chapter 1 was prepared by Atif Iqbal with help from Haitham Abu-Rub. Chapters 2, 4, and 8 were mainly prepared by Haitham Abu-Rub with help from Jaroslaw Guzinski. Chapter 6 was prepared by Zbigniew Krzeminski and Haitham Abu-Rub. Chapters 3 and 7 are the responsibility of Atif Iqbal, whereas Section 3.9 was written by Arkadiusz Lewicki and Section 3.10 was written by Arkadiusz Lewicki and Marcin Morawiec.

Chapter 5 was written by Truc Pham-dinh, and Chapter 9 was written by Jaroslaw Guzinski with help from Haitham Abu-Rub. Chapter 10 was written by Haitham Abu-Rub, Sertac Bayham, Shaikh Moinoddin, and Mariusz Malinowski. Chapter 11 was written by Marcin Morawiec and Arkadiusz Lewicki. All the chapters were revised by an English expert from the research team of Haitham Abu-Rub.

References

1. http://www.sparkmuseum.com/MOTORS.HTM (accessed 26 November 2020).
2. https://patents.google.com/patent/US381968A/en
3. Mecrow, B. C. and Jack, A. G. (2008) Efficiency trends in electric machines and drives. *Energy Policy*, 36, 4336–4341.
4. Zhu, Z. Q. (2011) Recent advances on permanent magnet machines including IPM technology. *Keynote Lect. IEEE IEMDC*, Niagara Falls, Canada, 15–18 May2011.
5. Jahns, T. M. and Owen, E. L. (2001) AC adjustable-speed drives at the millennium: How did we get here? *IEEE Trans. Power Electron.*, 6(1), 17–25.
6. Rahman, M. A. (1993) Modern electric motors in electronic world. 0-7803-0891-3/93, pp. 644–648.
7. Lorenz, R. D. (1999) Advances in electric drive control. *Proc. Int. Conf. on Elec. Mach. Drives IEMD*, Seattle, Washington, USA; 9–12 May 1999, pp. 9–16.
8. Rahman, M. A. (2005) Recent advances of IPM motor drives in power electronics world. Proc. IEEE Int. Conf. Power Elect. Drives Syst., PEDES, Kuala Lumpur, Malaysia, 28 November to 1 December 2005, pp. 24–31.
9. De Doncker, R. W. (2006) Modern electrical drives: Design and future trends. Proc. IPEMC-2006, Shanghai, 14–16 August, pp. 1–8.
10. Finch, J. W. and Giaouris, D. (2008) Controlled AC electrical drives. *IEEE Trans. Ind. Electron.*, 55(2), 481–491.
11. Toliyat, H. A. (2008) Recent advances and applications of power electronics and motor drives: Electric machines and motor drives. Proc. 34th IEEE Ind. Elect. Conf., IECON, Orlando, FL; 10–13 November 2008, pp. 34–36.
12. Bose, B. K. (1998) Advances in power electronics and drives: Their impact on energy and environment. Proc. Int. Conf. on Power Elect. Drives Ener. Syst. Ind. Growth, PEDES, vol. 1, Pretoria, South Africa; 7–10 July 1998.
13. Jahns, T. M. and Blasko, V. (2001) Recent advances in power electronics technology for industrial and traction machine drives. *Proc. IEEE*, 89(6), 963–975.
14. Bose, B. K. (2005) Power electronics and motor drives: Technology advances, trends and applications. Proc. IEEE Int. Conf. on Ind. Tech., ICIT, Bled, Slovenia; 12–16 July 1999, pp. 20–26.
15. Bose, B. K. (2008) Recent advances and applications of power electronics and motor drives: Introduction and perspective. Proc. IEEE Ind. Elect. Conf., Orlando, FL, USA; 10–13 November 2008, pp. 25–27.
16. Capilino, G. A. (2008) Recent advances and applications of power electronics and motor drives: Advanced and intelligent control techniques. Proc. IEEE Ind. Elect. Conf., IECON, Orlando, FL, USA; 10–13 November 2008, pp. 37–39.
17. Lin, F-J., Hwang, J-C., Tan, K-H., Lu, Z-H., and Chang, Y-R. (2010) Control of double-fed induction generator system using PIDNNs. 9th Int. Conf. Mach. Learn. Appl., Washington, DC, USA; 12–14 December 2010, pp. 675–680.
18. Muller, S., Deicke, M., and De Doncker, R. W. (2002) Doubly-fed induction generator systems for wind turbines. *IEEE Ind. Appl. Mag.*, 8(3), 26–33.
19. Bogalecka, E. (1993) Power control of a non-linear induction generator without speed or position sensor. *Conf. Rec. EPE*, 377(8), 224–228.
20. Jain, A. K. and Ranganathan, V. T. (2008) Wound rotor induction generator with sensorless control and integrated active filter for feeding non-linear loads in a stand-alone grid. *IEEE Trans. Ind. Electron.*, 55 (1), 218–228.

21. Iwanski, G. and Koczara, W. (2008) DFIG-based power generation system with UPS function for variable-speed applications. *IEEE Trans. Ind. Electron.*, 55(8), 3047–3054.
22. Pena, R., J. Clare, J. C., and Asher, G. M. (1996) Doubly fed induction generator using back-to-back PWM converters and its application to variable-speed wind-energy generation. *IEEE Proc. Elect. Power Appl.*, 143(3), 231–241.
23. Forchetti, D., Garcia, G., and Valla, M. I. (2002) Vector control strategy for a doubly-fed stand-alone induction generator. Proc. 28th IEEE Int. Conf., IECON., vol. 2, Sevilla, Spain; 5–8 Nov. 2002, pp. 991–995.
24. Pena, R., Clare, J. C., and Asher, G. M. (1996) A doubly fed induction generator using back-to-back PWM converters supplying an isolated load from a variable speed wind turbine. *IEEE Proc. Elect. Power Appl.*, 143(5), 380–387.
25. Wilamowski, B. M. and Irwin, J. D. (2011) The Industrial Electronics Handbook. CRC Press, Taylor & Francis Group, New York.
26. Forest, F., Labourı, E., Meynard, T. A., and Smet, V. A. (2009) Design and comparison of inductors and inter cell transformers for filtering of PWM inverter output. *IEEE Trans. Power Electron.*, 24(3), 812–821.
27. Shen, G., Xu, D., Cao, L., and Zhu, X. (2008) An improved control strategy for grid-connected voltage source inverters with an LCL filter. *IEEE Trans. Power Electron.*, 23(4), 1899–1906.
28. Gabe, I. J., Montagner, V. F., and Pinheiro, H. (2009) Design and implementation of a robust current controller for VSI connected to the grid through an LCL filter. *IEEE Trans. Power Electron.*, 24(6), 1444–1452.
29. Kojima, M., Hirabayashi, K., Kawabata, Y., Ejiogu, E. C., and Kawabata, T. (2004) Novel vector control system using deadbeat-controlled PWM inverter with output LC filter. *IEEE Trans. Ind. Appl.*, 40(1), 162–169.
30. Pasterczyk, R. J., Guichon, J-M., Schanen, J-L., and Atienza, E. (2009) PWM inverter output filter cost-to-losses trade off and optimal design. *IEEE Trans. Ind. Appl.*, 45(2), 887–897.
31. Bhattacharya, S. S., Deprettere, F., Leupers, R., and Takala, J. (2010) Handbook of Signal Processing Systems. Springer.
32. https://www.ti.com/product/TMS320C25?keyMatch=TMS32010&tisearch=Search-EN-everything (accessed 1 December 2020).
33. Rich, N. (2010) 'Xilinx puts ARM core into its FPGAs.' EE times. Available from, http://www. http://eetimes.com/electronics-products/processors/4115523/Xilinx-puts-ARM-core-into-its-FPGAs (accessed 27 April 2010).
34. Leonhard, W. (1996) Control of Electrical Drives, 2nd edn. Springer-Verlag.
35. Vas, P. (1998) Sensorless Vector and Direct Torque Control. Oxford University Press, London.
36. Vas, P. (1999) Artificial Intelligence Based Electric Machine and Drives: Application of Fuzzy, Neural, Fuzzy-Neural and Genetic Algorithm Based Techniques. Oxford University Press, Oxford.
37. Daryabeigi, E., Markadeh, G. R. A., and Lucas, C. (2010) Emotional controller (BELBIC) for electric drives: A review. Proc. IEEE IECON-2010, Glendale, AZ, USA; 7–10 November 2010, pp. 2901–2907.
38. Krzeminski, Z. (1987). Non-linear control of induction motor. IFAC 10th World Congr. Auto. Cont., Munich, 27–31 July 1987, pp. 349–354.
39. Abu-Rub, H., Krzemisnki, Z., and Guzinski, J. (2000) Non-linear control of induction motor: Idea and application. EPE–PEMC (9th Int. Power Elect. Mot. Cont. Conf.), Kosice/Slovac Republic, vol. 6, pp. 213–218.
40. Holtz, J. (2006) Sensorless control of induction machines: With or without signal injection? *IEEE Trans. Ind. Electron.*, 53(1), 7–30.
41. Holtz, J. (2002) Sensorless control of induction motor drives. *Proc. IEEE*, 90(8), 1359–1394.

42. Acarnley, P. P. and Watson, J. F. (2006) Review of position-sensorless operation of brushless permanent-magnet machines. *IEEE Trans. Ind. Electron.*, 53(2), 352–362.
43. Ward, E. E. and Harer, H. (1969) Preliminary investigation of an inverter-fed 5-phase induction motor. *Proc. IEEE*, 116(6), 980–984.
44. Levi, E. (2008) Multiphase electric machines for variable-speed applications. *IEEE Trans. Ind. Electron.*, 55(5), 1893–1909.
45. Abu-Rub, H., Iqbal, A., and Guzinski, J. (2010) Medium voltage drives: Challenges and requirements. IEEE Int. Symp. on Ind. Elect., ISIE 2010, 4–7 July, Bari, Italy, pp. 1372–1376.
46. Guzinski, J., Abu-Rub, H., and Strankowski, P. (2015) Variable Speed AC Drives with Inverter Output Filters. Wiley, Hoboken, NJ.
47. Bindra, A. (2015) Wide-bandgap-based power devices: Reshaping the power electronics landscape. *IEEE Power Electron. Mag.*, 2(1), 42–47.
48. Passmore, B. and O'Neal, C. Wolfspeed High-voltage SiC power modules for 10–25 kV applications. Available from, http://www.powermag.com/pdf/feature_pdf/1461163294_Woifspeed_Feature.pdf.
49. Mohamed, A., EL-Refaie, F., and Robert, D. K. (2012) Lowinductance, high-efficiency induction machine and method of making same, US Patent 20120126741 A1, May 24, 2012.
50. Gerada, D., Mebarki, A., Brown, N. L., Gerada, C., Cavagnino, A., and Boglietti, A. (2014) High-speed electrical machines: Technologies, trends, and developments. *IEEE Trans. Ind. Electron.*, 61(6), 2946–2959.
51. Hornberger, J., Lostetter, A. B., Olejniczak, K. J., McNutt, T., Lal, S. M., and Mantooth, A. (2004) Silicon-carbide (SiC) semiconductor power electronics for extreme high-temperature environments. Proc. 2004 IEEE Aerosp. Conf. (IEEE Cat. No.04TH8720), March 2004, Big Sky, MT, vol. 4, pp. 2538–2555.
52. Wrzecionko, B., Bortis D., and Kolar, J. W. (2014) A 120 °C ambient temperature forced air-cooled normally-off SiC JFET automotive inverter system. *IEEE Trans. Power Electron.*, 29(5), 2345–2358.
53. Frank, E. and CTO of ThinGap (2016) Efficient control of low inductance permananet magnet motors. UCSB IEE Conference 2016. Available from, http://iee.ucsb.edu/sites/iee.ucsb.edu/files/evan_frank.pdf.
54. Thin gap Mortos. https://www.thingap.com/standard-products/ (accessed 1 December 2020).
55. Krah, J. O. and Holtz, J. (1999) High-performance current regulation and efficient PWM implementation for low-inductance servo motors. *IEEE Trans. Ind. Appl.*, 35(5), 1039–1049.
56. De, S., Rajne, M., Poosapati, S., Patel, C., and Gopakumar, K. (2012) Low inductance axial flux BLDC motor drive for more electric aircraft. *IET Power Electron.*, 5(1), 124–133.
57. Mitsubishi electric develops EV motor drive system with built-in silicon carbide inverter, Mitsubishi Electric Corporation [Online]. Available from, http://www.mitsubishielectric.com/news/2014/pdf/0213-d.pdf.
58. KSB Motor Prototype demonstrates potential. Available from, https://www.expresswater.in/news/ksb-motor-prototype-demonstratespotential.
59. John Deere, Power America project SiC inverter for heavy-duty vehicles. Available from, https://www.poweramericainstitute.org/wpcontent uploads/2017/02/John-Deere.pdf.
60. U.S. Department of Energy (DOE), Office of Energy Efficiency and Renewable Energy (EERE) (2017) Medium voltage integrated drive and motor. Available from, https://www.energy.gov/sites/prod/files/2017/03/f34/Medium%20Voltage%20Integrated%20Drive%20and%20Motor_0.pdf.
61. PntPower.com (2017) About the SiC MOSFET modules in Tesla model 3. Available from, https://www.pntpower.com/tesla-model-3-powered-by-stmicroelectronics-sic-mosfets (accessed 9 July 2017).
62. Rohm Semiconductor (2013) SiC power devices and modules. Application note 13103EAY
63. Rujas, A., López, V. M., Mir, L., and Nieva, T. (2018) Gate driver for high power SiC modules: Design considerations, development and experimental validation. *IET Power Electron.*, 11(6), 977–983.

2

Mathematical and Simulation Models of AC Machines

2.1 Preliminary Remarks

This chapter describes the basic modeling procedures of power electronic converters and AC machines. The mathematical model thus obtained will be used to develop simulation models using MATLAB/Simulink. The standard approach of mathematical modeling will be described in addition to advanced modeling procedures such as the signal flow graph.

The modeling and simulation of AC machines will be presented. DC motors will be briefly modeled and discussed for educational benefit; those machines are currently rarely used and analyzed, and therefore will not be discussed in depth. The AC machines that will be discussed are the most popular three-phase induction machine (squirrel cage and rotor-wound) and three-phase permanent magnet synchronous machine. Machine modeling is a first important step in better understanding of the topic. The modeling of these machines will introduce the reader to the basics for understanding and analyzing the advanced control techniques of electric drives, such as 'field-oriented control,' 'direct torque control,' 'feedback linearization control,' 'predictive control,' and 'sensorless operation.'

The modeling of multiphase machines and IM with inverter output filters will be discussed in later chapters. This also applies to non-linear models of AC machines.

2.2 DC Motors

This section presents the modeling of separately excited and series DC motors, to provide educational benefits rather than for practical use. The mathematical models AC and DC machines can be found in [1–5].

High Performance Control of AC Drives With MATLAB®/Simulink, Second Edition.
Haitham Abu-Rub, Atif Iqbal, and Jaroslaw Guzinski.

Companion website: www.wiley.com/go/aburubcontrol2e

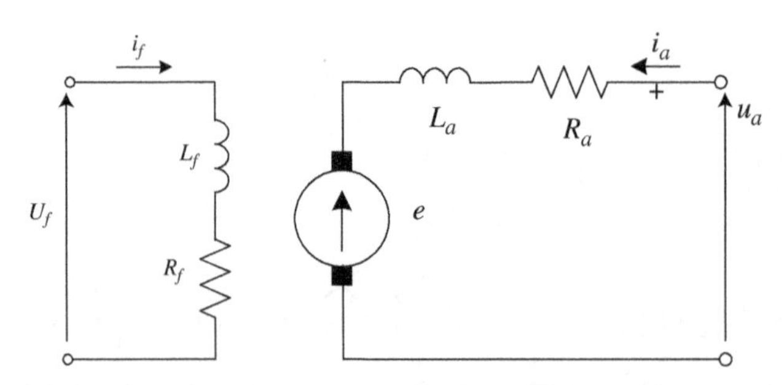

Figure 2.1 An equivalent circuit of a separately excited DC motor

2.2.1 Separately Excited DC Motor Control

An equivalent circuit of a separately excited DC motor is shown on Figure 2.1. The armature side of the motor is modeled using an ideal voltage source (back EMF) and an armature resistance. The excitation circuit is represented by a field resistor and an inductor. The two circuits are fed by separate voltage sources.

The mathematical model of the separately excited DC motor can be represented as [1–5]

$$u_a = R_a \cdot i_a + L_a \cdot \frac{di_a}{dt} + e \tag{2.1}$$

$$u_f = R_f \cdot i_f + \frac{d\Psi_f}{dt} \tag{2.2}$$

$$J \cdot \frac{d\omega_r}{dt} = t_e - t_l \tag{2.3}$$

where u_a, u_f, i_a, and i_f are armature voltage, field voltage, armature current, and field current, respectively; R_a and R_f are armature and field resistances, respectively; L_a is the armature inductance; J is the machine inertia; ω_r is the rotor angular speed; Ψ_f is the field flux; e is the electromagnetic force induced in the armature; and t_l is the load torque.

The induced voltage e and motor torque (t_e) in the motor are given by

$$e = c_E \cdot i_f \cdot \omega_r \tag{2.4}$$

$$t_e = c_M \cdot \Psi_f \cdot i_a \tag{2.5}$$

Assuming a linear magnetic path, the flux is

$$\Psi_f = L_f \cdot i_f \tag{2.6}$$

Using the above equations, the mathematical model presented as differential equations of state is described as

$$\frac{di_a}{dt} = \frac{1}{L_a} \cdot u_a - \frac{R_a}{L_a} \cdot i_a - \frac{c_e \cdot i_f \cdot \omega_r}{L_a} \tag{2.7}$$

$$\frac{di_f}{dt} = \frac{u_f - R_f \cdot i_f}{L_f} \tag{2.8}$$

$$\frac{d\omega_r}{dt} \frac{1}{J} \left(c_M \cdot L_f \cdot i_f \cdot i_a - t_l \right) \tag{2.9}$$

Let us now assume the next nominal values:

$$T_m = \frac{J \cdot \omega_{rn}}{t_n} \tag{2.10}$$

$$T_a = \frac{L_a}{R_a} \tag{2.11}$$

$$T_f = \frac{L_f}{R_f} \tag{2.12}$$

$$K_1 = \frac{u_{na}}{i_{na}} \cdot \frac{1}{R_a} \tag{2.13}$$

$$K_2 = \frac{u_{nf}}{i_{nf}} \cdot \frac{1}{R_f} \tag{2.14}$$

$$K_3 = \frac{E_n}{i_n \cdot R_a} \tag{2.15}$$

where u_{na}, u_{nf}, i_{na}, and i_{nf} are rated values of armature voltage, field voltage, armature current, and field current, respectively; t_n is the rated torque; ω_{rn} is the rated speed; and the rated induced voltage is

$$E_n = c_E \cdot i_{nf} \cdot \omega_{rn} \tag{2.16}$$

This voltage can be also presented as

$$E_n = u_{na} - R_a \cdot i_{na} \tag{2.17}$$

The per unit model of the motor is

$$\frac{di_a}{dt} = K_1 \cdot \frac{1}{T_a} \cdot u_a - \frac{1}{T_a} \cdot i_a - \frac{K_a \cdot i_f \cdot \omega_r}{T_a} \tag{2.18}$$

$$\frac{di_f}{dt} = K_2 \cdot \frac{u_f}{T_f} - \frac{i_f}{T_f} \tag{2.19}$$

$$\frac{d\omega_r}{dt} = \frac{1}{T_m} \left(i_f \cdot i_a - t_l \right) \tag{2.20}$$

The Simulink sub-blocks of the motor model are shown in Figures 2.2–2.4.

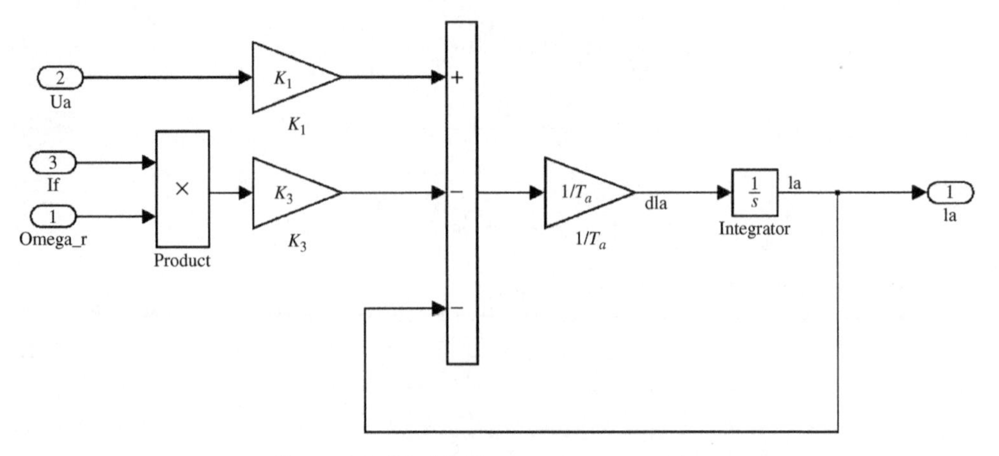

Figure 2.2 Model of an armature current loop

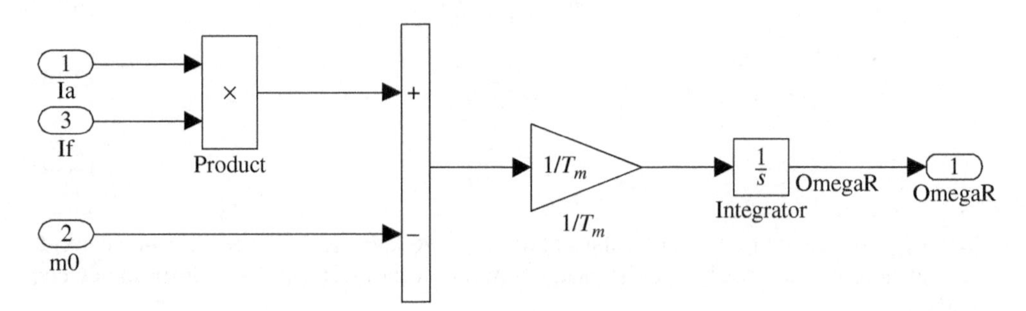

Figure 2.3 Model of a rotor speed loop

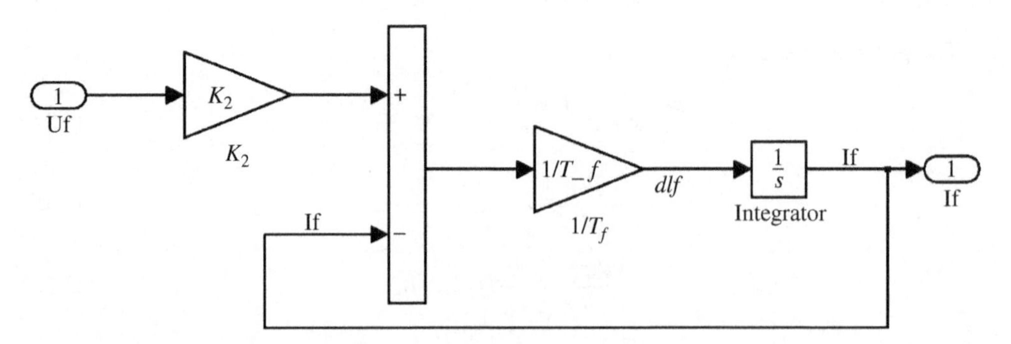

Figure 2.4 Model of a field current loop

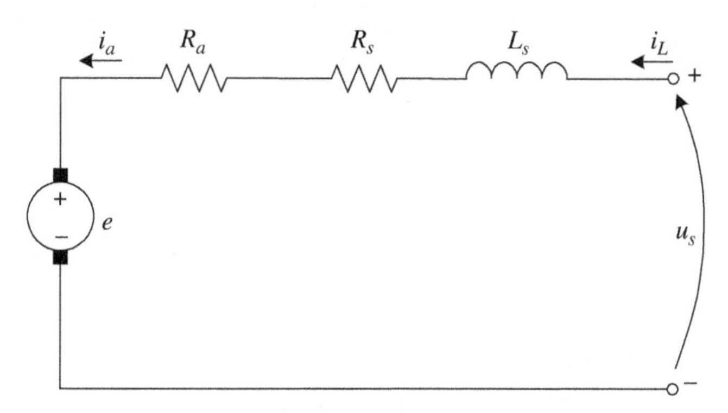

Figure 2.5 An equivalent circuit of a series DC motor

2.2.2 *Series DC Motor Control*

An equivalent circuit of a series DC motor is shown in Figure 2.5. The armature side of the motor is connected in series with the field side.

The mathematical model of a series DC motor is

$$\frac{di}{dt} = \frac{1}{L}(u_s - i_a \cdot R - e) \tag{2.21}$$

$$\frac{d\omega_r}{dt} = \frac{1}{J}(t_e - t_l - t_t) \tag{2.22}$$

where u_s is the voltage source, i_a is the armature current, t_l is the load torque, and J is the moment of inertia. The electromagnetic torque t_e is

$$t_e = c \cdot i^2 \tag{2.23}$$

where c is the motor constant and the t_t friction torque is

$$t_t = B \cdot \omega_r \tag{2.24}$$

The back EMF is

$$e = c \cdot i \cdot \omega_r \tag{2.25}$$

The resistance R is

$$R = R_f + R_a \tag{2.26}$$

The inductance L is

$$L = L_f + L_a \tag{2.27}$$

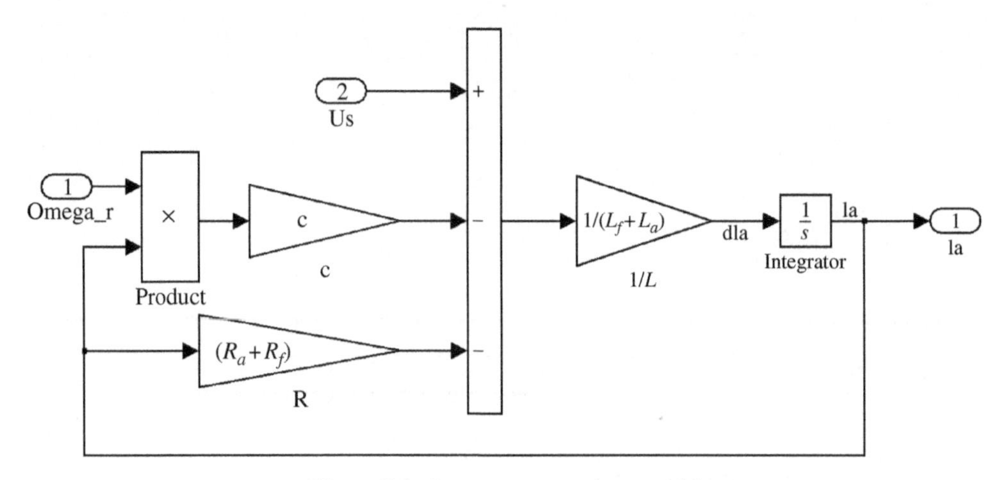

Figure 2.6 Armature current loop model

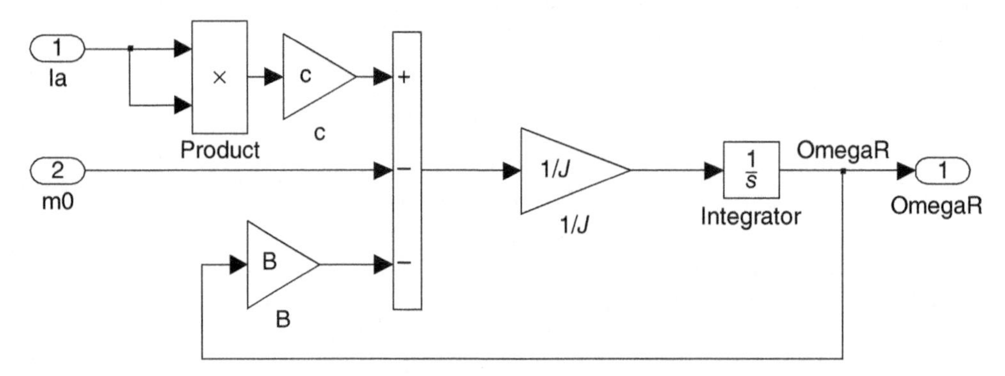

Figure 2.7 Rotor speed loop model

The DC series motor model, consisting of an armature current subsystem and a rotor speed subsystem, is shown in Figures 2.6 and 2.7.

2.3 Squirrel Cage Induction Motor

2.3.1 Space Vector Representation

A three-phase AC machine may be described using the space vector method [1–9], as was shown by Kovacs and Racz [6]. For this reason, the AC motor variables $K_A(t)$, $K_B(t)$, $K_C(t)$ for a symmetrical machine fulfill the condition:

$$K_A(t) + K_B(t) + K_C(t) = 0 \tag{2.28}$$

The summation of these variables gives the space vector:

$$K = \frac{2}{3}\left[K_A(t) + aK_B(t) + a^2K_C(t)\right] \tag{2.29}$$

Where $a = e^{j\frac{2}{3}\pi}$ and $a^2 = e^{j\frac{4}{3}\pi}$

Figure 2.8 shows the stator current complex space vector.

The space vector $\boldsymbol{K}$ may represent the motor variables (e.g. current, voltage, and flux). The vector control principle on AC machines takes advantages of transforming the variables from the physical three-phase ***ABC*** system to a stationary frame $\boldsymbol{\alpha\beta}$, or rotating frame $\boldsymbol{dq}$ [3, 6].

Let us analyze, as an example, the stator sinusoidal currents $[i_{SA}, i_{SB}, i_{SC}]$ transformation. The first transformation can be done to a stationary frame $\alpha\beta$ (Clark transformation) and then to a rotating frame dq (Park transformation) [1–9].

2.3.2 Clarke Transformation (ABC to $\alpha\beta$)

The sinusoidal three-phase variables can be represented as a space vector expressed on a two-orthogonal axis ($\alpha\beta$), as shown in Figure 2.9.

The stator current as a vector is described by the complex form:

$$\bar{i}_s = i_{s\alpha} + ji_{s\beta} \tag{2.30}$$

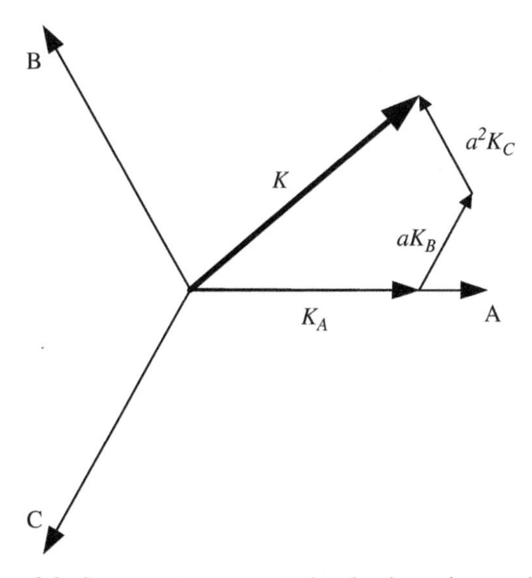

Figure 2.8 Space vector representation for three-phase variables

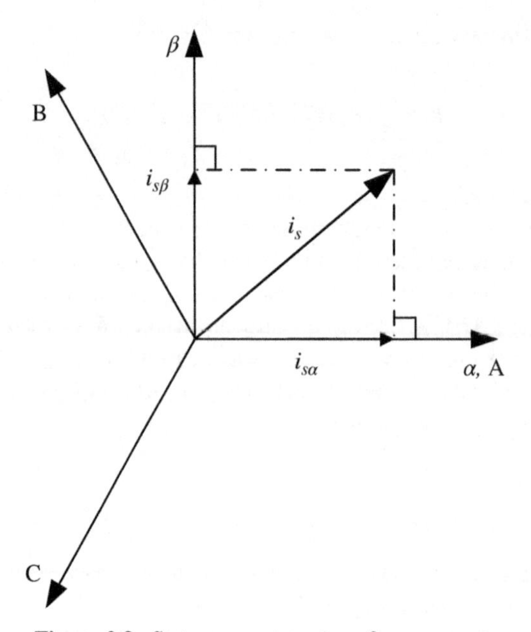

Figure 2.9 Stator current vector $\alpha\beta$ components

where

$$i_{s\alpha} = \text{Re}\left\{\frac{2}{3}\left[i_{sA} + ai_{sB} + a^2 i_{sC}\right]\right\} \tag{2.31}$$

$$i_{s\beta} = \text{Im}\left\{\frac{2}{3}\left[i_{sA} + ai_{sB} + a^2 i_{sC}\right]\right\} \tag{2.32}$$

then

$$i_{s\alpha} = i_{sA} \tag{2.33}$$

$$i_{s\beta} = \frac{1}{\sqrt{3}}(i_{sA} + 2i_{sB}) \tag{2.34}$$

This is equivalent to

$$\begin{bmatrix} i_{s\alpha} \\ i_{s\beta} \end{bmatrix} = \begin{bmatrix} 1 & 0 & 0 \\ \frac{1}{\sqrt{3}} & \frac{2}{\sqrt{3}} & 0 \end{bmatrix} \begin{bmatrix} i_{sA} \\ i_{sB} \\ i_{sC} \end{bmatrix} \tag{2.35}$$

The inverse Clarke transformation from $\alpha\beta$ to ABC is

$$\begin{bmatrix} i_{sA} \\ i_{sB} \\ i_{sC} \end{bmatrix} = \begin{bmatrix} 1 & 0 \\ -\frac{1}{2} & -\frac{\sqrt{3}}{2} \\ -\frac{1}{2} & \frac{\sqrt{3}}{2} \end{bmatrix} \begin{bmatrix} i_{s\alpha} \\ i_{s\beta} \end{bmatrix} \tag{2.36}$$

The above transformation does not take into account zero components, and it concedes maintaining variable vector length. It is also possible to have a transformation that maintains system power.

The matrix of variable transformation from the three-phase system ABC to the stationary $\alpha\beta 0$ system with retaining vector length is

$$A_W = \begin{bmatrix} \frac{1}{3} & \frac{1}{3} & \frac{1}{3} \\ \frac{2}{3} & -\frac{1}{3} & -\frac{1}{3} \\ 0 & \frac{1}{\sqrt{3}} & -\frac{1}{\sqrt{3}} \end{bmatrix} \tag{2.37}$$

where subscript W denotes transformation with retaining vector length.

The matrix of variable transformation from $\alpha\beta 0$ to ABC with retaining vector length is

$$\mathbf{A}_W^{-1} = \begin{bmatrix} 1 & 1 & 0 \\ 1 & -\frac{1}{2} & \frac{\sqrt{3}}{2} \\ 1 & -\frac{1}{2} & -\frac{\sqrt{3}}{2} \end{bmatrix} \tag{2.38}$$

The matrix of variable transformation from the three-phase system ABC to the stationary $\alpha\beta 0$ system with retaining system power is

$$\mathbf{A}_P = \begin{bmatrix} \frac{1}{\sqrt{3}} & \frac{1}{\sqrt{3}} & \frac{1}{\sqrt{3}} \\ \frac{\sqrt{2}}{\sqrt{3}} & -\frac{1}{\sqrt{6}} & -\frac{1}{\sqrt{6}} \\ 0 & \frac{1}{\sqrt{2}} & -\frac{1}{\sqrt{2}} \end{bmatrix} \tag{2.39}$$

where $\mathbf{A}_P$ denotes transformation with retaining system power.

The matrix of variable transformation from $\boldsymbol{\alpha\beta 0}$ to $\boldsymbol{ABC}$ with retaining system power is

$$\mathbf{A}_P^{-1} = \begin{bmatrix} \frac{1}{\sqrt{3}} & \frac{\sqrt{2}}{\sqrt{3}} & 0 \\ \frac{1}{\sqrt{3}} & -\frac{1}{\sqrt{6}} & \frac{\sqrt{2}}{2} \\ \frac{1}{\sqrt{3}} & -\frac{1}{\sqrt{6}} & -\frac{\sqrt{2}}{2} \end{bmatrix} \tag{2.40}$$

Variable transformation from one reference frame to another is done according to the procedures:

- For transformation with retaining vector length:

$$\begin{bmatrix} x_0 \\ x_\alpha \\ x_\beta \end{bmatrix} = \mathbf{A}_W \begin{bmatrix} x_A \\ x_B \\ x_C \end{bmatrix} \tag{2.41}$$

$$\begin{bmatrix} x_A \\ x_B \\ x_C \end{bmatrix} = \mathbf{A}_W^{-1} \begin{bmatrix} x_0 \\ x_\alpha \\ x_\beta \end{bmatrix} \tag{2.42}$$

- For transformation with retaining system power:

$$\begin{bmatrix} x_0 \\ x_\alpha \\ x_\beta \end{bmatrix} = \mathbf{A}_P \begin{bmatrix} x_A \\ x_B \\ x_C \end{bmatrix} \tag{2.43}$$

$$\begin{bmatrix} x_A \\ x_B \\ x_C \end{bmatrix} = \mathbf{A}_P^{-1} \begin{bmatrix} x_0 \\ x_\alpha \\ x_\beta \end{bmatrix} \tag{2.44}$$

where x denotes arbitrary variable in specific reference system.

2.3.3 *Park Transformation ($\alpha\beta$ to dq)*

This transformation aims to project the two phase components $\boldsymbol{\alpha\beta}$ in stationary frame to the $\boldsymbol{dq}$ rotating reference frame rotating with angular speed ω_k. If we consider the $\boldsymbol{d}$-axis aligned with the rotor flux in axis d, then we speak about rotor flux-oriented system. As an example, stator current space vector and its component in (α,β) and in the (d,q) are shown in Figure 2.10.

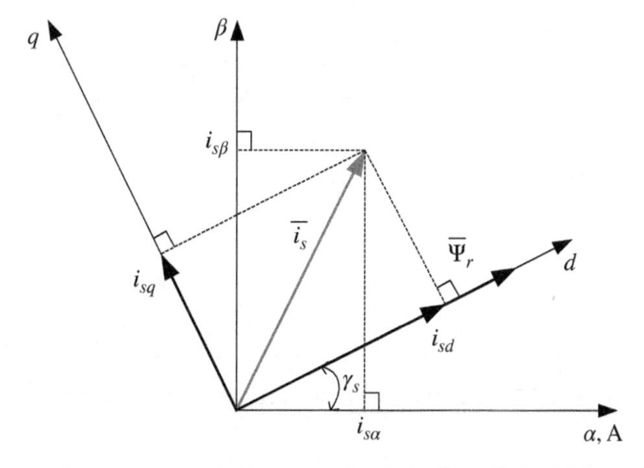

Figure 2.10 Stator current space vector and its component in (α, β) and in the ($d\,q$) rotating reference frame

The current vector in dq frame is

$$\overline{i}_s = i_{sd} + ji_{sq} \tag{2.45}$$

which is equivalent to

$$\overline{i}_s = \left(i_{s\alpha}\cos(\gamma_s) + i_{s\beta}\sin(\gamma_s)\right) + j\left(i_{s\beta}\cos(\gamma_s) - i_{s\alpha}\sin(\gamma_s)\right) \tag{2.46}$$

Following this:

$$\overline{i}_s = \begin{bmatrix} i_{sd} \\ i_{sq} \end{bmatrix} = \begin{bmatrix} \cos(\gamma_s) & \sin(\gamma_s) \\ -\sin(\gamma_s) & \cos(\gamma_s) \end{bmatrix} \begin{bmatrix} i_{s\alpha} \\ i_{s\beta} \end{bmatrix} \tag{2.47}$$

The inverse Park transformation is

$$\overline{i}_s = \begin{bmatrix} i_{s\alpha*} \\ i_{s\beta*} \end{bmatrix} = \begin{bmatrix} \cos(\gamma_s) & -\sin(\gamma_s) \\ \sin(\gamma_s) & \cos(\gamma_s) \end{bmatrix} \begin{bmatrix} i_{sd*} \\ i_{sq*} \end{bmatrix} \tag{2.48}$$

The transformation may be applied to all space vectors (e.g. stator voltage, rotor flux, etc.).

2.3.4 *Per Unit Model of Induction Motor*

A per-phase equivalent circuit of an induction motor is shown in Figure 2.11. It consists of stator side resistance and leakage inductance, mutual inductance, rotor side resistance, inductance, and induced voltage.

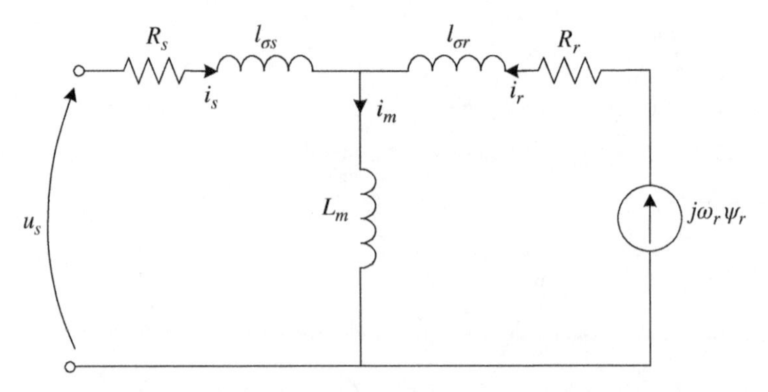

Figure 2.11 Per-phase equivalent circuit of induction motor

The induction motor in arbitrary reference frame ***K*** can be presented per unit rotating angular speed ω_k as

$$u_{sk} = R_{sk} i_{sk} + \frac{d\psi_{sk}}{dt} + j\omega_k \psi_{sk} \tag{2.49}$$

$$u_{rk} = R_{rk} i_{rk} + \frac{d\psi_{rk}}{dt} + j(\omega_k - \omega_r)\psi_{sk} \tag{2.50}$$

$$\psi_{sk} = L_s i_{sk} + L_m i_{rk} \tag{2.51}$$

$$\psi_{rk} = L_r i_{rk} + L_m i_{sk} \tag{2.52}$$

$$\frac{d\omega_r}{d\tau} = \frac{L_m}{JL_r}[\mathrm{Im}(\psi_{rk} * i_{sk}) - t_0]. \tag{2.53}$$

where T_M is the mechanical time constant; where u_s, i_s, i_r, ψ_s, ψ_r are voltage, current, and flux (stator and rotor) vectors, respectively; R_s and R_r are stator and rotor resistances; ω_r is the rotor angular speed; ω_a is an angular speed of reference frame; J is a moment of inertia; and t_0 is the load torque; L_m is the mutual inductance.

The currents relations are:

$$i_s = \frac{1}{L_s}\Psi_s - \frac{L_m}{L_s} i_r, \tag{2.54}$$

$$i_r = \frac{1}{L_r}\Psi_r - \frac{L_m}{L_r} i_s, \tag{2.55}$$

The induction motor (IM) model presented per unit in the rotating frame with arbitrary speed is given by [10, 11]

$$\frac{dx_{sx}}{d\tau} = -\frac{R_s L_r^2 + R_r L_m^2}{L_r \omega_\sigma} \cdot i_{sx} + \frac{R_r L_m}{L_r \omega_\sigma} \cdot \psi_{rx} + \omega_k \cdot i_{sy} + \omega_r \frac{L_m}{\omega_\sigma} \cdot \psi_{ry} + \frac{L_r}{\omega_\sigma} u_{sx} \tag{2.56}$$

$$\frac{dx_{sy}}{d\tau} = -\frac{R_s L_r^2 + R_r L_m^2}{L_r \omega_\sigma} \cdot i_{sy} + \frac{R_r L_m}{L_r \omega_\sigma} \cdot \psi_{ry} - \omega_k \cdot i_{sx} - \omega_r \frac{L_m}{\omega_\sigma} \cdot \psi_{rx} + \frac{L_r}{\omega_\sigma} u_{sy} \tag{2.57}$$

$$\frac{dx_{rx}}{d\tau} = -\frac{R_r}{L_r} \cdot \psi_{rx} + (\omega_k - \omega_r) \cdot \psi_{ry} + \frac{R_r L_m}{L_r} \cdot i_{srx} \tag{2.58}$$

$$\frac{dx_{ry}}{d\tau} = -\frac{R_r}{L_r} \cdot \psi_{ry} - (\omega_k - \omega_r) \cdot \psi_{rx} + \frac{R_r L_m}{L_r} \cdot i_{sy} \tag{2.59}$$

$$\frac{d\omega_r}{d\tau} = \frac{L_m}{JL_r} \left(\Psi_{rx} i_{sy} - \Psi_{ry} i_{sx} \right) - \frac{1}{J} t_o \tag{2.60}$$

Here, u_s, i_s, i_r, ψ_s, ψ_r are voltage, current, and flux (stator and rotor) vectors, respectively; R_s and R_r are stator and rotor resistances, respectively; ω_r is the rotor angular speed; ω_k is the angular speed of reference frame; J is the moment of inertia; and t_0 is the load torque.

The mathematical model of IM as differential equations of state variables presented in $(\alpha\beta)$ stationary reference frame and $(\omega_k = 0)$ is [10]

$$\frac{di_{s\alpha}}{dt} = a_1 \cdot i_{s\alpha} + a_2 \cdot \Psi_{r\alpha} + \omega_r \cdot a_3 \cdot \Psi_{r\beta} + a_4 \cdot u_{s\alpha} \tag{2.61}$$

$$\frac{di_{s\beta}}{dt} = a_1 \cdot i_{s\beta} + a_2 \cdot \Psi_{r\beta} - \omega_r \cdot a_3 \cdot \Psi_{r\alpha} + a_4 \cdot u_{s\beta} \tag{2.62}$$

$$\frac{d\Psi_{r\alpha}}{dt} = a_5 \cdot \Psi_{r\alpha} + (-\omega_r) \cdot \Psi_{r\beta} + a_6 \cdot i_{s\alpha} \tag{2.63}$$

$$\frac{d\Psi_{r\beta}}{d\tau} = a_5 \cdot \Psi_{r\beta} + \omega_r \cdot \Psi_{r\alpha} + a_6 \cdot i_{r\beta} \tag{2.64}$$

$$\frac{d\omega_r}{d\tau} = \frac{L_m}{L_r J} \left(\Psi_{r\alpha} i_{s\beta} - \Psi_{r\beta} \mathrm{i}_{s\alpha} \right) - \frac{1}{J} t_o \tag{2.65}$$

where:

$$a_1 = -\frac{R_s L_r^2 + R_r L_m^2}{L_r w} \tag{2.66}$$

$$a_2 = \frac{R_r L_m}{L_r w} \tag{2.67}$$

$$a_3 = \frac{L_m}{w} \tag{2.68}$$

$$a_4 = \frac{L_r}{w} \tag{2.69}$$

$$a_5 = -\frac{R_r}{L_r} \tag{2.70}$$

$$a_6 = R_r \frac{L_m}{L_r} \tag{2.71}$$

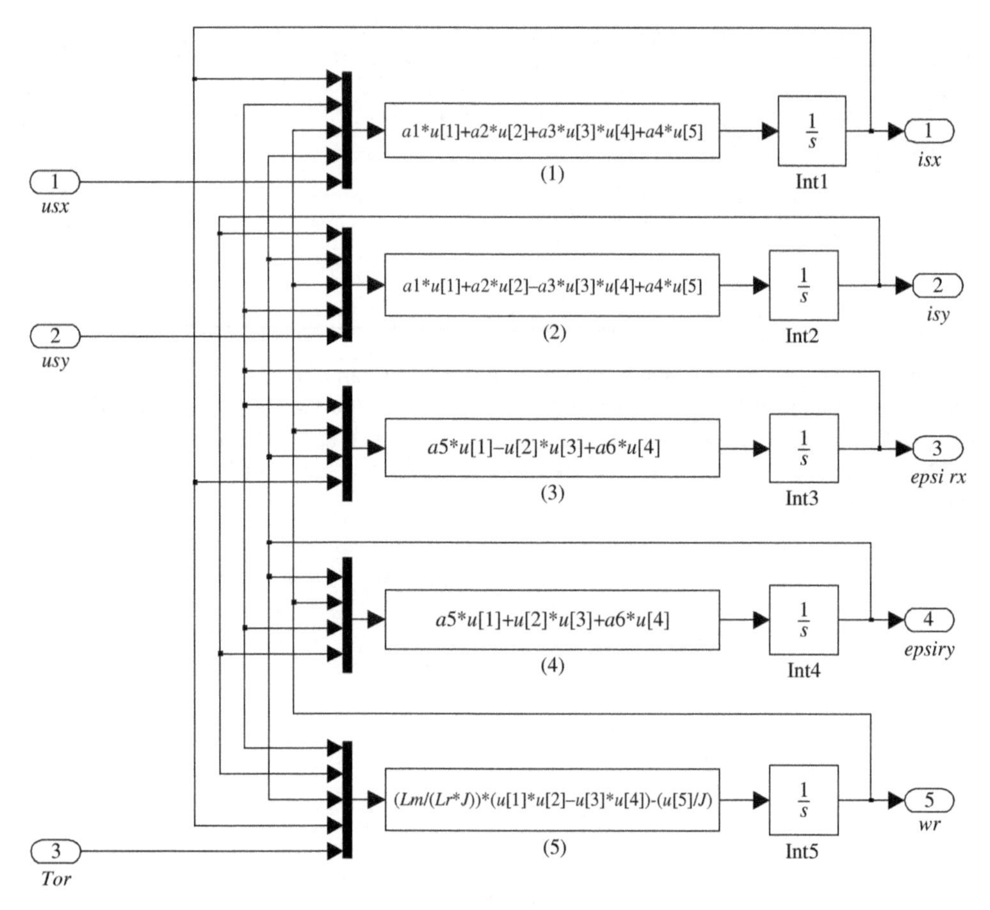

Figure 2.12 Induction motor model

$$w = \sigma L_r L_s = L_r L_s - L_m^2 \tag{2.72}$$

$$\sigma = 1 - \frac{L_m^2}{L_r L_s} \tag{2.73}$$

The above equations are modeled in Figure 2.12.

2.3.5 Double Fed Induction Generator (DFIG)

The mathematical model of DFIG is similar to the squirrel cage induction machine, the only difference being that the rotor voltage is not zero [10, 12–16]. The equations representing the DFIG [10, 12–16] are

$$\vec{v}_s = R_s \vec{i}_s + \frac{d\vec{\psi}_s}{d\tau} + j\omega_s \cdot \vec{\psi}_s \tag{2.74}$$

$$\vec{v}_r = R_r \vec{i}_r + \frac{d\vec{\psi}_r}{d\tau} + j(\omega_s - \omega_r)\cdot\vec{\psi}_r \tag{2.75}$$

$$\vec{\psi}_s = L_s \vec{i}_s + L_m \vec{i}_r \tag{2.76}$$

$$\vec{\psi}_r = L_r \vec{i}_r + L_m \vec{i}_s \tag{2.77}$$

where v_r and v_s are rotor and stator voltages, respectively; ψ_r and ψ_s are rotor and stator fluxes, respectively; R_r, R_s, L_r, and L_s are rotor and stator resistances and inductances, respectively; L_m represents mutual inductance; and ω_r and ω_s are rotor and synchronous angular speeds, respectively.

In the rotating *dq* reference frame, the machine model is

$$v_{ds} = R_s i_{ds} - \omega_s \psi_{qs} + \frac{d\psi_{ds}}{dt} \tag{2.78}$$

$$v_{qs} = R_s \cdot i_{qs} + \omega_s \cdot \psi_{ds} + \frac{d\psi_{qs}}{d\tau} \tag{2.79}$$

$$v_{dr} = R_r i_{dr} - (\omega_s - \omega_r)\psi_{qr} + \frac{d\psi_{dr}}{dt} \tag{2.80}$$

$$v_{qr} = R_r i_{qr} + (\omega_s - \omega_r)\psi_{dr} + \frac{d\psi_{qr}}{dt} \tag{2.81}$$

$$\psi_{ds} = L_s i_{ds} + L_m i_{dr} \tag{2.82}$$

$$\psi_{qs} = L_s i_{qs} + L_m i_{qr} \tag{2.83}$$

$$\psi_{dr} = L_r i_{dr} + L_m i_{ds} \tag{2.84}$$

$$\psi_{qr} = L_r i_{qr} + L_m i_{qs} \tag{2.85}$$

The produced torque is

$$T_e = \frac{3}{2} p\left(\psi_{dr} i_{qs} - \psi_{qr} i_{ds}\right) = \frac{3}{2} p L_m \left(i_{dr} i_{qs} - i_{qr} i_{ds}\right) \tag{2.86}$$

and the mechanical part of the model is

$$\frac{d\omega_m}{dt} = \frac{p}{J}(T_e - T_L) \tag{2.87}$$

The base values of the machine used to derive the above per unit model are shown in Table 2.1.

The $\alpha\beta$ (*xy*) per unit model of DFIG, connected with the grid, is described by following differential equations of state variables (stator flux components and rotor current referred to the rotor side) as

Table 2.1 Base values of DFIG

Base values	
Voltage	$V_b = V_n$
Power	$S_b = S_n$
Current	$I_b = \frac{S_b}{\sqrt{3}V_b}$
Impedance	$Z_b = \frac{V_b^2}{S_n}$
Speed	$\omega_b = 2\pi f_n$
Torque	$T_b = \frac{S_b}{\omega_b/p}$
Flux	$\psi_b = \frac{V_b}{\omega_b}$
Torque	$T_b = \frac{S_b}{\omega_b^3/p}$

$$\frac{d\phi_{sx}}{d\tau} = a_{11}\phi_{sx} + a_{12}i_{rx} + u_{sx} \tag{2.88}$$

$$\frac{d\phi_{sy}}{d\tau} = a_{11}\phi_{sy} + a_{12}i_{ry} + u_{sy} \tag{2.89}$$

$$\frac{di_{rxR}}{dt} = a_{21}i_{rxR} + a_{22}\phi_{sxR} - a_{23}\omega_r\phi_{syR} + b_{21}u_{rx} - b_{22}u_{sxR} \tag{2.90}$$

$$\frac{di_{ryR}}{dt} = a_{21}i_{ryR} + a_{22}\phi_{syR} - a_{23}\omega_r\phi_{sxR} + b_{21}u_{ry} - b_{22}u_{syR} \tag{2.91}$$

$$\frac{d\gamma_{fir}}{dt} = \omega_r \tag{2.92}$$

$$\frac{d\gamma_{ksi}}{dt} = \omega_s - \omega_r \tag{2.93}$$

The system constants are defined as:

$$a_{11} = -\frac{R_s}{L_s}, \quad a_{12} = \frac{R_s L_m}{L_s} \tag{2.94}$$

$$a_{21} = -\frac{L_s^2 R_r + L_m^2 R_s}{L_s\left(L_s L_r - L_m^2\right)}, \quad a_{22} = \frac{L_m R_s}{L_s\left(L_s L_r - L_m^2\right)}, \quad a_{23} = \frac{L_m}{L_s L_r - L_m^2} \tag{2.95}$$

$$b_{21} = -\frac{L_s}{L_s L_r - L_m^2}, \quad b_{22} = -\frac{L_m}{L_s L_r - L_m^2} \tag{2.96}$$

In these formulas, u_s and i_s are stator voltage and current vectors, respectively; R_s and L_s are stator resistances and inductances, respectively; and ω_r and ω_s are rotor and stator angular speed, respectively. The angle γ_{fir} is in the rotor angular position (fir in MATLAB/Simulink),

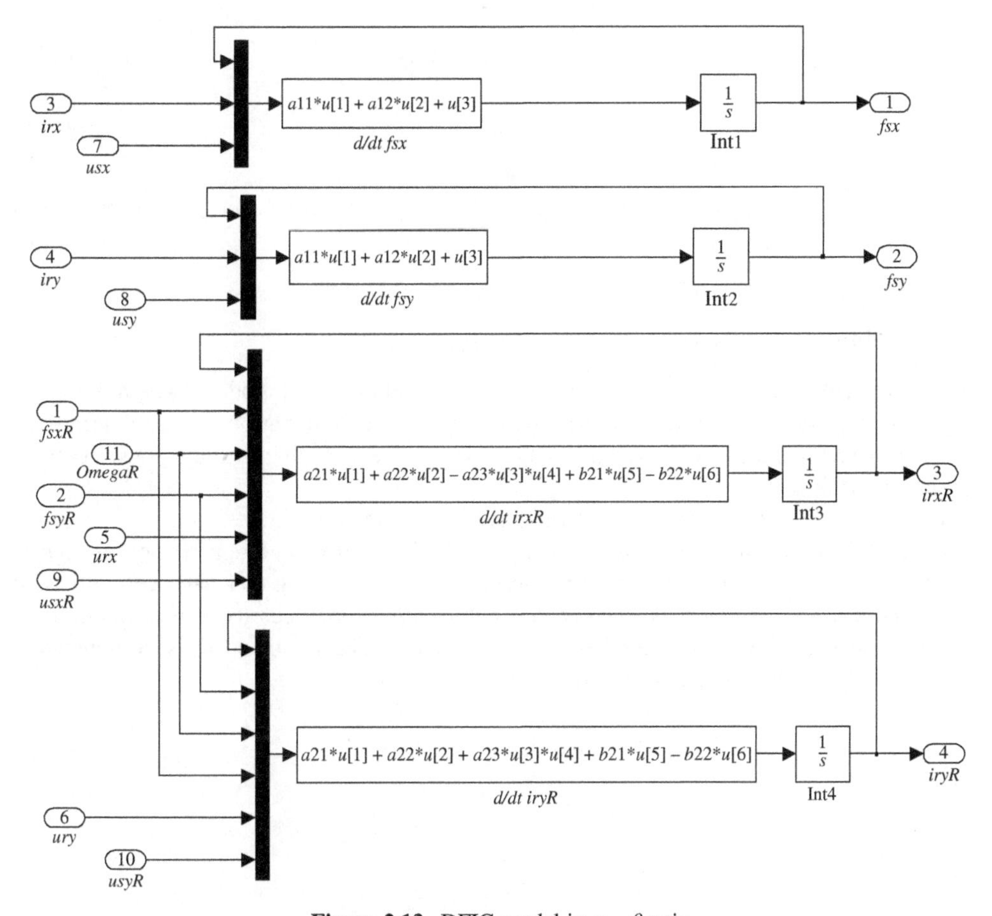

Figure 2.13 DFIG model in $\alpha - \beta$-axis

and γ_{ksi} is the angle between stator flux and rotor speed (ksi in MATLAB/Simulink). The above equations are modeled using MATLAB/Simulink, as shown in Figure 2.13.

2.4 Mathematical Model of Permanent Magnet Synchronous Motor

The mathematical model of PMSM per unit in an arbitrary reference frame rotating with ω_k is [17–22]

$$\vec{u}_s = R_s \vec{i}_s + \frac{d\vec{\psi}_s}{dt} + j\omega_k \vec{\psi}_s, \tag{2.97}$$

$$\vec{\psi}_s = L_s \vec{i}_s + \vec{\psi}_f, \tag{2.98}$$

$$\frac{d\omega_r}{dt} = \frac{1}{J}(t_e - t_l), \tag{2.99}$$

$$\frac{d\theta}{dt} = \omega_r, \tag{2.100}$$

where ω_r is the rotor angular speed; θ is the position of the rotor angle; and ψ_f is the permanent magnet flux.

2.4.1 Motor Model in dq Rotating Frame

It is convenient to design the control scheme and represent the motor model in a *dq* frame rotating with rotor speed. The variable transformation from/to the stationary frame ($\alpha\beta$) to/from the rotating one (*dq*) is realized according to space vector theory, using the equations presented earlier in this chapter.

The schematic of the PMSM for the *dq*-axis is represented in Figure 2.14.

In the PMSM, the main magnetic field is produced by permanent magnets. Those magnets are placed on the rotor. The result flux is constant in time, assuming that stator current has no effect (no armature reaction). In reality, the stator current produces its own magnetic field affecting the original, which is called the armature reaction. The net stator flux is the summation of the two fluxes (Figure 2.15) as [20]

$$\psi_d = \psi_{ad} + \psi_f = L_d i_d + \psi_f, \tag{2.101}$$

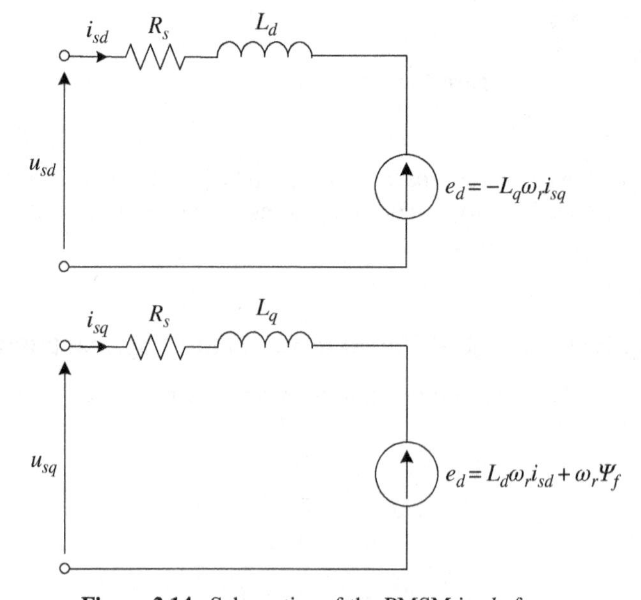

Figure 2.14 Schematics of the PMSM in *dq* frame

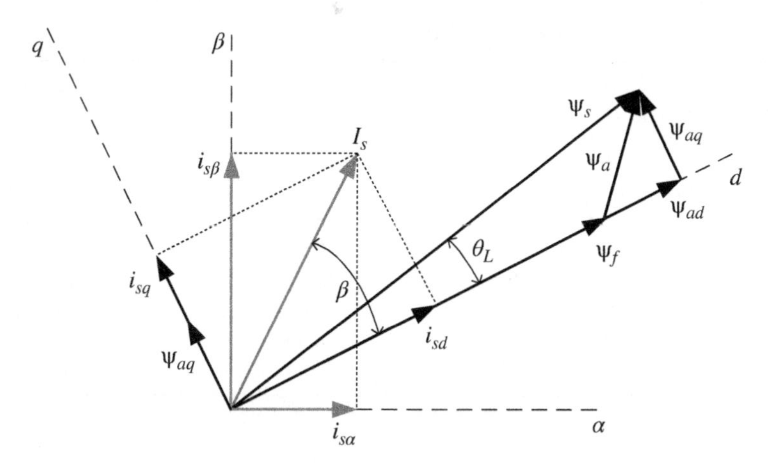

Figure 2.15 Phasor diagram of stator flux and stator current

$$\psi_q = \psi_{aq} = L_q i_q. \tag{2.102}$$

Where ψ_{ad} and ψ_{aq} are the stator leakage fluxes following the d and q, respectively, as shown in Figure 2.15.

From Figure 2.15, it is evident that the distortion of the magnetic field depends on the stator current position and magnitude. At no load condition, the armature reaction is assumed to be zero because of negligible amounts of stator current.

Back EMF (e) induced in the motor coils can be presented on the dq-axis as follows:

$$e_d = -L_q \omega_r i_q \tag{2.103}$$

$$e_q = L_d \omega_r i_d + \omega_r \psi_f \tag{2.104}$$

The produced torque of the machine can be presented as a function of the back EMF in the $\alpha\beta$ reference frame as

$$t_e = \frac{\left(e_\alpha i_\alpha + e_\beta i_\beta\right)}{\omega_r} \tag{2.105}$$

The maximum rotor speed can be identified from the relationship:

$$\omega_{max} = \frac{E_{s\,max}}{\psi_s} \tag{2.106}$$

where $E_{s\,max}$ is the maximum induced phase voltage in the machine and ω_{max} is the maximum motor speed for the rated flux. Higher speed can be obtained by weakening the flux.

Therefore, for a motor with $E_{s\,max}$ = 248 V and ω_{max} = 209.4 rad/sec, the flux is

$$\psi_s = \frac{E_{S\,max}}{\omega_{max}} = \frac{248}{209.4} = 1.18[\text{Wb}]$$

Table 2.2 Base values for PMSM motor model

Variable	Value	Unit	Description
U_b	429.5	[V]	Voltage
I_b	467.7	[A]	Current
Z_b	0.918518519	[Ohm]	Impedance
Ω_{mb}	868.7	[rad/s]	Mechanical speed
ψ_b	0.514020925	[Wb]	Flux
L_b	0.001099149	[H]	Inductance
T_b	961.5	[Nm]	Torque
Ω_o	835.7	[rad/s]	Electrical pulsation
J_b	0.001324577	[kg*m^2]	Inertia

Table 2.3 Motor parameters in per unit system [11]

Parameter	Value	Unit	Description
R_s	0.021774	[p.u.]	for 20[C]
R_s	0.032988	[p.u.]	for 150[C]
I_f	0.24564	[p.u.]	
$L_{ad} = L_m$	0.61866	[p.u.]	
L_d	0.86431	[p.u.]	
L_q	0.864305343	[p.u.]	
ψ_e	2.304	[p.u.]	
J	189	[p.u.]	For the motor only
T_n	0.65	[p.u.]	Rated torque

Source: Based on Krzemiński [11].

2.4.2 Example of Motor Parameters for Simulation

Motor parameters are presented in Tables 2.2 and 2.3 for rated values and in per unit [18, 19].

2.4.3 PMSM Model in Per Unit System

Mathematical models of the PMSM can be presented as differential equations of state variables. This can be presented either on the dq-axis, which is the preferred solution, or in a stationary frame. The motor model in the dq-axis rotating with rotor speed is [19, 21]

$$\frac{di_d}{dt} = -\frac{R_s}{L_d} i_d + \frac{L_q}{L_d} \omega_r i_q + \frac{1}{L_d} u_d \tag{2.107}$$

$$\frac{di_q}{dt} = -\frac{R_S}{L_q} i_q - \frac{L_d}{L_q} \omega_r i_d - \frac{I}{L_q} \omega_r \psi_f + \frac{1}{L_q} u_q \tag{2.108}$$

$$\frac{d\omega_r}{d\tau} = \frac{1}{J}\left[\psi_f i_q + \left(L_d - L_q\right) i_d i_q - t_0\right]. \tag{2.109}$$

$$\frac{d\theta_r}{dt} = \omega_r \tag{2.110}$$

where u_s and i_s are stator voltage and current vectors, respectively; R_s is the stator resistances; R_s, R_q, L_d, and L_q are rotor resistances and inductances when referring to d and q axes; ω_r is the rotor angular speed; J is the moment of inertia; t_l is the load torque, and T_M is the mechanical time constant.

The above equations are modeled in MATLAB/Simulink, as shown in Figure 2.16.

The parameters of the interior type PMSM to be simulated are

$R_s = 0.032988$;
$L_d = 0.86431$;
$L_q = 0.86431$;
$\psi_f = \text{fe} = 2.3036$;
$J = 0.00529$. Rated speed is 2000 rpm.

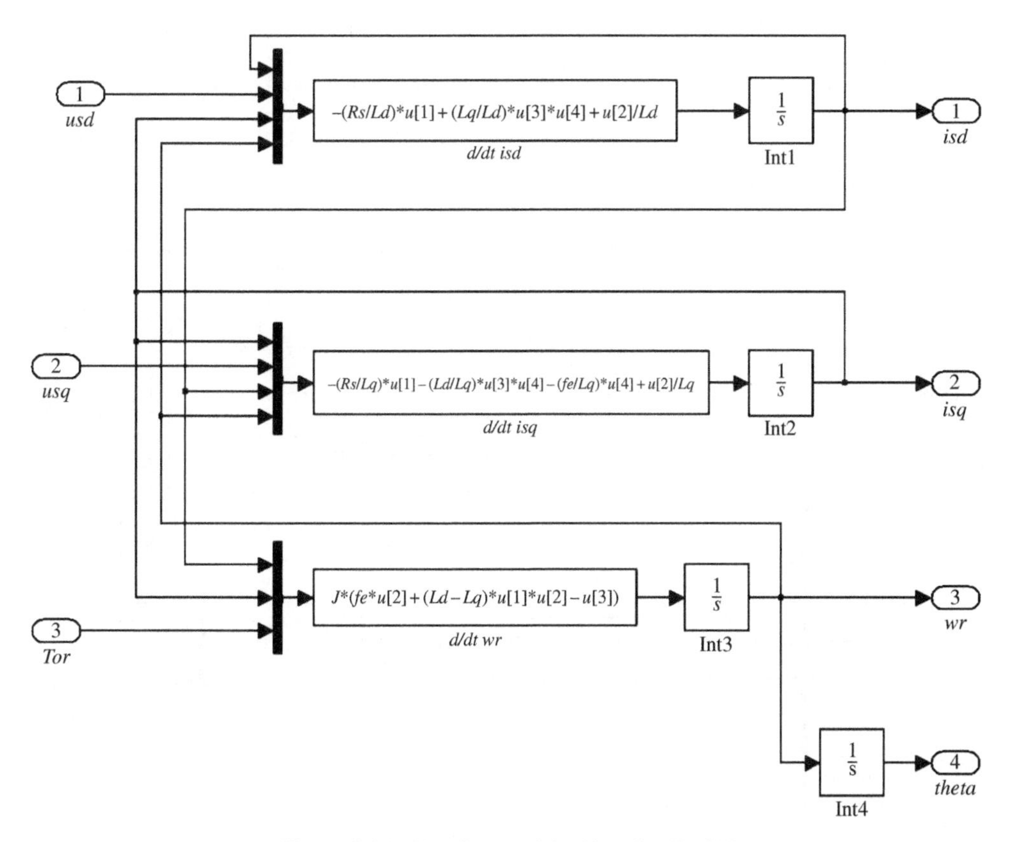

Figure 2.16 Simulink model of interior PMSM

2.4.4 PMSM Model in $\alpha - \beta$ $(x - y)$-Axis

The Permanent Magnet Synchronous Machine (PMSM) model presented in per unit $\alpha - \beta$ reference frame is given by [22]

$$\frac{di_{sx}}{dt} = \frac{1}{L_d}\omega_r\phi_{sy} + (u_{sx} - R_s i_{sx})\left(\frac{1}{L_d}\cos^2\theta + \frac{1}{L_q}\sin^2\theta\right) + 0.5\left(u_{sy} - R_s i_{sy}\right)\sin 2\theta\left(\frac{1}{L_d} - \frac{1}{L_q}\right) \tag{2.111}$$

$$\frac{di_{sy}}{dt} = \frac{1}{L_d}\omega_r\phi_{sy} + \left(u_{sy} - R_s i_{sy}\right)\left(\frac{1}{L_d}\sin^2\theta + \frac{1}{L_q}\cos^2\theta\right) + 0.5(u_{sx} - R_s i_{sx})\sin 2\theta\left(\frac{1}{L_d} - \frac{1}{L_q}\right) \tag{2.112}$$

$$\frac{d\omega_r}{dt} = J\left[\phi_{sx} i_{sy} - \phi_{sy} i_{sx} - t_l\right] \tag{2.113}$$

$$\frac{d\theta}{dt} = \omega_r \tag{2.114}$$

where

$$\phi_{sx} = \frac{L_d}{L_q}F_f\cos\theta - \left(1 - \frac{L_d}{L_q}\right)L_0\left(i_{sx}\cos 2\theta + i_{sy}\sin a2\theta\right) + L_2 i_{sx} \tag{2.115}$$

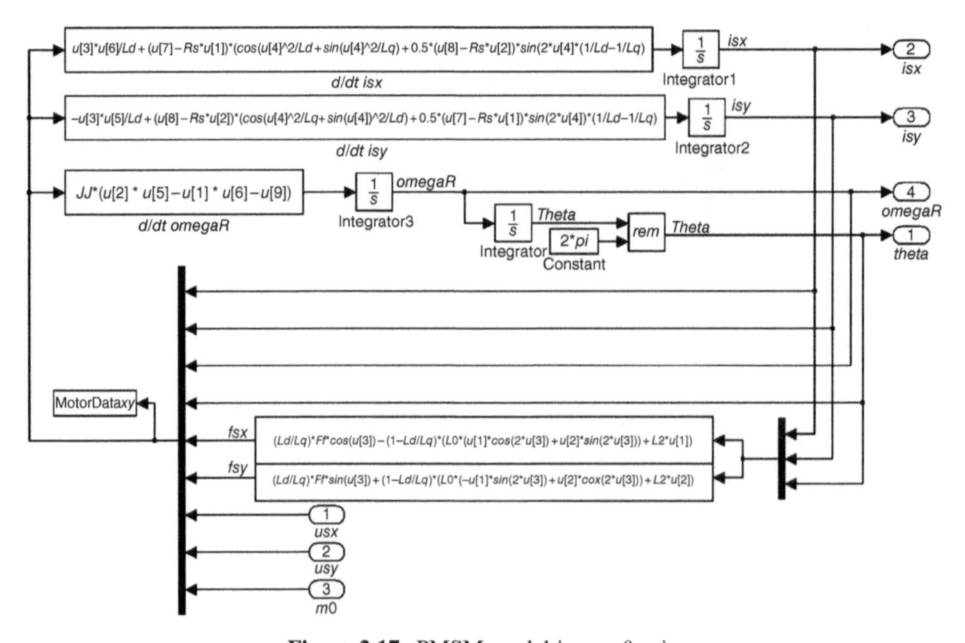

Figure 2.17 PMSM model in $\alpha - \beta$-axis

$$\phi_{sy} = \frac{L_d}{L_q} F_f \sin\theta - \left(1 - \frac{L_d}{L_q}\right) L_0 \left(i_{sy} \cos 2\theta + i_{sx} \sin a2\theta\right) + L_2 i_{sy} \tag{2.116}$$

$$L_0 = \frac{L_d + L_q}{2} \text{ and } L_2 = \frac{L_d - L_q}{2} \tag{2.117}$$

where, u_s and i_s are stator voltage and current vectors, respectively; R_s is the stator resistances, which are rotor inductances when referred to on the d and q axes; ω_r is the rotor angular speed; J is the moment of inertia; and t_l is the load torque. The above equations are modeled in Figure 2.17.

2.5 Problems

Design the models for IM, DFIG, and PMSM, by working with sinusoidal input signals.

References

1. Mohan, N. (2001) *Electric Drives an Integrative Approach*, 2nd edn. MNPERE, Minneapolis.
2. Bose, B. K. (2006) *Power Electronics and Motor Drives: Advances and Trends.* Academic Press, Elsevier, Amsterdam.
3. Leonhard, W. (2001) *Control of Electrical Drives.* Springer, Berlin.
4. Krishnan, R. (2001) *Electric Motor Drives: Modeling, Analysis and Control.* Prenctice Hall, London, UK.
5. Chiasson, J. (2005) *Modeling and High Performance Control of Electric Machines.* John Wiley & Sons, Chichester.
6. Kovacs, K. P. and Racz, I. (1959) *Transiente Vorgange in Wechselstrommachinen.* Akad. Kiado, Budapest.
7. Blaschke, F. (1971) Das Prinzip der Feldorientierung, die Grundlage fur Transvector-regelung von Drehfeld-maschine. *Siemens Z*, 45, 757–760.
8. Texas Instruments Europe (1998) *Field Orientated Control of 3-Phase AC-Motors.* Texas Instruments, Inc., Texas, USA.
9. Mohan, N. (2001) *Advanced Electric Drives Analysis, Control, and Modeling using Simulink.* MNPERE, Minneapolis.
10. Bogdan, M., Wilamowski, J., and Irwin, D. (2011) *The Industrial Electronics Handbook: Power Electronics and Motor Drives.* Taylor and Francis Group, LLC, UK.
11. Krzemiński, Z. (1987) Nonlinear control of induction motor. 10th World Cong. Auto. Cont., IFAC '87, Monachium, Niemcy, pp. 27–31.
12. Bogalecka, E. (199a3) Power control of a double fed induction generator without speed or position sensor. Conf. Rec. EPE, pt. 8, vol. 377, ch. 50, Munich, Germany, pp. 224–228.
13. Iwanski, G. and Koczara, W. (2008) DFIG-based power generation system with UPS function for variable-speed applications. *IEEE Trans. Ind. Electn.*, 55(8), 3047–3054.
14. Lin, F-J., Hwang, J-C., Tan, K-H., Lu, Z-H., and Chang, Y-R. (2010) Control of doubly-fed induction generator system using PIDNNs. 2010 9th Int. Conf. Mach. Learn. Appl., Washington, DC, USA; 12–14 December 2010, pp. 675–680.

15. Pena, R., Clare, J. C., and Asher, G. M. (1996) Doubly fed induction generator using back-to-back PWM converters and its application to variable-speed wind-energy generation. *IEEE Proc. -Electr. Power Appl.*, 143(3), 231–241.
16. Muller, M. and De Doncker, R. W. (2002) Doubly-fed induction generator systems for wind turbines. *IEEE IAS Mag.*, 8(3), 26–33.
17. Pillay, P. and Knshnan, R. (1988) Modeling of permanent magnet motor drives. *IEEE Trans. Ind. Elect.*, 35(4), 537–541.
18. Burgos, R. P., Kshirsagar, P., Lidozzi, A., Wang, F., and Boroyevich, D. (2006) Mathematical model and control design for sensorless vector control of permanent magnet synchronous machines. IEEE COMPEL Workshop, Rensselaer Polytech. Inst., July 16–19, Troy, NY.
19. Pajchrowski, T. and Zawirski, K. (2005) Robust speed control of servodrive based on ANN. IEEE ISIE 2005, June 20–23, Dubrovnik.
20. Stulrajter, M., Hrabovcova, V., and Franko, M. (2007) Permanent magnets synchronous motor control theory. *J. Elect. Eng.*, 58(2), 79–84.
21. Zawirski, K. (2005) *Control of Permanent Magnet Synchronous Motors (in Polish)*. Poznan, Poland.
22. Morawiec, M., Krzemiński, Z., and Lewicki, A. (2009) Control of PMSM with rotor speed observer: Sterowanie silnikiem o magnesach trwałych PMSM z obserwatorem prędko⊠ci kątowej wirnika. *Przegląd Elektrotechniczny (Electrical Review) (in Polish)*, 85.

3

Pulse-Width Modulation of Power Electronic DC–AC Converter

Atif Iqbal[1], Arkadiusz Lewicki[2], and Marcin Morawiec[2]
[1]*Department of Electrical Engineering, Qatar University, Doha, Qatar*
[2]*Electrical and Control Engineering Department, Gdansk University of Technology, Gdansk, Poland*

Lewicki and Morawiec have contributed Sections 3.8 and 3.9.
Iqbal has contributed the rest of the sections of this chapter.

3.1 Preliminary Remarks

The pulse-width modulation (PWM) technique is applied in the inverter (DC–AC converter) to output an AC waveform with variable voltage and variable frequency for use in mostly variable speed motor drives. The input to the inverter is DC, obtained from either a controlled or uncontrolled rectifier. Hence, an inverter is a two-stage power converter that transforms the grid AC to DC and then DC to AC. PWM and the control of the power electronic DC–AC converters have attracted much attention in the last three decades. Research is still active in this area and several control schemes have been suggested in the literature [1–6]. The basic idea is to modulate the duration of the pulses or duty ratio to achieve controlled voltage/current/power and frequency, satisfying the criteria of equal area. The implementation of the complex PWM algorithms has been made easier due to the advent of fast digital signal processors, microcontrollers, and field programmable gate arrays (FPGAs). The PWM is the basic energy processing technique used in power electronic converters initially implemented with analog technology using discrete electronic components. Nowadays, they are digitally implemented through modern signal processing devices. This chapter gives an overview of the PWM techniques based on the most basic and classical sinusoidal carrier-based scheme to modern space vector PWM

High Performance Control of AC Drives With MATLAB®/Simulink, Second Edition.
Haitham Abu-Rub, Atif Iqbal, and Jaroslaw Guzinski.

Companion website: www.wiley.com/go/aburubcontrol2e

(SVPWM). The literature shows an implicit relationship between the sinusoidal carrier-based PWM (SPWM) and SVPWM technique. A section is devoted to understanding the underlying relationship between these two techniques. The linear modulation range is mostly covered by PWM techniques and a special section is devoted to overmodulation methods (Section 3.4.8). The analytical approach is validated using a MATLAB/Simulink-based simulation. An artificial neural network (ANN) offers a good nonlinear mapping tool used to produce PWM for a three-phase voltage source inverter and elaborated on in Section 3.4.11. A comprehensive relationship between the SVPWM and SPWM is discussed in Section 3.5. For medium-voltage high-power applications, multilevel inverters are employed. Low-switching frequency modulation techniques such as selective harmonic elimination (SHE) method is employed high-power converters. Low-order harmonics are selectively eliminated from voltage waveforms by computing the exact switching angle of the power switches. However, computing exact switching angle or instant is not an easy task. Set of transcendental equations are formulated and solved to obtain the switching instant of the power switches. There are large number of mathematical tools and methods developed to solve transcendental equation. A brief summary of the available techniques are presented in this chapter in Section 3.6. Special inverters based on an impedance source network to incorporate the boost function, called Z-source and quasi Z-source inverters, are described in Sections 3.7 and 3.8. These inverters perform the boosting and inversion in a single stage and are considered an alternative to two-stage converters using a DC–DC converter and an inverter. ZSI and qZSI are popular in solar PV applications because of their single-stage structure and simple control. Basic topologies and PWM for multilevel (ML) inverters are given in Section 3.9. Three most commonly used ML inverter topologies namely neutral-point-clamped (NPC), flying capacitor (FLC), and cascaded H-bridge (CHB) inverters are elaborated. SVPWM applied to a three-level NPC and CHB topologies are discussed considering capacitor voltage balancing. The effect of using the dead time between the switching on and off of one inverter leg is presented in Section 3.9, and the summary is given in Section 3.10.

3.2 Classification of PWM Schemes for Voltage Source Inverters

Different types of PWM techniques are available in the literature, but the most basic PWM is the SPWM. The high-frequency carrier wave is compared with the sinusoidal modulating signals to generate the appropriate gating signals for the inverters. The other PWM techniques are evolved from this basic PWM technique. SVPWM, although appearing different to the SPWM, has a strong implicit relationship with SPWM [1–5]. The gating time of each power switch is directly calculated from the analytical time equations in SVPWM [6]. The power switches are then switched according to the predefined switching patterns. The achievable output in the case of SVPWM is higher when compared to the SPWM. The main aim of the modulation techniques is to attain the maximum voltage with the lowest total harmonic distortion (THD) in the output voltages. The PWM techniques can be broadly classified as:

- Continuous PWM;
- Discontinuous PWM.

In discontinuous PWM techniques, the power switches are not switched at regular intervals. Some of the switches do not change states in a sampling or switching period. This technique is employed to reduce switching losses. The SVPWM method is modified to achieve discontinuous switching. A detailed discussion is provided in Section 3.4.6.

3.3 Pulse-Width Modulated Inverters

This section elaborates on some basic inverter circuit topologies. The square-wave and PWM operation are described along with the harmonic spectrum of the output voltages.

3.3.1 Single-Phase Half-Bridge Inverters

The power circuit topology of a single-phase half-bridge voltage source inverter is shown in Figure 3.1. As shown in Figure 3.1a, the inverter consists of two power semiconductor switching devices. Each switch is composed of a transistor (BJT, MOSFET, IGBTs, etc.) and a freewheeling diode that provides an alternating path for the current. In this circuit topology, either S_1 is on or S_1' is on. Both switches of the same leg cannot be on simultaneously, otherwise the DC source will be short-circuited. Hence the two switches are complimentary in operation. The freewheeling diodes conduct when the load is inductive and a sudden change in the output voltage polarity does not change the direction of the current simultaneously. Indeed, for resistive loads, the diodes do not operate. When switching device S_1 is conducting, the input DC voltage of 0.5 V_{dc} appears across the load. On the other hand, when transistors S_1' is turned on, the voltage across the load is reversed and is (−0.5 V_{dc}). The switching states, the path of current flow, and the output voltage polarity are illustrated in Figure 3.2, and switching signals along with the output voltage waveform are depicted in Figure 3.3. The operation of the inverter can be well understood from Figures 3.2 and 3.3. The output fundamental frequency of the inverter can be varied by altering the frequency of the reference or modulating signal, and the voltage magnitude can be changed by changing the value of the DC-link voltage. The output voltage magnitude can be also varied by changing the modulation index value which is discussed in the below section.

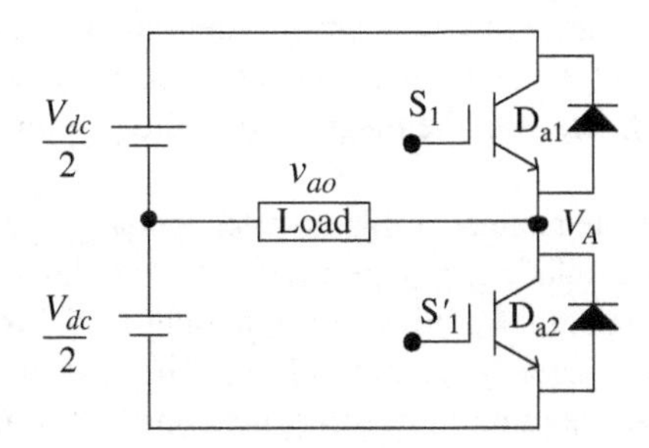

Figure 3.1 Power circuit of a half-wave bridge inverter

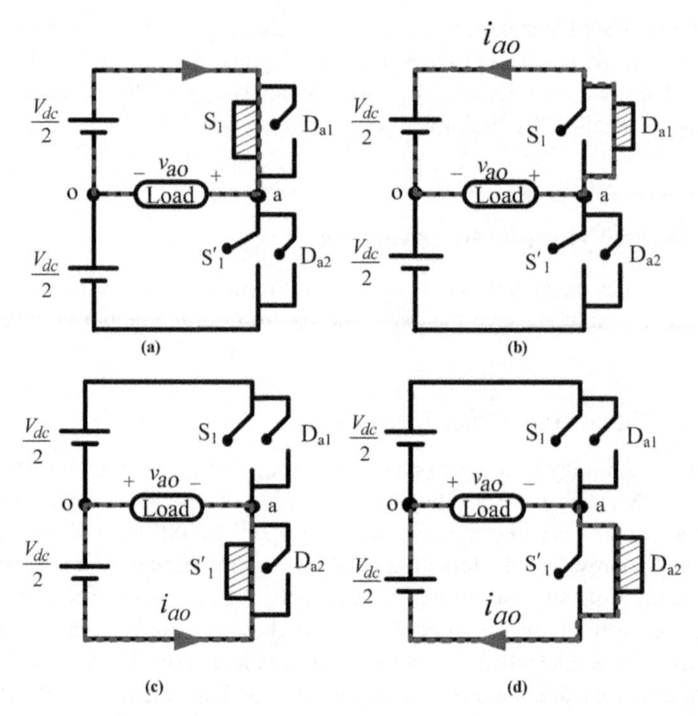

Figure 3.2 Switching states in half-bridge inverter: (a) and (c) $i_{ao} > 0$; (b) and (d) $i_{ao} < 0$

The output voltage is a square wave, as shown in Figure 3.3, and the Fourier analysis gives

$$v_{ao}(t) = \sum_{n=1}^{\infty} \frac{4}{n\pi}(0.5V_{dc}) \sin(n\,\omega t) \tag{3.1}$$

where $\omega = 2\pi f = 2\pi/T$; T is the output period and f is the output fundamental frequency.

The fundamental component of v_{ao} is obtained as $v_{ao1} = \frac{4}{\pi}(0.5\ V_{dc}) \sin(\omega t)$; the peak value of the fundamental is $V_{ao} = \frac{4}{\pi}(0.5V_{dc})$; and the rms value is $V_{ao,rms} = \frac{4}{\pi\sqrt{2}}(0.5V_{dc})$. A graphical view of the harmonic content in the output phase voltage is shown in Figure 3.4. The figure shows that the output contains a considerable amount of low-order harmonics such as 3rd, 5th, 7th, etc. and the magnitude of the harmonics varies as the inverse of its order.

An SPWM scheme is implemented in this inverter topology by comparing a sinusoidally varying reference/control/modulating voltage $v_m(t)$ with amplitude V_m and frequency f_m (the frequency of the modulating signal is low corresponding to the motor speed, if motor load is assumed) with a high-frequency carrier signal. The carrier frequency *is* f_c and the shape of the carrier wave is triangular in most applications. If a triangular shape of the carrier wave is chosen, it is called a double-edge modulation. A simpler shape is a saw-tooth carrier and the modulation using this type of carrier is called a single-edge modulation. Other

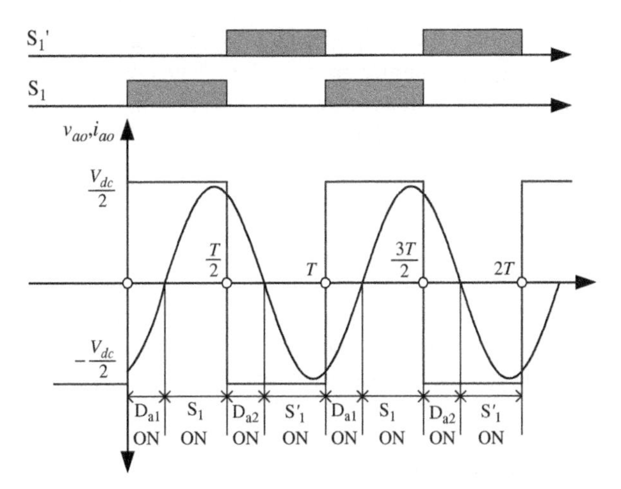

Figure 3.3 Switching signal and the output voltage and current in a half-bridge inverter

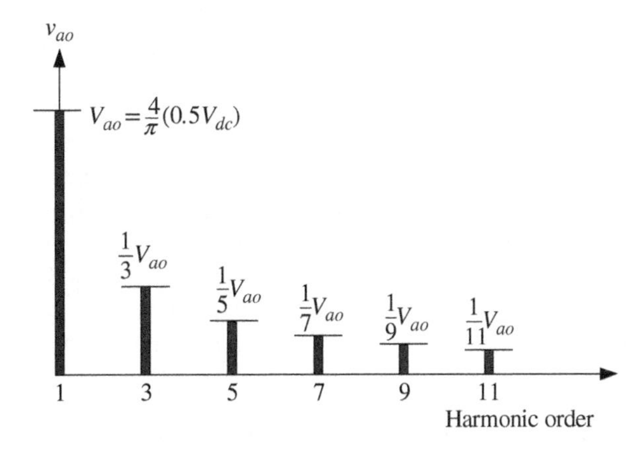

Figure 3.4 A typical harmonic spectrum of output voltage in a half-bridge inverter

shapes of the carrier wave, such as the inverted rectified sine wave, are also reported in the literature [6, 7]. Nevertheless, the triangular carrier is the most popular, as it offers good harmonic performance.

The ratio of the amplitude of the carrier signal and the control signal (or reference/modulating signal) is called 'amplitude modulation ratio' or 'modulation index':

$$m = \frac{|v_m|}{|v_c|} = \frac{V_m}{V_c} \tag{3.2}$$

When the modulating signal is less than or equal to the carrier signal, it is called the 'linear modulation region'. When the amplitude of the modulating signal becomes more than the carrier signal, it is called the 'pulse dropping mode' or the 'over modulation region'. The name is given because some of the edges of the modulating signal will not intersect with the carrier. The modulating signal is then modified accordingly. The output voltage increases or the modulation range increases, but low-order harmonics are introduced. Details on overmodulation are discussed separately in Section 3.4.8. The discussion here is limited to the linear modulation region. Another parameter defined by the SPWM is called the 'frequency modulation ratio,' and is given as:

$$m_f = \frac{f_c}{f_m} \tag{3.3}$$

This is also an important parameter that decides the harmonic performance of the output voltage. The choice of this ratio is discussed later in this chapter.

The inverter leg switching frequency is equal to the frequency of the triangular carrier signal, and the switching period is $T_c = \frac{1}{f_c}$.

The gating signals/switching signals are generated at the instant of intersection of the carrier signal and control or modulating signal. If the control or modulating signal amplitude V_m is more than the carrier signal amplitude (V_c), the upper switch S_1 is ON and $v_{ao} = \frac{V_{dc}}{2}$.

If the control or modulating signal amplitude (V_m) is less than the carrier signal amplitude (V_c), the lower switch S_1' is ON and $v_{ao} = -\frac{V_{dc}}{2}$.

With this type of switching of the inverter leg, the output voltage varies between $V_{dc}/2$ and $-V_{dc}/2$, as shown in Figure 3.5 and is called the bipolar PWM. The name 'bipolar' is given since the output leg voltage has both positive and negative values. The value of the average leg voltage V_{AO} during a switching period T_c can be determined from Figure 3.6, which shows one period of the triangular waveform. During an inverter switching period T_c, the amplitude of the modulating signal can be assumed to be nearly constant, as the switching frequency of the triangular carrier wave is high compared to modulating signal. The duration of S_1 being

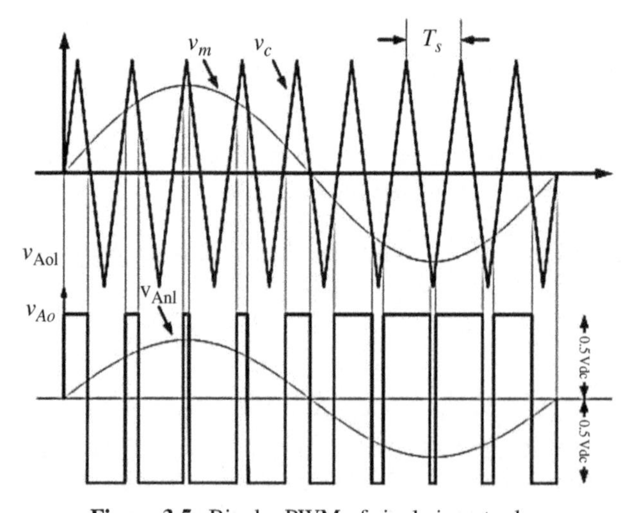

Figure 3.5 Bipolar PWM of single inverter leg

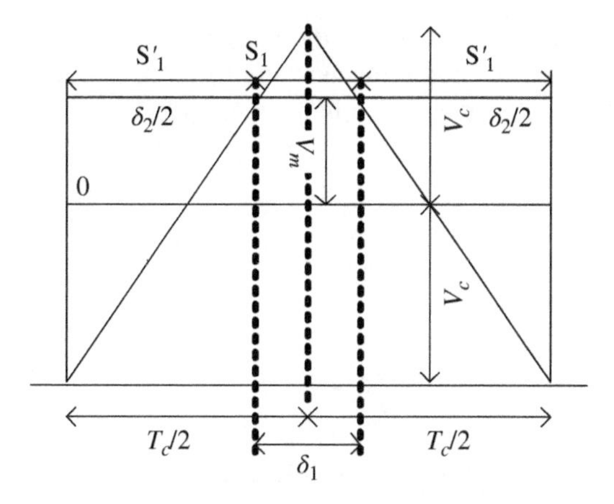

Figure 3.6 One switching cycle in SPWM

ON is denoted by δ_1 and the ON period of S_1' is denoted by δ_1'.Using equivalence of triangles, the following relation can be written:

$$\frac{\delta_1'/2}{T_c/2} = \frac{V_c + V_m}{V_c + V_c}$$
$$\delta_1' = T_c\left(\frac{V_c + V_m}{2V_c}\right) \tag{3.4}$$

and,

$$\delta_1 = T_c - \delta_1' = T_c\left(1 - \left(\frac{V_c + V_m}{2V_c}\right)\right) \tag{3.5}$$

To determine the average output voltage during one switching period, the equal volt-second principle (the product of voltage and time should be equal for the two switches and the average output) is applied:

$$\delta_1 \frac{V_{dc}}{2} - \delta_1' \frac{V_{dc}}{2} = T_c V_{Ao} \tag{3.6}$$

$$V_{Ao} = \frac{1}{T_c}\frac{V_{dc}}{2}(\delta_1 - \delta_1' =)\frac{V_m}{V_c}\frac{V_{dc}}{2} \tag{3.7}$$

$$V_{Ao} = m\frac{V_{dc}}{2} \tag{3.8}$$

The average leg voltage of the inverter is proportional to the modulation index (m) and the maximum value is $V_{dc}/2$ when the modulation index is 1. When the modulation signal is sinusoidal, the output voltage varies sinusoidally with a peak value of $V_{dc}/2$ for $m = 1$.

In the SPWM method, the harmonics in the inverter output voltage waveform appear as sidebands centered on the switching frequency and its multiples, i.e. around f_c, $2f_c$, $3f_c$, etc. A general form for the frequency at which harmonics appear is [3]

$$f_h = \left(j\frac{f_c}{f_m} \pm k \right) f_m \tag{3.9}$$

and the harmonic order h is

$$h = \frac{f_h}{f_m} = \left(j\frac{f_c}{f_m} \pm k \right) \tag{3.10}$$

It is important to note that for all odd j, the harmonics will appear for even values of k and vice versa. Normally, the frequency modulation ratio is chosen to be an odd integer, to avoid even harmonics appearing in the output voltage. The choice of switching frequency of an inverter depends upon the application. Ideally, the switching frequency should be as low as possible to avoid losses due to the switching of the power switches. However, when the switching frequency is low, the output voltage quality is poor. Hence, the switching frequency for the medium power motor drive system is kept between 4 and 10 kHz.

For low values of the frequency modulation ratio, usually for $m_f \leq 21$, the modulating or control signal and carrier signal are synchronized with each other, called synchronous PWM. The frequency modulation ratio is chosen as an integer number to facilitate proper synchronization. If the frequency modulation ratio is a non-integer, subharmonics (lower than the fundamental frequency) appears in the output voltage waveform. Hence, in synchronous PWM, the switching frequency should be changed with the change in the fundamental output frequency, to keep the ratio an integer and subsequently to avoid the production of subharmonics.

For higher values of the frequency modulation ratio, $m_f > 21$, the synchronization of the carrier wave and modulating signals are not necessary and the result is asynchronous PWM. Since the amplitude of the subharmonics will be small, asynchronous PWM can be used for medium-power applications.

Nevertheless, in high-power applications, even small subharmonics may lead to appreciable losses. Thus, it is recommended to use synchronous PWM for high-power applications.

3.3.1.1 MATLAB/Simulink Model of Half-Bridge Inverter

The Simulink model shown in this section uses ‘SimPowerSystem’ blocksets of the Simulink library. The IGBT switches are chosen from the built-in library. The gating signals are generated using the SPWM principle. The modulating signal is assumed to be a sinusoidal wave of unit amplitude and 50-Hz fundamental frequency. The switching frequency of the inverter is chosen as 1250 Hz ($m_f = 25$) (this can be changed by varying the frequency of the carrier wave), and hence the frequency of the triangular carrier wave is 1250 Hz. The gating signal generated is split into two complementary switching signals using a NOT gate (note: the deadband is not used in the model). The amplitude of the modulating signal is kept at 0.475 p.u. (m = 0.95) (note: the maximum output voltage is 0.5 V_{dc}, hence the maximum amplitude of the modulating signal is limited to 0.5 V_{dc}). The DC-link voltage is assumed at 1 p.u. The load used in the

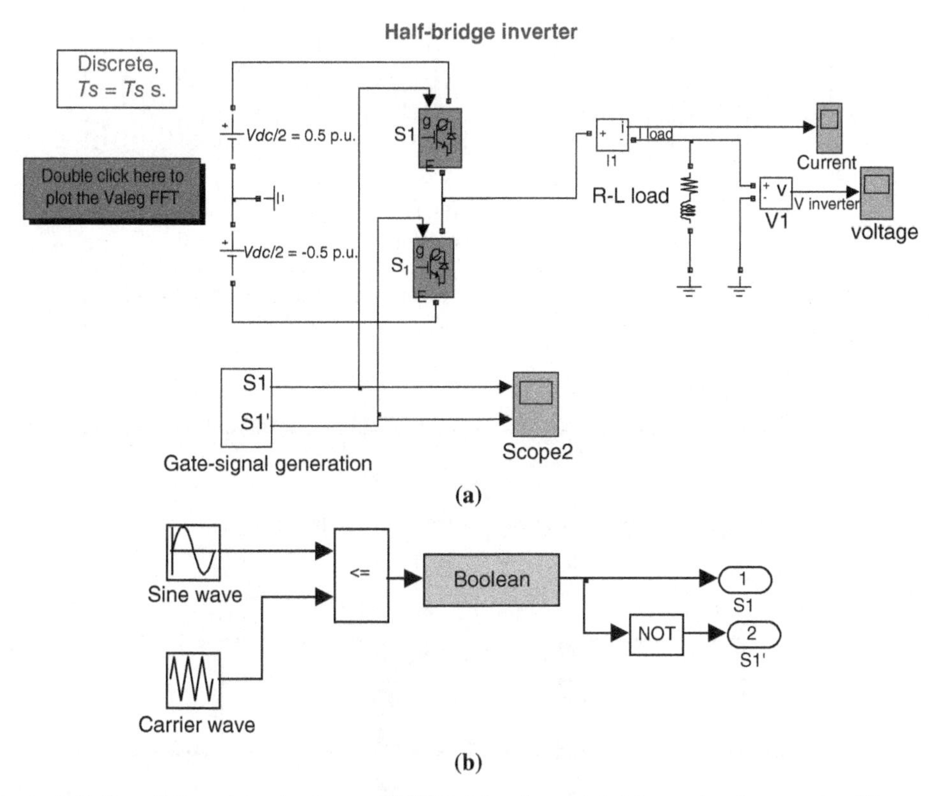

Figure 3.7 Simulink model to implement SPWM: (a) main model; (b) gate signal generation (file name: *Half_bridge_PWM.mdl)*

simulation model is RL, with values $R = 10\ \Omega$ and $L = 100$ mH. The simulation model is shown in Figure 3.7 and the resulting spectrum of the output voltage waveforms is shown in Figure 3.8. The harmonic components are marked in the output voltage spectrum.

Practically, the switching signals or gate drive signals are complementary but have some deadband that is introduced intentionally to prevent the two power switches of the same leg conducting simultaneously. An example of the deadband is shown in Figure 3.9a. To observe the effect of introducing a deadband (the two switches are not ON at the same time for safety purpose) in the switching signal, the deadband circuit shown in Figure 3.9b can be added to the gate signals of each switch. The deadband period can be defined and adjusted in the dialog box of the edge detector 'sample time'. Adding the deadband is left to the reader.

3.3.2 Single-Phase Full-Bridge or H-Bridge Inverters

The power circuit topology of a single-phase full-bridge or H-bridge inverter is depicted in Figure 3.10. There are four power semiconductor switches and two legs in this topology. This is also called the 'H-bridge' inverter due to its shape. Each switch is composed of a controlled switch (IBGTs, MOSFETs, BJT, etc.) and a freewheeling diode, which provides an alternating

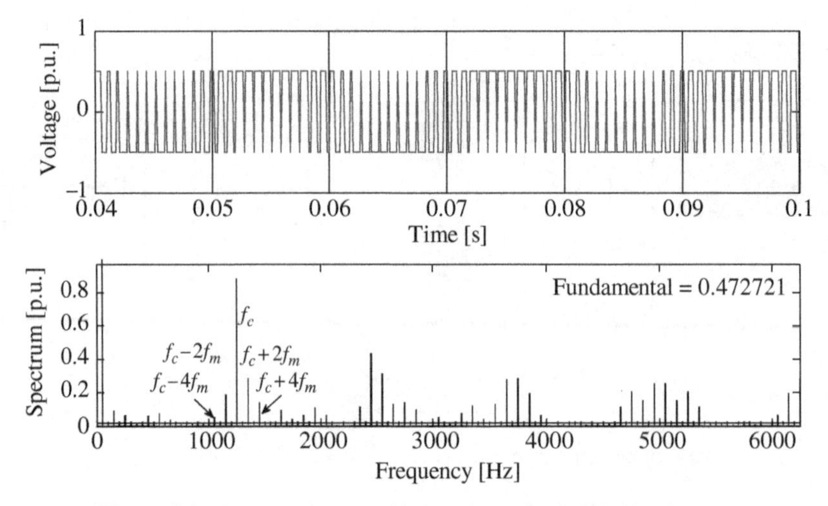

Figure 3.8 Output voltage and its spectrum for half-bridge inverter

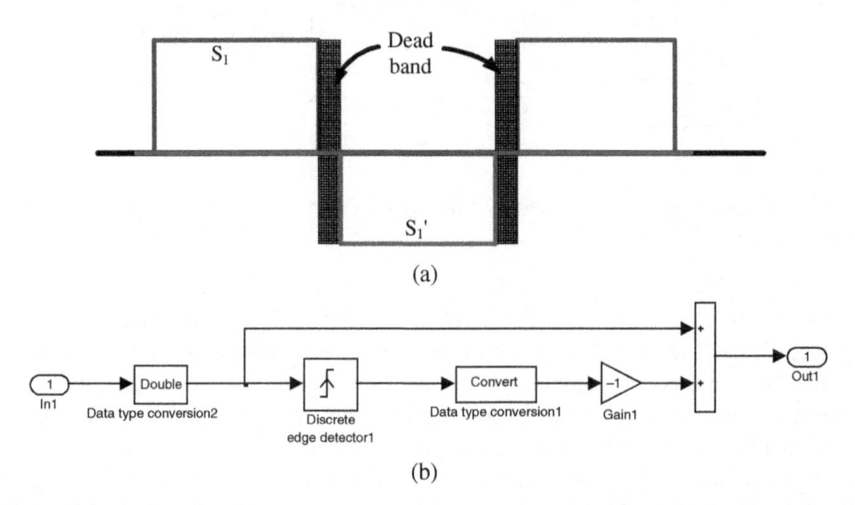

Figure 3.9 (a) Deadband between upper and lower gating signals and (b) deadband circuit

path for the current. When the switches S_1 and S_2' are conducting simultaneously, the input voltage V_{dc} appears across the load. On the other hand, when switches S_2 and S_1' are turned ON at the same time, the voltage across the load is reversed and is ($-V_{dc}$) and therefore, an alternating voltage and current are obtained at the output side while the input side is DC. The output voltage and frequency can be varied by varying the modulation index and fundamental frequency of the modulating signal, respectively. The switching states, the path of the current flow, and the output voltage polarity are shown in Figure 3.10. The alteration in the direction of the voltage and current is evident from this figure.

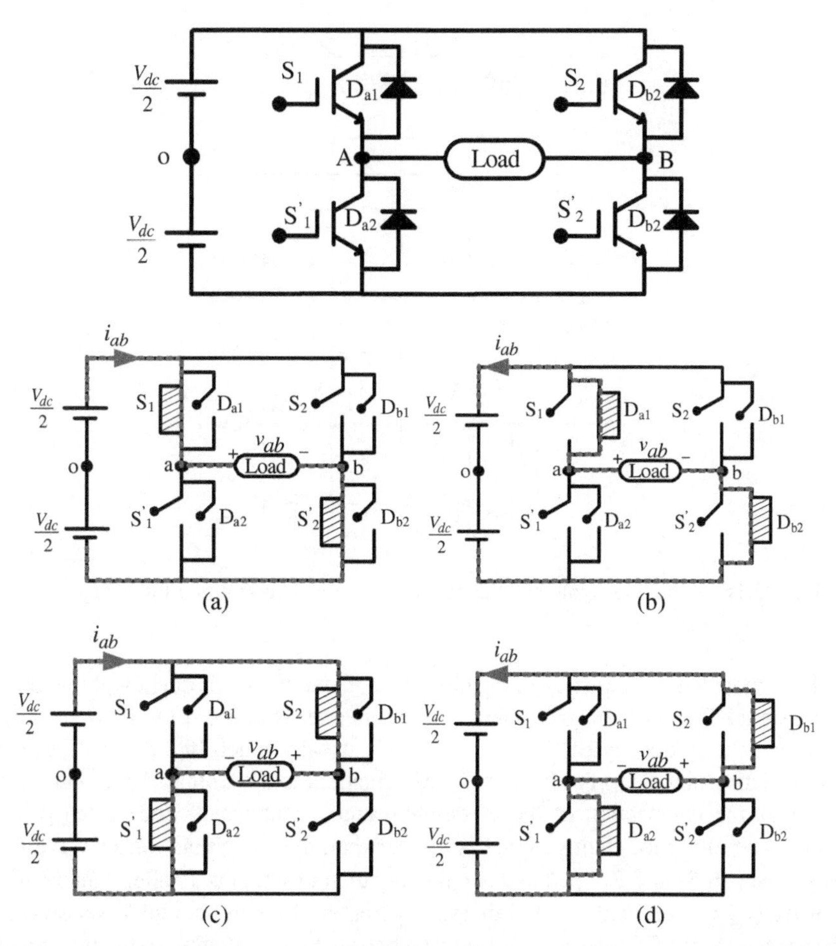

Figure 3.10 Power circuit topology of a single-phase full-bridge inverter: (b) switching states in full-bridge inverter; (a) and (c) $i_{ab} > 0$; (b) and (d) $i_{ab} < 0$

The output voltage and current waveforms for a fictitious RL load are shown in Figure 3.11. The currents do not change direction instantaneously due to the inductive load. The pole voltages vary between 0.5 V_{dc} and −0.5 V_{dc}, while the voltages across the load vary between V_{dc} and $-V_{dc}$. The switching signals are shown to be applied for 180 degrees but in practice, some deadband is given between the application of gating pulses for the upper and lower power switch. The output voltage is the same as that of the half-wave bridge inverter, except that the magnitude of the rms phase voltage, which is doubled in this case. Hence, the harmonic spectrum remains the same as that of a half-bridge inverter output, except with double fundamental voltage magnitude.

The SPWM technique is implemented in a full-bridge inverter by comparing either one or two modulating or control signals (180-degree phase-shifted) with the high-frequency triangular carrier wave. The former is called the bipolar PWM and the latter is called the unipolar

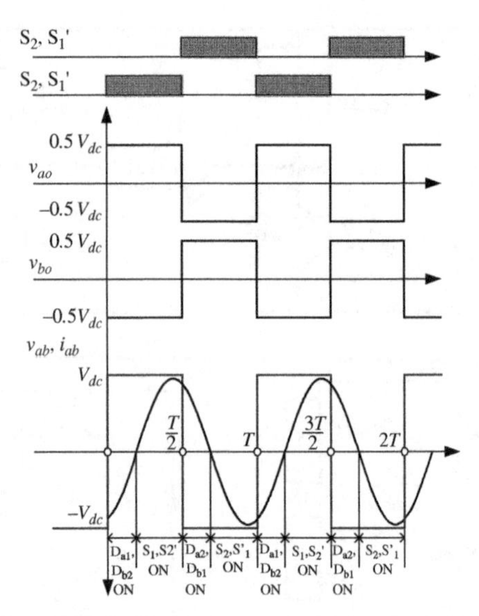

Figure 3.11 Switching signal and the output voltage and current in a full-bridge inverter

PWM. The fundamental frequency of the output is decided by the frequency of the modulating or control signal. The switching frequency of the inverter is dictated by the frequency of the carrier wave. The frequency of the carrier wave is usually much higher than the frequency of the modulating signal. In the bipolar PWM scheme, the switching of the power switches is decided by the intersection of the modulating and carrier wave. The switch pair S_1 and S_2' operate when the modulating signal is greater than the triangular carrier wave ($v_m > v_c$) and the switch pair S_2 and S_1' operate when the modulating signal is smaller than the triangular carrier wave ($v_m < v_c$). It is also noted that the operation of the upper and lower switch is still complimentary. A small deadband is introduced between the switching of the upper and lower switches in real-time implementation. The switching scheme for the bipolar PWM is not shown, as it is similar to that shown for the half-bridge inverter.

In the case of a unipolar PWM scheme, the two pairs of switches (S_1 and S_2') and (S_2 and S_1') operate simultaneously, as in the previous case. The switching signal for the upper switch S_1 is produced by comparing the positive modulating signal with the carrier signal and its complimentary signal (NOT signal) is given to the lower switch of the same leg, i.e. S_1'. The switching signal for the upper switch S_2 is produced by comparing the phase-shifted or negative modulating signal with the carrier signal and its complimentary signal (NOT signal) is given to the lower switch of the same leg, i.e. S_2'. The switching is done at the instant where $v_m > v_c$ for both S_1 and S_2 switches. The switching of the switches S_1 and S_1' and switches S_2 and S_2' are complimentary (i.e. when one is ON the other is OFF and vice versa). To obtain a better harmonic spectrum with reduced ripples in the current, a unipolar switching scheme is suggested. The switching scheme and the voltages are depicted in Figure 3.12. The leg voltages vary between $+V_{dc}/2$ and $-V_{dc}/2$ for both legs A and B. The output voltages across the load vary between $+V_{dc}$ and 0 in the positive half-cycle and $-V_{dc}$ and 0 for the negative half-cycle, thus it is called the 'unipolar PWM.'

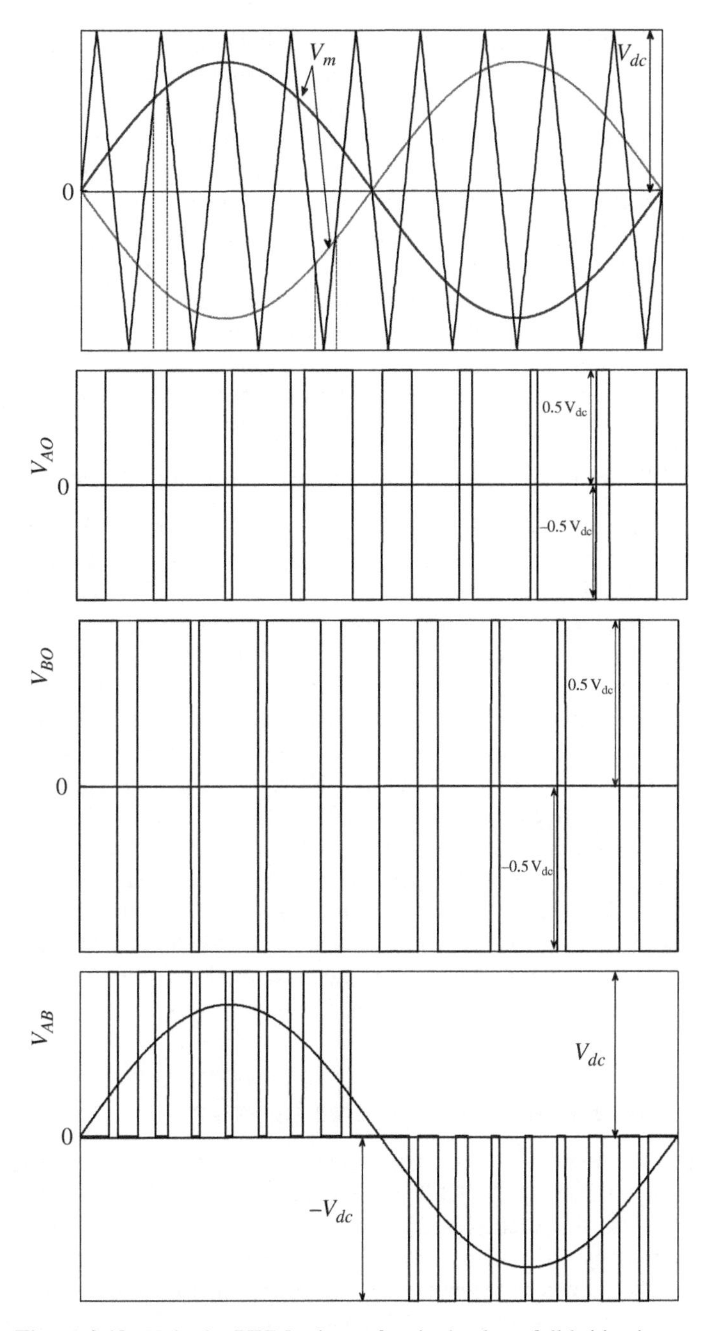

Figure 3.12 Unipolar PWM scheme for single-phase full-bridge inverter

The maximum fundamental output voltage is V_{dc} for both unipolar and bipolar PWM. In the case of bipolar PWM, the harmonic components in the output voltage are centered on m_f and multiples of m_f, i.e. $2m_f$, $3m_f$, and so on, similar to the half-bridge inverter.

In the case of a unipolar PWM scheme, the ratio between the modulating signal and the triangular carrier signal is chosen as even. The harmonic components at the output voltage is once again given by equation (3.9), but the harmonic at the switching frequency, three times the switching frequency, five times the switching frequency, etc. do not appear, i.e. to say that j is only even and k is odd. The harmonic at the carrier signal frequency is eliminated due to the fact that the output voltage is the difference between the leg voltages ($V_{AB} = V_{AO} - V_{BO}$). The harmonic in the leg voltages appears at the multiple of the switching frequency or the carrier signal frequency. Since the modulation signals are out of phase by 180 degrees, the harmonics at the switching will cancel out during subtraction at the load side.

3.3.2.1 MATLAB/Simulink Model of Single-Phase Full-Bridge Inverter

The Simulink model uses 'SimPowerSystem' blocksets of the Simulink library. The IGBT switches are chosen from the built-in library. The gating signals are generated using the unipolar SPWM principle. The implementation of bipolar PWM is left as an exercise. The modulating signals are assumed as two sinusoidal waves of unit amplitude and have a 180-degree phase shift and 50-Hz fundamental frequency. The switching frequency of the inverter is chosen as 1250 Hz (this can be changed by varying the frequency of the carrier wave), and hence the frequency of the triangular carrier wave is 1250 Hz. The gating signal generated is split into two complementary switching signals using a NOT gate (Note: the deadband is not used in the model). The load used in the simulation model is RL, with values $R = 10\,\Omega$ and $L = 100$ mH. The modulating signal is given as sinusoidal waves with a 180-degree phase shift and a peak value of 0.95 p.u. The simulation model is shown in Figure 3.13 and the resulting voltage spectrum is shown in Figure 3.14. The spectrum shows the peak magnitude of the fundamental frequency. The effect of introducing a deadband can be observed by adding the deadband circuit of Figure 3.9. This PWM can also be implemented using two 180-degree phase-shifted triangular carrier signals and one sinusoidal modulating signal. The resulting waveform will have the same quality as that of the method elaborated in this section.

3.4 Three-Phase PWM Voltage Source Inverter

The power circuit topology of a three-phase voltage source inverter is shown in Figure 3.15. Each power switch is a transistor or IGBT with antiparallel diodes. The pole or the leg voltages are denoted by a capital suffix letter V_A, V_B, and V_C and can attain the value +0.5 V_{dc} when the upper switch is operating and −0.5 V_{dc} when the lower switch is operating. The phase voltage applied to the load is denoted by the letters v_{an}, v_{bn}, and v_{cn}. The operation of the upper and the lower switches are complimentary (a small deadband is provided in real-time implementation).

The relationship between the leg voltage and switching signals are

$$V_k = S_k\left(\frac{V_{dc}}{2}\right); \quad k \in A, B, C \tag{3.11}$$

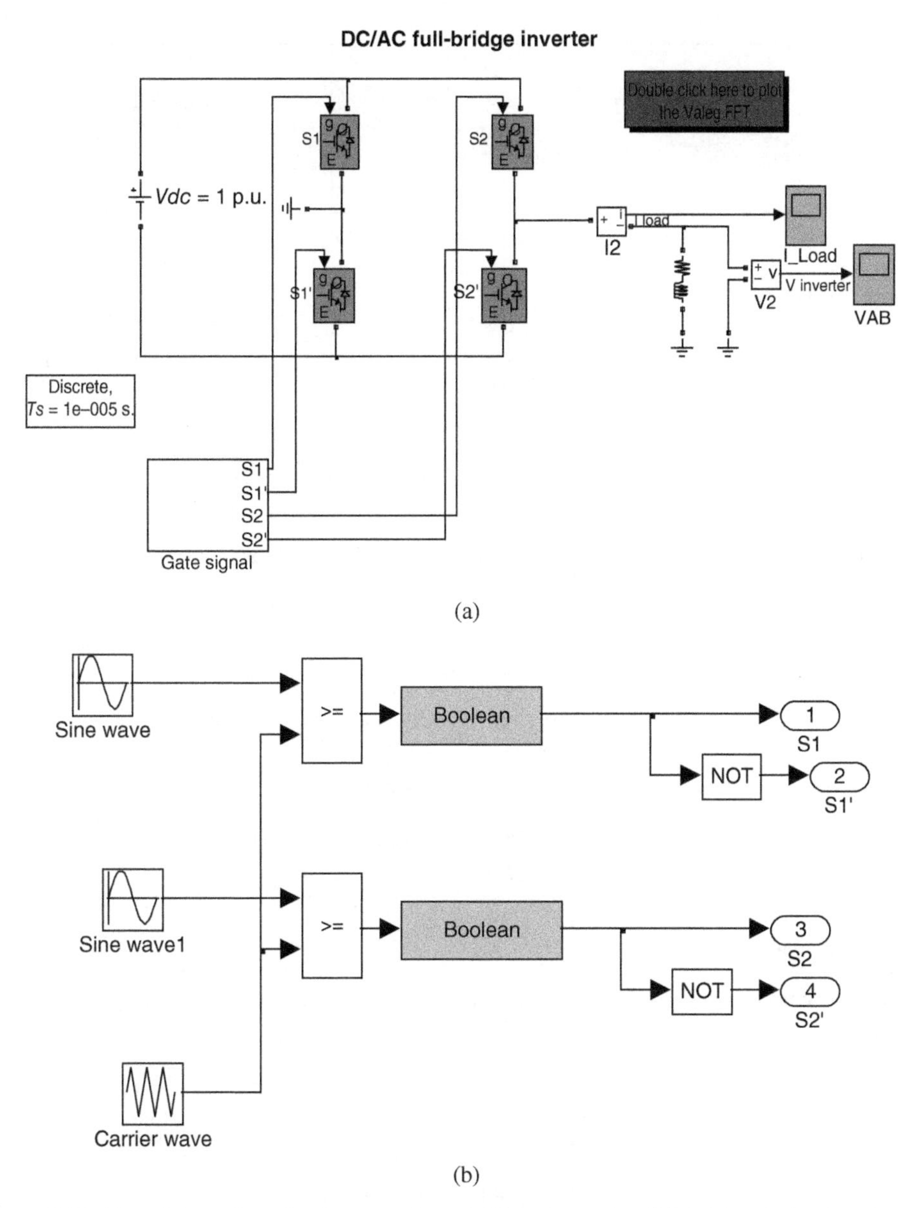

Figure 3.13 Simulink model to implement unipolar PWM scheme in a full-bridge single-phase inverter (file name: *Full_bridge_PWM.mdl*). (a) System block. (b) Gate signal generation.

where $S_k = 1$ when the upper power switch is 'ON' and $S_k = 0$ when the lower switch is 'ON.' If the load is assumed to be a star-connected three-phase, then the relation between the phase-to-neutral load voltage and the leg voltages can be written as:

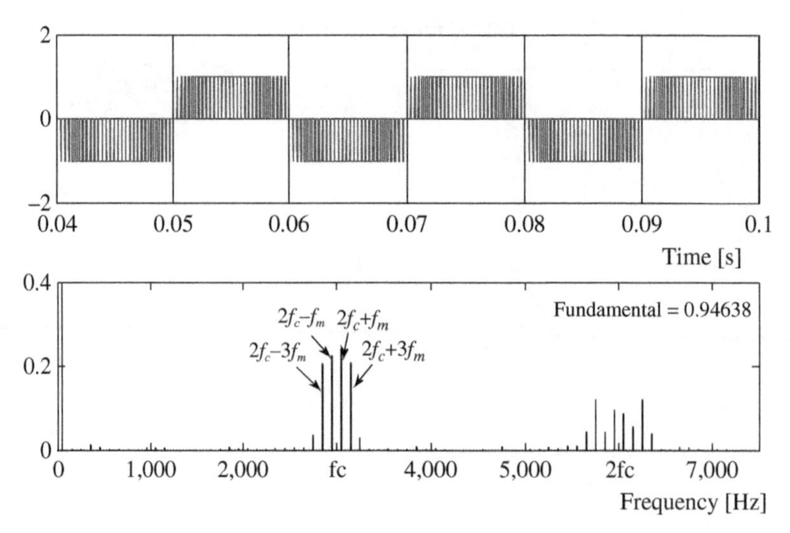

Figure 3.14 Voltage (V_{AB}) and its spectrum for unipolar PWM scheme in a single-phase inverter

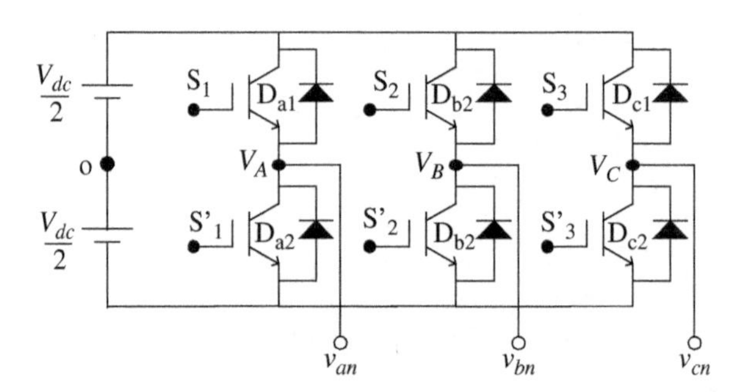

Figure 3.15 Power circuit topology of a three-phase voltage source inverter (*Source:* Based on Murai et al. [8])

$$
\begin{aligned}
V_A(t) &= v_{an}(t) + v_{no}(t) \\
V_B(t) &= v_{bn}(t) + v_{no}(t) \\
V_C(t) &= v_{cn}(t) + v_{no}(t)
\end{aligned}
\tag{3.12}
$$

where v_{no} is the voltage difference between the star point n of the load and the midpoint of the DC bus o, called the 'Common Mode Voltage'. This common mode voltage or neutral voltage is responsible for leakage bearing currents and their subsequent failure. This issue is taken up in more detail in Chapter 9.

By adding each term of the equation (3.12), and setting the sum of phase-to-neutral voltage to zero (assuming a balanced three-phase voltage whose instantaneous sum is always zero), the following is obtained:

$$v_{no}(t) = \frac{1}{3}[V_A(t) + V_B(t) + V_C(t)] \tag{3.13}$$

Substituting equation (3.13) back into equation (3.12), the following expressions for the phase-to-neutral voltages are obtained:

$$\begin{aligned} v_{an}(t) &= \frac{2}{3}V_A(t) - \frac{1}{3}[V_B(t) + V_C(t)] \\ v_{bn}(t) &= \frac{2}{3}V_B(t) - \frac{1}{3}[V_A(t) + V_C(t)] \\ v_{cn}(t) &= \frac{2}{3}V_C(t) - \frac{1}{3}[V_B(t) + V_A(t)] \end{aligned} \tag{3.14}$$

Equation (3.14) can also be written using the switching function definition of equation (3.11):

$$\begin{aligned} v_{an}(t) &= \left(\frac{V_{dc}}{3}\right)[2S_A - S_B - S_C] \\ v_{bn}(t) &= \left(\frac{V_{dc}}{3}\right)[2S_B - S_A - S_C] \\ v_{cn}(t) &= \left(\frac{V_{dc}}{3}\right)[4S_C - S_B - S_A] \end{aligned} \tag{3.15}$$

Equation (3.15) can be used to model a three-phase inverter in MATLAB/Simulink. The switching signals can be generated by analog circuits or by a digital signal processor/microprocessor for operation in six-step or PWM modes. The following section describes the operation of the inverter in both square-wave/six-step mode and PWM mode.

For inverter operation in square-wave or six-step mode, switching/gate signals are provided in such a way that the power switches change the state only twice in one fundamental cycle (OFF to ON and then ON to OFF). Each leg receives the gating signal with a phase shift of 120 degrees, so as to maintain the same phase shift among three output voltages. The output rms voltage magnitude in this case is the highest and the switching losses are minimal; however, the output voltages contain strong low-order harmonics, especially 5th and 7th. With the advent of the fast signal processing devices, it is easier to implement a PWM operation, and hence a normal step operation is avoided. The associated waveforms for six-step modes are shown in Figure 3.16. The leg voltage takes on the values +0.5 V_{dc} and –0.5 V_{dc}, and the phase voltage has six steps in one fundamental cycle. The steps in the phase voltages have amplitudes of ±1/3 V_{dc} and ±2/3 V_{dc} and the line voltage varies between $+V_{dc}$ and $-V_{dc}$, while the common mode voltage varies between +1/6 V_{dc} and −1/6 V_{dc}. During the six-step operation of the inverter, the values of the leg voltages are shown in Table 3.1.

To determine the phase-to-neutral voltages for the six-step mode, the leg voltages from Table 3.1 are substituted into equation (3.14), and the corresponding values are listed in

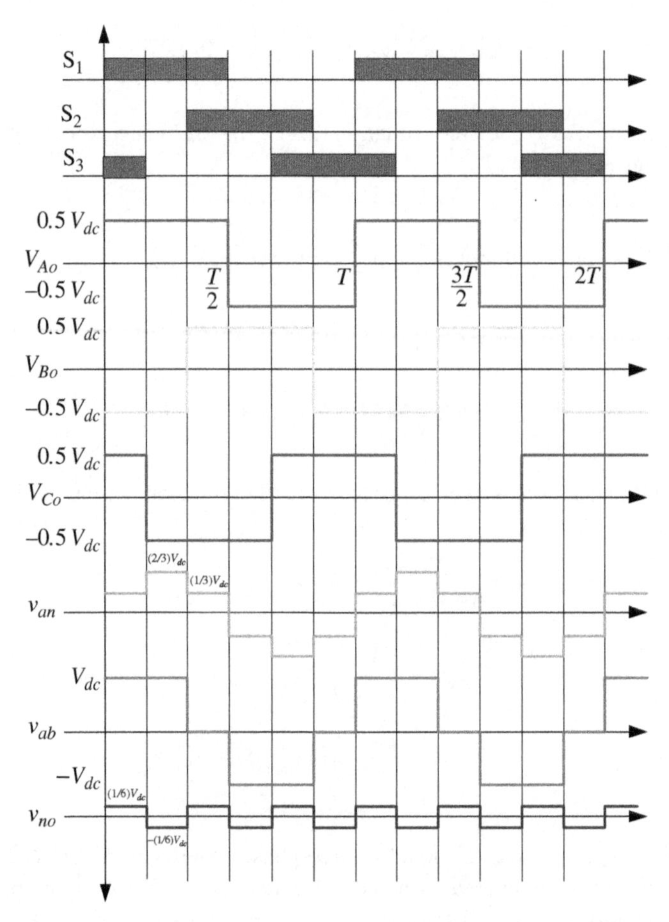

Figure 3.16 Waveforms for square-wave/six-step mode of operation of a three-phase inverter

Table 3.1 Leg/pole voltages of a three-phase VSI during six-step mode of operation

Switching mode	Switches ON	Leg voltage V_A	Leg voltage V_B	Leg voltage V_C
1	S_1, S_2', S_3	$0.5\ V_{dc}$	$-0.5\ V_{dc}$	$0.5\ V_{dc}$
2	S_1, S_2', S_3'	$0.5\ V_{dc}$	$-0.5\ V_{dc}$	$-0.5\ V_{dc}$
3	S_1', S_2, S_3'	$-0.5\ V_{dc}$	$0.5\ V_{dc}$	$-0.5\ V_{dc}$
4	S_1', S_2, S_3'	$-0.5\ V_{dc}$	$0.5\ V_{dc}$	$-0.5\ V_{dc}$
5	S_1', S_2, S_3	$-0.5\ V_{dc}$	$0.5\ V_{dc}$	$0.5\ V_{dc}$
6	S_1', S_2', S_3	$-0.5\ V_{dc}$	$-0.5\ V_{dc}$	$0.5\ V_{dc}$

Table 3.2 Phase-to-neutral voltages for six-step mode of operation

Switching mode	Switches ON	Phase voltage v_{an}	Phase voltage v_{bn}	Phase voltage v_{cn}
1	S_1, S'_2, S_3	1/3 V_{dc}	−2/3 V_{dc}	1/3 V_{dc}
2	S_1, S'_2, S'_3	2/3 V_{dc}	−1/3 V_{dc}	−1/3 V_{dc}
3	S_1, S_2, S'_3	1/3 V_{dc}	1/3 V_{dc}	−2/3 V_{dc}
4	S'_1, S_2, S'_3	−1/3 V_{dc}	2/3 V_{dc}	−1/3 V_{dc}
5	S'_1, S_2, S_3	−2/3 V_{dc}	1/3 V_{dc}	1/3 V_{dc}
6	S'_1, S'_2, S_3	−1/3 V_{dc}	−1/3 V_{dc}	2/3 V_{dc}

Table 3.2 for a star-connected load. The line voltages are obtained using equation (3.16) and are listed in Table 3.3:

$$\begin{aligned} v_{ab} &= v_{an} - v_{bn} \\ v_{bc} &= v_{bn} - v_{cn} \\ v_{ca} &= v_{cn} - v_{an} \end{aligned} \tag{3.16}$$

The maximum output peak phase-to-neutral voltage in the six-step mode is 0.6367 V_{dc} or $(2/\pi)$ V_{dc} and that of the line-to-line voltage is 1.1 V_{dc}. The Fourier series of phase-to-neutral voltage and line-to-line voltage can be obtained as:

$$v_{an}(t) = \frac{2}{\pi} V_{DC}\left[\sin \omega t + \frac{1}{5}\sin 5\omega t + \frac{1}{7}\sin 7\omega t + \frac{1}{11}\sin 11\omega t + \frac{1}{13}\sin 13\omega t + \cdots\right] \tag{3.17}$$

$$v_{ab}(t) = \frac{2\sqrt{3}}{\pi} V_{DC}\left[\sin\left(\omega t - \frac{\pi}{6}\right) + \frac{1}{5}\sin 5\left(\omega t - \frac{\pi}{6}\right) + \frac{1}{7}\sin 7\left(\omega t - \frac{\pi}{6}\right) + \cdots\right] \tag{3.18}$$

The other two phase voltages will be 120 degree phase- shifted from v_{an}. Similarly, the two line voltages will be 120 degree phase-shifted from v_{ab}

The Fourier series of a phase-to-neutral voltage and the line-to-line voltage shows that the triplen harmonic (multiple of three) does not appear (since the load is assumed with isolated

Table 3.3 Line voltages for six-step mode of operation

Switching mode	Switches ON	Line voltage v_{ab}	Line voltage v_{bc}	Line voltage v_{ca}
1	S_1, S'_2, S_3	V_{dc}	$-V_{dc}$	0
2	S_1, S'_2, S'_3	V_{dc}	0	$-V_{dc}$
3	S_1, S_2, S'_3	0	V_{dc}	$-V_{dc}$
4	S'_1, S_2, S'_3	$-V_{dc}$	V_{dc}	0
5	S'_1, S_2, S_3	$-V_{dc}$	0	V_{dc}
6	S'_1, S'_2, S_3	0	$-V_{dc}$	V_{dc}

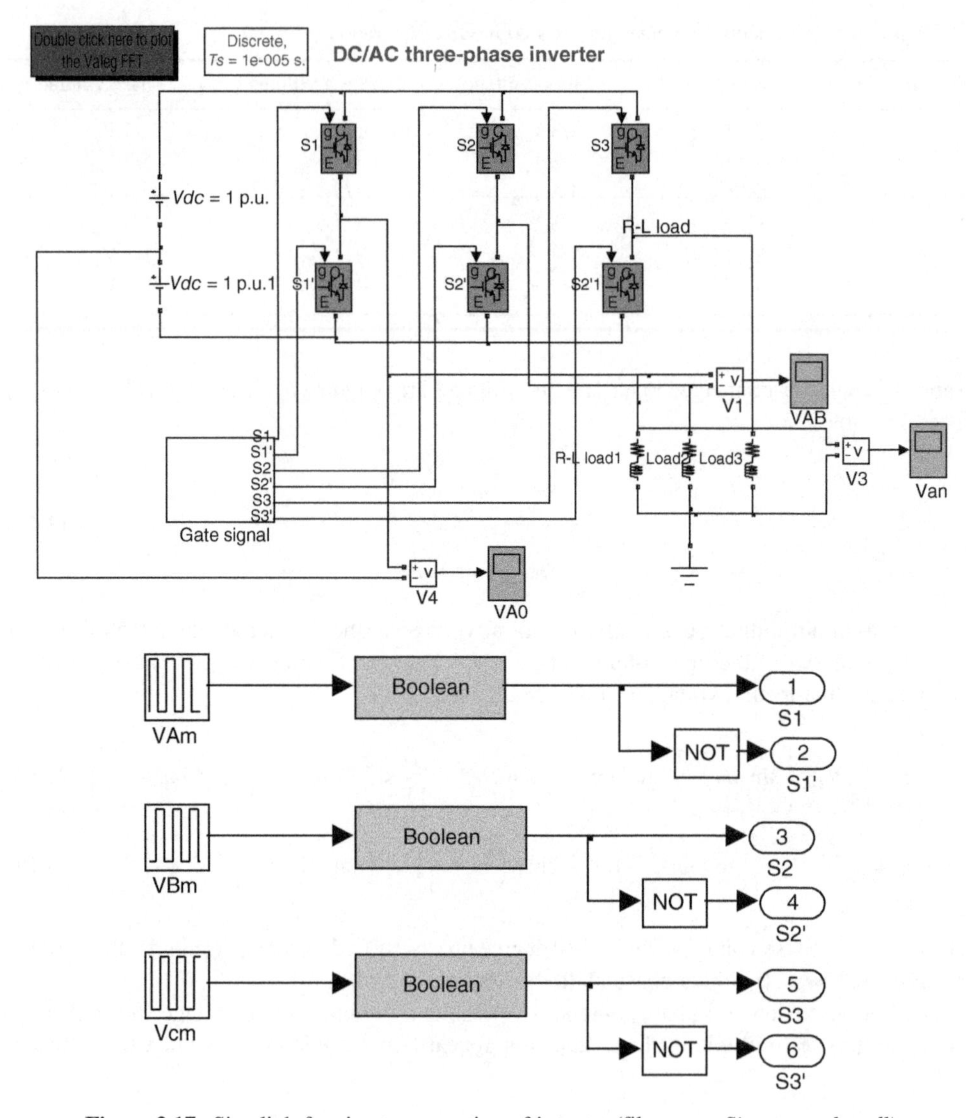

Figure 3.17 Simulink for six-step operation of inverter (file name: *Six-stepmode.mdl)*

neutral). The other harmonics have the inverse amplitude of their order. The output voltage magnitude can be controlled by controlling the DC-link voltage V_{dc}.

The MATLAB/Simulink model for generating the six-step waveform is given in Figure 3.17. The gating signal is generated using the repetitive block of the Simulink where the switching signal values are stored. In the six-step mode, the switching is done at the fundamental frequency, and hence the period of the gate drive signal depends upon the fundamental frequency of the output.

3.4.1 Carrier-Based Sinusoidal PWM

In the carrier-based SPWM, three-phase sinusoidal wave:

$$\begin{aligned} v_{Am} &= V_m \sin(\omega t) \\ v_{Bm} &= V_m \sin\left(\omega t - 2\frac{\pi}{3}\right) \\ v_{Cm} &= V_m \sin\left(\omega t + 2\frac{\pi}{3}\right) \end{aligned} \tag{3.19}$$

are used for the modulating or control signal and compared with a high-frequency triangular wave. One common triangular wave is used for the comparison of all three phases. The upper switch of leg A is ON when the modulating signal of phase A is greater than the amplitude of the triangular carrier wave, i.e. $v_{Am} \geq V_c$. The lower switch has a complementary operation. A small deadband is provided between the switching OFF of upper switch and switching ON of the lower switch and vice versa. The waveforms associated with the SPWM are given in Figure 3.18. The leg voltage, phase voltage, and line voltage are illustrated. The waveform is shown for $m_f = 9$.

The MATLAB/Simulink model for implementing a SPWM is shown in Figure 3.19a (power circuit is the same as that in Figure 3.17). The modulating index chosen is 0.95 (the reference modulating signal is given as $0.95 \times 0.5\ V_{dc}$, where $V_{dc} = 1$ p.u.). The switching frequency or the carrier frequency is chosen as 1500 Hz (hence, $m_f = 1500/50 = 30$). No deadband is shown in the simulation. The resulting phase-to-neutral voltage waveform and its spectrum are given in Figure 3.19b. The harmonics appear as the sideband of the switching frequency.

The choice of m_f is important from the harmonic spectrum point of view. The harmonic spectrum for six-step phase voltage is shown in Figure 3.20. Normally, this frequency ration m_f is chosen to be an odd multiple of three. This will ensure that the triple harmonic (multiple of three) will be eliminated from the motor phase currents since for the three-phase inverter, these harmonics are in phase and cancel out.

In an adjustable speed drive system with constant *v/f* control, variable frequency output is required. If the frequency modulation ration m_f is kept constant, the switching frequency will be constant and high at a high fundamental output frequency ($f_c = m_f f_m$). Therefore, the frequency modulation ratio m_f is varied for different frequency operations of the drive system. Changing the frequency modulation ratio in essence changes the slope of the line between f_c and f_{m}. There is a sudden jump in the slope of the line and this may cause jittering at the transition points. A typical relationship is shown in Figure 3.21. To avoid this, usually a hysteresis block is used. For a low range of fundamental output frequencies up to say f_L, asynchronous PWM is employed. The synchronous PWM method is used until it reaches a frequency value f_M and then a six-step operation is initiated. The typical value of f_L and f_M are 10 and 50 Hz, respectively.

3.4.2 Third-Harmonic Injection Carrier-Based PWM

The output voltage using the SPWM technique is limited to 0.5 V_{dc}. If the SPWM technique is used in motor drive applications, the available voltage may not be sufficient to run the motor at

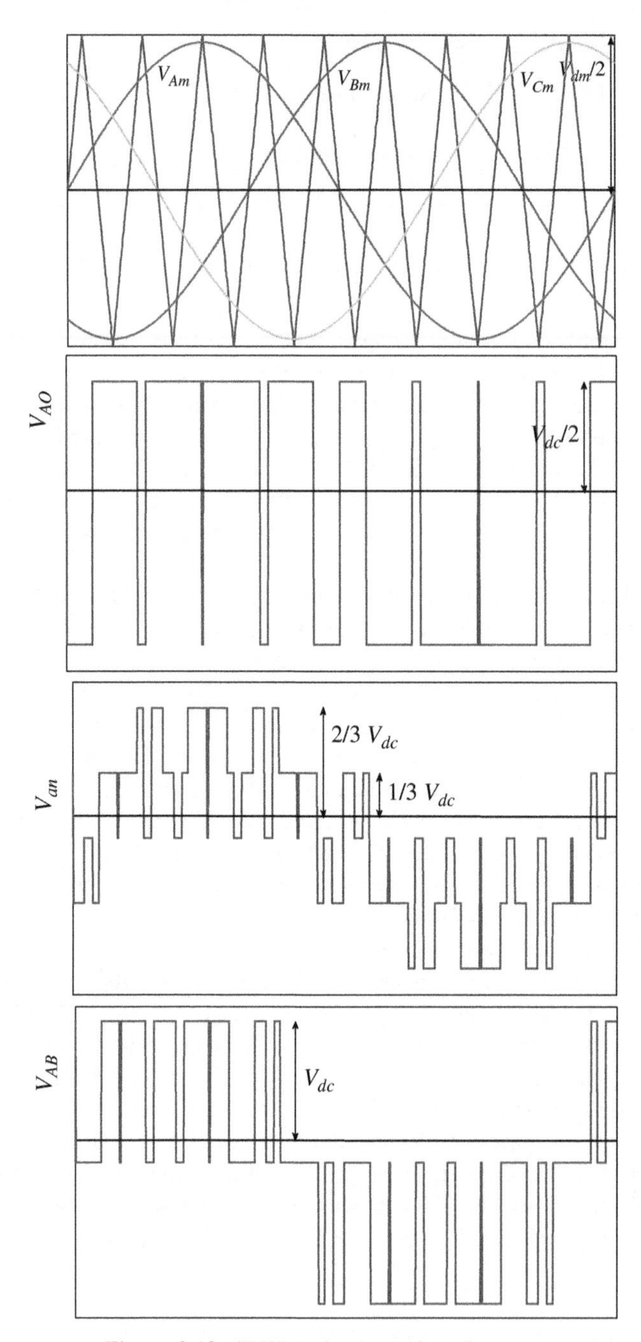

Figure 3.18 SPWM of a three-phase inverter

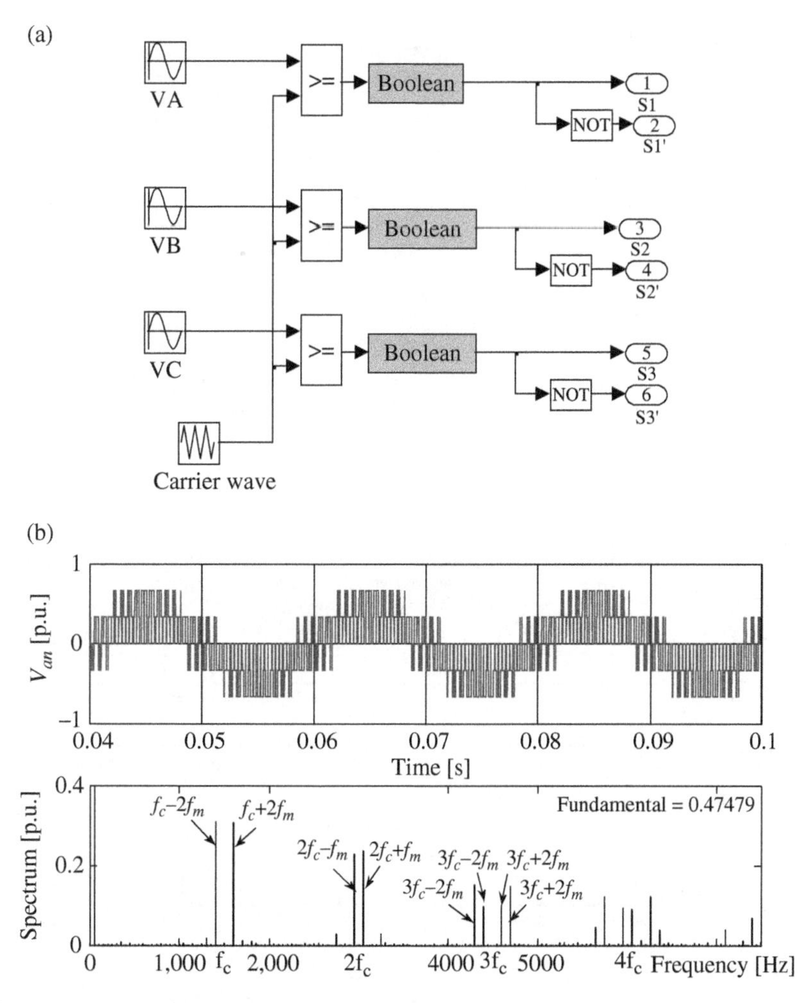

Figure 3.19 (a) Gate signal generation in MATLAB for three-phase inverter (File name: *PWM_Three_phase_CB.mdl)* and (b) phase-to-neutral voltage and spectrum for SPWM

rated condition. In this situation, the machine needs to be derated and a reduced torque is produced. To enhance the output voltage from the PWM inverter using carrier-based scheme, third-harmonic injection in the modulating signal is done. It is shown that by adding an appropriate third-harmonic component of the modulating signal in the fundamental modulating signal leads to a reduction in the peak of the resultant modulating signal. Hence, the peak value of the resulting modulated signals can be increased beyond 0.5 V_{dc} and that leads to the higher inverter output voltage. The injected third-harmonic component in the modulated signal or reference leg voltages cancels out in the legs and does not appear in the output phase voltages. Thus, the output voltage does not contain the undesired low-order harmonics. The optimal level of the third-harmonic injection can be determined by considering the modulating signals:

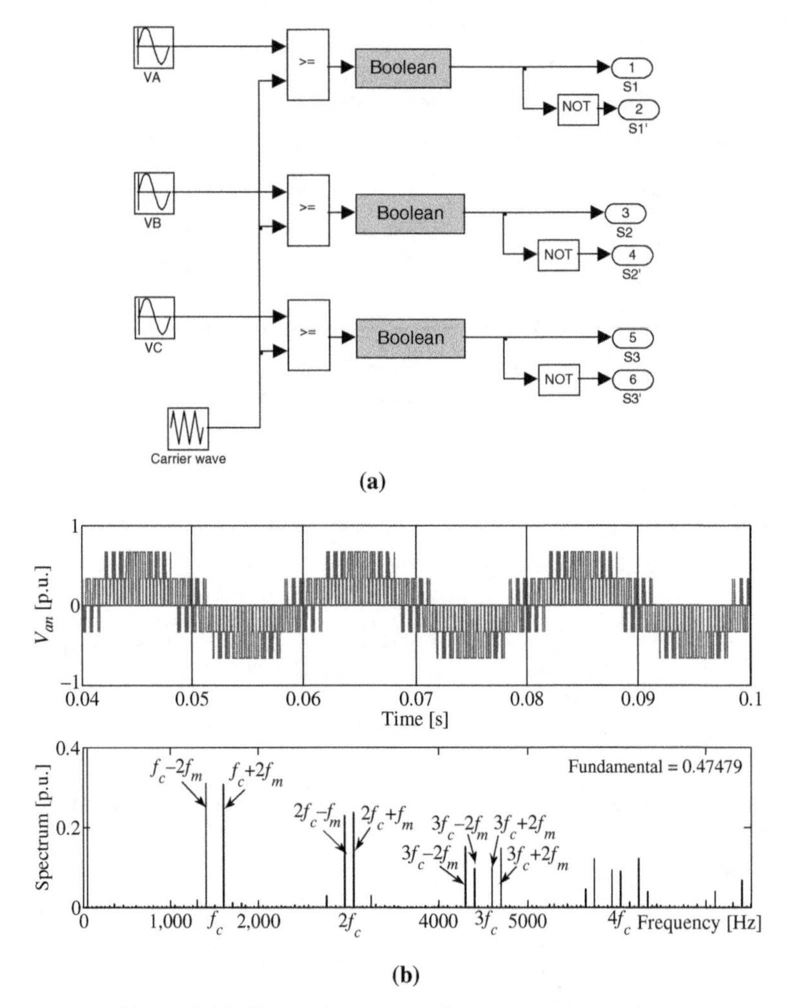

Figure 3.20 Harmonic spectrum for six-step phase voltage

$$
\begin{aligned}
v_{Am} &= V_{m1}\sin(\omega t) + V_{m3}\sin(3\omega t)\\
v_{Bm} &= V_m \sin\left(\omega t - 2\frac{\pi}{3}\right) + V_{m3}\sin(3\omega t)\\
v_{Cm} &= V_m \sin\left(\omega t + 2\frac{\pi}{3}\right) + V_{m3}\sin(3\omega t)
\end{aligned}
\tag{3.20}
$$

For the SPWM without harmonic injection, the fundamental peak magnitude of the output voltage is 0.5 V_{dc}. It is noted that the third harmonic has no effect on the value of the reference/modulating waveform when $\omega t = (2k+1)\frac{\pi}{3}$, since $\sin\left(3(2k+1)\frac{\pi}{3}\right) = 0$ for all odd k. Thus, the third harmonic is chosen to make the peak magnitude of the reference of equation (3.20)

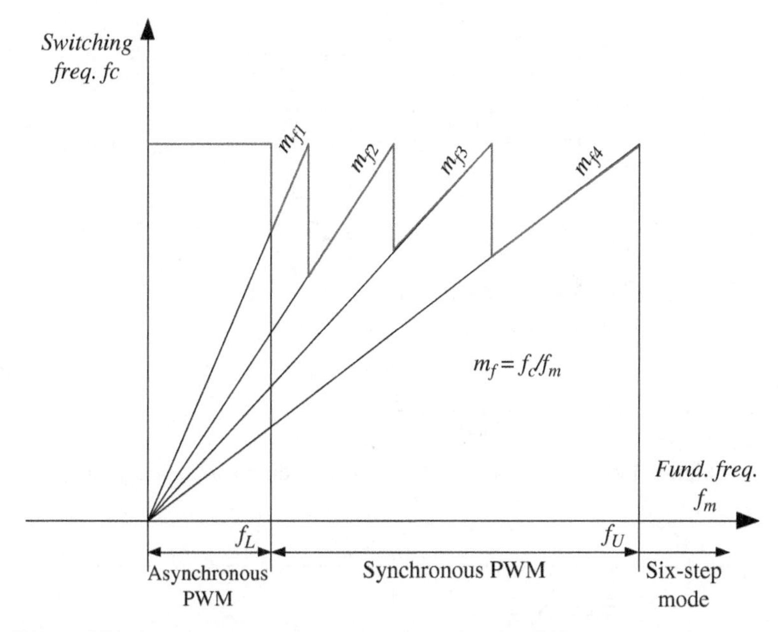

Figure 3.21 Varying frequency modulation ratios for different output frequency

occur where the third harmonic is zero. This ensures the maximum possible value of the fundamental component. The reference voltage reaches a maximum when

$$\frac{dv_{am}}{d\omega t} = V_{m1}\cos(\omega t) + 3V_{m3}\cos(3\omega t) = 0 \tag{3.21}$$

This yields

$$V_{m3} = -\frac{1}{3}V_{m1}\cos\left(\frac{\pi}{3}\right) \quad \text{for } \omega t = \frac{\pi}{3} \tag{3.22}$$

So the maximum modulation index can be

$|v_{am}| = \left|V_{m1}\sin(\omega t) - \frac{1}{3}V_{m1}\cos\left(\frac{\pi}{3}\right)\sin(3\omega t)\right| = 0.5\ V_{dc}$, The above equation gives

$$V_{m1} = \frac{0.5\ V_{dc}}{\sin\left(\frac{\pi}{3}\right)} \quad \text{for } \omega t = \frac{\pi}{3} \tag{3.23}$$

Thus, the output fundamental voltage is increased by 15.47% of the value obtainable using simple SPWM by injecting 1/6th third harmonic in the fundamental.

The implementation block diagram of the proposed method of PWM is shown in Figure 3.22. The reference modulating signal, the third-harmonic signal, and the resultant modulating signal after injecting third harmonic are shown in Figure 3.23.

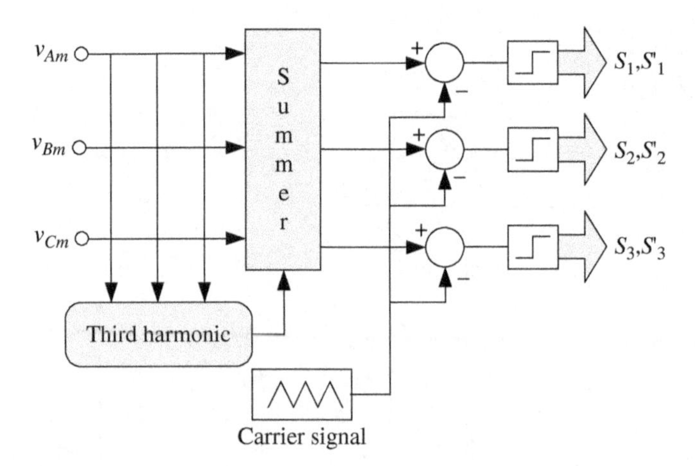

Figure 3.22 Block diagram of SPWM with third-harmonic injection

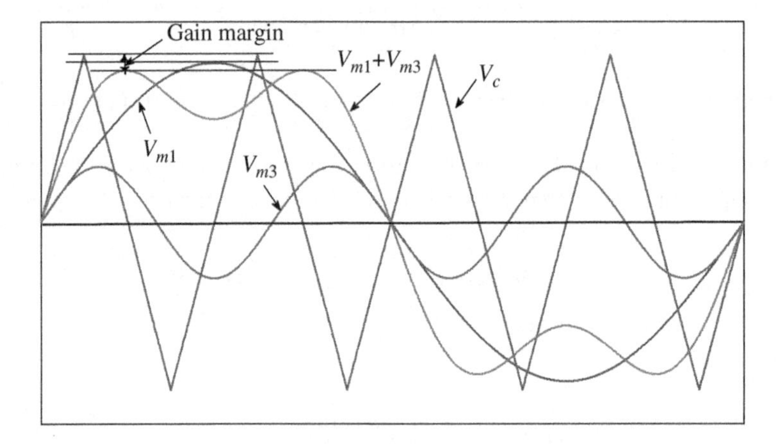

Figure 3.23 Modulating signals and carrier wave for third-harmonic injection PWM

3.4.3 MATLAB/Simulink Model for Third-Harmonic Injection PWM

The MATLAB/Simulink model of third-harmonic injection-based PWM is shown in Figure 3.24. The power circuit remains the same as that of Figure 3.17 and the gate signal generation is shown in Figure 3.24. The output voltage spectrum is shown in Figure 3.25, where the output is increased by 15.47% when compared to the simple SPWM scheme. The reference of the modulating signal is increased to 0.575 V_{dc}.

3.4.4 Carrier-Based PWM with Offset Addition

The output voltage magnitude from a three-phase voltage source inverter can be enhanced by adding an offset to the sinusoidal modulating signal. The offset addition PWM scheme follows

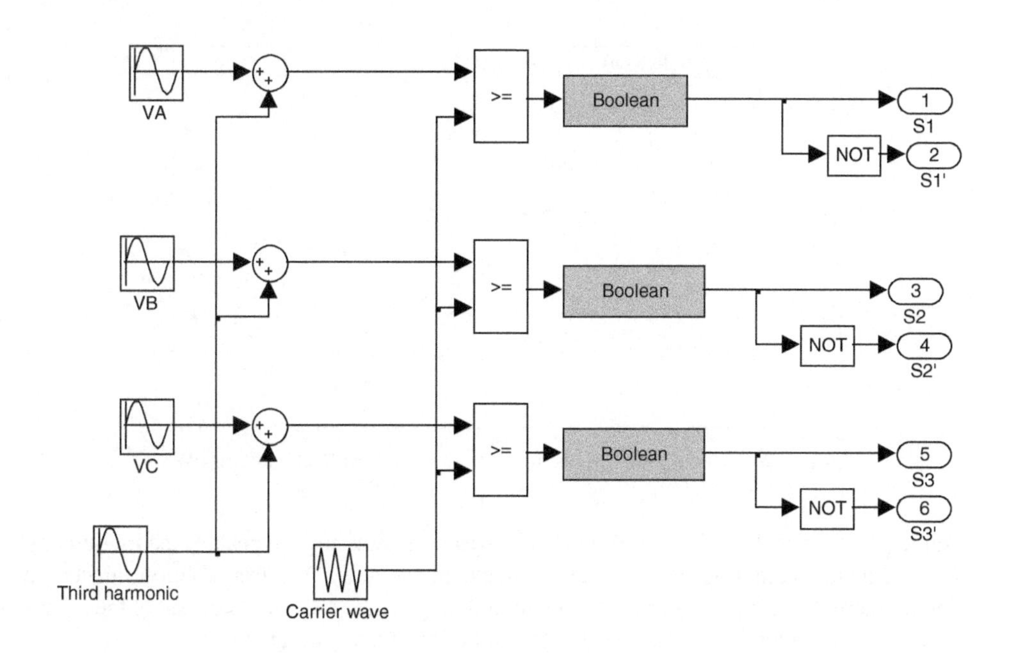

Figure 3.24 Gate signal generation for third-harmonic injection PWM (file name: *PWM_three-phase_CB_3rd_harmonic.mdl)*

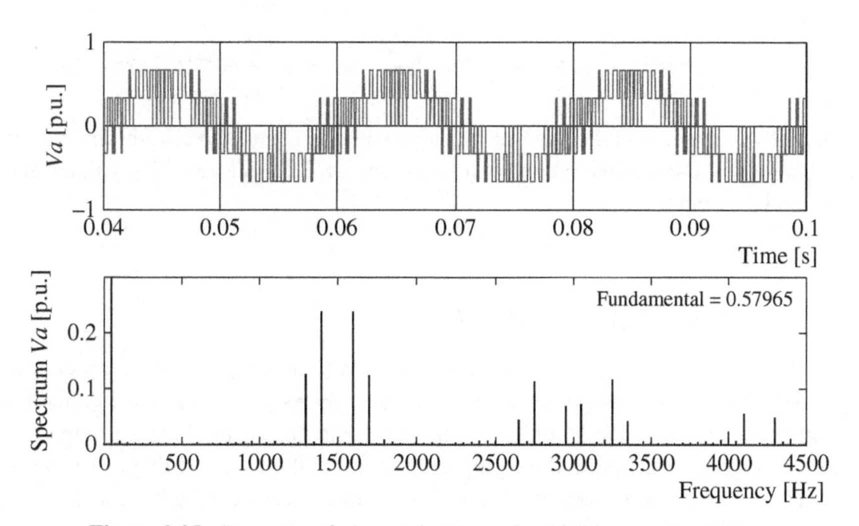

Figure 3.25 Spectrum of phase ‘a’ voltage for third-harmonic injection

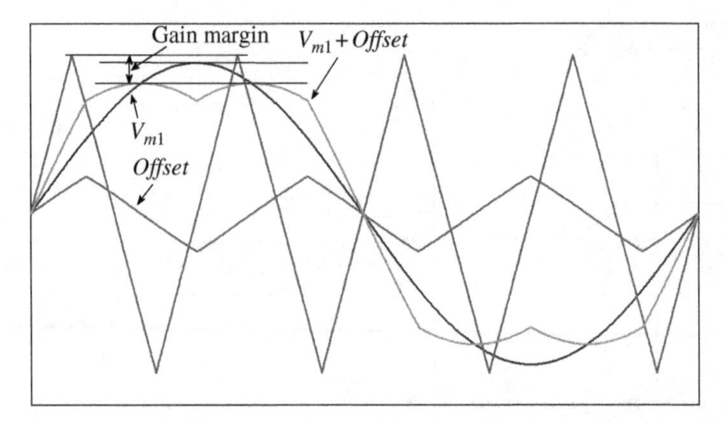

Figure 3.26 Modulating signals and carrier wave for offset addition PWM

the same principle as that of the third-harmonic injection, reducing the peak of the modulating signal and hence increasing the modulation index. In essence, the offset addition injects the triplen harmonics (multiple of three) into the modulating signal, reducing the peak of the resultant, as shown in Figure 3.26. The resultant modulating signals are given as:

$$\begin{aligned}
v_{Am} &= V_{m1}\sin(\omega t) + offset\\
v_{Bm} &= V_m \sin\left(\omega t - 2\frac{\pi}{3}\right) + offset\\
v_{Cm} &= V_m \sin\left(\omega t + 2\frac{\pi}{3}\right) + offset
\end{aligned} \tag{3.24}$$

where the offset is given as:

$$Offset = -\frac{V_{max}+V_{min}}{2};\, V_{max} = max\{v_{Am}, v_{Bm}, v_{Cm}\},\, V_{min} = min\{v_{Am}, v_{Bm}, v_{Cm}\} \tag{3.25}$$

The output voltage magnitude reaches the same value as that of the third-harmonic injection PWM. The MATLAB/Simulink model is shown in Figure 3.27. The resulting voltage and spectrum are shown in Figure 3.28.

3.4.5 Space Vector PWM (SVPWM)

The SVPWM technique is one of the most popular PWM techniques due to a higher DC bus voltage use (higher output voltage when compared with the SPWM) and easy digital realization [9]. The concept of the SVPWM relies on the representation of the inverter output as space vectors or space phasors. Space vector representation of the output voltages of the inverter is realized for the implementation of SVPWM. Space vector simultaneously represents three-phase quantities as one rotating vector; hence, each phase is not considered separately. The three phases are assumed as only one quantity. The space vector representation is valid for

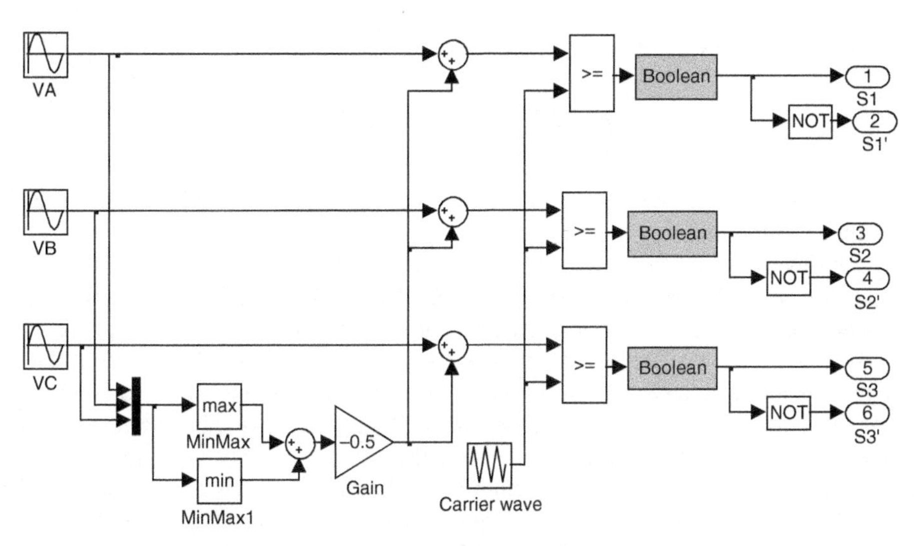

Figure 3.27 MATLAB/Simulink for offset addition PWM (file name: *PWM_3_phase_CB_offset.mdl*)

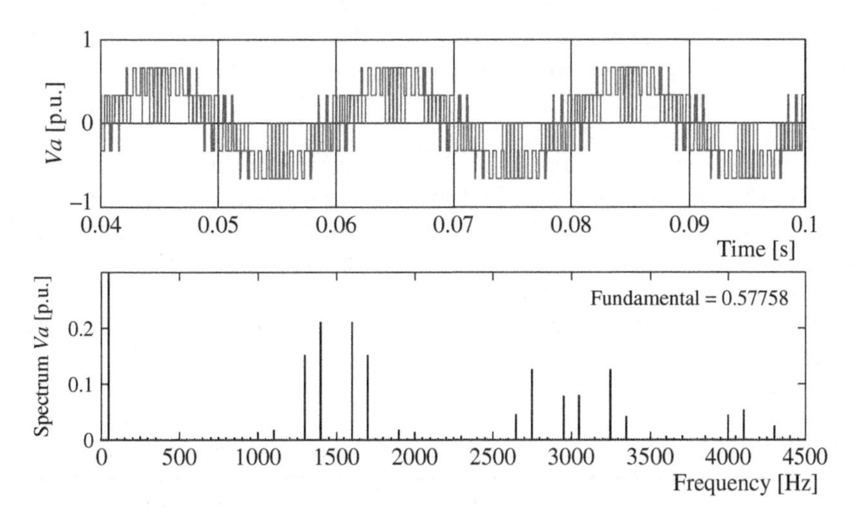

Figure 3.28 Output voltage and voltage spectrum for offset addition PWM

both transient and steady-state conditions in contrast to phasor representation, which is valid only for steady-state conditions. Space vector representation is valid for sinusoidal and non-sinusoidal signals, while phasor only represents sinusoidal signals. The concept of the space vector arises from the rotating air-gap MMF in a three-phase induction machine. By supplying balanced three-phase voltages to the three-phase balanced winding of a three-phase induction machine, rotating MMF is produced, which rotates at the same speed as that of individual voltages, with an amplitude of 1.5 times the individual voltage amplitude.

The space vector is defined as:

$$\underline{f}_s = \frac{2}{3}\left[f_a + e^{j2\frac{\pi}{3}} f_b + e^{j4\frac{\pi}{3}} f_c \right] \tag{3.26}$$

where f_a, f_b, and f_c are the three-phase quantities of voltages, currents, and fluxes. In the inverter PWM, the voltage space vectors are considered.

Since the inverter can attain either +0.5 V_{dc} or −0.5 V_{dc} (if the DC bus has midpoint) or V_{dc}, 0, i.e. only two states, the total possible outputs are 2^3 = 8 (000, 001, 010, 011, 100, 101, 110, 111). Here, 0 indicates the upper switch is 'OFF' and 1 represents the upper switch is 'ON'. Thus, there are six active switching states (power flows from the input/DC-link side of the inverter to the output/load side of the inverter) and two zero-switching states (no power transfer from input/DC link to the load/load side). During zero-switching states, all the three upper switches or three lower switches are ON at the same time. Thus, the output voltages are zero, hence called zero states. The operation of the lower switches is complimentary. Using equation (3.26) and Table 3.2, the possible space vectors are computed and listed in Table 3.4.

The space vectors are shown graphically in Figure 3.29. The tips of the space vectors, when joined together, form a hexagon. The hexagon consists of six distinct sectors spanning over 360 degrees (one sinusoidal wave cycle corresponds to one rotation of the hexagon) with each sector of 60 degrees. Space vectors 1, 2, … 6 are called active state vectors, and 7 and 8 are called zero-state vectors. The zero-state vectors are redundant but they are used to minimize the switching frequency. The space vectors are stationary, while the reference vector v_s^* is rotating at the speed of the fundamental frequency of the inverter output voltage. It circles once for one cycle of the fundamental frequency.

The reference voltage follows a circular trajectory in a linear modulation range and the output is sinusoidal. The reference trajectory will change in overmodulation and the trajectory will be a hexagon boundary when the inverter is operating in the six-step mode. In implementing the SVPWM, the reference voltage is synthesized using the nearest two neighboring active vectors and zero vectors. The choice of the active vectors depends upon the sector number in which the reference is located. Hence, it is important to locate the position of the reference voltage. Once the reference vector is located, the vectors to be used for the SVPWM implementation to be

Table 3.4 Phase-to-neutral space vector for a three-phase voltage source inverter

Switching states	Space vector number	Phase-to-neutral voltage space vectors
000	7	0
001	5	$\frac{2}{3}V_{dc}e^{j\frac{4\pi}{3}}$
010	3	$\frac{2}{3}V_{dc}e^{j\frac{2\pi}{3}}$
011	4	$\frac{2}{3}V_{dc}e^{j\pi}$
100	1	$\frac{2}{3}V_{dc}e^{j0}$
101	6	$\frac{2}{3}V_{dc}e^{j\frac{5\pi}{3}}$
110	2	$\frac{2}{3}V_{dc}e^{j\frac{\pi}{3}}$
111	8	0

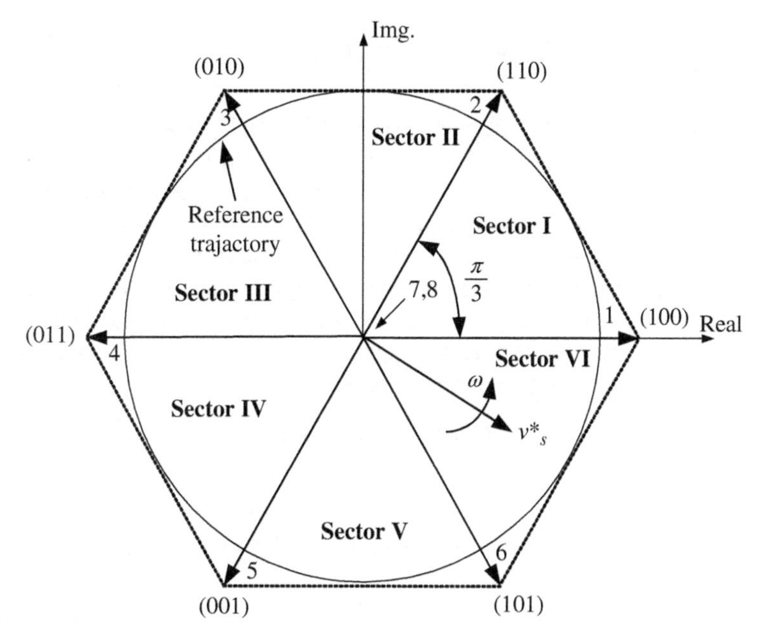

Figure 3.29 Voltage space vector locations corresponding to different switching states

identified. After identifying the vectors to be used, the next task is to find the time of application of each vector, called the 'dwell time'. The output voltage frequency of the inverter is the same as the speed of the reference voltage and the output voltage magnitude is the same as the magnitude of the reference voltage.

The maximum modulation index or the maximum output voltage that is achievable using the SVPWM technique is the radius of the largest circle that can be inscribed within the hexagon. This circle is tangential to the midpoints of the lines joining the ends of the active space vector. Thus, the maximum obtainable fundamental output voltage is calculated from the right-angled triangle (Figure 3.30) as:

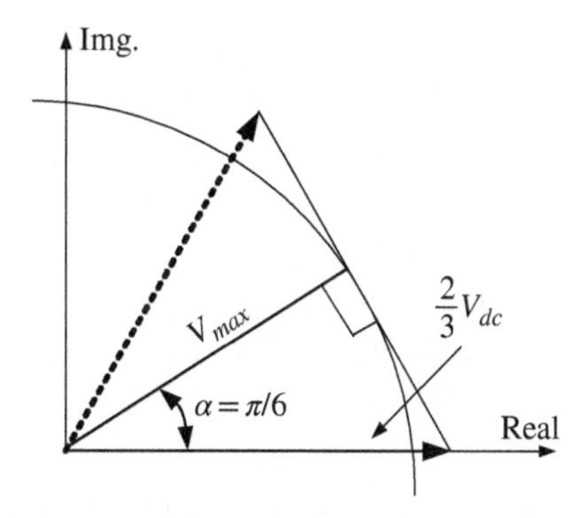

Figure 3.30 Determining the maximum possible output using SVPWM

$$V_{max} = \left(\frac{2}{3}\right) V_{dc} \cos\left(\frac{\pi}{6}\right) = \frac{1}{\sqrt{3}} V_{dc} \tag{3.27}$$

The maximum possible output voltage using SPWM is 0.5 V_{dc}, and hence the increase in the output, when using SVPWM, is (2/3 V_{dc})/(0.5 V_{dc}) = 1.154 which is 15.4%.

The time of applications of the different space vectors are calculated using the 'equal volt-second principle'. According to this principle, the product of the reference voltage and the sampling/switching time (T_s) must be equal to the product of the applied voltage vectors and their time of applications, assuming that the reference voltage remains fixed during the switching interval. When the reference voltage is in sector I, the reference voltage can be synthesized using the vectors V_1, V_2, and V_o (zero vector), applied for time t_a, t_b, and t_o, respectively. Hence, using the equal volt-second principle, for sector I:

$$v_s^* T_s = V_1 t_a + V_2 t_b + V_o t_o \tag{3.28a}$$

where

$$T_s = t_a + t_b + t_o \tag{3.28b}$$

The space vectors are given as:

$$v_s^* = \left|v_s^*\right| e^{j\alpha} \quad V_1 = \frac{2}{3} V_{dc} e^{jo} \quad V_2 = \frac{2}{3} V_{dc} e^{j\frac{\pi}{3}} \quad V_o = 0 \tag{3.29}$$

Substituting equation (3.29) into equation (3.28a) and separating the real (a-axis) and imaginary (β-axis) components:

$$\left|v_s^*\right| \cos(\alpha) T_s = \frac{2}{3} V_{dc} t_a + \frac{2}{3} V_{dc} \cos\left(\frac{\pi}{3}\right) t_b \tag{3.30a}$$

$$\left|v_s^*\right| \sin(\alpha) T_s = \frac{2}{3} V_{dc} \sin\left(\frac{\pi}{3}\right) t_b \tag{3.30b}$$

where $\alpha = \pi/3$.

Solving equations (3.30a) and (3.30b) for the time of applications t_a and t_b:

$$\begin{aligned} t_a &= \frac{\sqrt{3}\left|v_s^*\right|}{V_{dc}} \sin\left(\frac{\pi}{3} - \alpha\right) T_s \\ t_b &= \frac{\sqrt{3}\left|v_s^*\right|}{V_{dc}} \sin(\alpha) T_s \\ t_o &= T_s - t_a - t_b \end{aligned} \tag{3.31}$$

Generalizing the above equation for six sectors gives the following, where $k = 1, 2, \dots 6$ is the sector number:

$$\begin{aligned} t_a &= \frac{\sqrt{3}\left|v_s^*\right|}{V_{dc}} \sin\left(k\frac{\pi}{3} - \alpha\right) T_s \\ t_b &= \frac{\sqrt{3}\left|v_s^*\right|}{V_{dc}} \sin\left(\alpha - (k-1)\frac{\pi}{3}\right) T_s \\ t_o &= T_s - t_a - t_b \end{aligned} \tag{3.32}$$

After locating the reference location and calculating the dwell time, the next step in SVPWM implementation is the determination of the switching sequence. The requirement is the minimum number of switchings to reduce switching loss, ideally one power switch should turn 'ON' and turn 'OFF' in one switching period.

To obtain a fixed switching frequency and optimum harmonic performance from SVPWM, each leg should change its state only once in each switching period. This is achieved by applying the zero-state vector followed by two adjacent active state vectors in a half-switching period. The next half of the switching period is the mirror image of the first half. The total switching period is divided into seven parts, the zero vector (000) is applied for (1/4)th of the total zero vector time, followed by the application of active vectors for half of their application time and then again zero vector (111) is applied for (1/4)th of the zero vector time. This is then repeated in the next half of the switching period. This is how symmetrical SVPWM is obtained. The switching patterns showing the leg voltages in one switching period is depicted in Figure 3.31 for sectors I–VI.

Average leg voltages can be determined from the switching patterns (using equal volt-second principle) of Figure 3.31, for example, for sector I:

$$
\begin{aligned}
V_{A,avg} &= \frac{(V_{dc}/2)}{T_s}[t_0 + t_a + t_b - t_0] \\
V_{B,avg} &= \frac{(V_{dc}/2)}{T_s}[t_0 - t_a + t_b - t_0] \\
V_{C,avg} &= \frac{(V_{dc}/2)}{T_s}[t_0 - t_a - t_b - t_0]
\end{aligned}
\tag{3.33}
$$

Substituting the equation of time of applications from equation (3.31) into equation (3.33), the following is obtained for sector I and similarly can be derived for other sectors:

$$
\begin{aligned}
&V_{A,avg} = \frac{\sqrt{3}}{2}\left|v_s^*\right| \sin\left(\alpha + \frac{\pi}{3}\right), \quad V_{B,avg} = \frac{3}{2}\left|v_s^*\right| \sin\left(\alpha - \frac{\pi}{6}\right), \\
&V_{C,avg} = -\frac{\sqrt{3}}{2}\left|v_s^*\right| \sin\left(\alpha + \frac{\pi}{3}\right)
\end{aligned}
\tag{3.34}
$$

3.4.6 Discontinuous Space Vector PWM

Discontinuous SVPWM results when one of the two zero vectors is not used in the implementation of the SVPWM. One leg of the inverter does not switch during the whole switching period and remains tied to either the positive or negative DC bus [10, 11]. This is known as discontinuous SVPWM, since the switching is not continuous. Due to the manipulation of the zero space vector application in a switching period, one branch of the inverter remains unmodulated during one switching interval. Switching takes place in two branches: one branch is either tied to the positive DC bus or the negative DC bus, as shown in Figure 3.32 (when zero voltage (000) is eliminated, the leg voltage is tied to positive DC bus 0.5 V_{dc} and when zero voltage (111) is eliminated, the leg voltage is tied to the negative bus voltage –0.5 V_{dc}). The number of switching is thus reduced to two-thirds compared to the continuous SVPWM, and

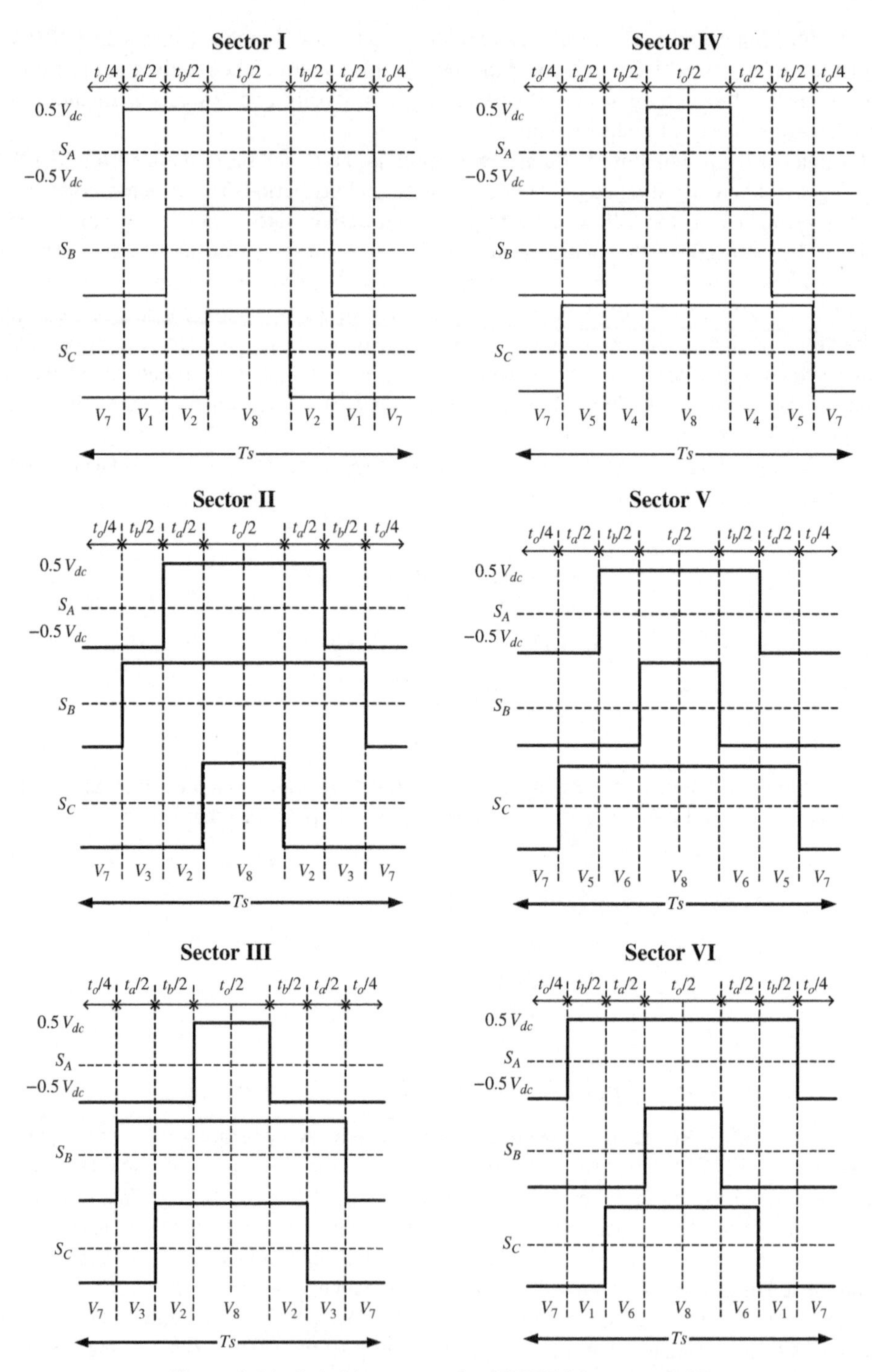

Figure 3.31 Switching pattern for SVPWM for sector I–VI

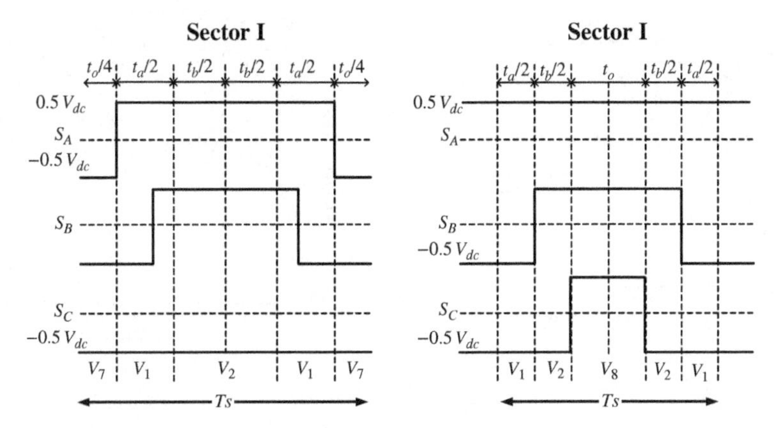

Figure 3.32 Leg voltage (switching pattern) for discontinuous SVPWM

hence switching losses are reduced significantly. This type of PWM is especially useful when switching losses are to be reduced. There are several schemes of discontinuous SVPWM; however, all essentially merely rearrange the placement of the zero output voltage pulse within each half-carrier or carrier interval. Discontinuous SVPWM is classified according to the location of the zero space vector and for how long it is not applied. The nine different discontinuous SVPWM techniques are

1. $t_7 = 0$ for all sectors, known as DPWMMAX;
2. $t_8 = 0$ for all sectors, known as DPWMMIN; zero vectors are made inoperative alternatively in a different sector:
3. Discontinuous modulation DPWM 0 (zero vector (000), i.e. t_7 eliminated in odd sectors and zero vector (111), i.e. t_8 is eliminated in even sectors);
4. Discontinuous modulation DPWM 1(zero vector (000), i.e. t_7 eliminated in even sectors and zero vector (111), i.e. t_8 is eliminated in odd sectors);
5. Discontinuous modulation DPWM 2 (each sector is subdivided into parts and zero vector (000), i.e. t_7 eliminated in odd sectors and zero vector (111), i.e. t_8 is eliminated in even sectors);
6. Discontinuous modulation DPWM 3 (each sector is subdivided into parts and zero vector (000), i.e. t_7 eliminated in even sectors and zero vector (111), i.e. t_8 is eliminated in odd sectors);
7. Discontinuous modulation DPWM 4 (the region is divided into four sectors each spanning 90 degrees and zero vector (000), i.e. t_7 eliminated in odd sectors and zero vector (111), i.e. t_8 is eliminated in even sectors);
8. Discontinuous modulation DPWM 5 (zero vector (000), i.e. t_7 eliminated in sectors 1, 2, 3 and zero vector (111), i.e. t_8 is eliminated in sectors, 4, 5, 6);
9. Discontinuous modulation DPWM 6 (the whole region is divided into 8 sectors of 45 degrees each, zero vector (000), i.e. t_7 eliminated in odd sectors and zero vector (111), i.e. t_8 is eliminated in even sectors).

The placement of the zero vectors and corresponding voltages (V_{avg} (leg voltage), V_a (phase voltage), and V_{nN} (voltage between neutral point)) for the discontinuous SVPWM are shown in Figure 3.33.

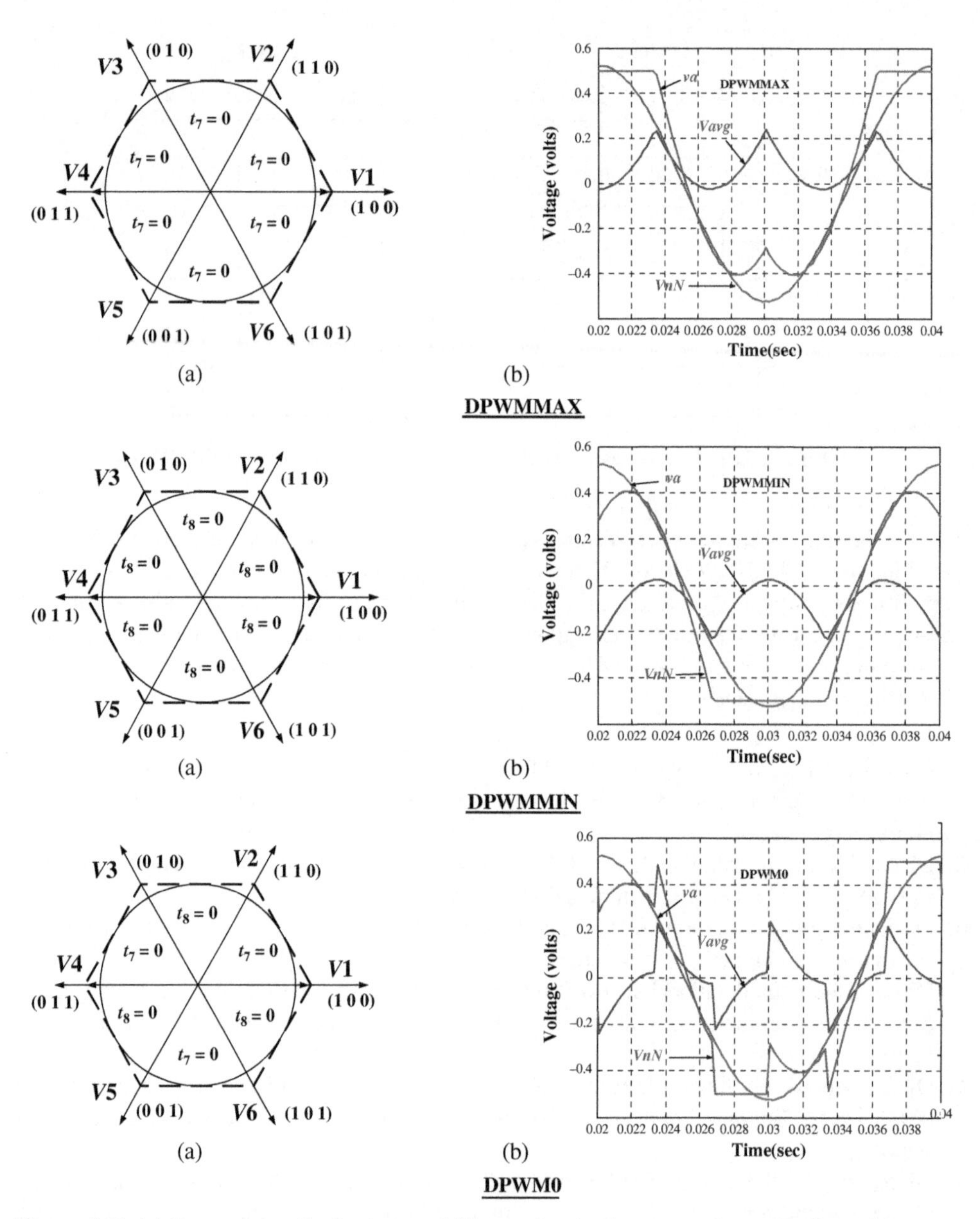

Figure 3.33 (a) Zero-voltage distributions and (b) associated voltage waveforms for discontinuous SVPWM

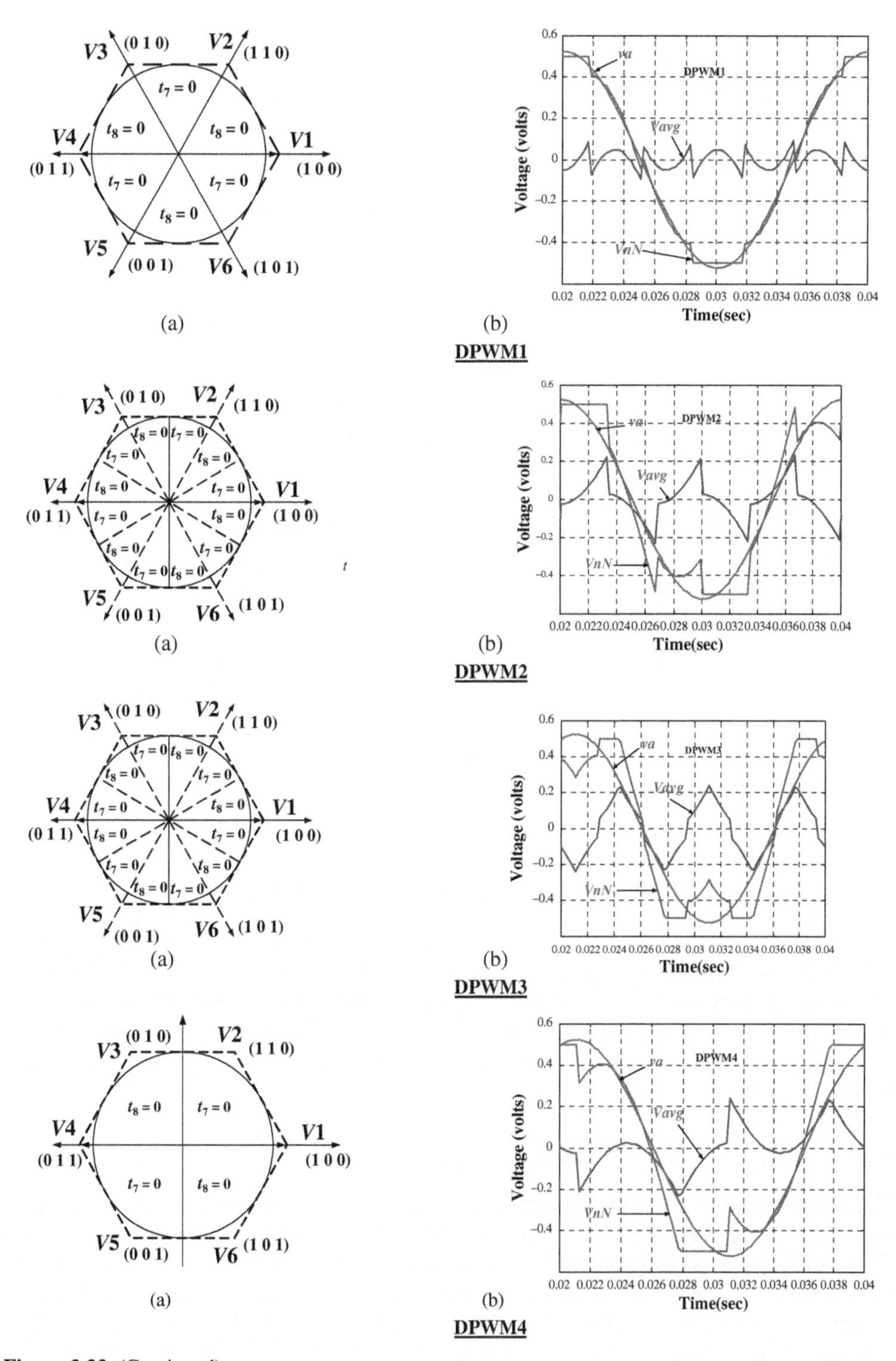

Figure 3.33 (Continued)

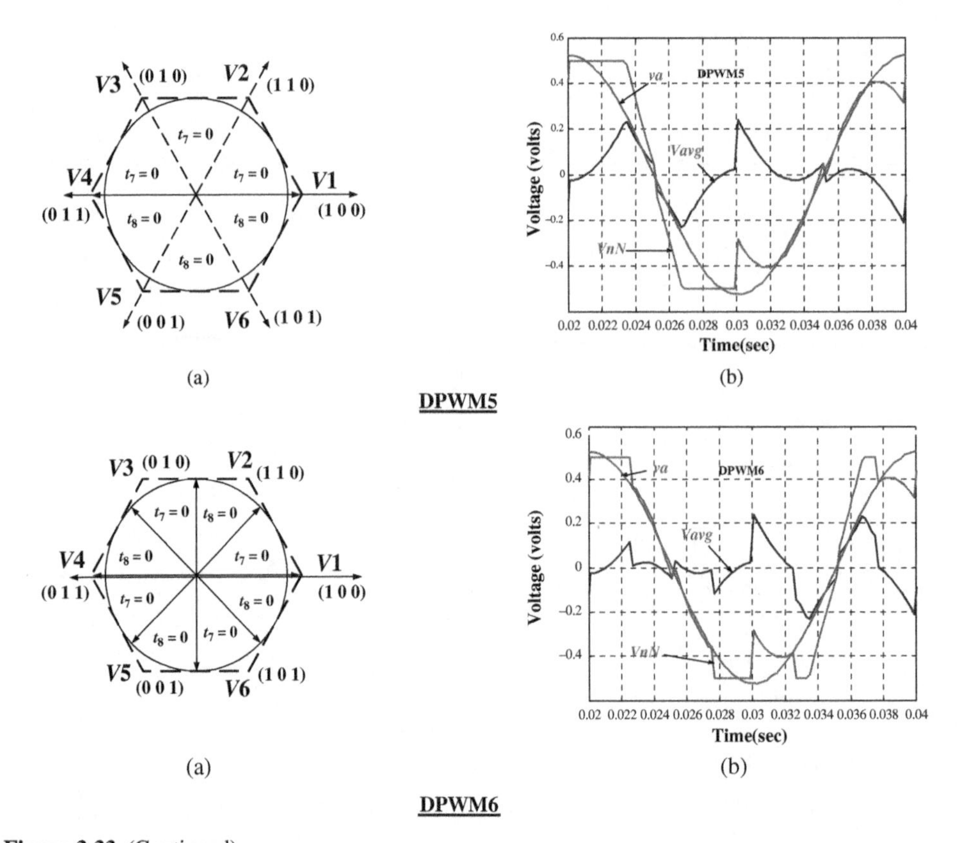

Figure 3.33 (Continued)

3.4.7 *MATLAB/Simulink Model for Space Vector PWM*

The MATLAB/Simulink model can be developed using different approaches, using all Simulink blocks or all MATLAB coding [12, 13]. This section shows a model developed using the MATLAB function block and Simulink blocks. The model shown is flexible in nature and can be used to model continuous as well as discontinuous SVPWM by changing in the MATLAB function command. The simulation model is shown in Figure 3.34. Each block is further elaborated on in Figure 3.35.

3.4.7.1 Reference Voltage Generation Block

This block is used to simulate the balanced three-phase input reference. The three-phase input sinusoidal voltage is generated using the "'function' block from the 'Functions & Tables' sublibrary of Simulink. This is then converted into a two-phase equivalent using Clark's transformation equations. This is once again implemented using the 'function' blocks. Furthermore, the two-phase equivalent is transformed into the polar form using the 'Cartesian to polar' block from the 'Simulink extras' sublibrary. The output of this block is the magnitude of the reference

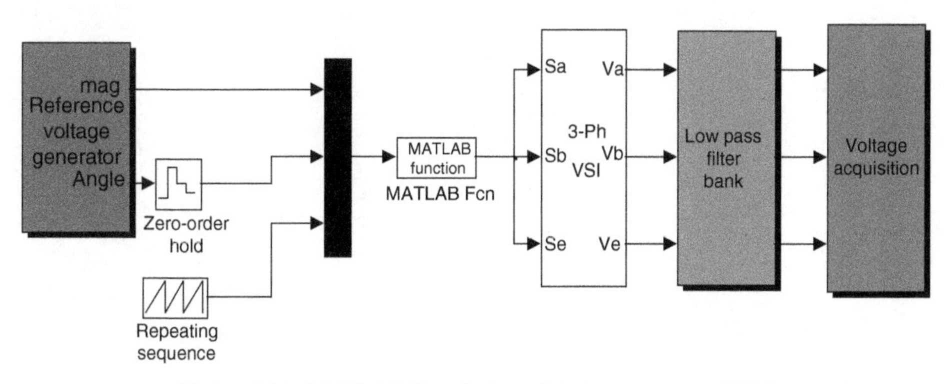

Figure 3.34 MATLAB/Simulink model of space vector PWM

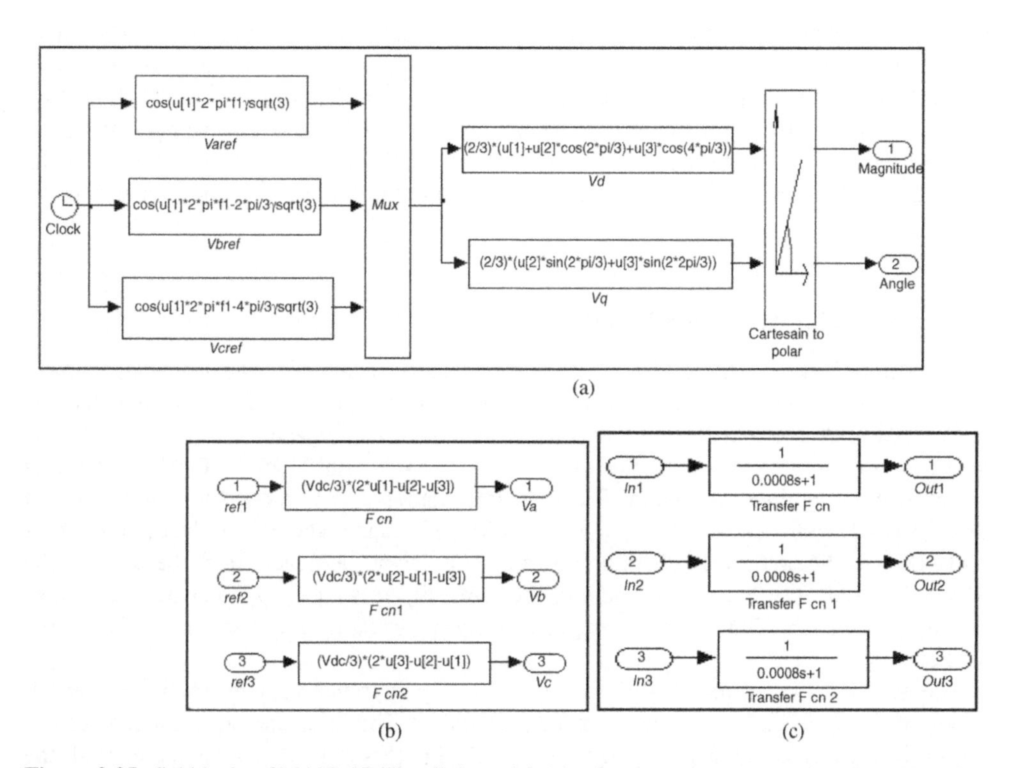

Figure 3.35 Subblocks of MATLAB/Simulink model: (a) reference voltage generation, (b) VSI, and (c) filters (file name: *Space3.mdl and aaa.m*)

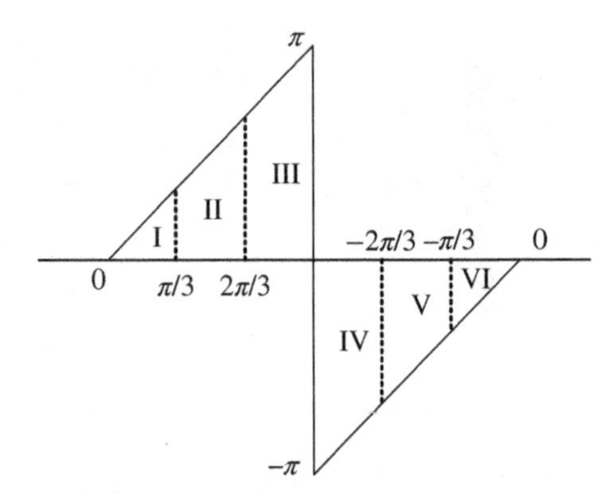

Figure 3.36 Sector identification logic

as the first output and the corresponding angle of the reference as the second output. The waveform of the magnitude is simply a constant line, as its value remains fixed in an open-loop mode. In the closed-loop mode, the reference voltage can be dictated by the closed-loop controller. The waveform of the angle is shown in Figure 3.36. It is a saw-tooth signal with a peak value of $\pm\pi$. Sector identification is done using the comparison of the angle waveform with the predefined values, as shown in Figure 3.36. The number inside the waveform represents the sector number.

3.4.7.2 Switching Time Calculation

The switching time and the corresponding switch state for each power switch are calculated in MATLAB function block 'sf.' The MATLAB code requires a magnitude of the reference, the angle of the reference, and the timer signal for comparison. The angle of the reference voltage is kept constant for one sampling period using the 'zero order hold' block, so that its value does not change during time calculation. The angle information is used for sector identification in the MATLAB code 'aaa_OM1' for continuous SVPWM and code 'a1' for the discontinuous SVPWM (DPWMMAX). Furthermore, a ramp time signal is generated to be used in the MATLAB code. The height and width of the ramp signal is equal to the switching time of the inverter branch. This ramp is generated using a 'repeating sequence' from the source sublibrary.

The MATLAB code first identifies the sector of the reference voltage. The time of application of the active and zero vectors are then calculated. The times are then arranged according to the predefined switching pattern (Figure 3.31). This time is then compared with the ramp timer signal. The height and width of the ramp are equal to the switching period of the inverter branch. Depending upon the location of the time signal within the switching period, the switch state is defined. This switch state is then passed on to the inverter block for further calculation.

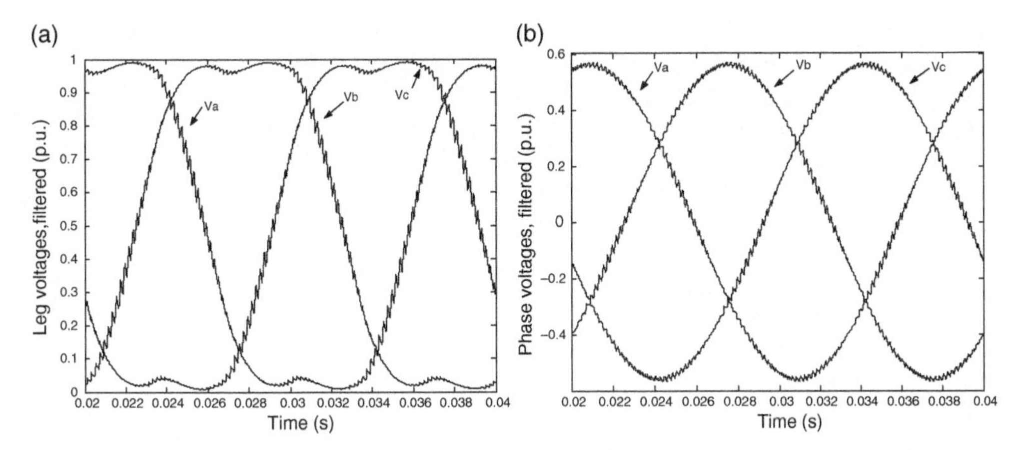

Figure 3.37 (a) Filtered leg voltages for continuous SVPWM and (b) filtered phase voltages for DSVPWM

3.4.7.3 Three-Phase Inverter Block

This block is built to simulate a voltage source inverter assuming a constant DC-link voltage. The inputs to the inverter block are the switching signals and the outputs are the PWM phase-to-neutral voltages. The inverter model is built using 'function' blocks.

3.4.7.4 Filter Blocks

To visualize the actual output of the inverter block, the filtering of the PWM ripple is required. The PWM voltage signal is filtered here using the first-order filter. This is implemented using the 'Transfer function' block from 'Continuous' sublibrary. The time constant of the first-order filter is chosen as 0.8 ms.

3.4.7.5 Voltage Acquisition

This block stores the output results. The results are stored in a 'workspace' taken from the 'sink' library of the Simulink. The simulation waveform is shown in Figure 3.37. The input reference fundamental frequency is kept at 50 Hz and the switching frequency of the inverter in chosen as 5 kHz. The output is seen to be sinusoidal without any ripple except the switching harmonics. By simply changing the input reference's magnitude and the frequency, the output voltage magnitude and frequency can be varied.

3.4.7.6 Continuous SVPWM – % MATLAB Code to Generate Switching Functions

```
% Inputs are magnitude u1(:),angle u2(:), and
% ramp time signal for comparison u3(:)
function [sf]=aaa(u)
ts=0.0002;vdc=1;peak_phase_max= vdc/sqrt(3);
x=u(2); y=u(3);
mag=(u(1)/peak_phase_max) * ts;
```

```
%sector I
if (x>=0) & (x<pi/3)
ta = mag * sin(pi/3-x);
tb = mag * sin(x);
t0 =(ts-ta-tb);
t1=[t0/4 ta/2 tb/2 t0/2 tb/2 ta/2 t0/4];
t1=cumsum(t1);
   v1=[0 1 1 1 1 1 0];
   v2=[0 0 1 1 1 0 0];
 v3=[0 0 0 1 0 0 0];
     for j=1:7
     if(y<t1(j))
        break
      end
   end
sa=v1(j);
sb=v2(j);
sc=v3(j);
end

% sector II
if (x>=pi/3) & (x<2*pi/3)
adv= x-pi/3;
tb = mag * sin(pi/3-adv);
ta = mag * sin(adv);
t0 =(ts-ta-tb);
t1=[t0/4 ta/2 tb/2 t0/2 tb/2 ta/2 t0/4];
t1=cumsum(t1);
v1=[0 0 1 1 1 0 0];
v2=[0 1 1 1 1 1 0];
v3=[0 0 0 1 0 0 0];
     for j=1:7
        if(y<t1(j))
         break
       end
end
sa=v1(j);
sb=v2(j);
sc=v3(j);
end

%sector III
if (x>=2*pi/3) & (x<pi)
adv=x-2*pi/3;
ta = mag * sin(pi/3-adv);
tb = mag * sin(adv);
t0 =(ts-ta-tb);
```

```
t1=[t0/4 ta/2 tb/2 t0/2 tb/2 ta/2 t0/4];
t1=cumsum(t1);
v1=[0 0 0 1 0 0 0];
v2=[0 1 1 1 1 1 0];
v3=[0 0 1 1 1 0 0];
     for j=1:7
        if(y<t1(j))
         break
        end
         end
sa=v1(j);
sb=v2(j);
sc=v3(j);
end

%sector IV
if (x>=-pi) & (x<-2*pi/3)
adv = x + pi;
tb= mag * sin(pi/3 - adv);
ta = mag * sin(adv);
t0 =(ts-ta-tb);
t1=[t0/4 ta/2 tb/2 t0/2 tb/2 ta/2 t0/4];
t1=cumsum(t1);
v1=[0 0 0 1 0 0 0];
v2=[0 0 1 1 1 0 0];
v3=[0 1 1 1 1 1 0];
     for j=1:7
        if(y>t1(j))
         break
       end
     end
sa=v1(j);
sb=v2(j);
sc=v3(j);
end

% sector V
if (x>=-2*pi/3) & (x<-pi/3)
adv = x+2*pi/3;
ta = mag * sin(pi/3-adv);
tb = mag * sin(adv);
t0 =(ts-ta-tb);
t1=[t0/4 ta/2 tb/2 t0/2 tb/2 ta/2 t0/4];
t1=cumsum(t1);
v1=[0 0 1 1 1 0 0];
v2=[0 0 0 1 0 0 0];
```

```
v3=[0 1 1 1 1 1 0];
        for j=1:7
          if(y<t1(j))
           break
          end
        end
sa=v1(j);
sb=v2(j);
sc=v3(j);
end
```

```
%Sector VI
if (x>=-pi/3) & (x<0)
adv = x+pi/3;
tb = mag * sin(pi/3-adv);
ta = mag * sin(adv);
t0 =(ts-ta-tb);
t1=[t0/4 ta/2 tb/2 t0/2 tb/2 ta/2 t0/4];
t1=cumsum(t1);
v1=[0 1 1 1 1 1 0];
v2=[0 0 0 1 0 0 0];
v3=[0 0 1 1 1 0 0];
 for j=1:7
          if(y<t1(j))
           break
          end
        end
sa=v1(j);
sb=v2(j);
sc=v3(j);
     end
sf=[sa, sb, sc];
```

3.4.7.7 Discontinuous SVPWM – % MATLAB Code

```
function [sf]=a1(u)
f=50;
ts=0.0002;
vdc=1;
peak_phase_max = vdc/sqrt(3);
x=u(2); y=u(3);
mag=(u(1)/peak_phase_max) * ts;
```

```
%sector I
if (x>=0) & (x<pi/3)
     ta = mag * sin(pi/3-x);
tb = mag * sin(x);
   t0 = (ts-ta-tb);
```

```
   t1 = [ta/2 tb/2 t0 tb/2 ta/2];
   t1 = cumsum(t1);
   v1 = [ 1 1 1 1 1];
   v2 = [ 0 1 1 1 0];
v3 = [ 0 0 1 0 0];
    for j = 1:5
      if(y<t1(j))
        break
      end
    end
   sa=v1(j);
   sb=v2(j);
   sc=v3(j);
end

% sector II
if (x>=pi/3) & (x<2*pi/3)
   adv= x-pi/3;
tb = mag * sin(pi/3-adv);
   ta = mag * sin(adv);
   t0 =(ts-ta-tb);
   t1=[ta/2 tb/2 t0 tb/2 ta/2 ];
   t1=cumsum(t1);
   v1=[0 1 1 1 0];
   v2=[1 1 1 1 1];
v3=[ 0 0 1 0 0];
    for j=1:5
      if(y<t1(j))
        break
      end
    end
   sa=v1(j);
   sb=v2(j);
   sc=v3(j);
    end

%sector III
if (x>=2*pi/3) & (x<pi)
   adv=x-2*pi/3;
ta = mag * sin(pi/3-adv);
   tb = mag * sin(adv);
   t0 =(ts-ta-tb);
   t1=[ta/2 tb/2 t0 tb/2 ta/2];
   t1=cumsum(t1);
   v1=[ 0 0 1 0 0];
   v2=[ 1 1 1 1 1];
v3=[ 0 1 1 1 0];
  for j=1:5
    if(y<t1(j))
```

```
      break
     end
   end
   sa=v1(j);
   sb=v2(j);
   sc=v3(j);
end

%sector IV
if (x>=-pi) & (x<-2*pi/3)
    adv = x + pi;
tb= mag * sin(pi/3 - adv);
    ta = mag * sin(adv);
    t0 =(ts-ta-tb);
    t1=[ta/2 tb/2 t0 tb/2 ta/2];
    t1=cumsum(t1);
    v1=[ 0 0 1 0 0];
    v2=[ 0 1 1 1 0];
v3=[ 1 1 1 1 1];
  for j=1:5
    if(y<t1(j))
      break
    end
  end
  sa=v1(j);
  sb=v2(j);
  sc=v3(j);
end

% sector V
if (x>=-2*pi/3) & (x<-pi/3)
    adv = x+2*pi/3;
ta = mag * sin(pi/3-adv);
    tb = mag * sin(adv);
    t0 =(ts-ta-tb);
    t1=[ta/2 tb/2 t0 tb/2 ta/2];
t1=cumsum(t1);
    v1=[0 1 1 1 0];
    v2=[0 0 1 0 0];
    v3=[1 1 1 1 1];
    for j=1:5
      if(y<t1(j))
        break
      end
    end
    sa=v1(j);
    sb=v2(j);
sc=v3(j);
end
```

```
%Sector VI
if (x>=-pi/3) & (x<0)
   adv = x+pi/3;
tb = mag * sin(pi/3-adv);
   ta = mag * sin(adv);
   t0 =(ts-ta-tb);
   t1=[ta/2 tb/2 t0 tb/2 ta/2];
   t1=cumsum(t1);
   v1=[1 1 1 1 1];
   v2=[0 0 1 0 0];
v3=[0 1 1 1 0];
    for j=1:5
    if(y<t1(j))
       break
     end
   end
   sa=v1(j);
   sb=v2(j); sc=v3(j); end sf=[sa, sb, sc]
```

3.4.8 Space Vector PWM in Overmodulation Region

In the SVPWM method, the maximum sinusoidal output is obtained when the reference is $|v_s^*| = 1/\sqrt{3}V_{dc}$ and the trajectory is a circle inscribed inside the hexagon, called the linear modulation range. If the references increase more than this value, then the output cannot be realized using the linear modulation technique and this is called the overmodulation region [14–17]. Since the time of application of the zero space vector becomes negative, it does not make any physical sense. Considering equation (3.31) and assuming $|v_s^*| = 1/\sqrt{3}V_{dc}$, $V_{dc} = 1$ p.u. and $T_s = 1$ second, the time of application of the zero space vector in sector I is calculated (Figure 3.38). It is seen that the zero vector time of application just touches zero in the middle of the sector at 30 degrees. Increasing further the length of the reference voltage space vector will lead to a negative value of time of application. This is true for the other sectors

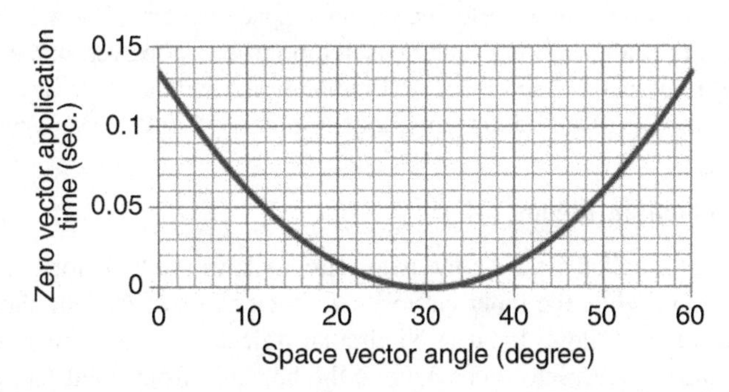

Figure 3.38 Zero vector time of application in linear modulation range in sector I

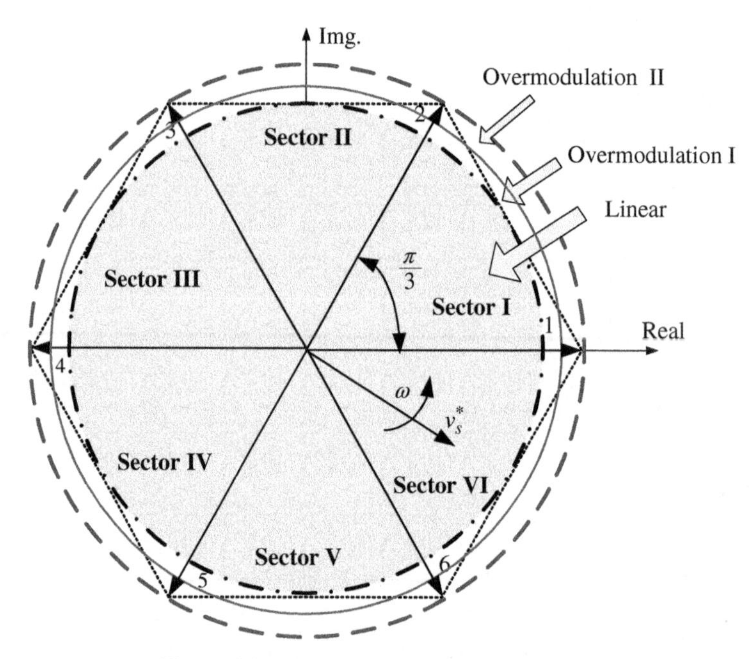

Figure 3.39 Linear and overmodulation range

too. Hence, for reference voltage vectors exceeding this limit, the zero voltage vectors cannot be used as such.

When the reference voltage vector is more than this limiting value of $|v_s^*| = 1/\sqrt{3}V_{dc}$, the inverter moves into the overmodulation region for $\frac{1}{\sqrt{3}}V_{dc} < |v_s^*| < \frac{2}{\pi}V_{dc}$. When the reference voltage vector is $|v_s^*| = \frac{2}{\pi}V_{dc}$, the inverter operates in the square-wave mode or six-step mode, as shown in Figure 3.39.

The overmodulation region is further divided into two submodes, the overmodulation region I $(0.5773 < |v_s^*| < 0.6061)$ and the overmodulation region II $(0.6061 < |v_s^*| < 0.6366)$. In overmodulation region I, the reference voltage space vector is modified in such a way that it becomes a 'distorted continuous reference voltage space vector.' The magnitude of the reference voltage vector is altered, but the angle is not modified. However, in the overmodulation region II, both the reference voltage space vector magnitude and the angles are modified in such a way that it becomes a 'distorted discontinuous reference voltage space vector.'

3.4.8.1 Overmodulation Region I

Consider Figure 3.40 (called Case 1): the outer circle shows the trajectory of the desired reference voltage vector, while the inner circle shows the end of linear modulation region. The reference voltage vector follows the desired circular trajectory from point x to point y. When the reference voltage space vector goes beyond the hexagon from point y to point p, the reference voltage space vector trajectory slides along the hexagonal straight line. After the end of the straight line (at point p), the trajectory once again follows the circular path from point

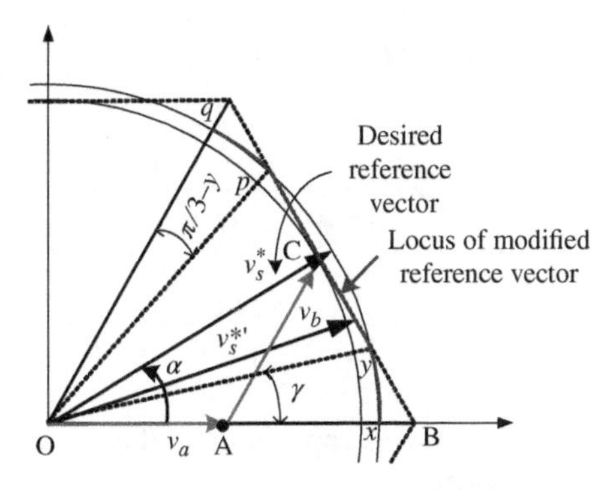

Figure 3.40 Overmodulation I in SVPWM-Case 1

p to q; this repeats in other sectors too. In this way, the reference voltage vector is modified for overmodulation region I.

The reference voltage space vector follows a circular trajectory for some time and then a straight line trajectory. The time of the application of the active space vectors for the circular trajectory is given by equation (3.31). The time of application of the active space vectors for the region y to p (straight line) can be obtained as follows:

Let the modified space vector be $|v_s^{*\prime}|$ and be decomposed into a real and an imaginary part:

$$\begin{gathered} |v_s^{*\prime}|\cos(\alpha) = |v_a| + |v_b|\cos\left(\frac{\pi}{3}\right) \\ \text{for} \quad \gamma < \alpha < \frac{\pi}{3} - \gamma \\ |v_s^{*\prime}|\sin(\alpha) = |v_b|\sin\left(\frac{\pi}{3}\right) \end{gathered} \tag{3.35}$$

Solving equation (3.35) for v_a:

$$|v_a| = |v_s^{*\prime}|\left(\cos(\alpha) - \frac{1}{\sqrt{3}}\sin(\alpha)\right) \tag{3.36}$$

From Figure 3.40, ABC forms an equilateral triangle, hence:

$$\text{OB} = \text{OA} + \text{AB} = |v_a| + |v_b| = \frac{2}{3}V_{dc} \tag{3.37}$$

Using equations (3.35)–(3.37) to determine the length of the modified space vector:

$$|v_s'| = \frac{\frac{2}{3}V_{dc}}{\left(\cos(\alpha) + \frac{1}{\sqrt{3}}\sin(\alpha)\right)} \tag{3.38}$$

The length of the component vectors are obtained for $\gamma < \alpha < \frac{\pi}{3} - \gamma$:

$$|v_a| = \frac{2}{3} V_{dc} \left[\frac{\sqrt{3}\cos(\alpha) - \sin(\alpha)}{\sqrt{3}\cos(\alpha) + \sin(\alpha)} \right] \tag{3.39a}$$

$$|v_b| = \frac{2}{3} V_{dc} \left[\frac{2\sin(\alpha)}{\sqrt{3}\cos(\alpha) + \sin(\alpha)} \right] \tag{3.39b}$$

To obtain the expression for the time of application of the active space voltage vectors, the equal volt-second principle is applied:

$$v_a t_a + v_b t_b = v_s^* T_s \tag{3.40}$$

$$\begin{aligned} t_a &= \left(\frac{\sqrt{3}\cos(\alpha) - \sin(\alpha)}{\sqrt{3}\cos(\alpha) + \sin(\alpha)} \right) T_s \\ t_b &= T_s - t_a \\ t_o &= 0 \end{aligned} \tag{3.41}$$

These equations are valid for the first sector; nevertheless, a similar relationship can be derived for other sectors. Implementation of the SVPWM follows the same procedure as that of the linear mode, except the time of application is now governed by equation (3.41).

From Figure 3.41 (called Case 2), it is seen that by modifying the trajectory of the reference space vector, the actual output becomes smaller than the desired reference. This is due to the fact that the volt-second balance is not met and the loss in the output corresponds to the volt-second area A_2. To compensate for this loss and to maintain the volt-second balance, the reference voltage vector is further modified. The trajectory of the reference voltage vector follows the outer circle for a small distance when the hexagon boundary is met at point r. It then slides along the hexagon straight line up to point s, where it meets the circular trajectory and then

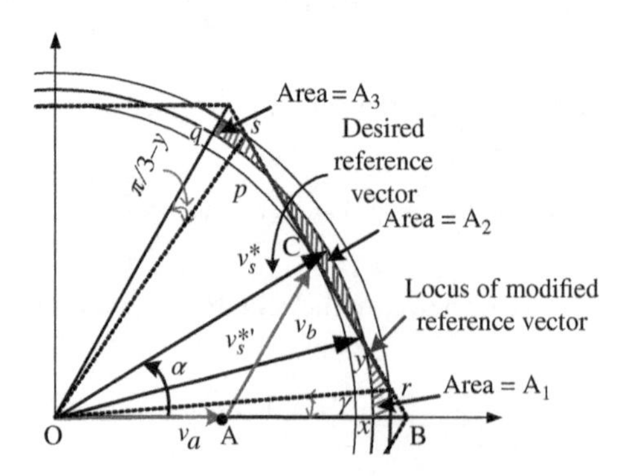

Figure 3.41 Overmodulation I in SVPWM-Case 2

moves along this circle. In this way, the volt-second lost in A_2 is gained in area A_1 and A_3. A more precise calculation of the modified trajectory is needed in this case. The modified reference voltage expression is derived for different parts of the sectors. Once the voltage modified voltage reference is known, the time of application of the vector is calculated using equation (3.31) and accordingly the switching pattern is obtained and further used for the SVPWM implementation. The modified space voltage vector expression is obtained as [15]:

$$\left|v_s^*\right| = \frac{1}{\sqrt{3}} V_{dc} \tan(\alpha) \quad \text{for } 0 < \alpha < \left(\frac{\pi}{6} - \gamma\right) \tag{3.42a}$$

$$\left|v_s^{*\prime}\right| = \frac{1}{\sqrt{3}} V_{dc} \frac{1}{\cos\left(\frac{\pi}{6} - \gamma\right)} \quad \text{for} \left(\frac{\pi}{6} - \gamma\right) < \alpha < \left(\frac{\pi}{6} + \gamma\right) \tag{3.42b}$$

$$\left|v_s^{*\prime}\right| = \frac{1}{\sqrt{3}} V_{dc} \frac{1}{\cos\left(\frac{\pi}{3} - \alpha\right)} \quad \text{for} \left(\frac{\pi}{6} + \gamma\right) < \alpha < \left(\frac{\pi}{2} - \gamma\right) \tag{3.42c}$$

$$\left|v_s^{*\prime}\right| = \frac{1}{\sqrt{3}} V_{dc} \frac{1}{\cos\left(\frac{\pi}{6} - \gamma\right)} \quad \text{for} \left(\frac{\pi}{2} - \gamma\right) < \alpha < \frac{\pi}{2} \tag{3.43}$$

A linearized relationship between the angle j and the modulation index is obtained as [15]:

$$\gamma = \begin{cases} -30.23MI + 27.94 & \text{for } 0.09068 \le MI < 0.9095 \\ -8.58MI + 8.23 & \text{for } 0.9095 \le MI < 0.9485 \\ -26.43MI + 25.15 & \text{for } 0.9485 \le MI < 0.9517 \end{cases} \tag{3.44}$$

where $MI = \frac{\left|v_s^*\right|}{\frac{2}{3}V_{dc}}$. After determining the angle γ, the modified reference voltage vectors length is calculated from equation (3.43), which is then used to calculate the time of application of the space vector using equation (3.31).

3.4.8.2 Overmodulation Region-II

In the overmodulation region I, the angular velocity of the modified reference voltage space vector and desired reference voltage space vector is the same and constant for each fundamental cycle. The reference voltage and actual output voltage are equal during the overmodulation region I, due to the fact that the loss in the area A_2 (due to reduced length of the reference vector, since it cannot go beyond the hexagon boundary) is compensated for by increasing the length of the reference in area A_1 and area A_3. A situation arises when the desired reference voltage vector reaches 0.6061 V_{dc}, where no more room exists in area A_1 and area A_3 to compensate for the loss in area A_2. At this point, overmodulation region II starts. During this mode of overmodulation, both the reference voltage space vector length and angle are modified. For the region x to y and p to q (Figure 3.42), the modified space vector is held constant at the vertex of the hexagon. Hence, the desired reference voltage moves while the modified voltage vector does not.

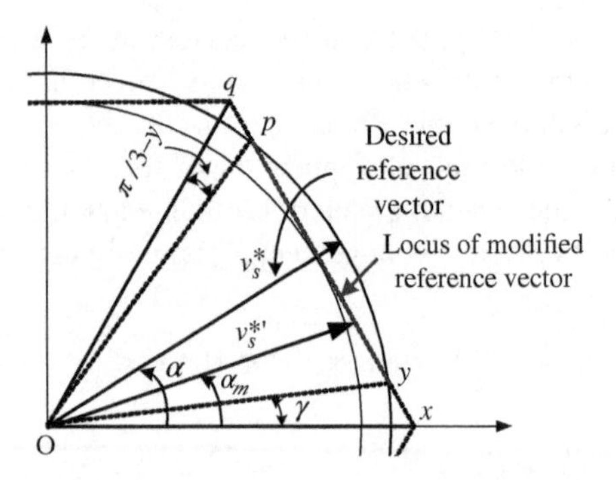

Figure 3.42 Overmodulation II in SVPWM

Once the desired reference reaches point y, the modified vector starts to slide along the side of the hexagon. In this way, the reference becomes discontinuous. The angle of the modified reference voltage vector is

$$\alpha_m = \begin{cases} 0 & 0 \le \alpha \le \gamma \\ \dfrac{\pi}{6}\dfrac{\alpha-\gamma}{\pi/6-\gamma} & \gamma \le \alpha \le \pi/3-\gamma \\ \pi/6 & \pi/3-\gamma \le \alpha \le \pi/3 \end{cases} \tag{3.45}$$

The expression for the modified reference voltage vector is [15]

$$\left|v_s^{*\prime}\right| = \frac{1}{\sqrt{3}}V_{dc}\tan(\alpha) \quad \text{for } 0 < \alpha < \left(\frac{\pi}{6}-\gamma\right) \tag{3.46a}$$

$$\left|v_s^{*\prime}\right| = \frac{1}{3}V_{dc} \quad \text{for}\left(\frac{\pi}{6}-\gamma\right) < \alpha < \left(\frac{\pi}{6}+\gamma\right) \tag{3.46b}$$

$$\left|v_s^{*\prime}\right| = \frac{1}{\sqrt{3}}V_{dc}\frac{1}{\cos\left(\frac{\pi}{3}-\alpha_m\right)} \quad \text{for } \left(\frac{\pi}{6}+\gamma\right) < \alpha < \left(\frac{\pi}{2}-\gamma\right) \tag{3.46c}$$

$$\left|v_s^{*\prime}\right| = \frac{2}{3}V_{dc} \quad \text{for } \left(\frac{\pi}{2}-\gamma\right) < \alpha < \frac{\pi}{2} \tag{3.46d}$$

The linearized equation is found for the modified angle is

$$\gamma = \begin{cases} 6.4MI - 6.09 & \text{for } 0.95 \le MI < 0.98 \\ 11.75MI - 11.34 & \text{for } 0.98 \le MI < 0.9975 \\ 48.96MI - 48.3 & \text{for } 0.9975 \le MI < 1.0 \end{cases} \tag{3.47}$$

The SVPWM in the overmodulation region II is implemented using equations (3.31), (3.46a)–(3.46d), and (3.47).

3.4.9 MATLAB/Simulink Model to Implement Space Vector PWM in Overmodulation Regions

The MATLAB/Simulink model for implementing overmodulation using SVPWM is the same as that in Figure 3.34. The MATLAB function block is modified in order to account for the overmodulation. The time of application of the space vector is calculated using equation (3.31) for the period when the reference space vector lies between the region ($0 < \alpha \leq \gamma$ and $\pi/3 - \gamma \leq \alpha < \pi/3$) and for the remaining period, time of application is computed using equation (3.41). The angle g is calculated using equation (3.44). The MATLAB code requires the magnitude of the reference, the angle of the reference, and the timer signal for comparison. The angle of the reference voltage is on hold for each switching period, so that its value does not change during time calculation. The angle information is used for sector identification in the MATLAB code 'aaa_OM1' for overmodulation I and 'aaa_OM2' for overmodulation II and subsequent generation of gating signals for the inverter. For the 50-Hz fundamental, filtered output voltages for different modulation indices are shown in Figure 3.43.

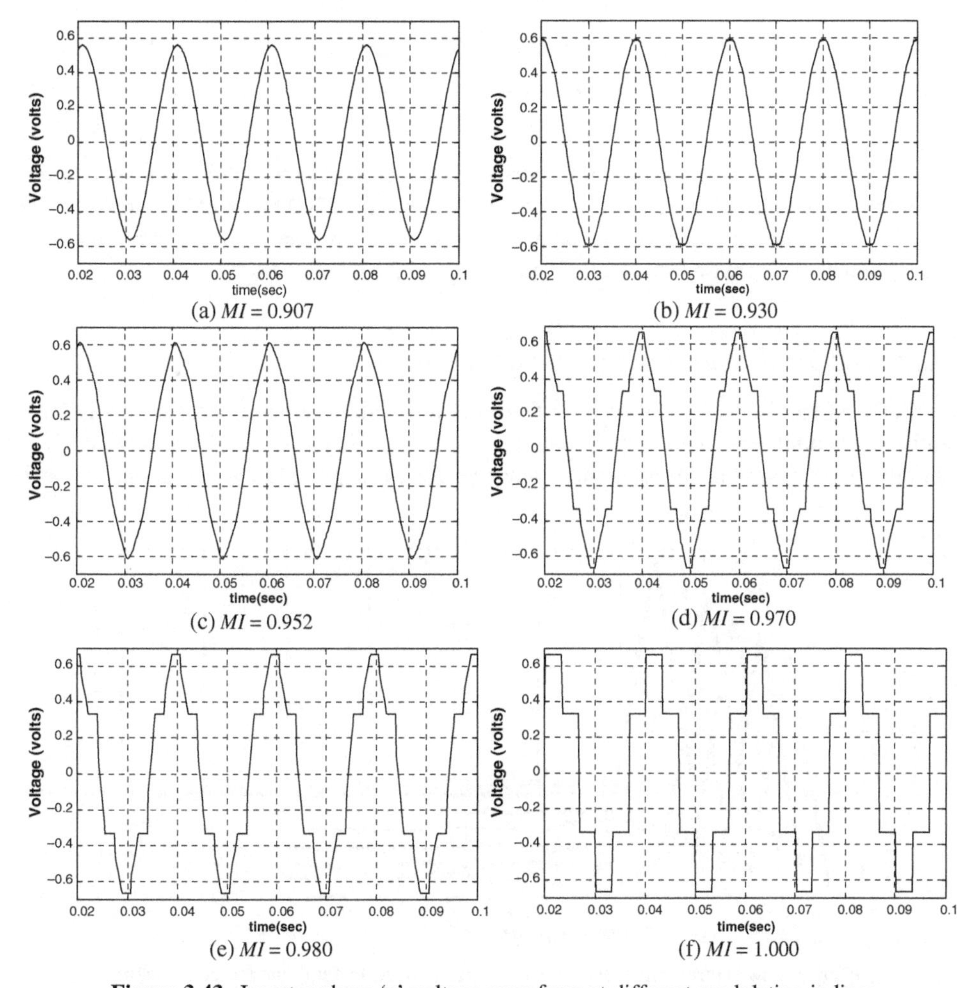

Figure 3.43 Inverter phase 'a' voltage waveform at different modulation indices

3.4.10 Harmonic Analysis

Harmonic components of the output voltage are given, using the Fourier series expression as:

$$F_n(\theta) = \frac{4}{\pi}\left[\int_0^{\pi/2} f(\theta)\sin(n\theta)d\theta\right] \tag{3.48}$$

where $f(\theta)$ is given by equations (3.42a)–(3.42c) and (3.43) in the overmodulation mode I and equation (3.46a)–(3.46d) in the overmodulation mode II. Numerical integration of equation (3.48) shows that even-ordered harmonics and triplen harmonics are eliminated in the output voltages. The six lowest harmonic components (3rd, 5th, 7th, 9th, 11th, and 13th) versus the modulation index (MI) are shown in Figure 3.44. The low-order harmonic components increase with the increase in the modulation index.

The THD factor is defined as:

$$THD = \frac{\sqrt{\left(V_r^2 - V_1^2\right)}}{V_1^2} \tag{3.49}$$

where V_r and V_1 are the rms values of the rth harmonic components and fundamental component of the phase voltage, respectively. Figure 3.45 shows the *THD* factor of the output voltage up to the 25th harmonic. As the modulation index increases, especially in overmodulation mode II, the THD increases steeply and culminates to 31.1% at $MI = 1$.

3.4.11 Artificial Neural Network-Based PWM

The SVPWM can be implemented by a feed-forward neural network because the SVPWM algorithm is a nonlinear input/output mapping and the ANN is a powerful nonlinear mapping technique. This means that the reference voltage vector $V*$ magnitude, and the angle, $\alpha*$, can be

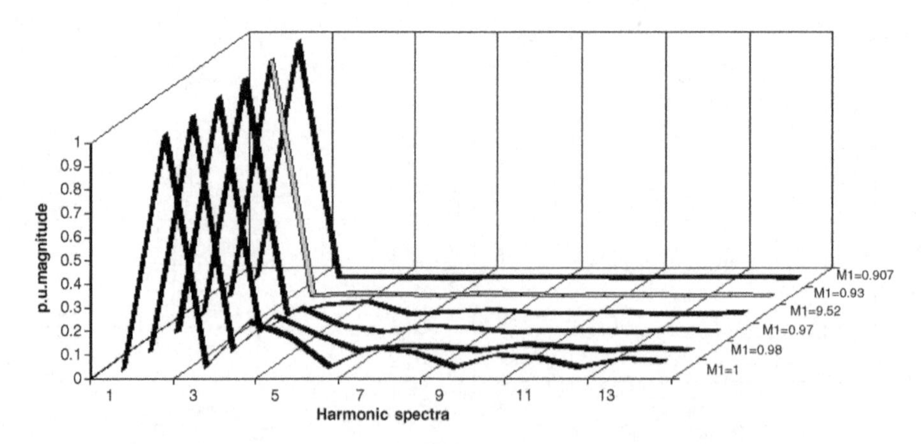

Figure 3.44 Harmonic spectra by FFT, normalized to fundamental component

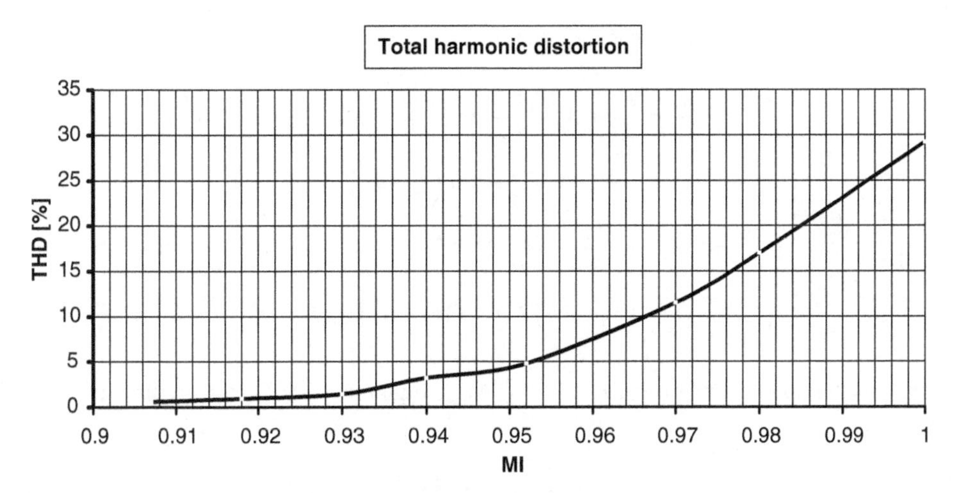

Figure 3.45 THD in overmodulation region

impressed at the input of the network and the corresponding pulse-width pattern of the three phases can be generated at the output. There are two approaches in implementing the SVPWM, using the ANN called 'direct method' and the second is the 'indirect method' [1]. In the so-called 'Direct method,' the feed-forward back-propagation ANN directly replaces the conventional SVPWM algorithm. Since the feed-forward ANN network can map only one input pattern onto only one output pattern, the switching period T_s is divided into n subintervals. Thus, each subinterval includes only one output switching pattern for every input pattern, so requires a huge data set for proper training of the network. Therefore, this approach is limited in use [18–22]. The 'indirect method' uses two separate feed-forward back-propagation ANN, one for the magnitude of the reference voltage and other for the reference voltage position. The magnitude network yields a voltage magnitude scaling function, which is linear in the linear modulation region and is a nonlinear function of V_{DC} in the overmodulation region. The reference position network yield turn-on pulse-width function at unit voltage magnitude. This pulse-width functions are then multiplied by a suitable bias signal and the product is compared with the up/down counter to generate the appropriate switching signals for the inverter. The complete implementation block is shown in Figure 3.46.

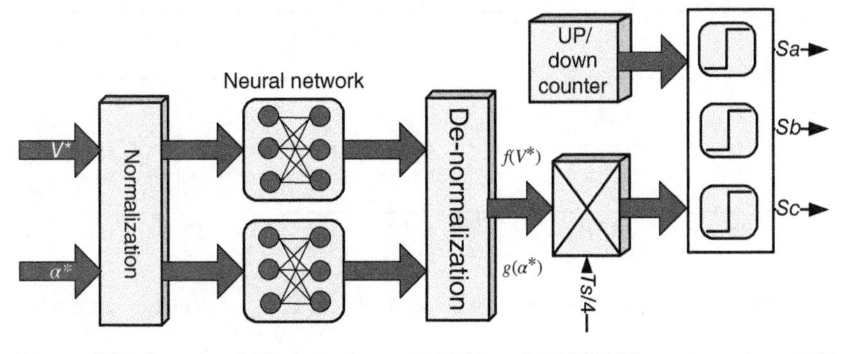

Figure 3.46 Functional block diagram of ANN-based SVPWM for a three-phase VSI

The phase 'A' turn on time can be given as:

$$T_{A=ON} = \begin{cases} \dfrac{t_0}{4} = \dfrac{T_s}{4} + K.V^*\left[-\sin\left(\dfrac{\pi}{3}\right) - \sin\left(\alpha^*\right)\right], S = 1,6 \\ \dfrac{t_0}{2} + t_b = \dfrac{T_s}{4} + K.V^*\left[-\sin\left(\dfrac{\pi}{3} - \alpha^*\right) + \sin\left(\alpha^*\right)\right], S = 2 \\ \dfrac{t_0}{2} + t_0 + t_b = \dfrac{T_s}{4} + K.V^*\left[\sin\left(\dfrac{\pi}{3} - \alpha^*\right) + \sin\left(\alpha^*\right)\right], S = 3,4 \\ \dfrac{t_0}{2} + t_a = \dfrac{T_s}{4} + K.V^*\left[\sin\left(\dfrac{\pi}{3} - \alpha^*\right) - \sin\left(\alpha^*\right)\right], S = 5 \end{cases} \tag{3.50}$$

where S is the sector number and $K = \sqrt{3}T_s/4V_{dc}$. Equation (3.50) can be written in the general form:

$$T_{A-ON} = T_s/4 + f(V^*)g_A(\alpha^*) \tag{3.51}$$

where $f(V*)$ is the voltage amplitude scale factor, $T_s/4$ is the bias time, and $g(\alpha*)$ is called the turn-on signal at unit voltage and it is given as (Figure 3.47):

$$g_A(\alpha^*) = \begin{pmatrix} K\left[-\sin\left(\dfrac{\pi}{3} - \alpha^*\right) - \sin\left(\alpha^*\right)\right], S = 1,6 \\ K\left[-\sin\left(\dfrac{\pi}{3} - \alpha^*\right) + \sin\left(\alpha^*\right)\right], S = 2 \\ K\left[\sin\left(\dfrac{\pi}{3} - \alpha^*\right) + \sin\left(\alpha^*\right)\right], S = 3,4 \\ K\left[\sin\left(\dfrac{\pi}{3} - \alpha^*\right) - \sin\left(\alpha^*\right)\right], S = 5 \end{pmatrix} \tag{3.52}$$

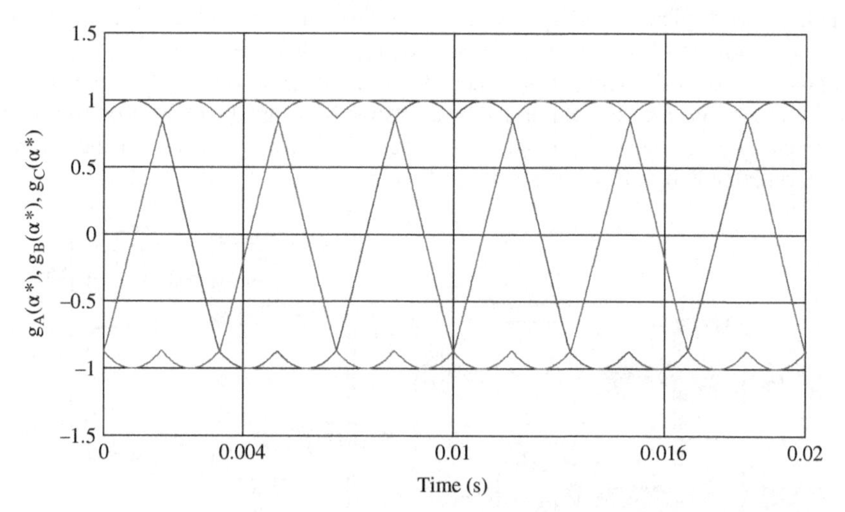

Figure 3.47 Turn-on pulse-width function of phase A, B, and C as a function of angle a in different sectors

3.4.12 MATLAB/Simulink Model of Implementing ANN-Based SVPWM

The MATLAB/Simulink model for implementing the ANN-based SVPWM is shown in Figure 3.48. Initially, a neural network is trained with the reference voltage input and the turn-on time as output using equations (3.50)–(3.52). The ANN model is then generated using the neural network toolbox, and the generated model is placed in the Simulink, as shown in Figure 3.48. The input signal to the neural network is the reference voltage position a∗. The model uses a multilayer function in the first and second layers. The number of nodes can be set at the training stage. The number of hidden nodes can be chosen between 10 and 20. The weights are generated during the training mode. The digital words corresponding to turn-on time are generated by multiplying the output of the neural network with $V{*}T_S$ and then

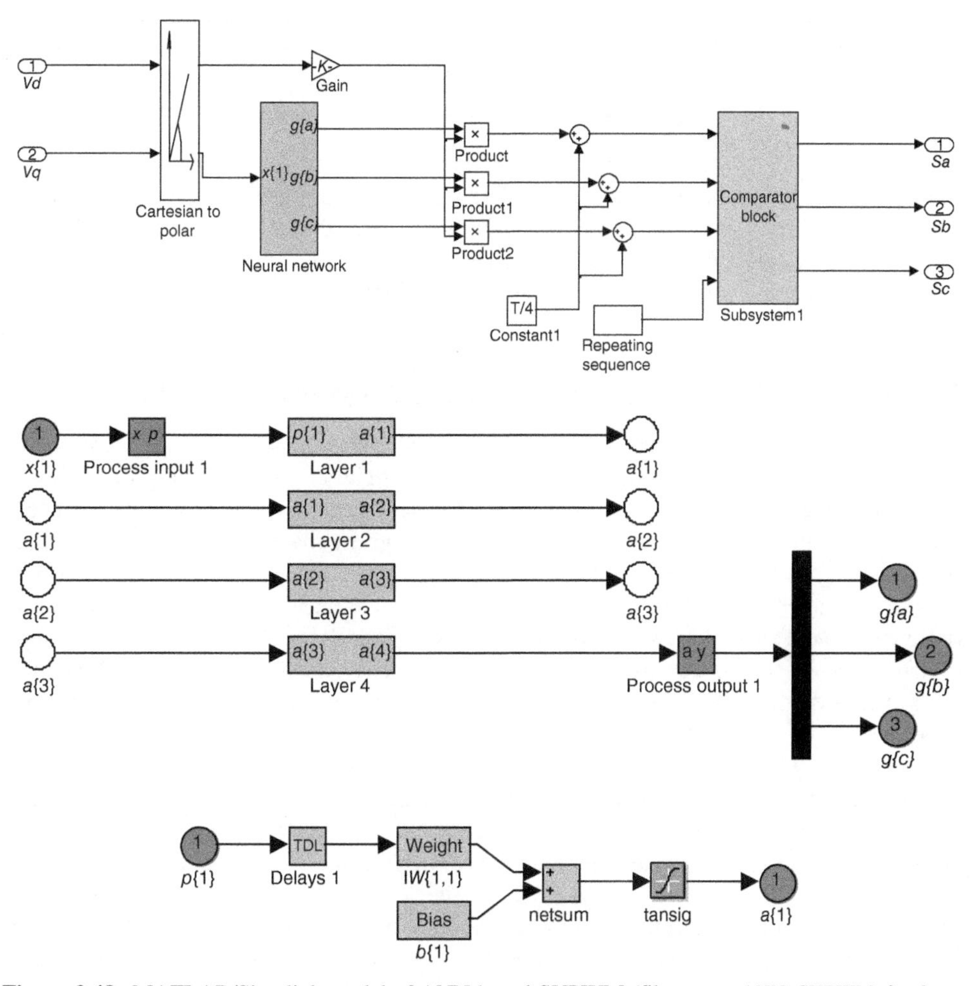

Figure 3.48 MATLAB/Simulink model of ANN-based SVPWM (file name: *ANN_SVPWM_3_phase.mdl*)

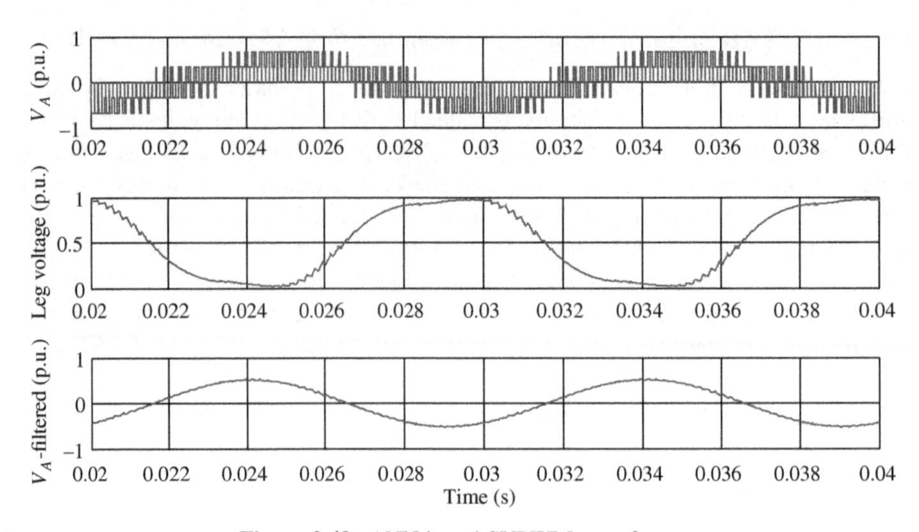

Figure 3.49 ANN-based SVPWM waveforms

adding $T_s/4$, as shown in the figure. The PWM signals are then generated by comparing the turn-on time with a triangular reference having a time period of T_s and amplitude $T_s/2$, and the PWM signals thus obtained are then applied to the inverter.

The resulting simulation waveform is presented in Figure 3.49.

3.5 Relationship Between Carrier-Based PWM and SVPWM

Although the two PWM schemes, carrier-based and SVPWM, appear different, there exists a similarity between the two approaches [23]. In the SPWM approach, the three-phase modulating waves are compared with a triangular high-frequency carrier to determine the switching instants of the three phases. The modulating wave of a given phase is the average leg voltage corresponding to that phase. The most commonly used modulating signals are sinusoidal waves. The triplen (multiple of three) frequency components can be added as a zero-sequence component to the three-phase sinusoidal waves or modulating signals. Hence, the choice of the triplen frequency components to be added to the fundamental signals lead to different modulating waveforms with different harmonic spectrum. The choice of these triplen frequency components is thus a degree of freedom in the SPWM technique.

In the SVPWM technique, the voltage reference is provided in terms of a rotating space vector. The magnitude and the frequency of the fundamental component of the output are specified by the magnitude and frequency of the reference vector. The reference vector is sampled once in every switching period. The inverter is switched such that the average voltage vector is equal to the sampled reference vector in every switching period. The inverter switching is done according to the reference voltage vector location. Two active and two zero vectors are used in every switching period. The time of application of each vector is obtained from the relationship obtained on the basis of the 'equal volt-second' principle. The degree of freedom used in the SVPWM is the division of the time between the two zero states (000 and 111). The

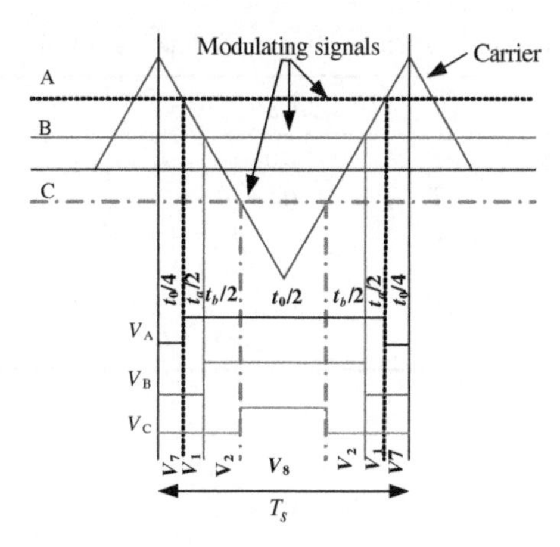

Figure 3.50 Relationship between carrier-based and SVPWM in sector I

distribution of the time duration between the two zero states 7(000) and 8(111) is equivalent to adding the zero-sequence component to the three-phase average leg voltages. Hence, the two PWM approaches are equivalent in this sense.

3.5.1 Modulating Signals and Space Vectors

The switching pattern generated due to the interaction between a triangular carrier signal and the modulating signal is shown in Figure 3.50; the magnitude of the modulating signal is assumed constant during the sampling period. The switching pattern shown is the leg voltage with magnitude ±0.5 V_{dc}. By comparing Figures 3.50 and 3.31, it is seen that the same switching pattern and space vector disposition are available in the SVPWM in sector I.

The same similarity also exists in other sectors. The relationship between the modulating signals and space vector in different sectors can be found from Figure 3.31 as:

$$\begin{aligned} V_A T_s &= 0.5V_{dc}\left(V_A^* + v_{nN}\right)T_s = 0.5V_{dc}(t_a + t_b + t_8 - t_7) \\ V_B T_s &= 0.5V_{dc}\left(V_B^* + v_{nN}\right)T_s = 0.5V_{dc}(-t_a + t_b + t_8 - t_7) \\ V_A T_s &= 0.5V_{dc}\left(V_A^* + v_{nN}\right)T_s = 0.5V_{dc}(t_a + t_b + t_8 - t_7) \end{aligned} \tag{3.53}$$

The relationship between the modulating signal and the space vector in sector I is

$$\begin{aligned} U_A &= (t_a + t_b + t_8 - t_7)/T_s \\ U_B &= (-t_a + t_b + t_8 - t_7)/T_s \\ U_C &= (-t_a - t_b + t_8 - t_7)/T_s \end{aligned} \tag{3.54}$$

where t_8 is the time of application of the zero vector 111, t_7 is the time of application of the zero vector 000, and U_A, U_B, and U_C represent the three-phase modulating signals for the SPWM. The modulating signals for SPWM are

Table 3.5 Relationship between the space vector and modulating signal

Sector no.	U_A	U_B	U_C
I	$(t_a + t_b + t_8 - t_7)/T_s$	$(-t_a + t_b + t_8 - t_7)/T_s$	$(-t_a - t_b + t_8 - t_7)/T_s$
II	$(t_a - t_b + t_8 - t_7)/T_s$	$(t_a + t_b + t_8 - t_7)/T_s$	$(-t_a - t_b + t_8 - t_7)/T_s$
III	$(-t_a - t_b + t_8 - t_7)/T_s$	$(t_a + t_b + t_8 - t_7)/T_s$	$(-t_a + t_{b'} + t_8 - t_7)/T_s$
IV	$(-t_a - t_b + t_8 - t_7)/T_s$	$(t_a - t_b + t_8 - t_7)/T_s$	$(t_a + t_b + t_8 - t_7)/T_s$
V	$(-t_a + t_b + t_8 - t_7)/T_s$	$(-t_a - t_b + t_8 - t_7)/T_s$	$(t_a + t_b + t_8 - t_7)/T_s$
VI	$(t_a + t_b + t_8 - t_7)/T_s$	$(-t_a - t_b + t_8 - t_7)/T_s$	$(t_a - t_b + t_8 - t_7)/T_s$

$$\begin{aligned} U_A &= V_A^* + v_{nN} \\ U_B &= V_B^* + v_{nN} \\ U_C &= V_C^* + v_{nN} \end{aligned} \tag{3.55}$$

where

$$v_{nN} = \frac{1}{3}(U_A + U_B + U_C) \tag{3.56}$$

is the zero-sequence signal.

The relationship between the modulating signals and the space vectors in the other five sectors can be obtained in a similar fashion using Figure 3.31 and are listed in Table 3.5.

The relation between the zero-sequence signal and the space vector time of applications can be determined by substituting equation (3.54) into equation (3.56). The resulting relationships are given for odd sectors (sectors I, III, and V) and for even sectors (sectors II, IV, and VI) in equations (3.57) and (3.58), respectively:

$$v_{nN} = \frac{1}{3T_s}[-t_a + t_b - 3t_7 + 3t_8] \tag{3.57}$$

$$v_{nN} = \frac{1}{3T_s}[t_a - t_b - 3t_7 + 3t_8] \tag{3.58}$$

3.5.2 Relationship Between Line-to-Line Voltages and Space Vectors

A relationship between the line-to-line voltage and space vectors can be obtained using the equations of Table 3.5 and equation (3.54), now listed in Table 3.6:

$$U_{ij}T_s = (U_i - U_j)T_s = \frac{V_{dc}}{2}\left(U_i^* - U_j^*\right)T_s, \quad \text{Where } i,j = a,b,c \text{ and } i \neq j \tag{3.59}$$

A direct relationship between the line-to-line voltages and space vectors exist. It can be noted that the line-to-line voltages are determined by only active vectors, there being no role of zero vectors.

Table 3.6 Line-to-line voltages and space vectors

Sector no.	U_{AB}	U_{BC}	U_{CA}
I	$2t_a/T_s$	$2t_b/T_s$	$-2(t_a + t_b)/T_s$
II	$-2t_b/T_s$	$2(t_a + t_b)/T_s$	$-2t_a/T_s$
III	$-2(t_a + t_b)/T_s$	$2t_a/T_s$	$2t_b/T_s$
IV	$-2t_a/T_s$	$-2t_b/T_s$	$2(t_a + t_b)/T_s$
V	$2t_b/T_s$	$-2(t_a + t_b)/T_s$	$2t_a/T_s$
VI	$2(t_a + t_b)/T_s$	$-2t_a/T_s$	$-2t_b/T_s$

3.5.3 Modulating Signals and Space Vector Sectors

The relationship between the modulating signals and the sectors of the space vector can be obtained by plotting the three-phase sinusoidal waveform and identifying the region of 0–60 degrees (sector I), 60–120 degrees (sector II), 120–180 degrees (sector III), 180–240 degrees (sector IV), 240–300 degrees (sector V), and 300–360 degrees (sector VI). The relationship is shown in Figure 3.51.

3.6 Low-Switching Frequency PWM

In high-power application of power electronics converters, the switching losses caused during the commutation is considerably high which results in high switching loss and hence lower converter efficiency. Therefore, high-switching frequency PWM techniques such as carrier-based SPWM and SVPWM are not suitable in such applications. As a solution to operate power electronics switches in high motor power applications, the SHE method has been developed [24]. In this method, the information about the harmonics profile is available and thus can be limited to desired magnitude and desirable fundamental component by having the switching angles at the known instants, obtained after computations. The Fourier series analysis of the

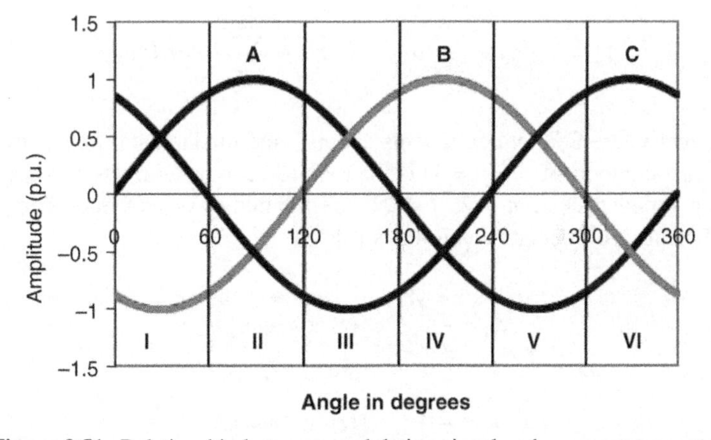

Figure 3.51 Relationship between modulating signal and space vector sectors

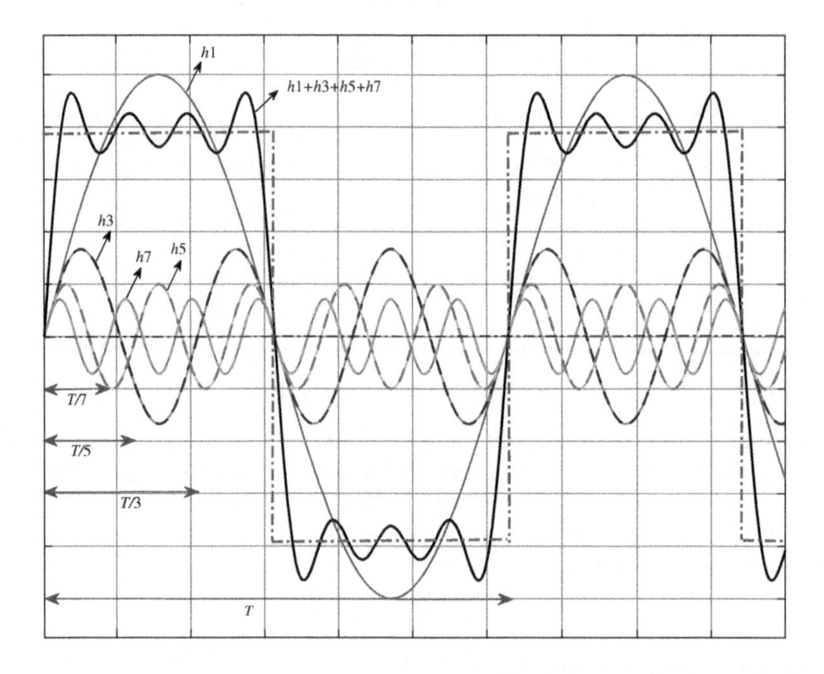

Figure 3.52 Decomposition of non-sinusoidal waveform in fundamental and harmonics components

predefined output waveform is utilized to eliminate the non-desired harmonics components and to have the desired fundamental component defined from the modulation index obtained from the reference value. Any periodic waveform can be analyzed using Fourier series of trigonometric functions having fundamental components as well as harmonics component. Figure 3.52 shows the fundamental components and up to 7th order harmonics components of a square wave.

Any general periodic function in mathematical notation as follows can be written as:

$$f(t) = \frac{1}{2}a_0 + \sum_{n=1}^{\infty} a_n \cos(n\omega t) + \sum_{n=1}^{\infty} b_n \sin(n\omega t) \tag{3.60}$$

where a_0, a_n, and b_n are Fourier coefficients and ω is the fundamental frequency component. The fundamental component (i.e. $n = 1$) is the required component, whereas the multiples of fundamental components (i.e. $n = 2, 3, 4\ldots,$) are harmonics or unwanted components. The Fourier coefficients are calculated as follows [2]:

$$\begin{aligned} a_n &= \frac{2}{T}\int_{-T/2}^{T/2} f(t)\cos(n\omega t)dt \\ b_n &= \frac{2}{T}\int_{-T/2}^{T/2} f(t)\sin(n\omega t)dt \end{aligned} \tag{3.61}$$

3.6.1 Types of Symmetries and Fourier Analysis

Following five types of symmetries are found in power converters output waveform *f*(*t*) with period T, which is useful in the simplification of Fourier series analysis. Different waveform symmetries are presented in Figure. 3.53 and the corresponding coefficients are summarized in Table 3.7:

1. Even-function symmetry

$$f(t) = f(-t) \tag{3.62}$$

2. Odd-function symmetry

$$f(t) = -f(-t) \tag{3.63}$$

3. Half-wave symmetry

$$f(t) = -f(-t + T/2) \tag{3.64}$$

4. Even quarter-wave symmetry

$$f(t) = f(-t) \tag{3.65}$$

$$f(t) = -f(-t + T/2) \tag{3.66}$$

5. Odd quarter-wave symmetry

$$f(t) = -f(-t) \tag{3.67}$$

$$f(t) = -f(-t + T/2) \tag{3.68}$$

3.6.2 Selective Harmonics Elimination in a two-Level VSI

The three-phase VSI is created by three inverter legs similar to three half-H-bridges having each its own binary control as shown in Figure 3.54. In each leg, the power electronics switches are controlled with complementary signal in order to avoid the direct short circuit [2].

The bipolar waveform from a leg of the VSI is shown in Figure 3.55. It is either $+V_{dc}$ or $-V_{dc}$ at the output (say R). The line voltage is obtained by $V_{RY} = V_{Rn} - V_{Yn}$ and it will not contain triplen odd order harmonics for balanced system as it will be canceled out in the final waveform.

The Fourier series of the quarter symmetric SHE output PWM waveform shown in Figure 3.54, and the expression is given in equation (3.69):

$$v(\omega t) = \sum_{n=1}^{\infty} b_n \sin(n\omega t) \tag{3.69}$$

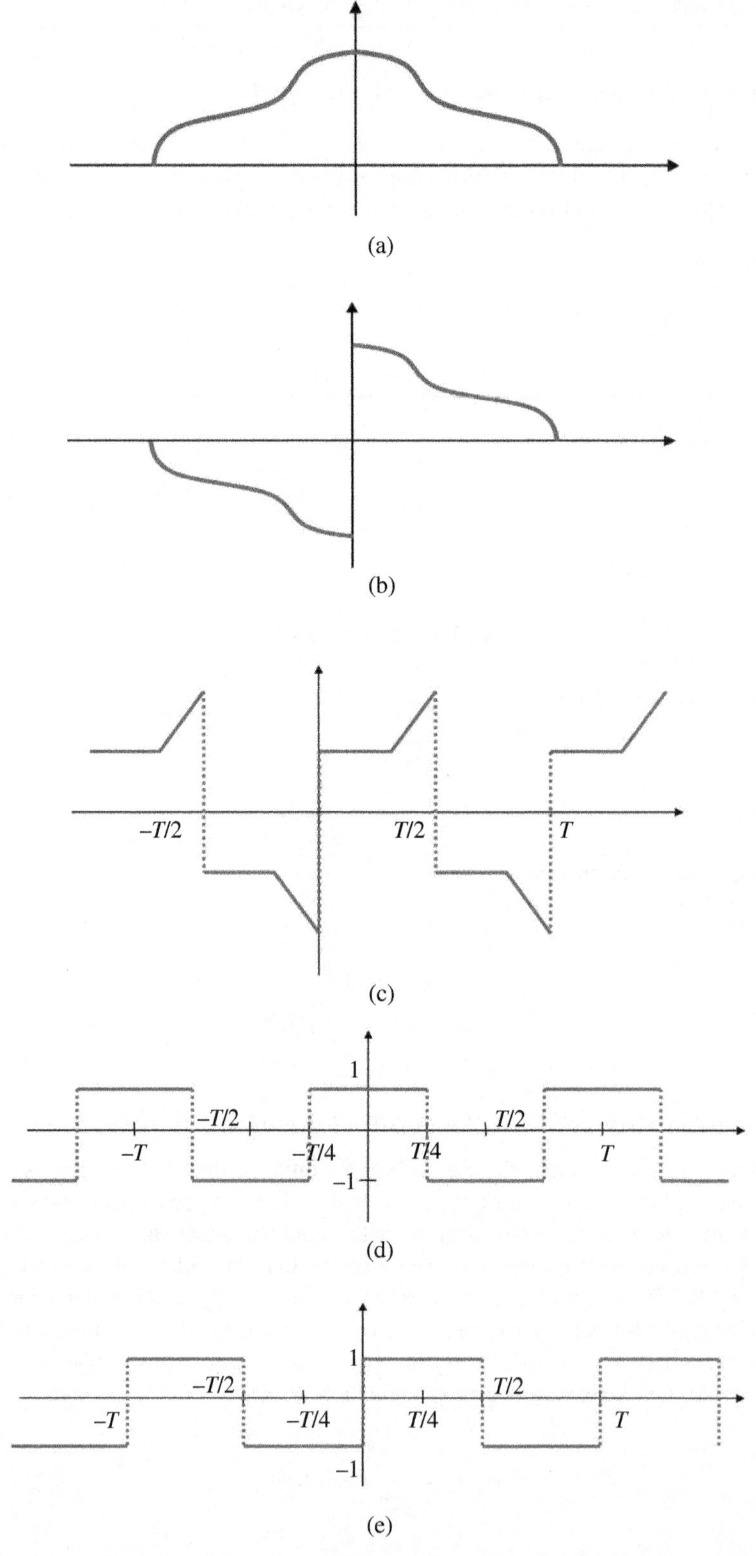

Figure 3.53 Different type of waveform symmetries: (a) even-function symmetry, (b) odd-function symmetry, (c) half-wave symmetry, (d) even-quarter way symmetry, and (e) odd quarter-wave symmetry

Table 3.7 Waveform symmetry and corresponding Fourier coefficient [2]

Symmetry type	Definition	Nonzero Fourier coefficients	Zero Fourier coefficients
Even	$f(t) = f(-t)$	a_n	b_n
Odd	$f(t) = -f(-t)$	b_n	a_n
Half-wave	$f(t) = -f(-t + T/2)$	a_{2n-1}, b_{2n-1}	a_{2n}, b_{2n}
Even, quarter wave	$f(t) = f(-t)$ and $f(t) = -f(-t + T/2)$	a_{2n-1}	a_{2n}, b_n
Odd, quarter wave	$f(t) = -f(-t)$ and $f(t) = -f(-t + T/2)$	b_{2n-1}	b_{2n}, a_n

Source: Holmes and Lipo [2].

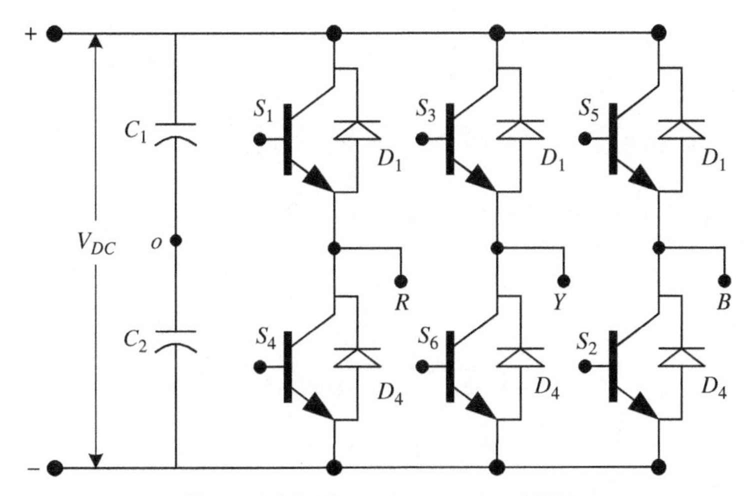

Figure 3.54 Three-phase two-level VSI

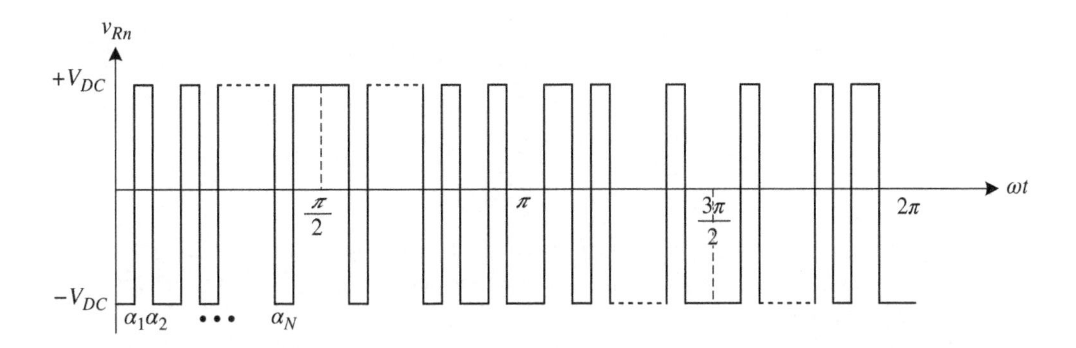

Figure 3.55 Bipolar two-level waveform

where n is the harmonic order and b_n is the Fourier coefficient signifies the harmonics of nth order harmonics component. For a bipolar waveform shown in Figure 3.54, b_n in general can be expressed by equation (3.70):

$$b_n = \frac{4}{\pi}\int_0^{\pi/2} f(\omega t)\sin(n\omega t)d\omega t$$

$$b_n = \frac{4}{\pi}\left[\int_0^{\alpha_1} f(\omega t)\sin(n\omega t)d\omega t + \int_{\alpha_1}^{\alpha_2} f(\omega t)\sin(n\omega t)d\omega t + \cdots + \int_{\alpha_N}^{\pi/2} f(\omega t)\sin(n\omega t)d\omega t\right]$$

$$b_n = \frac{4}{\pi}\left[\int_0^{\alpha_1} (-V_{dc})\sin(n\omega t)d\omega t + \int_{\alpha_1}^{\alpha_2} (V_{dc})\sin(n\omega t)d\omega t + \cdots + \int_{\alpha_N}^{\pi/2} (V_{dc})\sin(n\omega t)d\omega t\right]$$

$$b_n = \frac{4V_{dc}}{n\pi}\left[\cos(n\alpha)|_0^{\alpha_1} - \cos(n\alpha)|_{\alpha_1}^{\alpha_2} + \cdots - \cos(n\alpha)|_{\alpha_N}^{\pi/2}\right]$$

$$b_n = \frac{4V_{dc}}{n\pi}\left[-1 + 2\cos(n\alpha_1) - 2\cos(n\alpha_2) + \cdots + 2(-1)^N\cos(n\alpha_N)\right]$$

or

$$b_n = \begin{cases} \dfrac{4V_{dc}}{n\pi}\left[-1-2\sum_{k=1}^{N}(-1)^k\cos(n\alpha_k)\right], & \text{for odd n} \\ 0 & \text{for even n} \end{cases} \tag{3.70}$$

The magnitude of nth harmonics components for single and three phases can be given by equations (3.71) and (3.72), respectively:

$$V_n(\alpha) = \frac{4V_{dc}}{(2N-1)\pi}\left[-1-2\sum_{k=1}^{N}(-1)^k\cos((2N-1)\alpha_k)\right] \tag{3.71}$$

$$V_n(\alpha) = \frac{4V_{dc}}{(3N-2)\pi}\left[-1-2\sum_{k=1}^{N}(-1)^k\cos((3N-2)\alpha_k)\right] \tag{3.72}$$

where V_{dc} is DC source voltage and N is the number of switching angles. By having N switching angles, $N-1$ harmonics can be eliminated from the output waveform along with control of desired fundamental component.

For a balanced three-phase system, the triplen harmonics need not to be considered for elimination because the triplen harmonics will get canceled automatically from the output waveform.

Table 3.8 Switching angles in different quadrants

Quadrant	Switching angles (radians)
1st, $(0, \pi/2)$	$\alpha_1, \alpha_2, \alpha_3 \cdots \alpha_{N-1}, \alpha_N$
2nd, $(0, \pi)$	$\pi - \alpha_N, \pi - \alpha_{N-1}, \cdots \pi - \alpha_2, \pi - \alpha_1$
3rd, $(0, 3\pi/2)$	$\pi + \alpha_1, \pi + \alpha_2, \cdots \pi + \alpha_{N-1}, \pi + \alpha_N$
4th, $(0, 2\pi)$	$2\pi - \alpha_N, 2\pi - \alpha_{N-1}, \cdots 2\pi - \alpha_2, 2\pi - \alpha_1$

For quarter wave symmetric waveform, the switching angles are computed only for quadrant and for remaining quadrants it can be found as shown in Table 3.8.

Let the system of nonlinear equations in the SHE problem is defined as equation (3.73):

$$f(\alpha_1, \alpha_2, \alpha_3, \ldots, \alpha_n) = \begin{bmatrix} f_1(\alpha_1, \alpha_2, \alpha_3, \ldots, \alpha_n) \\ f_2(\alpha_1, \alpha_2, \alpha_3, \ldots, \alpha_n) \\ \vdots \\ f_n(\alpha_1, \alpha_2, \alpha_3, \ldots, \alpha_n) \end{bmatrix} \tag{3.73}$$

where $f_k : R^n \rightarrow R$.

Also let $\alpha \in R^n$. Then, α represents the design variable or vector:

$$\alpha = \begin{bmatrix} \alpha_1 \\ \alpha_2 \\ \vdots \\ \alpha_n \end{bmatrix} \tag{3.74}$$

where $\alpha_k \in R$ and $k = 1, 2, \ldots$ n and calculate the Jacobian matrix $J(\alpha)^{-1}$ as (3.75):

$$J(\alpha)^{-1} = \begin{bmatrix} \frac{\partial f_1(\alpha)}{\partial \alpha_1} & \frac{\partial f_1(\alpha)}{\partial \alpha_2} \cdots & \frac{\partial f_1(\alpha)}{\partial \alpha_n} \\ \frac{\partial f_2(\alpha)}{\partial \alpha_1} & \frac{\partial f_2(\alpha)}{\partial \alpha_2} \cdots & \frac{\partial f_2(\alpha)}{\partial \alpha_n} \\ \vdots & \vdots & \vdots \\ \frac{\partial f_n(\alpha)}{\partial \alpha_1} & \frac{\partial f_n(\alpha)}{\partial \alpha_2} \cdots & \frac{\partial f_n(\alpha)}{\partial \alpha_n} \end{bmatrix}^{-1} \tag{3.75}$$

The steps of Newton's method are as follows:

1. Assume any random initial of switching angles. Let it be $\alpha^j = \left[\alpha_1^j, \alpha_2^j, \alpha_3^j, \alpha_4^j, \alpha_5^j\right]^T$ with $j = 0$
2. Start modulation index $m = 0$. Take an increment of 0.0001 for next cases.

3. Calculate the nonlinear system matrix f and inverse of Jacobian matrix J^{-1} at α^j.
4. Compute $d\alpha^j = -J^{-1}/F$ and then the updated switching angles will become $\alpha^{j+1} = \alpha^j + d\alpha^j$.
5. Perform $\alpha^{j+1} = \text{mod}(\alpha^{j+1}, \pi/2)$ and $\text{sort}(\alpha^{j+1})$ to ensure $0 < \alpha_1 < \alpha_2 < \alpha_3 < \alpha_4 < \alpha_5 < \pi/2$.
6. Update $j = j + 1$ and repeat the process from 3 to 5 until $d\alpha^j$ satisfied to the desired accuracy or maximum number of iterations reached.
7. If solution found plot the switching angles against m.
8. Take another value of $m = m + 0.0001$.
9. Repeat the process from 1 to 8.

3.6.3 MATLAB Code

```
clear all; close all; clc
Tol=1;
MaxIter=1000;
M=0.7;
x0=sort(rand(5,1)*(pi()/2));
x=x0;
 i=1;
  delf=1;
    while (delf>1e-6 && i<MaxIter)
    f=[-1+2*cos(x(1))-2*cos(x(2))+2*cos(x(3))-2*cos(x(4))+2*cos
(x(5))-M;
   -1+2*cos(5*x(1))-2*cos(5*x(2))+2*cos(5*x(3))-2*cos(5*x(4))
+2*cos(5*x(5));
   -1+2*cos(7*x(1))-2*cos(7*x(2))+2*cos(7*x(3))-2*cos(7*x(4))
+2*cos(7*x(5));
   -1+2*cos(11*x(1))-2*cos(11*x(2))+2*cos(11*x(3))-2*cos(11*x
(4))+2*cos(11*x(5));
   -1+2*cos(13*x(1))-2*cos(13*x(2))+2*cos(13*x(3))-2*cos(13*x
(4))+2*cos(13*x(5))];
J=[-2*sin(x(1)) 2*sin(x(2)) -2*sin(x(3)) 2*sin(x(4)) -2*sin(x(5));
   -10*sin(5*x(1)) 10*sin(5*x(2)) -10*sin(5*x(3)) 10*sin(5*x
(4)) -10*sin(5*x(5));
   -14*sin(7*x(1)) 14*sin(7*x(2)) -14*sin(7*x(3)) 14*sin(7*x
(4)) -14*sin(7*x(5));
   -22*sin(11*x(1)) 22*sin(11*x(2)) -22*sin(11*x(3)) 22*sin
(11*x(4)) -22*sin(11*x(5));
   -26*sin(13*x(1)) 26*sin(13*x(2)) -26*sin(13*x(3)) 26*sin
(13*x(4)) -26*sin(13*x(5))];
   w=-inv(J);
   delx=w*f;
   X=x+delx;
   X=mod(X,pi/2);
   Tol=abs(X-x);
   delf=max(abs(f));
```

```
x=sort(X);
i=i+1;
S=x';
if i<MaxIter;
     F=max(f );
     disp([i M S F]);    end; end
```

Figure 3.56 shows the switching angles trajectories for $N = 5$ (five angles) and $N = 7$ (seven angles) as function of modulation index (M).The harmonics spectrum is shown in Figure 3.57 clearly depicts that the targeted harmonics for elimination are absent from the output waveforms.

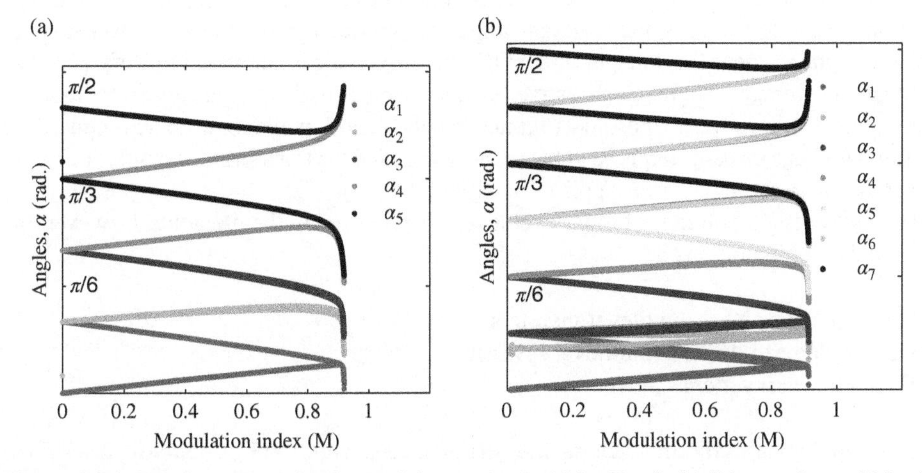

Figure 3.56 Solution trajectories for (a) firing angles vs M for $N = 5$, (b) firing angles vs M for $N = 7$

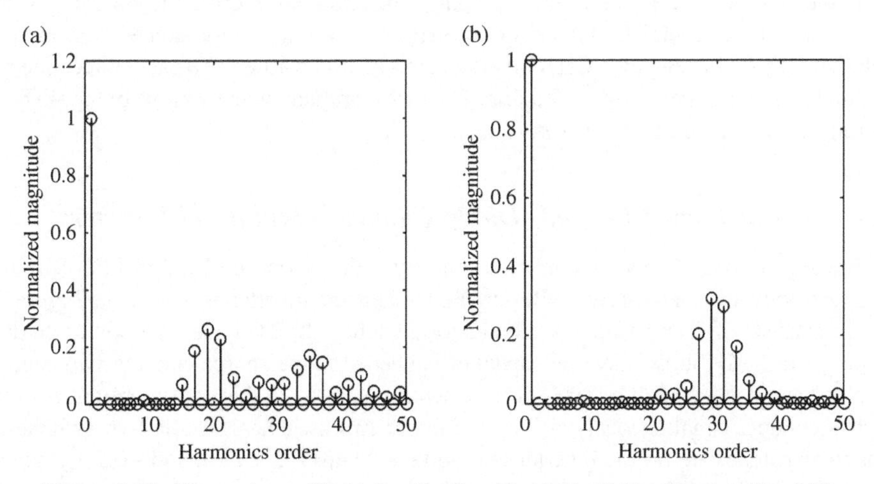

Figure 3.57 Harmonic profile (a) for $N = 5$ at $M = 0.91$ (b) For $N = 5$ at $M = 0.91$

3.7 Multilevel Inverters

The voltage source inverters discussed so far produces only two levels of output, either ±0.5 V_{dc} or 0 and $V_{dc,}$ hence called the two-level inverter. Two-level inverters pose several problems in high-power medium-voltage applications due to limited voltage blocking capability of the power semiconductor switches. Moreover, d*v*/d*t* is considerably higher, causing high stress on the switching devices. The solution to this problem lies in using a series connection of several power switching devices or alternatively using multilevel inverters [25–31]. The former solution may lead to unequal voltage sharing among the devices. Hence, the later solution is preferred. The output voltage waveform of the multilevel inverter is significantly improved when compared to a two-level output. The THD is low and the output quality is acceptable, even at low-switching frequency. In high-power drives, switching losses constitute a considerable portion of the total losses, and hence the reduction of this is a major goal. Nevertheless, the number of power switching devices used in a multilevel inverter is high, and additional diodes and capacitors are needed. Another important shortcoming of the multilevel inverter is imbalance in the capacitor voltages at the DC link. This issue is addressed in Refs. [32–35].

Multilevel inverters have many attractive features, such as high voltage capability, reduced common mode voltages near sinusoidal outputs, low d*v*/d*t*, and smaller or even no output filter; sometimes no transformer is required at the input side, called the transformerless solution, making them suitable for high-power applications [36–39].

There are several topologies of multilevel inverters proposed in the literature. However, the most popular configurations are

- Diode-clamped or NPC multilevel inverters;
- Capacitor-clamped or FLC multilevel inverters;
- CHB multilevel inverters.

The NPC topology uses diodes to clamp the voltage levels, while the FLC topology uses floating capacitors to clamp the voltage levels. The major problems associated with these two topologies are the balancing of capacitors voltages. Additional problems associated with the FLC type is the starting procedure, as the clamping capacitors are required to be charged to the appropriate voltage level. The balancing problem becomes more pronounced with increasing number of levels. However, the CHB inverter is modular in nature, with each unit connected in series and there is no problem of balancing. The major problem with this type of topology is the requirement of complex transformers [5].

3.7.1 Neutral-Point-Clamped (Diode-Clamped) Multilevel Inverters

The three-level NPC or diode-clamped inverter was first proposed by Ref. [40]. The power circuit topology of a three-level NPC or diode-clamped inverter is shown in Figure 3.58. One leg consists of four power semiconductor switches (IGBTs) and two power diodes for clamping. At the DC-link side, one capacitor is needed at the source end and two capacitors are halved and connected in series to form a neutral point *N*. In this configuration, two power switches conduct simultaneously. When the upper two switches, middle two switches, and lower two switches are on the inverter leg, voltage attains 0.5 V_{dc}, 0, and −0.5 V_{dc} voltages, respectively. The switches S_a and S'_a, S_b and S'_b, and S_c, and S'_c are complimentary in operation.

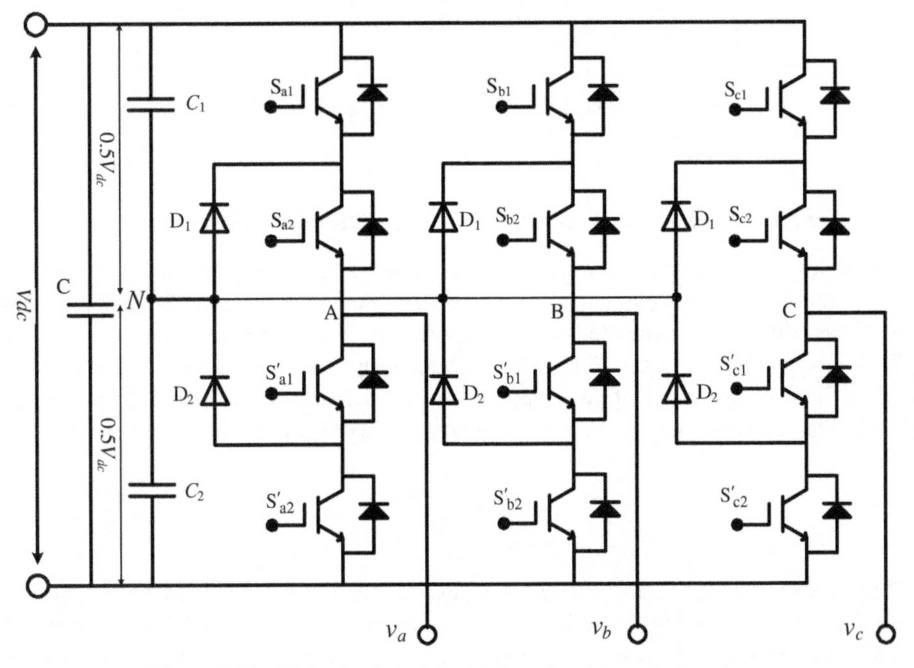

Figure 3.58 Neutral-point-clamped three-level inverter topology

Defining the switching function S_{k1}, S_{kO}, and S_{k2} ($k = A, B, C$) (zero when switch is off and one when switch is on, with the constraint: $S_{k1} + S_{kO} + S_{k2} = 0$) when the upper two switches are on, the middle two switches are on, and the lower two switches are on, respectively. The leg voltages can then be written as:

$$\begin{aligned} V_{AN} &= (S_{A1} - S_{A2})V_{dc} \\ V_{BN} &= (S_{B1} - S_{B2})V_{dc} \\ V_{CN} &= (S_{C1} - S_{C2})V_{dc} \end{aligned} \tag{3.76}$$

The line voltage relation can be obtained as:

$$\begin{aligned} v_{ab} &= V_{AN} - V_{BN} = (S_{A1} - S_{A2} - S_{B1} + S_{B3})V_{dc} \\ v_{bc} &= V_{BN} - V_{CN} = (S_{B1} - S_{B2} - S_{C1} + S_{C3})V_{dc} \\ v_{ca} &= V_{CN} - V_{AN} = (S_{C1} - S_{C2} - S_{A1} + S_{A3})V_{dc} \end{aligned} \tag{3.77}$$

The phase-to-load neutral voltage can be obtained as:

$$\begin{aligned} v_{an} &= \frac{2}{3}V_{dc}(S_{A1} - S_{A3} - 0.5(S_{B1} - S_{B3} + S_{C1} - S_{C3})) \\ v_{bn} &= \frac{2}{3}V_{dc}(S_{B1} - S_{B3} - 0.5(S_{A1} - S_{A3} + S_{C1} - S_{C3})) \\ v_{cn} &= \frac{2}{3}V_{dc}(S_{C1} - S_{C3} - 0.5(S_{A1} - S_{A3} + S_{B1} - S_{B3})) \end{aligned} \tag{3.78}$$

The common mode voltage (voltage between the load neutral and inverter neutral) is

$$v_{nN} = \frac{1}{3} V_{dc}(S_{A1} + S_{B1} + S_{C1} - S_{A1} - S_{B1} - S_{C3}) \tag{3.79}$$

3.7.1.1 Carrier-Based PWM Technique for NPC Three-Level Inverter

A SPWM scheme of a two-level inverter can be readily extended to a multilevel inverter. However, the number of triangular carrier waves is equal to L – 1, where L is the level number. Multicarrier signals are generated and compared with the sinusoidal modulating signal of SPWM fundamental frequency. The switching/gating signals are thus generated. Broadly classified, a SPWM scheme is of two types, phase-shifted and level-shifted. In phase-shifted SPWM schemes, the required phase shifts among different carriers and is given as:

$$\phi = \frac{2\pi}{(L-1)} \tag{3.80}$$

Thus, in a three-level phase-shifted SPWM, two carrier waves are needed, with a 180-degree phase shift. In a five-level inverter, four carrier waves are required, with a 90-degree phase shift. As the number of level increases, the number of carrier waves increases and the phase shift angle decreases.

Level-shifted SPWM schemes are further classified as:

- In-phase disposition (IPD), in which all triangular signals are in phase;
- Phase opposition disposition (POD), in which all the carrier above reference zero are in opposite phase to those below zero reference;
- Alternate phase opposition disposition (APOD), in which each carrier pair is in the opposite phase.

The principle of different types of SPWM schemes, as applied to a five-level diode-clamped inverter, is shown in Figure 3.59. In a five-level inverter, four carrier signals are required, as shown in Figure 3.59 and are compared with the three-phase sinusoidal modulating wave (in the figure, only the phase 'a' modulating signal is shown). The gating signals are generated at the point of intersection of the modulating wave and the carrier wave. The operation of the upper and third from the top power switches are complimentary. Similarly, the second from the top and the bottom-most switches are complimentary in operation.

3.7.1.2 MATLAB/Simulink Model of Carrier-Based PWM Scheme for Three-Level NPC

MATLAB/Simulink model of a three-level NPC inverter is shown in Figure 3.60. The simulation model is developed using the mathematical model of the inverter. The DC-link voltage is assumed as unity, the fundamental output frequency is chosen as 50 Hz, and the switching frequency is kept at 2 kHz. The carrier signals are chosen such that they are IPD.

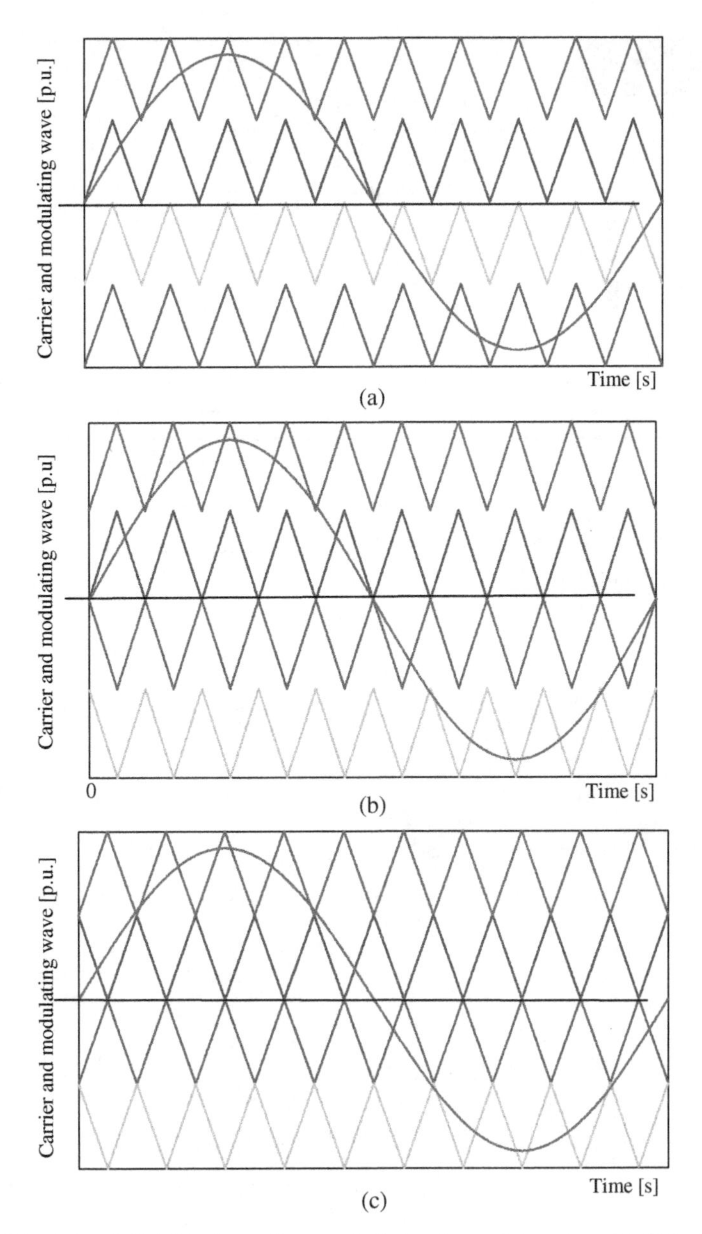

Figure 3.59 Principle of SPWM for a five-level diode-clamped inverter: (a) IPD; (b) POD; and (c) APOD

The resulting waveforms obtained from the simulation are shown in Figure 3.61. Seven-level phase voltages are obtained, while three-level line voltages are generated. More number of step leads to lower THD. Spectrum of phase voltage is also shown in the figure.

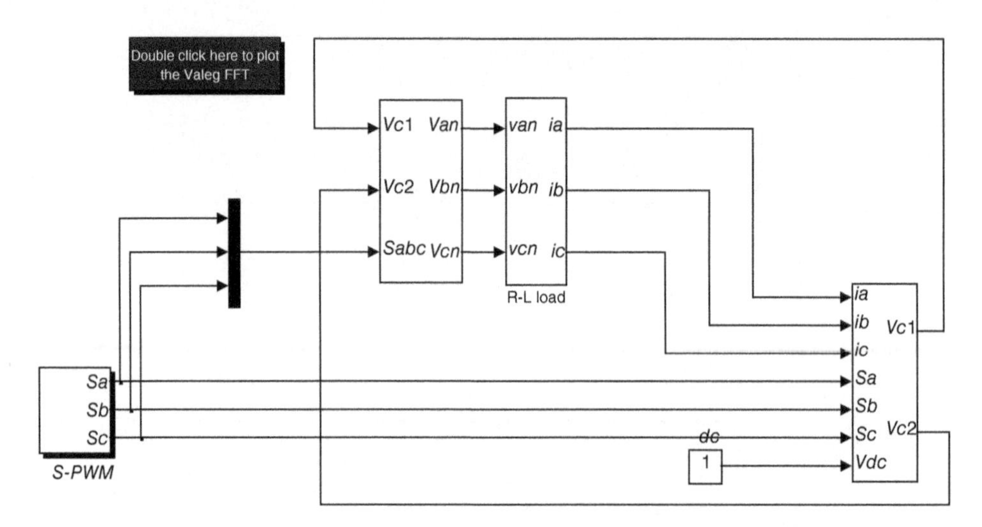

Figure 3.60 MATLAB/Simulink model of SPWM for a three-level NPC (*File name: NPC_3_phase_3_level.mdl*)

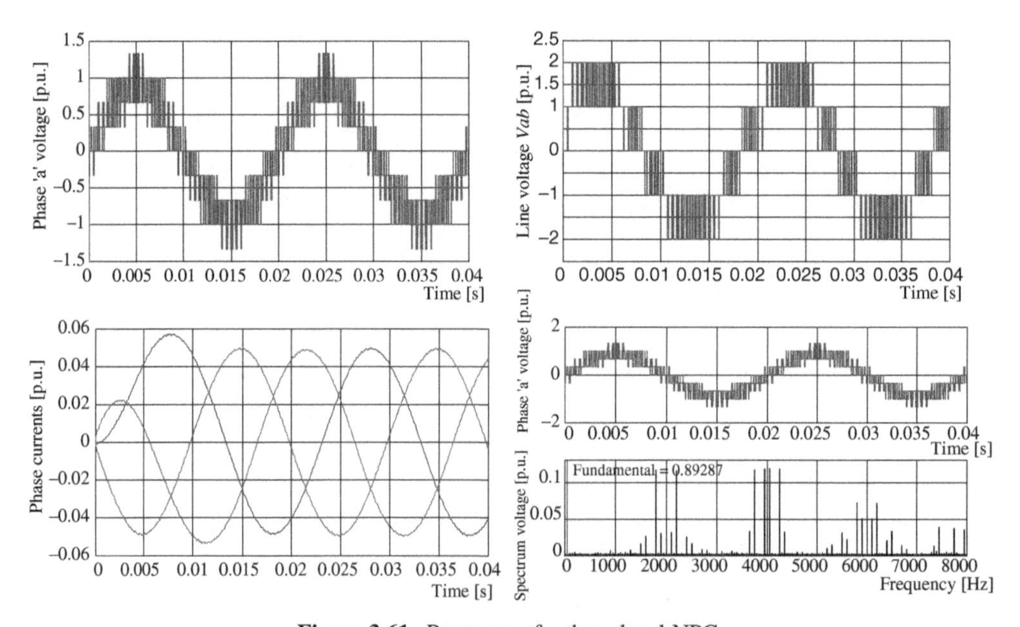

Figure 3.61 Response of a three-level NPC

3.7.2 Flying Capacitor-Type Multilevel Inverter

The FLC or capacitor-clamped topology has some distinct advantages over other topologies of multilevel inverters, for example, the redundancies of the voltage levels, which means that more than one switch combination can create the same output voltage and hence freedom exists in the choice of the appropriate switching combination. The flying-level capacitor inverter has phase redundancies, unlike the NPC that has only line-to-line redundancies [41–49]. This feature allows the charging and discharging of particular capacitors in a way that may be integrated within the control algorithm for balancing the voltages. Another advantage of this inverter is that the active and reactive power flow can be controlled. Also, a large number of capacitors enable an operation through short duration outages and significant voltage sags.

FLC multilevel topology is a good approach, particularly for more than three voltage levels. A negative aspect of using a FLC is that the output current changes the voltage of the capacitors. In general, the capacitance must be selected as high to make the changes as low as possible. Therefore, the size of the capacitors increases as an inverse of the switching frequency. This topology is therefore not practical for low-switching frequency applications [41]. Another important shortcoming of this topology is the starting/excitation of the inverter. The capacitor must be charged to the appropriate voltage level before operating the inverter. Hence, a special starting procedure is required for this type of multilevel inverter.

The FLC voltage balancing is a tedious task in the FLC topology [41–43]. Numerous techniques have been presented in the literature to balance the voltage across the clamping capacitors [42–49]. These include the SPWM, as well as the SVPWM, using switching vector redundancies. Among the carrier-based techniques, three methods are most popular, namely phase-shifted PWM, modified carrier redistribution PWM, and saw-tooth rotation PWM. A comprehensive comparison of these methods is presented in Ref. [47] for a three-phase system. It is concluded that the modified carrier redistribution PWM offers the optimum performance. This technique is further used in Refs. [15, 18]. Hence, the same technique is elaborated on in this chapter.

3.7.2.1 Operation of Three-Level Flying Capacitor Inverter

The power circuit topology of a five-phase three-level FLC-type voltage source inverter is shown in Figure 3.62. In this topology of a multilevel inverter, the floating capacitors are

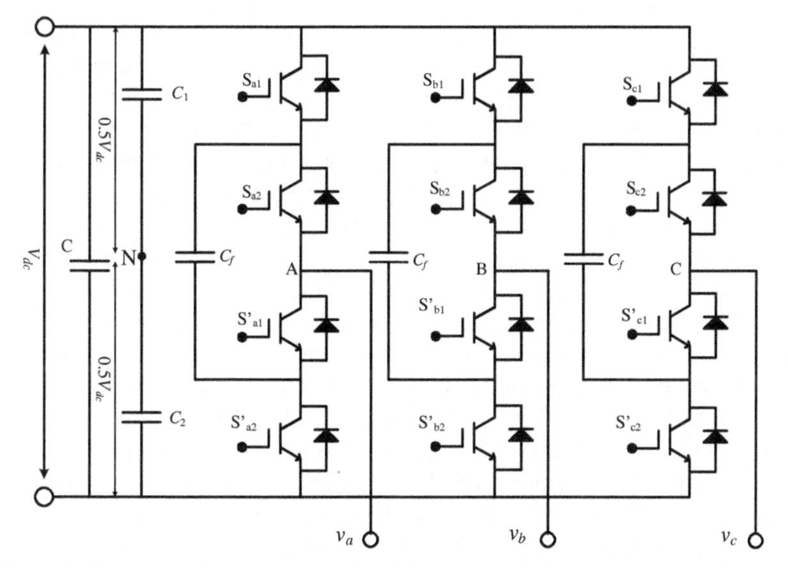

Figure 3.62 Power circuit of a three-level three-phase flying capacitor-type inverter

employed to clamp the node voltages of the series connected power switching devices. Four power switching devices are connected in series to form one leg of the inverter. The complimentary switches are indicated in the Figure 3.62 $((s_{n1}, s'_{n1}), (s_{n2}, s'_2); n \in a, b, c)$; the top and the bottom switches and the two middle switches are complementary. Two power switching devices are turned on simultaneously to provide three different voltage levels at the output phase, 0.5 V_{dc}, –05 V_{dc}, and 0. Since the voltage across the FLCs is limited to 0.5 V_{dc}, the same voltage stress will be borne by each switch. As the number of levels increases the required voltage, the blocking value reduces, thus lower rating switches can be used or alternatively higher voltage can be achieved. The relationship between the pole voltage and the output phase voltage remains the same as that of an NPC three-level inverter.

One cycle of the operation of one leg of the FLC inverter is depicted in Figure 3.63a–h. The path of the current is shown as a dotted line. The pole voltage is 0.5 V_{dc} and the current either flows to or from the load in Figure 3.63a and b. The capacitor state remains unchanged.

The FLC charges during the switching state of Figure 3.63c and f. The charging current is equal to the load current. Hence, the design of FLC should take into account the maximum load current. The pole voltage is zero during this operation. The FLC discharges during the switching state of Figure 3.63d and e. Once again the current flow through the capacitor is equal to the load current. The pole voltage remains zero during this switching state. Figure 3.63g and h show the switching state when the pole voltage attains the value of –0.5 V_{dc} and the FLC conditions remain unchanged.

It is evident from the above discussion that the capacitor charging state changes throughout Figure 3.63c–f. Thus, to balance the FLC voltage, these switching states need special attention. Thus, the PWM technique described in the next section optimally exploits these redundant switching states.

3.7.2.2 PWM Technique to Balance Flying Capacitor Voltage [45, 47]

The major goal of the PWM scheme is to balance the FLC voltage and keep it constant at half of the DC-link voltage, irrespective of the load condition. The basic idea behind voltage balancing is equal charging and discharging time of the FLC. From the switching diagram (Figure 3.63), it is clear that for switch states of Figure 3.63c and f, the output pole voltage remains the same (zero), thus these states are redundant states. However, the FLC charges in the state of Figure 3.63c and f and discharges to the load in state of Figure 3.63d and e. Therefore, states (Figure 3.63c and f) and (Figure 3.63d and e) should persist for the same time for equal charging and discharging of the FLC. To accomplish this, the carrier signal should be formulated, as shown in Figures 3.64 and 3.65, for switch S_{n1} and S_{n2}, respectively. The upper carrier is used when the reference voltage lies between 0.5 V_{dc} and V_{dc}, while the lower carrier is used when reference voltage lies between 0 and 0.5 V_{dc}. The gate signals are shown in Figures 3.66 and 3.67 for the two different values of the reference voltages. The carrier signals for four sampling periods are shown. It is evident that the redundant switching state is active in the first sample and then in the third sample in Figure 3.66. Hence, the FLC charges in the first sampling time and discharges in the third sampling time, keeping the same charging and discharging time in four sample periods. The switching time of each power switch will be equal to $2T_s$ as evident from Figures 3.66 and 3.67. A similar observation can be made from Figure 3.67. Thus, it is seen that for all the values of the reference voltages, the charge and discharge time of the FLCs

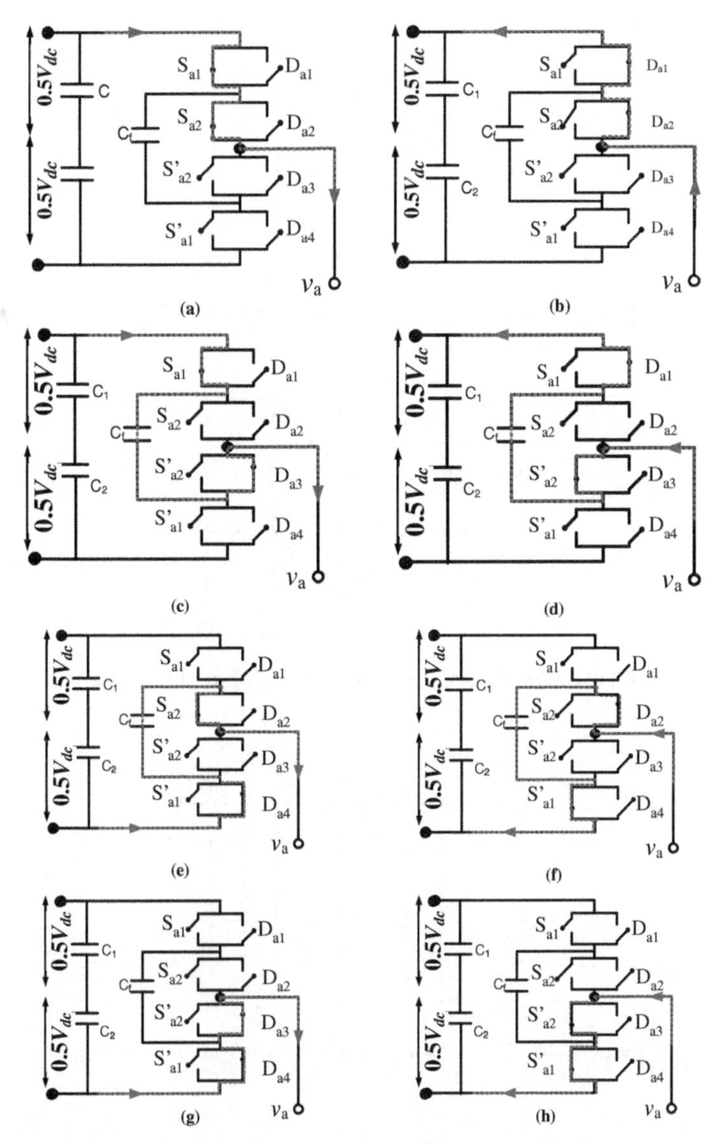

Figure 3.63 (a) Switch state 1100 with positive current flow; (b) switch state 1100 with negative current flow; (c) switch state 1010 with positive current flow (flying capacitor charges); (d) switch state 1010 with negative current flow (flying capacitor discharges); (e) switch state 0101 with positive current flow (flying capacitor discharges); (f) switch state 0101 with negative current flow (flying capacitor charges); (g) switch state 0011 with positive current flow; and (h) switch state 0011 with negative current flow

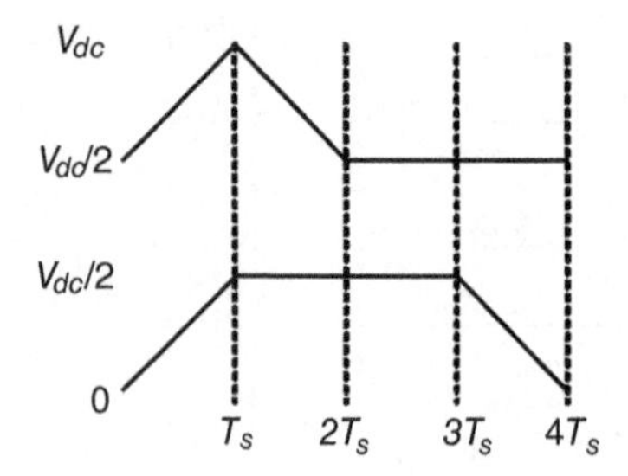

Figure 3.64 Carrier wave for S_{n1} and S_{n1}' in FLC inverter

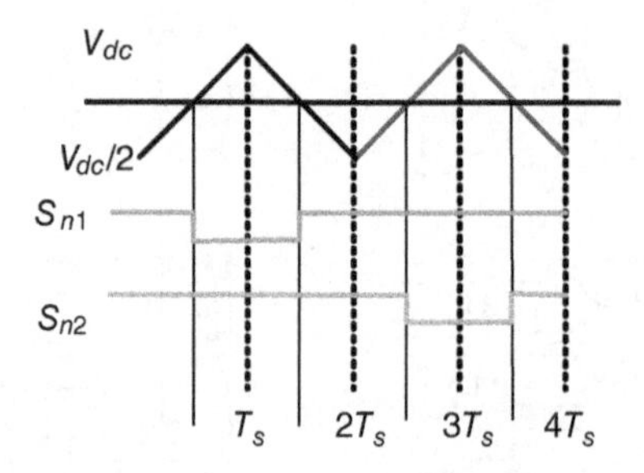

Figure 3.65 Carrier wave for S_{n2} and S_{n2}' in FLC inverter

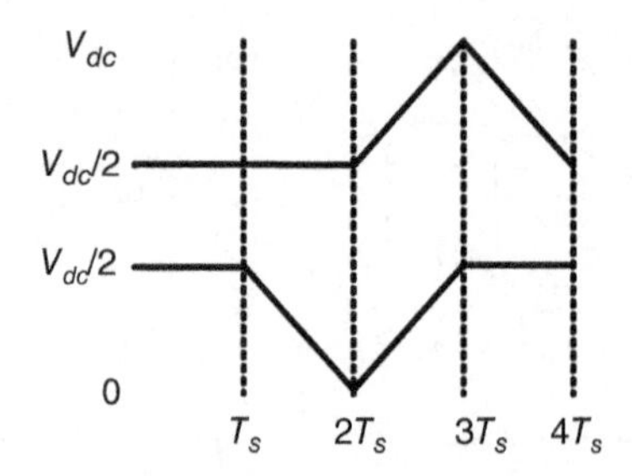

Figure 3.66 Gate signal generation when $V_{dc}/2 \leq |v_{ref}| \leq V_{dc}$

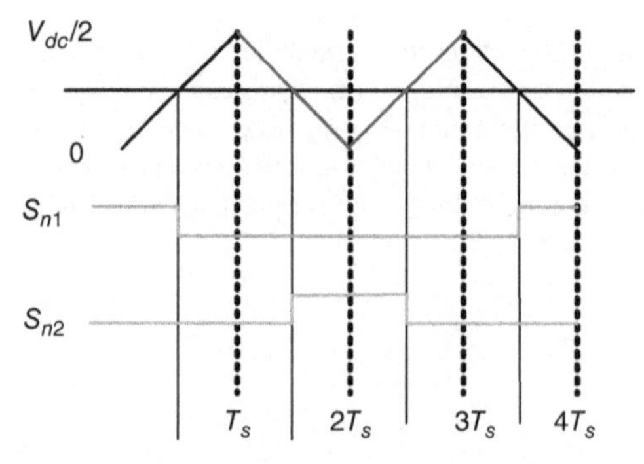

Figure 3.67 Gate signal generation when $0 \leq |v_{ref}| \leq V_{dc}/2$

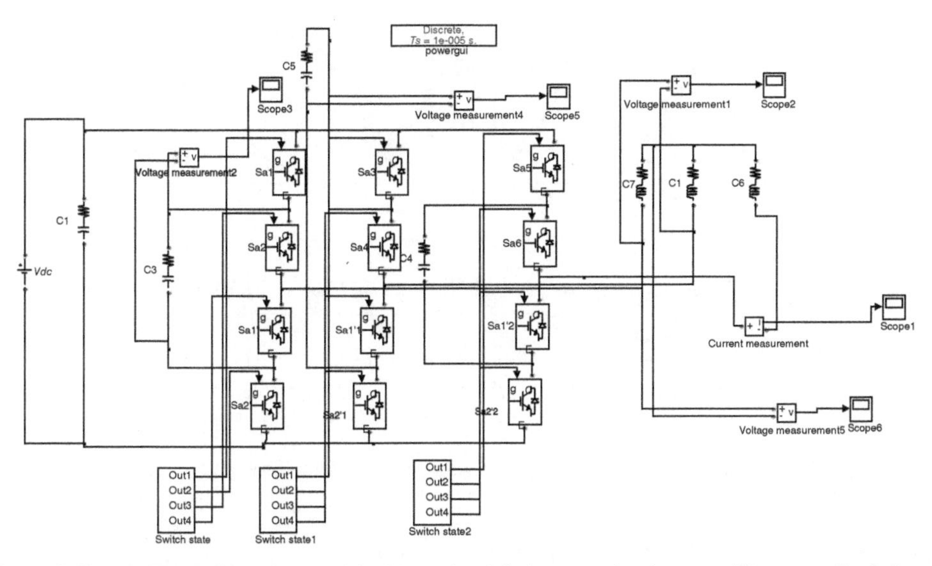

Figure 3.68 MATLAB/Simulink model of three-level flying capacitor inverter (file name: *flc_3ph.mdl*)

remain the same. Thus, the voltage is balanced at the FLC terminals as an average in four sample time. This average balancing time will be higher for higher-level numbers.

3.7.2.3 Start-Up/Excitation Procedure of FLC

One of the important requirements of a FLC inverter is its start-up procedure. This problem still persists in a five-phase inverter as in a three-phase inverter. At the start, the FLC needs to be charged equal to 0.5 V_{dc}. Thus, the gating signal is provided in such a way that the switch S_{n1} and S'_{n1} remain 'ON' to charge the capacitor and the switches S_{n2} and S'_{n2} remain 'OFF.' Once the voltage of the FLC builds up to the required voltage level (i.e. 0.5 V_{dc}), all the switches are kept off so that the FLC voltage floats at 0.5 V_{dc}. Then, the normal PWM inverter operation is implemented.

3.7.2.4 MATLAB/Simulink Model of Three-Level Capacitor-Clamped or FLC Inverter

The MATLAB/Simulink model for implementing the three-level FLC inverter is shown in Figure 3.68. The model is shown using the SimPowerSystem blocksets. SPWM is further illustrated in Figures 3.66 and 3.67, which are used in this model.

DC-link voltage of 100 V is retained and the switching frequency remains at 2.5 kHz, the FLC is chosen as 100 μF, and the load parameters are $R = 5$ W and $L = 0.5$ mH. The resulting output line and phase voltages are shown in Figure 3.69. The output line voltage assumes three levels and the phase voltage has seven levels.

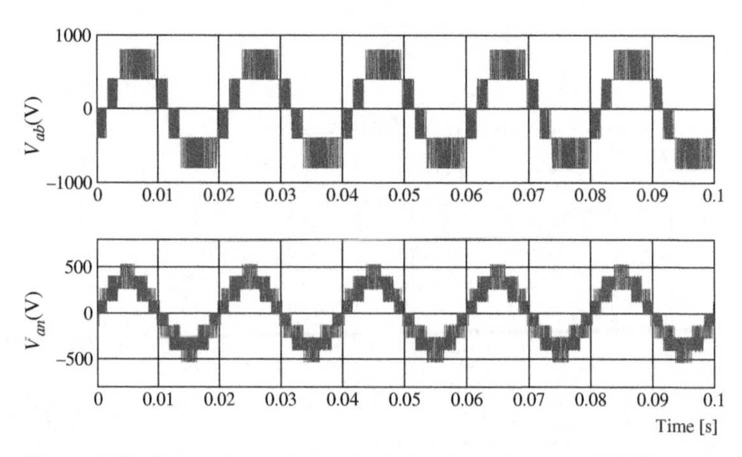

Figure 3.69 Output line and phase voltages from three-level FLC inverter

3.7.3 Cascaded H-Bridge Multilevel Inverter

Several units of the H-bridge inverters are connected in series to form a CHB topology that outputs multilevel voltages. This is a highly modular configuration, where by adding each H-bridge cell increases the level of the output voltages. Each H-bridge unit requires isolated DC supplies. The number of units to be connected in series to obtain an L level (odd) of the output voltages is given as:

$$N = \frac{L-1}{2} \tag{3.81}$$

where L is the number of output levels and N is the number of H-bridge unit. The number of output voltage levels can only be odd. Hence, to obtain a five-level output, two H-bridge units are connected in series.

If two units are connected in series, they form a five-level output, as shown in Figure 3.70. The switching states and the possible output voltages for this configuration are given in Table 3.9. The output phase voltages can acquire four different voltage levels, $-2\ V_{dc}$, $-V_{dc}$, $0\ V_{dc}$, and $2\ V_{dc}$. It is further noted that the switching redundancy exists in the same way as the output voltage level is obtained from different switching combination. This redundancy can be exploited in a SVPWM to choose the best possible vectors to generate the reduced common mode voltages.

3.7.3.1 Carrier-Based PWM Technique for a Multilevel CHB Inverter

The SPWM technique can be broadly classified into two types, phase-shifted and level-shifted. In the phase-shifted SPWM technique, the phase shift angle between two adjacent carrier signals is given by equation (3.80). Hence, in a five-level CHB inverter, four carrier waves are required with the phase shift between the two carriers at 90 degrees. The modulating signals are three-phase sinusoids. In a five-level inverter, the carrier with a 0-degree phase shift will control the switching of the cell-1 upper power switch S_{a11} and the 90-degree phase-shifted

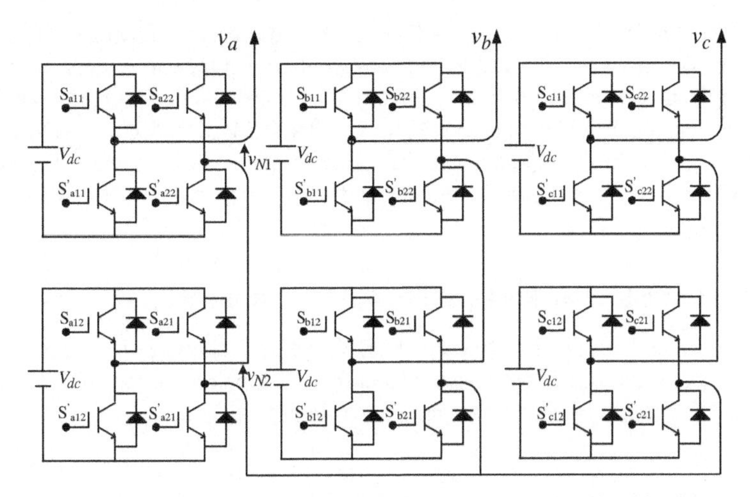

Figure 3.70 Power circuit topology of a five-level cascaded H-bridge inverter

Table 3.9 Switching states and output voltages in a five-level CHB inverter

S_{a11}	S_{a22}	S_{a12}	S_{a21}	v_{N1}	v_{N2}	V_a
1	0	1	0	V_{dc}	V_{dc}	$2V_{dc}$
1	0	1	1	V_{dc}	0	V_{dc}
1	0	0	0	V_{dc}	0	V_{dc}
1	1	1	0	0	V_{dc}	V_{dc}
0	0	1	0	0	V_{dc}	V_{dc}
0	0	0	0	0	0	0
0	0	1	1	0	0	0
1	1	0	0	0	0	0
1	1	1	1	0	0	0
1	0	0	1	V_{dc}	$-V_{dc}$	0
0	1	1	0	$-V_{dc}$	V_{dc}	0
0	1	1	1	$-V_{dc}$	0	$-V_{dc}$
0	1	0	0	$-V_{dc}$	0	$-V_{dc}$
1	1	0	1	0	$-V_{dc}$	$-V_{dc}$
0	0	0	1	0	$-V_{dc}$	$-V_{dc}$
0	1	0	1	$-V_{dc}$	$-V_{dc}$	$-2V_{dc}$

carrier signal controls the switching of the cell-2 upper power switch S_{a12}. The 180-degree phase-shifted carrier signal controls the cell-1 switch S_{a22} and the 270-degree phase-shifted carrier controls the switching of S_{a21} of cell-2 (Figure 3.70). The gating signals are generated at the intersection of the triangular carrier signals and the modulating signals.The inverter switching frequency and the device switching frequency is related by:

$$f_{s,inverter} = (L-1)f_{s,device} \tag{3.82}$$

In level-shifted SPWM technique, (L – 1)-level-shifted carriers are required. The carriers are arranged vertically between $-V_{dc}$ and V_{dc} and the amplitude of the carrier depend on their number, for example, in the case of a five-level inverter, four carrier signals are required. The amplitude of the carrier signals will then be 0.5 V_{dc} to V_{dc}, 0 to 0.5 V_{dc}, –0.5 V_{dc} to $-V_{dc}$, and –0.5 V_{dc} to 0. The inverter switching frequency and the carrier frequency is the same in the case of level-shifted PWM. The device switching frequency and the inverter switching frequency are related, as shown in equation (3.82).

3.7.3.2 MATLAB/Simulink Model of Five-Level CHB Inverter

A MATLAB/Simulink model for a five-level CHB inverter is shown in Figure 3.71. The model is formulated using the SimPowerSystem blocksets. The phase-shifted SPWM technique is employed and the resulting waveforms are shown in Figure 3.72.

3.8 Space Vector Modulation and DC-Link Voltage Balancing in Three-Level Neutral-Point-Clamped Inverters

The topology of three-level NPC inverter consists of 2 DC-link capacitors, 12 transistors with antiparallel diodes, and 6 diodes connected to the DC-link midpoint (Figure 3.58). Such structure allows connecting the inverter output terminals to one of the DC-link potentials: positive, negative, or potential of the DC-link midpoint.

Because the three-level NPC inverters are used to generate higher output voltages using lower rating devices, the DC-link voltages have to be balanced. The DC-link voltage unbalance causes an increase in voltage stress on switching devices. Additionally, the additional harmonic distortion occur in the inverter output voltage. The voltages of the series-connected DC-link capacitors should be equalized. If the DC-link voltage asymmetry is allowed, it is necessary to ensure that it does not affect the inverter output voltage (Figure 3.73).

3.8.1 The Output Voltage of Three-Level NPC Inverter in the Case of the DC-Link Voltage Unbalance

The potentials on the inverter terminals depend on the activated transistors and not depend on the phase current direction (Figure 3.74a and b). The voltages between the inverter terminals and the DC-link midpoint can be calculated as:

$$\begin{aligned} v_{aN} &= v_{DCu} \cdot (S_{a1} \cdot S_{a2}) - v_{DCl} \cdot (S'_{a1} \cdot S'_{a2}), \\ v_{bN} &= v_{DCu} \cdot (S_{b1} \cdot S_{b2}) - v_{DCl} \cdot (S'_{b1} \cdot S'_{b2}), \\ v_{cN} &= v_{DCu} \cdot (S_{c1} \cdot S_{c2}) - v_{DCl} \cdot (S'_{c1} \cdot S'_{c2}), \end{aligned} \tag{3.83}$$

where v_{DCu} and v_{DCl} are the DC-link voltages, S_{x1}, S_{xa2}, S'_{x1}, and S'_{x2} ($x = a$, b or c) are the gate signals for upper (u) and lower (l) transistors with the values ‘1’ for the switched-on and ‘0’ for switched-off transistors.

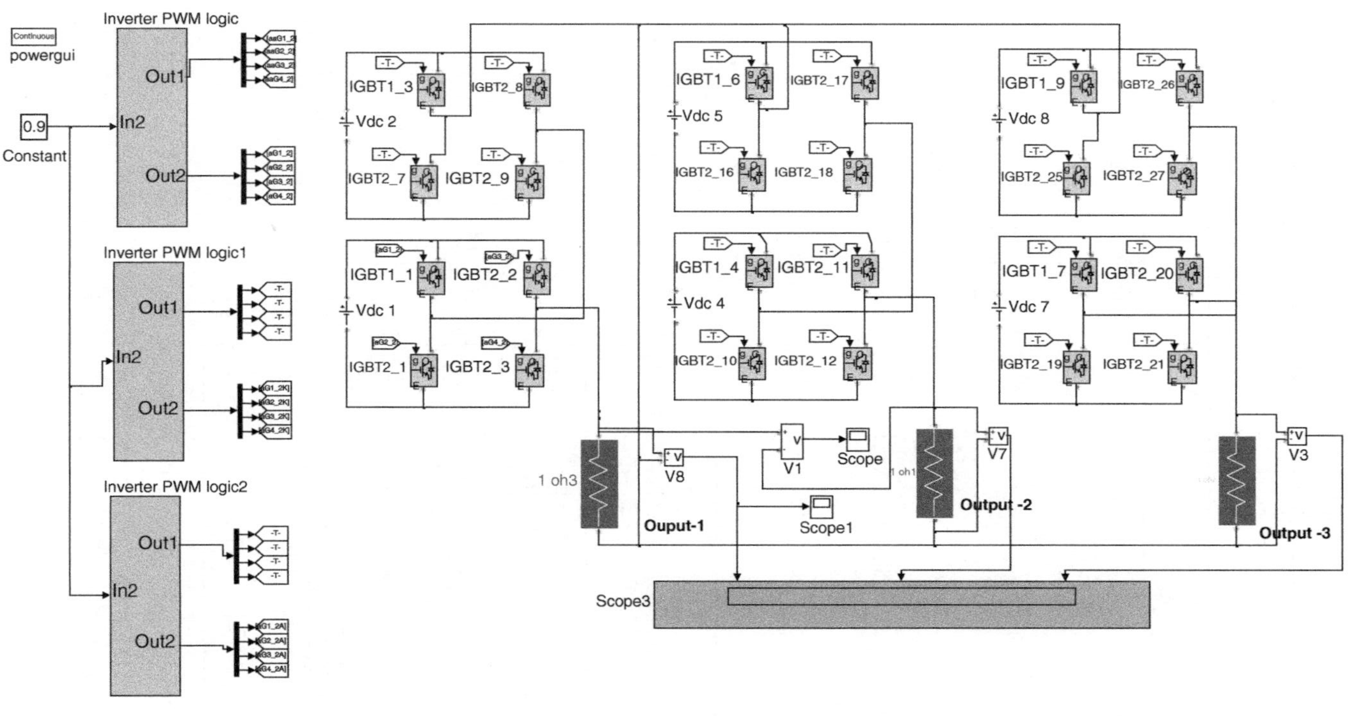

Figure 3.71 MATLAB/Simulink of five-level three-phase CHB (file Name: *CHB_5_level.mdl*)

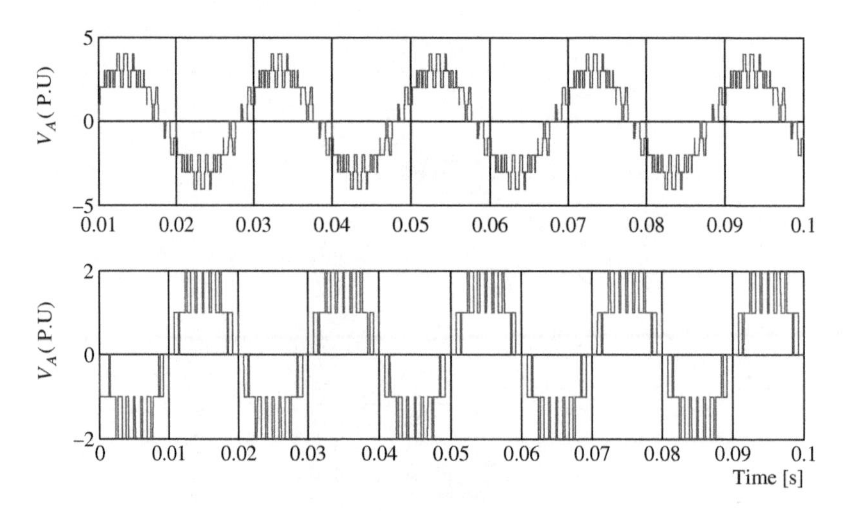

Figure 3.72 Output line and phase voltages from five-level CHB inverter

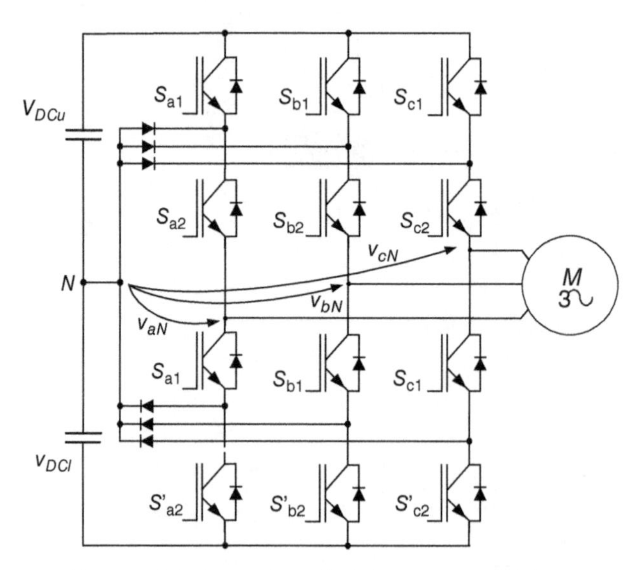

Figure 3.73 Three-phase NPC inverter-based drives

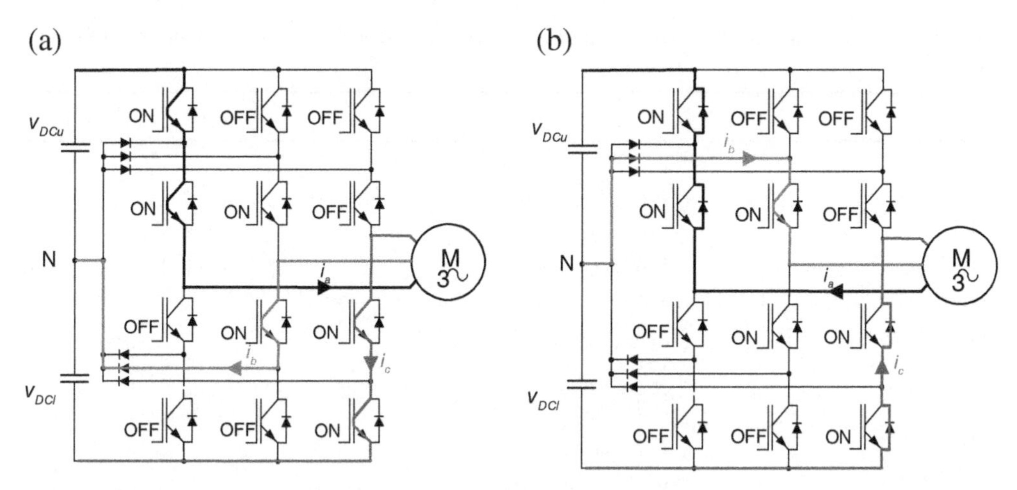

Figure 3.74 The flow of the three-phase NPC inverter currents for allowable switching-on combinations of transistors and phase currents: $i_a > 0$, $i_b < 0$, $i_c < 0$ (a), and $i_a < 0$, $i_b > 0$, $i_c > 0$ (b)

The voltages on the inverter terminals are listed in Table 3.10.

The output voltage of NPC inverter depends on the activated transistors. The components of switching vectors (active and zero vectors) can be determined using Clarke's transformation as follows:

$$v_\alpha = \sqrt{\frac{1}{6}} \cdot \begin{pmatrix} v_{DCu} \cdot (2 \cdot (S_{a1} \cdot S_{a2}) - (S_{b1} \cdot S_{b2}) - (S_{c1} \cdot S_{c2})) \\ - v_{DCl} \cdot (2 \cdot (S'_{a1} \cdot S'_{a2}) - (S'_{b1} \cdot S'_{b2}) - (S'_{c1} \cdot S'_{c2})) \end{pmatrix},$$
$$v_\beta = \sqrt{\frac{1}{2}} \cdot \begin{pmatrix} v_{DCu} \cdot ((S_{b1} \cdot S_{b2}) - (S_{c1} \cdot S_{c2})) \\ - v_{DCl} \cdot ((S'_{b1} \cdot S'_{b2}) - (S'_{c1} \cdot S'_{c2})) \end{pmatrix}. \tag{3.84}$$

Assuming the following gate signal markings:

$$\begin{aligned} &Sx = -1 \Leftrightarrow S_{x1} = 0;\ \ S_{x2} = 0;\ \ S'_{x1} = 1;\ \ S'_{x2} = 1; \\ &Sx = 0 \Leftrightarrow S_{x1} = 0;\ \ S_{x2} = 1;\ \ S'_{x1} = 1;\ \ S'_{x2} = 0; \\ &Sx = 1 \Leftrightarrow S_{x1} = 1;\ \ S_{x2} = 0;\ \ S'_{x1} = 0;\ \ S'_{x2} = 0; \\ &x = a, b \text{ or } c \end{aligned} \tag{3.85}$$

The components of active and zero vectors of three-level three-phase NPC inverter can be determined as shown in Table 3.10.

In three-level three-phase NPC inverter, there is possible to generate $3^3 = 27$ switching vectors including three zero vectors: (1,1,1), (0,0,0), and (−1,−1,−1). All the active vectors are shown in Figure 3.75.

The active vectors can be divided into three groups: the large active vectors (1,−1,−1), (1,1,−1), (−1,1,−1), (−1,1,1), (−1,−1,1), and (1,−1,1), the medium vectors: (1,0,−1), (0,1,

Table 3.10 The components of active and zero vectors in orthogonal space $\alpha\beta$ for unbalanced DC-link voltages

(S_a, S_b, S_c)	u_α	u_β
(1,1,1)	0	0
(1,1,–1)	$\sqrt{1/6}\cdot(v_{DCu}+v_{DCl})$	$\sqrt{1/2}\cdot(v_{DCu}+v_{DCl})$
(1,1,0)	$\sqrt{1/6}\cdot v_{DCu}$	$\sqrt{1/2}\cdot v_{DCu}$
(1,–1,1)	$\sqrt{1/6}\cdot(v_{DCu}+v_{DCl})$	$-\sqrt{1/2}\cdot(v_{DCu}+v_{DCl})$
(1,–1,–1)	$\sqrt{2/3}\cdot(v_{DCu}+v_{DCl})$	0
(1,–1,0)	$\sqrt{2/3}\cdot(v_{DCu}+0.5\cdot v_{DCl})$	$-\sqrt{1/2}\cdot v_{DCl}$
(1,0,1)	$\sqrt{1/6}\cdot v_{DCu}$	$-\sqrt{1/2}\cdot v_{DCu}$
(1,0,–1)	$\sqrt{2/3}\cdot(v_{DCu}+0.5\cdot v_{DCl})$	$\sqrt{1/2}\cdot v_{DCl}$
(1,0,0)	$\sqrt{2/3}\cdot v_{DCu}$	0
(–1,1,1)	$-\sqrt{2/3}\cdot(v_{DCu}+v_{DCl})$	0
(–1,1,–1)	$-\sqrt{1/6}\cdot(v_{DCu}+v_{DCl})$	$\sqrt{1/2}\cdot(v_{DCu}+v_{DCl})$
(–1,1,0)	$-\sqrt{2/3}\cdot(0.5\cdot v_{DCu}+v_{DCl})$	$\sqrt{1/2}\cdot v_{DCu}$
(–1,–1,1)	$-\sqrt{1/6}\cdot(v_{DCu}+v_{DCl})$	$-\sqrt{1/2}\cdot(v_{DCu}+v_{DCl})$
(–1,–1,–1)	0	0
(–1,–1,0)	$-\sqrt{1/6}\cdot v_{DCl}$	$-\sqrt{1/2}\cdot v_{DCl}$
(–1,0,1)	$-\sqrt{2/3}\cdot(0.5\cdot v_{DCu}+v_{DCl})$	$-\sqrt{1/2}\cdot v_{DCu}$
(–1,0,–1)	$-\sqrt{1/6}\cdot v_{DCl}$	$\sqrt{1/2}\cdot v_{DCl}$
(–1,0,0)	$-\sqrt{2/3}\cdot v_{DCl}$	0
(0,1,1)	$-\sqrt{2/3}\cdot v_{DCu}$	0
(0,1,–1)	$\sqrt{1/6}\cdot(-v_{DCu}+v_{DCl})$	$\sqrt{1/2}\cdot(v_{DCu}+v_{DCl})$
(0,1,0)	$-\sqrt{1/6}\cdot v_{DCu}$	$\sqrt{1/2}\cdot v_{DCu}$
(0,–1,1)	$-\sqrt{1/6}\cdot(v_{DCu}-v_{DCl})$	$\sqrt{1/2}\cdot(-v_{DCu}-v_{DCl})$
(0,–1,–1)	$\sqrt{2/3}\cdot v_{DCl}$	0
(0,–1,0)	$\sqrt{1/6}\cdot v_{DCl}$	$-\sqrt{1/2}\cdot v_{DCl}$
(0,0,1)	$-\sqrt{1/6}\cdot v_{DCu}$	$-\sqrt{1/2}\cdot v_{DCu}$
(0,0,–1)	$\sqrt{1/6}\cdot v_{DCl}$	$\sqrt{1/2}\cdot v_{DCl}$
(0,0,0)	0	0

–1), (–1,1,0), (–1,0,1), (0,–1,1), and (1,–1,0), and short vectors (1,0,0), (0, –1,–1), (1,1,0), (0,0, –1), (0,1,0), (–1,0,1), (0,1,1), (–1,0,0), (0,0,1), (–1,–1,0), (1,0,1), and (0,–1,0). The length and position of large active vectors do not depend on the DC-link voltage unbalance. Activating these vectors has no influence on the DC-link unbalance occurrence. The position and length of medium active vectors depend on the DC-link voltage unbalance, while only the lengths of the short vectors depend on the DC-link voltage unbalance. For the short vectors, the lengths of six of them: (1,0,0), (1,1,0), (0,1,0), (0,1,1), (0,0,1), and (1,0,1) depend on the upper capacitor voltage, while the lengths of the further six vectors (0,–1,–1), (0,0,–1), (–1,0,–1), (–1,0,0), and

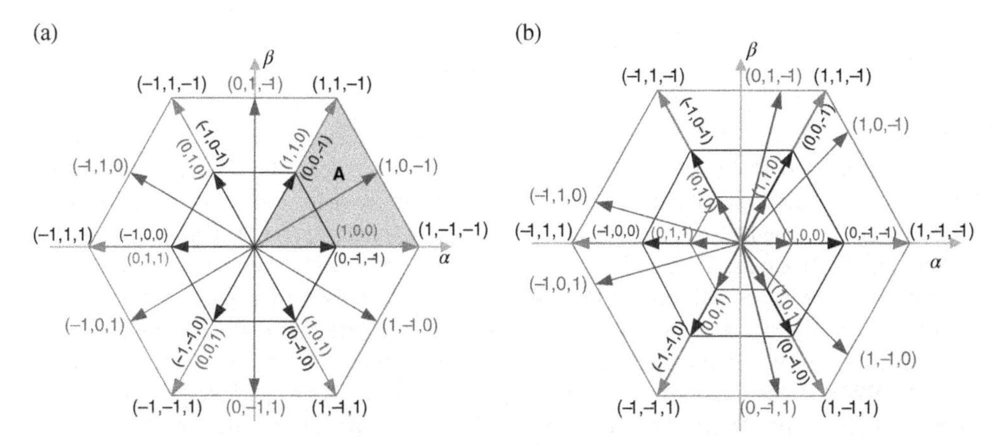

Figure 3.75 Space vectors of three-level NPC inverter in the case of balanced ($v_{DCu} = v_{DCl}$) (a) and unbalanced ($v_{DCu} < v_{DCl}$) (b) DC link

(−1,−1,0), (0,−1,0) depend on the lower capacitor voltage. The pairs of these vectors have the same positions, and they are often called redundant vectors.

The switching vectors of NPC inverter are easy to identify – the zero vectors connect all the inverter terminals to the same DC-link potential, the large active vectors connect the inverter terminals to both positive and negative DC-link potentials, and the medium vectors connect the inverter terminals to all DC-link potentials (one terminal to the positive potential, one terminal to the N point, and one terminal to the negative potential). The short vectors dependent on the voltage on upper capacitor connect the inverter terminals to positive and N potentials, while the short vectors dependent on the lower capacitor voltage connects the terminals to the negative and N potentials.

The short vector can be grouped in pairs of redundant vectors. Each pair of redundant vectors affects the DC-link voltages in the opposite way (for example, (0,−1,−1) and (1,0,0) (Figure 3.76). Their activation, as well as the activation of medium active vectors, connects the motor terminals to the DC-link neutral point. The changes of voltages on upper and lower DC-link capacitors can be estimated from:

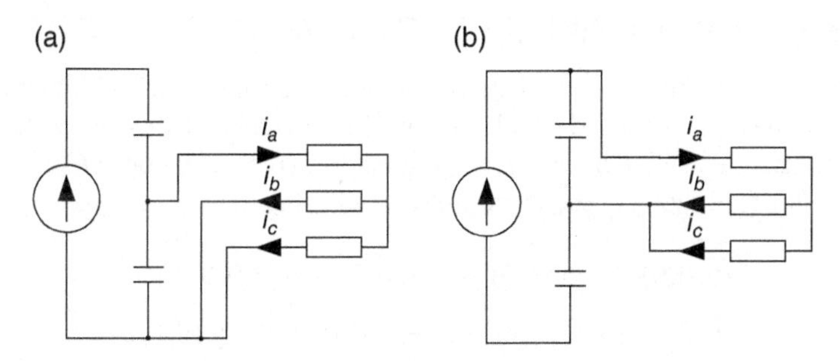

Figure 3.76 Influence of redundant vectors: (0,−1,−1) (a) and (1,0,0) (b) on DC-link voltages

$$v_{DCu}(t) = \frac{1}{2C}\int_{0}^{t} i(\tau)d\tau + u_c(0) \tag{3.86}$$

where C is the DC-link capacity ($C = C_u = C_l$), v_{DCu} is the voltage on the upper capacitor of DC link, $i(\tau)$ is the neutral point current, and τ is the duration of redundant or medium vectors.

In the case of large DC-link capacitance, the neutral point current has a negligible impact on the DC-link voltage unbalancing. The lengths of active vectors with an assumption of balanced DC-link voltages can be calculated. This approach allows to divide the space vector area into six equal sectors (Figure 3.75a). Each of them can be divided into equal subsectors. This assumption is correct as long as the DC-link voltage remains balanced. In many solutions of the SVPWM strategies, one or two switching sequences are strictly assigned to specific subsectors [50, 51]. The control strategies of DC-link voltage balance are based on the change of switching sequences depending on the DC-link voltage unbalance [52–55]. Some of the PWM solutions are based on using controllers to control neutral-point voltages [56–60]. The duration of redundant vectors in the switching sequence is changed in order to reduce the DC-link voltage unbalance [52–54, 61–64]. A similar solution was proposed in Ref. [65]. The voltage balance control strategy for one-phase NPC converter is based on the adjusting of effective time of the positive and the negative small vectors in response to the DC-link voltage unbalance and the direction of the load current. The method proposed is based on virtual vectors to control the DC-link voltage unbalance.

In the case of small DC-link capacitance, the DC-link voltage is usually unbalanced. In this case, the approach based on assumption of equal voltages in the DC-link may be a reason of incorrect output voltage vector generation. A corrective method of vector duration dependent on the DC-link voltage unbalance was proposed in Ref. [66]. The PWM strategy, proposed in Ref. [66], takes into consideration the influence of DC-link voltage unbalance on the length and position of active vectors in the space vector area. The DC-link voltage unbalance is reduced by appropriate choice of redundant vectors due to direction of neutral point current and voltage unbalance on the DC-link capacitors. The calculation of active vectors duration considers the actual difference between voltages on upper and lower DC-link capacitors. The method proposed in Ref. [67] is based on choosing one of several alternative vector sequences and changing the activation times of redundant vectors to balance DC-link voltages.

3.8.2 The Space Vector PWM for NPC Inverters

The SVPWM presented below is based on the solution proposed in Ref. [67]. In this algorithm, the vector durations are calculated simultaneously for four alternative sequences. If an output voltage vector takes a position in 'A' sector (Figure 3.75a), the durations of switching vectors are calculated for the following sequences [68]:

$$(0,0,0) \rightarrow (1,1,0) \rightarrow (1,0,0) \rightarrow (0,0,0) \tag{3.87}$$

$$(0,0,0) \rightarrow (0,-1,-1) \rightarrow (0,0,-1) \rightarrow (0,0,0) \tag{3.88}$$

$$(0,0,0) \rightarrow (1,-1,-1) \rightarrow (1,0,-1) \rightarrow (0,0,0) \tag{3.89}$$

$$(0,0,0) \rightarrow (1,-1,-1) \rightarrow (1,1,-1) \rightarrow (0,0,0) \tag{3.90}$$

The switching sequence (3.90) utilizes two long and two zero vectors which have no influence on the DC-link voltage balance. Sequences (3.87) and (3.88) utilize two short active and two zero vectors. The length of active vectors, utilized in (3.87), depends on the voltage on upper capacitor, while the length of active vectors in sequence (3.88) depends on the voltage on lower capacitor. The maximum length of the output voltage vector generated using these two sequences depends on the unbalance of the DC-link voltages and in unrelated to its location (Figure 3.75b). This permits choosing one of these sequences in the case of voltage unbalance occurrence, if the amplitude of inverter output voltage is higher than half of the DC-link voltage. These sequences may be used only in case if the inverter output voltage vector can be inscribed into hexagon, bounded by ends of vectors (1,1,0), (1,0,0),(1,0,1), (0,0,1), (0,1,1), and (0,1,0) (for seq.(3.87)) and vectors (0,0,–1), (0,–1,–1), (0,–1,0), (–1,–1,0), (–1,0,0), and (–1,0,–1) (for seq. (3.88)) (Figure 3.75b). The sequence (3.89) contains also two active vectors, but only one of them, the vector (1,0,–1), has an influence on the DC-link voltage balance.

The selection of one of the sequences has to ensure proper generation of the output voltage vector as well as the minimization of the DC-link voltage unbalance. The SV-PWM algorithm utilizes the information about the unbalance criterion, which is determined from (3.86):

$$x_{act} = i_x \cdot t_x = 2 \cdot C \cdot \Delta v \tag{3.91}$$

where t_x is the time to balance the voltages in DC link when the neutral-point current is equal to i_x. Δv is DC-link voltage unbalance:

$$\Delta v = v_{DCu} - v_{DCl} \tag{3.92}$$

Activating of one of the sequences (3.87)–(3.90) will change the DC-link voltage unbalance:

$$x = x_{pred} + x_{act}, \tag{3.93}$$

$$x_{pred} = \sum_{k=1}^{2} t_k \cdot i_k, \tag{3.94}$$

and i_k is the neutral-point current specified for k-th active vector in switching sequences (3.87)–(3.90), and t_k is the duration of k-th active vector.

The DC-link unbalance (3.93) is calculated for all active vectors and all predefined switching sequences (3.87)–(3.90). The sequences (3.87) and (3.88) contain only zero and short vectors; these sequences can be activated interchangeably and they are not modified in any way. The sequences (3.89) and (3.90) contain at least one long active vector that can be replaced by a short vector (the long and short vectors have the same position regardless of the DC-link voltage unbalance (Figure 3.75b). For example, the sequence (3.89) can be replaced by the following:

$$(0,0,0) \rightarrow (0,-1,-1) \rightarrow (1,-1,-1) \rightarrow (1,0,-1) \rightarrow (0,0,0) \tag{3.95}$$

or

$$(0,0,0) \rightarrow (1,-1,-1) \rightarrow (1,0,-1) \rightarrow (1,0,0) \rightarrow (0,0,0) \tag{3.96}$$

The active vectors (0,–1,–1) and (1,0,0) in (3.95) and (3.96) have the opposite effect on the DC-link voltage unbalance. The choice of the sequence (3.95) or (3.96) depends on the previously calculated DC-link unbalance criterion:

$$\begin{aligned} &x > 0, i_{(1,0,0)} < 0 \Rightarrow (1,0,0) \\ &x > 0, i_{(0,-1,-1)} < 0 \Rightarrow (0,-1,-1) \\ &x < 0, i_{(1,0,0)} > 0 \Rightarrow (1,0,0) \\ &x < 0, i_{(0,-1,-1)} > 0 \Rightarrow (0,-1,-1) \end{aligned} \tag{3.97}$$

where $i_{(0,-1,-1)}$ and $i_{(1,0,0)}$ are neutral-point currents during the activation of (0,–1,–1) and (1,0,0) redundant vector, respectively.The duration of redundant vector can be calculated as:

$$t_{(0,-1,-1),(1,0,0)} = abs\left(\frac{x}{i_{(0,-1,-1),(1,0,0)}}\right) \tag{3.98}$$

where $t_{(0,-1,-1)}$ and $t_{(1,0,0)}$ are the durations of redundant vectors (0,–1,–1) or (1,0,0), respectively. Activation of the redundant vector with the duration (3.94) will reduce the unbalance of the DC-link voltages to zero.

Because the medium active vector replaces the large vector, its duration have to be modified according to:

$$t'_{(1,-1,-1)} = t_{(1,-1,-1)} - t_{(0,-1,-1),(1,0,0)} \cdot \left(\frac{v_{\alpha(0,-1,-1),(1,0,0)}}{v_{\alpha(1,-1,-1)}}\right) \tag{3.99}$$

If

$$t'_{(1,-1,-1)} < 0 \tag{3.100}$$

the activation time of redundant vector should be reduced to:

$$t_{(0,-1,-1),(1,0,0)} = t_{(1,-1,-1)} \frac{v_{\alpha(1,-1,-1)}}{v_{\alpha(0,-1,-1),(1,0,0)}} \tag{3.101}$$

and the duration of vector (1,–1,–1) will be set to zero. In this case, the unbalance of the DC-link voltages will be reduced during a few pulse periods.

The redundant vector can also be introduced to (3.90) sequence. The algorithm for duration calculation is performed in the same manner as that proposed for the switching sequence (3.89).

The choice of final switching sequence from (including introduced redundant vectors) is based on minimization of the unbalance criterion (3.93) calculated for all alternative switching sequences. Only the sequences that enable correct generation of the output voltage are analyzed; for the highest values of reference output vector length, the sum of active vector durations calculated for some of the alternative vector sequences will be greater than a pulse period; in this case, this sequence (or these sequences) is not taken into account. The calculation of unbalance criterion (3.93) takes into consideration all the redundant vectors introduced to the switching sequences and their calculated durations. The sequence of vectors for which the lowest DC-link voltage unbalance is predicted will be utilized to generate the output voltage.

3.8.3 MATLAB/Simulink of SVPWM

The Simulink model of NPC inverter with SVPWM algorithm is shown in Figures 3.77 and 3.78, respectively. The inverter is loaded with RL system. The waveforms of DC-link voltages, phase currents, phase-to-phase voltage, and phase voltage are shown in Figure 3.79.

The SVPWM algorithm was prepared as an S-Function Block (Figure 3.80). It determines the switching pattern (the vector sequence and durations). The determined switching sequence is realized in the block 'Gate Signal Generator' based on the timer signal. The active and zero vectors are visualized on the scope (Figure 3.81). The vector numbering is according to the Table 3.10.

The following input signals are provided to the SVPWM algorithm:

u – the reverence voltage amplitude,
freq – the reference voltage frequency.
Udc – the measured DC-link voltages;
Ref_DC_imbalance – the reference DC-link voltage unbalance (0 – for balanced system).

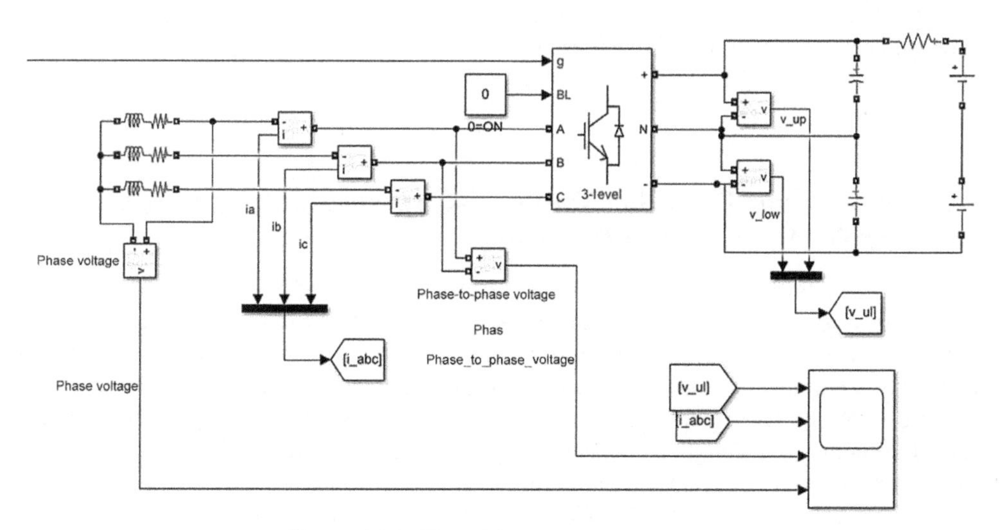

Figure 3.77 The NPC inverter and RL load

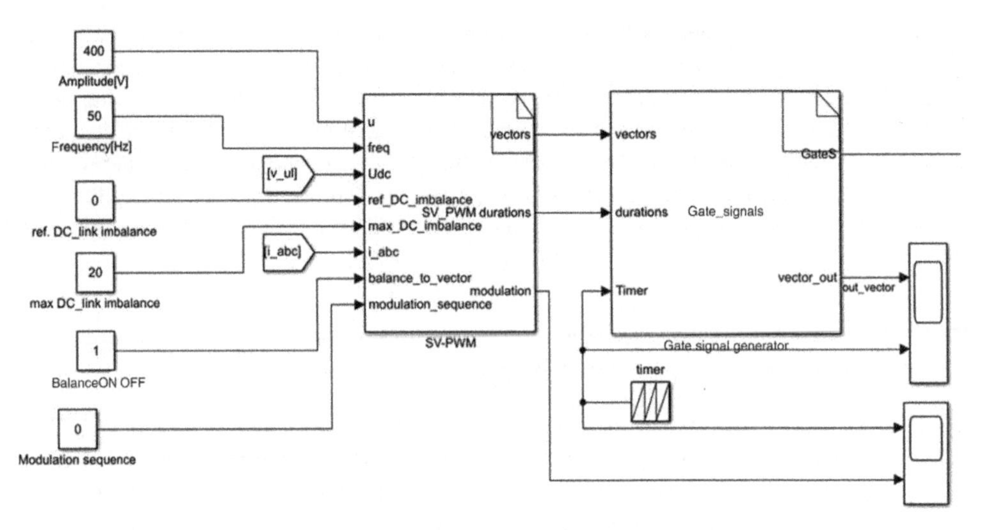

Figure 3.78 The blocks with SV-PWM algorithm and gate signal generator

Max DC_link unbalance – maximum value of DC-link voltage unbalance (if the DC-link voltage unbalance exceeds this value, the output voltage is generated based on the sequence (3.90) with introduced redundant vectors).

balance_to _vector – if '1' the DC-link voltage unbalance is taken into account, if '0' the DC-link voltage unbalance is ignored and the DC-link voltages are not balanced.

Modulation_sequence – manual selection of one of the sequences: 1 utilizes the active vectors with the lengths dependent on the voltage of upper DC-link capacitor (3.87), 2 utilizes the active vectors with the lengths dependent on the voltage on lower DC-link capacitor (3.88), 3 utilizes the sequences with medium active vector (3.89), and 4 utilizes the sequence with the long active vectors (3.90). If '0' the switching sequence is chosen automatically, depending on the predicted DC-link voltage unbalance.

3.9 Space Vector PWM for Multilevel-Cascaded H-Bridge Converter with DC-Link Voltage Balancing

The main advantages of ML inverters are the improved quality of voltage waveforms and an increase in the output voltage for a given blocking voltage capacity of the semiconductors. One of the most interesting constructions has the CHB converters [69, 70]. They are composed of low-voltage H-bridges connected serially, with the semiconductor blocking voltages much lower than the nominal output voltage of the converter. The inverter output voltage is the sum of the output voltages of individual H-bridges. Increasing the value of inverter output voltage requires only increase in the amount of H-bridges connected in series.

The CHB converters require galvanically isolated power sources (Figure 3.82a) [71]. The DC-link circuits of the H-bridges can be supplied by rectifiers connected to a multi-pulse

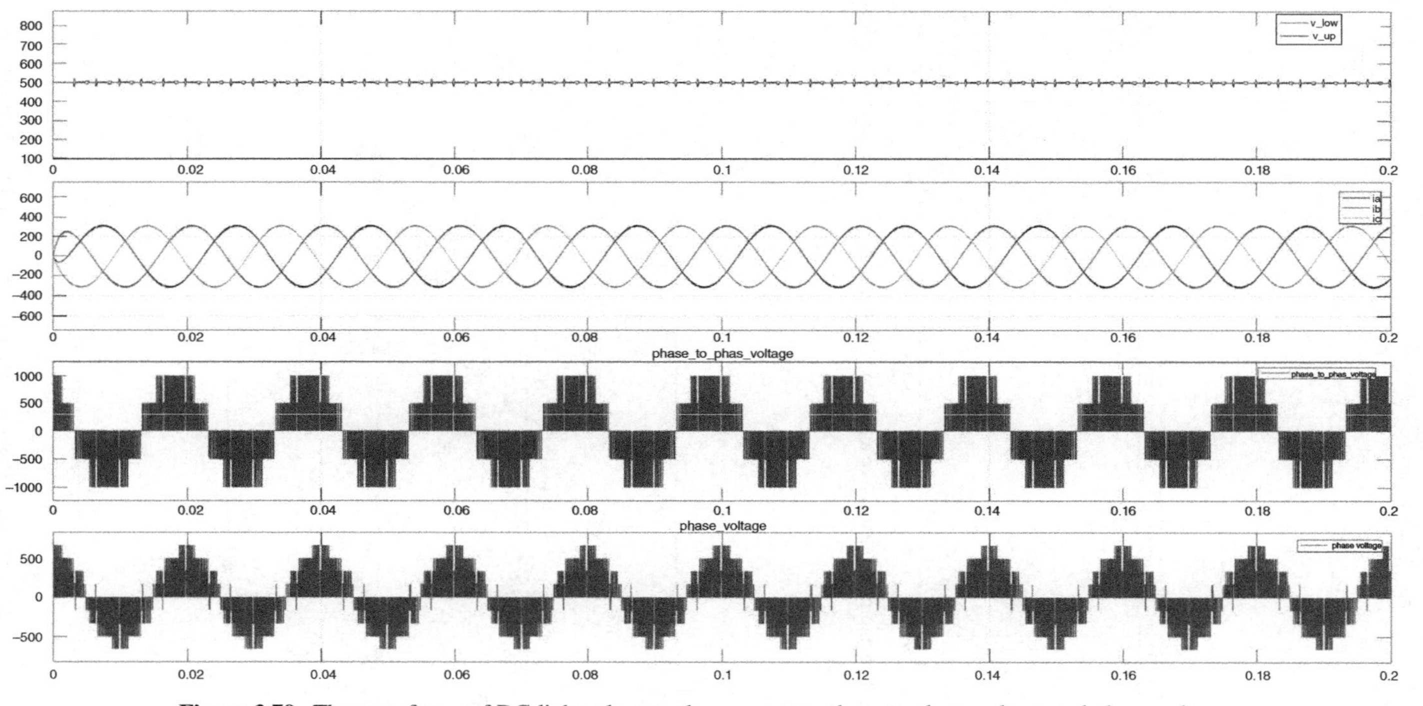

Figure 3.79 The waveforms of DC-link voltages, phase currents, phase-to-phase voltage and phase voltage

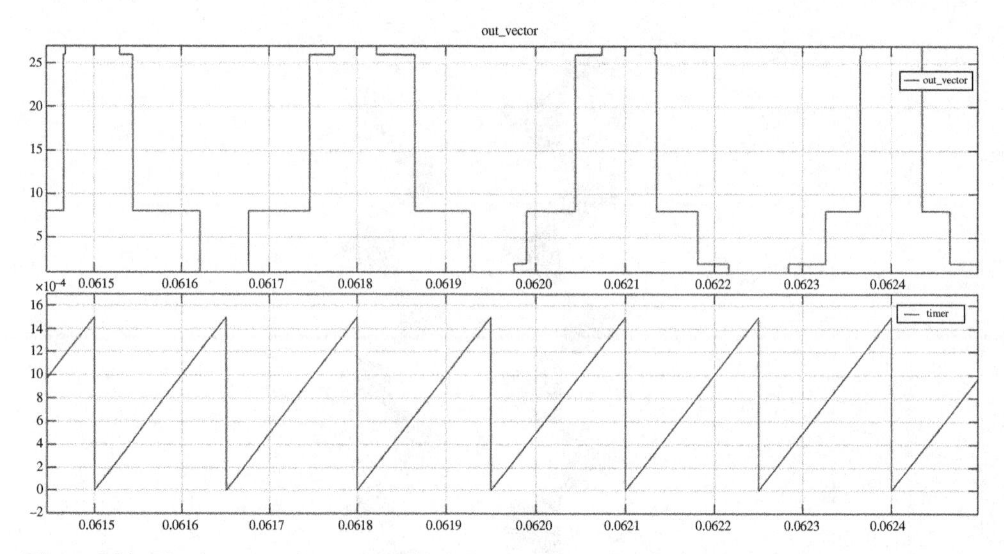

Figure 3.80 The timer waveform and generated active and zero vectors (vector numbering is according to the Table 3.10)

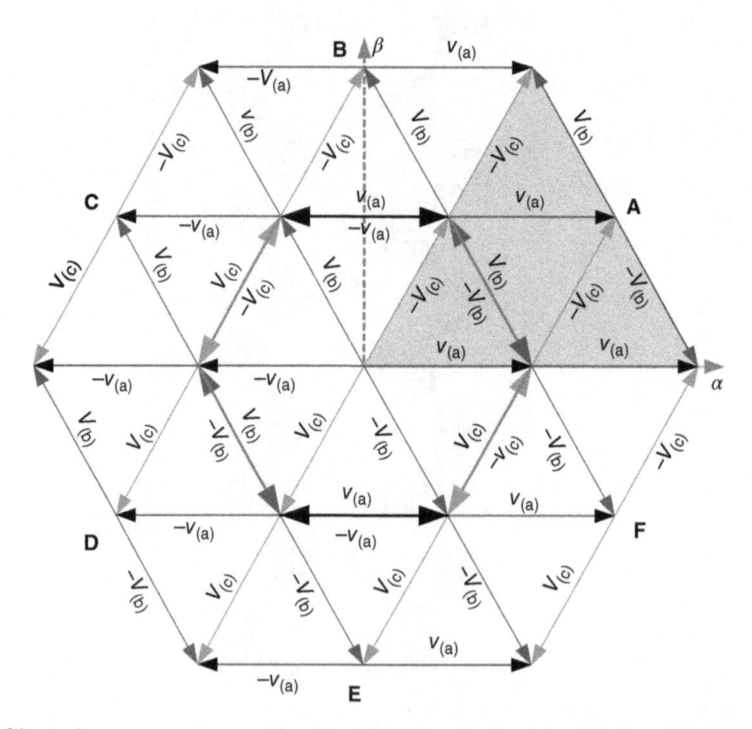

Figure 3.81 Active vectors generated by three H-bridges in the case of balanced DC-link voltages

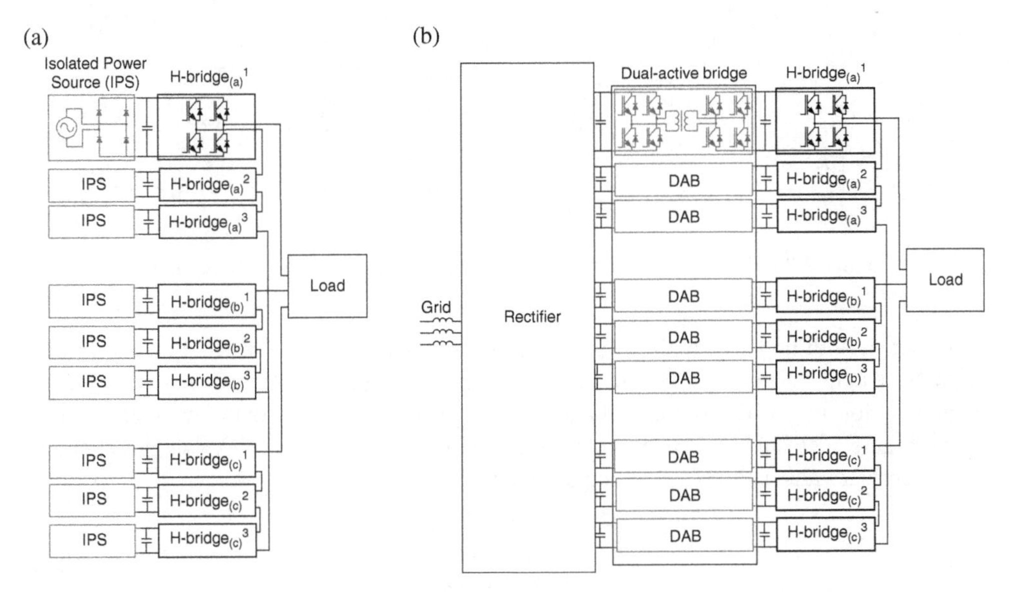

Figure 3.82 The seven-level CHB inverter with isolated power sources (a) and isolated dual-active bridges (b)

transformer [70] or to single-phase transformers [72]. The energy can be also delivered to the DC links using isolated dual-active bridges (DABs) [73] (Figure 3.82b).

The most popular modulation strategies applied to CHB converters are carrier-based SPWM [74, 75] and noncarrier-based selective-harmonic elimination pulse-width modulation (SHE-PWM) [76]. This is mainly due to the simplicity of implementation of these algorithms in multilevel inverters – the SVPWM strategies for more than three-level inverters are more complex due to the large amount of space vectors [77, 78]. The SVPWM strategies usually utilize three space vectors nearest to the sector, where the reference voltage vector is located [79, 80]. Some of the algorithms concentrate on generating the output voltage vector in the CHB inverters, wherein the DC-link voltages are not equalized [81].

The DC-link voltages of CHB inverter depend on delivered energy and the load. Because the energy is not uniformly delivered to all DC-link capacitors, the DC-link voltages are changing and must be controlled by additional balancing algorithms. The most popular balancing methods utilize the management of the H-bridges used to generate the inverter output voltages [76], and they also based on the modification of the individual reference voltages for individual H-bridges as well as on changing the modulation indexes of the H-bridges [77, 82, 83]. The DC-link voltages can be balanced using controllers [77, 84] or by rotation the carrier signals for every modulation cycle between H-bridges [85].

3.9.1 Control of a Multilevel CHB Converter

Each of the H-bridges of a CHB inverter can be configured to operate in two states: the zero (bypass) state (Figure 3.83a and b) and the active state (Figure 3.83c and d). While the zero state

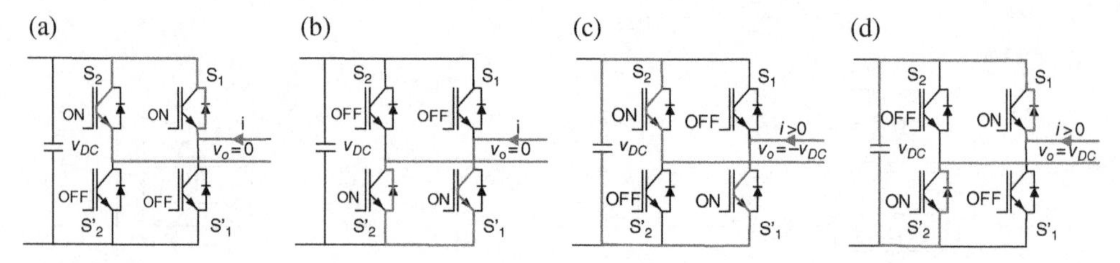

Figure 3.83 The H-bridge in zero state (a) and (b) and in active state (c) and (d)

is activated, both upper or both lower transistors are switched-on. The DC-link capacitor is bypassed and its voltage does not change. When the active state is switched on, two transistors S_1 and S_2' (for the positive output voltage) or S_2, S_1' (for the negative output voltage) are activated (Figure 3.83c and d). The current flow causes a change in the DC-link voltage depending on the current value and H-bridge output voltage:

$$\begin{aligned} &\text{if } i \cdot v_o > 0 \Rightarrow v_{dc} \uparrow\uparrow, \\ &\text{if } i \cdot v_o < 0 \Rightarrow v_{dc} \downarrow\downarrow, \end{aligned} \tag{3.102}$$

where i is the H-bridge output current and v_o is H-bridge output voltage (Figure 3.83b). The equation (3.102) does not take into account the supply current.

It is worth noting that the same current flows through all H-bridges in a single inverter phase regardless of the activated state.

3.9.2 *The Output Voltage of a Single H-Bridge*

The output voltage of a single H-bridge can be adjusted by the duty cycle. The durations of active and zero states can be determined as:

$$\begin{aligned} t_{active} &= \gamma \cdot T_s, \\ t_{zero} &= T_s - t_{active}, \end{aligned} \tag{3.103}$$

where t_{active} and t_{zero} are the durations of active and zero states, and T_s is a switching period.

The duty cycle γ is equal to:

$$\gamma = \frac{|v_o|}{v_{DC}}, \tag{3.104}$$

where v_{DC} is a DC-link voltage.The instantaneous value of H-bridge output voltage can be calculated as:

$$v_o = v_{DC} \cdot (S_1 - S_2), \tag{3.105}$$

where S_1 and S_2 are the gate signals for the upper transistors (Figure 3.83) with the values: 1 – upper transistor is switched on, 0 – lower transistor is activated.Hence, the output voltage of H-bridge is equal to:

$$\begin{aligned} v_o &= u_{DC} \Leftrightarrow S_1 = 1,\ \ S_2 = 0, \\ v_o &= -u_{DC} \Leftrightarrow S_1 = 0,\ \ S_2 = 1, \\ v_o &= 0 \Leftrightarrow S_1 = 0,\ \ S_2 = 0 \ \text{ or } \ S_1 = 1,\ \ S_2 = 1. \end{aligned} \tag{3.106}$$

The mean value of the output voltage is equal to:

$$v_o = \pm \gamma \cdot u_{DC}. \tag{3.107}$$

3.9.3 Three-Level CHB Inverter

Let's consider the case, when only three of the H-bridges (one in each phase) are in active states. The other H-bridges are bypassed (their duty cycles are equal to zero).

The components of output voltage vectors, generated in these three H-bridges, can be determined using (3.105) and the Clarke's transformation. The components of the voltage vectors generated in the subsequent H-bridges can be calculated as:

- the H-bridge in phase a:

$$\begin{aligned} v_{\alpha(a)} &= \sqrt{\frac{2}{3}} \cdot v_{DC(a)} \cdot \left(S_{1(a)} - S_{2(a)}\right), \\ v_{\beta(a)} &= 0, \end{aligned} \tag{3.108}$$

- the H-bridge in phase b:

$$\begin{aligned} v_{\alpha(b)} &= \sqrt{\frac{2}{3}} \cdot v_{DC(b)} \cdot \cos\left(\frac{2\pi}{3}\right) \cdot \left(S_{1(b)} - S_{2(b)}\right), \\ v_{\beta(b)} &= \sqrt{\frac{2}{3}} \cdot v_{DC(b)} \cdot \sin\left(\frac{2\pi}{3}\right) \cdot \left(S_{1(b)} - S_{2(b)}\right), \end{aligned} \tag{3.109}$$

- the H-bridge in phase c:

$$\begin{aligned} v_{\alpha(c)} &= \sqrt{\frac{2}{3}} \cdot v_{DC(c)} \cdot \cos\left(\frac{4\pi}{3}\right) \cdot \left(S_{1(c)} - S_{2(c)}\right), \\ v_{\beta(c)} &= \sqrt{\frac{2}{3}} \cdot v_{DC(c)} \cdot \sin\left(\frac{4\pi}{3}\right) \cdot \left(S_{1(c)} - S_{2(c)}\right), \end{aligned} \tag{3.110}$$

where $v_{\alpha(p)}$ and $v_{\beta(p)}$ are the components of output voltage vector generated in the 'p' – phase H-bridge ($p = a$, b or c), and $u_{DC(p)}$ are the DC-link voltages.If these H-bridges are in actives states ($S_{1(p)} - S_{2(p)} = \pm 1$), the active voltage vector components can be rewritten as:

$$\begin{aligned}
v_{\alpha(a)} &= \pm\sqrt{\frac{2}{3}} \cdot v_{DC(a)}, \\
v_{\beta(a)} &= 0, \\
v_{\alpha(b)} &= \pm\sqrt{\frac{2}{3}} \cdot v_{DC(b)} \cdot \cos\left(\frac{2\pi}{3}\right), \\
v_{\beta(b)} &= \pm\sqrt{\frac{2}{3}} \cdot v_{DC(b)} \cdot \sin\left(\frac{2\pi}{3}\right), \\
v_{\alpha(c)} &= \pm\sqrt{\frac{2}{3}} \cdot v_{DC(c)} \cdot \cos\left(\frac{4\pi}{3}\right), \\
v_{\beta(c)} &= \pm\sqrt{\frac{2}{3}} \cdot v_{DC(c)} \cdot \sin\left(\frac{4\pi}{3}\right).
\end{aligned} \tag{3.111}$$

The positions of obtained active vectors are shown in Figure 3.84.

The output voltages can be simultaneously generated in three H-bridges giving the active voltage vectors shown in Figure 3.81.

The CHB inverter with three H-bridges (one in each phase of the CHB inverter) can be treated as a three-level CHB inverter, where the output voltage is the sum of the voltages generated be these H-bridges:

$$\begin{aligned}
v_{\alpha(3L)} &= v_{\alpha(a)} + v_{\alpha(b)} + v_{\alpha(c)} \\
v_{\beta(3L)} &= v_{\beta(a)} + v_{\beta(b)} + v_{\beta(c)}
\end{aligned} \tag{3.112}$$

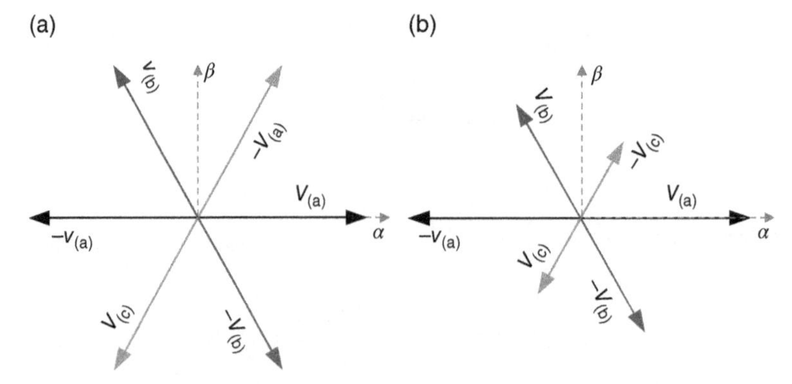

Figure 3.84 Output voltage vectors generated by three H-bridges in the case of balanced ($v_{DC(c)} = v_{DC(b)} = v_{DC(a)}$) (a) and unbalanced ($v_{DC(c)} < v_{DC(b)} < v_{DC(a)}$), (b) DC-link voltages

The additional H-bridges in the individual phases of multilevel CHB inverter also form the three-level inverters (each group of three H-bridges – one in each phase of the CHB inverter). The number of output voltage levels of the multilevel CHB inverter can be defined as:

$$n = 2k + 1 \tag{3.113}$$

where k is the number of H-bridges in a single CHB inverter phase.

Thus, the CHB inverter with 1 H-bridge in each phase gives 3-level output voltage, 2 H-bridges connected in series give the 5-level inverter, 3 H-bridges give the 7-level inverter, 6 H-bridges give the 13-level CHB inverter. The output voltage of the n-level CHB inverter is the sum of the voltages generated in three-level CHB inverters (with one H-bridge in each phase):

$$\begin{aligned} v_{o\alpha(ML)} &= \sum_{j=1}^{k} v_{o\alpha(3L)}{}^{j}, \\ v_{o\beta(ML)} &= \sum_{j=1}^{k} v_{o\beta(3L)}{}^{j}, \end{aligned} \tag{3.114}$$

where k is the number of three-level CHB inverters (number of the H-bridges in a single CHB inverter phase).

Generating a nonzero voltage state in a single H-bridge changes its DC-link voltage. Thus, the H-bridges used to construct the three-level CHB inverters can be selected arbitrarily only in the case if all the DC-link voltages are equal. Otherwise, the impact of the H-bridge exploitation on its DC-link voltage has to be analyzed. The H-bridges with lowest DC-link voltages (considered for each phase separately) should be used before all others to compose the first three-level CHB inverter if the following condition is fulfilled [39]:

$$i_{(p)} \cdot v_{ref(p)} > 0, \tag{3.115}$$

where $i_{(p)}$ is the 'p'- phase current ($p = a$, b or c) and $v_{ref(p)}$ is the reference phase voltage of a multilevel CHB inverter.

If the condition (3.115) is not satisfied, the three-level CHB inverter should be constructed using the H-bridges with highest DC-link voltages. As a result, the voltage unbalance will be reduced in the first place in the H-bridges with the highest or lowest DC-link voltages.

The next three-level inverters will be constructed in the same manner, utilizing previously unused H-bridges and taking into consideration their DC-link voltages and the condition (3.115).

3.9.4 The Space Vector Modulation for Three-Level CHB Inverter

The SVPWM algorithm proposed in Ref. [69] assumes that the same output voltage vector can be generated in a three-level CHB inverter using one of three alternative methods (Figure 3.85). For all of them, the choice of the sector (and the choice of active voltage vectors) depends on the

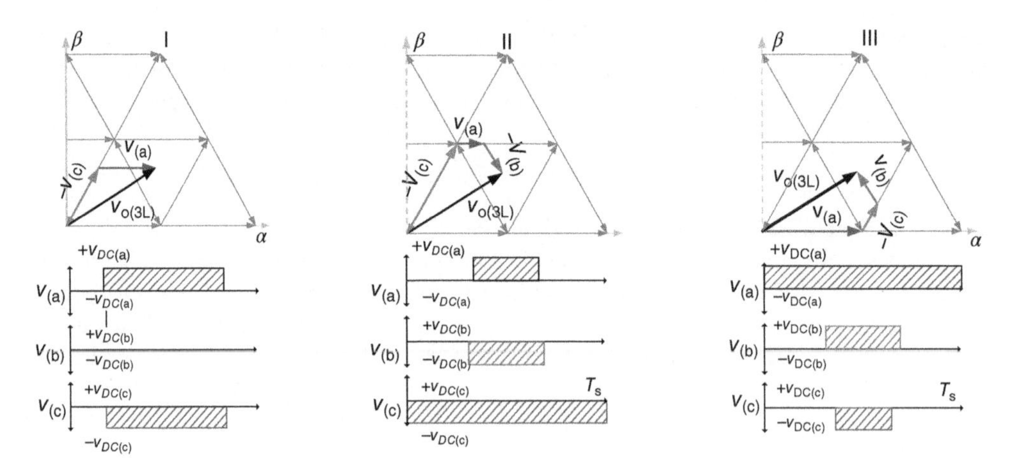

Figure 3.85 Generating of an output voltage vector in a three-level CHB converter using alternative strategies of the corresponding waveforms of H-bridge output voltages

angular position of the reference voltage of multilevel CHB inverter. Because the three-level inverter is considered to generate the reference voltage vector assigned to the multilevel CHB inverter, the end of the reference vector may be located outside the area shown in Figure 3.81. All the modulation strategies are calculated at the same time and the most suitable is selected to generate the output voltage vector.

It is worth noting that all the methods shown in Figure 3.85 ensure zero-switching number of the transistors in one of the inverter H-bridges. The method I ensures zero-switching number in phase *b*, while the methods II and III reduces to zero the switching number in the phases *c* and *a*, respectively. This is particularly important, because the CHB inverters are mainly applied to medium- and high-power applications, where the electrical energy has to be converted with maximum efficiency. It should be also noted that each of the sequences shown in Figure 3.85 affects the DC-link voltage distribution in different way. The predicted DC-link voltage distribution for each of the predefined modulation methods should be analyzed before generating the inverter output voltage.

If the position of the reference voltage vector is in the range of 0–$\frac{2\pi}{6}$ rad, the reference voltage vector is assigned to sector A (Figure 3.81). In the first of the strategies shown in Figure 3.85, the active states are in the H-bridges in phase ‘a’ (positive output voltage) and phase ‘c’ (negative output voltage). The duty cycles can be calculated as:

$$\begin{aligned}
\gamma_{(a)} &= \frac{v_{ref\alpha} \cdot v_{\beta(c)} - v_{ref\beta} \cdot v_{\alpha(c)}}{v_{\beta(c)} \cdot v_{\alpha(a)} - v_{\alpha(c)} \cdot v_{\beta(a)}}, \\
\gamma_{(b)} &= 0, \\
\gamma_{(c)} &= \frac{v_{ref\beta} \cdot v_{\alpha(a)} - v_{ref\alpha} \cdot v_{\beta(a)}}{v_{\beta(c)} \cdot v_{\alpha(a)} - v_{\alpha(c)} \cdot v_{\beta(a)}},
\end{aligned} \tag{3.116}$$

where $v_{ref\alpha}$ and $v_{ref\beta}$ are the reference voltage vector components of the multilevel CHB inverter.

The equation (3.115) allow to determine the duty cycles for three-level CHB inverter, while the reference voltage is defined for multilevel CHB inverter. As a result, the calculated duty cycles can be greater than 1. They should be limited to 1, which means that not all reference voltage will be generated:

$$\text{if } \gamma_{(a,b,c)} > 1 \Rightarrow \gamma_{(a,b,c)} = 1. \tag{3.117}$$

The second strategy is determined simultaneously with the first. It based on the duty cycles determined in (3.116). If the duty cycle for the H-bridge in phase ‘c’ is less than 1 (Figure 3.86a), it is assumed that it is equal to ‘1’ (Figure 3.86b). Thus, this duty cycle is increased by the value:

$$\Delta\gamma_{(c)} = 1 - \gamma_{(c)}. \tag{3.118}$$

This operation affects the length and position of the output voltage vector. The change in the β-component of the output voltage vector (Figure 3.86b) can be determined as:

$$\Delta v_\beta = \left(\Delta\gamma_{(c)} \cdot \left|v_{(c)}\right|\right) \cdot \sin\left(\frac{\pi}{3}\right) \tag{3.119}$$

and can be compensated by the H-bridge in phase ‘b’ (Figure 3.86c). The duty cycle for this H-bridge can be calculated as:

$$\gamma_{(b)} = \frac{\Delta v_\beta}{\left|v_{(b)}\right| \cdot \sin\left(\frac{\pi}{3}\right)} \tag{3.120}$$

The next step is to compensate the α-component of the output voltage. It is increased by the value (Figure 3.86c):

$$\Delta v_\alpha = \left(\Delta\gamma_{(c)} \cdot \left|v_{(c)}\right| + \gamma_{(b)} \cdot \left|v_{(b)}\right|\right) \cdot \cos\left(\frac{\pi}{3}\right) \tag{3.121}$$

an can be compensated by modifying the duty cycle for the H-bridge in phase ‘a.’ The new duty cycle can be calculated as:

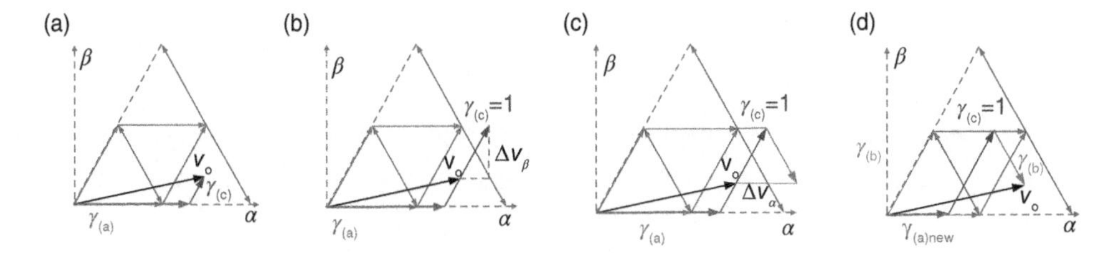

Figure 3.86 The steps of duty cycle selection

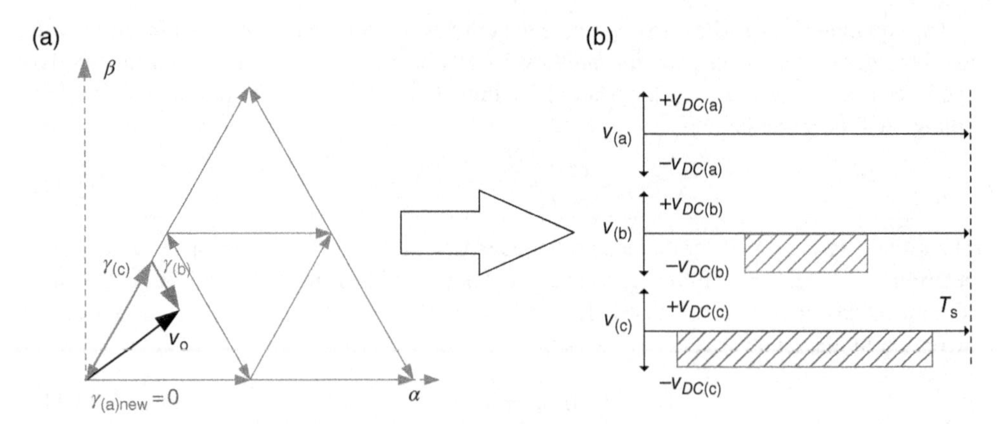

Figure 3.87 The duty cycles (a) and corresponding phase voltage waveforms (b) in the case of zero duty cycle for phase 'a' H-bridge

$$\gamma_{(a)new} = \gamma_{(a)} - \frac{\Delta v_\alpha}{|v_{(a)}|} \tag{3.122}$$

where $\gamma_{(a)new}$ is the updated duty cycle (Figure 3.86d).

If the new duty cycle $\gamma_{(a)new}$ is negative, the output voltage vector is generated using two active voltage vectors (H-bridges in phases 'b' and 'c'), while the H-bridge in phase 'a' is bypassed (the transistors are not switched) (Figure 3.87):

$$\begin{aligned}
\gamma_{(a)new} &= 0, \\
\gamma_{(b)} &= \frac{v_{ref\alpha} \cdot v_{\beta(c)} - v_{ref\beta} \cdot v_{\alpha(c)}}{v_{\beta(c)} \cdot v_{\alpha(b)} - v_{\alpha(c)} \cdot v_{\beta(b)}}, \\
\gamma_{(c)} &= \frac{v_{ref\beta} \cdot v_{\alpha(b)} - v_{ref\alpha} \cdot v_{\beta(b)}}{v_{\beta(c)} \cdot v_{\alpha(b)} - v_{\alpha(c)} \cdot v_{\beta(b)}}.
\end{aligned} \tag{3.123}$$

If the calculated duty cycles are greater than 1, they should be limited to 1.

The third method is realized simultaneously with the first and second methods. It also based on the duty cycles determined in (3.116). Unlike the second method, the third method increases the duty cycle for the H-bridge in phase a. The other operations are identical to the third method (Figure 3.88).

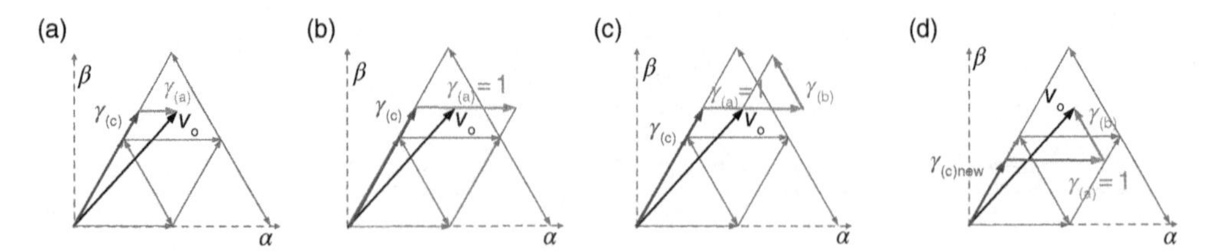

Figure 3.88 The steps of duty cycle selection

3.9.5 The Space Vector Modulation for Multilevel CHB Inverter

For all modulation methods (Figure 3.85), the components of the output voltage vector are calculated using limited duty cycles:

$$\begin{aligned} v_{o\alpha(3L)} &= \gamma_{(a)} \cdot v_{\alpha(a)} + \gamma_{(b)} \cdot v_{\alpha(b)} + \gamma_{(c)} \cdot v_{\alpha(c)}, \\ v_{o\beta(3L)} &= \gamma_{(a)} \cdot v_{\beta(a)} + \gamma_{(b)} \cdot v_{\beta(b)} + \gamma_{(c)} \cdot v_{\beta(c)}, \end{aligned} \tag{3.124}$$

where $v_{o\alpha(3L)}$ and $v_{o\beta(3L)}$ are the components of the output voltage vector of the three-level CHB inverter.

If all considered modulation strategies operate on limited duty cycles (the precalculated duty cycles were greater than 1), the choice of the modulation strategy used to generate the output voltage should provide the output voltage vector of the three-level CHB inverter as close as possible to the reference voltage vector of the multilevel CHB inverter. In this case, the output voltage vector should be generated using this modulation strategy, which ensures the minimum value of the function:

$$f\left(v_{o\alpha(3L)}, v_{o\beta(3L)}\right) = \left(v_{ref\alpha} - v_{o\alpha(3L)}\right)^2 + \left(v_{ref\beta} - v_{o\beta(3L)}\right)^2, \tag{3.125}$$

where $v_{o\alpha(3L)}$ and $v_{o\beta(3L)}$ are the components of the output voltage vector of a three-level inverter, calculated using (3.124) for all switching patterns, and $v_{ref\alpha}$ and $v_{ref\beta}$ are the reference voltage vector components of multilevel CHB inverter.

If the voltage vector obtained in the three-level CHB inverter is not equal to the reference voltage vector (of a multilevel CHB inverter), the next three H-bridges are activated (the next three-level CHB inverter). The new reference voltage for the next three-level CHB inverter can be calculated as:

$$\begin{aligned} v'_{ref\alpha} &= v_{ref\alpha} - v_{o\alpha(3L)}, \\ v'_{ref\beta} &= v_{ref\beta} - v_{o\beta(3L)}. \end{aligned} \tag{3.126}$$

where $v'_{ref\alpha}$ and $v'_{ref\beta}$ are the new reference voltage vector components used to determine the duty cycles for the next three-level CHB inverter.

The algorithm is repeated until the output voltage vector is determined for at least one of the methods (Figure 3.85) and the reference voltage vectors will be equal and at least one of the methods will give unlimited duty cycles. In this case, the modulation strategy is selected in a different way. The output voltage is generated using this method which operates on unlimited duty cycles and ensures minimization of the predicted DC-link voltage unbalance:

$$\begin{aligned} f\left(v_{DC(a)}, v_{DC(b)}, v_{DC(c)}\right) &= \left(v_{DC(a)}(k+1) - v_{DC(AV)}\right)^2 \\ &+ \left(v_{DC(b)}(k+1) - v_{DC(AV)}\right)^2 + \left(v_{DC(c)}(k+1) - v_{DC(AV)}\right)^2, \end{aligned} \tag{3.127}$$

where

$$v_{DC(AV)} = \frac{v_{DC(a)}(k+1) + v_{DC(b)}(k+1) + v_{DC(c)}(k+1)}{3}, \tag{3.128}$$

and

$$v_{DC(p)}(k+1) = v_{DC(p)}(k) + \frac{1}{C} \cdot \gamma_{(p)} \cdot T_s \cdot i_{(p)}, \tag{3.129}$$

where $i_{(p)}$ is a phase current, p is the phase of CHB inverter ($p = a$, b or c), C is the DC-link capacitance, and (k) and $(k+1)$ denote actual and predicted DC-link voltages, respectively.

3.9.6 MATLAB/Simulink Simulation of SVPWM

The Simulink model of two seven-level CHB rectifiers is shown in Figure 3.89. Both inverters cooperate with various supply networks and load each other through coupling in DC links.

Any phase of the CHB inverter contains three H-bridges connected in series (Figure 3.90). The simulation is based on simplified H-bridge model, where the transistor gate signals are not used and the currents and voltages are calculated in relation to the calculated duty cycles (Figure 3.91). This simplification significantly speeds up the simulations and has a negligible impact on the results. The output voltage of a single H-bridge is calculated as:

$$v_{o(p)} = v_{DC(p)} \cdot \gamma_{(p)}, \tag{3.130}$$

The DC-link current is calculated as:

$$i_{DC(p)} = i_{(p)} \cdot \gamma_{(p)} - i_{load}, \tag{3.131}$$

where $i_{(p)}$ is a phase current and i_{load} is a load current of the coupled H-bridge.The results of the simulation test during the step change of the DC-link voltage of the seven-level CHB inverter with DC-balancing method are shown in Figure 3.92.

3.10 Impedance Source or Z-source Inverter

The impedance source or Z-source inverters are special types of inverters that provide the voltage boost capability in conventional inverters. The conventional inverters work as a buck converter only because the output voltage is always lower than the DC input voltage. Moreover, the upper and lower power switches cannot conduct simultaneously, otherwise the DC source will short-circuit. Hence, a deadband is provided intentionally between the switching on and switching off of the complimentary power switches of the same leg. This deadband causes distortion in the output current. These shortcomings are overcome in the Z-source inverter. Two inductors of equal values and two capacitors of equal values, arranged in the form of a letter X with series inductors and diagonal capacitors, are inserted between the DC source and inverter of six power switches. This intermediate stage offers the advantage of boosting the DC source voltage, which is a highly attractive solution for renewable energy interface, such as photovoltaic system, fuel cells, etc., where the generated voltage is low and the load demand voltage is high. Detailed discussions on the Z-source inverters are given in Refs. [86–95]. The concept of

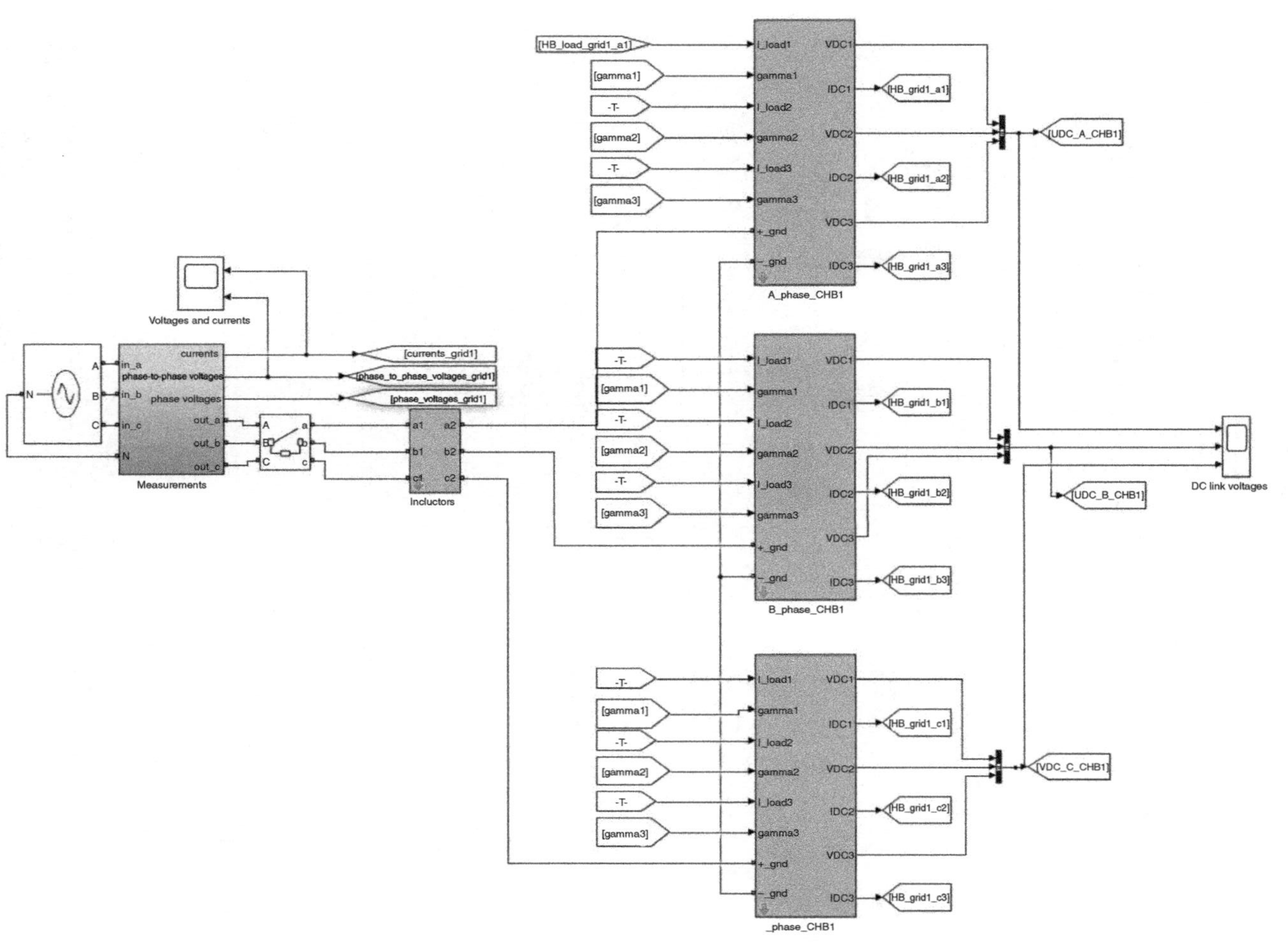

Figure 3.89 The structure of a seven-level CHB inverter

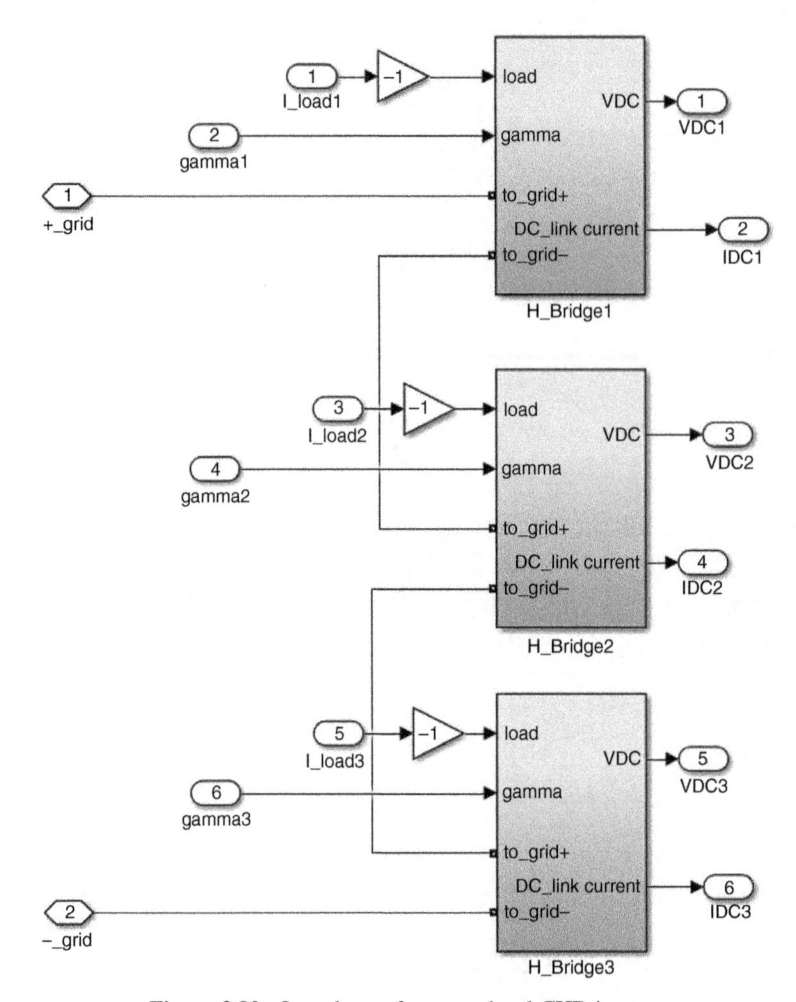

Figure 3.90 One phase of a seven-level CHB inverter

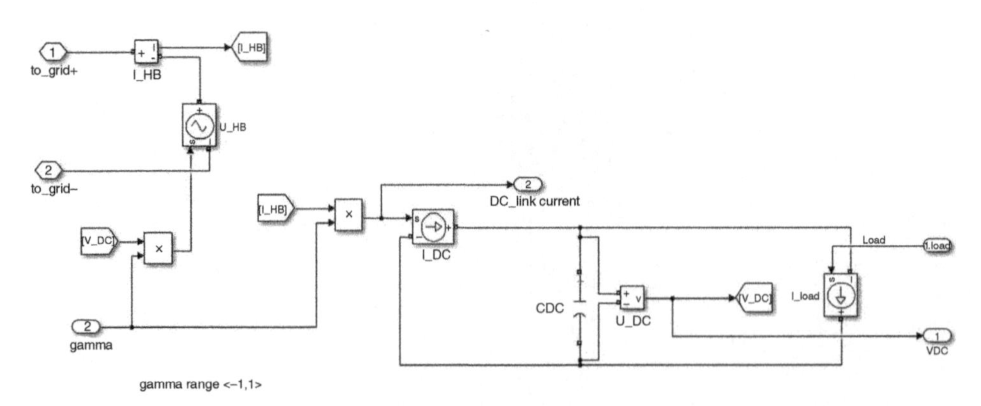

Figure 3.91 The H-bridge model

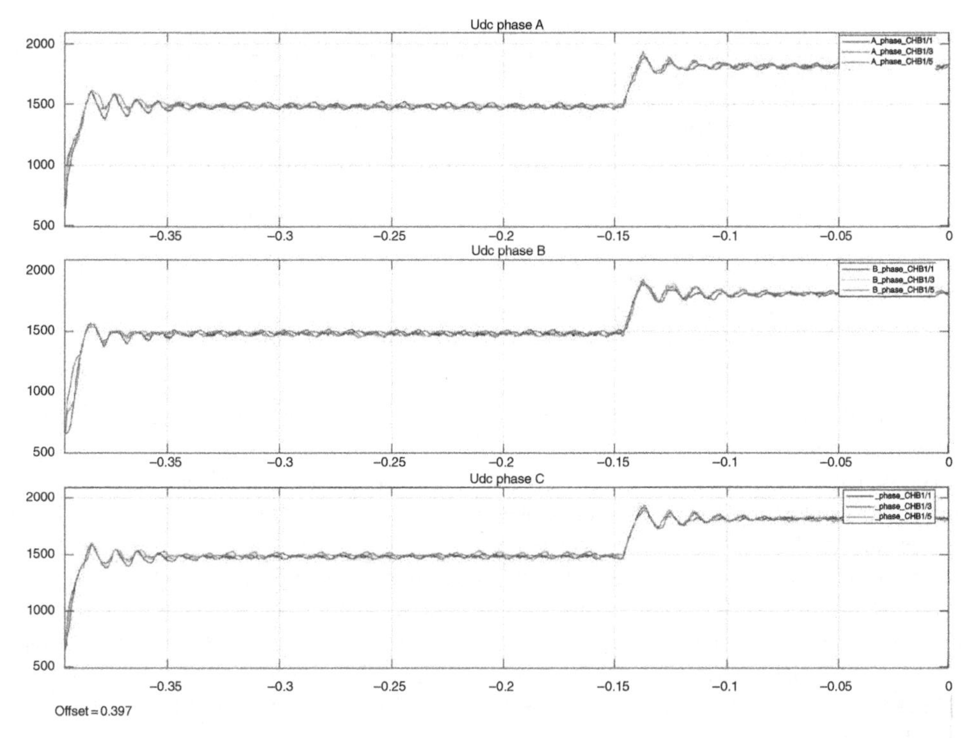

Figure 3.92 The change in DC-link voltages in the seven-level CHB inverter. $u_{DC(p)i}$ is the DC-link voltages in phase 'p' (p = a, b, c) and 'i' is the H-bridge number (i = 1, 2, 3). Simulation results

impedance source is equally applicable to AC–DC, DC–DC, AC–AC, and DC–AC power converter topology. This section introduces this new class of inverters (DC–AC converter) and will elaborate the basic principle of operation and control power circuit topology of a Z-source inverter, as shown in Figure 3.93. The DC source obtained from the renewable energy sources

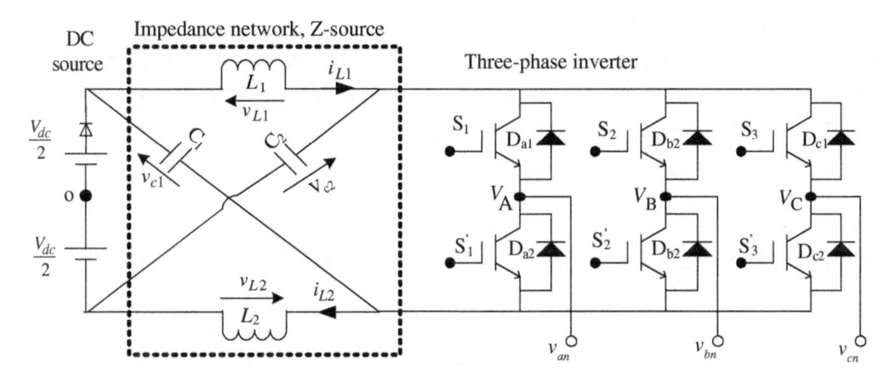

Figure 3.93 Power circuit topology of a Z-source inverter

Table 3.11 Shoot-through states in a three-phase Z-source inverter

Switching state name	Conducting switches
Shoot-through 1 (E1)	S_1, S'_1
Shoot-through 2 (E2)	S_2, S'_2
Shoot-through 3 (E3)	S_3, S'_3
Shoot-through 4 (E4)	S_1, S'_1, S_2, S'_2
Shoot-through 5 (E5)	S_1, S'_1, S_3, S'_3
Shoot-through 6 (E6)	S_2, S'_2, S_3, S'_3
Shoot-through 7 (E7)	S_1, S'_1, S_2, S'_2, S_3, S'_3

or a diode rectifier is connected to the conventional three-leg inverter through an intermediate impedance network.

The switching of a Z-source inverter produces seven additional states when compared with a conventional inverter where only eight states are possible (six active and two zero). This is because the operation of the power switch of the same leg is permitted in this type of inverters. When one or more leg is conducting with both power switches being 'ON' simultaneously, this leads to these additional switching states called 'shoot-through' states and operation of the inverter in these states causes voltage boost. During 'shoot-through' states, the output voltage at the load terminal is zero, while the voltage at the input of the inverter or output DC from the impedance network increases. The switching table, with additional shoot-through states, is presented in Table 3.11.

3.10.1 Circuit Analysis

The equivalent circuit of a Z-source inverter is shown in Figure 3.94. The inverter behaves as a constant current source when looked at from the impedance source to the inverter side, as shown in Figure 3.94a, where the inverter is operating in normal 'non-shoot-through' mode. The inverter is a short-circuit and the equivalent diode at the DC side is off when the inverter operates in the 'shoot-through' mode, as shown in Figure 3.94b.

For the circuit analysis, the following assumptions are made; the inductance values are assumed to be the same, i.e. $L_1 = L_2 = L$, as are the capacitor values, i.e. $C_1 = C_2 = C$. The inductor current is assumed to be large and constant. If the switching period is T_s, the shoot-through period is T_{sh} and non-shoot-through period is T_{nsh}, where

$$T_s = T_{sh} + T_{nsh} \tag{3.132}$$

Due to the symmetrical impedance network, the following relation holds true:

$$v_{c1} = v_{c2} = V_c \tag{3.133}$$

$$v_{L1} = v_{L2} = V_L \tag{3.134}$$

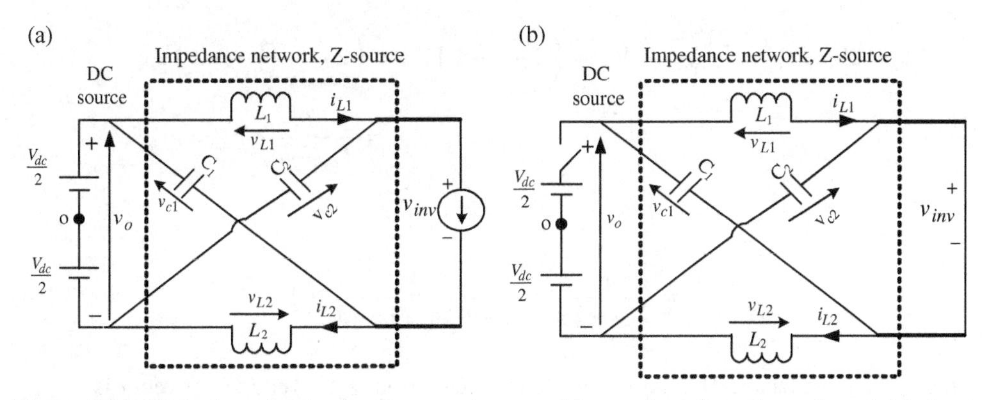

Figure 3.94 Equivalent circuit of a Z-source inverter: (a) non-shoot-through mode and (b) shoot-through mode

During the non-shoot-through state, the following relations hold true from the equivalent circuit (Figure 3.94a):

$$v_{L-nsh} = V_L = V_{dc} - V_c \tag{3.135}$$

$$v_{inv} = V_c - V_L = 2V_c - v_o \tag{3.136}$$

$$v_o = V_{dc} \tag{3.137}$$

During the shoot-through period, the following relation is obtained from Figure 3.94b:

$$v_o = 2V_c \tag{3.138}$$

$$v_{L-sh} = V_L = V_c \tag{3.139}$$

$$v_{inv} = 0 \tag{3.140}$$

The average voltage across the inductor over one switching period should be zero in steady state and hence, from volt-second principle, the following relation can be written as:

$$\begin{aligned} V_L T &= v_{L-sh} T_{sh} + v_{L-nsh} T_{snh} = 0 \\ V_L T &= v_c T_{sh} + (V_{dc} - V_c) T_{snh} = 0 \end{aligned} \tag{3.141}$$

From equation (3.141), the following relation is obtained:

$$\frac{V_c}{V_{dc}} = \frac{T_{nsh}}{T_{nsh} - T_{sh}} \tag{3.142}$$

The peak DC-link voltage across the inverter bridge is obtained using equations (3.136) and (3.131):

$$\hat{v}_{inv} = 2V_c - V_{dc} = \frac{2V_c - v_{dc}}{V_{dc}} V_{dc} = \left(2\frac{V_c}{V_{dc}} - 1\right) V_{dc} = \left(2\frac{T_{nsh}}{T_{nsh} - T_{sh}} - 1\right) V_{dc}$$
$$= \frac{T_s}{T_{nsh} - T_{sh}} V_{dc} = gV_{dc} \quad (3.143)$$

where

$$g = \frac{T_s}{T_{nsh} - T_{sh}} = \frac{1}{1 - 2\frac{T_{sh}}{T_s}} = \frac{1}{1 - 2\delta} \geq 1 \quad (3.144)$$

where g is boost factor. The output peak AC voltage from a conventional inverter is

$$\hat{v} = M\frac{V_{dc}}{2} \quad (3.145)$$

The peak output voltage in a Z-source inverter is

$$\hat{v} = M\hat{v}_{inv}/2 = Mg\frac{V_{dc}}{2} \quad (3.146)$$

where M is the modulation index. It is thus seen that the output AC peak voltage from a Z-source inverter is g (boost factor) times the conventional inverter output. The value g is always more than unity and hence gain in the output voltage is achieved. It is seen from equation (3.146) that the boost factor g is zero if the shoot-through period $T_{sh} = 0$. Hence, it can be said that the boost in the output voltage is obtained due to the shoot-through in the switching state.

3.10.2 Carrier-Based Simple Boost PWM Control of a Z-source Inverter

Conventional PWM schemes, such as SPWM and SVPWM, are equally applicable to a Z-source inverter. The only difference is seen when voltage boost is required and shoot-through state operation is needed. In the conventional PWM, two zero states and two active states are used. The shoot-through state is inserted within the conventional zero states, as shown in Figure 3.95. The triangular carrier signal is shown along with the sampled modulating signals. Two additional constant signals are used, namely V_P^*, V_N^*, to modulate the shoot-through duty ratio. Hence, by controlling the amplitude of the two additional signals, the output voltage from the Z-source inverter is controlled. It is seen that the zero vector time only altered, while the active vector time remains the same.

The shoot-through duty ratio (δ) decreases with the increase in the modulation index M. The maximum duty ratio is $1 - M$ and hence reaches zero for modulation index of unity. The voltage gain reduces with the increase in the modulation index. The voltage gain due to the boost becomes zero for $M = 1$. The gain is high for a low modulation index, hence to obtain a higher gain the modulation index must be kept low.

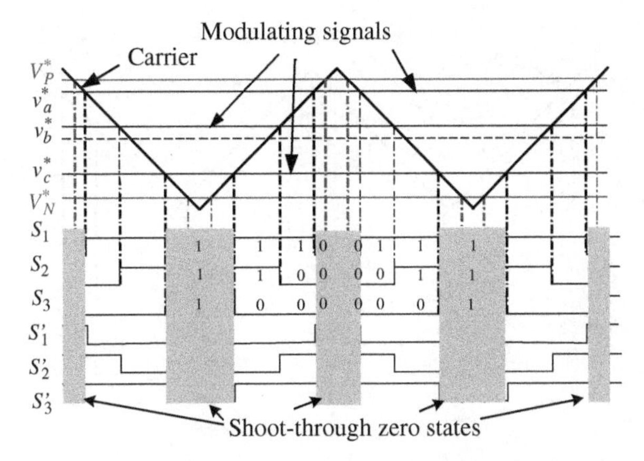

Figure 3.95 Principle of carrier-based simple boost PWM for a Z-source inverter

The overall gain G in the output voltage using the simple boost PWM method is

$$G = M \cdot g = \frac{\hat{v}}{V_{dc}/2} = \frac{M}{2M-1} \tag{3.147}$$

The maximum modulation for given g is

$$M = \frac{G}{2G-1} \tag{3.148}$$

The voltage stress across the power semiconductor switch using the carrier-based simple boost PWM method is

$$V_{stress} = gV_{dc} = (2G-1)V_{dc} \tag{3.149}$$

Higher gain at the output of inverter leads to high stress on the power semiconductor power switch.

3.10.3 Carrier-Based Maximum Boost PWM Control of a Z-source Inverter

The limitation of the simple boost control method is that the stress across the power switches is higher for higher voltage gain, since the voltage gain is given as $M \cdot g$ and the voltage stress across the switch is $g \cdot V_{dc}$. In order to minimize the stress of the power switch, the boost factor g should be minimized and do simultaneously maximize the modulation index M, keeping the output voltage gain at a desired value. However, in order to increase the voltage gain, the boost factor g should be enhanced. To meet these conflicting requirements, the zero states are all converted into the shoot-through period. The active states remain unchanged, as shown in Figure 3.96.

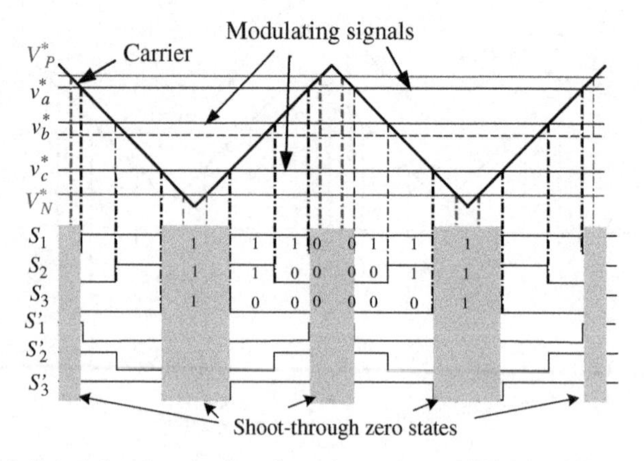

Figure 3.96 Principle of carrier-based maximum boost PWM for a Z-source inverter

The average duty ratio of the shoot-through is [87]

$$\delta = \frac{T_{sh}}{T_s} = \frac{2\pi - 3\sqrt{3}M}{2\pi} \tag{3.150}$$

The boost factor is given by:

$$g = \frac{1}{1-2\delta} = \frac{\pi}{3\sqrt{3}M - \pi} \tag{3.151}$$

The output voltage gain is given by:

$$G = Mg = \frac{\pi M}{3\sqrt{3}M - \pi} \tag{3.152}$$

The voltage stress across the power switches are

$$V_{stress} = gV_{dc} = \frac{3\sqrt{3}G - \pi}{\pi} V_{dc} \tag{3.153}$$

The voltage stress in a PWM scheme is considerably lower than the simple boost PWM method. Hence, a higher gain can be achieved with this maximum constant boost PWM method.

The variation of the voltage gain G versus the modulation index for simple boost PWM and maximum constant boost PWM is shown in Figure 3.97.

To increase the modulation index M to $1/\sqrt{3}$, the third-harmonic injection method can be adopted. The modulating signal consists of a sinusoidal signal plus the one-fourth or one-sixth third harmonic.

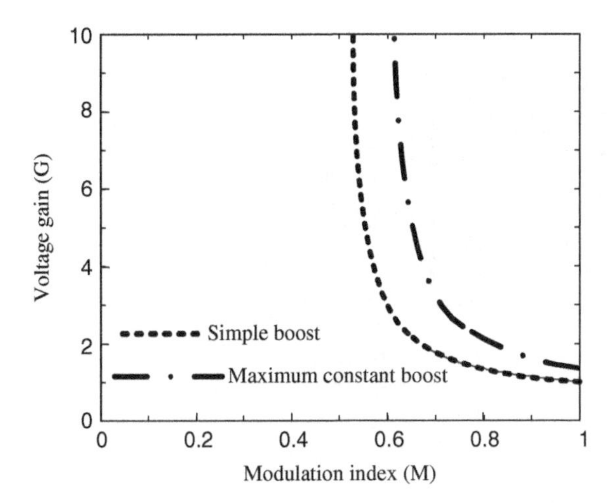

Figure 3.97 Voltage gain versus modulation index for Z-source inverter

3.10.4 MATLAB/Simulink Model of Z-source Inverter

The MATLAB/Simulink model of a Z-source inverter controlled using a simple SPWM scheme is shown in Figure 3.98. The switching frequency of the inverter is kept at 2 kHz, the fundamental frequency is chosen as 50 Hz, the impedance network parameters are $L = 500$ μH and $C = 400$ μF, and the output filter parameters are $L_v = 0.01$ H and $C_v = 110$ μF. The load is chosen as a variable with different requirements of active and reactive powers. The load is changed at the instant of $t = 0.5$ s and $t = 0.8$ s, the modulation index is kept at 0.75, the output phase voltage requirement is 249 Vrms, and the DC-link voltage is kept at 400 V. The resulting waveforms are shown in Figures 3.99 and 3.100.

3.11 Quasi Impedance Source or qZSI Inverter

The impedance source or Z-source inverter, described in Section 3.9, suffers from the drawback of discontinuous input (DC) current during boost mode, high voltages across the capacitors, and higher stress on power switches. These shortcomings are overcome by a different topology of inverter called qZSI [8, 96–98]. The qZSI topology is derived from the original Z-source inverter. The advantages of a qZSI are

- draws continuous current from DC source;
- the voltage across the capacitor C2 is reduced;
- lower component count and hence higher reliability and higher efficiency; and
- lower voltage stress on the power switches.

There are several topologies of qZSI reported in the literature [97]. This section elaborates on the voltage fed qZSI with continuous input current, as shown in Figure 3.101. The two operating modes of the inverter are the shoot-through and non-shoot-through modes, and the

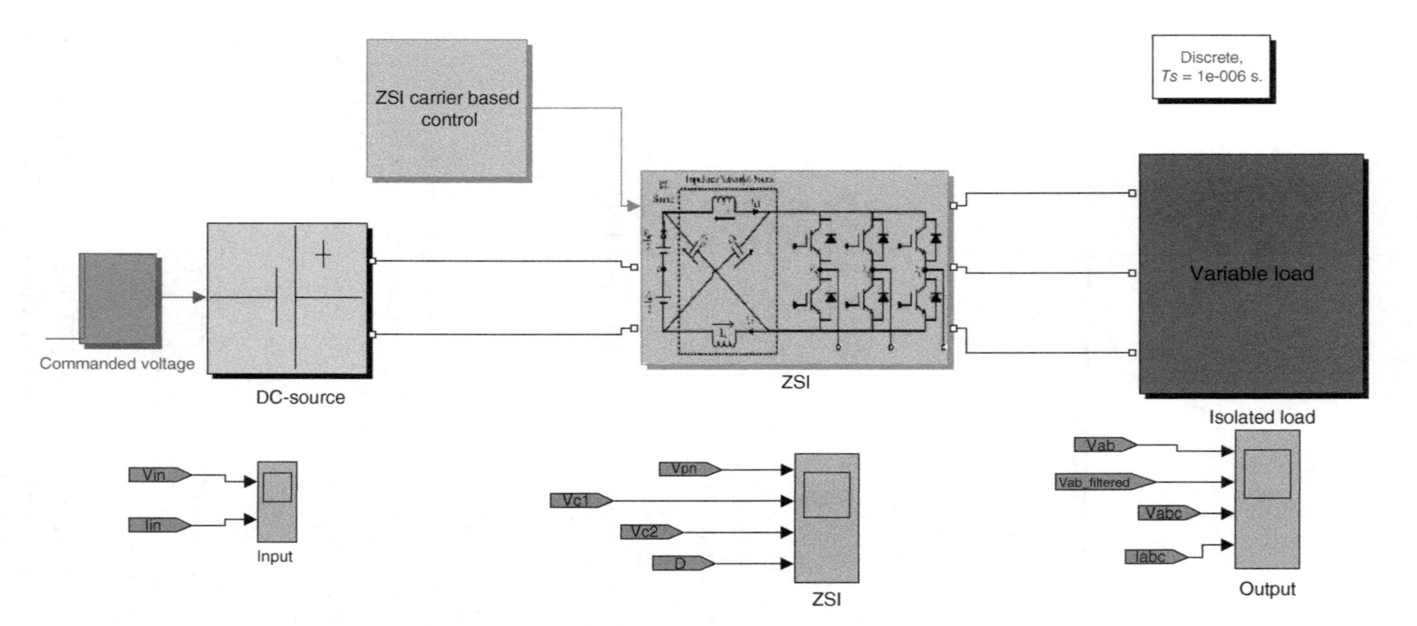

Figure 3.98 MATLAB/Simulink model of Z-source inverter

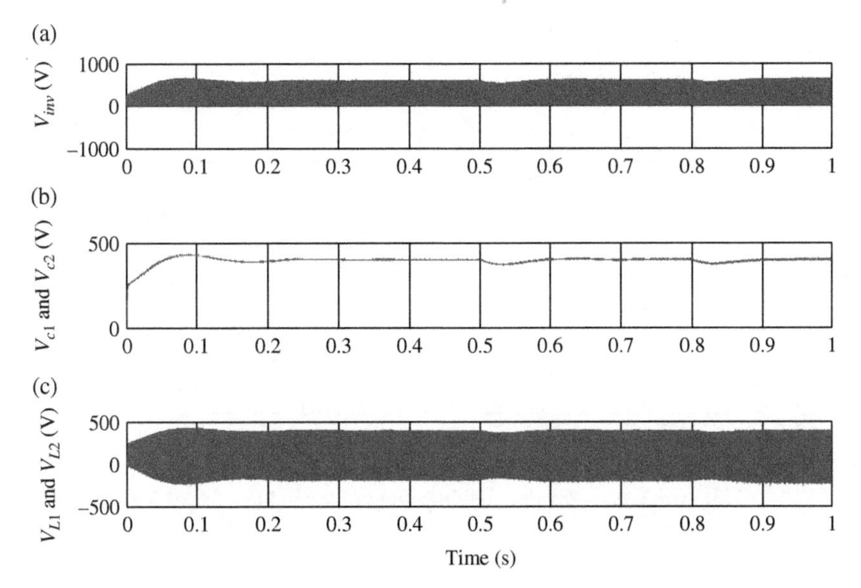

Figure 3.99 Voltages at the source side: (a) at the input of the bridge inverter; (b) across two capacitors; and (c) across two inductors

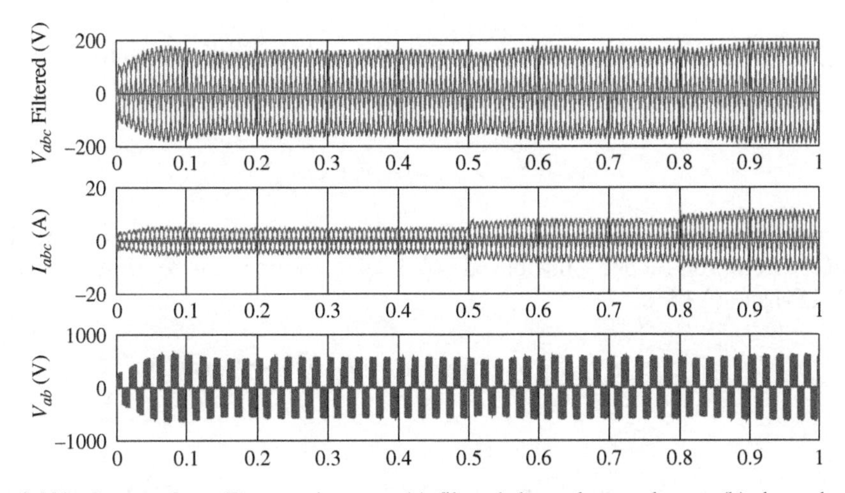

Figure 3.100 Outputs from Z-source inverter: (a) filtered three-phase voltages, (b) three-phase load current, and (c) line voltage

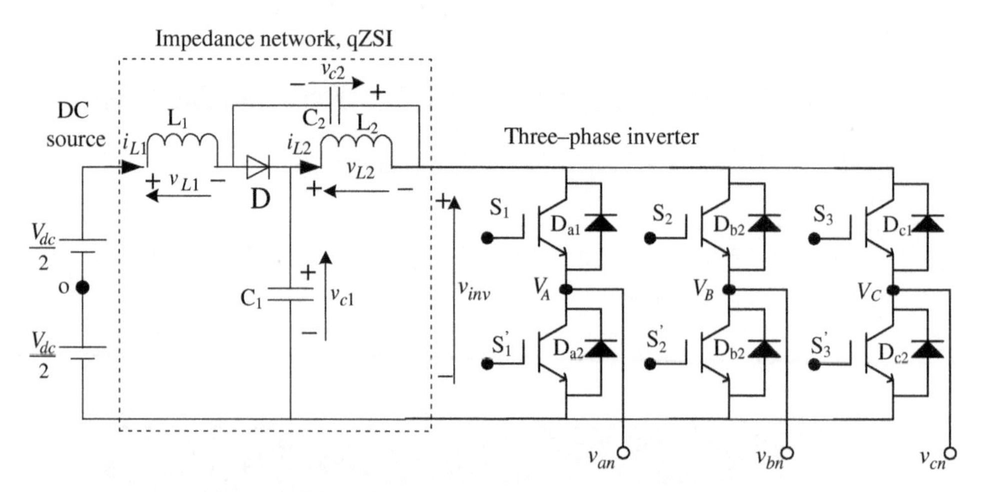

Figure 3.101 Power circuit topology of the voltage fed qZSI

equivalent circuits during the modes of operation are shown in Figure 3.101. From the circuit (Figure 3.102a), the voltage relations during the non-shoot-through period (T_{nsh}) are

$$v_{L1} = V_{dc} - v_{C1}, \quad v_{L2} = -v_{C2}, \tag{3.154}$$

$$v_{inv} = v_{C1} - v_{L2} = v_{C1} + v_{C2}, \quad v_{diode} = 0. \tag{3.155}$$

From the circuit (Figure 3.102b), the voltage relations during the shoot-through period (T_{sh}) are

$$v_{L1} = v_{C2} + V_{in}, \quad v_{L2} = v_{C1} \tag{3.156}$$

$$v_{inv} = 0, \quad v_{diode} = -(v_{C1} + v_{C2}) \tag{3.157}$$

The average voltage across inductor over one switching period should be zero in steady state and hence, from volt-second principle, the following relation can be written from equations (3.154) and (3.156):

$$V_{L1} = \overline{v}_{L1} T_s = T_{sh}(V_{C2} + V_{dc}) + T_{nsh}(V_{dc} - V_{C1}) = 0 \tag{3.158}$$

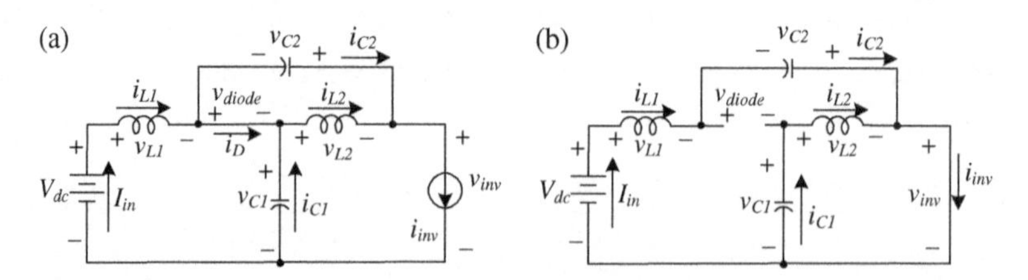

Figure 3.102 Equivalent circuit of the qZSI: (a) non-shoot-through state and (b) shoot-through state

$$V_{L2} = \bar{v}_{L2} T_s = T_{sh}(V_{C1}) + T_{nsh}(-V_{C2}) = 0 \tag{3.159}$$

Thus, the following relations are

$$V_{C1} = \frac{1-\delta}{1-2\delta} V_{dc}, \quad V_{C2} = \frac{\delta}{1-2\delta} V_{dc} \tag{3.160}$$

From equations (3.155), (3.157), and (3.160), the peak DC-link voltage across the inverter bridge is

$$v_{inv} = V_{C1} + V_{C2} = \frac{T_s}{T_{nsh} + T_{sh}} v_{dc} = \frac{1}{1-2\delta} V_{dc} = g V_{dc} \tag{3.161}$$

where g is the boost factor of the qZSI. The average currents through the two inductors L_1 and L_2 can be calculated by the system power rating P:

$$I_{L1} = I_{L2} = I_{in} = P/V_{dc} \tag{3.162}$$

According to Kirchhoff's current law and equation (3.162), the following current relations are obtained:

$$I_{C1} = I_{C2} = I_{inv} - I_{L1}, \quad I_D = 2I_{L1} - I_{inv} \tag{3.163}$$

Defining the following terms:

1. M is the modulation index; $\hat{v}$ is the ac peak phase voltage or output peak voltage of the inverter;
2. $p = (1-\delta)/(1-2\delta)$; $k = \delta/(1-2\delta)$; $g = 1/(1-2\delta)$.

A comparison between the Z-source and quasi Z-source inverter average parameters are given in Table 3.12.

The qZSI and Z-source inverter have the same features. The advantage of boosting the input DC voltage in a single-stage configuration is also available with the qZSI. There is no need to provide dead time between the switching on and off of the power switches of the same leg.

Table 3.12 Comparison of average quantities of ZSI and qZSI [8]

	$v_{L1} = v_{L2}$		v_{inv}		v_{diode}							
	T_{sh}	T_{nsh}	T_{sh}	T_{nsh}	T_{sh}	T_{nsh}	V_{C1}	V_{C2}	$\hat{v}$	$I_{in} = I_{L1} = I_{L2}$	$I_{C1} = I_{C2}$	I_D
ZSI	pV_{dc}	$-kV_{dc}$	0	gV_{dc}	gV_{dc}	0	pV_{dc}	pV_{dc}	$GV_{dc}/2$	P/V_{dc}	$I_{inv} = I_{L1}$	$2I_{L1} = I_{inv}$
qZSI	pV_{dc}	$-kV_{dc}$	0	gV_{dc}	gV_{dc}	0	pV_{dc}	kV_{dc}	$GV_{dc}/2$	P/V_{dc}	$I_{inv} = I_{L1}$	$2I_{L1} = I_{inv}$

3.11.1 MATLAB/Simulink Model of qZ-source Inverter

The MATLAB/Simulink model of the qZSI is shown in Figure 3.103. The inverter is controlled using a simple SPWM technique. The simulation parameters are the same as that of the ZSI. The resulting waveforms are shown in Figures 3.104 and 3.105.

3.12 Dead Time Effect in a Multiphase Inverter

This section further elaborates on the effect of introducing dead time (time lag between the switching 'ON' and switching 'OFF' of the two power semiconductor switches of the same leg of an inverter). The two power switches of the same leg should not be turned on simultaneously in order to avoid the source-side short circuit and uncontrolled flow of current through inverter switches and subsequent failure of the inverter; the time delay thus to be introduced between the turning on and turning off of the two switches of the same leg. The power semiconductor switch turn-on and turn-off time depend on several factors, such as the power rating of the switch, the operating temperature, and gate drive current [99]. The required dead time is a function of the power rating of the device, considering one leg of a multiphase inverter with a gate signal supplied from a gate drive circuit.

Figure 3.106 shows one of the legs of the basic multiphase PWM inverter with RL load. V_c(t) and V_i(t) are carrier and modulating signals, respectively, which are compared in a comparator.

The waveforms associated with Figure 3.106 are shown in Figure 3.107.

Output of the comparator is $V_1(t)$. This output signal and its complement signal ($V_2(t)$) are fed to the gates of upper (T_1) and lower (T_2) IGBTs of one leg through a time-delay circuit. Time delay between upper and lower IGBTs switching is required to avoid shoot-through conditions and subsequently to avoid the source short circuit. The waveforms of the PWM inverter in

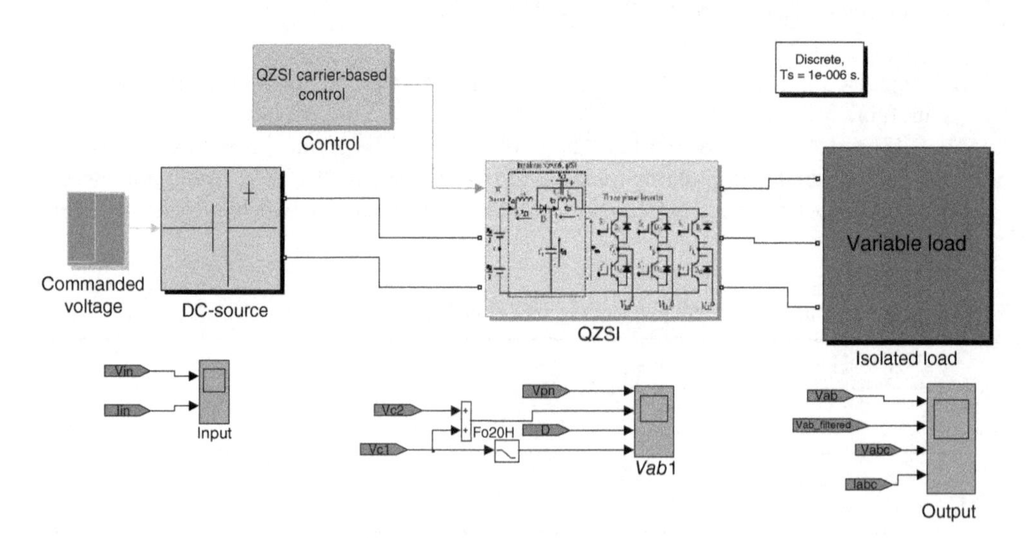

Figure 3.103 MATLAB/Simulink model of Z-source inverter

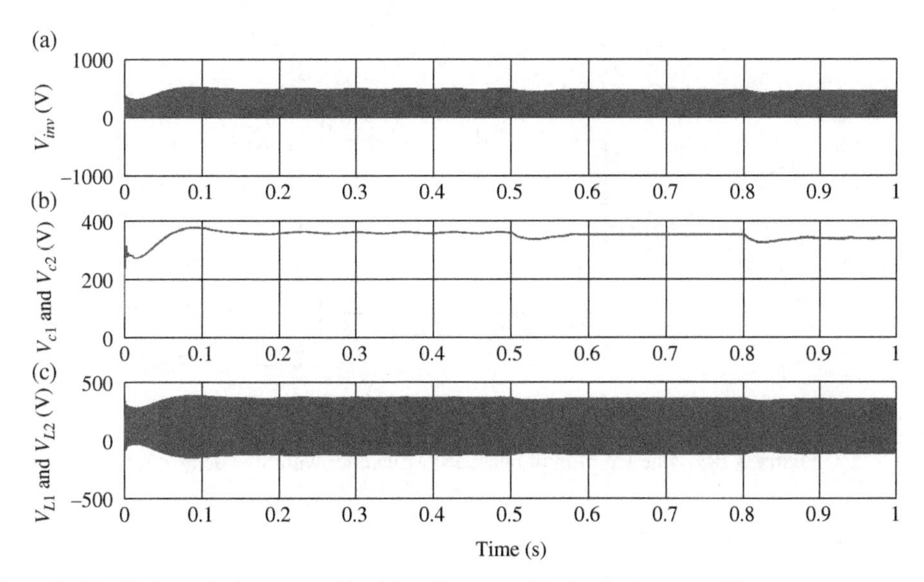

Figure 3.104 Voltages at the source side: (a) at the input of the bridge inverter; (b) across two capacitors; and (c) across two inductors

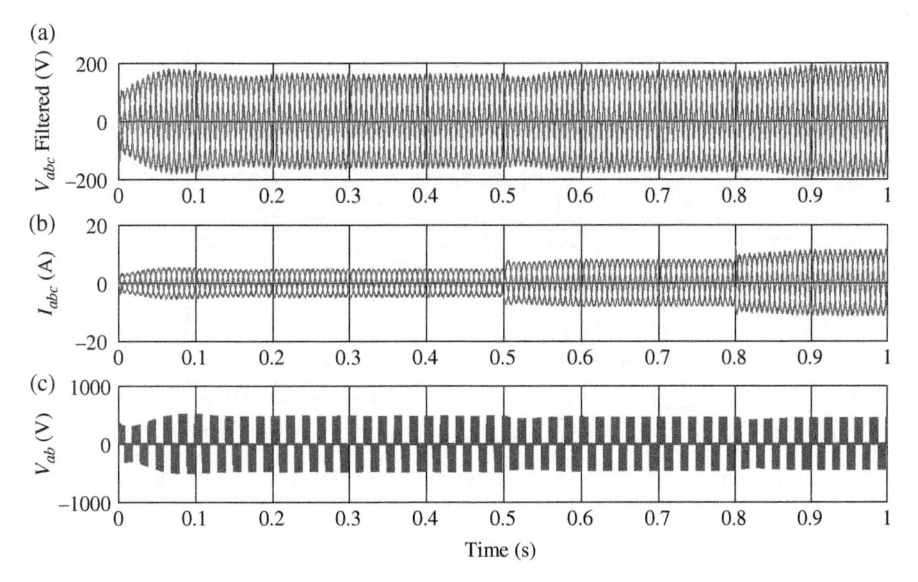

Figure 3.105 Outputs from qZ-source inverter: (a) filtered three-phase voltages; (b) three-phase load current; and (c) line voltage

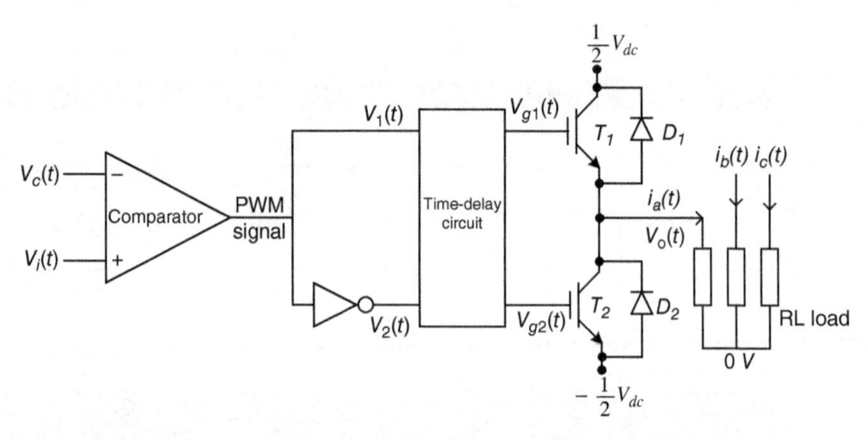

Figure 3.106 One leg of multiphase PWM inverter with time-delay circuit

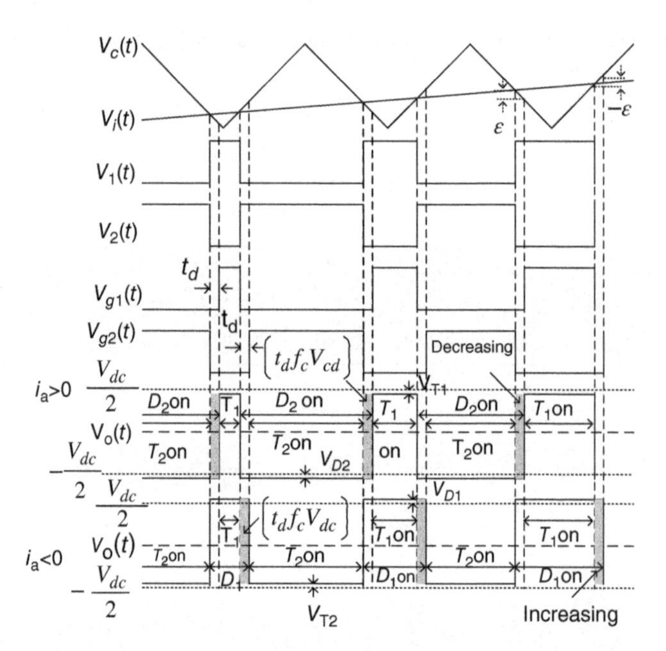

Figure 3.107 Gate control signals V_{g1} and V_{g2}, and output voltages at $i_a > 0$ and $i_a < 0$, showing IGBTs (T_1, T_2) and diode (D_1, D_2) ON conditions

Figure 3.107 are shown in Figure 3.108 for positive phase current $i_a > 0$ and negative phase current $i_a < 0$. Figure 3.107 also shows the duration at which D_1, D_2, T_1, and T_2 are conducting. V_{D1}, V_{D2}, V_{T1}, and V_{T2} are the voltage drops across D_1, D_2, T_1, and T_2, respectively. D_1 and D_2 are the freewheeling diodes. Here, it is assumed that the delay time t_d includes t_{on} and t_{off} of the

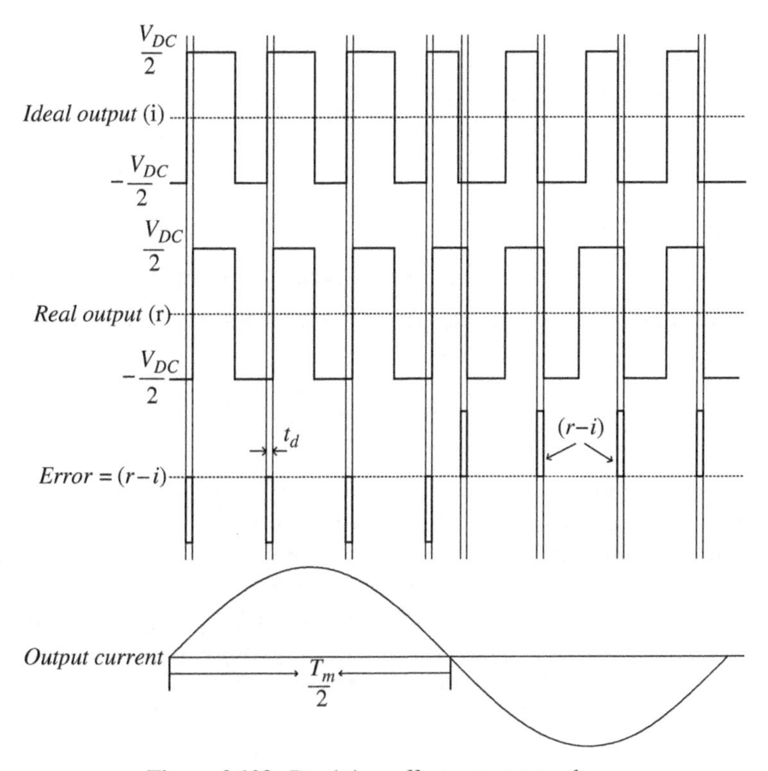

Figure 3.108 Dead time effect on output voltage

devices. The shaded portion for $i_a > 0$ is responsible for the decrease in output voltage, whereas the shaded portion for $i_a < 0$ is responsible for the increase in output voltage.

The voltage deviation ε depends on the delay time t_d and slope of the carrier wave $V_c(t)$. Therefore,

$$\frac{-\varepsilon}{t_d} = \frac{-2V_c}{(T_c/2)} \text{ or } \varepsilon = t_d \frac{2V_c}{(T_c/2)} = 4f_c t_d V_c \tag{3.164}$$

where V_c, f_c, and T_c are the amplitude, frequency, and time period of the carrier signal, respectively.

Figure 3.108 depicts ideal, real, and error output voltages waveforms. Here, it is assumed that the IGBTs switch instantly. The error pulses, whose height and width are equal to $\frac{V_{DC}}{2}$ and t_d, respectively, are in opposition to the output current.

The area of each pulse is

$$\Delta e = \frac{V_{DC}}{2} \times t_d \tag{3.165}$$

where the number of pulses is

$$n_p = \frac{\left(\frac{T_m}{2}\right)}{T_c} \tag{3.166}$$

The pulses due to dead time effect can be summed to form a square wave whose height and width are 'h' volt and $\frac{T_m}{2}$ s, respectively, and T_c is the period of the triangular wave or the inverter switching period. Therefore, the area of the resulting square wave is

$$h \times \frac{T_m}{2} = n_p \times \frac{V_{DC}}{2} \times t_d \text{ or } h = \Delta V = n_p \times \frac{V_{DC}}{T_m} \times t_d = f_c \times \frac{V_{DC}}{2} \times t_d \tag{3.167}$$

where ΔV is the amplitude.

In the ideal PWM waveform, the predominant harmonics are shifted to the switching frequency and the low-order harmonics are either suppressed or canceled. However, due to switching dead time, t_d, the presence of the error voltage can introduce low-frequency components such as the 5th and 7th, especially at very high-switching frequencies. The rectangular pulses can be approximated to a square wave with a frequency equal to that of the output fundamental frequency. If 'h' is the height of the square wave, then

$$h \times \frac{T_m}{2} = n_p \times V_{DC} \times t_d \tag{3.168}$$

i.e.

$$h\frac{T_m}{2} = \frac{\left(\frac{T_m}{2}\right)}{T_c} \times V_{DC} \times t_d \tag{3.169}$$

$$h = f_c \times V_{DC} \times t_d \tag{3.170}$$

where f_c is the frequency of the carrier wave or the inverter switching frequency. Using the Fourier series expression for the square wave, the nth harmonic of the error voltage is

$$V_n = \frac{4h}{\pi n} = \frac{4f_c \times V_{DC} \times t_d}{\pi n} \tag{3.171}$$

The above equation shows that the more the f_c increases, the more the V_n rises. Therefore, the upper limitation of carrier frequency exists when using the high-frequency carrier waveform and the product of f_c and t_d should be selected to a minimum value to avoid waveform distortion due to low-frequency harmonics introduced by switching lag times.

The effect of dead time is compensated in real-time implementation using several open-loop and closed-loop techniques. Some of the important references related to dead time compensation are Refs. [100–102].

3.13 Summary

This chapter discusses the PWM control of a three-phase two-level and multilevel voltage source inverter. The basic structure of the inverter is discussed starting from half-bridge to full-bridge and then three-phase inverter is elaborated on. A simple SPWM scheme is discussed followed by a third-harmonic injection and offset addition-based PWM. SVPWM is presented and a general simulation model is given that can be modified to implement both continuous and discontinuous SVPWM. An overmodulation region is also presented along with its MATLAB/Simulink implementation. Low-switching frequency PWM called SHE is presented and its MATLAB model is given. The use of the artificial intelligence technique for implementing SVPWM is described. SVPWM for NPC and CHB inverter is discussed. The effect of dead-banding is also discussed. Basic topologies of the multilevel inverter are presented and their SPWM is elaborated on. SVPWM for multilevel inverter is elaborated. A comparison between phase-shifted and level-shifted PWM technique is presented. Special inverters (Z-source and quasi Z-source) that have the capability of boosting the voltage are discussed in detail. Design of impedance source inverter is discussed in the chapter.

Problems

3.1: A single-phase half-bridge voltage source inverter is supplied from a center-tapped DC input voltage 120 V. The inverter supplies a resistive load of 20 Ω. Calculate;
(a) The fundamental output peak and rms voltage
(b) First three harmonics of the output voltage
(c) Power consumed by the load

3.2: A single-phase full-bridge voltage source inverter is supplied from a DC input voltage of 300 V. The inverter supplies a resistive load of 20 Ω. Calculate;
(a) The fundamental output peak and rms voltage
(b) First three harmonics of the output voltage
(c) Power consumed by the load

3.3: A single-phase half-bridge inverter is controlled using bipolar PWM method and supplied from a center-tapped 120 V DC supply. The fundamental output voltage frequency is 50 Hz. The triangular carrier signal's frequency is 5 kHz. For a modulation index of 0.9, determine;
(a) Frequency modulation ratio
(b) Fundamental output peak and rms voltage
(c) First three significant harmonic order produced in the output voltage

3.4: A single-phase full-bridge voltage source is supplied by a DC source of 300 V and controlled using unipolar PWM technique. The modulation index is 0.9, the fundamental output frequency is 50 Hz, and the carrier signal frequency is 2 kHz. Compute the following:
(a) Fundamental output voltage peak value
(b) First three dominant harmonic components

3.5: A three-phase voltage source inverter is operating in 180° conduction mode and supplied from a 600-V DC-link voltage. Calculate;

(a) The peak and rms value of the output fundamental phase and line voltages
(b) Three most significant harmonic component of phase peak value

3.6: A three-phase voltage source inverter is controlled using carrier-based sinusoidal PWM technique. The frequency of the fundamental output voltage is to be kept at 50 Hz and the switching frequency of the inverter is 2 kHZ. The inverter is supplied from a DC-link voltage of 600 V. The modulation index is 0.9. Compute the fundamental output phase RMS voltage value.

To enhance the output voltage, third-harmonic components are injected. Compute the value of the peak value of the third-harmonic component required to raise the output voltage by 15.47%.

3.7: A three-phase induction machine is supplied from a three-phase voltage source inverter. The three-phase VSI is controlled using space vector PWM technique and sinusoidal supply is to be given to the motor. The inverter switching frequency is 5 kHz and the DC bus voltage at the inverter input is 300 V. The inverter output frequency is 50 Hz. The reference phase voltage rms is 100 V. Identify the space vectors that will be applied and their time of application for the time instants are 1, 5, and 10.5 ms.

References

1. Kazmierkowaski, M. P., Krishnan, R., and Blaabjerg, F. (2002) *Control in Power Electronics: Selected Problems*. Academic Press, USA.
2. Holmes, D. G. and Lipo, T. A. (2003) *Pulse Width Modulation for Power Converters: Principles and Practice*. IEEE Power Engineering Series, Wiley Interscience, USA.
3. Mohan, N., Undeland, T. M., and Robbins, W. P. (1995) *Power Electronics-Converters, Applications and Design*, 2nd edn. John Wiley & Sons, Ltd., USA.
4. Holtz, J. (1992) Pulse width modulation: A survey. *IEEE Trans. Ind. Electron.*, 39(5), 410–420.
5. Wu, B. (2006) *High Power Converters and AC Drives*. IEEE Power Engineering Series, Wiley Interscience, USA.
6. Chen, Y. M. and Cheng, Y. M. (1999) PWM control using a modified triangle signal. *IEEE Trans. Ind. Electron.*, 1(29), 312–317.
7. Boost, M. A. and Ziogas, P. D. (1998) State-of-art-carrier PWM techniques: A critical evaluation. *IEEE Trans. Ind. Appl.*, 24(2), 271–280.
8. Murai, Y., Watanabe, T., and Iwasaki, H. (1987) Waveform distortion and correction circuit for PWM inverters with switching lag-times. *IEEE Trans. Ind. Appl.*, IA–23, 881–886.
9. Van der Broeck, H. W., Skudenly, H., and Stanke, G. (1988) Analysis and realization of a pulse width modulator based on voltage space vectors. *IEEE Trans. Ind. Electron.*, 24(1), 142–150.
10. Hava, A. M., Kerkman, R. S., and Lipo, T. A. (1997) A high performance generalized discontinuous PWM algorithm. *IEEE APEC*, 2, 886–891.
11. Ojo, O. (2004) The generalized discontinuous PWM scheme for three-phase voltage source inverters. *IEEE Trans. Ind. Electron.*, 51(6), 1280–1289.
12. Iqbal, A., Lamine, A., Ashraf, I., and Mohibullah (2006) MATLAB/Simulink model for space vector modulation of three-phase VSI. *IEEE UPEC 2006*, 6–8 September, Newcastle, UK, vol. **3**, pp. 1096–1100.
13. Iqbal, A., Ahmed, M., Khan, M. A., and Abu-Rub, H. (2010) Generalised simulation and experimental implementation of space vector PWM techniques of a three-phase voltage source inverter. *Int. J. Eng. Sci. Technol.*, 2(1), 1–12.

14. Holtz, J., Khamabdkone, A. M., and Lotzkat, W. (1993) On continuous control of PWM inverters in the over modulation range including the six-step mode. *IEEE Trans. Power Electron.*, 8(4), 546–553.
15. Lee, D. C. and Lee, G. M. (1998) A novel over-modulation technique for space vector PWM inverter. *IEEE Trans. Power Electron.*, 13(6), 1144–1151.
16. Have, A. M. (1998) Carrier-based PWM voltage source inverter in the over-modulation range. PhD thesis, University of Wisconsin.
17. Kerkman, R. J., Leggate, D., Seibel, B. J., and Rowan, T. M. (1996) Operation of PWM voltage source-inverters in the over-modulation region. *IEEE Trans. Ind. Electron.*, 43(1), 132–140.
18. Bose, B. K. (2007) Neural network applications in power electronics and motor drives: an introduction and perspective. *IEEE Trans. Ind. Electron.*, 54(1), 14–33.
19. Yuan, Z. and Cheng, J. (2013) Research on space vector PWM inverter based on artificial neural network. *2013 Sixth International Symposium on Computational Intelligence and Design*, Hangzhou, 2013, pp. 80–83, doi: 10.1109/ISCID.2013.134.
20. Kashif, S. A. R., Saqib, M. A., Zia, S., and Kaleem, A. (2009) Implementation of neural network based space vector pulse width modulation inverter-induction motor drive system. *Proc. 3rd Int. Conf. Eletc. Eng., ICEE*, Lahore, 2009, pp. 1–6, doi: 10.1109/ICEE.2009.5173177.
21. Mondal, S. K., Pinto, O. P., and Bose, B. K. (2002) A neural-network-based space-vector PWM controller for a three-level voltage-fed inverter induction motor drive. *IEEE Trans. Ind. Appl.*, 38(3), 660–669.
22. Xiang, G. and Huang, D. (2007) A neural network based space vector PWM for permanent magnet synchronous motor drive. *Proc. Int. Conf. Mech. Autom., ICMA*, Harbin, 2007, pp. 3673–3678, doi: 10.1109/ICMA.2007.4304157.
23. Zhou, K. and Wang, D. (2002) Relationship between space vector modulation and three-phase carrier-based PWM: A comprehensive analysis. *IEEE Trans. Ind. Electron.*, 49(1), 186–196.
24. Abu-Rub, H., Holtz, J., Rodriguez, J., and Ge, B. (2010) Medium-voltage multilevel converters-state of the art, challenges and requirements in industrial applications. *IEEE Trans. Ind. Electron.*, 57(8), 2581–2596.
25. Meynard, T. A., Fadel, M., and Aouda, N. (1997) Modeling of multilevel converters. *IEEE Trans. Ind. Electron.*, 4(3), 356–364.
26. Tolbert, L. M., Peng, F. Z., and Habetler, T. (1999) Multilevel converters for large electric drives. *IEEE Trans. Ind. Appl.*, 35, 36–44.
27. Rodriguez, J., Lai, J-S., and Peng, R. Z. (2002) Multilevel inverters: A survey of topologies, controls, and applications. *IEEE Trans. Ind. Electron.*, 49(4), 724–738.
28. Abu-Rub, H., Lewiscki, A., Iqbal, A., and Guzinski, J. (2010) Medium voltage drives-challenges and requirements. *Proc. IEEE ISIE-2010*, Bari, Italy; 4–7 July 2010, pp. 1372–1377.
29. Abu-Rub, H., Holtz, J., Rodriguez, J., and Boaming, G. (2010) Medium voltage multilevel converter-state of the art, challenges and requirements in industrial applications. *IEEE Trans. Ind. Electron.*, 57(8), 2581–2596.
30. Krug, D., Bernet, S., Fazel, S. S., Jalili, K., and Malinowski, M. (2007) Comparison of 2.3-kV medium-voltage multi-level converters for industrial medium-voltage drives. *IEEE Trans. Ind. Electron.*, 54(6), 2979–2992.
31. Franquelo, L. G., Rodriguez, J., Leon, J. I., Kouro, S., Portillo, R., and Prats, M. A. M. (2008) The age of multi-level converters arrives. *IEEE Ind. Electron. Mag.*, 2(2), 28–39.
32. Lin, L., Zou, Y., Wang, Z., and Jin, H. (2005) Modeling and control of neutral point voltage balancing problem in three-level NPC PWM inverters. *Proc. 36th IEEE PESC*, Recife, Brazil, 16 June 2005, pp. 861–866.

33. Ojo, O. and Konduru, S. (2005) A discontinuous carrier-based PWM modulation method for the control of neutral point voltage of three phase three-level diode clamped converters. *Proc. 36th IEEE Annu. Power Electron. Spec. Conf.*, Recife, Brazil, 16 June 2005, pp. 1652–1658.
34. Bernet, S. (2006) State of the art and developments of medium voltage converters: an overview. *Prz Elektrotech. (Elect. Rev.)*, 82(5), 1–10.
35. Peng, F. (2001) A generalized multilevel inverter topology with self voltage balancing. *IEEE Trans. Ind. Appl.*, 37(2), 611–618.
36. Fracchia, M., Ghiara, T., Marchesoni, M., and Mazzucchelli, M. (1992) Optimized modulation techniques for the generalized N-level converter. *Proc. IEEE Power Electron. Spec. Conf.*, vol. **2**, Toledo, Spain; 29 June to 3 July 1992, pp. 1205–1213.
37. Mukherjee, S. and Poddar, G. (2010) Series-connected three-level inverter topology for medium-voltage squirrel-cage motor drive applications. *IEEE Trans. Ind. Appl.*, 46(1), 179–186.
38. Du, Z., Tolbert, L. M., Chiasson, J. N., Ozpineci, B., Li, H., and Huang, A. Q. (2006) Hybrid cascaded H-bridges multilevel motor drive control for electric vehicles. *Proc. 37th IEEE Power Elect. Spec. Conf.*, South Korea; 18–22 June 2006, pp. 1–6.
39. Sirisukprasert, S., Xu, Z., Zhang, B., Lai, J. S., and Huang, A. Q. (2002) A high frequency 1.5 MVA H-bridge building block for cascaded multilevel converters using emitter turn-off thyrister. *Proc. IEEE Appl. Power Elect. Conf.*, Dalas, TX, USA; 10–14 March 2002, pp. 27–32.
40. Nabae, A., Takahashi, I., and Akagi, H. (1981) A new neutral point clamped PWM inverter. *IEEE Trans. Ind. Appl.*, IA-17(5), 518–523.
41. Escalante, M. F., Vannier, J. C., and Arzande, A. (2002) Flying capacitor multilevel inverters and DTC motor drive applications. *IEEE Trans. Ind. Electron.*, 49(4), 809–815.
42. Fazel, S. S., Bernet, S., Krug, D., and Jalili, K. (2007) Design and comparison of 4 kV neutral-point-clamped, flying capacitor and series-connectd H-bridge multilevel converters. *IEEE Trans. Ind. Appl.*, 43(4), 1032–1040.
43. Shukla, A., Ghosh, A., and Joshi, A. (2007) Capacitor voltage balancing schemes in flying capacitor multilevel inverters. *Proc. IEEE Power Elect. Spec. Conf. PESC-2007*, 17–21 June, Orlando, FL, pp. 2367–2372.
44. Jin, B. S., Lee, W. K. Kim, T. J., Kang, D. W., and Hyun, D. S. (2005) A study on the multi-carrier PWM methods for voltage balancing of flying capacitor in the flying capacitor multilevel inverter. *Proc. 31st IEEE Ind. Elect. Conf. IECON*, 6–10 November, North Carolina, pp. 721–726.
45. Lee, S. G., Kang, D. W., Lee, W. H., and Hyun, D. S. (2001) The carrier-based PWM method for voltage balance of flying capacitor multilevel inverter. *Proc. 3rd IEEE Power Elect. Spec. Conf.*, vol. **1**, Vancouver, BC, Canada; 17–21 June 2001, pp. 126–131.
46. Shukla, A., Ghosh, A., and Joshi, A. (2008) Improved multilevel hysteresis current regulation and capacitor voltage balancing schemes for flying capacitor multilevel inverter. *IEEE Trans. Power Electron.*, 23(2), 518–529.
47. Lee, W. K., Kim, S. Y., Yoon, J. S., and Baek, D. H. (2006) A comparison of the carrier-base PWM techniques for voltage balance of flying capacitor in the flying capacitor multilevel inverter. *Proc. 21st IEEE Appl. Power Elect. Conf. Exp.*, 19–23 March, Dallas, TX, pp. 1653–1659.
48. Xin, M. C., Ping, S. L., Xu, W. T., and Bao, C. C. (2009) Flying capacitor multilevel inverters with novel PWM methods. *Procedia Earth Planet. Sci.*, 1, 1554–1560.
49. Khazraei, M., Sapahvand, H., Corzine, K., and Fredowsi, M. (2010) A generalized capacitor voltage balancing scheme for flying capacitor multilevel converters. *Proc. 25th Appl. Power Elect. Conf. Exp.*, 6–10 March, Dallas, TX, pp. 58–62.

50. Lai, Y., Chou, Y., and Pai, S. (2007) Simple PWM technique of capacitor voltage balance for three-level inverter with DC-link voltage sensor only. *The 33rd Annu. Conf. IEEE Ind. Electron. Soc. IECON.*, Taipei, Taiwan, 5–8 November 2007.
51. Bhalodi, K. H. and Agrawal, P. (2006) Space vector modulation with DC-link voltage balancing control for three-level inverters. *Int. Conf. Power Electron., Drives Energy Syst., PEDES.*, New Delhi, India; 12–15 December 2006.
52. Suh, J., Choi, C., and Hyun, D. (1999) A new simplified space-vector PWM method for three-level inverters. *14 Annu. Appl. Power Electron. Conf. Expo., APEC.*, Dallas, TX, USA; 14–18 March 1999.
53. Holtz, J. and Oikonomou, N. (2005) Neutral point potential balancing algorithm at low modulation index for three-level inverter medium voltage drives. *Ind. Appl. Conf. IAS.*, Hong Kong, China; 2–6 October 2005.
54. Holtz, J. and Oikonomou, N. (2007) Neutral point potential balancing algorithm at low modulation index for three-level inverter medium voltage drives. *IEEE Trans. Ind. Electron.*, 43(3), 761–768.
55. Malinowski, M., Stynski, S., Kolomyjski, W., and Kazmierkowski, M. P. (2009) Control of three-level PWM converter applied to variable-speed-type turbines. *IEEE Trans. Ind. Electron.*, 56(1), 69–77.
56. Portillo, R. C., Prats, M. M., León, J. I., Sánchez, J. A., Carrasco, J. M., Galván, E., and Franquelo, L. G. (2006) Modeling strategy for back-to-back three-level converters applied to high-power wind turbines. *IEEE Trans. Ind. Electron.*, 53(5), 1483–1491.
57. Pereira, I. and Martins, A. (2009) Neutral-point voltage balancing in three-phase NPC converters using multicarrier PWM control. *Int. Conf. Power Eng., Energy Electr. Drives.*, Lisbon, Portugal; 18–20 March 2009.
58. Pereira, I. and Martins, A. (2009) Multicarrier and space vector modulation for three-phase NPC converters: A comparative analysis. *13th Eur. Conf. Power Electron. Appl. EPE.*, Barcelona, Spain; 8–10 September 2009.
59. Umbría, F., Gordillo, F., Salas, F., and Vázquez, S. (2010) Voltages balance control in three phase three-level NPC rectifiers. *IEEE Int. Symp. Ind. Electron., ISIE.*, Bari, Italy; 4–7 July 2010.
60. Alloui, H., Berkani, A., and Rezine, H. (2010) A three level NPC inverter with neutral point voltage balancing for induction motors direct torque control. *XIX Int. Conf. Electr. Mach. ICEM.*, Rome, Italy; 6–8 September 2010.
61. Seo, J. H and Choi, C. H (2001) Compensation for the neutral-point potential variation in three-level space vector PWM. *Sixteenth Annu. IEEE Appl. Power Electron. Conf. Expo., APEC.*, Anaheim, CA, USA; 4–8 March 2001.
62. Seo, J. H., Choi, C. H., and Hyun, D. S. (2001) A new simplified space–vector PWM method for three-level inverters. *IEEE Trans. Power Electron.*, 16(4), 545–550.
63. Patel, P. J., Patel, R. A., Patel, V., and Tekwani, P. N. (2008) Implementation of self balancing space vector switching modulator for three-level inverter. *IEEE Region 10 and the Third Int. Conf. Ind. Inf. Syst., ICIIS.*, Kharagpur, India; 8–10 December 2008.
64. Zhu, R.-W., Wu, X.-J., Xiao-yan, J., and Peng, D. (2010) An improved neutral-point-potential balance control strategy for three-level PWM rectifier. *Asia Pac. Conf. Postgrad. Res. Microelectron. Electron.*, Shanghai, China; 22–24 September 2010.
65. Zhi, Z., Xie, Y.-X., Wei-ping, H., Jiang-yuan, L., and Lin, C. (2009) A new SVPWM method for single-phase three-level NPC inverter and the control method of neutral point voltage balance. *Int. Conf. Electr. Mach. Sys. ICEMS.*, Tokyo, Japan; 15–18 November 2009.
66. Pou, J., Boroyevich D., and Pindado, R. (2002) New feedforward space-vector PWM method to obtain balanced AC output voltages in a three-level neutral-point-clamped inverter. *IEEE Trans. Ind. Electron.*, 49(5), 1026–1034.

67. Lewicki, A., Krzeminski, Z., and Abu-Rub, H. (2011) Space-vector pulsewidth modulation for three-level NPC converter with the neutral point voltage control. *IEEE Trans. Ind. Electron.*, 58(11), 5076–5086.
68. Guo, F., Yang, T., Bozhko, S., and Wheeler, P. (2018) 3L-NPC AC-DC power converter using virtual space vector PWM with optimal switching sequence based on g-h coordinate. *IEEE Int. Conf. Electr. Sys. Aircr., Railw., Ship Propul. Road Veh. Int. Transp. Electrification Conf. (ESARS-ITEC)*, Orlando, FL, USA; 17–21 June 2007.
69. Lewicki, A. and Morawiec, M. (2017) Space-vector pulsewidth modulation for a seven-level cascaded H-bridge inverter with the control of DC-link voltages. *Bull. Polish Acad. Sci.: Tech. Sci.*, 65(5), 619–628.
70. Rodriguez, J., Pontt, J., Silva, E., Espinoza, J., and Perez, M. (2003) Topologies for regenerative cascaded multilevel inverters. *Proc. 34th Annu. Power Electron. Spec. Conf. IEEE PESC*, vol. **2**, United States; 15–19 June 2003, pp. 519–524.
71. Islam, M.R., Guo, Y., and Zhu, J. (2014) A high-frequency link multilevel cascaded medium-voltage converter for direct grid integration of renewable energy systems. *IEEE Trans. Power Electron.*, 29(8), 4167–4182.
72. Rodriguez, J., Morán, L., Pontt, J., Hernández, J. L., Silva, L., Silva, C., and Lezana, P. (2002) High-voltage multilevel converter with regeneration capability. *IEEE Trans. Ind. Electron.*, 49(4), 839–846.
73. Goodman, A., Watson, A., Dey, A., Clare, J., Wheeler, P., and Zushi, Y. (2014) DC side ripple cancellation in a cascaded multi-level topology for automotive applications. *Proc. IEEE ECCE*, pp. 5916–5922.
74. Vahedi, H., Al-Haddad, K., Labbe, P.-A., and Rahmani, S. (2014) Cascaded multilevel inverter with multi-carrier PWM technique and voltage balancing feature. *Proc. 23rd Int. Symp. Ind. Electron. IEEE ISIE*, Istanbul, Turkey; 1–4 June 2014, pp. 2155–2160.
75. Kumar, A., Kumar, D., and Meena, D. R. (2014) SRF based modeling and control of cascaded multilevel active rectifier with uniform DC-buses. *Proc. Recent Adv. Eng. Comp. Scien. RAECS*, Chandigarh, India; 6–8 March 2014, pp. 1–5.
76. Marzoughi, A. and Imaneini, H. (2014) Optimal selective harmonic elimination for cascaded H-bridge-based multilevel rectifiers. *IET Power Electron.*, 7(2), 350–356.
77. Taha, O. A. and Pacas, M. (2014) Hardware implementation of balance control for three-phase grid connection 5-level Cascaded H-Bridge converter using DSP. *Proc. 23rd Int. Symp. Ind. Electron. IEEE ISIE*, Istanbul, Turkey; 1–4 June 2014, pp. 1366–1371.
78. Ahmed, I. and Borghate, V. B. (2014) Simplified space vector modulation technique for seven-level cascaded H-bridge inverter. *IET Power Electron.*, 7(3), 604–613.
79. Correa Vasquez, P. I. (2006) Fault tolerant operation of series connected H-bridge multilevel inverters. Vigo, Spain; 4–7 June 2007.
80. Gholinezhad, J. and Noroozian, R. (2012) Application of cascaded H-bridge multilevel inverter in DTC-SVM based induction motor drive. *Proc. 3rd Power Electron. Drive Sys. Techn. PEDSTC*, Tehran, Iran; 15–16 February 2012, pp. 127–132.
81. Nowicki, E. P. and Roodsari, B. N. (2013) Fast space vector modulation algorithm for multilevel inverters and its extension for operation of the cascaded H-bridge inverter with non-constant DC sources. *IET Power Electron.*, 6(7), 1288–1298.
82. Marzoughi, A., Neyshabouri, Y., and Imaneini, H. (2014) Control scheme for cascaded H-bridge converter-based distribution network static compensator. *IET Power Electron.*, 7(11), 2837–2845.
83. Sun, Y., Zhao, J., and Ji, Z. (2013) An improved CPS-PWM method for cascaded multilevel STATCOM under unequal losses. *Proc. 39th Annu. Conf. IEEE IECON*, Vienna, Austria; 10–14 November 2013, pp. 418–423.

84. Chavarria, J., Biel, D., Guinjoan, F., Meza, C., and Negroni, J. J. (2013) Energy-balance control of PV cascaded multilevel grid-connected inverters under level-shifted and phase-shifted PWMs. *IEEE Trans. Ind. Electron.*, 60(1), 98–111.
85. Angulo, M., Lezana, P., Kouro, S., Rodriguez, J., and Wu, B. (2007) Level-shifted PWM for cascaded multilevel inverters with even power distribution. *Proc. Power Electron. Spec. Conf. IEEE PESC*, Orlando, FL, USA; 17–21 June 2007, pp. 2373–2378.
86. Peng, F. Z. (2003) Z-source inverter. *IEEE Trans. Ind. Appl.*, 39(2), 504–510.
87. Peng, F. Z., Shen, M., and Qian, Z. (2005) Maximum boost control of the Z-source inverter. *IEEE Trans. Power Electron.*, 20(4), 833–838.
88. Shen, M., Wang, J., Joseph, A., Peng, F. Z., Tolbert, L. M., and Adams, D. J. (2006) Constant boost control of the Z-source inverter to minimize current ripple and voltage stress. *IEEE Trans. Ind. Appl.*, 42(3), 770–778.
89. Badin, R., Huang, Y., Peng, F. Z., and Kim, H. G. (2007) Grid interconnected Z-source PV system. *Proc. IEEE PESC'07*, June, Orlando, FL, pp. 2328–2333.
90. Shen, M. and Peng, F. Z. (2008) Operation modes and characteristics of the Z-source inverter with small inductance or low power factor. *IEEE Trans. Ind. Electron.*, 55(1), 89–96.
91. Peng, F. Z. (2008) Z-source networks for power conversion. *23rd Ann. IEEE App. Power Elect. Conf. Exp., APEC2008*, 24–28 February, Austin, TX, pp. 1258–1265.
92. Peng, F. Z. (2008) Recent advances and applications of power electronics and motor drives: Power converters. *34th Ann. Conf. IEEE Ind. Elect., IECON*, 10–13 November, Orlando, FL, pp. 10–13.
93. Peng, F. Z., Shen, M., and Holland, K. (2007) Application of Z-source inverter for traction drive of fuel cell-battery hybrid electric vehicles. *IEEE Trans. Power Electron.*, 22(3), 1054–1061.
94. Peng, F. Z., Yuan, X., Fang, X., and Qian, Z. (2003) Z-source inverter for adjustable speed drives. *IEEE Power Electron. Lett.*, 1(2), 33–35.
95. Peng, F. Z., Joseph, A., Wang, J. et al. (2005) Z-source inverter for motor drives. *IEEE Trans. Power Electron.*, 20(4), 857–863.
96. Anderson, J. and Peng, F. Z. (2008) A class of quasi-Z-source inverters. *IEEE Ind. Appl. Soc. Ann. Mtg., IAS'08*, 5–9 October., Edmonton, Alta, pp. 1–7.
97. Li, Y., Anderson, J., Peng, F. Z., and Liu, D. (2009) Quasi-Z-source inverter for photovoltaic power generation systems. *24th Ann. IEEE Appl. Power Elect. Conf. Exp., APEC 2009*, 15–19 February, Washington, DC, pp. 918–924.
98. Park, J., Kim, H., Nho, E., Chun, T., and Choi, J. (2009) Grid-connected PV system using a quasi-Z-source inverter. *24th Ann. IEEE AplL. Power Elect. Conf. Exp., APEC 2009*, 15–19 February, Washington, DC, pp. 925–929.
99. Jeong, S. G. and Park, M. H. (1991) The analysis and compensation of dead-time effects in PWM inverters. *IEEE Trans. Ind. Electron.*, 38, 108–114.
100. Lin, J. L. (2002) A new approach of dead time compensation for PWM voltage inverters. *IEEE Trans. Circ. Syst. I-Fund. Theory Appl.*, 49(4), 476–483.
101. Munoz, A. R. and Lipo, T. A. (1999) On-line dead time compensation technique for open-loop PWM VSI drives. *IEEE Trans. Power Electron.*, 14(4), 683–689.
102. Victor, G., Cgrdenas, M., Horta, S. M., and Echavarria, S. R. (2006) Elimination of dead time effects in three-phase inverters. *Proc. CIEP*, 14–17 October, Mexico, pp. 258–262.

4

Field-Oriented Control of AC Machines

4.1 Introduction

This chapter focuses on vector control of AC machines, particularly induction motors (IM), permanent magnet synchronous motors (PMSM), and double fed induction generator (DFIG). We consider these machines to be the most interesting for modern electric drives, and so the focus will be on their control and simulation. The control of AC machines can be classified into "scalar" and "vector" controls. Scalar controls are simple to implement and offer good steady-state response; however, the dynamics are slow because the transients are not controlled. To obtain high precision and good dynamics, vector control schemes have been invented for use with closed-loop feedback controls. Thus, this chapter focuses on the "vector control" schemes of AC machines after a brief description of the scalar control method.

Adjustable speed drives offer significant energy savings and fast and precise responses in industrial applications. At the beginning of the 1970s [1–9], the principles of torque and flux control were introduced and called "field-oriented control" or "vector control" for squirrel cage induction machines and later for synchronous machines. The vector control idea relies on the control of stator current space vectors in a similar, but more complicated, way to a DC machine. The advancement in adjustable speed AC drives was slow until the 1980s, when the microprocessor revolution made it possible to implement complex control algorithms, which made AC machines the dominating machine in the drives market.

The research on vector control of AC machines is active and up to date. The new advancements are high precision and include the use of novel solutions, such as the sensorless operation.

High Performance Control of AC Drives With MATLAB®/Simulink, Second Edition.
Haitham Abu-Rub, Atif Iqbal, and Jaroslaw Guzinski.

Companion website: www.wiley.com/go/aburubcontrol2e

4.2 Induction Machines Control

Squirrel cage types of IM have been considered for a long time as the workhorse in industry [2]. It has been claimed that 90% of installed motors are of this type. Among the reasons for their popularity are robustness, reliability, low price, and relatively high efficiency.

Different control methods are popular in the industry (e.g. scalar method [named *V/f*], vector control, direct torque control, etc.). In this chapter, we will first describe the *V/f* method and then focus on the vector control approach.

4.2.1 *Control of Induction Motor Using V/f Methods*

The aim of this section is to model a scalar control method for IM, named *V/f* = constant. This method is the simplest way for controlling this type of machine.

In this type of control, a constant ratio between the voltage magnitude and the frequency should be maintained. This is to keep the flux constant in the machine.

A well-known machine model is presented in Chapter 2 and can be rewritten as [1–8]:

$$u_{sx} = R_s i_{sx} + \frac{d\psi_{sx}}{d\tau} - \omega_a \psi_{sy} \tag{4.1}$$

$$u_{sy} = R_s i_{sy} + \frac{d\psi_{sy}}{d\tau} - \omega_a \psi_{sx} \tag{4.2}$$

It is assumed that the co-ordinate system is connected with stator current, hence:

$$\psi_{sx} = |\psi_s| \tag{4.3}$$

$$\psi_{sy} = 0 \tag{4.4}$$

which in a steady state would be

$$\frac{d\psi_{sx}}{d\tau} = 0 \tag{4.5}$$

Therefore,

$$u_{sx} = R_s i_{sx} \tag{4.6}$$

$$u_{sy} = R_s i_{sy} + \omega_a \psi_{sx} \tag{4.7}$$

The voltage vector magnitude is

$$|u_s| = \sqrt{u_{sx}^2 + u_{sy}^2} \tag{4.8}$$

$$|u_s| = \sqrt{(R_s i_{sx})^2 + (R_s i_{sy})^2 + (\omega_a \psi_{sx})^2 + 2 \cdot R_s i_{sy} \cdot \omega_a \psi_{sx}} \tag{4.9}$$

Neglecting stator resistance, the following is achieved:

$$|u_s| = \sqrt{(\omega_a \psi_{sx})^2} \tag{4.10}$$

Keeping constant motor flux magnitude at rated value (in per unit, this is 1), the following is correct:

$$|\psi_s| = 1 \text{ (rated value in per unit)} \tag{4.11}$$

Therefore, equation (4.10) becomes

$$|u_s| = \sqrt{(\omega_a)^2} \tag{4.12}$$

Where

$$\omega_a = 2\pi f \tag{4.13}$$

Thus, in *V/f* control, a constant ratio between the voltage magnitude and frequency is maintained. This is to keep constant and optimal flux in the machine. The *V/f* control of the IM model is shown in Figure 4.1. It consists of the motor model and the control system model, as shown in Figure 4.2. All the models are described per unit unless otherwise specified. The Simulink software file of *V/f* control of the IM model is [IM_Vbyf_control]. The motor parameters are located in the [IM_Vbyf_param], which should be first executed.

4.2.1.1 Current and Flux Computation

The magnitude of current and flux can be calculated from the real and imaginary parts, as follows, and are also shown in Figure 4.3:

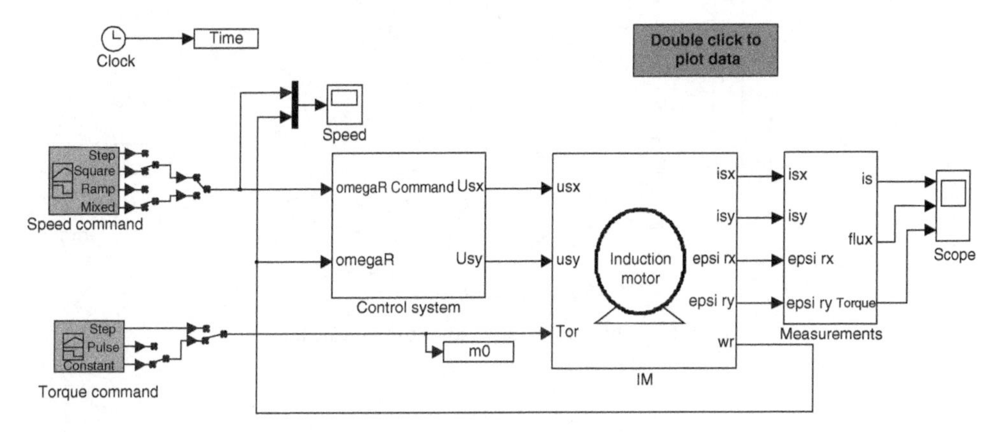

Figure 4.1 *V/f* control of an induction motor

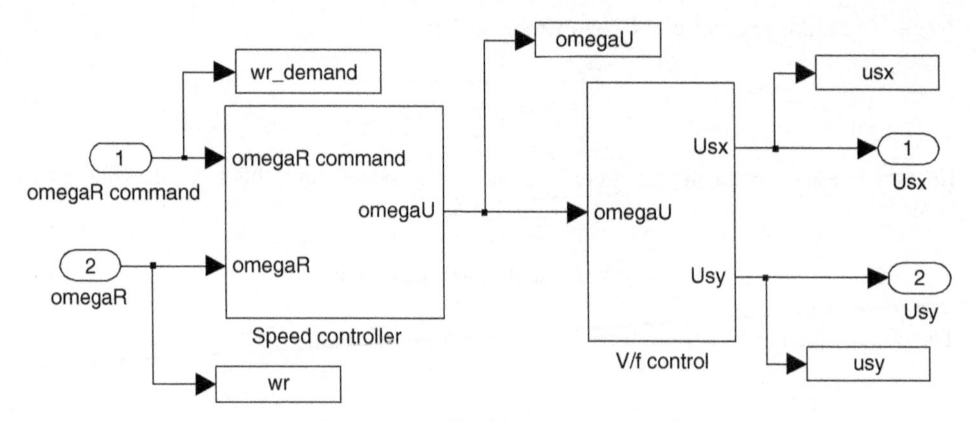

Figure 4.2 *V/f* control system model

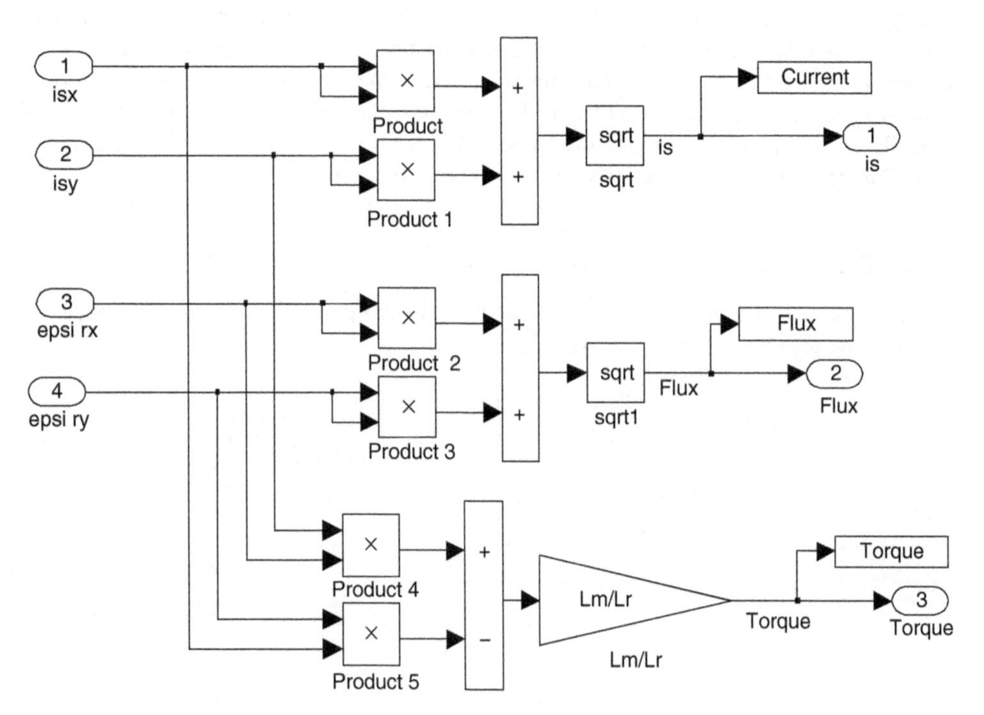

Figure 4.3 Current flux and motor torque calculations

$$|I_s| = \sqrt{I_{s\alpha}^2 + I_{s\beta}^2} \tag{4.14}$$

$$|\psi_s| = \sqrt{\psi_{s\alpha}^2 + \psi_{s\beta}^2} \tag{4.15}$$

The motor torque is given by [1–7]

$$t_e = \frac{L_m}{L_r}\left(\psi_{r\alpha} i_{s\beta} - \psi_{r\beta} i_{s\alpha}\right) \tag{4.16}$$

4.2.1.2 Control System

The control system model is a proportional integral (PI) speed controller that calculates the stator voltage. The PI speed controller is shown in Figure 4.4. The integrator is only active when the error is within 0.4 per unit.

The *V*/*f* control calculates the real and imaginary parts of the stator voltage, based on constant *V*/*f* formulation (Figure 4.5). The magnitude of the stator voltage is calculated from the absolute value of the PI controller output, and the angle is calculated by integrating the PI controller output.

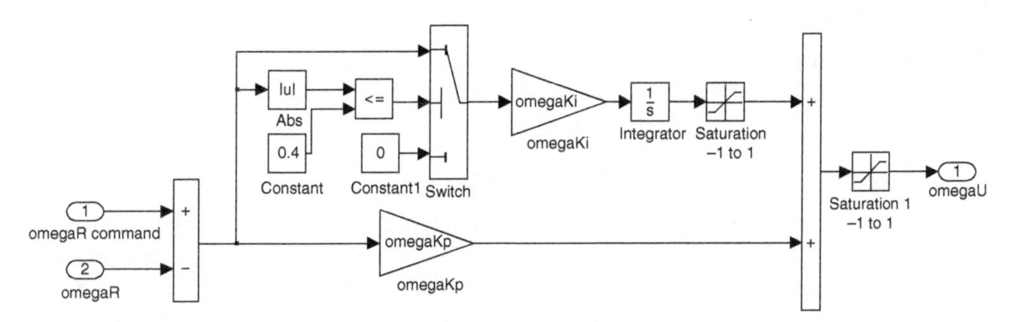

Figure 4.4 Limited authority PI control systems model

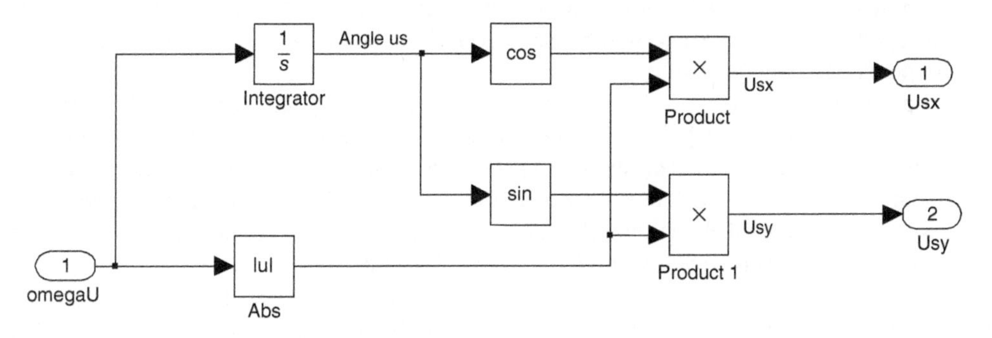

Figure 4.5 *V*/*f* control

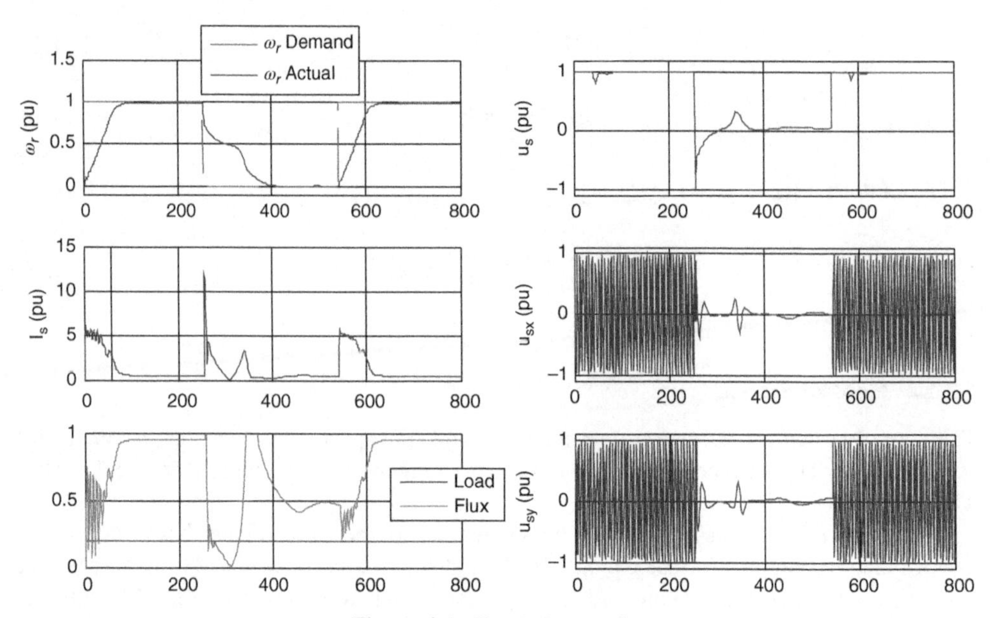

Figure 4.6 Simulation results

4.2.1.3 Simulation Results

The simulation result for square pulse command in speed is shown in Figure 4.6. The given results are selected examples of machine response.

4.2.2 Vector Control of Induction Motor

Squirrel cage IM (and other types of AC machines) cannot be controlled as easily as they can be in the case of separately excited DC motor [1–16]. For all types of control, the magnetic flux and produced torque should be decoupled for maintaining linearity between input and output and for achieving high dynamic drive. In the case of AC machines, the dynamic models are nonlinear and more complex than those in DC machines.

Overcoming this problem became possible by using space vector representations of AC machines. The flux-oriented control methods allow representation of the mathematically complicated induction machine in a similar manner to DC machines for obtaining control linearity, decoupling, and high performance AC drives.

Proper control of motor speed and produced electromagnetic torque are needed in high performance adjustable speed drives. The torque production depends on the armature current and machine flux. The magnetic field should be kept at the optimal level (rated flux) for producing maximum torque, while protecting against entering the saturation level. Keeping constant flux allows for obtaining a linear relationship between motor torque and armature current. In this way, linear control is achieved with high machine dynamics.

Decoupled control between flux and torque is easily achieved in the case of separately excited DC motors, but it is not so simple for AC drives. In squirrel cage motors, the controlled signal is only the stator current because the rotor current is inaccessible. In such away, the torque equation is not linear, so linear control with maximum torque production is difficult to achieve.

The vector control approach was first formulated by Blaschke [9] to overcome this problem. The principle of field-oriented control was introduced and decoupled control between torque and flux was possible.

In summary, by using vector representation, it is possible to present the variables in an arbitrary co-ordinate system. If the co-ordinate system rotates together with a flux space vector, then we use different terminology: flux-oriented control, rotor-oriented, etc.

Relationships of motor models presented in Chapter 2 can be described in frame $d-q$, rotating with rotor flux on the d axis (the q flux component is zero $\psi_{sq} = 0$), as

$$\frac{dx_{sd}}{d\tau} = a_1.i_{sd} + a_2.\psi_r + \omega_a.i_{sq} + \omega_r.a_3.0 + a_4.u_{sd} \tag{4.17}$$

$$\frac{dx_{sq}}{d\tau} = a_1.i_{sq} + a_2.0 - \omega_a.i_{sd} - \omega_r.a_3.\psi_r + a_4.u_{sq} \tag{4.18}$$

$$\frac{d\psi_r}{d\tau} = a_5.\psi_r + (\omega_a - \omega_r).0 + a_6.i_{sd} \tag{4.19}$$

$$0 = a_5.0 - (\omega_a - \omega_r).\psi_r + a_6.i_{sq} \tag{4.20}$$

$$\frac{d\omega_r}{dt} = \frac{L_m}{JL_r}\left(.\psi_r i_{sq} - 0.i_{sd}\right) - \frac{1}{J}t_0 \tag{4.21}$$

Hence:

$$t_e = \frac{L_m}{L_r}\left(.\psi_r i_{sq}\right) \tag{4.22}$$

$$a_1 = -\frac{R_s L_r^2 + R_r L_m^2}{L_r \omega_\sigma} \quad a_2 = \frac{R_r L_m}{L_r \omega_\sigma} \quad a_3 = \frac{L_m}{\omega_\sigma} \quad a_4 = \frac{L_r}{\omega_\sigma} \quad a_5 = -\frac{R_r}{L_r} \quad a_6 = \frac{R_r L_m}{L_r} \tag{4.23}$$

$$w_\sigma = L_s L_s - L_m^2 \tag{4.24}$$

R_s, R_r are motor resistances; L_s, L_r, L_m are motor inductances; ω_r is the rotor angular speed.

In this way, it was possible to represent the electromagnetic torque as a product of a flux-producing current and a torque-producing current. Keeping constant flux, an induction machine may be controlled, like a separately excited DC machine.

Assuming that the flux is kept constant (at rated level), equation (4.19) can be written as follows:

$$0 = a_5.\psi_r + a_6.i_{sd} \tag{4.25}$$

Which gives

$$0 = a_5.\psi_r + a_6.i_{sd} = -\frac{a_5}{a_6}.\psi_r = \frac{\psi_r}{L_m} \tag{4.26}$$

From the above equations, it is evident that in a quadrant co-ordinate system *dq*, the axis *d* is consistent with the direction of the rotor flux linkage vector, and the component on axis *q* affects the quantity of electromagnetic torque.

4.2.2.1 Basic Control Scheme

Figure 4.7 describes the basic scheme of vector control of an AC machine.

In typical AC drive systems, two motor phase currents and the DC bus voltage are measured. The measured currents are transformed using the Clarke transformation block into a stationary frame ($i_{s\alpha}$ and $i_{s\beta}$). These last two components are further transformed, using the Park transformation, into rotating components (*dq*). The PI controllers compare the command values with the measured components (after transformation) and command proper values to establish the desired condition.

The outputs of the controllers are transformed from a rotating to a stationary frame using the Park transformation. The commanded signals of the stator voltage are sent to the pulse width modulation (PWM) block.

4.2.2.2 MATLAB/Simulink Model for Induction Motor Vector Control

The control scheme of an IM is shown in Figure 4.8. The control depends on vector transformation using Park/Clark transformations. The idea is to control this type of machine in a similar

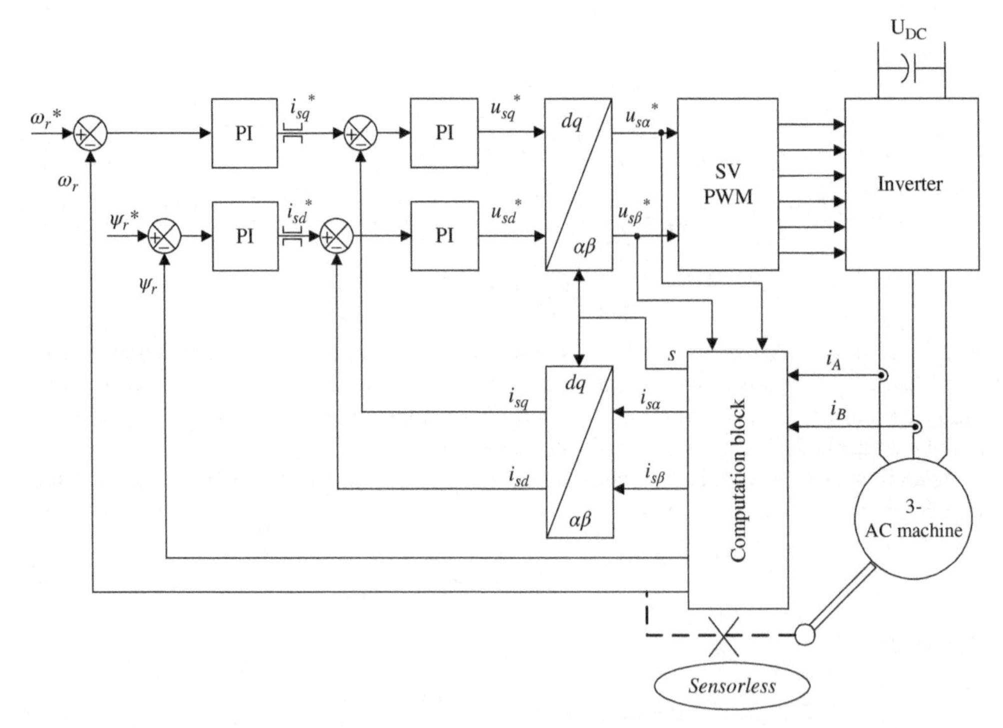

Figure 4.7 Basic scheme of field oriented control (FOC) for the three-phase AC machine

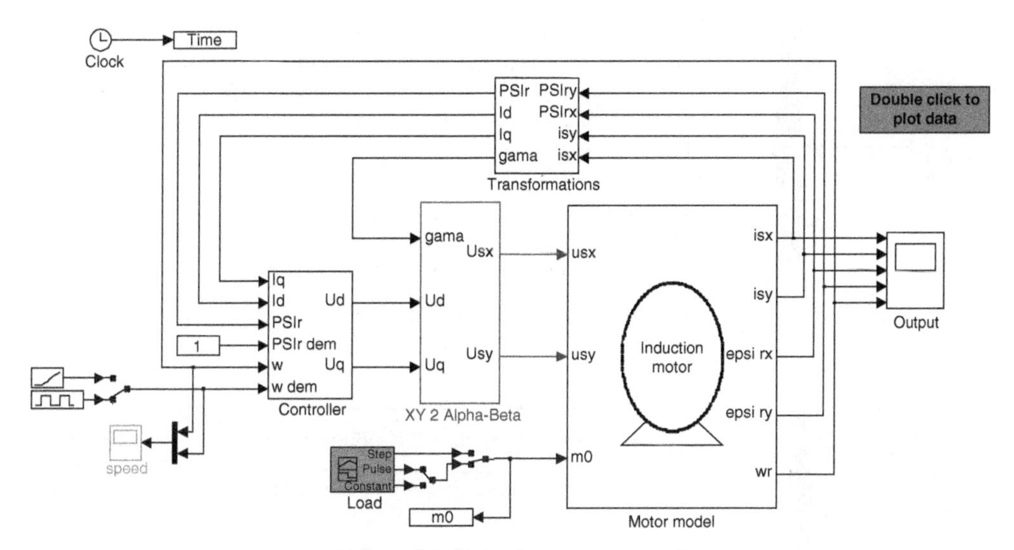

Figure 4.8 Induction motor control

way to a separately excited DC motor, in which the torque is controlled separately from the flux. In this way, decoupled control between the two subsystems is achieved when the flux is kept constant.

For control purposes, the motor model is first transformed into the $d-q$ axis model. It consists of a motor model, various transformations, and the control system model. All the models are in per unit unless otherwise specified. The Simulink software file of vector control of IM model in [IM_vector_control]. The motor parameters are located in the [IM_param], which should be first executed.

The IM model in Simulink is described in Chapter 2. The motor current and flux magnitude are calculated in the following:

$$|I_s| = \sqrt{I_{s\alpha}^2 + I_{s\beta}^2} \tag{4.27}$$

$$|\psi_s| = \sqrt{\psi_{s\alpha}^2 + \psi_{s\beta}^2} \tag{4.28}$$

4.2.2.3 Variables Transformation

The machine variables can be transformed into dq components by using the "$\alpha-\beta$ to $x-y$" transformation, as shown in Figure 4.9. The block in Figure 4.9 calculates the voltages to be applied in the $d-q$ axis. The $d-q$ axis voltages are converted into the x-y components by "$x-y$ to $\alpha-\beta$" transformations, as shown in Figure 4.10. Both of these transformations require the rotor flux angle (gamma) and the rotor flux magnitude, which are calculated by using the transformation in Figure 4.11.

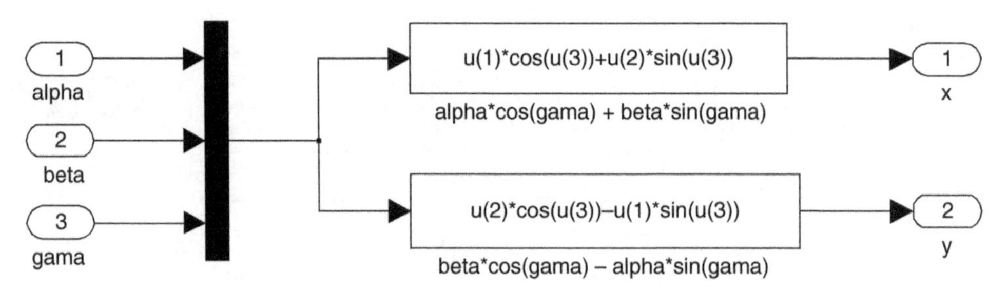

Figure 4.9 α–β to x–y transformation

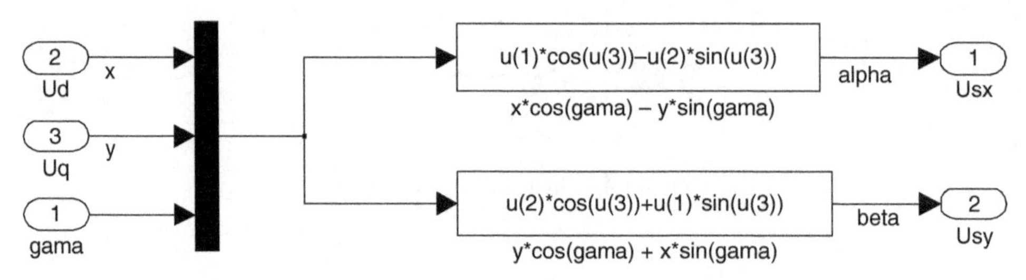

Figure 4.10 x–y to α–β transformation

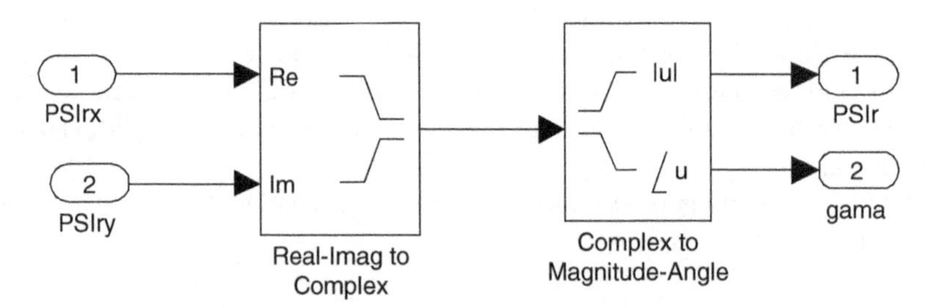

Figure 4.11 Real-imaginary to magnitude-angle transformation

4.2.2.4 Control Scheme

The inner and outer loop control system model is shown in Figure 4.12. The inner loop controller works on the *d-q* axis model. The outer loop controller works on the flux and speed to calculate the demand for I_d and I_q currents, respectively. The inner loop controller then controls the I_d and I_q currents.

4.2.2.5 Simulation Results

The simulation results for square pulse demand in speed are shown in Figure 4.13.

The given simulation results are selected examples of machine response.

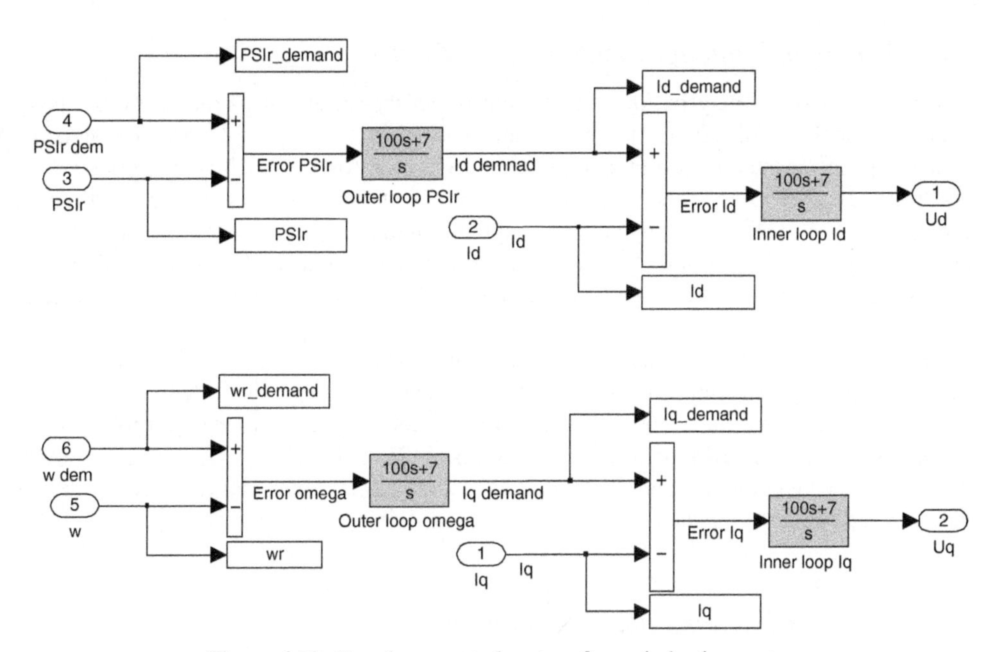

Figure 4.12 Two-loop control system for an induction motor

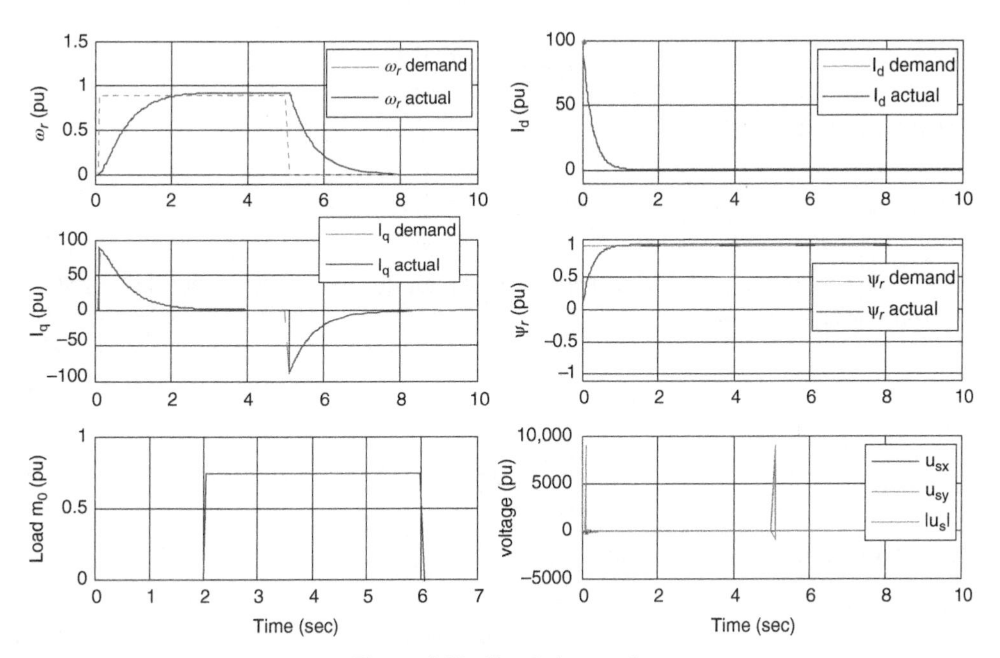

Figure 4.13 Simulation results

4.2.3 Direct and Indirect Field-Oriented Control

Motor state variables may be identified using direct measurement or indirectly by using machine models or observers. In most control schemes, the optimal variables selected as state space are stator current components and rotor (or stator) flux components. In the direct, as well as indirect, control methods, the stator current is measured directly. In direct control schemes, the machine's magnetic field is measured using special sensors located in the machine air gap. This method is not practical and can be used for laboratory investigations; however, for practicality, the machine flux is computed using special methods.

These two basic advanced methods are used for flux computation and control purpose. A first method is based on modeling the AC machine by its state space equations. A sinusoidal magnetic field in the machine is assumed. The machine models are defined as open-loop (e.g. stator voltage model or as closed-loop adaptive observers [1–9, 11–16]). The adaptive observers are receiving more attention than open-loop models because of their robustness against parameters variation and higher computation precision.

4.2.4 Rotor and Stator Flux Computation

The stator flux vector could be computed from the next well-known voltage models [1–9, 13, 14]:

$$\overline{\psi}_s = \int \left(\overline{u}_s - R_s\overline{i}_s\right) dt \tag{4.29}$$

$$\overline{\psi}_s = L_s\overline{i}_s + L_m\overline{i}_r \tag{4.30}$$

The rotor flux vector is

$$\overline{\psi}_r = L_r\overline{i}_r + L_m\overline{i}_s \tag{4.31}$$

After some algebra, we get the next expressions of rotor flux vector components in the stationary co-ordinate system ($\alpha\beta$):

$$\psi_{r\alpha} = \frac{L_r}{L_m}\left(\psi_{s\alpha} - \delta L_s i_{s\alpha}\right) \tag{4.32}$$

$$\psi_{r\beta} = \frac{L_r}{L_m}\left(\psi_{s\beta} - \delta L_s i_{s\beta}\right) \tag{4.33}$$

The rotor flux magnitude is computed as

$$|\psi_r| = \sqrt{\psi_{r\alpha}^2 + \psi_{r\beta}^2} \tag{4.34}$$

The angle of the rotor flux position γ_s is

$$\gamma_s = \tan^{-1}\left(\frac{\psi_{r\beta}}{\psi_{r\alpha}}\right) \tag{4.35}$$

The disadvantage of this method is that it is operationally limited at lower frequencies (below 3%) because of problems with integrating small signals at low frequencies.

An increase of lower speed operation range is possible using the current model [3, 6]:

$$\hat{\tau}'_\sigma \frac{d\hat{i}_s}{d\tau} = \frac{k_r}{\hat{r}_\sigma \hat{\tau}_r}(1 - \mathrm{j}\hat{\omega}_m \hat{\tau}_r)\hat{\psi}_r + \frac{1}{\hat{r}_\sigma} u_s \tag{4.36}$$

The model fails in operation at higher speeds because of differentiating higher signals.

The control systems designed using those open models are sensitive to stator and rotor resistances. For the voltage model, the effect of stator resistance is particularly visible at lower speed range. The method is simple but needs significant modifications in order to increase the operation precision and cancel the DC drift for lower speed operation [12–15].

4.2.5 Adaptive Flux Observers

In this section, we will discuss the basic description of flux closed-loop observers. More examples will be given in Chapter 9.

The full-order observer used for estimating state variables (mostly stator current and rotor flux components) is described as [12]

$$\frac{d}{dt}\hat{x} = A\hat{x} + Bu_s + G\left(i_s - \hat{i}_s\right) \tag{4.37}$$

$\hat{x}$ and $\hat{i}_s$ are the estimated values, and G is the feedback observer gain matrix.

One of the commonly used observers is the Luenberger observer [16], which is mainly used for flux and stator current computation. The equations representing this type of observer are development of the machine model discussed in Chapter 2. The equations may be presented as [15]

$$\frac{d\hat{i}_{s\alpha}}{dt} = a_1 \cdot \hat{i}_{s\alpha} + a_2 \cdot \hat{\Psi}_{r\alpha} + \omega_r \cdot a_3 \cdot \hat{\Psi}_{r\beta} + a_4 \cdot u_{s\alpha} + k_i\left(i_{s\alpha} - \hat{i}_{s\alpha}\right) \tag{4.38}$$

$$\frac{d\hat{\psi}_{s\beta}}{d\tau} = a_1 . \hat{i}_{s\beta} + a_2 . \hat{\psi}_{s\beta} - \omega_r . a_3 . \hat{\psi}_{s\alpha} + a_4 . \hat{i}_{s\beta} + k_i\left(i_{s\beta} - \hat{i}_{s\beta}\right) \tag{4.39}$$

$$\frac{d\hat{\psi}_{r\alpha}}{d\tau} = a_5 . \hat{\psi}_{r\alpha} - \omega_r . \hat{\psi}_{r\beta} + a_6 . \hat{i}_{s\alpha} + k_{f1}\left(i_{s\alpha} - \hat{i}_{s\alpha}\right) - k_{f2}\left(i_{s\beta} - \hat{i}_{s\beta}\right) \tag{4.40}$$

$$\frac{d\hat{\psi}_{r\beta}}{d\tau} = a_5 . \hat{\psi}_{r\beta} + \omega_r . \hat{\psi}_{r\alpha} + a_6 . \hat{i}_{s\beta} + k_{f2}\left(i_{s\alpha} - \hat{i}_{s\alpha}\right) + k_{f1}\left(i_{s\beta} - \hat{i}_{s\beta}\right) \tag{4.41}$$

Here, $u_{s\alpha}$, $u_{s\beta}$ are voltage components; $i_{s\alpha}$, $i_{s\beta}$ are measured current components; $\hat{i}_{s\alpha}, \hat{i}_{s\beta}$ are estimated current components, with $\hat{\Psi}_{r\alpha}, \hat{\Psi}_{r\beta}$ as estimated rotor flux components; k_i, k_{f1}, k_{f2} are the observer gains. Based on the above differential equations, the Luenberger observer is shown in Figure 4.14.

The Simulink software file of vector control of IM model is [IM_VC_Luenberger]. The motor parameters are located in the model initialization function.

4.2.6 *Stator Flux Orientation*

Stator flux-oriented control (Figure 4.15) is less sensitive to machine parameters than rotor field orientation [11, 13, 17]. The usual method for computing stator flux is the open-loop voltage model. Such equations are sensitive to stator resistance, which is dominant only at lower speeds [6]. The method enables producing more torque in the field weakening region than rotor field oriented control (RFO), as will be discussed in the next section.

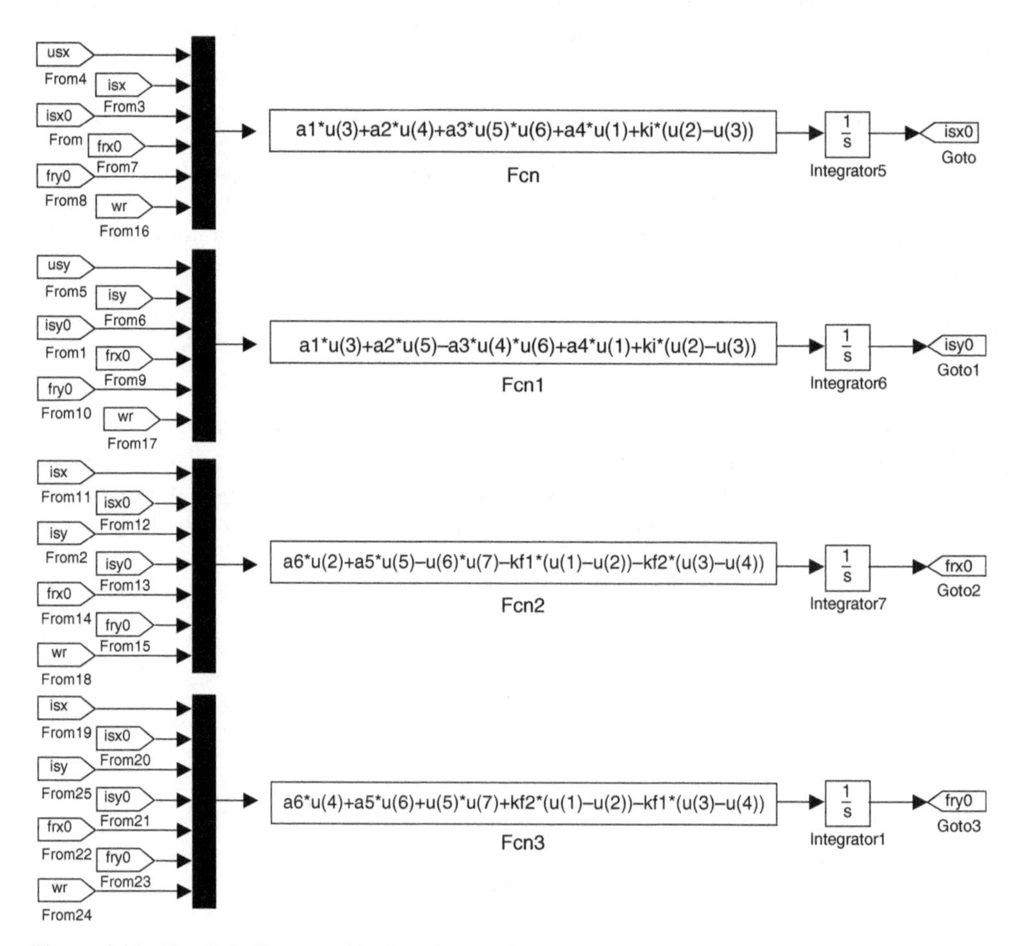

Figure 4.14 Simulink diagram of the Luenberger observer for stator currents and rotor fluxes estimator

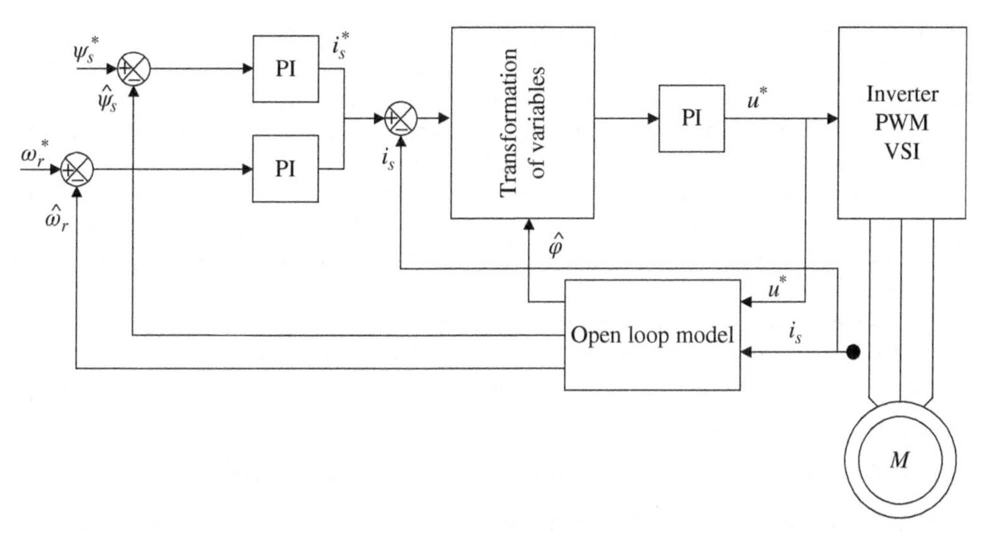

Figure 4.15 Stator flux-oriented vector control

The use of the stator field oriented control (SFO) scheme in field weakening is an excellent solution for many existing problems. This control principle gives a solution for decoupled control at field weakening. The operation at the absolute voltage limit of the inverter for maximum torque production eliminates the voltage margin required by the current controllers to adjust the respective current components i_d and i_q. Decoupled control of torque and flux would then become difficult in a classical vector control scheme [17, 18].

4.2.7 Field Weakening Control

One of the important applications of the machine is its running at the field weakening region, in which it operates at speeds higher than the rated one [17–24]. Such an operation is required in specific applications, such as spindle and traction drives.

The reason for flux weakening (decreasing) is that the magnetic field of the machine operating above rated speed level would be decreased due to the limit in the machine's voltage capability, which is imposed by a stator winding voltage limit and the DC link voltage [17–24].

To increase the produced torque to a maximum level in the field weakening region, it is essential to properly adjust the machine's magnetic field by maintaining the maximum voltage and maximum current. The loss of torque and power in the case of not properly adjusting the machine flux (e.g. in the case of $1/\omega$ method) is up to 35% [18–23].

Therefore, the machine flux should be weakened to such a level that it would guarantee a maximum possible torque at the whole speed range.

Field weakening control can be categorized into three methods [17–24], including:

1. adjustment of the machine flux in inverse proportion to speed ($1/\omega$);
2. forward control of the flux based on simplified machine equations; and
3. closed-loop control of the stator voltages to keep a maximum level.

For the conventional control method ($1/\omega$), the flux is established inversely proportional to the motor speed. The DC link voltage cannot be fully utilized for maximum torque production, which means this method is not able to provide the maximum torque in the field weakening region. The reason is that enough of the voltage margin needed to control the stator current cannot be achieved [17, 18].

The second method is based on simplified machine equations, making it parameter dependent. It gives reasonable results, assuming that machine parameters are known. Indeed, this type of control is not a preferable and optimal way for maximum torque control at the field weakening region.

The voltage control field weakening method ensures maximum torque production at the whole field weakening region when in a steady state. The method provides maximum required torque during a steady state, and is not dependent on motor parameters and DC link voltage.

4.2.7.1 Current and Voltage Limits for Maximum Torque Production

Current limiting during an overload condition in the field weakening must be controlled, in order to limit the torque producing current component i_q, while giving the priority to the flux producing component i_d. The i_q must be reduced to keep the total stator current within the maximum level imposed by the inverter and machine's capabilities. Hence, the optimum flux and maximum torque are achieved at field weakening [18].

The limit for the maximum voltage is reached when the system operates at maximum modulation index [18]. The inverter DC link voltage sets the maximum stator voltage limit. The priority in the voltage is for the d component $\boldsymbol{u}_d$, responsible for machine excitation to insure the optimal flux and hence the maximum possible torque. The limiting condition is $u_a \leq \sqrt{u_{s\,max}^2 - u_d^2}$, where $u_{s\,max}$ is limited voltage by inverter DC bus [18, 20, 22–24].

The block diagram for the control scheme at field weakening is shown in Figure 4.16 and is widely described in [18]. The system operates with a stator field or rotor field orientation. An open-loop voltage model is used for estimating the machine flux components and transformation angle and even the rotor angular speed ω_r.

4.3 Vector Control of Double Fed Induction Generator (DFIG)

4.3.1 Introduction

The DFIG is a rotor-wound three-phase induction machine that is connected to the AC supply from both stator and rotor terminals (Figure 4.17). The stator windings of the machine are connected to the utility grid without using power converters, and the rotor windings are fed by an active front-end converter. The machine can be fed by current or voltage source inverters with controlled voltage magnitude and frequency [12, 25–32].

In the control schemes of double fed induction motor (DFIM), two output variables in the stator side are generally defined and used for the control. These defined variables could be electromagnetic torque and reactive power, active and reactive powers, or stator voltage and its frequency. Different structures are used to control each pair of such variables.

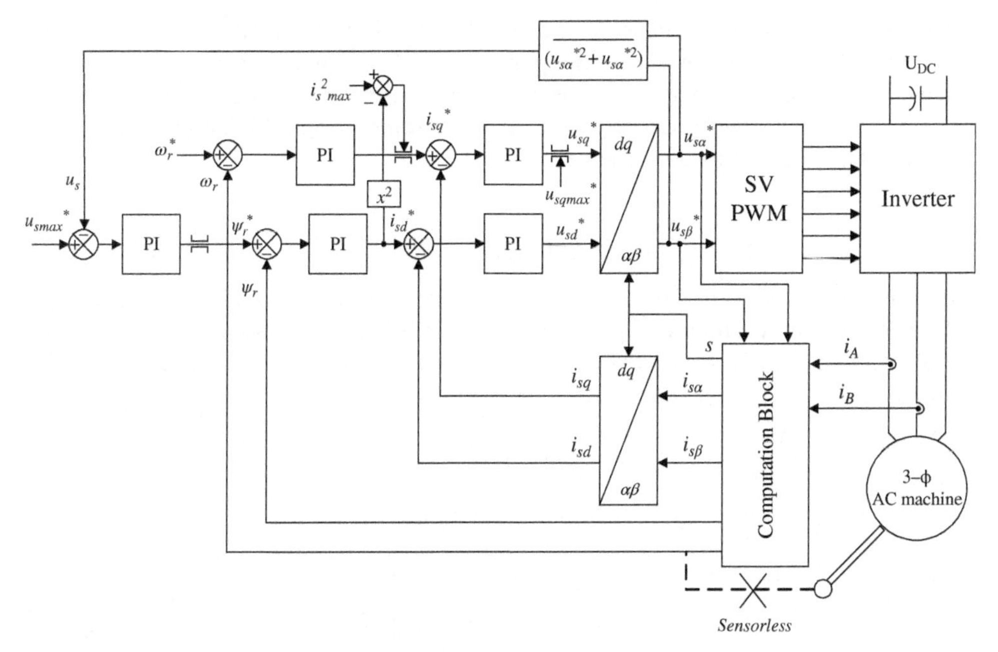

Figure 4.16 Induction motor field weakening system

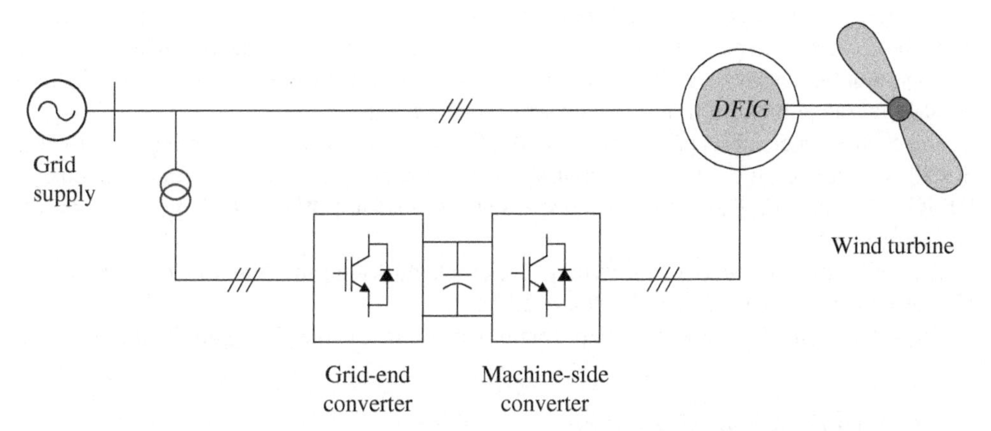

Figure 4.17 General view of DFIG connected to wind system and utility grid

The DFIM is popular and widely adopted for high power wind generation systems and other types of generators with similar variable speed high power sources (e.g. hydro systems). The advantage of using this type of machine is that the required converter capacity is much lower (up to three times lower) than those that connect the converter to the stator side [12, 25–27, 32]. Hence, the costs and losses in the conversion system are drastically reduced [1].

DFIG can be used either in an autonomous generation system (stand-alone) or in parallel with the grid. If the machine is working autonomously, the stator voltage and frequency are selected as the controlled signals. However, when the machine is connected to the infinite bus, the stator voltage and frequency are dictated by the system. In such situations, the controlled variables are the active and reactive powers [12, 27–32]. Indeed, there are different types of control strategies for this type of machine; however, the most popular is vector control, which has different orientation frames similar to the squirrel cage IM, but the most popular of these is the stator orientation scheme.

4.3.2 Vector Control of DFIG Connected with the Grid (αβ Model)

The mathematical model of DFIM is similar to the squirrel cage induction machine, the only difference being that the rotor voltage is not zero. The machine model is presented in Chapter 2.

The active and reactive powers of stator and rotor circuits are needed in the control of DFIG and may be described in per unit as [32, 33]

$$P_s = \left(v_{ds}i_{ds} + v_{qs}i_{qs}\right) \tag{4.42}$$

$$Q_s = \left(v_{qs}i_{ds} - v_{ds}i_{qs}\right) \tag{4.43}$$

$$P_r = \left(v_{dr}i_{dr} + v_{qr}i_{qr}\right) \tag{4.44}$$

$$Q_r = \left(v_{qr}i_{dr} - v_{dr}i_{qr}\right) \tag{4.45}$$

The DFIG system with vector control scheme is shown in Figure 4.18 and consists of a DFIG model presented in the $\alpha\beta$ frame and control system part. All variables and parameters are in per unit unless otherwise specified. In practice, vector control of this machine is similar to the vector control of the squirrel cage IM with a difference in controlled variables. The control depends on vectors transformation from three phase to rotating frame.

The control scheme was programmed in MATLAB/Simulink, as shown in Figure 4.19. The Simulink software file of vector control of double fed induction generator scheme is [DFIG_VC_-model]. The scheme parameters are located in [DFIG_VC_init], which should be first executed.

The DFIG model consists of different blocks related to the machine model, computation blocks, and control part, as will be described below.

The machine model in the stationary frame is given in Chapter 2 (Figure 2.10).

4.3.3 Variables Transformation

4.3.3.1 Stator to Rotor Transformation

The stator parameters (flux, current, and voltage) are converted into the rotor reference frame by using the following transformation:

$$\psi_{sxR} = \psi_{sx}\cos\gamma_{fir} + \psi_{sy}\sin\gamma_{fir} \tag{4.46}$$

$$\psi_{syR} = \psi_{sy}\cos\gamma_{fir} - \psi_{sx}\sin\gamma_{fir} \tag{4.47}$$

These equations are modeled in Figure 4.20.

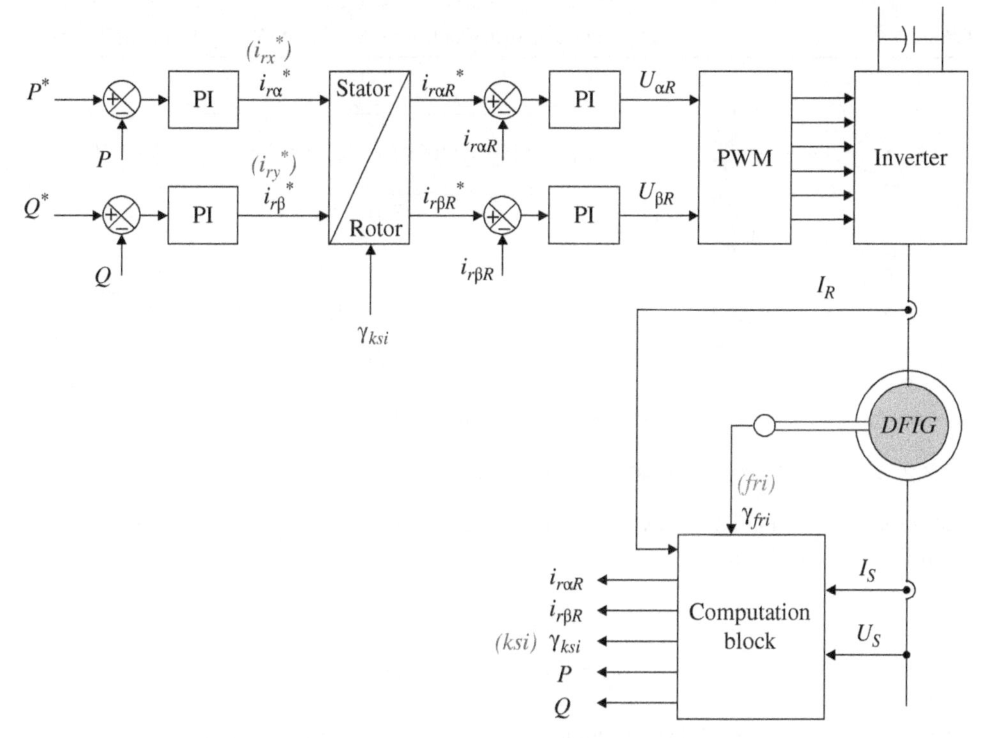

Figure 4.18 Block diagram of the DFIG control system

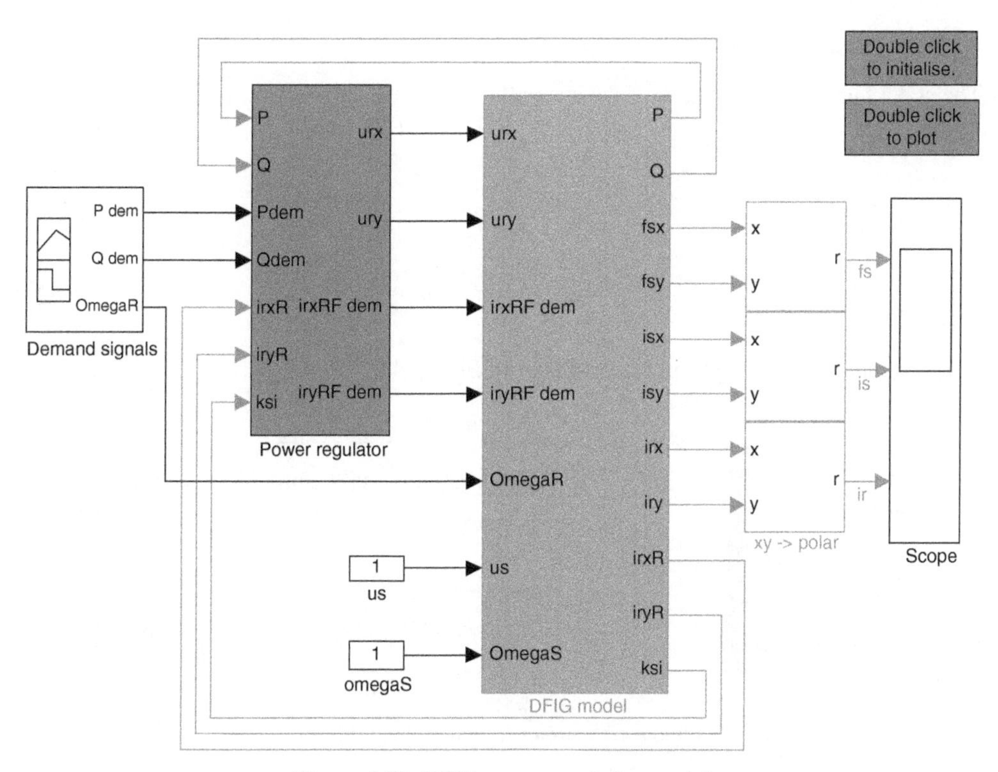

Figure 4.19 DFIG power regulation model

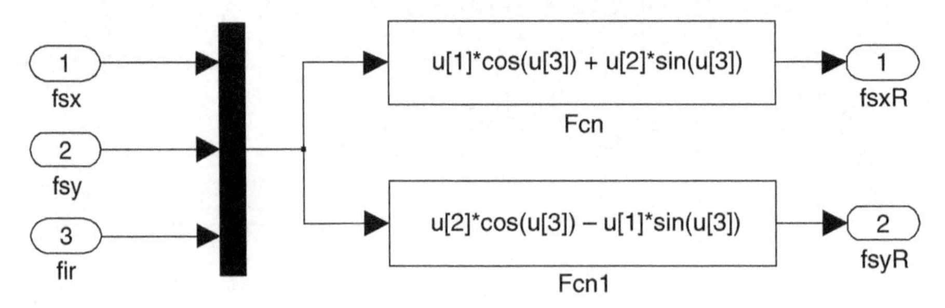

Figure 4.20 Stator to rotor transformation

4.3.3.2 Rotor to Stator Transformation

The rotor parameters (flux and current) are converted into the stator reference frame by using the following transformation:

$$i_{sxR} = i_{sx} \cos \gamma_{fir} - i_{sy} \sin \gamma_{fir} \tag{4.48}$$

$$i_{syR} = i_{sy} \cos \gamma_{fir} + i_{sx} \sin \gamma_{fir} \tag{4.49}$$

The above equations are shown in Figure 4.21. The angle γ_{fir} is in the rotor angular position (fir in MATLAB/Simulink), which is shown in Chapter 2.

4.3.3.3 Stator Current Relationship

The stator currents are function of the stator flux and rotor current following the two axes as follows:

$$i_{sx} = \frac{\phi_{sx}}{L_s} - \frac{L_m}{L_s} i_{rx} \tag{4.50}$$

$$i_{sy} = \frac{\phi_{sy}}{L_s} - \frac{L_m}{L_s} i_{ry} \tag{4.51}$$

These equations are programmed, as shown in Figure 4.22.

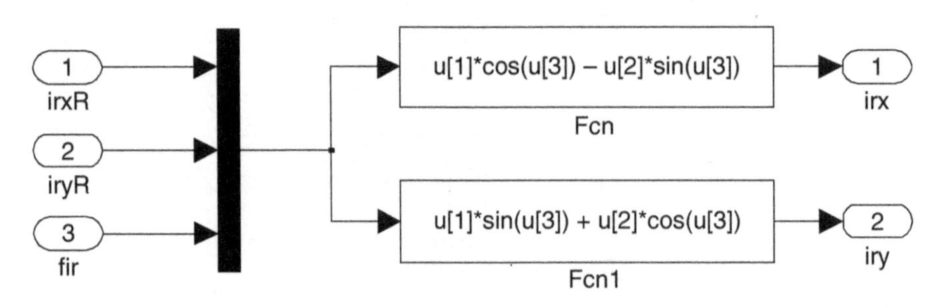

Figure 4.21 Rotor to stator transformation

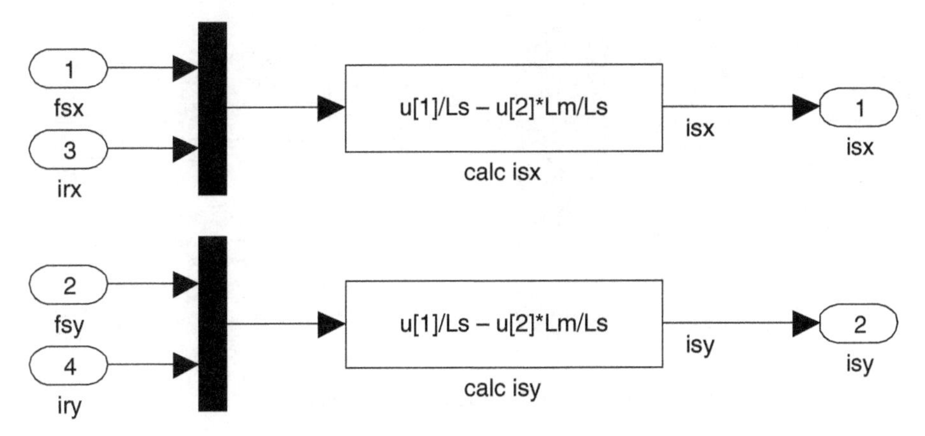

Figure 4.22 Stator current calculations

4.3.3.4 Power Calculations

The active and reactive power components are given as

$$P = u_{sx} i_{sxF} + u_{sy} i_{syF} \tag{4.52}$$

$$Q = u_{sy}.i_{sxF} - u_{sx}.i_{syF} \tag{4.53}$$

The index F denotes the filtered value. The filtration is not essential for the control; however, it provides smoother waveforms. Additional filters may be added to other variables if necessary. These equations are modeled, as shown in Figure 4.23.

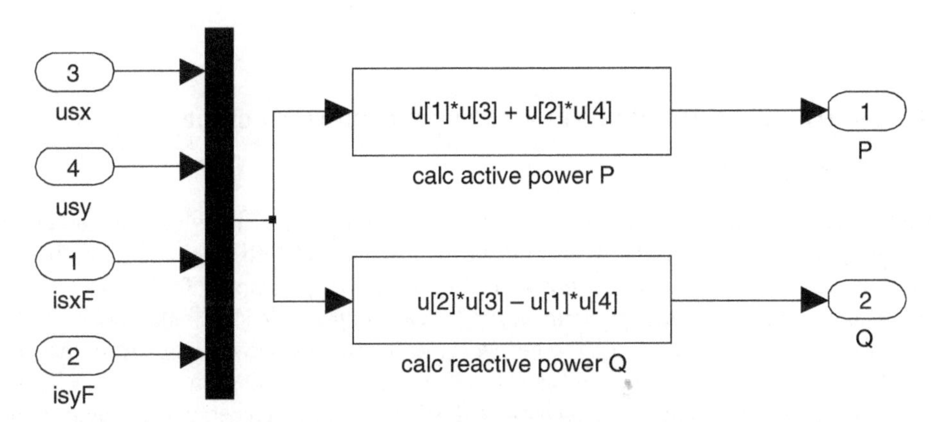

Figure 4.23 Active and reactive power calculations

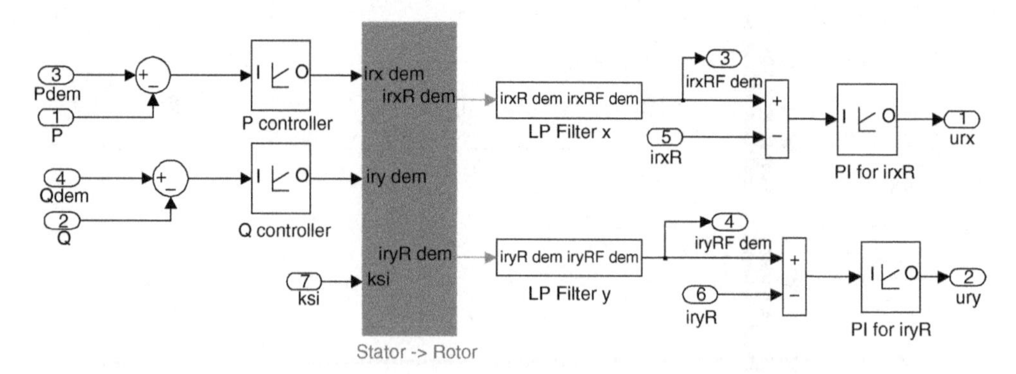

Figure 4.24 Cascaded control system model

4.3.3.5 Control System

Cascade control is used for power regulation. The PI control for active and reactive power provides the demand for an inner current loop, as shown in Figure 4.24. The active power controller commands the rotor current in the α frame (i_{rx}) in the stator reference frame. The reactive power controller commands the b component of this current. The angle γ_{ksi} is the angle between stator flux and rotor speed (*ksi* in MATLAB/Simulink), as shown in Chapter 2. After transformation from stator flux reference frame to rotor speed frame, the $\alpha\beta$ rotor current components i_{rxR} and i_{ryR} are obtained. Two additional PI controllers are used to generate the rotor voltage for the PWM voltage inverter. To simplify this, we neglect this inverter and command those two voltages directly to the DFIG model.

4.3.4 Simulation Results

The simulation result for power regulation with step changes in speed, which controls the active and reactive powers for different speeds, is shown in Figures 4.25 and 4.26.

4.4 Control of Permanent Magnet Synchronous Machine

4.4.1 Introduction

PMSM have received increased attention in industry due to their mature material and design technology. Their high torque to inertia ratio, power density, and high efficiency make them an attractive alternative to industry's workhorse IM for many applications (Table 4.1).

Depending on the dynamic requirements and speed control precision, either brushless DC motor (BLDC) or PMSM are used. The former is mostly used with lower amounts of magnets. The latter is used for high requirements and has higher amounts of magnets. The BLDC is also named the switched permanent magnet motor (SPMM) or trapezoidal permanent magnet motor (TPMM). This type of motor is fed using a square current waveform; hence, the character of the

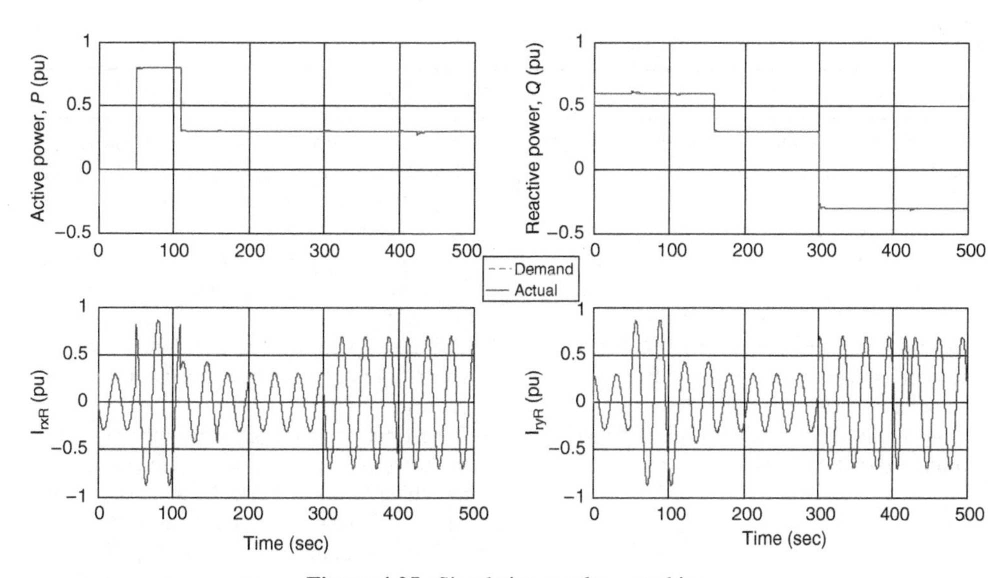

Figure 4.25 Simulation results – tracking

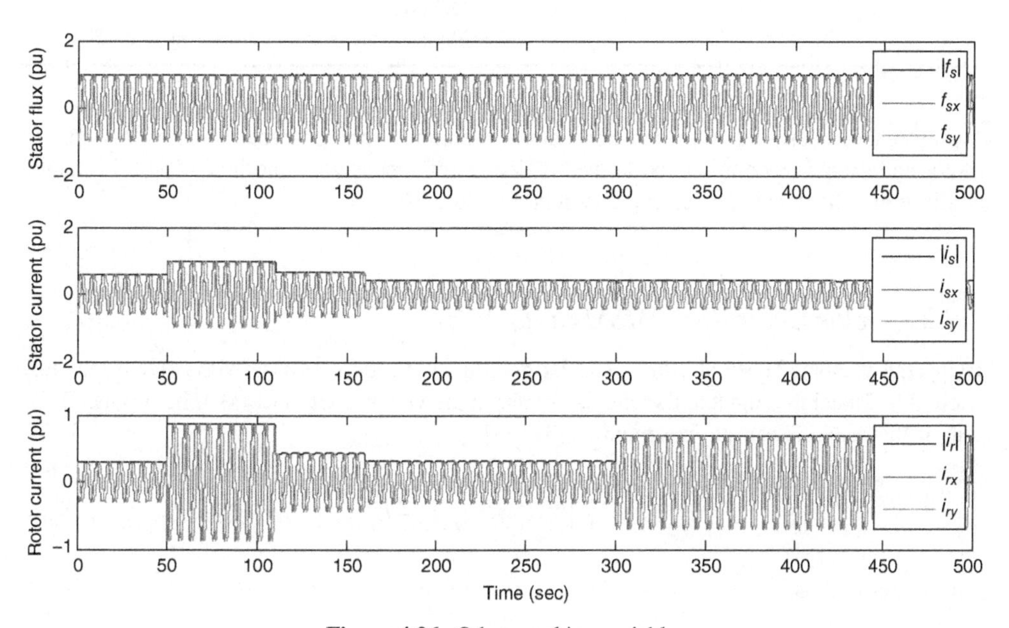

Figure 4.26 Other machine variables

back emf is trapezoidal. The PMSM is fed by a sinusoidal current waveform and is widely used in industry.

In this chapter, we will focus on controlling the motor without PWM. We will only be using PI controllers of the commanded currents.

Table 4.1 Advantages and disadvantages of PMSM comparing with induction motor

Advantages	Disadvantages
• Higher torque for the same dimension. For the same power, the dimension is lower by almost 25%; • Lower weight for the same power, around 25%; • Lower rotor losses, which results in higher efficiency of up to 3%; • Decreased motor noise by 3 dB, which causes that the rotor to run smoother. This helps in reducing the harmonics that result from an irregularity of the air gap; • Higher inductance inside the magnetic path, around 1.2[T].	• In the case of inverter faults, it is not possible to reduce the magnetic field by reducing the torque surge, which forces a use of switch between the motor and the inverter; • It is possible to connect only one motor to the inverter; therefore group motor work is not possible; • The use of permanent magnet forces using enclosed motor housing complicates the cooling process; • Importance of using high switching frequency of the power transistors, which increases the switching losses and also the inverter structure, should be changed, e.g. trapezoidal inverter; • An increase of the inverter voltage in the case of inverter blocking during motor running. This results in the necessity of using chopper in the DC link; • Higher price of the motor (around 10–20%).

Any reader interested in more details about the control and characteristics of this type of motor and drive is recommended to use the list of references given at the end of this chapter and in other chapters (particularly Chapters 2, 9, and 10).

4.4.2 Vector Control of PMSM in dq Axis

Using the vector of PMSM allows similar dealings with the motor as with a DC separately excited DC machine. Indeed, the idea is similar to the vector control of IM. The produced torque of the machine can be presented as [34–38]

$$t_e = \frac{3p}{2}\left[\Psi_f i_q + i_d i_q \left(L_d - L_q\right)\right] \tag{4.54}$$

and in per unit as

$$t_e = \psi_f i_q + \left(L_d - L_q\right) i_d i_q \tag{4.55}$$

For surface mounted PMSM, the d and q inductances are equal $(L_d = L_q)$. The produced torque in such a situation is

$$t_e = \psi_f i_q \tag{4.56}$$

Hence the torque can be expressed as

$$t_e = \psi_f i_s \sin\beta \tag{4.57}$$

The torque obtains maximum value for β (torque angle) equal to 90 degrees for a given value of stator current. This gives maximum torque per ampere and hence a higher efficiency [37].

A general scheme of the PMSM model in the *dq* axis is shown in Figure 4.27. It consists of a PMSM model and the control system in the *dq* axis connected with the rotor speed control loop. All the parameters and variables are presented in per unit unless otherwise specified.

Keeping zero i_{sd} is the most common control for permanent magnet machines. As explained earlier, this helps protect the machine against under- or over-excited conditions. Therefore, in the MATLAB/Simulink model, we start from commanding zero *d* axis stator current.

The Simulink model for the above scheme is shown in Figure 4.28. The Simulink software file of vector control of interior PMSM scheme is [PMSM_VC_dq]. The motor parameters are located in the model initialization function.

PMSM in stationary and rotating frame are given in Chapter 2.

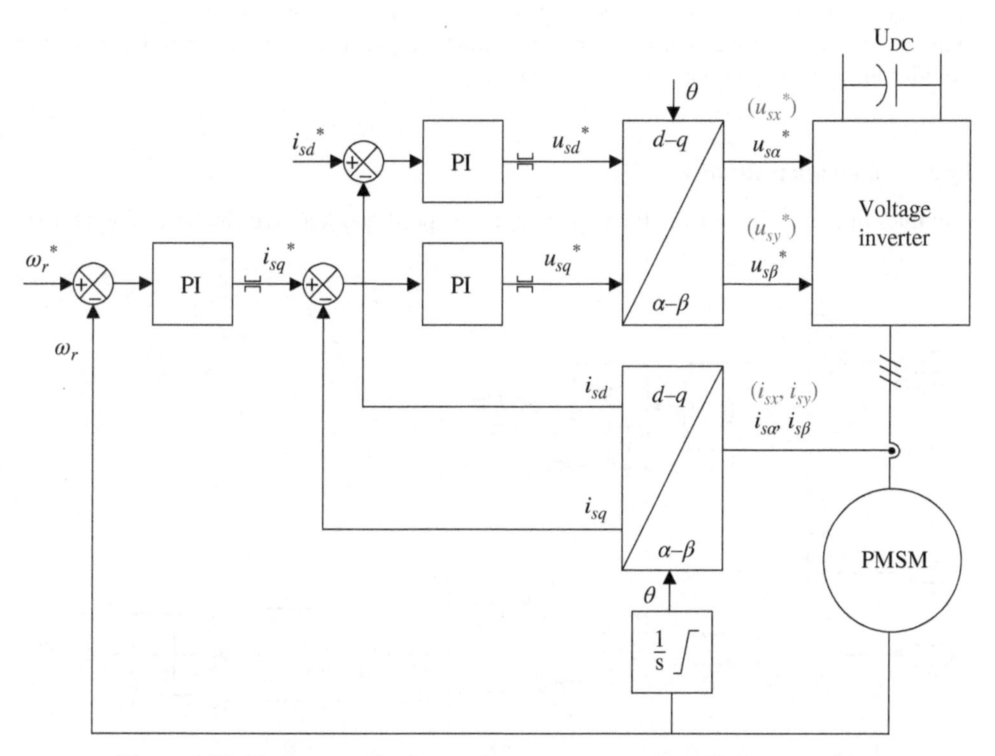

Figure 4.27 Vector control scheme of permanent magnet synchronous machine

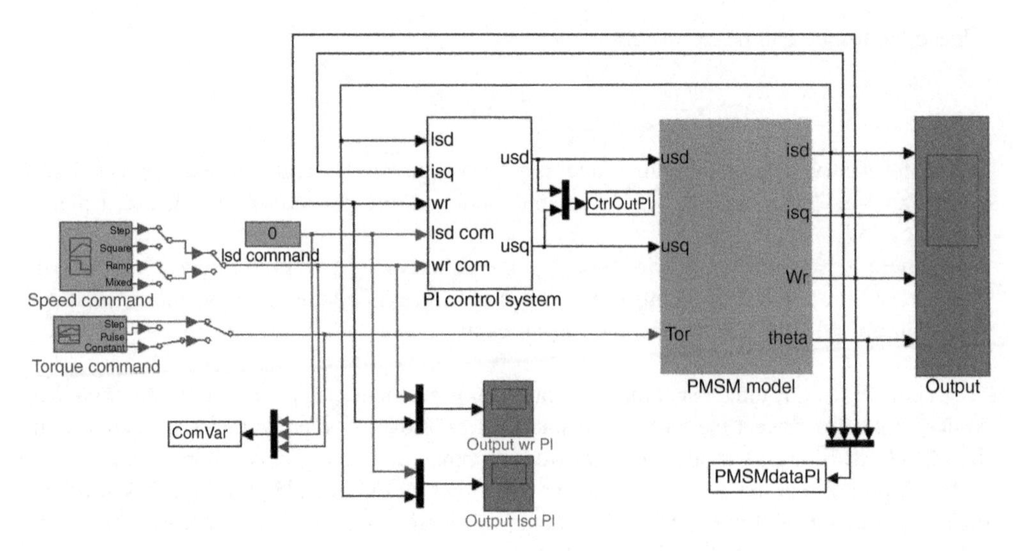

Figure 4.28 Vector control of PMSM

4.4.2.1 Control System with Cascaded PI Controllers

PI control loop for *d*-current is used to provide the control input voltage u_{sd}, as shown in Figure 4.29. The rotor speed loop controller calculates the demand for *q*-current, which in turn provides the voltage component in axis *q* axis u_{sq}.

4.4.2.2 Simulation Results

Examples of simulation results for step changes in speed and load are shown in Figure 4.30.

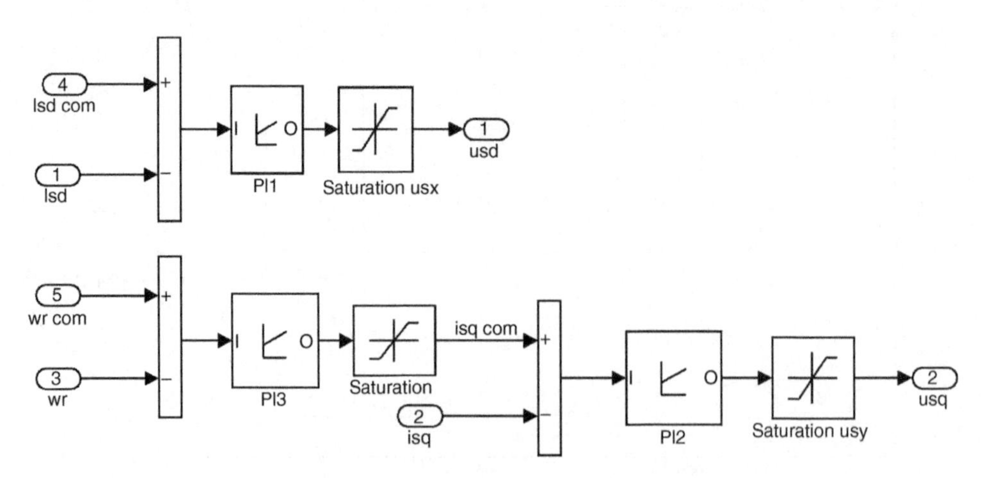

Figure 4.29 PI control system for PMSM

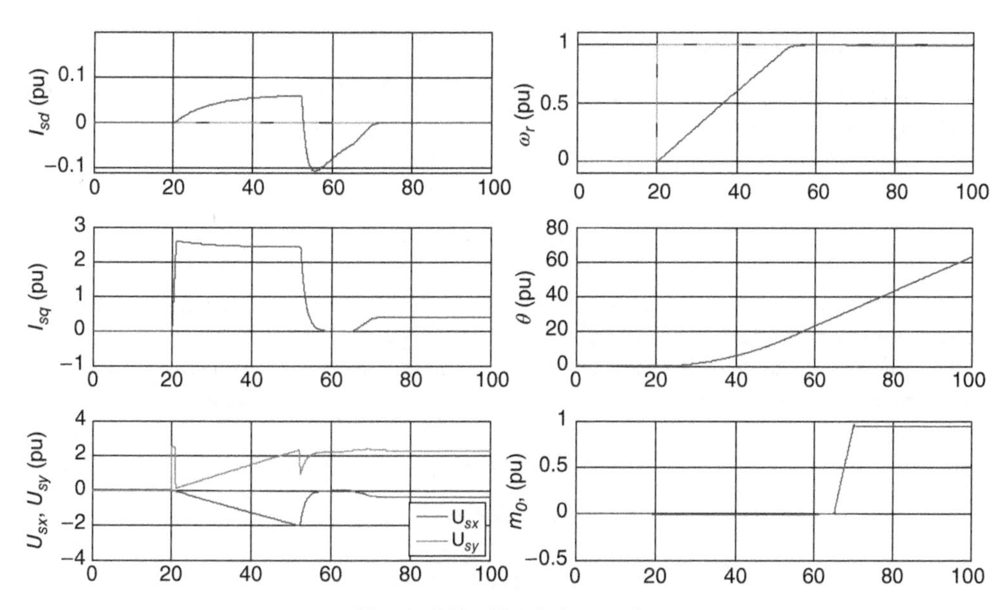

Figure 4.30 Simulation results

4.4.3 Vector Control of PMSM in α–β Axis Using PI Controller

The control of the PMSM model in the α–β (x–y) axis is shown in Figure 4.31. It consists of the PMSM model and the control system model. All the models are in per unit unless otherwise specified. The machine model in stationary α–β reference frame is given in Chapter 2.

The Simulink software file of vector control of interior PMSM scheme is [PMSM_VC_alpha_beta]. The scheme parameters are located in [PMSM_param], which should be first executed.

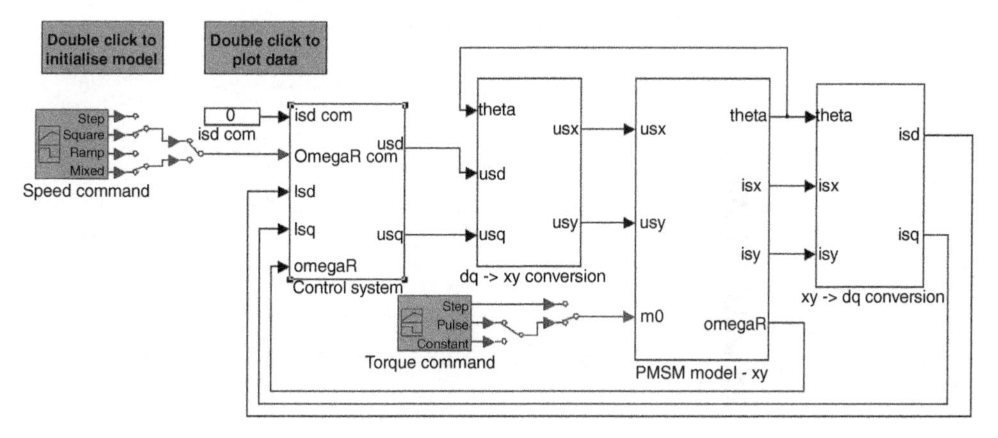

Figure 4.31 Vector control of PMSM

4.4.3.1 Variables Transformations

The currents in the x–y (α–β) axis are converted into the d–q frame of reference by using the following transformation:

$$i_{sd} = i_{sx} \cos\theta + i_{sy} \sin\theta \tag{4.58}$$

$$i_{sq} = -i_{sx} \sin\theta + i_{sy} \cos\theta \tag{4.59}$$

These equations are modeled, as shown in Figure 4.32. The currents in the d–q axis are used for control purposes. The controller provides control input in the d–q axis.

The control voltage in the d–q axis is then converted into the x–y axis by using the following transformation and then applied to the machine in the x–y reference frame:

$$u_{sx} = u_{sd} \cos\theta - u_{sq} \sin\theta \tag{4.60}$$

$$i_{sy} = u_{sd} \sin\theta + u_{sq} \cos\theta \tag{4.61}$$

These above equations are modeled, as shown in Figure 4.33.

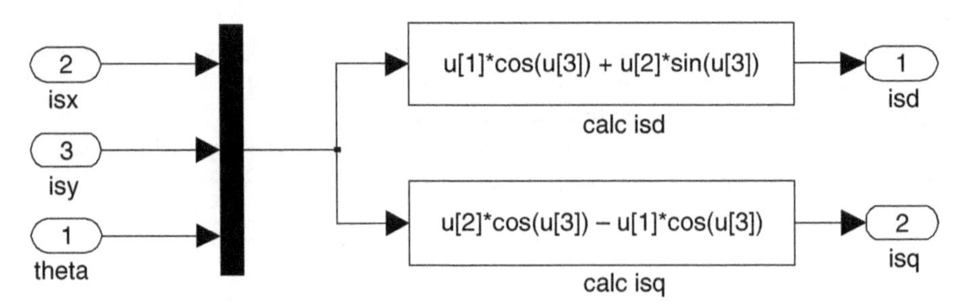

Figure 4.32 x–y to d–q transformation

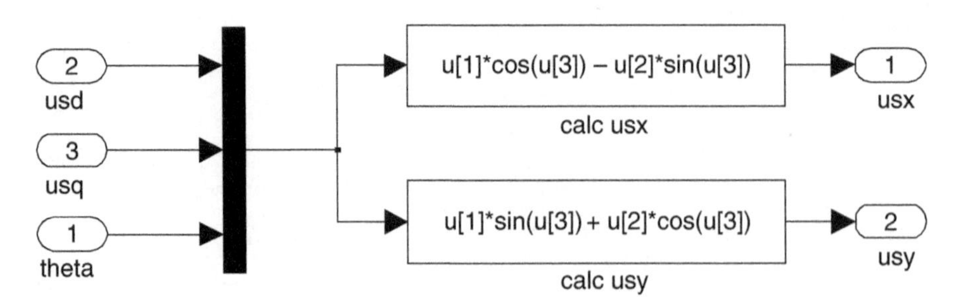

Figure 4.33 d–q to x–y transformation

4.4.3.2 Control System

The control scheme is shown in Figure 4.34. The PI controller for the *d*-current provides the control input u_{sx} and a cascaded PI control for the speed is shown in Figures 4.35–4.37. The speed loop controller calculates the demand for the *q*-current loop, which in turn provides the other control input u_{sy}.

The transformation and individual PI controllers are shown in Figures 4.32–4.35.

4.4.3.3 Simulation Results

The simulation results for step demand in speed are shown in Figures 4.38 and 4.39.

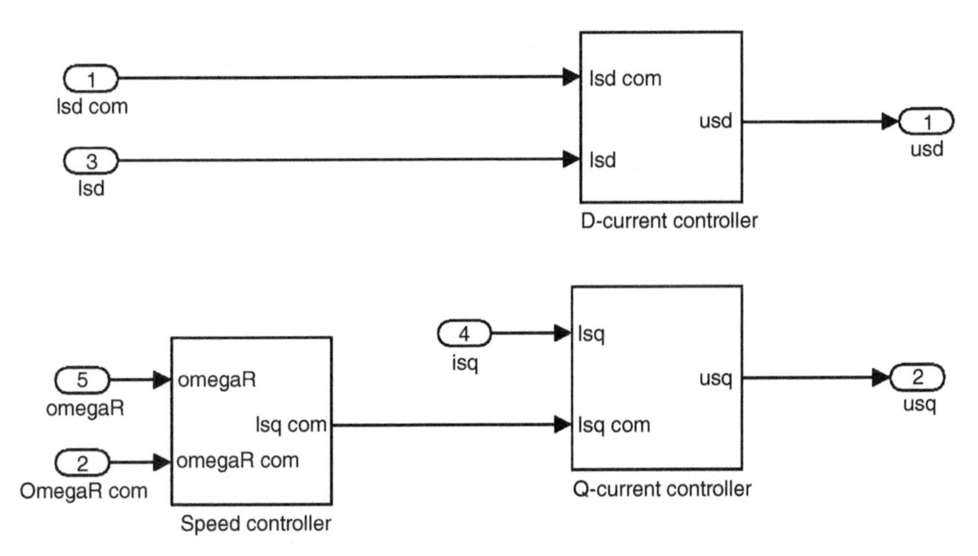

Figure 4.34 Control system model

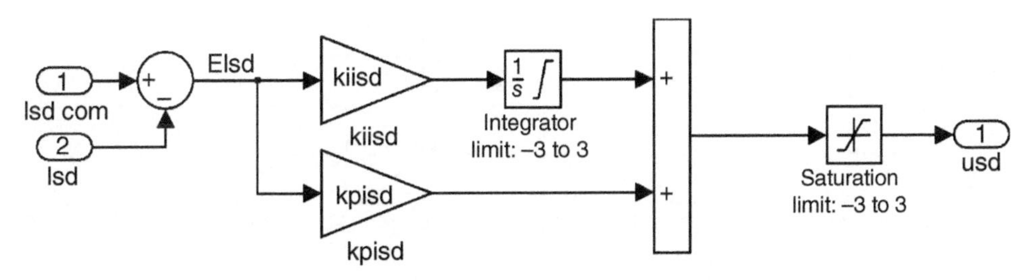

Figure 4.35 *d*-Channel PI controller

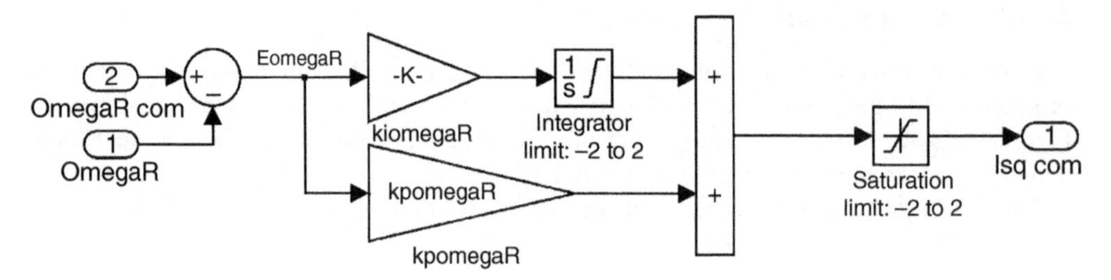

Figure 4.36 Outer loop PI controller for speed control

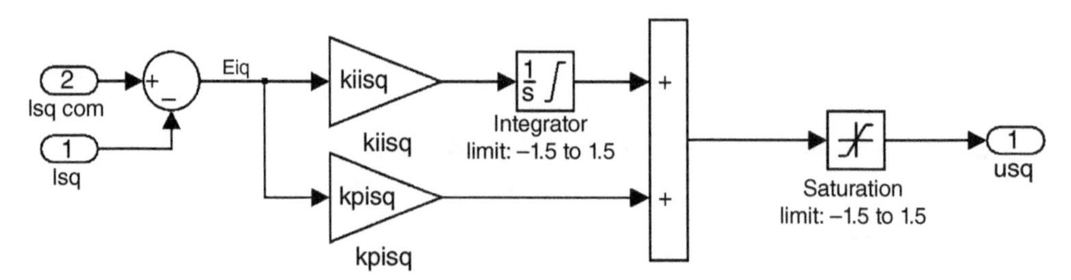

Figure 4.37 q-Channel PI controller

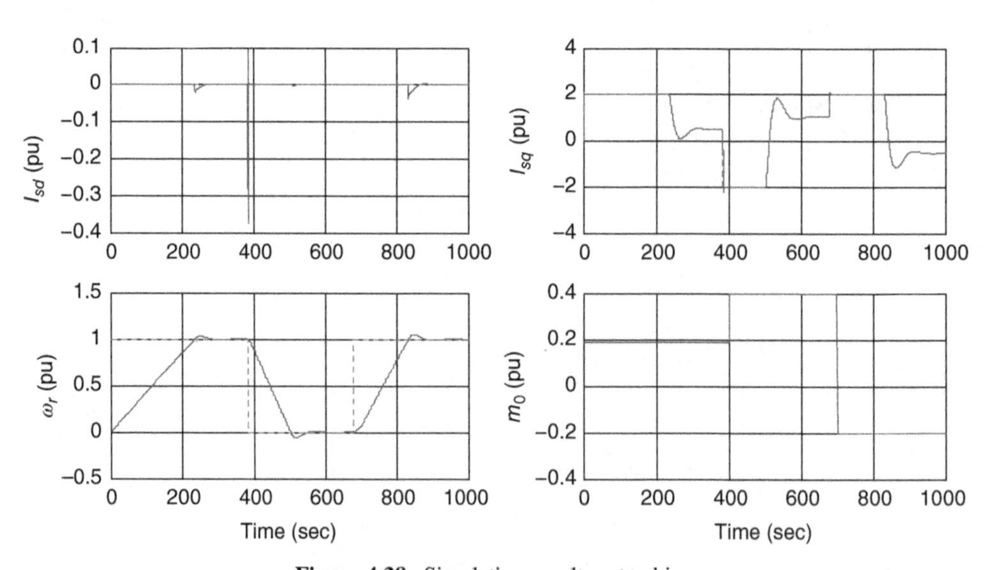

Figure 4.38 Simulation results – tracking

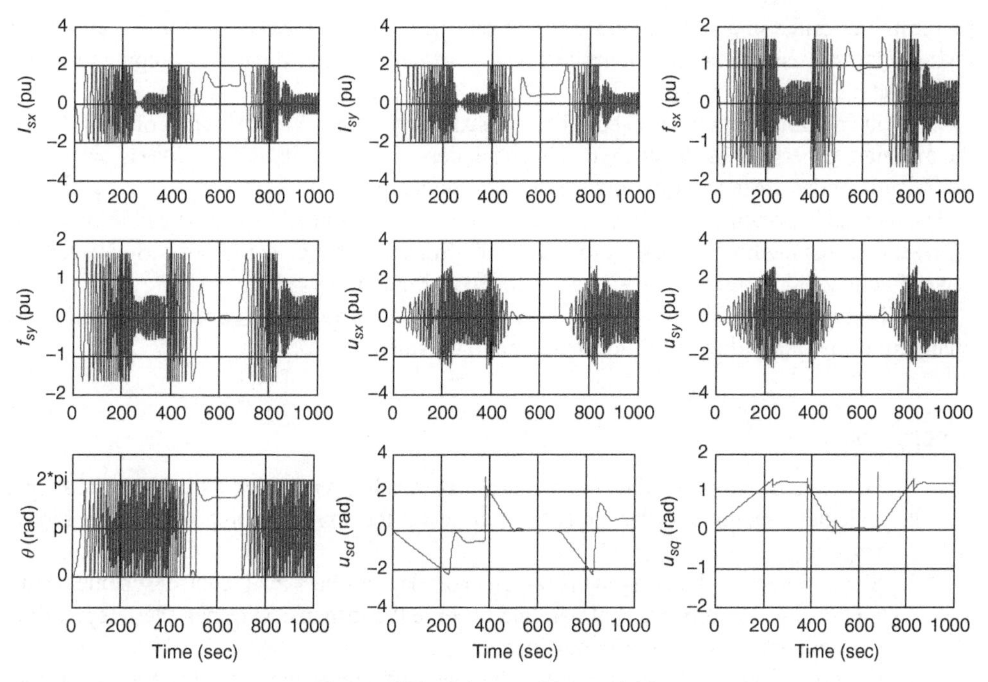

Figure 4.39 Other machine variables

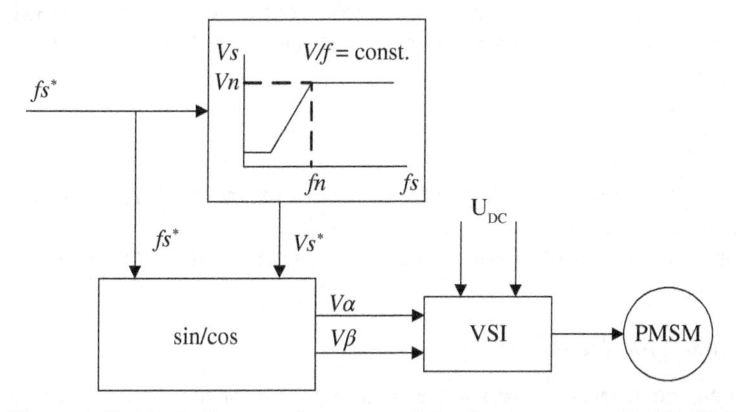

Figure 4.40 Block diagram of constant volt by frequency control of PMSM

4.4.4 Scalar Control of PMSM

Scalar control of PMSM is a simple way of controlling this type of machine; however, it is imprecise and has a slow response similar to the induction machine [34]. Hence, in this section, we will not go into detailed equations, but will focus on the main characteristics and control scheme. The method is named constant volt per hertz or *V* by *f* (*V*/*f*), which is an open-loop control method operating on the voltage magnitude rather than on the vectors. It is easy to perform, but results in a slow response and imprecise values. Nevertheless, this technique for PMSM offers a sensorless control scheme. The rotor angular speed can be computed using a stator voltage frequency [34].

The voltage to frequency ratio should be constant (V/f = const.) to maintain constant flux in the machine. Otherwise the machine may reach under- or over-excitation conditions, which are not recommended from stability and economical points of view.

The method in principle is similar to the IM, which was explained earlier. The block diagram representing this method is shown in Figure 4.40. Stator resistance could not be neglected for lower frequencies; therefore, the speed is limited in the low frequency range.

One disadvantage of using this method is that the stator current is not controlled. This allows changing of the main flux in the machine above the rated value.

Exercises

1. Program the equation of the torque in all AC machines and give its value to the scope.
2. Calculate motor speed reverse and observe the current components and torque response?
3. See the effect of changing the parameters of PI controllers in all control schemes (hint: first change the proportional part and observe the response and then the integral time constant and observe the steady state error).
4. Observe what will happen if you command different values of i_d in motors control.
5. Program the scalar control scheme with speed control loop.

The Simulink software file of V by f vector control of interior PMSM scheme is [PMSM_scalar]. The scheme parameters are located in [PMSM_param], which should be first executed.

Additional Tasks

1. Research a comparison study between stator flux and rotor flux field-oriented control schemes.
2. Make a broad test of motors operation in term of starting, breaking, loading, reversing, etc.

Possible Tasks for DFIG

1. Command different powers while changing the rotor speed around the rated value (up to 30% change). Observe the active and reactive powers, motor currents, voltages, and fluxes.
2. Obtain the unity power factor out of the DFIG (hint: the reactive power should be zero).
3. Observe the effect of changing controllers' parameters on the control performance. As an example, test the effect of changing the active power parameters.

Questions

Q1: What is scalar control of motors?
Q2: What are a few drawbacks of scalar control?

Q3 Why would one choose field-oriented control over vector control?
Q4 Classify field-oriented control (vector control) methods.
Q5 State an advantage of using vector control.
Q6 State some drawbacks of vector control.
Q7 What are different types of reference frames used in vector control?

References

1. Mohan, N. (2001) *Electric Drives an Integrative Approach*, 2nd edn. MNPERE, Minneapolis.
2. Bose, B. K. (2006) *Power Electronics and Motor Drives: Advances and Trends*. Academic Press, Elsevier.
3. Vas, P. (1998) *Sensorless Vector and Direct Torque Control*. Oxford University Press, Oxford.
4. Novotny, D. W. and Lipo, T. A. (1996) *Vector Control and Dynamics of AC Drive*. Oxford University Press, Oxford.
5. Kazmierkowski, M. P., Krishnan, R., and Blaabjerg, F. (2002) *Control in Power Electronics-Selected Problems*. Academic Press Elsevier.
6. Leonhard, W. (2001) *Control of Electrical Drives*. Springer.
7. Krishnan, R. (2001) *Electric Motor Drives: Modeling, Analysis and Control*. Prentice Hall.
8. Chiasson, J. (2005) *Modeling and High Performance Control of Electric Machines*. Wiley.
9. Blaschke, F. (1971) Das Prinzip der Feldorientierung, die Grundlage fur Transvector-regelung von Drehfeld-maschine. *Siemens Z.*, 45, 757–760.
10. Kovacs, K. P. and Racz, I. (1959) *Transiente Vorgange in Wechselstrommachinen*. Akad. Kiado, Budapest.
11. Leonhard, W. (1991) 30 Years of space vector, 20 years field orientation, 10 years digital signal processing with controlled AC drives: a review. *EPE J.*, 1, 13–19.
12. Bogdan, M. Wilamowski, J., and Irwin, D. (2011) *The Industrial Electronics Handbook – Power Electronics and Motor Drives*. Taylor and Francis Group, LLC.
13. Holtz, J. and Khambadkone, A. (1991) Vector controlled induction motor drive with a self commissioning scheme. *IEEE Trans. Ind. Electr.*, 38(5), 322–327.
14. Hu, J. and Wu, B. (1998) New integrations algorithms for estimating motor flux over a wide speed range. *IEEE Trans. Power Electron.*, 13(3), 969–977.
15. Krzeminski, Z. (2008) Observer of induction motor speed based on exact disturbance model. Power Elect. Mot. Cont. Conf., EPE-PEMC, pp. 2294–2299.
16. Khalil, H. K. (2002) *Nonlinear Systems*. Prentice Hall.
17. Xu, X. and Novotny, D. W. (1992) Selection of the flux reference for induction machine drive in the field weakening region. *IEEE Trans. Ind. Appl.*, 28(6), 1353–1358.
18. Abu-Rub, H., Schmirgel, H., and Holtz, J. (2006) Sensorless control of induction motors for maximum steady-state torque and fast dynamics at field weakening. IEEE/IAS 41st Ann. Mtg., Tampa, FL.
19. Grotstollen, H. and Wiesing, J. (1995) Torque capability and control of a saturated induction motor over a wide range of flux weakening. *IEEE Trans. Ind. Electr.*, 42(4), 374–381.
20. Briz, F., Diez, A., Degner, M. W., and Lorenz, R. D. (2001) Current and flux regulation in field weakening operation of induction motors. *IEEE Trans. Ind. Appl.*, 37(1), 42–50.
21. Harnefor, L., Pietilainen, K., and Gertmar, L. (2001) Torque-maximizing field-weakening control: design, analysis, and parameter selection. *IEEE Trans. Ind. Electr.*, 48(1), 161–168.

22. Bünte, A., Grotstollen, H., and Krafka, P. (1996) Field weakening of induction motors in a very wide region with regard to parameter uncertainties. 27th Ann. IEEE Power Elect. Spec. Conf., 23–27 June, vol. 1, pp. 944–950.
23. Bünte, A. (1998) Induction motor drive with a self-commissioning scheme and optimal torque adjustment (in German). PhD thesis, Paderborn University, Germany.
24. Kim, S. H. and Sul, S. K. (1997) Voltage control strategy for maximum torque operation of an induction machine in the field weakening region. *IEEE Trans. Ind. Electr.*, 44(4), 512–518.
25. Lin, F.-J., Hwang, J.-C., Tan, K.-H., Lu, Z.-H., and Chang, Y.-R. (2010) Control of doubly-fed induction generator system using PIDNNs. 2010 9th Int. Conf. Mach. Learn. Appl., pp. 675–680.
26. Pena, R., Clare, J. C., and Asher, G. M. (1996) Doubly fed induction generator using back-to-back PWM converters and its application to variable-speed wind-energy generation. *IEEE Proc. Electr. Power Appl.*, 143(3), 231–241.
27. Muller, S., Deicke, M., and De Doncker, R. W. (2002) Doubly-fed induction generator systems for wind turbines. *IEEE IAS Mag.*, 8(3), 26–33.
28. Pena, R., Clare, J. C., and Asher, G. M. (1996) A doubly fed induction generator using back-to-back PWM converters supplying an isolated load from a variable speed wind turbine. *IEEE Proc. Electr. Power Appl.*, 143(5), 380–387.
29. Forchetti, D., Garcia, G., and Valla, M. I. (2002) Vector control strategy for a doubly-fed stand-alone induction generator. Proc. 28th IEEE Int. Conf., IECON, vol. 2, pp. 991–995.
30. Jain, A. K. and Ranganathan, V. T. (2008) Wound rotor induction generator with sensorless control and integrated active filter for feeding nonlinear loads in a stand-alone grid. *IEEE Trans. Ind. Electr.*, 55(1), 218–228.
31. Iwanski, G. and Koczara, W. (2008) DFIG-based power generation system with UPS function for variable-speed applications. *IEEE Trans. Ind. Electr.*, 55(8), 3047–3054.
32. Bogalecka, E. (1993) Power control of a double fed induction generator without speed or position sensor. Conf. Rec. EPE, vol. pt. 8, ch. 50, pp. 224–228.
33. Akagi, H., Kanazawa, Y., and Nabae, A. (1998) *Theory of the Instantaneous Reactive Power in Three-Phase Circuits*. IPEC, Tokyo.
34. Stulrajter, M. Hrabovcova, V., and Franko, M. (2007) Permanent magnets synchronous motor control theory. *J. Electr. Eng.*, 58(2), 79–84.
35. Binns, K. J. and Shimmin, D. W. (1996) Relationship between rated torque and size of permanent magnet machines. *Electr. Mach. Drives IEE Proc. Electr. Power Appl.*, 143(6), 417–422.
36. Pillay, P. and Knshnan, R. (1988) Modeling of permanent magnet motor drives. *IEEE Trans. Ind. Electr.*, 35(4), 537–541.
37. Adamowicz, M. (2006) Observer of induction motor speed based on simplified dynamical equations of disturbance model. Proc. Doct. Sch. of Energy and Geotech., 16–21 January, Kuressaare, Estonia, pp. 63–67.
38. Pajchrowski, T. and Zawirski, K. (2005) Robust speed control of Servodrive based on ANN. IEEE ISIE, 20–23 June, Dubrovnik, Croatia.

5

Direct Torque Control of AC Machines

Truc Phamdinh
The Mind Lab, Auckland, New Zealand

5.1 Preliminary Remarks

Theoretical principles of direct torque control (DTC) for high-performance drives were introduced in the second half of the 1980s [1–3]. Compared with field-oriented control, with origins that date back to the beginning of the 1970s, DTC is a significantly newer concept. It took almost 20 years for vector control to gain acceptance by industry. In contrast, the concept of DTC has been taken on board by industry relatively quickly in only 10 years [4, 5]. While vector control predominantly relies on mathematical modeling of an induction machine, DTC makes direct use of physical interactions that take place within the integrated system of the machine and its supply. The DTC scheme adopts simple signal processing methods and relies entirely on the nonideal nature of the power source that is used to supply an induction machine within the variable speed drive system (two-level or three-level voltage source inverters, matrix converters, etc.). It can therefore be applied to power electronic converter-fed machines only. The on–off control of converter switches is used for the decoupling of the nonlinear structure of the induction machine [6]. The most frequently discussed and used power electronic converter in DTC drives is a voltage source inverter.

DTC takes a different look at the induction machine and the associated power electronic converter. First, it recognizes that regardless of how the inverter is controlled, it is by default a voltage source rather than a current source. Next, it dispenses with one of the main characteristics of the vector control indirect flux and torque control by means of two stator current components. In essence, DTC recognizes that if flux and torque can be controlled indirectly by these

High Performance Control of AC Drives With MATLAB®/Simulink, Second Edition.
Haitham Abu-Rub, Atif Iqbal, and Jaroslaw Guzinski.

Companion website: www.wiley.com/go/aburubcontrol2e

two current components, then there is no reason why it should not be possible to control flux and torque directly, without intermediate current control loops.

The control part of a DTC drive consists of two parallel branches, similar to those of vector control. The references are the flux set point and the torque set point, which may or may not be the output of the speed controller, depending on whether the drive is torque-controlled or speed-controlled. DTC asks for the estimation of stator flux and the motor torque, so that closed loop flux and torque control can be established. However, the errors between the set and the estimated flux and torque values are used in a completely different way when compared to the vector control. In vector control, these errors would be used as inputs to the PI controllers, whose outputs are then set points for the stator *d*-*q*-axis current references. The basic idea of DTC is that the existing errors in torque and flux can be used directly to drive the inverter without any intermediate current control loops or coordinate transformation. Flux and torque controllers are hysteresis types, and their outputs are used to determine which of the possible inverter states should be applied to the machine terminals, so that the errors in the flux and torque remain within the prescribed hysteresis bands.

DTC is inherently sensorless. Information about an actual rotor speed is not necessary in the torque mode of operation because of the absence of coordinate transformation. However, the correctness of the estimation of the stator flux and torque is important for the accurate operation of hysteresis controllers. An accurate mathematical model of an induction machine is therefore essential in DTC. Accuracy of DTC is also independent of variations of the rotor's parameters. Only variations of stator resistance, due to changes in thermal operating conditions, cause problems for high-performance DTC at low speed.

In summary, the main features of DTC and differences compared to vector control are [7]:

- direct control of flux and torque;
- indirect control of stator currents and voltages;
- absence of coordinate transformation;
- absence of separate voltage modulation block, usually required in vector drives;
- requirement to know only the sector in which the stator flux linkage space vector is positioned, rather than the exact position of it (necessary in vector drives for coordinate transformation);
- absence of current controllers;
- inherently sensorless control, since speed information is never required in the torque mode of operation;
- in its basic form, the DTC scheme is only sensitive to variation in stator resistance.

5.2 Basic Concept and Principles of DTC

5.2.1 Basic Concept

The DTC principle was introduced in the late 1980s [8–10]. In contrast to vector control, which became accepted by drive manufacturers after 20 years of extensive research, DTC needed only just over a decade to really take off. A direct torque-controlled induction motor drive has been manufactured commercially by ABB since the mid-1990s [5]. In the direct torque controller developed by ABB, the optimum inverter switching pattern is determined in every sampling period (25 μs). The core of the control system in DTC is the sub-system containing torque and flux hysteresis controllers and optimal inverter switching logic. An accurate machine model is also important, since estimation of the stator flux and motor torque is based on the machine model and the measurement of the machine input stator voltages and currents. The measurement of actual speed is not required [5]. A machine model in stationary reference frame is used to develop DTC theory. Detailed discussion regarding mathematical modeling of induction machincs is presented in Chapter 6.

In the stationary reference frame, the stator flux linkage is the integral of the stator emf. If the stator voltage drop on stator resistance can be neglected, then the stator flux is the integral of the applied voltage. Hence, in a short period of time, the increments of stator flux are proportional to the applied voltage. It therefore follows that the inverter output voltage space vector directly impresses on the stator flux and a desired locus of stator flux can be obtained by selecting the appropriate inverter output voltages [7]. The rotor time constant of induction machines is usually large; therefore, the rotor flux linkage changes slowly compared to the stator flux linkage and can be assumed as constant in magnitude and in speed of rotation during short transients [7]. When the forward active voltage space vectors are applied, the stator flux linkage vector is moved away from the rotor flux linkage vector. This will increase the machine's torque because the torque angle increases. However, if the zero or backward active voltage space vectors are applied, the torque angle is reduced. The torque is therefore reduced [6]. It can now be seen that torque can be controlled directly and increased or decreased almost instantly by moving the stator flux linkage space vector to the required position, being determined by torque demand. This in turn can be done quickly by selecting the appropriate voltage vector, while the stator flux linkage magnitude is kept within the hysteresis band. This is the reason why the control scheme is called DTC [7].

DTC requires set points of flux and torque as independent inputs. The estimated values of these quantities are needed to establish closed loop control of the flux and torque. However, the errors between the estimates of actual quantities and set points are used in a completely different way compared to vector control. There is no utilization of current controllers in DTC. The torque and flux controllers are two-level or three-level hysteresis controllers, which determine whether an increase or a decrease of flux and/or torque is required, depending on whether or not torque or flux errors fall outside the predefined ranges. From this information, together with the knowledge of the position of the stator flux linkage space vector, an appropriate voltage vector will be selected based on the switching strategy. An accurate knowledge of the magnitude of the stator flux linkage space vector in the machine is needed. However, knowledge of the precise value of the stator flux space vector instantaneous position is not required. The control system only needs to know in which sector of the voltage vector space, which is a two-dimensional complex plane, the flux linkage space vector is. In the case of the standard two-level voltage source inverter, there are six sectors in the space vector plane corresponding to the six active voltage space vectors, with the voltage vectors positioned at the centers of the sectors. Each sector expands 60 degrees so that all six of them cover the voltage vector complex plane. DTC is inherently sensorless. Different structures of speed estimators can be incorporated into a direct torque controller for sensorless drive operation with closed loop speed control.

5.2.2 Principle of DTC

5.2.2.1 Torque Production in a Direct Torque-Controlled Drive

In a direct torque-controlled induction motor drive supplied by a voltage source inverter, it is possible to directly control the stator flux linkage and the electromagnetic torque by the selection of the optimum stator voltage space vectors in the inverter. The selection of the most appropriate voltage vector is done in such a way that the flux and torque errors are restricted within the respective flux and torque hysteresis bands, fast torque response is obtained, and the inverter switching frequency is kept at the lowest possible level.

In the case of rotor flux-oriented control of an induction motor, the electromagnetic torque developed by the motor is described by

$$T_e = (3/2)P(L_m/L_r)\psi_r i_{qs} \tag{5.1}$$

where the stator q-axis current is the imaginary component of the stator current space vector in the coordinate system fixed to the rotor flux space vector, as shown Figure 4.6a.

The torque equation (5.1) can be written in terms of the amplitude and phase of the stator current space vector with respect to the d-axis of the reference frame as

$$T_e = K\psi_r|\underline{i}_s|\sin\lambda \tag{5.2}$$

Instantaneous change of the torque requires, according to equation (5.2), change in the amplitude and phase of the stator current space vector, such that the d-axis current component remains the same (so that rotor flux is constant), while the torque is stepped to the new appropriate value by the change in the stator q-axis current component.

An alternative expression for the torque uses the stator flux space vector and stator current space vector. Regardless of the applied method of control, the torque developed by the motor can be written as

$$T_e = \frac{3}{2}P\left|\underline{\psi}_s\right||\underline{i}_s|\sin\alpha \tag{5.3}$$

where the angle α is the instantaneous value of the angle between the stator current and stator flux space vectors. Figure 5.1b shows the stator current space vector and stator flux space vector's relative positions.

It can be shown [7] that at certain rotor speeds, if the amplitude of the stator flux is kept constant, an electromagnetic torque can be changed rapidly by altering its instantaneous position so that angle α in equation (5.3) is rapidly changed. In other words, if such stator voltages are imposed on the motor, which not only keep the stator flux constant (at the set value), but also quickly rotate the stator flux space vector into the required position (determined by the torque command), then fast torque control is obtained. It follows that if in the DTC drive, the developed torque is smaller than the reference, the torque should be increased by using the maximum possible rate of change of the stator flux space vector position ϕ_s. If the stator flux space vector is accelerated in the forward direction, an increase in torque is produced; however, when it is decelerated backwards, a decrease in torque results. The stator flux space vector

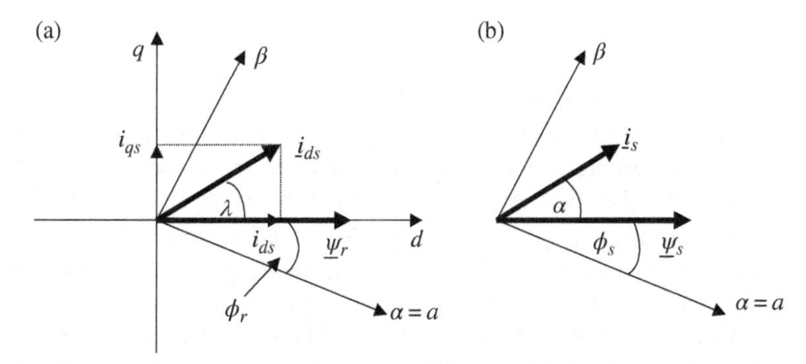

Figure 5.1 (a) Stator current and rotor flux space vectors in a rotor flux-oriented induction machine; (b) stator current and stator flux's relative position in an induction machine

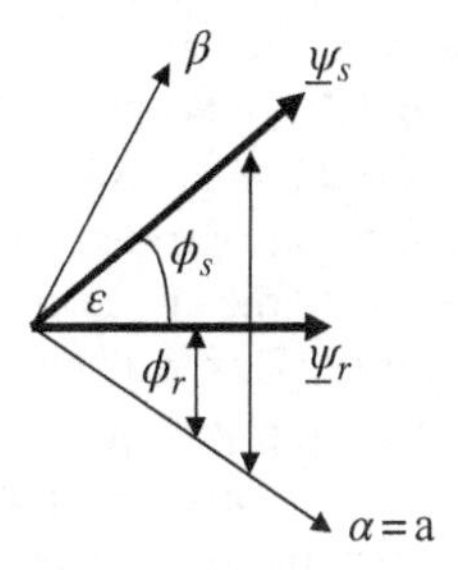

Figure 5.2 Relative positions of stator flux and rotor flux space vectors

can be adjusted by using the appropriate stator voltage space vectors, obtainable from the VSI operated in the PWM mode. Thus, there is a direct stator flux and torque control achieved by means of the voltage source, hence the name "DTC." Another form of the torque equation (Figure 5.2) is the one given in terms of stator and rotor flux:

$$T_e = \frac{3}{2} P \frac{L_m}{\sigma L_s L_r} \left|\underline{\psi}_r\right| \left|\underline{\psi}_s\right| \sin \varepsilon \tag{5.4}$$

where ε is the angle between stator flux and rotor flux space vectors.

Rotor flux changes slowly because its rate of change depends on a relatively large rotor time constant; therefore, it can be assumed to be constant in a short period of time. The stator flux amplitude is also kept constant in the DTC control scheme; hence, both the vectors in equation (5.4) have constant amplitudes. Rapid change of torque can be obtained if instantaneous positions of the stator flux space vector are changed quickly so that ε is quickly varied. This is the essence of DTC. The instantaneous change of ε can be obtained by switching on the appropriate stator voltage space vector of the VSI.

If stator resistance voltage drop is neglected, the stator voltage equation in the stationary reference frame is

$$\underline{v}_s = d\underline{\psi}_s / dt \tag{5.5}$$

Hence, the applied stator voltage directly impresses the stator flux. If the voltage abruptly changes, then stator flux will change accordingly to satisfy equation (5.5). The variation of the stator flux changes during a short period of time when the stator voltage is changed as $\Delta\underline{\psi}_s = \underline{v}_s \Delta t$. This shows that the stator flux space vector moves in the direction of the applied stator voltage space vector during this period. By selecting the appropriate stator voltage space vectors in subsequent time intervals, it is then possible to change the stator flux in the desired way. Decoupled control of the stator flux and torque is achieved by acting on the radial and tangential components of the stator flux space vector. These two components are directly proportional to the components of the stator voltage space vector in the same directions.

The angle ε between stator and rotor flux space vectors is important in torque production. Assume that the rotor flux space vector is traveling at a given speed, at a certain point in the steady-state operation; this speed is initially equal to the average one of the stator flux space vector. The induction motor is accelerated. Appropriate stator voltage vector is applied to increase torque, and quick rotation of stator flux space vector occurs. However, rotor flux space vector amplitude does not change appreciably because of the significant rotor time constant. The rotor flux space vector's speed of rotation is also not changed abruptly; recall that speed of rotation of this vector can be given in terms of rotor flux components and their derivatives; if rotor flux components do not change due to the large time constant, the angular speed of rotor flux space vector does not change either. This will result in an increase of ε that in return increases the motor's torque. If deceleration is required, an appropriate voltage vector will be applied to reduce the angle ε and therefore the decrease the torque developed by the motor. The application of zero voltage vectors will almost stop the rotation of the stator flux space vector. If the angle ε becomes negative, the torque will change signs and a braking process takes place.

5.2.2.2 Inverter Switching Table

A number of methods for the selection of the optimum voltage space vectors for DTC have already been mentioned [1, 8, 9, 11–13]. However, this book only discusses the classical method originally suggested by [1]. Space vectors of inverter output phase voltages are shown again in Figure 5.3 as further explanation. In addition, sections of the plane, identified with Roman numeral I–VI, are also included. The sectors are all of 60 degrees and are distributed ±30 degrees around the corresponding voltage space vector. If the stator flux space vector lies in the kth sector, where $k = 1, 2, 3, 4, 5, 6$, its magnitude can be increased by using the voltage vectors k, $k + 1$, $k - 1$. Its magnitude can be decreased by using $k + 2$, $k - 2$, and $k + 3$ vectors. In other words, stator flux will be increased if either the voltage vector belonging to the sector or any of the two adjacent voltage vectors is applied. It will be decreased if the remaining three active voltage vectors are applied. Vectors $k + 1$ and $k + 2$ are called active forward voltage vectors, whereas vectors $k - 1$ and $k - 2$ are called active backward voltage vectors.

However, the selected voltage vector will also affect the torque production of the induction motor. In addition, the switching frequency will also be affected. The idea is to always keep the switching frequency as low as possible so that the most appropriate voltage space vector is the one that requires the minimum number of switchings and simultaneously drives both the stator

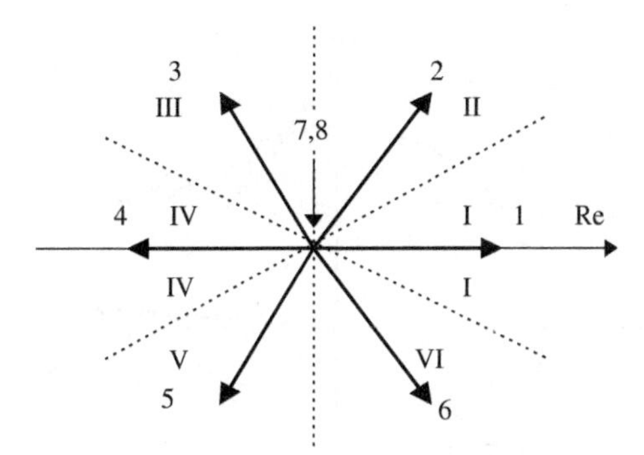

Figure 5.3 Phase voltage space vectors and appropriate sectors

Table 5.1 Phase to neutral voltages of three-phase voltage inverter

Switching state	Switches on	Space vector	Phase voltage v_a	Phase voltage v_b	Phase voltage v_c
1	1,4,6	$\underline{v}_{1\text{phase}}(100)$	$(2/3)V_{DC}$	$-(1/3)V_{DC}$	$-(1/3)V_{DC}$
2	1,3,6	$\underline{v}_{2\text{phase}}(110)$	$(1/3)V_{DC}$	$(1/3)V_{DC}$	$-(2/3)V_{DC}$
3	2,3,6	$\underline{v}_{3\text{phase}}(010)$	$-(1/3)V_{DC}$	$(2/3)V_{DC}$	$-(1/3)V_{DC}$
4	2,3,5	$\underline{v}_{4\text{phase}}(011)$	$-(2/3)V_{DC}$	$(1/3)V_{DC}$	$(1/3)V_{DC}$
5	2,4,5	$\underline{v}_{5\text{phase}}(001)$	$-(1/3)V_{DC}$	$-(1/3)V_{DC}$	$(2/3)V_{DC}$
6	1,4,5	$\underline{v}_{6\text{phase}}(101)$	$(1/3)V_{DC}$	$-(2/3)V_{DC}$	$(1/3)V_{DC}$
7	1,3,5	$\underline{v}_{7\text{phase}}(111)$	0	0	0
8	2,4,6	$\underline{v}_{8\text{phase}}(000)$	0	0	0

flux and the torque errors in the desired direction. From Table 5.1, where inverter switching states are defined as a set of binary signals for each of the six possible non-zero space vectors, we see that progressing from state one toward state six requires only switching one of three inverter legs. Hence, if the inverter operates with a vector whose state is 100, then the most appropriate subsequent vectors are 110, 101, and 000 (zero vector number 8), as these vectors require switching in a single inverter leg. Minimum switching frequency is achieved in this way. Which of the three possible vectors will be applied depends on the flux and torque errors.

An illustration of the switching process is given in Figure 5.4. Operation in the base speed region is assumed so that the stator flux reference is constant and is equal to the rated stator flux. A deviation by a hysteresis band equal to $\pm\Delta\psi_s$s is allowed for the actual flux with respect to the reference. It is assumed that in certain time instants, the stator flux is in the first sector and has just reached the outer boundary of the allowed deviation (point A). The inverter switching state therefore has to be changed, since the stator flux must be reduced. The direction of rotation is anticlockwise. If the stator flux is in sector one, then the application of voltage vectors 1, 2, and 6 would further increase the flux. Reduction in the flux can be obtained using vectors

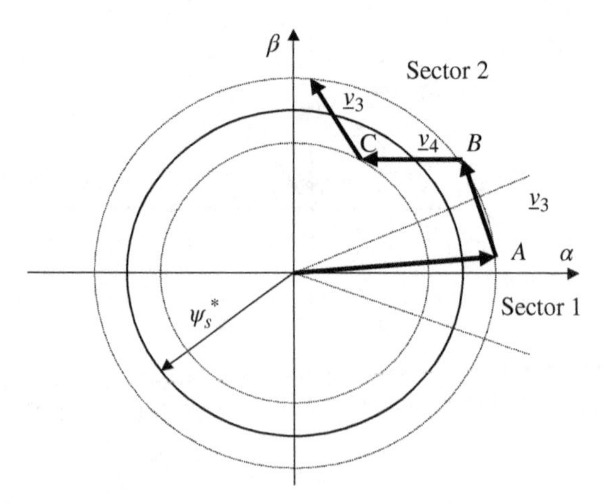

Figure 5.4 Control of stator flux space vector by means of appropriate voltage vector application (*Source:* Vas [7])

3, 4, and 5. Out of these three, vector 3 is the only one that asks for switching in only one of the inverter legs. Vector 3 is therefore applied and drives the stator flux toward point B, where the stator flux once again reaches the upper hysteresis band. Since the stator flux is now in sector 2 and once again needs to be reduced, the possible voltage vectors that will lead to the reduction in the stator flux are vectors 4, 5, and 6. Vector 4 is the only one that asks for switching in a single inverter leg. Voltage vector 4 is therefore applied, and it drives stator flux toward the lower boundary of the allowed deviation (point C). In point C, the stator flux is still in the second sector and needs to be increased. An increase can be obtained using vectors 1, 2, and 3. As vector 3 asks for a single switching, this vector will be applied again. The process continues over time.

Previous considerations did not take into account the impact of the torque error on the selection of the switching state. Stopping the rotation of the stator flux space vector corresponds to the case when the torque does not have to be changed (actual value of the torque is within the prescribed hysteresis bands). However, when the torque has to be changed in the clockwise or anticlockwise direction, corresponding to negative or positive torque variation, then the stator flux space vector has to be rotated in an appropriate direction. In general, if an increase in torque is required, then the torque is controlled by applying voltage vectors that advance the stator flux space vector in the direction of rotation. If a decrease is required, voltage vectors are applied to oppose the direction of rotation of stator flux space vector. If zero change in torque is required, then zero vectors are applied. It follows that the angle of the stator flux space vector is controlled indirectly through the flux amplitude and torque control. The torque demands are reduced to choices of increase, decrease, or zero. Similarly, the stator flux amplitude control is limited to the choice of increase and decrease. Figure 5.5 illustrates the selection of the optimum voltage vector as a function of the required change in flux and torque. The situation is shown for an anticlockwise direction of rotation of the stator flux space vector and for the initial position of the stator flux space vector in the first two sectors. Similar figures can be constructed for other sectors.

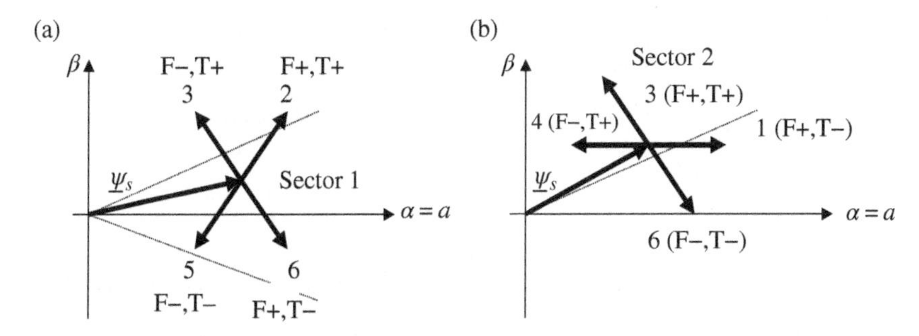

Figure 5.5 Selection of the appropriate voltage vector for required changes in stator flux and torque: (a) when the stator flux is in sector 1 and (b) when the stator flux is in sector 2 (F = flux, T = torque)

Table 5.2 Optimum voltage vector look-up table

$\Delta\psi_s$	ΔT_e	Sector 1	Sector 2	Sector 3	Sector 4	Sector 5	Sector 6
	1	$\underline{v}_2$	$\underline{v}_3$	$\underline{v}_4$	$\underline{v}_5$	$\underline{v}_6$	$\underline{v}_1$
1	0	$\underline{v}_7$	$\underline{v}_8$	$\underline{v}_7$	$\underline{v}_8$	$\underline{v}_7$	$\underline{v}_8$
	−1	$\underline{v}_6$	$\underline{v}_1$	$\underline{v}_2$	$\underline{v}_3$	$\underline{v}_4$	$\underline{v}_5$
	1	$\underline{v}_3$	$\underline{v}_4$	$\underline{v}_5$	$\underline{v}_6$	$\underline{v}_1$	$\underline{v}_2$
0	0	$\underline{v}_8$	$\underline{v}_7$	$\underline{v}_8$	$\underline{v}_7$	$\underline{v}_8$	$\underline{v}_7$
	−1	$\underline{v}_5$	$\underline{v}_6$	$\underline{v}_1$	$\underline{v}_2$	$\underline{v}_3$	$\underline{v}_4$

Active voltage vectors: 1 (100), 2 (110), 3 (010), 4 (011), 5 (001), 6 (101).
Zero voltage vectors: 7 (111); 8 (000).

The result of these considerations can be represented in the optimum voltage vector selection table (Table 5.2) or inverter switching table. This gives the optimum selection of the voltage vectors for all the possible stator flux space vector positions (in terms of sectors) and the desired control inputs, which are the output of the torque and flux hysteresis comparator, respectively. The outputs will give commands of increase or decrease for torque and stator flux, in order to keep both of them within the respective hysteresis band. If a stator flux increase is required, then $\Delta\psi_s = 1$, and if a stator flux decrease is required, then $\Delta\psi_s = 0$. This notation corresponds to the fact that the digital output signals of a two-level flux hysteresis comparator are

$$\begin{aligned} \Delta\psi_s &= 1 \text{ if } \quad \left|\underline{\psi}_s\right| \le \left|\underline{\psi}_s^*\right| - |\text{hysteresis band}| \\ \Delta\psi_s &= 0 \text{ if } \quad \left|\underline{\psi}_s\right| \ge \left|\underline{\psi}_s^*\right| + |\text{hysteresis band}| \end{aligned} \tag{5.6}$$

Similarly, if a torque increase is required, then $\Delta T_e = 1$; if a torque decrease is required, then $\Delta T_e = -1$; and if no change in torque is required, then $\Delta T_e = 0$. The notation corresponds to the fact that the digital outputs of a three-level hysteresis comparator are

$$\begin{aligned} \Delta T_e &= 1 && \text{if} \quad T_e \le T_e^* - |\text{hysteresis band}| \\ \Delta T_e &= 0 && \text{if} \quad T_e = T_e^* \\ \Delta T_e &= -1 && \text{if} \quad T_e \ge T_e^* + |\text{hysteresis band}| \end{aligned} \tag{5.7}$$

5.3 DTC of Induction Motor with Ideal Constant Machine Model

5.3.1 Ideal Constant Parameter Model of Induction Motors

5.3.1.1 Induction Machine's Dynamic Model in *dq* Rotating Reference Frame

An induction machine with a perfectly smoothed air gap is considered. The phase windings of the machine are assumed to be physically 120 degrees apart for both stator and rotor. Winding resistances and leakage inductances are assumed to be constant. All the parasitic phenomena, such as iron loss and main flux saturation, are ignored at this stage. The induction machine under consideration is therefore an ideal smooth air-gap machine with a sinusoidal distribution of windings, and all the effects of MMF harmonics are neglected [10, 14]. A schematic representation of the machine is shown in Figure 5.6.

The notations in the figure above are:

1. ω_r: angular rotor speed;
2. ω_s: angular speed of stator flux space vector;
3. θ: instantaneous rotor angular position with respect to phase a magnetic axis of stator;
4. θ_s: instantaneous angular position of common rotating reference frame with respect to phase a magnetic axis of stator.
5. θ_r: instantaneous angular position of common rotating reference frame with respect to phase a magnetic axis of rotor.

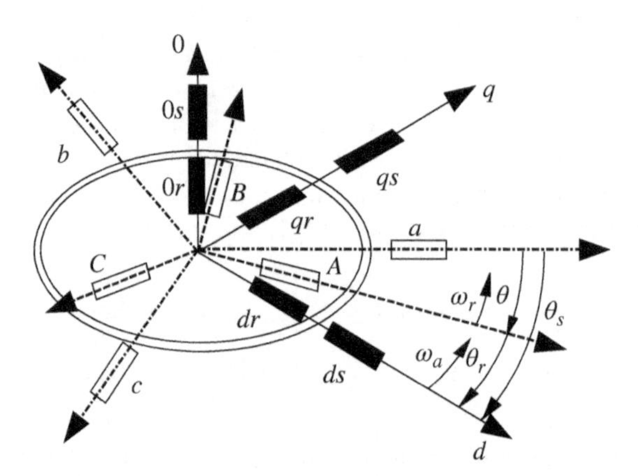

Figure 5.6 Schematic representation of induction machine with reference frames in three-phase domain and common rotating d–q reference frame

After the transformation from a three-phase domain to a common rotating reference frame with an arbitrary angular speed, the stator and rotor voltage equations of the induction machine are

$$\begin{aligned} v_{ds} &= R_s i_{ds} + \frac{d\psi_{ds}}{dt} - \omega_a \psi_{qs} \\ v_{qs} &= R_s i_{qs} + \frac{d\psi_{qs}}{dt} + \omega_a \psi_{ds} \end{aligned} \tag{5.8}$$

$$\begin{aligned} v_{dr} &= R_r i_{dr} + \frac{d\psi_{dr}}{dt} - (\omega_a - \omega_r)\psi_{qr} \\ v_{qr} &= R_r i_{qr} + \frac{d\psi_{qr}}{dt} + (\omega_a - \omega_r)\psi_{dr} \end{aligned} \tag{5.9}$$

The flux linkages are

$$\begin{aligned} \psi_{ds} &= L_s i_{ds} + L_m i_{dr} \\ \psi_{qs} &= L_s i_{qs} + L_m i_{qr} \end{aligned} \tag{5.10}$$

$$\begin{aligned} \psi_{dr} &= L_r i_{dr} + L_m i_{ds} \\ \psi_{qr} &= L_r i_{qr} + L_m i_{qs} \end{aligned} \tag{5.11}$$

Symbols L_s and L_r denote stator and rotor self-inductance, respectively, while L_m is the magnetizing inductance.

The relationship between the stator, rotor, and magnetizing inductances and the three-phase model self and mutual inductances is

$$\begin{aligned} L_s &= L_{\sigma s} + L_m \\ L_r &= L_{\sigma r} + L_m \end{aligned} \tag{5.12}$$

where $L_{\sigma s}$ and $L_{\sigma r}$ are stator and rotor leakage inductances, respectively. The equation of mechanical motion remains as

$$T_e - T_L = \frac{J}{P}\frac{d\omega_r}{dt} \tag{5.13}$$

where T_e is the electromagnetic torque, T_L is the load torque, J is the inertia of the induction machine, and P is the number of pole pairs.

Electromagnetic torque can be expressed in terms of $d{-}q$ components of stator flux and stator current as

$$T_e = \frac{3}{2}P\left(\psi_{ds} i_{qs} - \psi_{qs} i_{ds}\right) \tag{5.14}$$

The per phase equivalent circuit of the ideal constant parameter machine mode in the $d{-}q$ rotating reference frame is shown in Figure 5.7.

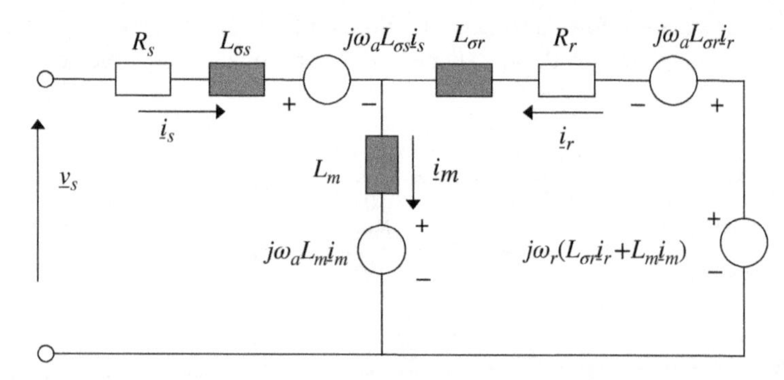

Figure 5.7 Dynamic equivalent circuit of an IM in an arbitrary rotating common reference frame

5.3.1.2 Induction Machine's Dynamic Model in α–β Stationary Reference Frame

The induction machine model that will be used later in this chapter for simulation purposes is the machine model in the stationary reference frame. Components of stator current, rotor current, and angular electrical speed ω_r are used as state-space variables in the fifth-order differential equation system. The stator and rotor voltage equations with stator current, rotor current, and rotor angular speed as state-space variables are obtained by using equations (5.8)–(5.11), and (5.13) as

$$\begin{aligned} v_{\alpha s} &= R_s i_{\alpha s} + L_s \frac{di_{\alpha s}}{dt} + L_m \frac{di_{\alpha r}}{dt} \\ v_{\beta s} &= R_s i_{\beta s} + L_s \frac{di_{\beta s}}{dt} + L_m \frac{di_{\beta r}}{dt} \end{aligned} \tag{5.15}$$

$$\begin{aligned} 0 &= L_m \frac{di_{\alpha s}}{dt} + \omega_r L_m i_{\beta s} + R_r i_{\alpha r} + L_r \frac{di_{\alpha r}}{dt} + \omega_r L_r i_{\beta r} \\ 0 &= -\omega_r L_m i_{\alpha s} + L_m \frac{di_{\beta s}}{dt} - \omega_r L_r i_{\alpha r} + R_r i_{\beta r} + L_r \frac{di_{\beta r}}{dt} \end{aligned} \tag{5.16}$$

$$\frac{J}{P}\frac{d\omega_r}{dt} = \frac{3}{2}P\left[i_{\beta s}(L_s i_{\alpha s} + L_m i_{\alpha r}) - i_{\alpha s}\left(L_s i_{\beta s} + L_m i_{\beta r}\right)\right] - T_L \tag{5.17}$$

These differential equations will be used to build the induction machine model in MATLAB/Simulink for the simulation of the performance of a DTC induction motor drive.

5.3.2 Direct Torque Control Scheme

5.3.2.1 Basic Control Scheme of Direct Torque-Controlled Induction Motor

Since $\Delta\underline{\psi}_s = \underline{v}_s \Delta t$, then the stator flux space vector will move fast if non-zero voltage vectors are applied to the motor. It will almost stop if zero voltage vectors are applied. In DTC drives, at every sampling period, the stator voltage vectors are selected on the basis of keeping the stator flux amplitude error and torque error within the prescribed hysteresis bands. The size of the hysteresis bands will significantly affect the inverter switching frequency. In general, the larger the hysteresis band, the lower the switching frequency and the poorer the response of the drive to change in reference. Because the stator flux space vector is the integral of the stator voltage vector, it will move in the direction of the stator voltage space vector for as long as this voltage vector is applied to the motor.

Basic control schemes of a DTC induction motor drive are shown in Figures 5.8 and 5.9 for torque mode and speed mode of operation, respectively. In each figure, there are two parallel branches, one for stator flux amplitude and the other for the torque control. The torque reference is either an independent input (for a torque-controlled drive, Figure 5.8) or the output of the speed controller (in a speed-controlled drive, Figure 5.9). Both the stator flux and the torque controllers are of the hysteresis type. The drive requires appropriate measurements that will enable the estimation of the stator flux and torque for closed loop control of these two quantities. Stator currents and measured or reconstructed stator voltage are used for the estimation, as given by equations (5.18) and (5.19):

$$\begin{aligned}
\psi_{\alpha s} &= \int (v_{\alpha s} - R_s i_{\alpha s})dt \\
\psi_{\beta s} &= \int (v_{\beta s} - R_s i_{\beta s})dt \\
\psi_s &= \sqrt{\psi_{\alpha s}^2 + \psi_{\beta s}^2} \quad \cos\phi_s = \psi_{\alpha s}/\psi_s \quad \sin\phi_s = \psi_{\beta s}/\psi
\end{aligned} \tag{5.18}$$

$$T_e = \frac{3}{2}P(\psi_{\alpha s} i_{\beta s} - \psi_{\beta s} i_{\alpha s}) \tag{5.19}$$

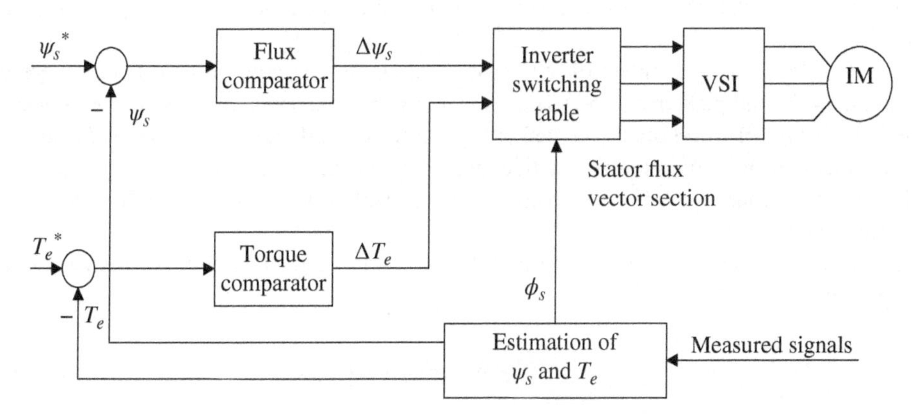

Figure 5.8 Control scheme of a DTC induction motor drive for torque mode of operation

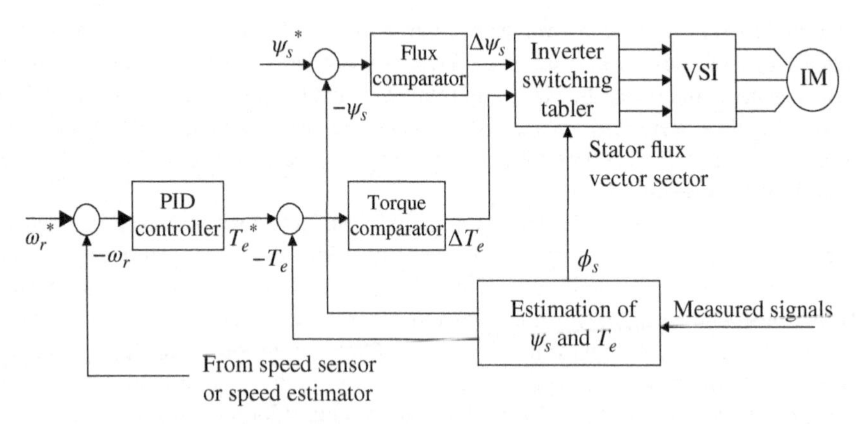

Figure 5.9 Control scheme of DTC induction motor drive for speed mode of operation with or without speed sensor

In addition, an estimation of the speed of rotation is required for closed loop speed control in sensorless DTC drives. Note that the DTC induction motor drive inherently lends itself to sensorless operation, since no coordinate transformation is involved and a speed signal is only needed for closing the speed loop.

5.3.2.2 Stator Flux and Torque Estimation

From the considerations of the previous sub-section, it follows that for successful operation of a DTC scheme, it is necessary to have accurate estimates of the stator flux amplitude and the electromagnetic torque. In addition, it is necessary to estimate in which sector of the complex plane the stator flux space vector is situated. As already shown in the sub-section above, the stator flux amplitude and torque can be obtained in a straightforward manner if stator currents are measured and stator voltages are reconstructed (or measured) according to equations (5.18) and (5.19).

The two problems encountered in the process of stator flux and torque estimation are the requirement for the pure integration and temperature-related variation in stator resistance in equation (5.18). Stator resistance-related voltage drop is significant at low speed of operation. The accuracy of the estimation at low frequency depends on the accuracy of stator resistance value. An open loop stator flux estimator can work well down to 1–2 Hz, but not below this frequency [7].

Equation (5.18) can also be used to find the location of the stator flux space vector in the complex plane:

$$\phi_s^e = \tan^{-1}\left(\psi_{\beta s}/\psi_{\alpha s}\right) \tag{5.20}$$

The exact position of the stator flux space vector is not needed in DTC induction motor drives. It is only necessary to know in which sector (out of six possible ones) of the complex plane the vector is located. Equation (5.20) will be used in the simulations for determining the sector containing the stator flux space vector. The determination of the sector of the complex plane containing the stator flux space vector can be done by using the algebraic signs of the stator flux α–β components obtained from equation (5.18) and the sign of the phase "*b*" stator flux, which is obtained by using $\psi_{bs} = -0.5\psi_{\alpha s} + 0.5\sqrt{3}\psi_{\beta s}$, as discussed in [7].

5.3.3 Speed Control with DTC

The general speed control scheme is shown in Figure 5.9. A PID controller is used as a speed controller whose input is the error between the speed command and feedback speed. Feedback speed can be from the speed sensor or speed estimator. A PI controller is commonly used instead of a PID controller.

For better performance of speed responses (lower overshoot, faster time response, reduced or zero steady-state error), a PI controller with anti-wind up is used as a speed controller. Figure 5.10 shows the structure of a PID controller with anti-wind up in MATLAB/Simulink.

Simulation of the performance of the DTC of the induction motor using MATLAB/Simulink will be discussed in the next sub-section. The discussion will focus on both the torque control mode and the speed control mode. Structures of the simulation programs are also presented.

5.3.4 MATLAB/Simulink Simulation of Torque Control and Speed Control with DTC

5.3.4.1 MATLAB/Simulink Simulation of Torque Control with DTC

This sub-section discusses the implementation of the simulation of the independent control of torque and stator flux with the DTC method suggested by [1]. The machine model does not consider iron losses, magnetic saturation, or stray losses. The machine's parameters are constant during the operation.

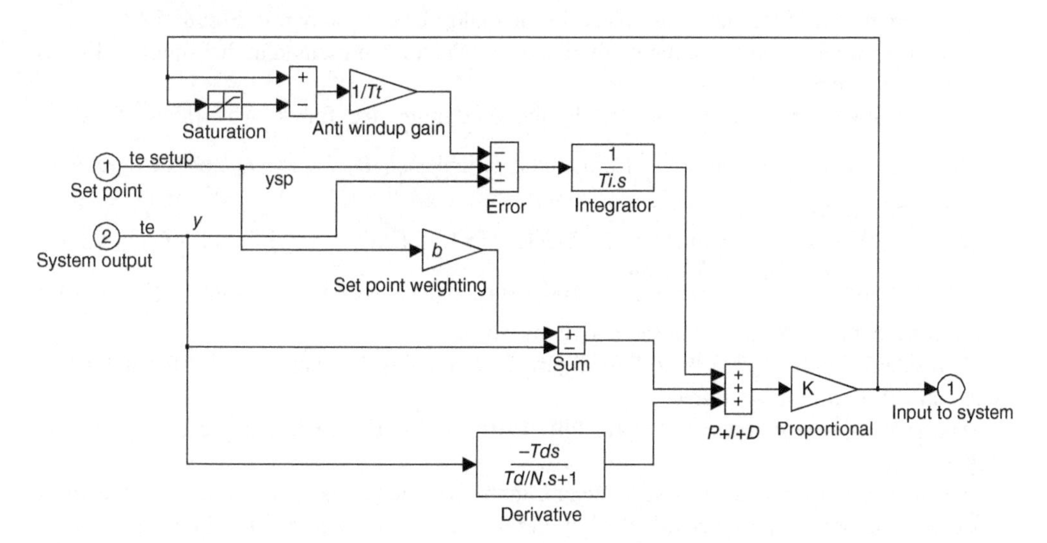

Figure 5.10 PID controller with anti-wind up

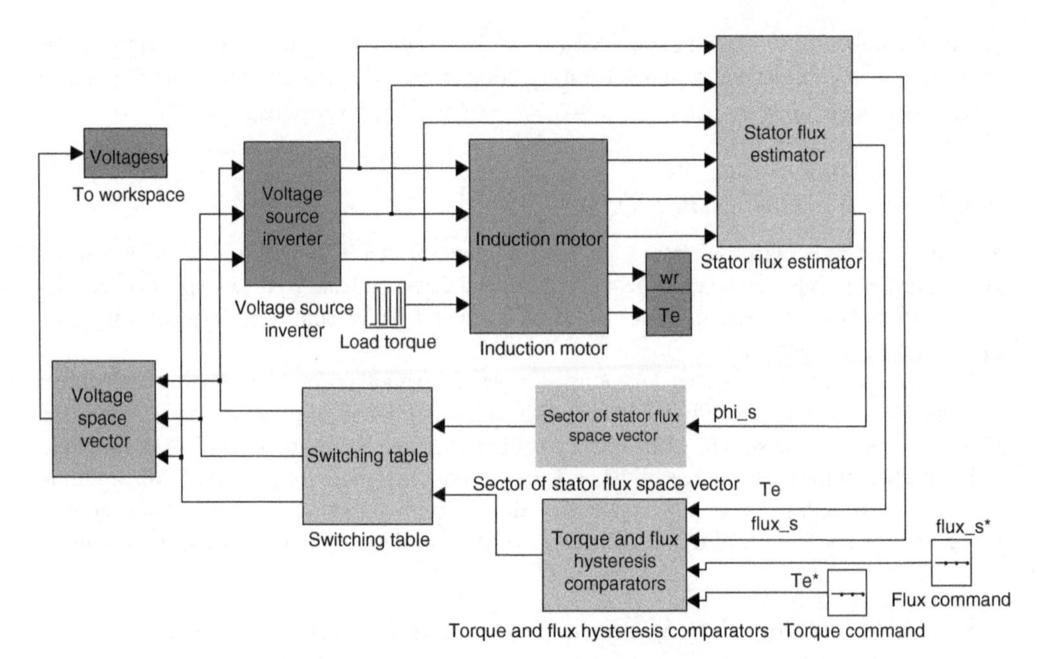

Figure 5.11 Simulation program of DTC of induction motor

MATLAB/Simulink is used for the simulation. Figure 5.11 presents the overall layout of the simulation program.

Figure 5.12 shows the structure of the block that represents the dynamic mathematical model of the induction motor with constant parameters. Because the model is ideal, iron losses and magnetic saturation are not considered.

The parameters of the motor are listed in the dialog box, as shown in Figure 5.13.

The differential equations of the mathematical model are represented in the Simulink file, as shown in Figure 5.14.

The machine model is represented in the α–β reference frame. The block "Fcn" in Figure 5.14 is the expression of $\frac{di_{\alpha s}}{dt}$ from all the variables $i_{\alpha s}$, $i_{\beta s}$, $i_{\alpha r}$, and $i_{\beta r}$ and the input voltage $v_{\alpha s}$ and $v_{\beta s}$, as

shown in Figure 5.15. Similarly, the block "Fcn1," "Fcn2," "Fcn3," and "Fcn4" are the expressions for $\frac{di_{\beta s}}{dt}, \frac{di_{\alpha r}}{dt}, \frac{di_{\beta r}}{dt}, \frac{d\omega_r}{dt}$, respectively. The block "Fcn5" is the expression for the electromagnetic torque from the above variables.

The subsystem "Flux Calculator" in Figure 5.14 contains the expression for the stator flux and rotor flux from the variables.

The layout of the sub-system "Stator Flux Estimator" in Figure 5.11 is presented below in Figure 5.16.

The location of the stator flux space vector in the complex plane is determined by the block "Sector of stator flux space vector" block in Figure 5.11. The structure of this block is shown in Figure 5.17.

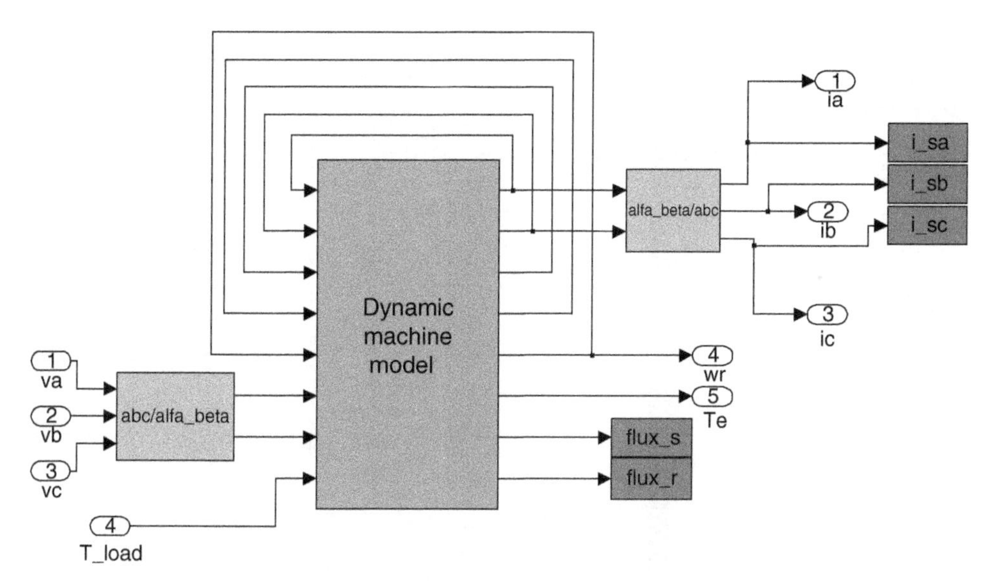

Figure 5.12 Simulink layout of the machine model

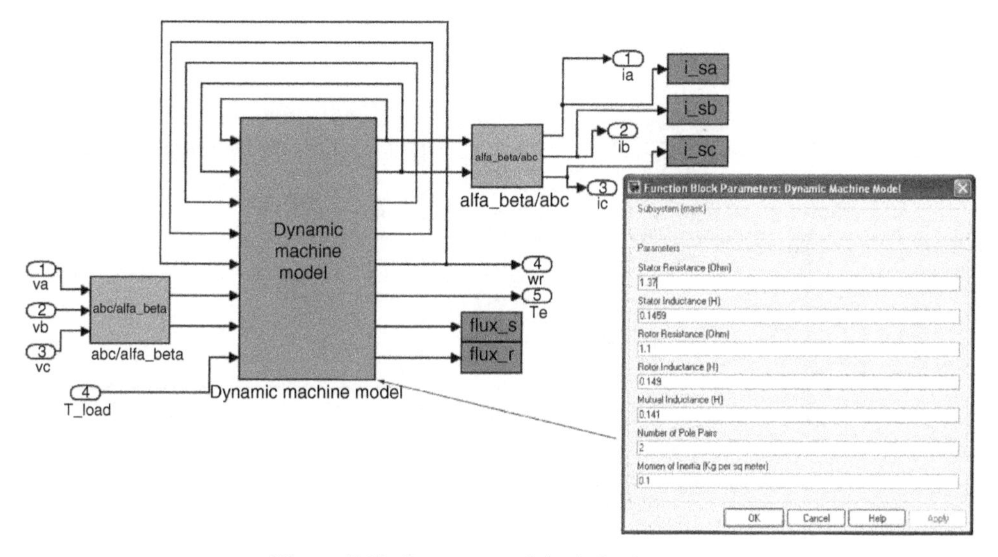

Figure 5.13 Parameters of the induction motor

The structure of the hysteresis comparators in the block "Torque and Flux Hysteresis Comparators" is shown in Figure 5.18. Flux and torque hysteresis comparators are shown in Figures 5.19 and 5.20, respectively.

The structure of the switching table suggested in [1] is shown in Figure 5.21. Each "Look-up Table" block represents the switching state of one leg of the three-phase voltage source inverter.

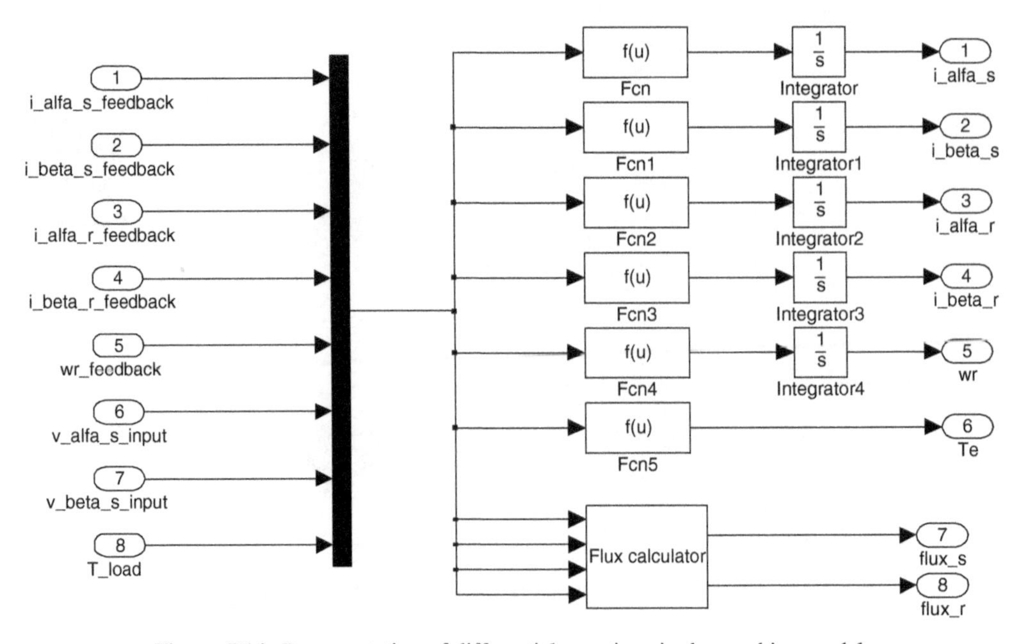

Figure 5.14 Representation of differential equations in the machine model

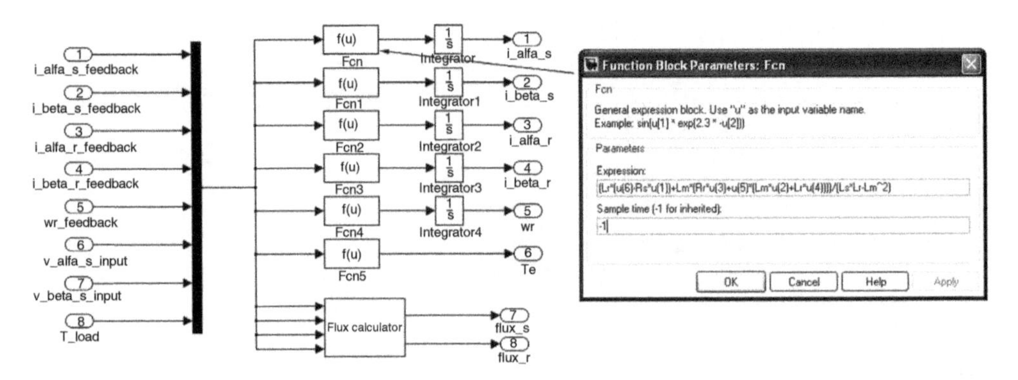

Figure 5.15 Expression of differential equation in block "Fcn"

The block used to observe the voltage space vector applied to the induction motor is presented in Figure 5.22. The principle of operation of the three-phase voltage source inverter is illustrated in Figure 5.23. Each "Switch" block represents a leg.

The values of the flux command, torque command, and load torque of the DTC system are illustrated in Figure 5.24. The flux command is the rated flux, and the torque command is one and a half times of the rated torque. The load torque is equal to the rated value and applied instantaneously at the moment the rotor speed reaches the rated speed.

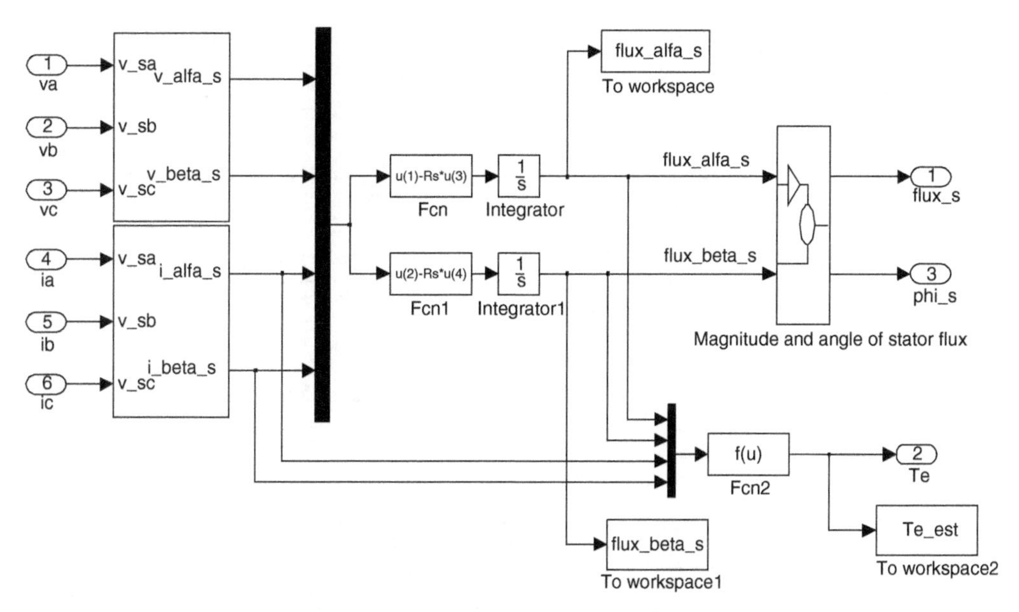

Figure 5.16 Schematic diagram of "Stator Flux Estimator" block

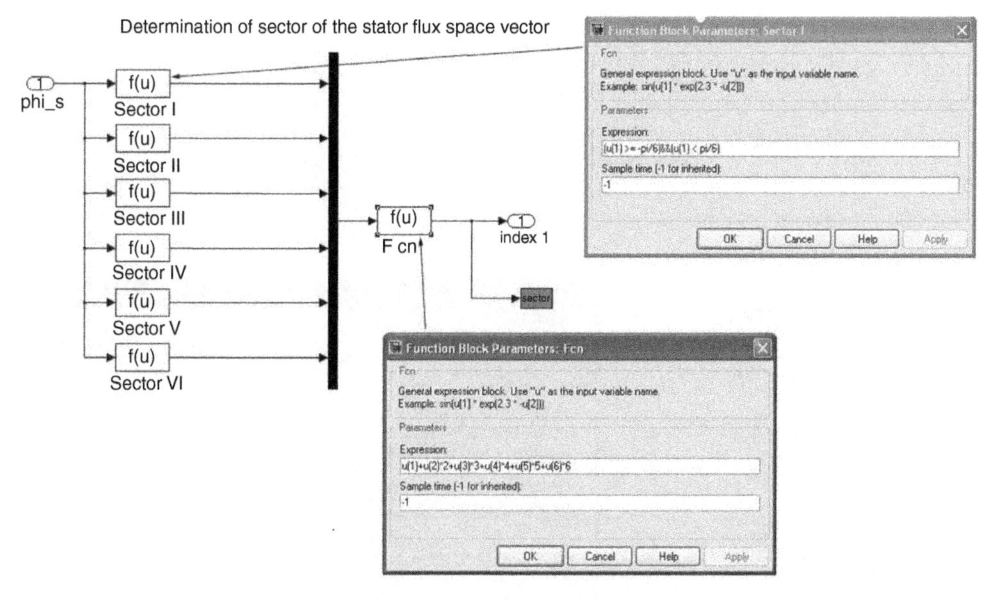

Figure 5.17 Structural diagram of "Sector of stator flux space vector" block

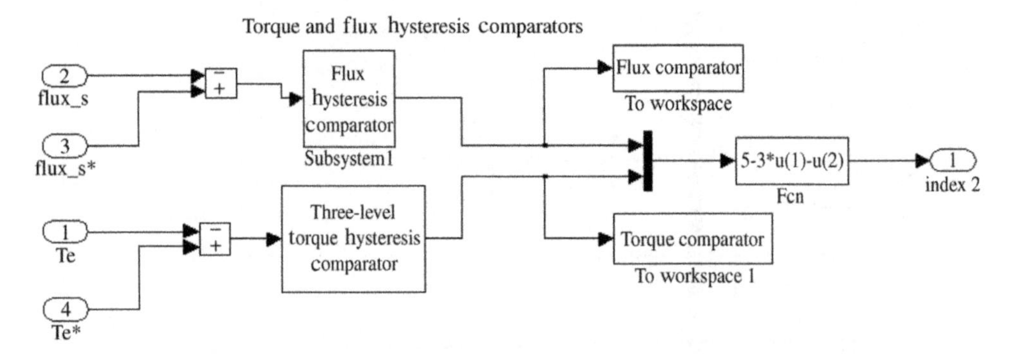

Figure 5.18 Structure of the "Flux and Torque Hysteresis Comparators" block

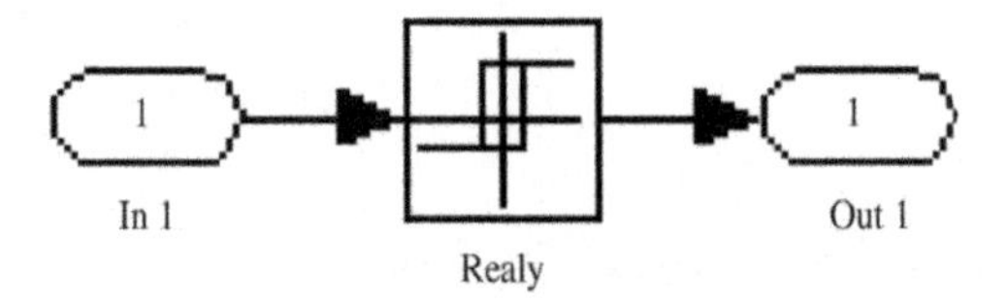

Figure 5.19 Flux hysteresis comparator

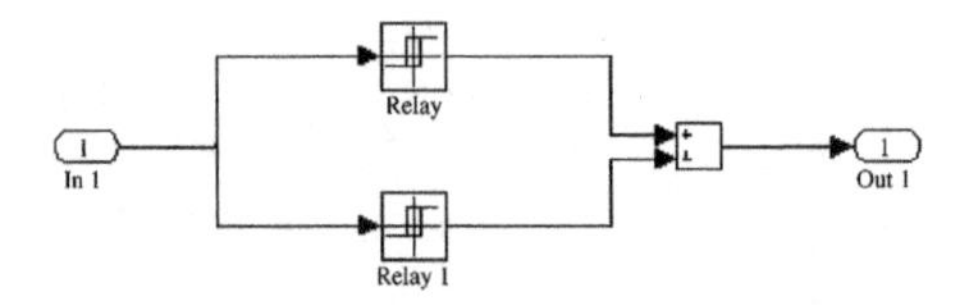

Figure 5.20 Torque hysteresis comparator

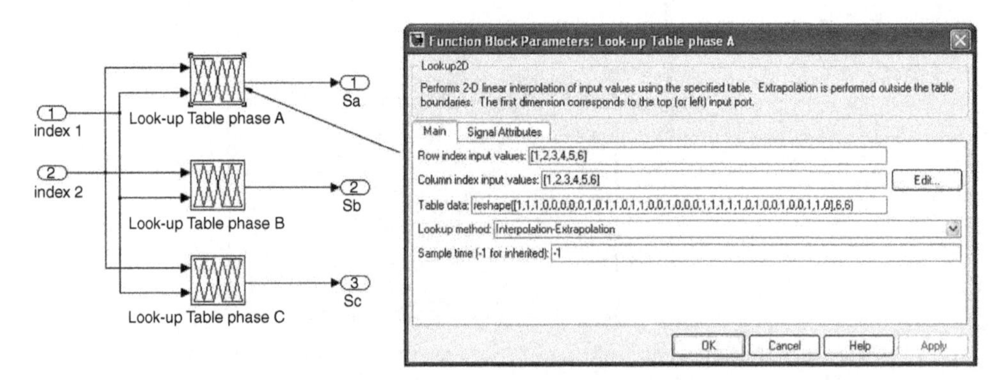

Figure 5.21 Simulink diagram of the DTC look-up table

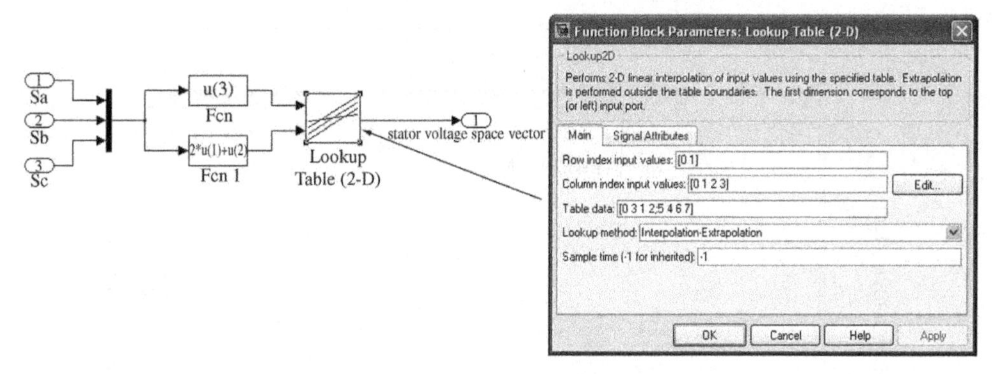

Figure 5.22 Simulink diagram for observation of the applied voltage space vectors

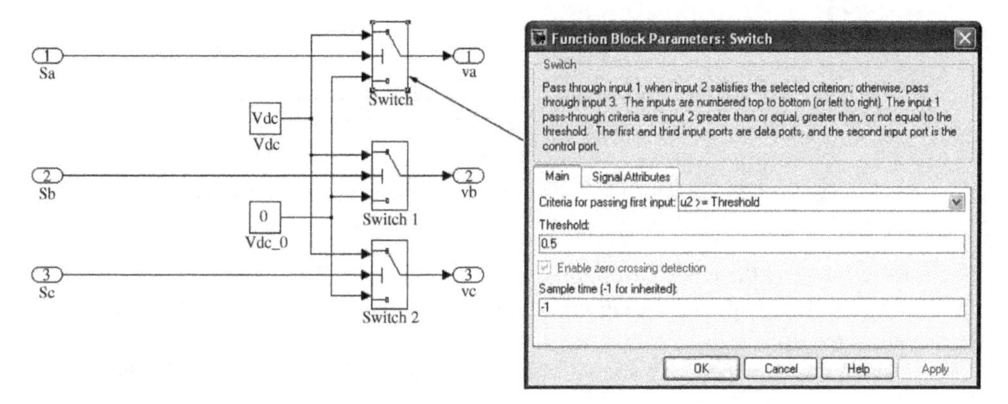

Figure 5.23 Simulink diagram of a three-phase voltage source inverter

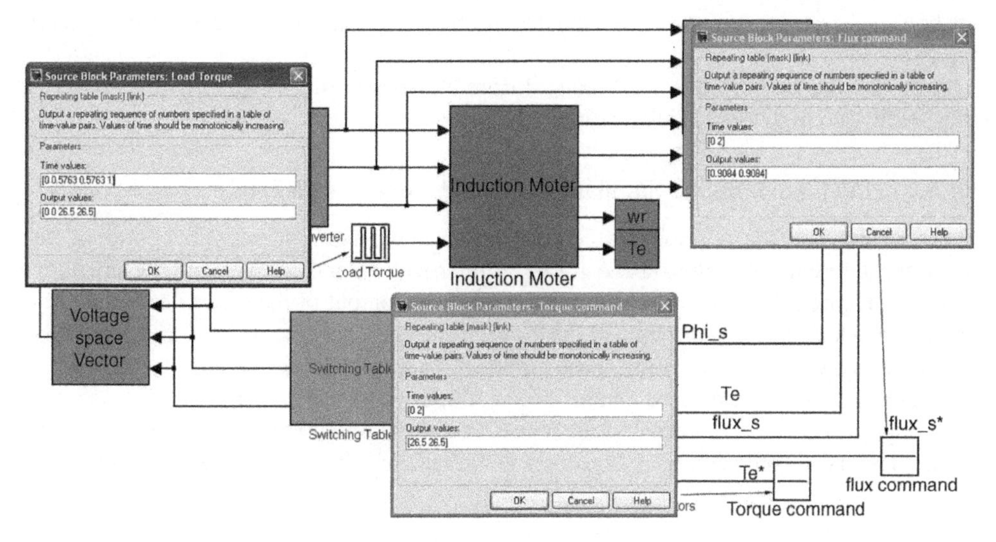

Figure 5.24 Values of flux command, torque command, and load torque

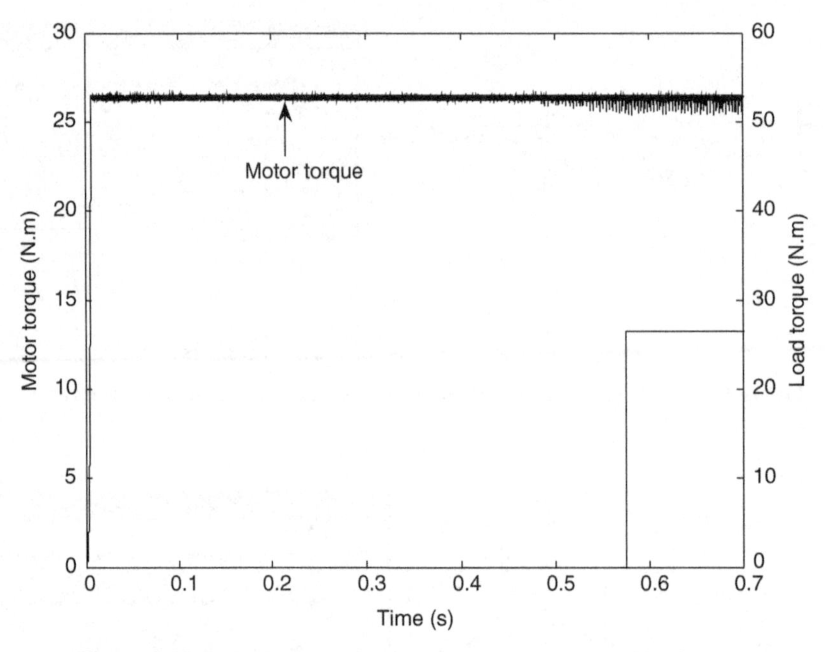

Figure 5.25 Induction motor's electromagnetic torque and load torque

Torque response in the control of torque and stator flux with DTC is shown in Figure 5.25. The load torque is not applied during the starting point, but the motor is fully loaded when it reaches the rated speed.

Stator flux and rotor speed are shown in Figure 5.26, and the stator current is illustrated in Figure 5.27.

The rotor speed in Figure 5.26 is slightly reduced with the progression of time. This is due to the fact that the average value of the motor's electromagnetic torque at high speed is slightly less than the load torque because of high torque ripples. The torque command and load torque are both at the rated values; therefore, the rate of change of speed with time is negative according to equation (5.13).

From the simulation results, the two significant weaknesses of DTC are high flux ripples at low speeds and high torque ripple at high speeds.

5.3.4.2 MATLAB/Simulink Simulation of Speed Control with DTC

Speed control can then be implemented after the development of independent control of electromagnetic torque and stator flux. A PID controller with an anti-wind up, whose structure is shown in Figure 5.10, is used for control of speed. The output of the controller is the torque command, which in turn is the input of the torque hysteresis comparator. The input of the speed controller is the error between the speed command and actual speed. Figure 5.28 shows the schematic structure of the speed control of an induction motor with DTC.

The speed control DTC system of the induction motor in Figure 5.28 is without load of the motor. The speed command goes from zero to the rated electrical speed. Details of the speed

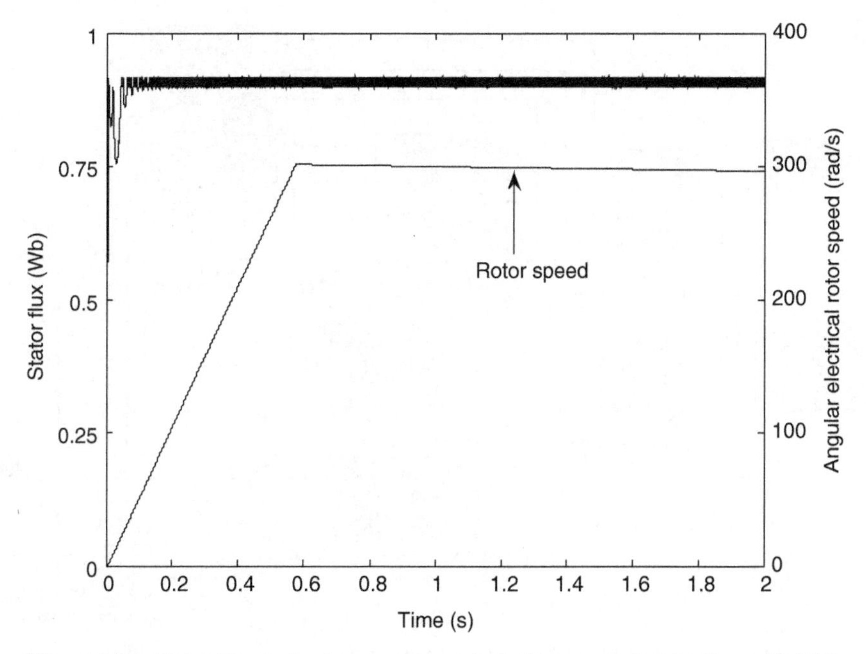

Figure 5.26 Stator flux and rotor speed in control of torque and stator flux with DTC

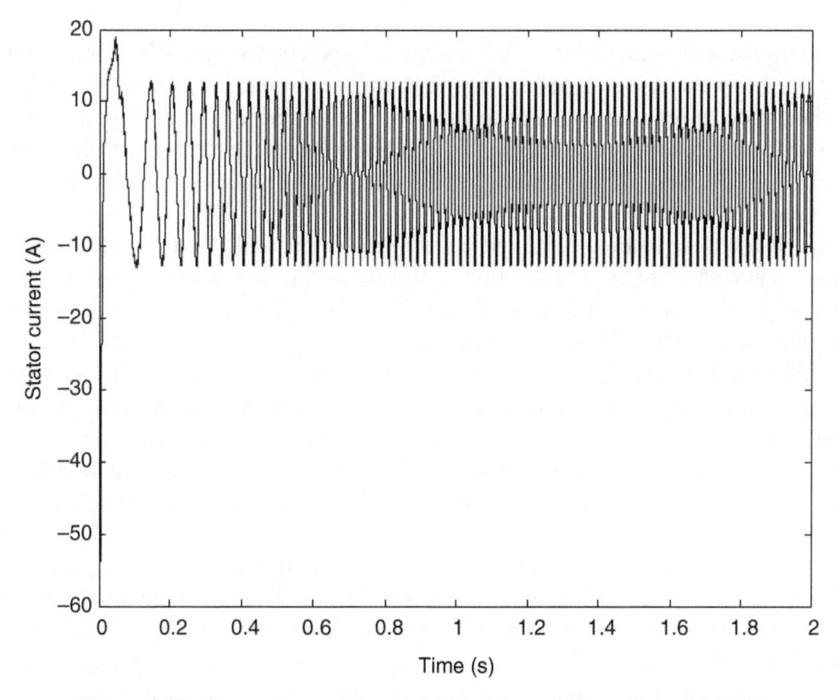

Figure 5.27 Stator current in control of torque and stator flux with DTC

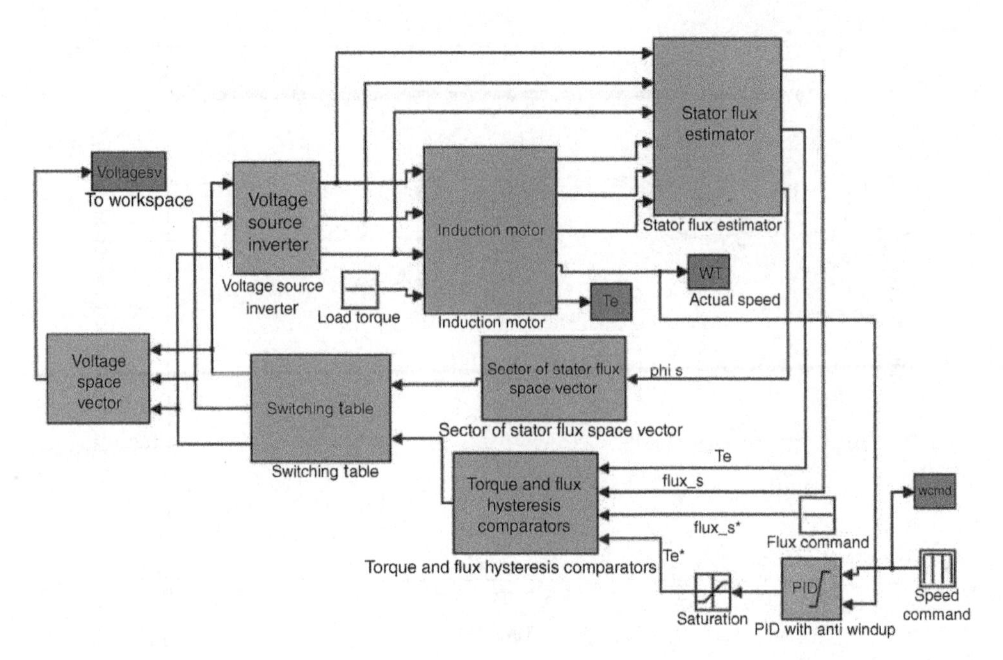

Figure 5.28 Simulation program of DTC of induction motor in speed control without load

command signal and values of the PID controller's parameters are shown in Figure 5.29. Figure 5.30 shows the value of the load torque during the simulation. All other configurations of the simulation program are similar to the ones in the torque control program, except the values of the speed command and PID controller. Responses of the rotor speed, electromagnetic torque of the motor, stator flux, and phase current obtained from the simulation will be analyzed and are presented in Figures 5.31–5.34.

The speed response in Figure 5.31 meets the requirement and is stable after reaching the command value. The obtained stator flux is similar to the one in the torque control and is not affected by the variation of speed. Figure 5.32 shows that the steady-state error during no-load operation of the DTC speed control is insignificant.

The torque has been increased up to a maximum allowable value during the acceleration of the motor from standstill to commanded value. When the speed reaches and remains at the required value, the torque is reduced to proximity of zero. High torque ripples are still present due to the high speed of the rotor. Figure 5.34 shows the similarities between the current response and the response obtained in torque control. In both cases, the starting current is still high.

The motor is run again in the speed control mode without a load at starting point. Then, a rated load torque is applied and removed stepwise; this is a highly hypothetical situation to test the noise removal capability of the controller. The load is suddenly applied after starting 1 second and also suddenly removed at the moment of 1.2 seconds from the starting point. Speed, torque, flux, and stator phase current responses are shown in Figures 5.35–5.38, respectively.

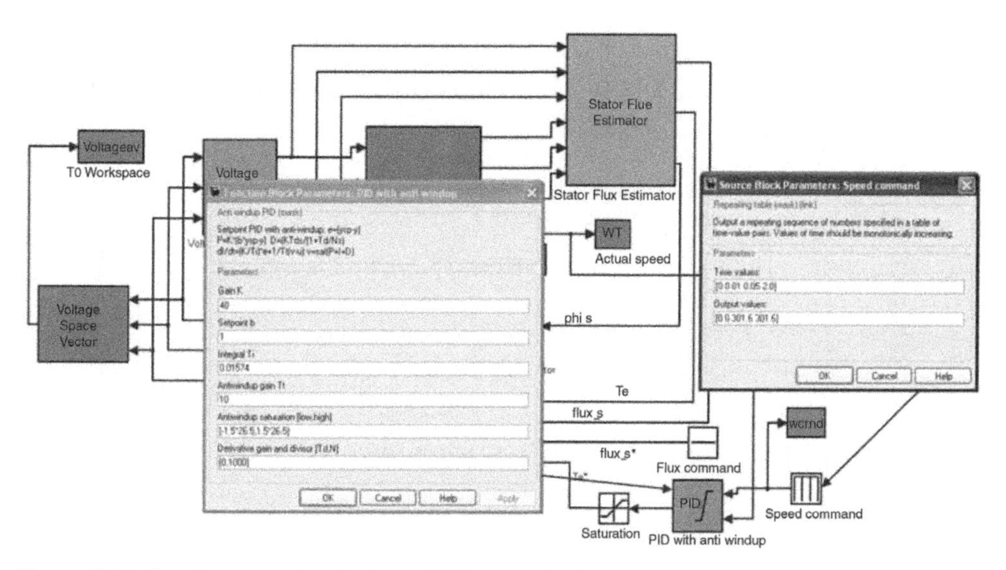

Figure 5.29 Speed command and values of PID controller's parameters of the induction motor control system in Figure 5.28

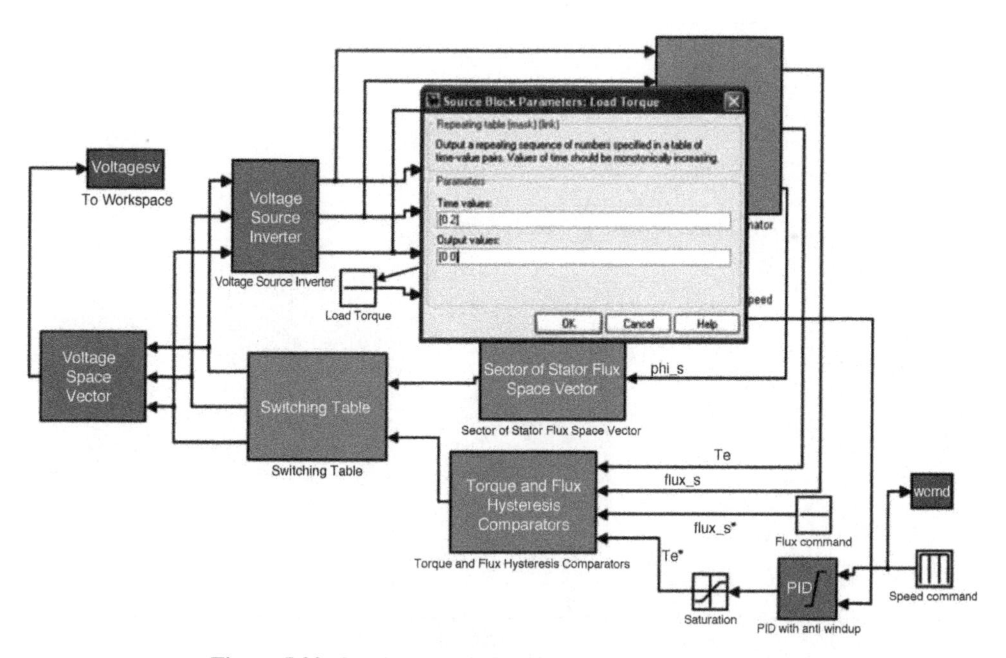

Figure 5.30 Load torque during the speed control simulation

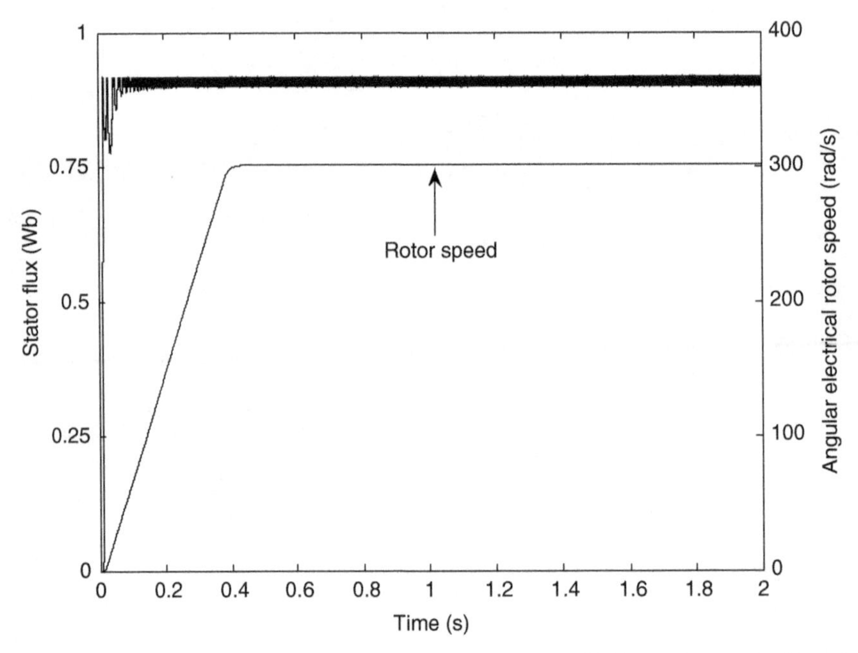

Figure 5.31 Stator flux and rotor speed of induction motor in DTC speed control without load

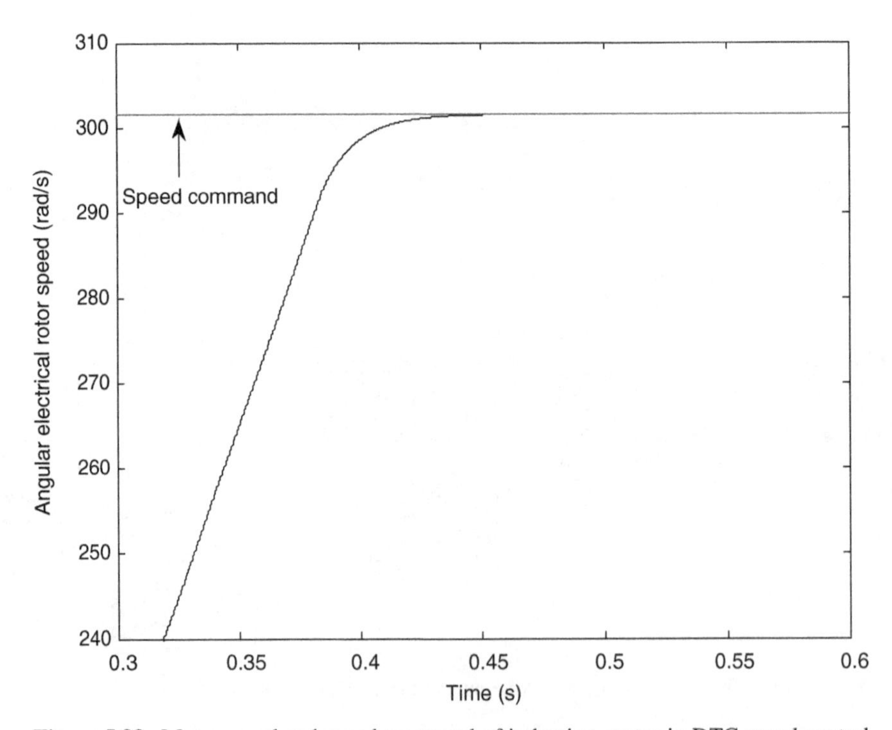

Figure 5.32 Motor speed and speed command of induction motor in DTC speed control

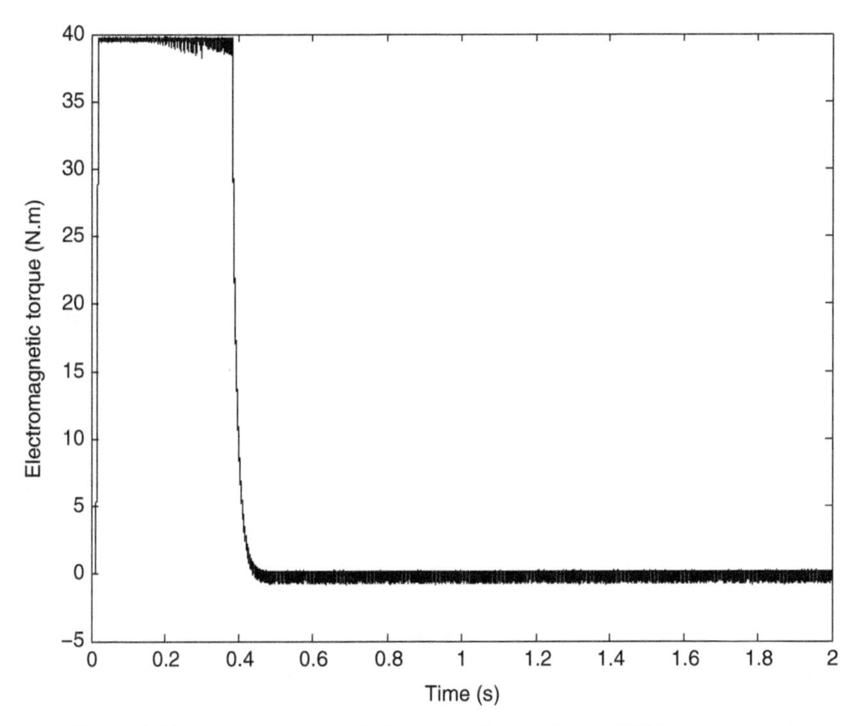

Figure 5.33 Torque response during no-load operation of DTC speed control

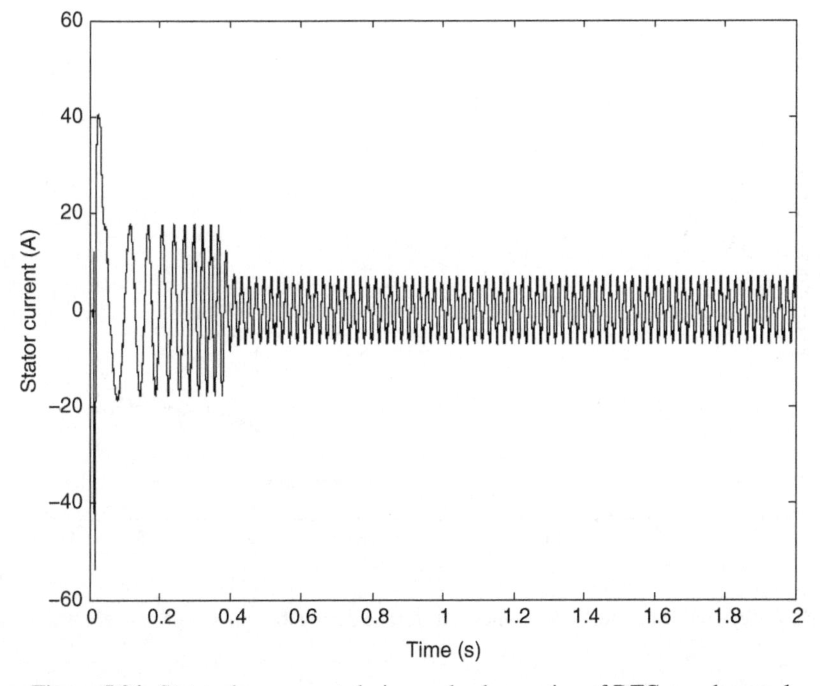

Figure 5.34 Stator phase current during no-load operation of DTC speed control

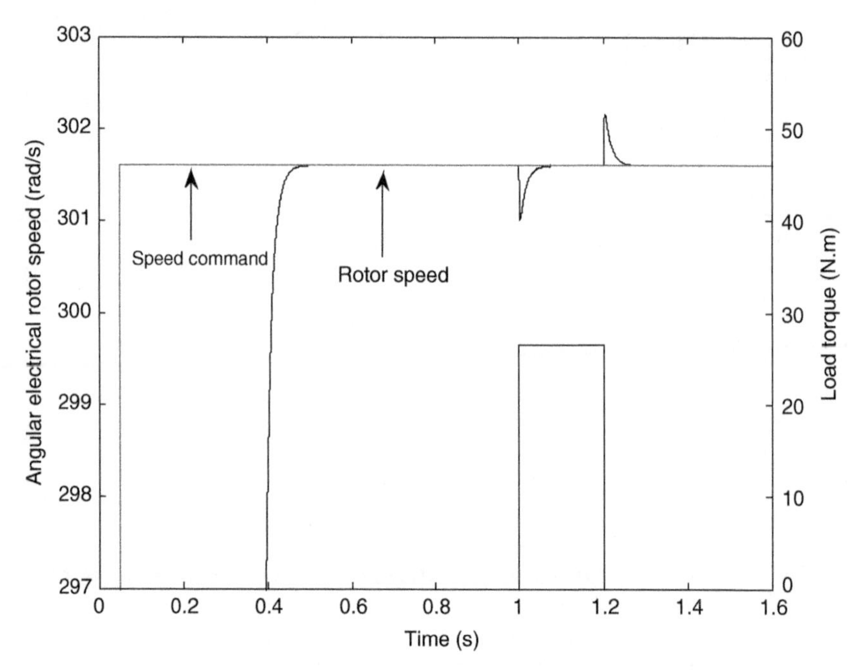

Figure 5.35 Speed response and load torque when load is applied and removed during DTC speed control

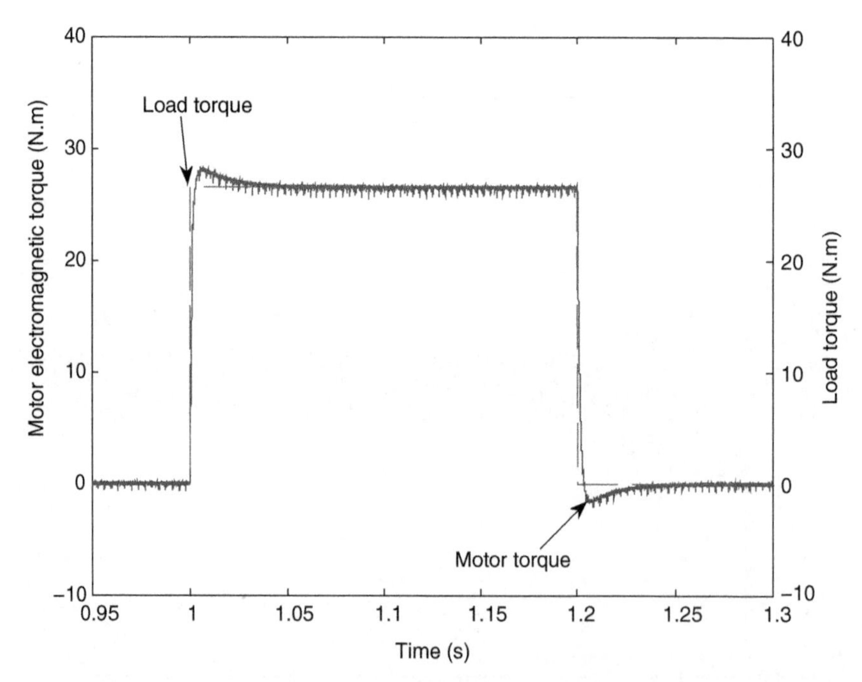

Figure 5.36 Dynamic of motor's electromagnetic torque during the application and removal of rated load torque

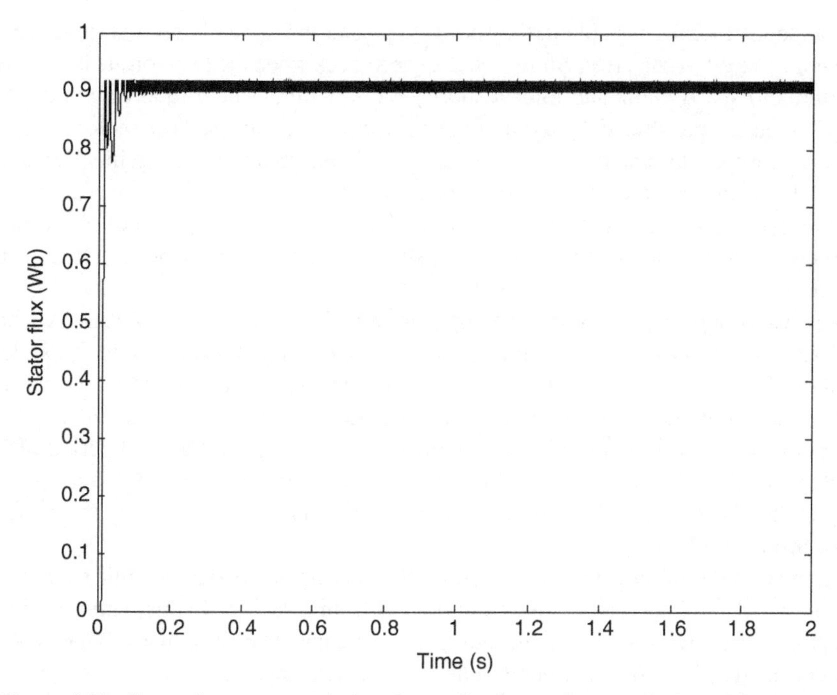

Figure 5.37 Stator flux response during the application and removal of rated load torque

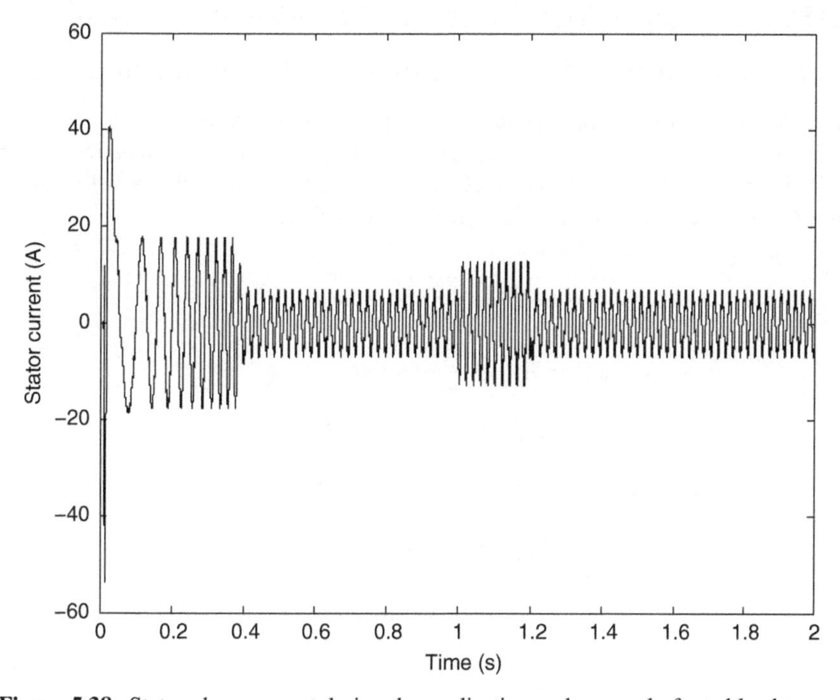

Figure 5.38 Stator phase current during the application and removal of rated load torque

When the load torque is applied, the speed drops and then quickly returns to the command value without steady-state error. Similarly, the speed rises when the load torque is removed and then returns to speed command after a short period of time. The torque response shown in Figure 5.36 has demonstrated the dynamic of electromagnetic torque. The motor's torque cannot change abruptly to match the load torque. This is the reason for a dip in speed shown in Figure 5.35. However, the motor torque increases rapidly to bring the speed back to the required value and stabilize the speed response. Similar dynamic response of electromagnetic torque occurs when the load torque is removed suddenly. This explains the variation of speed in Figure 5.35.

Stator flux is not affected during the application and removal of load torque, as shown in Figure 5.37, which causes the variation of the motor's electromagnetic torque. This illustrates the ability of DTC to independently control the torque and flux in the induction motor.

Stator phase current is increased during the loading period of the motor, as shown in Figure 5.38. However, high overshoots are not detected during the transient period of loading and unloading. This demonstrates that the current is well controlled in DTC, except during the starting of the motor. The problem of high starting current can be overcome by the combination of soft starting and DTC.

In summary, DTC of induction motor with ideal constant parameter dynamic machine model in both torque control and speed control modes has been discussed in this section and has also been demonstrated with MATLAB/Simulink simulations. The DTC with dynamic machine model considering iron loss will be presented in the next section.

5.4 DTC of Induction Motor with Consideration of Iron Loss

5.4.1 Induction Machine Model with Iron Loss Consideration

All the equations for this dynamic model are shown in an arbitrary, rotating common-reference frame. The equivalent iron loss resistance is placed in parallel to the magnetizing branch, as shown in Figure 5.39. The space vector equations for the new machine model in an arbitrary reference frame follow from Figure 5.39 in the form [15, 16]:

$$\begin{aligned} \underline{v}_s &= R_s \underline{i}_s + \frac{d\underline{\psi}_s}{dt} + j\omega_a \underline{\psi}_s \\ 0 &= R_r \underline{i}_r + \frac{d\underline{\psi}_r}{dt} + j(\omega_a - \omega_r)\underline{\psi}_r \end{aligned} \tag{5.21}$$

$$\begin{aligned} \underline{\psi}_s &= L_{\sigma s}\underline{i}_s + L_m \underline{i}_m = L_{\sigma s}\underline{i}_s + \underline{\psi}_m \\ \underline{\psi}_r &= L_{\sigma r}\underline{i}_r + L_m \underline{i}_m = L_{\sigma r}\underline{i}_r + \underline{\psi}_m \end{aligned} \tag{5.22}$$

$$R_{Fe}\underline{i}_{Fe} = L_m \frac{d\underline{i}_m}{dt} + j\omega_a L_m \underline{i}_m \tag{5.23}$$

$$\underline{i}_m + \underline{i}_{Fe} = \underline{i}_s + \underline{i}_r \tag{5.24}$$

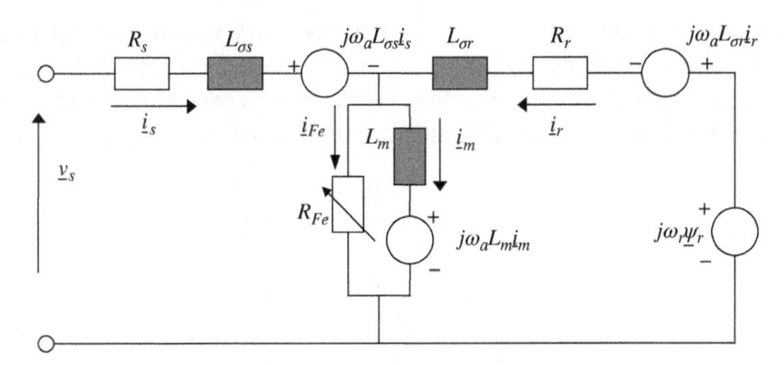

Figure 5.39 Equivalent circuit of an induction machine with iron loss consideration in space vector form in an arbitrary rotating reference frame

$$T_e = \frac{3}{2} P \frac{L_m}{L_{\sigma r}} \left(\psi_{dr} i_{qm} - \psi_{qr} i_{dm} \right) \tag{5.25}$$

The equivalent iron loss resistance is considered as a function of fundamental frequency only in the machine model above, where $R_{Fe} = f(f)$. In reality, equivalent iron loss resistance is a function of both flux density and frequency. However, the experimental method of its determination in [17] has already taken into account the dependency on flux density.

This dynamic model only takes into consideration the fundamental component of the actual iron loss because the equivalent iron loss resistance in Figure 5.39 models only the fundamental component of the iron loss. The value of the equivalent iron loss resistance is calculated using a variable-frequency, no-load test where voltage magnitude is varied using the V/f = const. control law. The source used in the experiment can be either an ideal sinusoidal voltage source (e.g. a synchronous generator) or a PWM voltage source inverter, where measurement of the fundamental component of the input power and current is required. In real operation with PWM voltage source inverter, there will be higher-order voltage harmonics in the supply voltage. These harmonics will "see" the fundamental equivalent iron loss resistance when the model of Figure 5.39 is used in dynamic simulation of a DTC induction motor drive, since the inverter model will be included. This is an inherent limitation of the model in equations (5.18)–(5.23), but is of no consequence when the vector control is considered. In simulation of the vector-controlled drives, the voltage source can be assumed to be ideally sinusoidal without any loss of generality, so that the fundamental equivalent iron loss resistance suffices. However, this is not the case in DTC, where the inverter model has to be included in the study. Since the use of the fundamental equivalent iron loss resistance for higher-order voltage harmonics is of questionable accuracy, simulation results obtained with the model of Figure 5.39 will exhibit a change in the torque ripple characteristic. Unfortunately, there appears to be no better solution to the problem. We would ideally need to use the theory of multiple reference frames, in which an appropriate circuit, similar to the one of Figure 5.39, would be devised for each voltage harmonic. Equivalent iron loss resistance values in these harmonic equivalent circuits could be obtained using the extension of the iron loss identification procedure proposed in [17]. However, since DTC relies on hysteresis control, voltage harmonics that will be present in the inverter output voltage cannot be determined in advance; moreover, output voltage spectrum is continuous, so application of the multiple reference frame theory is not possible.

The complete model in the stationary reference frame, convenient for dynamic simulation, is obtained from equations (5.19) to (5.25) in the following forms:

$$
\begin{aligned}
v_{\alpha s} &= R_s i_{\alpha s} + L_{\sigma s}\frac{di_{\alpha s}}{dt} + L_m\frac{di_{\alpha m}}{dt}\\
v_{\beta s} &= R_s i_{\beta s} + L_{\sigma s}\frac{di_{\beta s}}{dt} + L_m\frac{di_{\beta m}}{dt}
\end{aligned}
\tag{5.26}
$$

$$
\begin{aligned}
0 &= R_r i_{\alpha r} + L_{\sigma r}\frac{di_{\alpha r}}{dt} + L_m\frac{di_{\alpha m}}{dt} + \omega_r\left(L_{\sigma r} i_{\beta r} + L_m i_{\beta m}\right)\\
0 &= R_r i_{\beta r} + L_{\sigma r}\frac{di_{\beta r}}{dt} + L_m\frac{di_{\beta m}}{dt} - \omega_r\left(L_{\sigma r} i_{\alpha r} + L_m i_{\alpha m}\right)
\end{aligned}
\tag{5.27}
$$

$$
i_{\alpha m} + i_{\alpha Fe} = i_{\alpha s} + i_{\alpha r} \quad i_{\beta m} + i_{\beta Fe} = i_{\beta s} + i_{\beta r}
\tag{5.28}
$$

$$
R_{Fe} i_{\alpha Fe} = L_m\frac{di_{\alpha m}}{dt} \quad R_{Fe} i_{\beta Fe} = L_m\frac{di_{\beta m}}{dt}
\tag{5.29}
$$

$$
T_e = \frac{3}{2}P\frac{L_m}{L_{\sigma r}}\left[i_{\beta m}\left(L_{\sigma r} i_{\alpha r} + L_m i_{\alpha m}\right) - i_{\alpha m}\left(L_{\sigma r} i_{\beta r} + L_m i_{\beta m}\right)\right]
\tag{5.30}
$$

$$
T_e - T_L = \frac{J}{P}\frac{d\omega_r}{dt}
\tag{5.31}
$$

The state-space variables of the differential equations are selected as $\alpha-\beta$ components of stator current, rotor current, and magnetizing current, as well as angular electrical speed of rotor. The final set of differential equations is derived from equations (5.26) to (5.31) in a straightforward manner, by eliminating iron loss current components from equation (5.29) using equation (5.27).

Actual variation of the equivalent iron loss resistance with fundamental frequency is given in equation (5.32) for a three-phase squirrel cage, 4 kW, 380 V, 8.7 A, 4-pole, 50 Hz, 1440 rpm induction motor. The fundamental iron loss component was determined experimentally, using the procedure of [17]. The testing was performed in the frequency range from 10 Hz up to 100 Hz. The base speed region, relevant for the analysis and simulation in this chapter, encompasses frequencies up to 50 Hz. When the fundamental iron loss component, as a function of the operating frequency, is known, the equivalent iron loss resistance of Figure 5.39 can be easily calculated. Figure 5.40 presents the measurement results related to the fundamental iron loss component (Figure 5.40a) and corresponding equivalent iron loss resistance (Figure 5.40b). The appropriate analytical approximation of the equivalent iron loss resistance, required for the circuit of Figure 5.39, is also shown in Figure 5.40b. Analytical approximation of the fundamental iron loss, included in Figure 5.40a, will be required later for compensation purposes. These analytical functions, obtained using least squares fitting, are given with the following expressions, in terms of fundamental frequency:

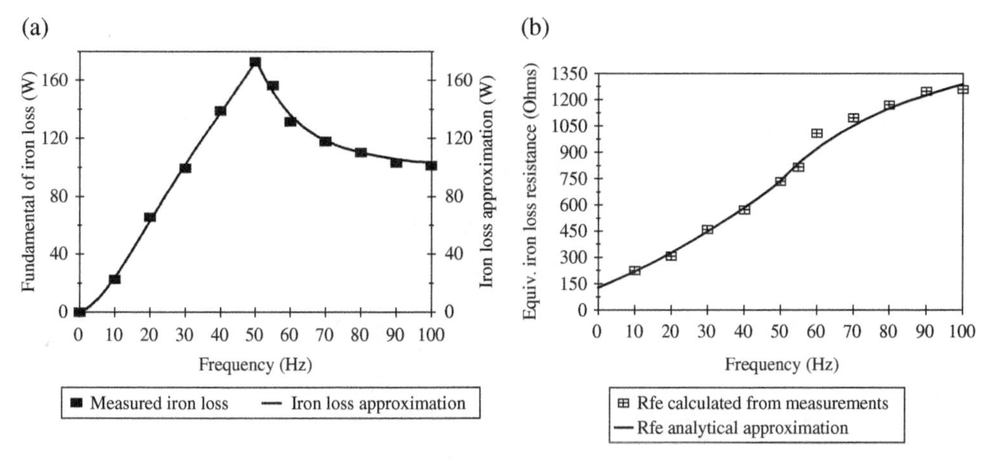

Figure 5.40 Fundamental iron loss component: (a) calculated equivalent iron loss resistance; and (b) experimentally identified points and corresponding analytical approximations

$$P_{Fe}(\mathrm{W}) = \begin{cases} 0.00003808f^4 - 0.004585f^3 + 0.183f^2 + 1.0254f - 0.2784 & f \leq 50\ \mathrm{Hz} \\ 0.00002087f^4 - 0.0073f^3 + 0.9658f^2 - 57.684f + 1468.3 & f > 50\ \mathrm{Hz} \end{cases}$$

$$R_{Fe}(\Omega) = \begin{cases} 128.92 + 8.242f + 0.0788f^2 & f \leq 50\ \mathrm{Hz} \\ 1841 - 55275/f & f > 50\ \mathrm{Hz} \end{cases} \tag{5.32}$$

5.4.2 MATLAB/SIMULINK Simulation of the Effects of Iron Losses in Torque Control and Speed Control

5.4.2.1 Dynamic Machine Model with Consideration of Iron Losses

A new MATLAB/SIMULINK model based on a dynamic machine model with iron loss consideration is developed. This new simulation model will then replace the model of ideal machine in simulation of DTC in torque control mode and speed control mode. Figure 5.41 below shows this new model for an induction machine with iron losses in a DTC torque control MATLAB/SIMULINK program. Figure 5.42 illustrates the block diagram of the model.

The details of the block "Induction Motor Dynamic Model" in Figure 5.42 are shown below in Figure 5.43. Details of the differential equations are demonstrated in Figures 5.44–5.46.

The variables in the differential equations in the above figures are components of stator current, rotor current, air-gap flux, and rotor speed.

The structural diagram of the block "Frequency Calculator" is shown in Figure 5.47. The frequency is calculated from the components of stator flux. The components of stator flux are determined from the components of stator current and air-gap flux. The equation for frequency determination is

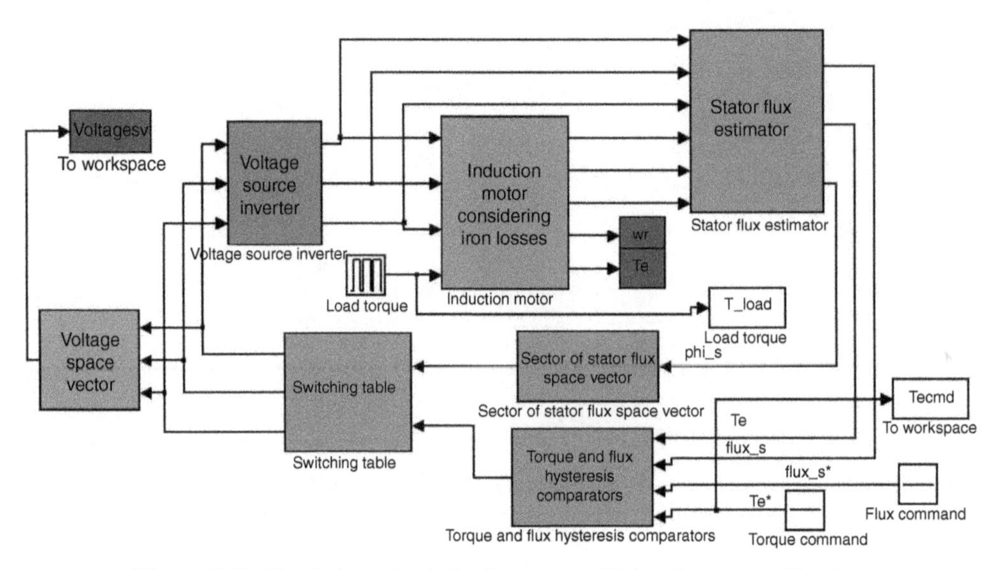

Figure 5.41 Simulation of an induction motor with iron losses consideration

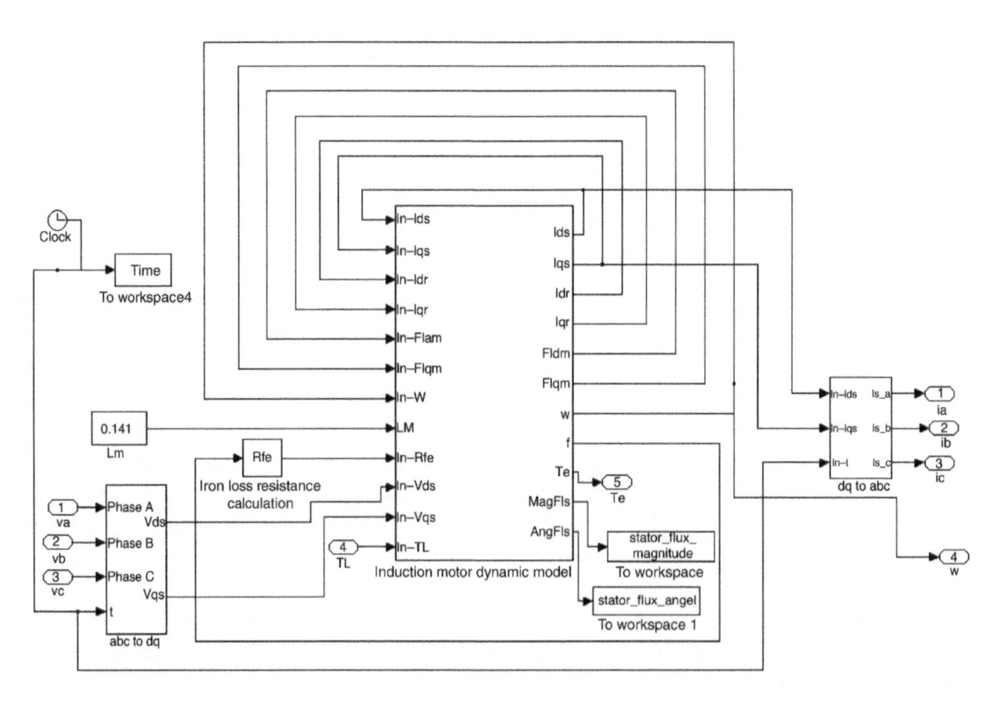

Figure 5.42 Block diagram model for an induction motor with iron losses

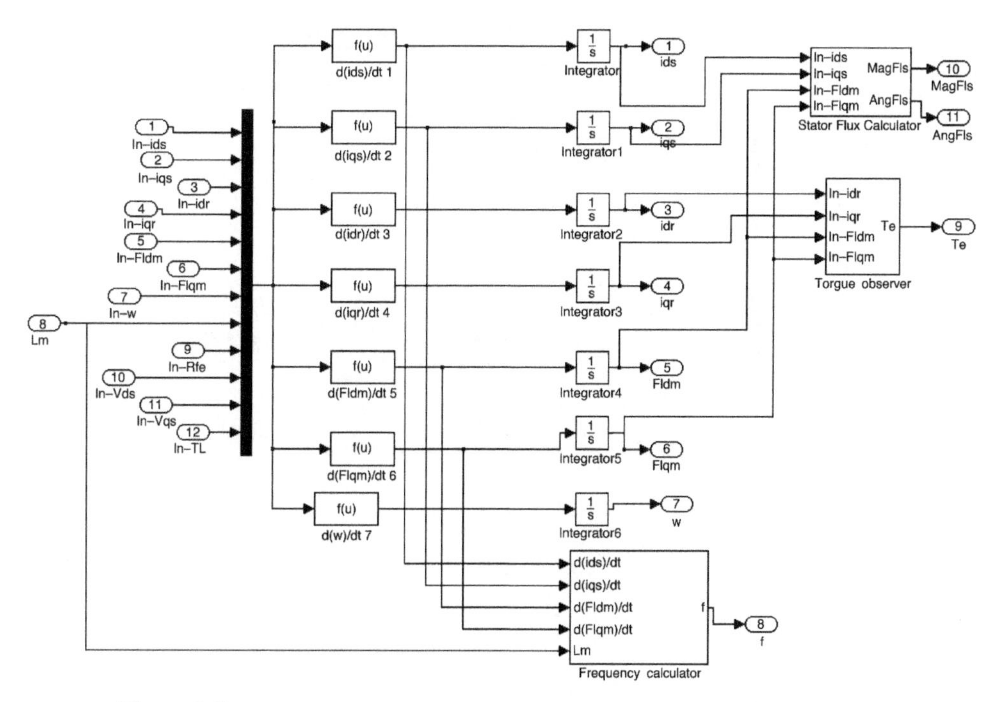

Figure 5.43 Representation of differential equations in a model with iron losses

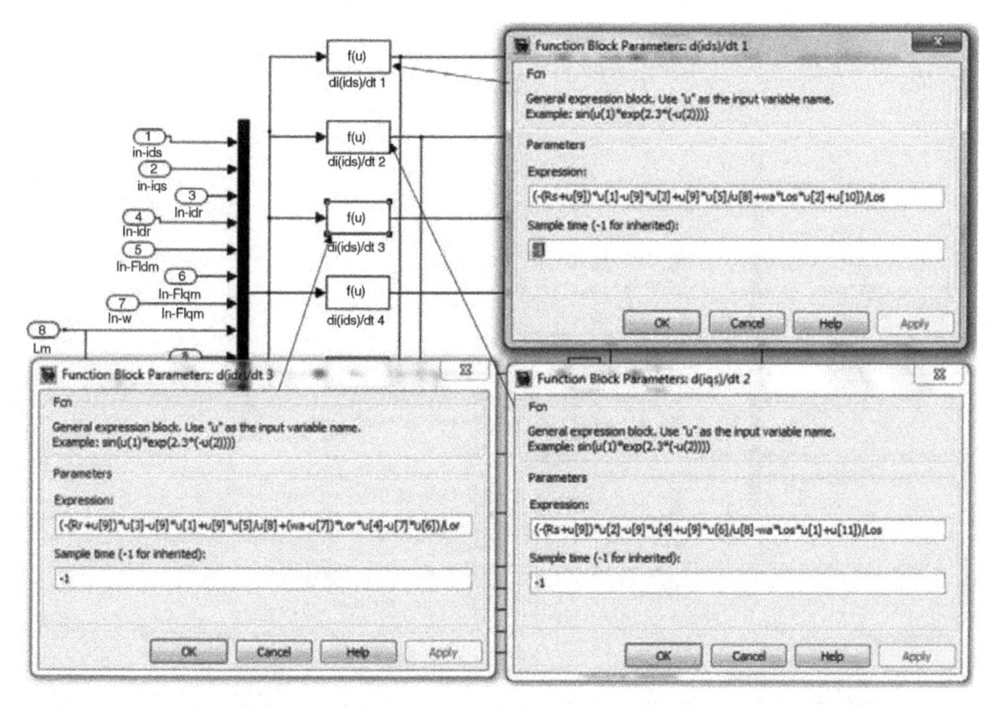

Figure 5.44 Expressions for the blocks "*d(ids)/dt*," "*d(iqs)/dt*," and "*d(idr)/dt*" (*Source:* SIMULINK software)

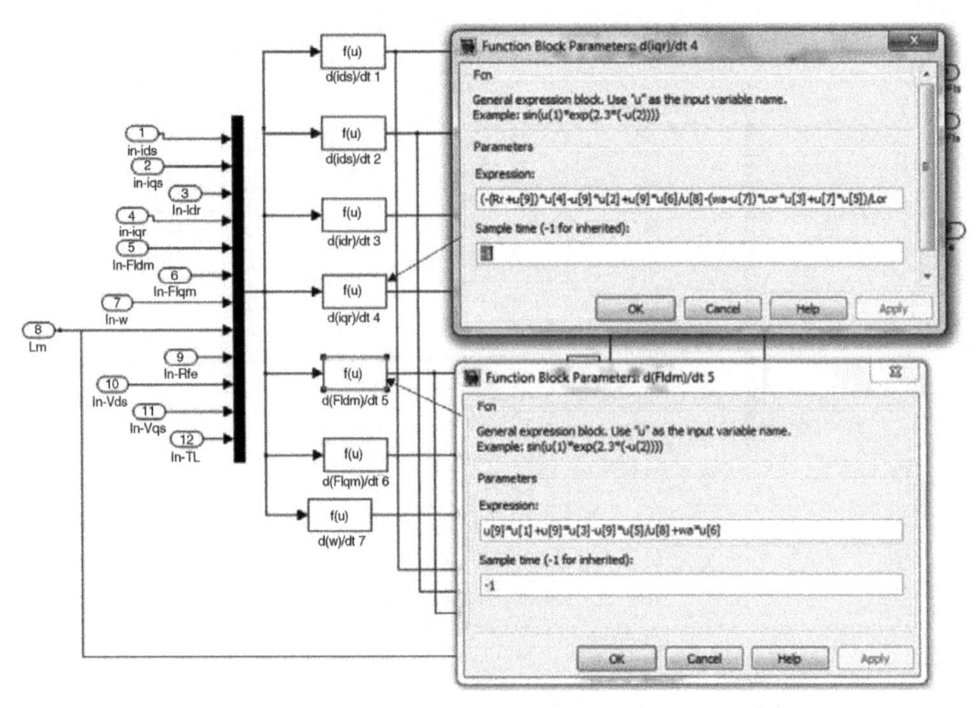

Figure 5.45 Expressions for the blocks "*d(iqr)/dt*" and "*d(Fldm)/dt*" (*Source:* SIMULINK software)

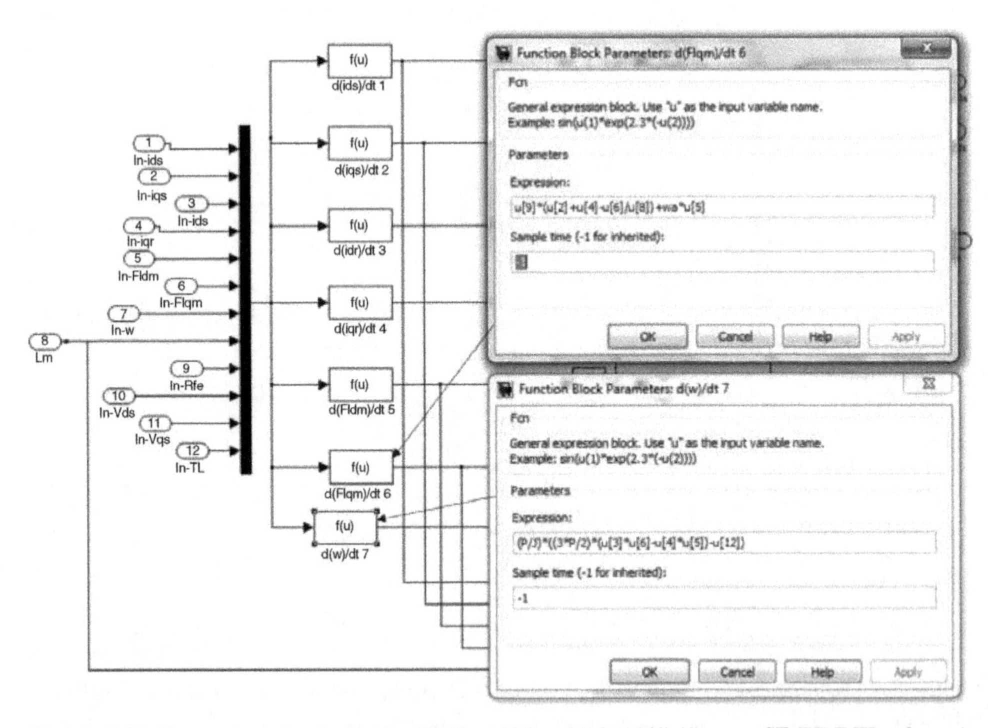

Figure 5.46 Expressions for the blocks "*d(Flqm)/dt*" and "*d(w)/dt*" (*Source:* SIMULINK software)

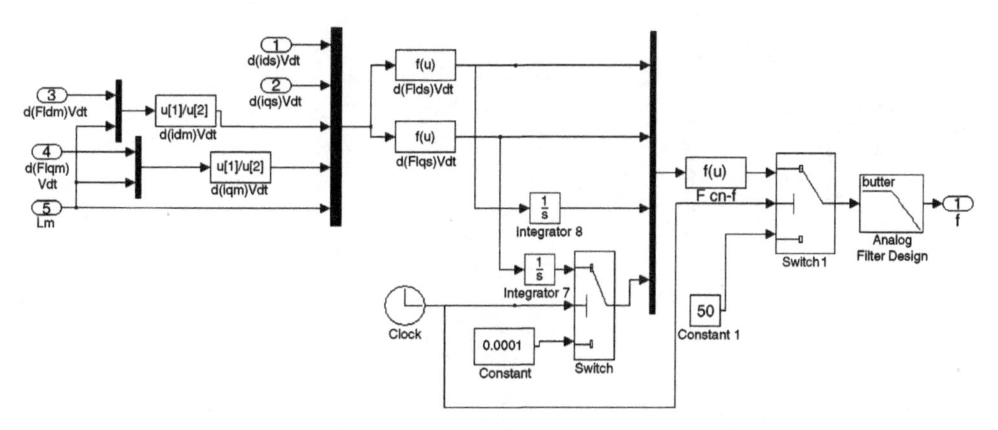

Figure 5.47 Structural diagram of the block "Frequency Calculator" (*Source:* SIMULINK software)

$$f = \frac{1}{2\pi}\frac{\psi_{\alpha s}d\psi_{\beta s}/dt - \psi_{\beta s}d\psi_{\alpha s}/dt}{\psi_{\alpha s}^2 + \psi_{\beta s}^2} \tag{5.33}$$

Because the values of the denominator in equation (5.33) are small when the simulation starts running, the values of the variables are artificially kept for the first few step sizes of the simulation to avoid the error "NaN" (not a number). The frequency is also kept at 50 Hz during this period. Due to the small step size of the simulation, this constant value does not affect the results of the simulation process. The calculated frequency then goes through a Butterworth low-pass filter before being used to determine the iron loss equivalent resistance.

Equations for calculating the derivatives of α and β components of stator flux to be used in equation (5.33) are shown in Figure 5.48. Figure 5.49 demonstrates implementation of equation (5.33). The parameters of the Butterworth filter are also presented in Figure 5.49.

The selecting value for the switches is 5 μs, as shown in Figure 5.50.

Figure 5.51 shows the block diagram and expression in "Stator Flux Calculator" in Figure 5.43. Figure 5.52 illustrates the diagram and expression for electromagnetic torque estimation in the "Torque Observer" block of Figure 5.43.

5.4.2.2 Simulation of DTC with Iron Losses Consideration in Torque Control Mode

The torque response obtained from the simulation of DTC in the torque control mode, when iron losses are considered, is illustrated in Figure 5.53. The torque command is still the same as the one in the simulations of Section 5.3 (26.5 Nm). Similarly, flux command remains at 0.9084. The induction motor is still the same, only changing the dynamic model from one with constant parameter and no iron loss consideration to the one that includes iron losses as a function of frequency.

As can be seen in Figure 5.53 that the electromagnetic torque of the motor cannot reach the commanded value as it does in the case of ideal machine model due to the iron losses. Therefore, the actual torque created by the motor is lower than the expected value from the ideal

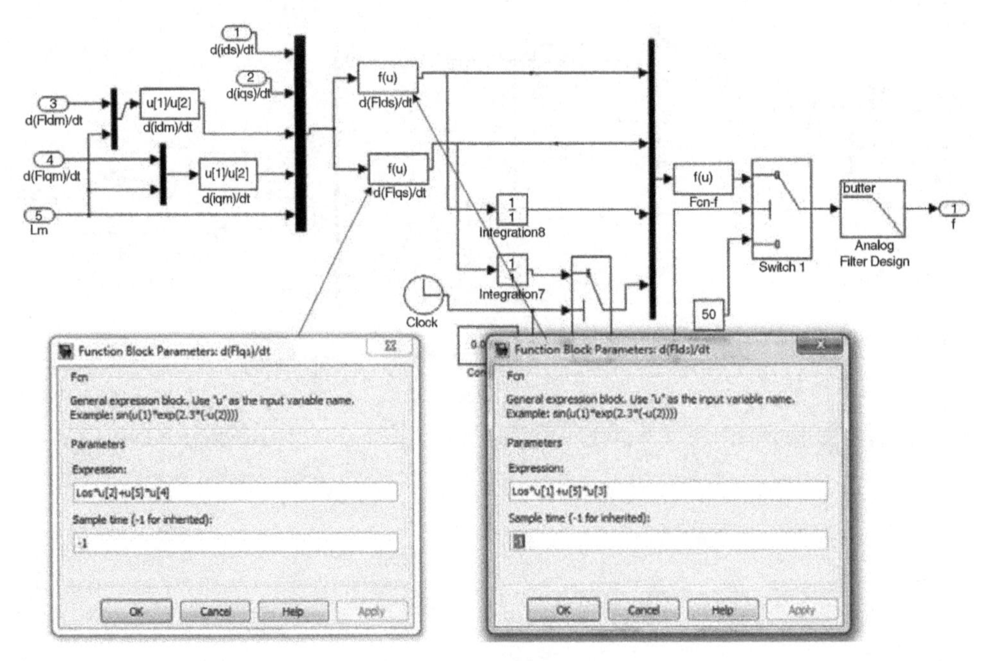

Figure 5.48 Equations for derivatives of α and β components of stator flux (*Source:* SIMULINK software)

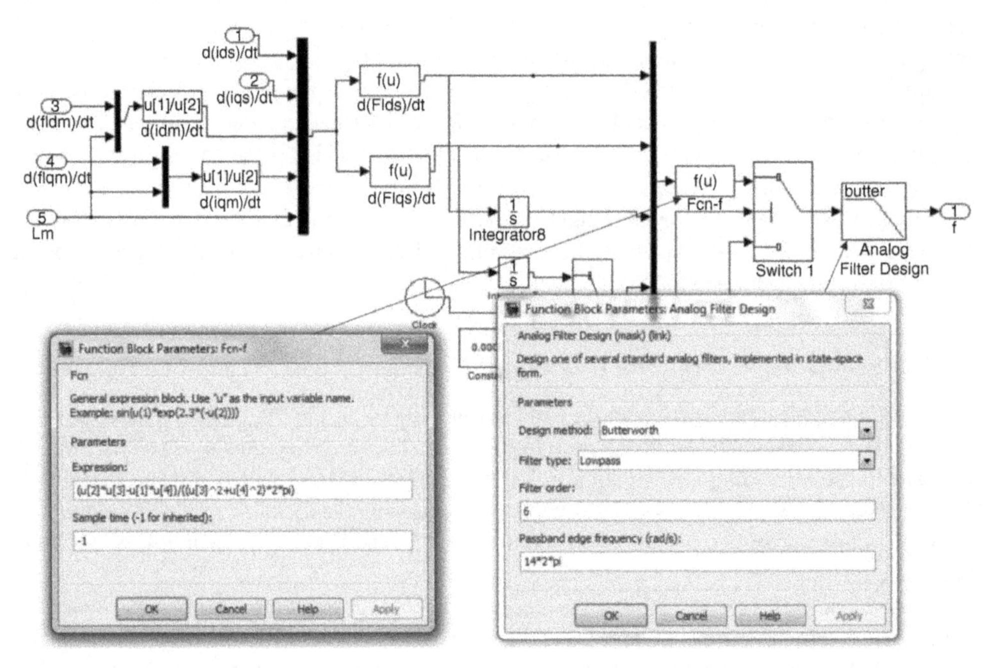

Figure 5.49 Expression in "Fcn-f" block and parameters for Butterworth filter (*Source:* SIMULINK software)

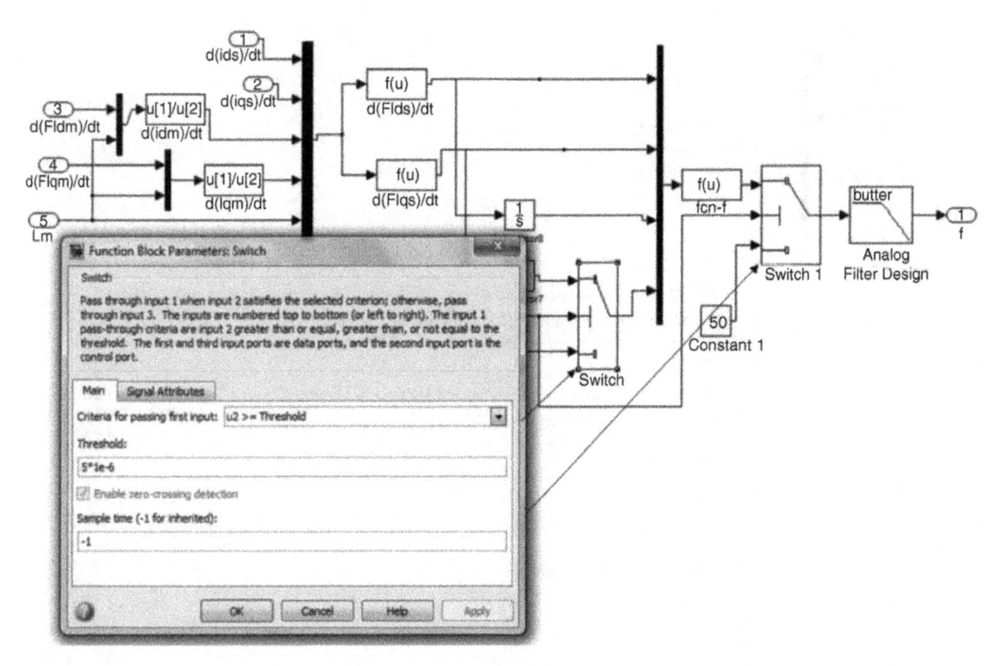

Figure 5.50 Selecting a value for the switches in Figure 5.47 (*Source:* SIMULINK software)

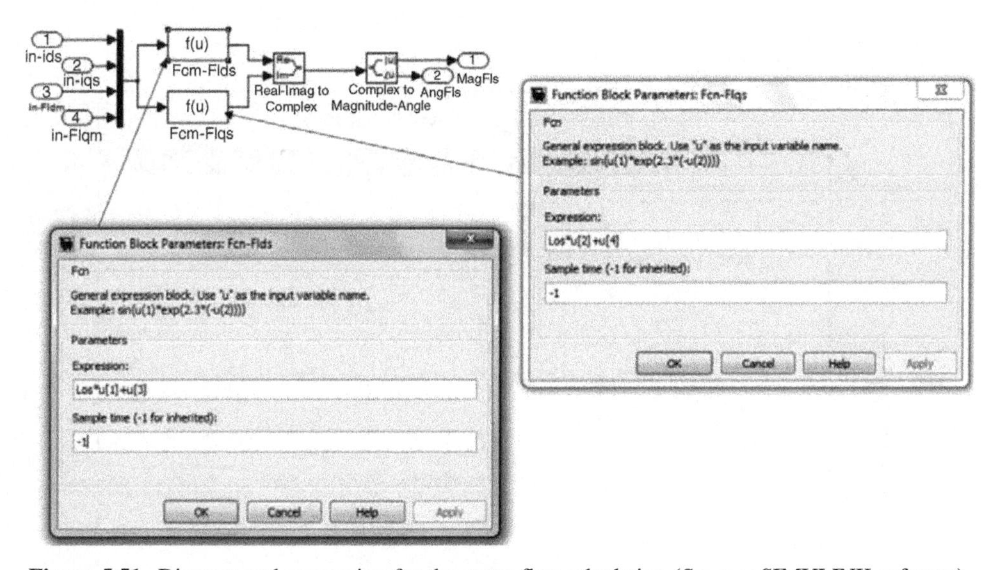

Figure 5.51 Diagram and expression for the stator flux calculation (*Source:* SIMULINK software)

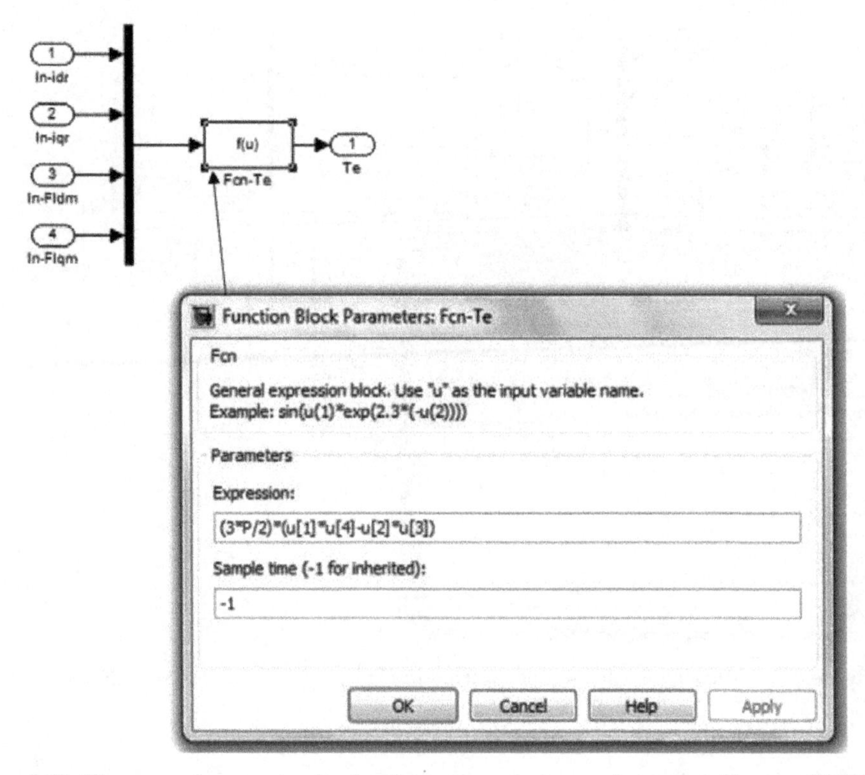

Figure 5.52 Diagram and expression for the electromagnetic torque observation (*Source:* SIMULINK software)

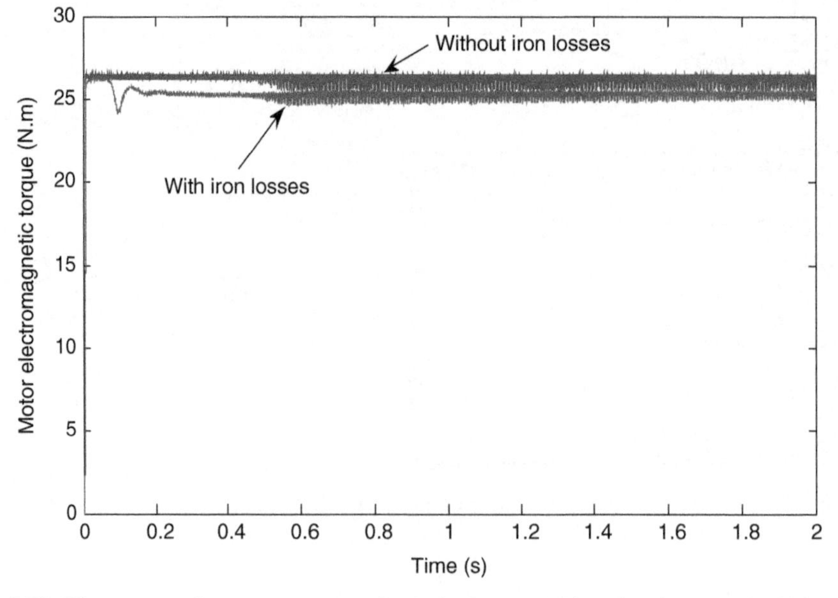

Figure 5.53 Electromagnetic torque responses for the both cases, without iron losses and with iron losses consideration

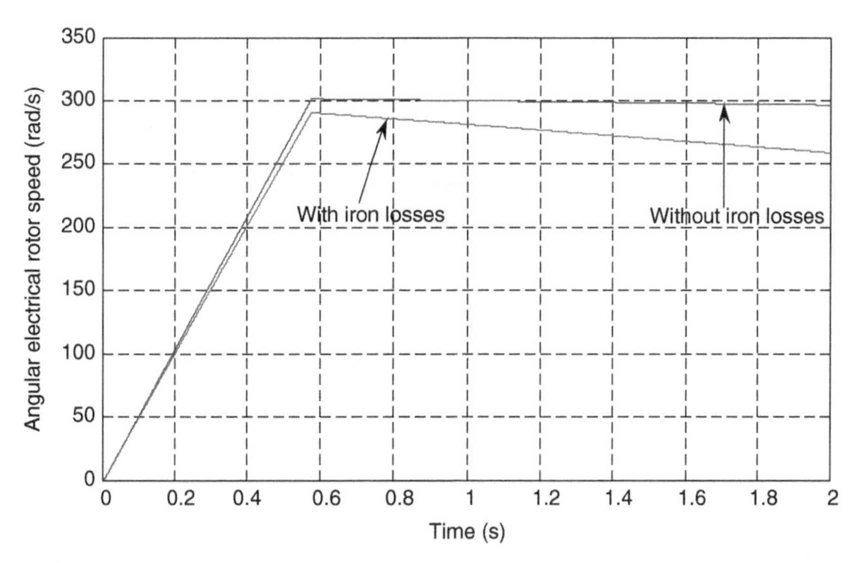

Figure 5.54 Rotor speed responses for the both cases, without iron losses and with iron losses consideration

machine model. This deficiency of the electromagnetic torque will cause inaccuracy in speed response, as shown in Figure 5.54. In this figure, the speed response of the ideal motor reaches the rated electrical speed when the rated load torque is applied. However, the rotor speed does not reach that rated value when iron losses are considered in the machine model because of the deficiency of torque, as explained above and shown in Figure 5.55. The difference in torque responses during the steady state in Figure 5.55 results in a quick decrease of rotor speed, in Figure 5.54, when iron losses are included. Stator flux and phase current responses are not affected by the inclusion of iron losses in the machine model, as demonstrated in Figures 5.56 and 5.57.

5.4.2.3 Simulation of DTC with Iron Losses Consideration in Speed Control Mode

The ideal constant parameter machine model in the MATLAB/Simulink program for speed control of DTC in Section 5.3.4.1 is replaced with the machine model considering iron loss, as presented in Section 5.4.2.1. The parameters of the PID speed controller are also adjusted to get a good response of the rotor speed. Figure 5.58 shows the response of the rotor speed during the start-up, application, and removal of load torque.

The speed response, when iron losses are included, is a little slower than the one obtained with an ideal constant machine model. This is due to the lower actual electromagnetic torque produced by the motor when output power is reduced by iron losses. However, the steady-state response of the speed is acceptable. The rotor speed reaches the commanded value.

When the load torque is applied instantaneously, the speed goes down deeper for the case with iron loss. The recovering period of the speed response is therefore slightly longer for the case with iron loss. The steady-state error of speed response is the same for both cases; rotor

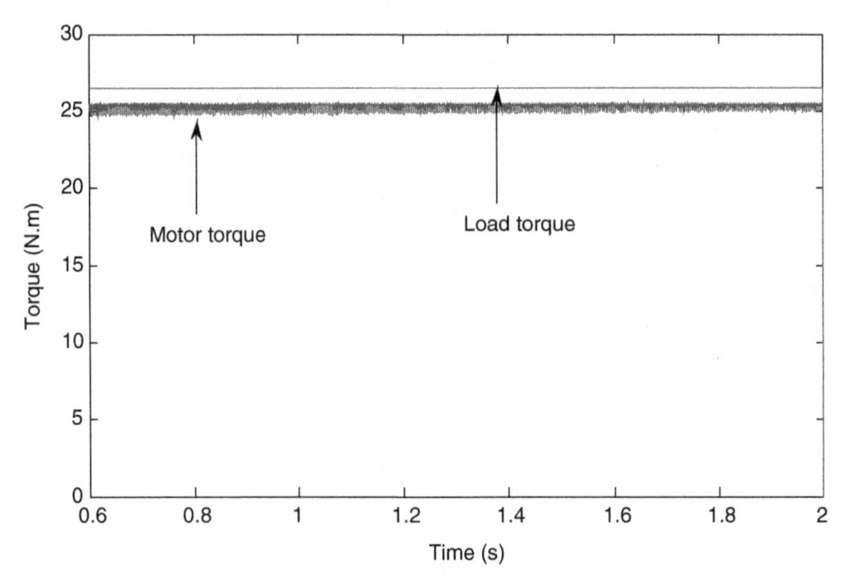

Figure 5.55 Load torque and electromagnetic torque of the induction motor with iron losses consideration during steady state

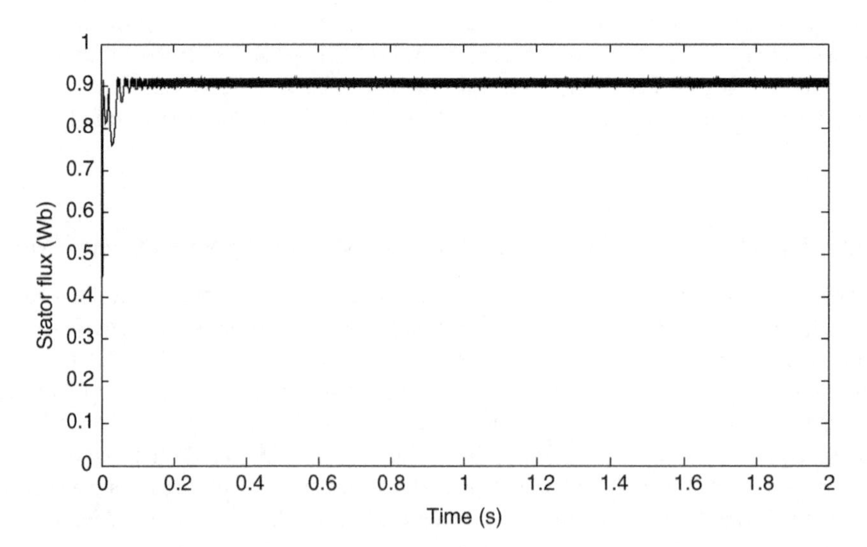

Figure 5.56 Stator flux response when iron losses are considered

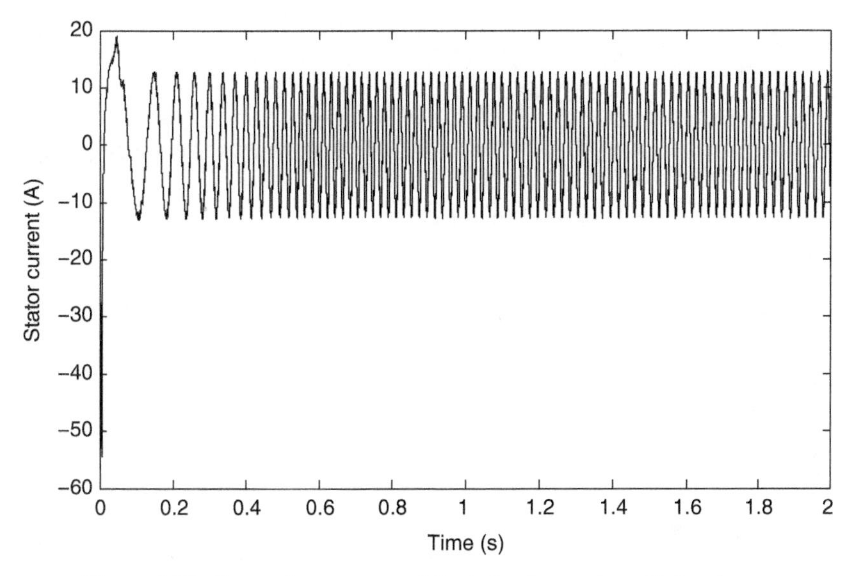

Figure 5.57 Stator phase current of the induction motor with iron losses

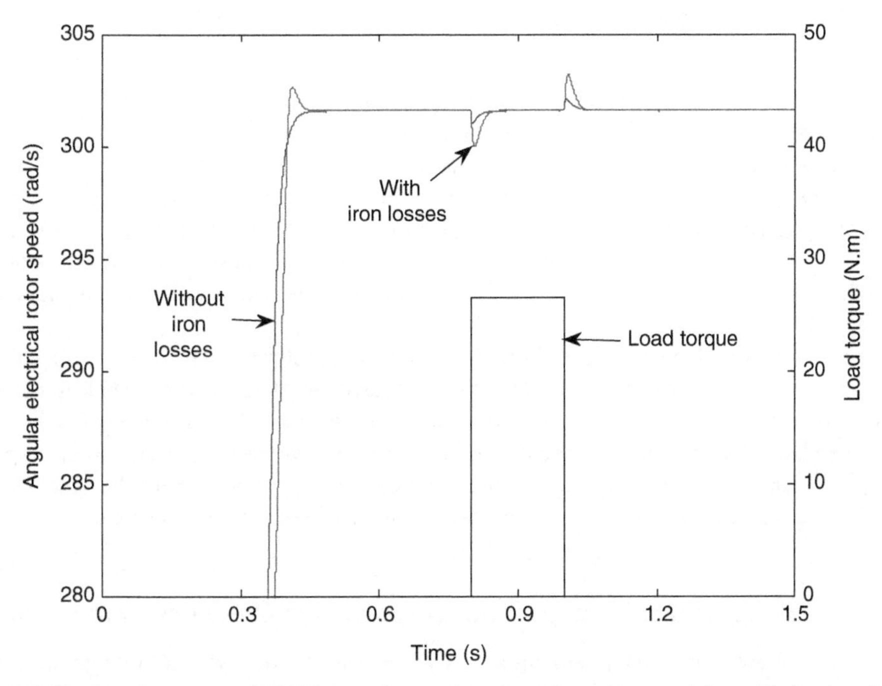

Figure 5.58 Speed response of DTC in speed control mode with machine model considering iron losses

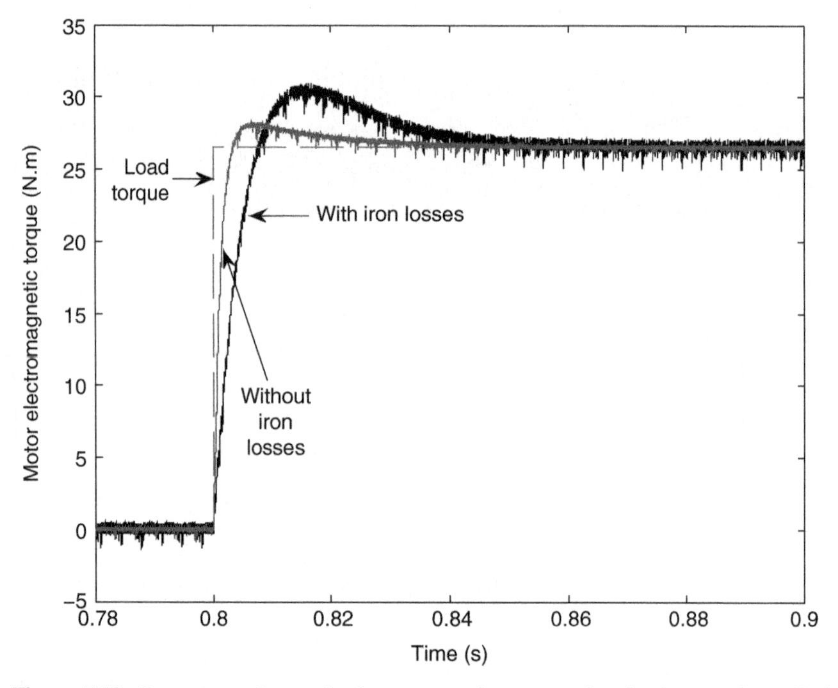

Figure 5.59 Responses of motor's electromagnetic torque when load torque is applied

speed returns to commanded value in both cases. A similar difference between the two cases is observed when load torque is removed instantaneously.

Figures 5.59 and 5.60 show the responses of the motor's electromagnetic torque when load torque is applied and removed, respectively. In both figures, the torque responses, when iron losses are ignored and included, are presented. From Figure 5.59, torque response in the case of iron losses has higher torque ripple in steady state and longer response time compared with the case when iron losses are not considered. Similar deficiency is also observed when load torque is removed.

The performance of speed control in DTC of induction motor is generally not affected seriously by the presence of iron losses. The transient response of speed is affected by the deficiency in torque response. Simple compensation schemes for torque deficiency are recommended. More complex compensation schemes can give more accurate torque response during transient state; however, they require more computing capabilities of the systems. The next sub-section will discuss a simple compensation scheme for torque response.

5.4.3 Modified Direct Torque Control Scheme for Iron Loss Compensation

It has been shown that iron losses affect only the performance of electromagnetic torque response of the motor. Stator flux response is unchanged when iron losses are considered. Therefore, only compensation for iron losses is discussed. Equation (5.19), used for torque estimation in DTC, is based on the ideal machine model; therefore, it does not realize the power

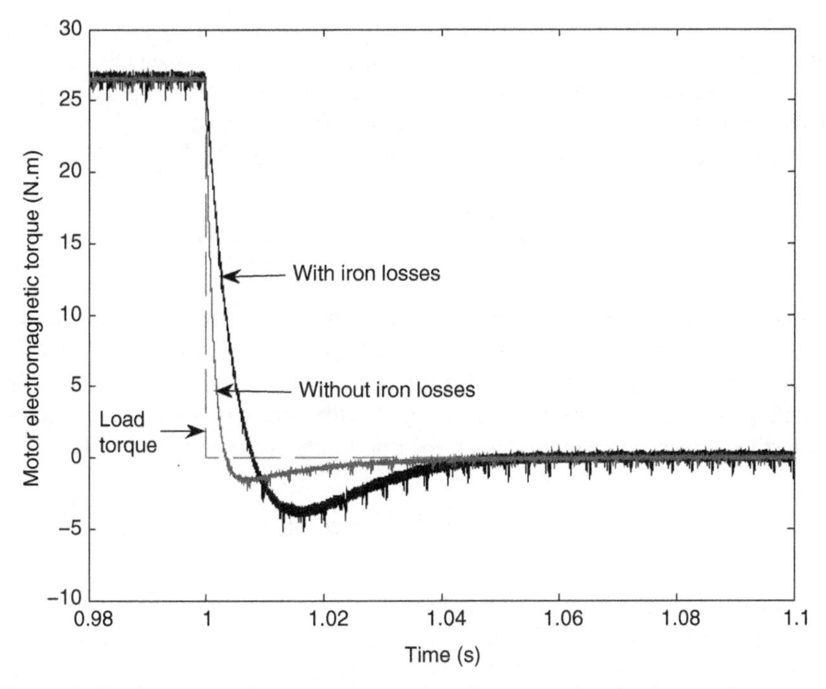

Figure 5.60 Responses of motor's electromagnetic torque when load torque is removed

deficiency by the existence of iron losses. The estimated value of electromagnetic torque is higher than the actual one. A reduction of estimated torque is introduced in [18] and is expressed by

$$T_e' = (3/2)P\left(\psi_{\alpha s} i_{\beta s} - \psi_{\beta s} i_{\alpha s}\right) - \Delta T_{Fe} \tag{5.34}$$

The mechanical power produced by the motor is theoretically assumed as $T_e\omega_m$, while the actual output power is $T_e'\omega_m$ [18], and ω_m is the mechanical angular speed of rotor rotation ($\omega_m = \omega_r/P$). The torque compensation ΔT_{Fe} in equation (5.34) can be obtained from the following equations [18]:

$$\begin{aligned} T_e\omega_m &= T_e'\omega_m + P_{Fe} \\ \Delta T_{Fe} &= P_{Fe}/\omega_m \\ P_{Fe} &= \Delta T_{Fe}\omega_m \end{aligned} \tag{5.35}$$

The symbol P_{Fe} stands for the fundamental component of the iron losses. Equation (5.34) holds true for both modes of operation of the motor, braking and motoring. The iron losses will lower the output torque in motoring, while the torque is higher in magnitude than the value obtained by equation (5.19) during braking.

The iron losses, as shown in equation (5.32), are a function of frequency. However, stator frequency is generally not available in DTC. The angular electrical speed of rotor can be used instead to approximate the iron losses. The difference between stator angular frequency and angular electrical speed of the rotor is a small slip frequency, and rotor speed is usually measured or estimated. The torque compensation now can be calculated as [19]

$$\Delta T_{Fe} = \frac{P_{Fe}(f)}{\omega_m} \quad \text{or} \quad \Delta T_{Fe} = \frac{P_{Fe}(P\omega_m/2\pi)}{\omega_m} \tag{5.36}$$

Experimentally identified fundamental component of iron losses often varies almost linearly with the stator frequency [20–22]. From Figure 5.40, the variation of the fundamental iron loss with frequency in the range from 0 to 50 Hz for the motor considered here is almost linear. By approximating the iron loss in the base speed region with a linear dependence on frequency, the simplest method of constant compensation for iron losses is suggested by [19]

$$\Delta T_{Fe} = \frac{P_{Fen}}{P_n} T_{en} = \text{const.} \tag{5.37}$$

As the fundamental iron loss at 50 Hz equals 173.4 W, while rated motor power and torque are 4 kW and 26.5 Nm, respectively, the constant torque compensation is 1.15 Nm.

The simulation results of DTC with constant compensation for iron losses in torque control mode are demonstrated in Figures 5.61 and 5.62.

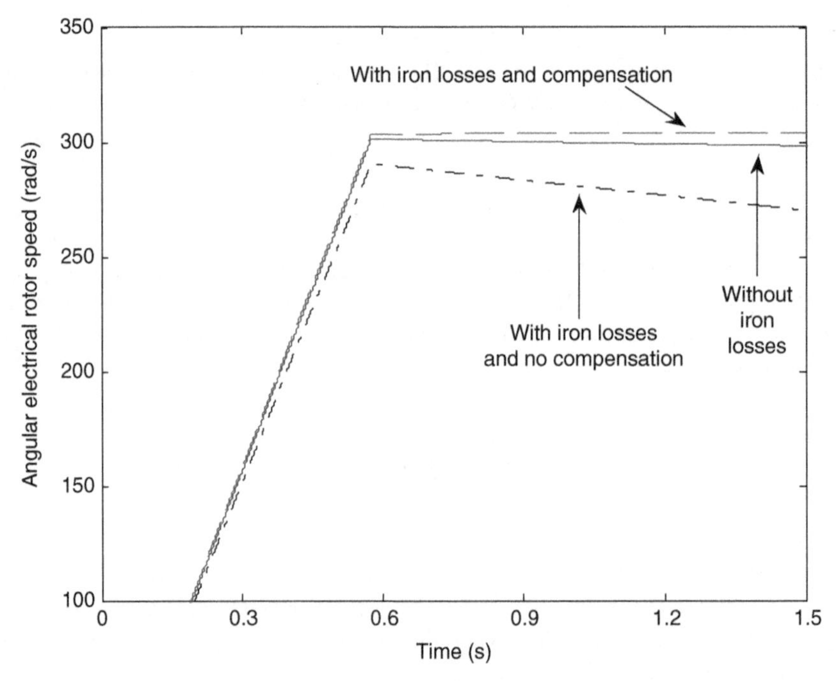

Figure 5.61 Rotor speed responses of DTC in torque control mode for the three cases: without iron losses; with iron losses but no compensation; with iron losses and constant compensation

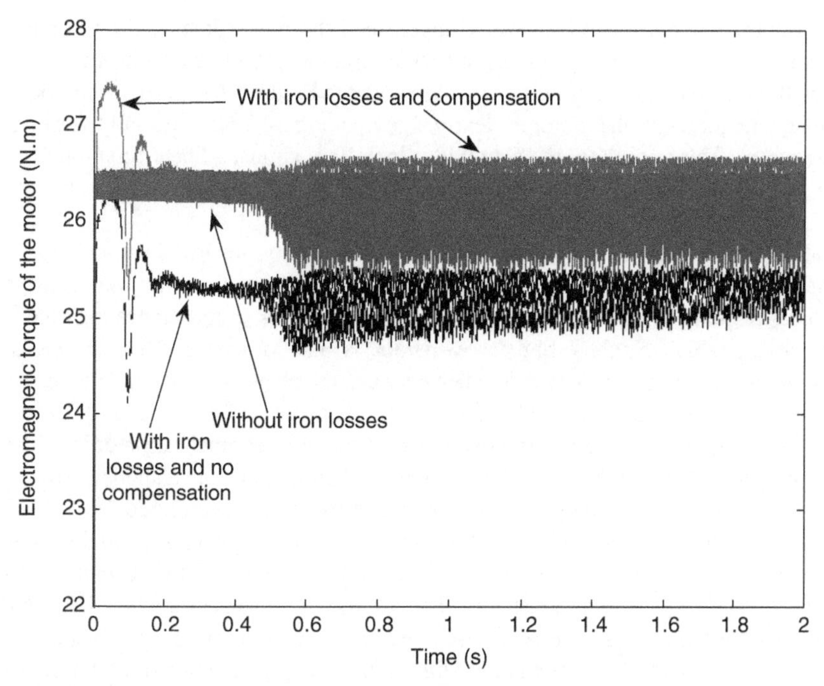

Figure 5.62 Torque responses of DTC in torque control mode for the three cases: without iron losses; with iron losses but no compensation; with iron losses and constant compensation

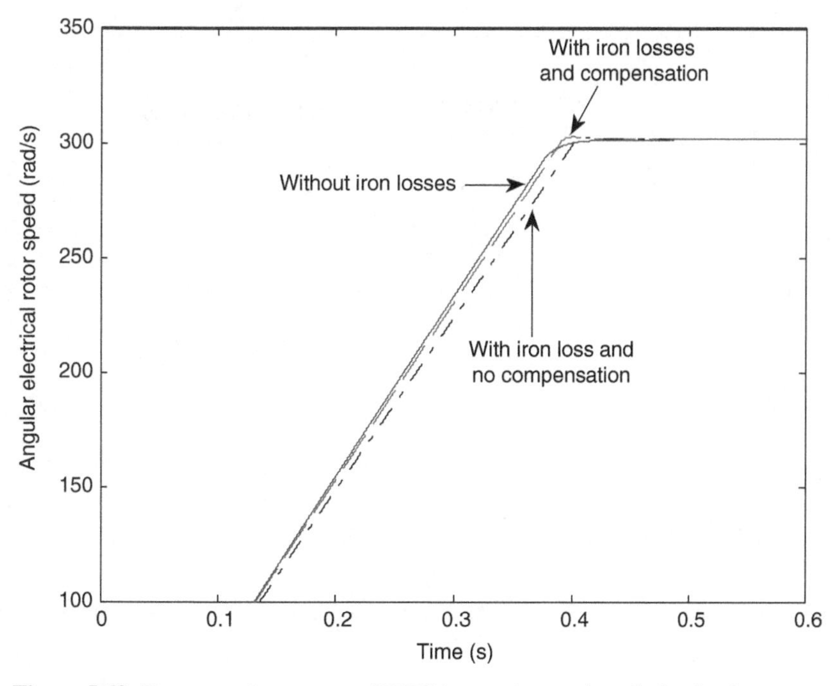

Figure 5.63 Rotor speed responses of DTC in speed control mode for the three cases

The proposed improvement has over-compensated the power lost by the presence of iron losses by a small amount. The speed response with iron losses compensation is even better than the one without consideration of iron losses, as shown in Figure 5.61. The overcompensation in electromagnetic torque of the motor is clearly shown in Figure 5.62, especially during the start-up of the motor. This is because the actual iron losses are almost a linear function of the stator frequency in the low speed region and are much less than the compensated value during the start-up.

The simulation results of DTC with constant compensation for iron losses in speed control mode are demonstrated in Figures 5.63 and 5.64. During start-up of the motor, as shown in Figure 5.63, response time of the speed with compensation is slightly more than the one with ideal machine model. Slightly higher overshoot is also observed when compensation is included. However, there is a clear difference between the responses with and without compensation when iron losses are included.

When load torque is applied and removed instantaneously, compensation of iron losses does not improve the performance of speed, as shown in Figure 5.64. The responses are similar for the cases with and without compensation when iron losses are considered.

In summary, iron losses consideration for DTC of induction motor has been discussed in this sub-section. Simulation of the effects of iron losses on performance of DTC in both torque and speed control mode has been demonstrated. The results have also been analyzed. From the analysis, compensation methods have been discussed and simulation of the simplest one has been presented. The next section of this chapter will discuss the consideration of both iron losses and magnetic saturation in DTC.

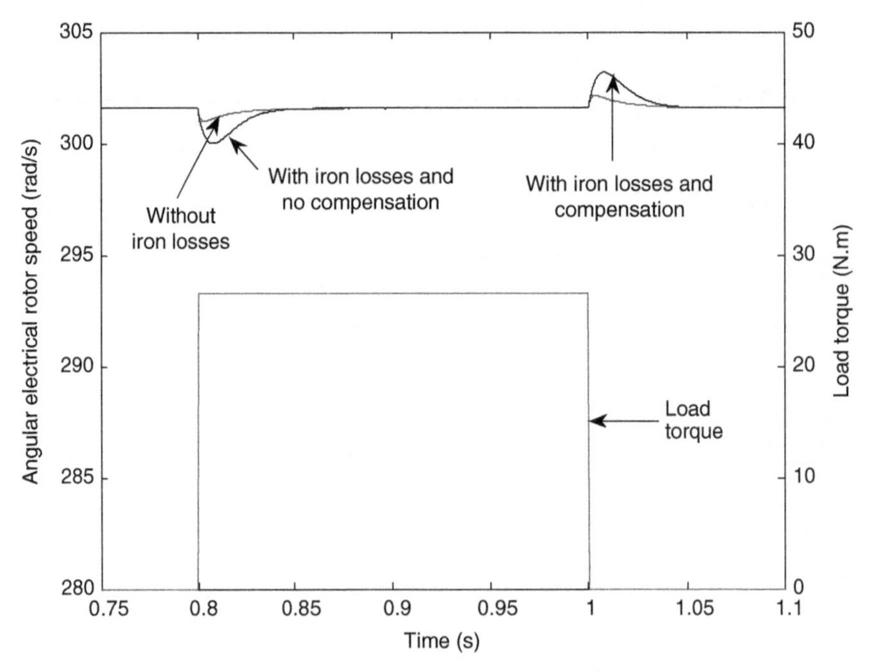

Figure 5.64 Speed responses of the three cases when load torque is applied and removed

5.5 DTC of Induction Motor with Consideration of Both Iron Losses and Magnetic Saturation

5.5.1 Induction Machine Model with Consideration of Iron Losses and Magnetic Saturation

Representations of iron losses and magnetic saturation in dynamic space vector equivalent circuit of induction machines in an arbitrary frame of reference rotating at angular speed ω_a are shown in Figure 5.65.

Differential equations of the machine model are [23].

$$
\begin{aligned}
v_{\alpha s} &= R_s i_{\alpha s} + L_{\sigma s}\frac{di_{\alpha s}}{dt}\frac{d\psi_{\alpha m}}{dt} \\
v_{\beta s} &= R_s i_{\beta s} + L_{\sigma s}\frac{di_{\beta s}}{dt}\frac{d\psi_{\beta m}}{dt}
\end{aligned}
\tag{5.38}
$$

$$
\begin{aligned}
0 &= R_r i_{\alpha r} + L_{\sigma r}\frac{di_{\alpha r}}{dt} + \frac{d\psi_{\alpha m}}{dt} + \omega\left(L_{\sigma r} i_{\beta r} + \psi_{\beta m}\right) \\
0 &= R_r i_{\beta r} + L_{\sigma r}\frac{di_{\beta r}}{dt} + \frac{d\psi_{\beta m}}{dt} + \omega\left(L_{\sigma r} i_{\alpha r} + \psi_{\alpha m}\right)
\end{aligned}
\tag{5.39}
$$

$$
\begin{aligned}
0 &= \frac{d\psi_{\alpha m}}{dt} - R_{Fe} i_{\alpha s} - R_{Fe} i_{\alpha r} + \frac{R_{Fe}}{L_m}\psi_{\alpha m} \\
0 &= \frac{d\psi_{\beta m}}{dt} - R_{Fe} i_{\beta s} - R_{Fe} i_{\beta r} + \frac{R_{Fe}}{L_m}\psi_{\beta m}
\end{aligned}
\tag{5.40}
$$

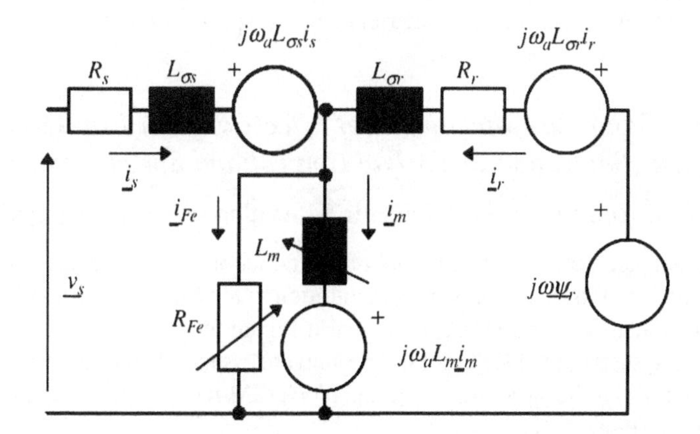

Figure 5.65 Space vector dynamic equivalent circuit of induction machine in an arbitrary reference frame (*Source:* Sokola and Levi [23])

$$\begin{aligned}\psi_{\alpha s} &= L_{\sigma s} i_{\alpha s} + L_m i_{\alpha m} \\ \psi_{\beta s} &= L_{\sigma s} i_{\beta s} + L_m i_{\beta m} \\ \psi_{\alpha r} &= L_{\sigma r} i_{\alpha r} + L_m i_{\alpha m} \\ \psi_{\beta r} &= L_{\sigma r} i_{\beta r} + L_m i_{\beta m}\end{aligned} \tag{5.41}$$

$$\frac{d\omega}{dt} = \frac{P}{J}\left[\frac{3}{2}P\left(i_{\alpha r}\psi_{\beta m} - i_{\beta r}\psi_{\beta m}\right) - T_L\right] \tag{5.42}$$

$$T_e = \frac{3}{2}P\left(i_{\alpha r}\psi_{\beta m} - i_{\beta r}\psi_{\beta m}\right) \tag{5.43}$$

The equivalent iron loss resistance as a function of stator frequency is [23]

$$R_{Fe} = \begin{bmatrix} 128.92 + 8.242f + 0.0788f^2 & (\Omega); f \le 50 \text{ Hz} \\ 1841 - 55272/f & (\Omega); f \ge 50 \text{ Hz} \end{bmatrix} \tag{5.44}$$

The equivalent magnetizing inductance is considered as a nonlinear function of main flux:

$$L_m = f(\psi_m); \text{ where } L_m = \frac{\psi_m}{i_m} \tag{5.45}$$

Magnetizing curve approximation (rms values) is [23].

$$\psi_m = \begin{bmatrix} 0.1964285 i_m; & i_m < 2.2 \text{ A} \\ 0.8374 + 0.0067 i_m - 0.924/i_m; & i_m > 2.2 \text{ A} \end{bmatrix} \tag{5.46}$$

The indices s and r denote stator and rotor parameters and variable, respectively; index σ stands for leakage inductances; index m defines parameters and variables associated with magnetizing flux, magnetizing current, and magnetizing inductances; P is the number of pole pairs; J is the inertia constant; and ω is the electrical angular speed of rotor.

5.5.2 *MATLAB/Simulink Simulation of Effects of Both Iron Losses and Magnetic Saturation in Torque Control and Speed Control*

5.5.2.1 Effects of Iron Losses and Magnetic Saturation in Torque Control Mode

The ideal constant parameter machine model of the induction motor is replaced by the machine model considering iron losses and magnetic saturation in the MATLAB/Simulink program for torque control simulation using DTC, as shown in Figure 5.66.

The other sub-systems of the DTC control system are essentially the same as those shown in Figures 5.11 and 5.41. The structure inside the block "IM model with iron losses and saturation" is shown in Figure 5.67.

In addition to the blocks for transformation of reference frame, there are also blocks for calculation of instantaneous values of magnetizing inductance and equivalent resistance of iron

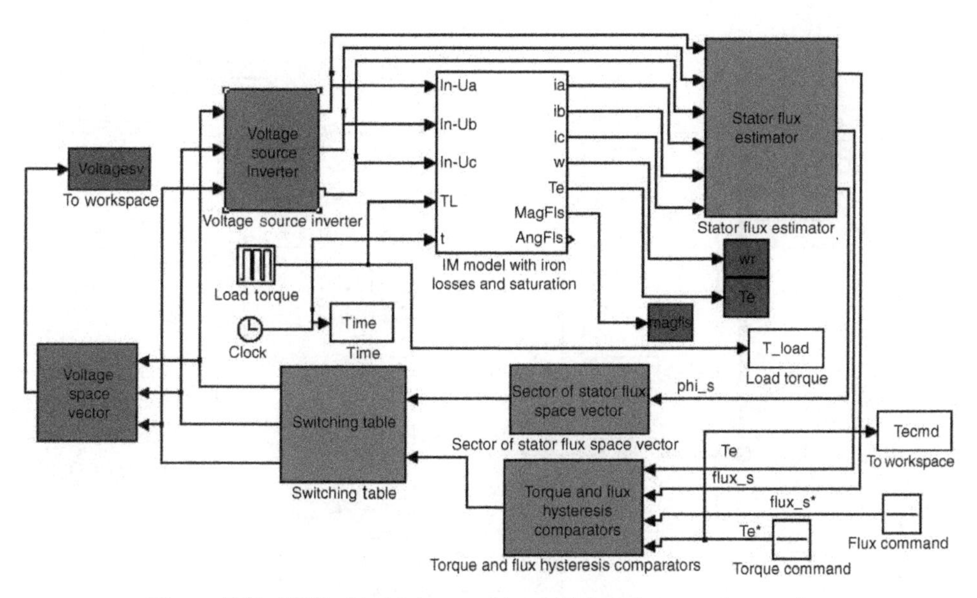

Figure 5.66 DTC of induction machine with iron losses and saturation

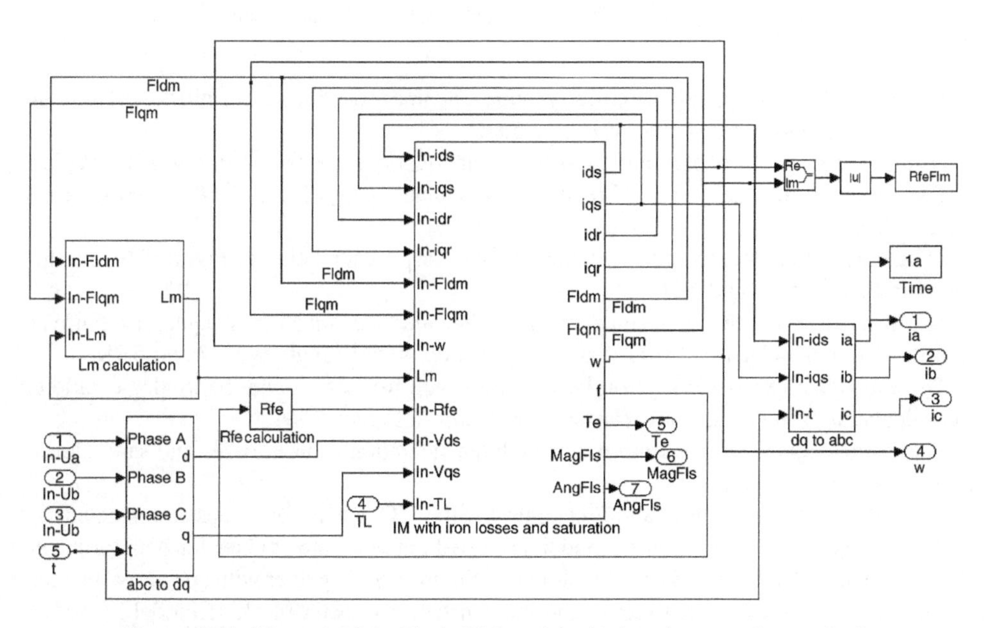

Figure 5.67 Diagram of the block "IM model with iron losses and saturation"

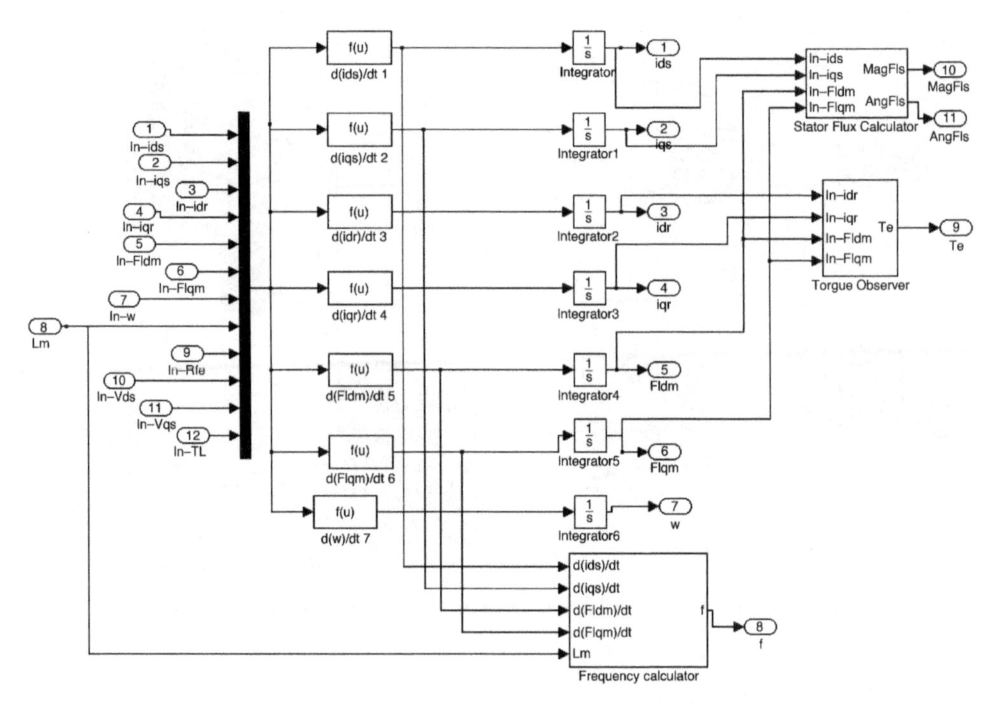

Figure 5.68 Blocks for differential equations of the dynamic machine model

losses. The differential equations of the dynamic machine model inside the block "IM with iron losses and saturation" are shown in Figure 5.68.

Details of the blocks expressing the differential equations are shown in Figures 5.69–5.71.

The structure of the block "Frequency Calculator" in Figure 5.68 is presented in Figures 5.72–5.74.

Structures of the blocks used for stator flux and torque calculation in Figure 5.68 are illustrated in Figures 5.75 and 5.76.

The instantaneous value of the magnetizing inductance is deduced from equations (5.45) and (5.46). The implementation of this calculation is presented in Figures 5.77 and 5.78.

The equivalent resistance for iron losses varies with frequency, and the implementation of the calculation of instantaneous resistance is shown in Figure 5.79.

The parameters of the induction motor with consideration of iron losses and saturation are presented in Figure 5.80.

Simulation of torque control and flux control with DTC, when iron losses and magnetic saturation are included in the machine model, is carried out with rated values for both torque and flux commands. The obtained speed is shown in Figure 5.81, together with speeds of the motor when only iron losses are considered and the constant parameter machine model is used.

The responses of speed with only iron losses and with both iron losses and saturation are almost identical. The speeds decrease when rated load torque is applied due to the high ripples in torque responses, which result in lower average value of electromagnetic torque of the motor. Magnetic saturation does not affect the speed response in this region of operation of the motor.

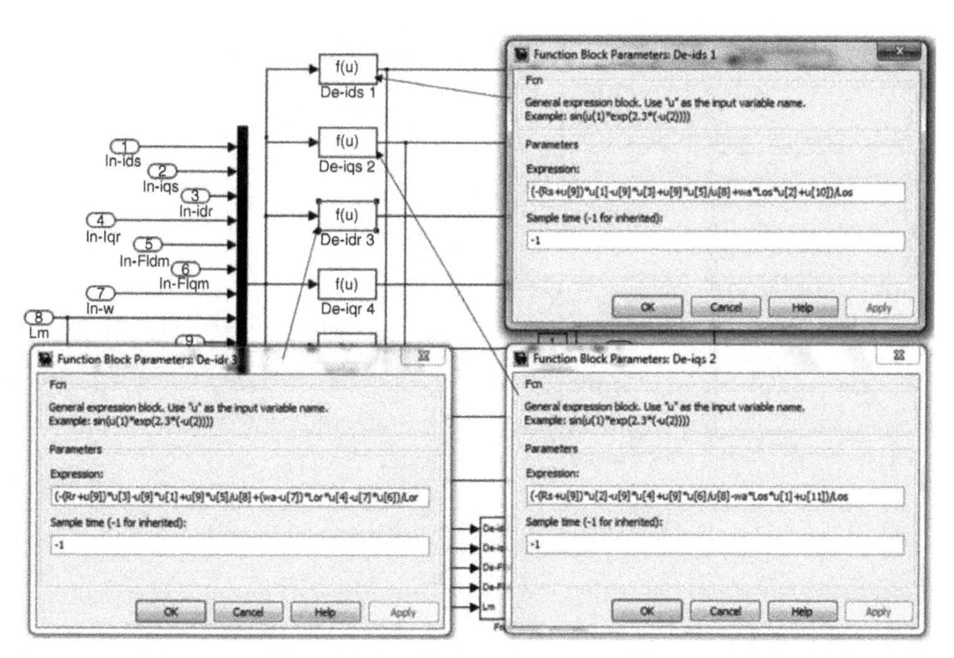

Figure 5.69 Expressions for differential equations of i_{ds}, i_{dr}, and i_{dr} (*Source:* SIMULINK software)

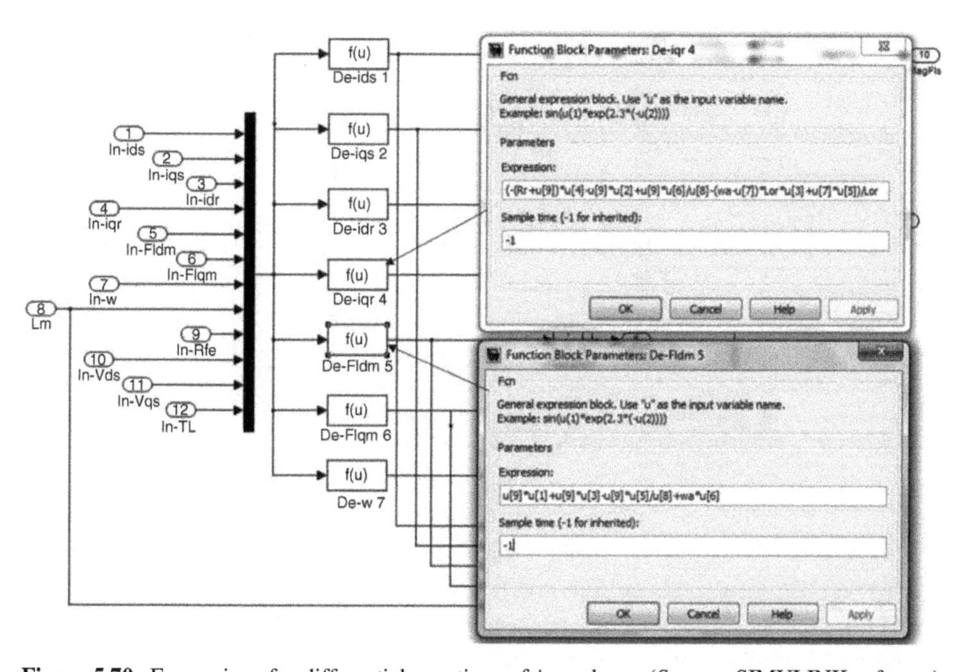

Figure 5.70 Expressions for differential equations of i_{qr} and ψ_{dm} (*Source:* SIMULINK software)

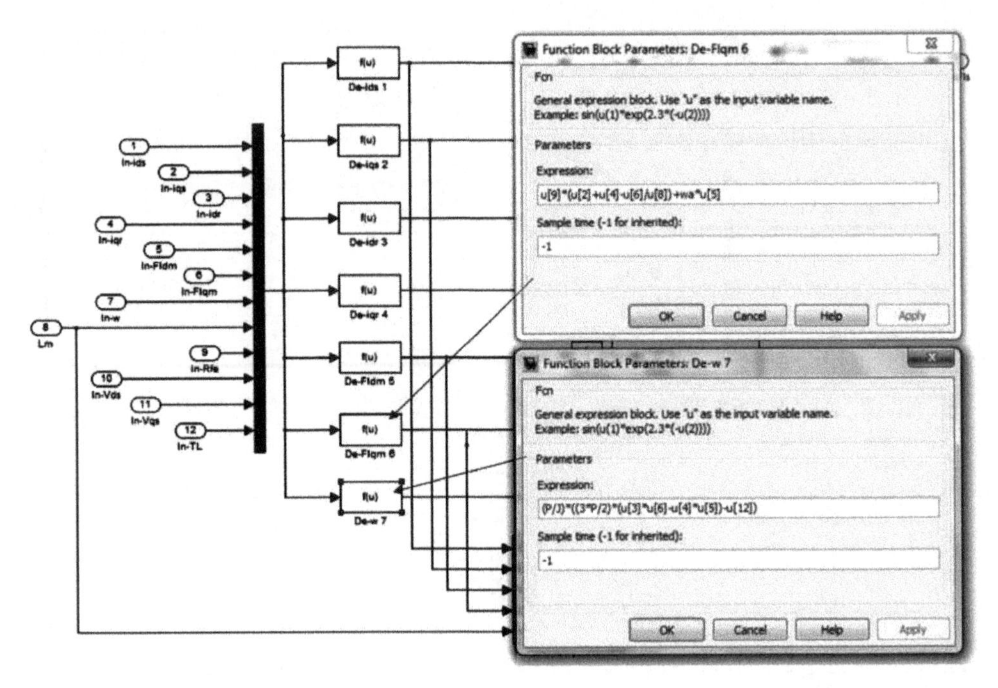

Figure 5.71 Expressions for differential equations of ψ_{dqm} and ω (*Source:* SIMULINK software)

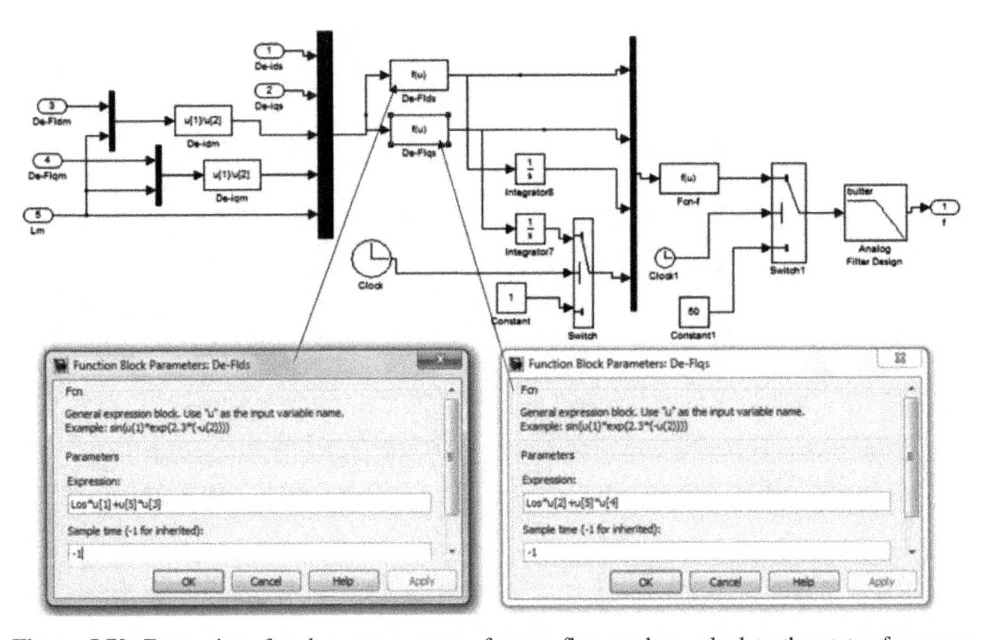

Figure 5.72 Expressions for d–q components of stator flux used to calculate the stator frequency (*Source:* SIMULINK software)

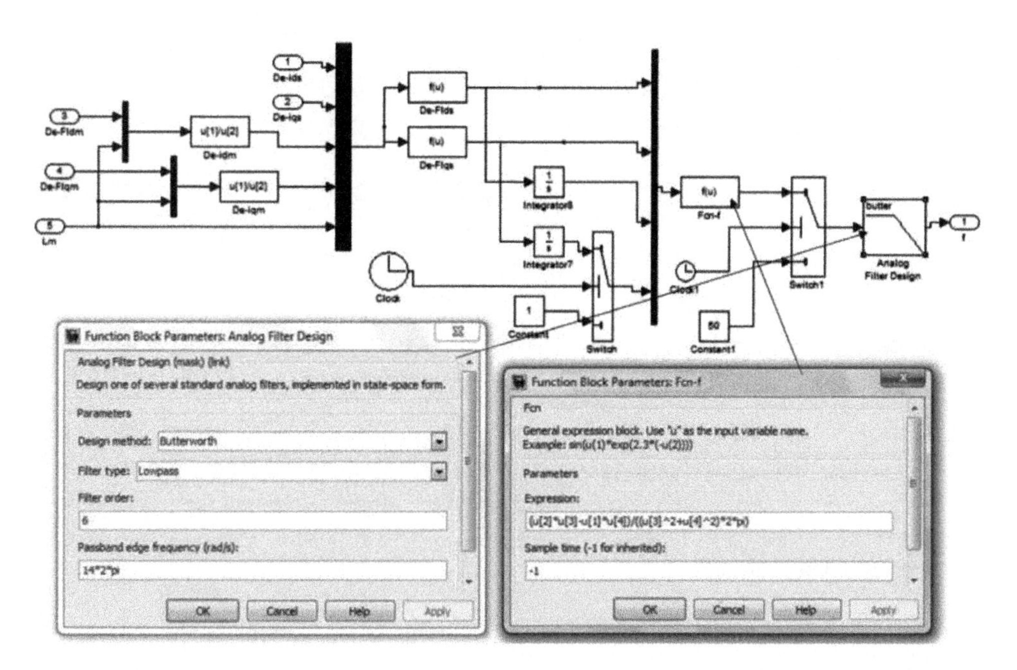

Figure 5.73 Expression for frequency calculation and design of Butterworth filter to obtain the fundamental component of stator frequency (*Source:* SIMULINK software)

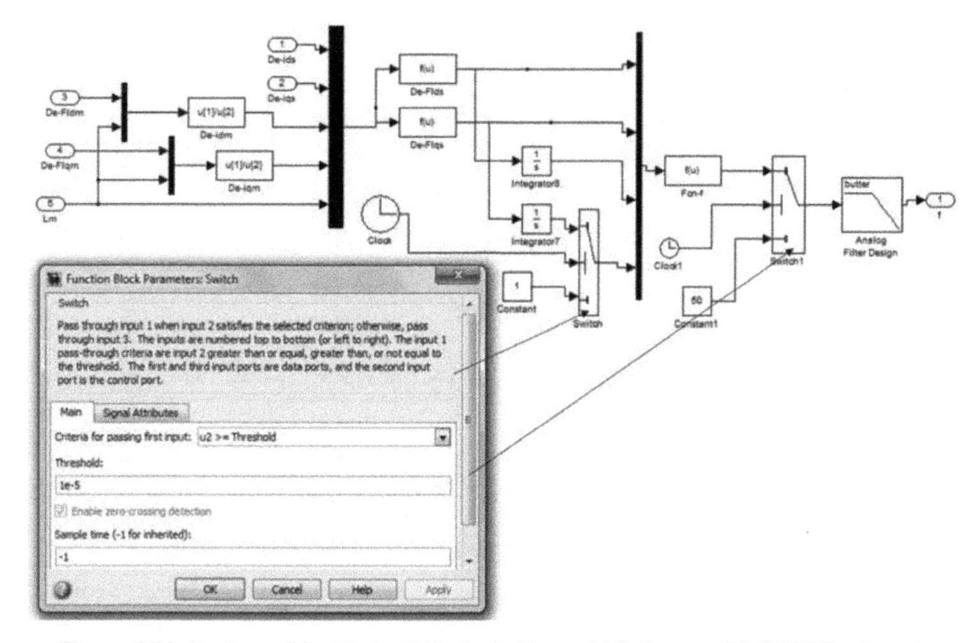

Figure 5.74 Designs of the "Switch" blocks in Figure 5.72 (*Source:* SIMULINK software)

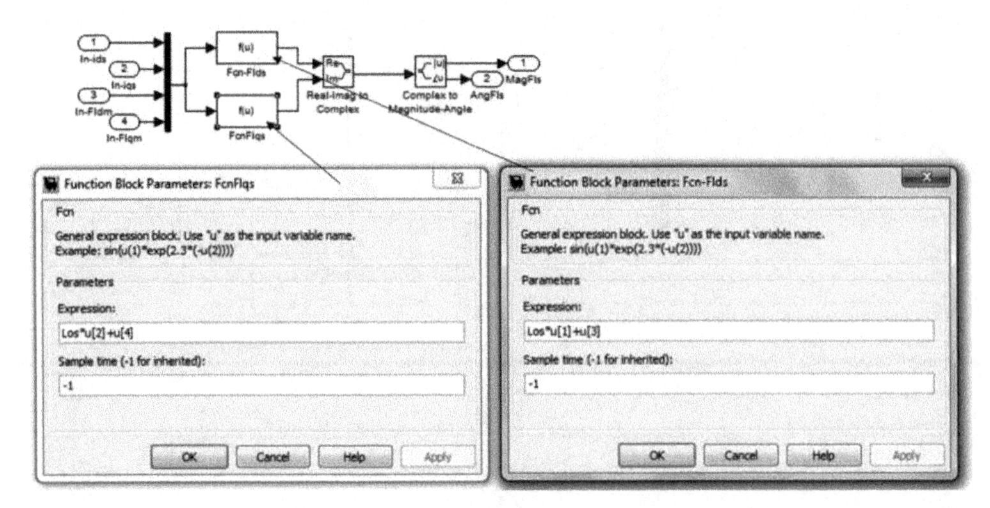

Figure 5.75 Stator flux calculation in induction machine model with iron losses and magnetic saturation (*Source:* SIMULINK software)

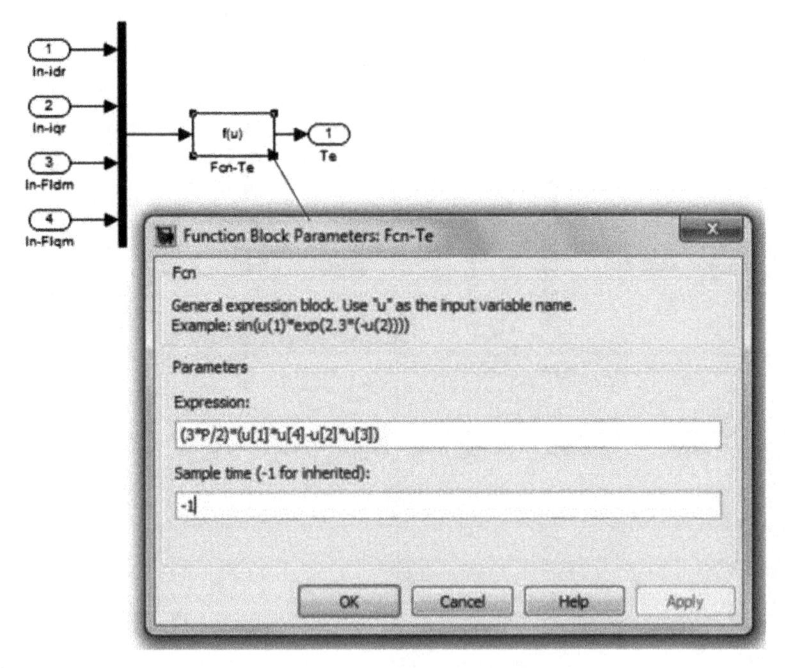

Figure 5.76 Electromagnetic torque calculation in the induction machine model (*Source:* SIMULINK software)

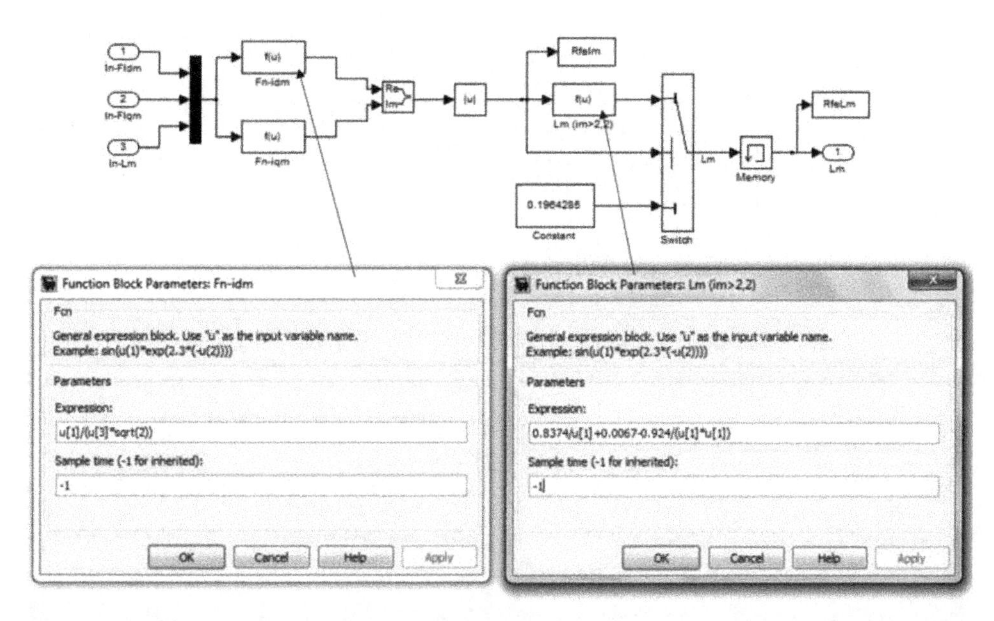

Figure 5.77 Expressions for i_{dm} and L_m (when i_{dm} is >2.2 A) in calculation of instantaneous value of the magnetizing inductance (*Source:* SIMULINK software)

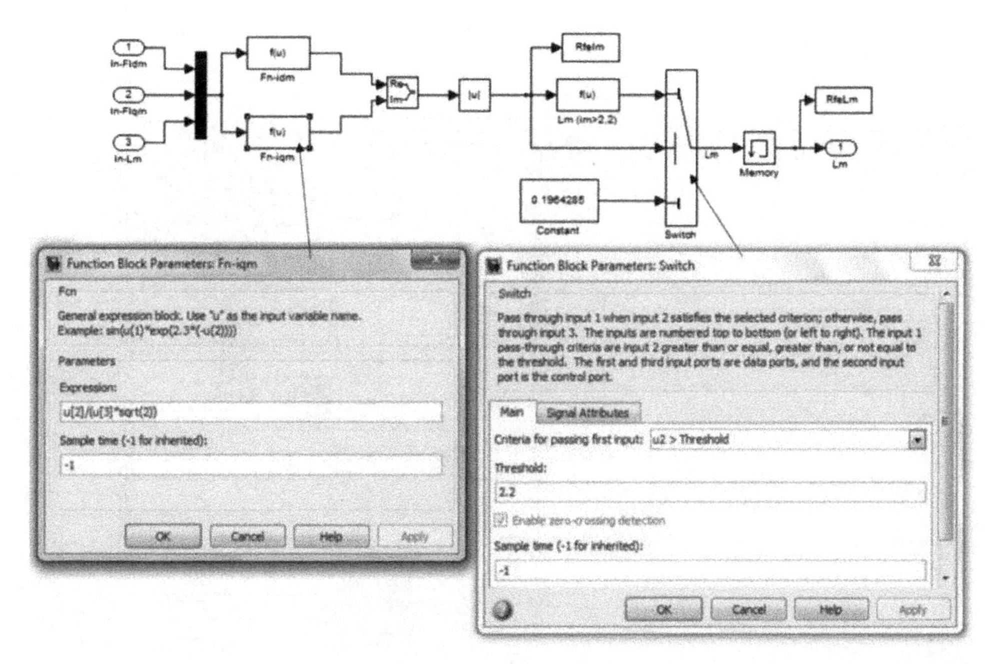

Figure 5.78 Expression of i_{qm} and values of "Switch" block in Figure 5.77 (*Source:* SIMULINK software)

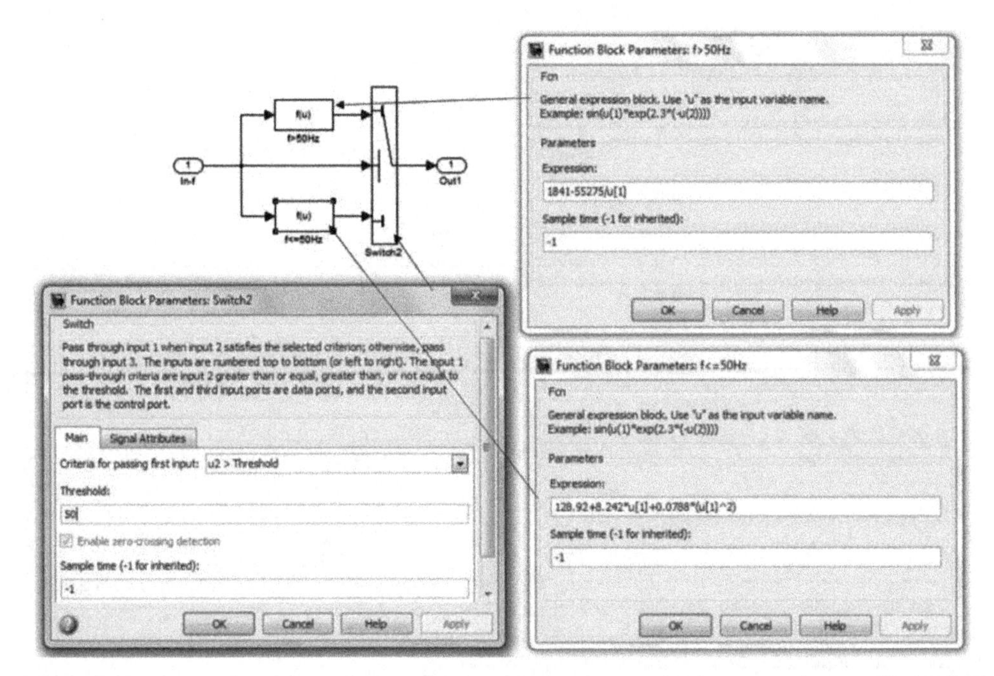

Figure 5.79 Calculation of the equivalent resistance for iron losses with frequency (*Source:* SIMULINK software)

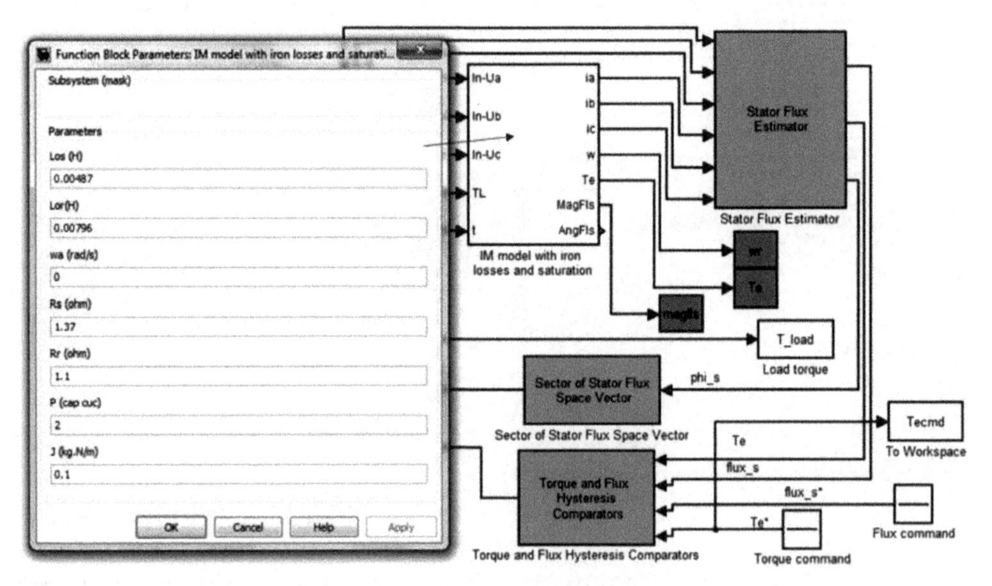

Figure 5.80 Parameters of the induction motor (*Source:* SIMULINK software)

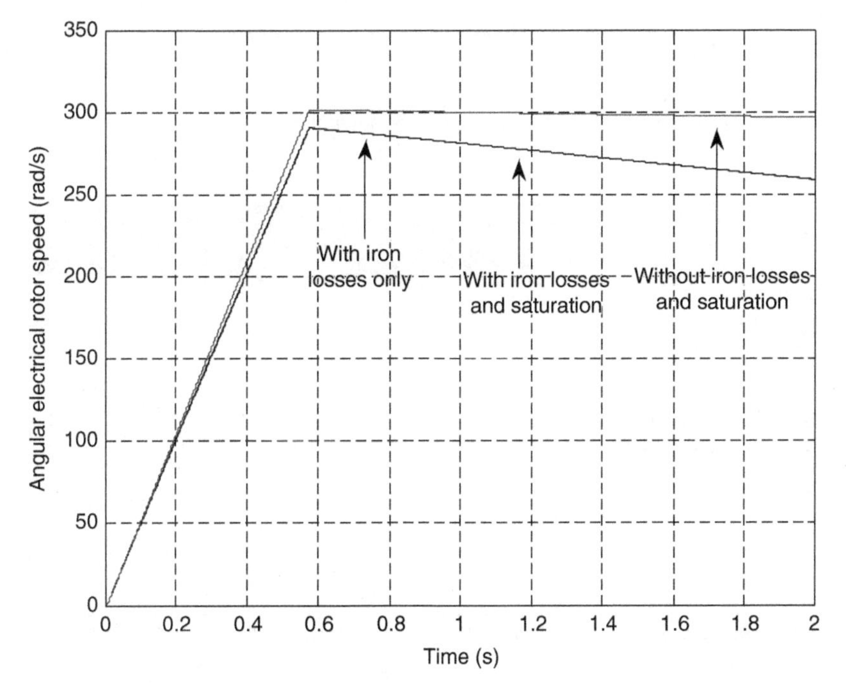

Figure 5.81 Speed responses in torque control mode of DTC

Torque responses are shown in Figure 5.82. Again, the additional inclusion of magnetic saturation does not change significantly torque response. Discrepancies between load torque and the actual torque of the motor in steady state are shown in Figure 5.83. This difference is the cause of the decrease in speed.

Flux and stator current responses are demonstrated in Figures 5.84 and 5.85. The inclusion of magnetic saturation and iron losses does not cause significant changes in the responses.

5.5.2.2 Effects of Iron Losses and Magnetic Saturation in Speed Control Mode

The simulation program of DTC in speed control mode is essentially the same as that shown in sub-section 5.3.4.1; only the Simulink block for the induction motor is changed to include iron losses and magnetic saturation. Figures 5.86 and 5.87 show the schematic diagram of the program and parameters of the PID controller, respectively.

Figure 5.88 shows the response of speed during the acceleration, applying and removing of rated load torque, and the speed response with the ideal machine model is also presented. Figure 5.89 shows the respective torque response when the iron losses and magnetic saturation are included.

The responses of the motor's electromagnetic torque during the application and removal of rated load torque are presented in Figures 5.90 and 5.91, respectively. The electromagnetic torque with the ideal constant parameter machine model is also demonstrated for comparison.

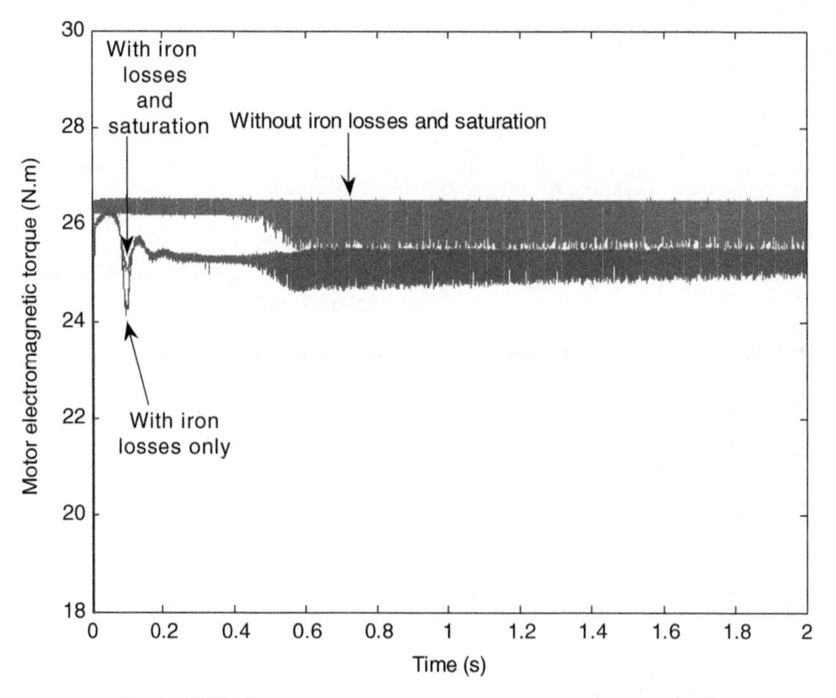

Figure 5.82 Torque responses in torque control mode of DTC

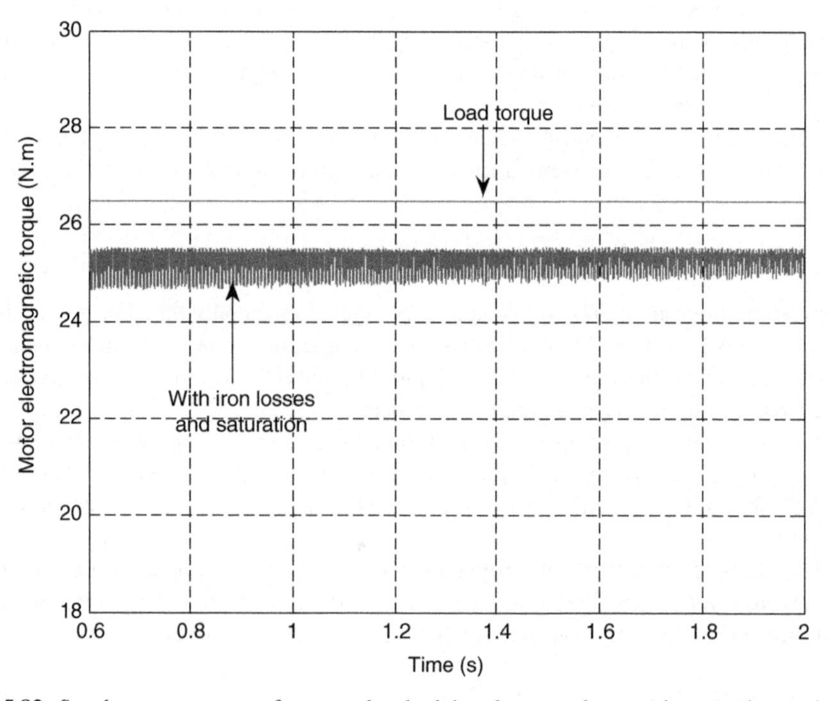

Figure 5.83 Steady-state response of torque when both iron losses and magnetic saturation are included

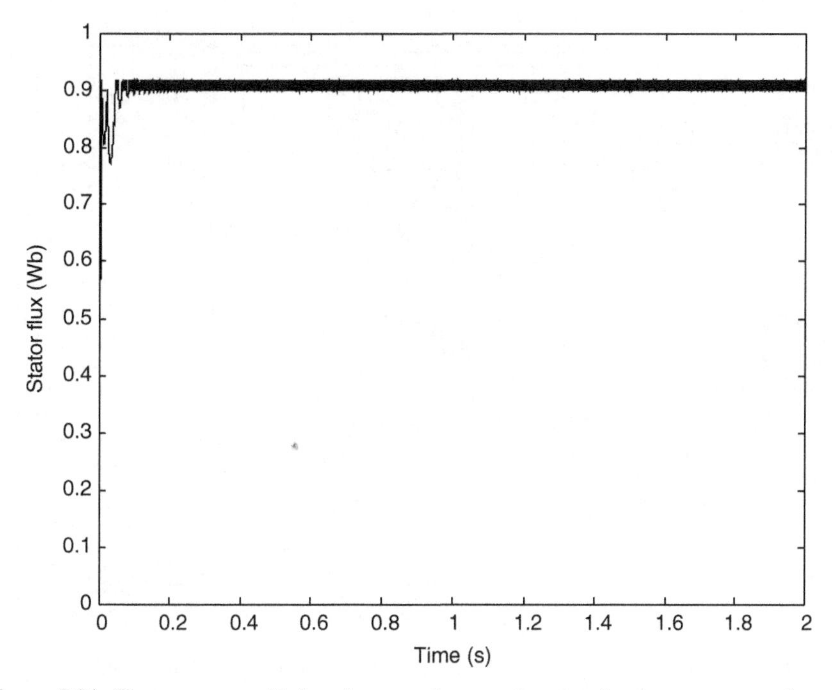

Figure 5.84 Flux response with iron losses and magnetic saturation in torque control mode

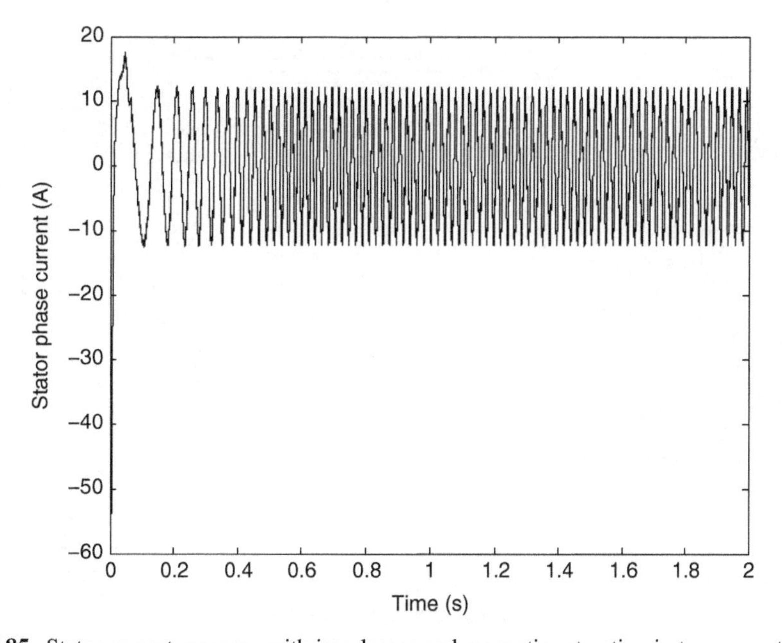

Figure 5.85 Stator current response with iron losses and magnetic saturation in torque control mode

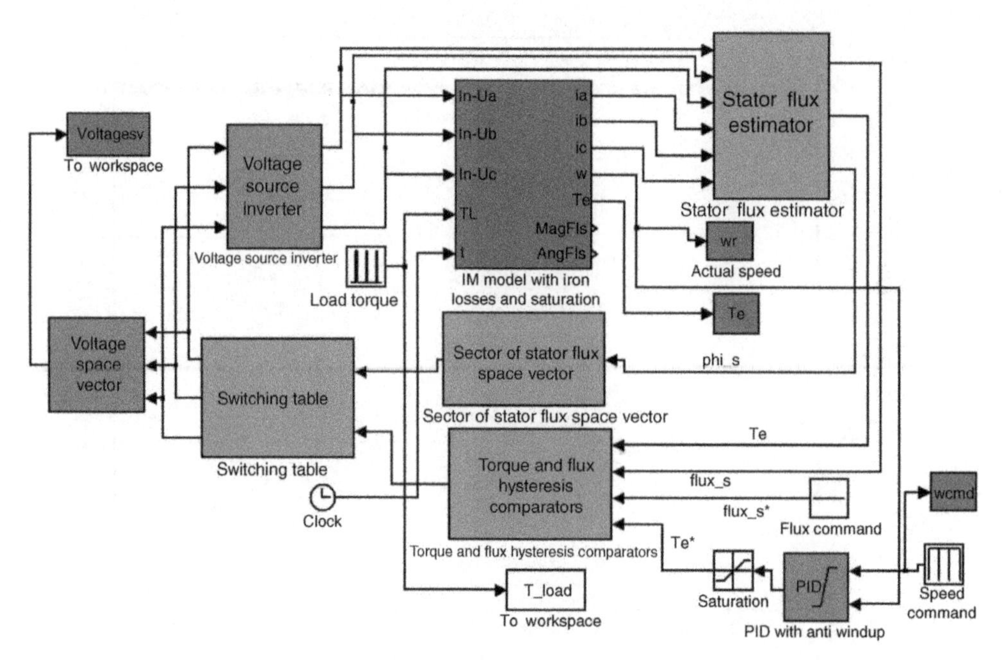

Figure 5.86 Simulation of DTC in speed control mode with the inclusion of both iron losses and magnetic saturation

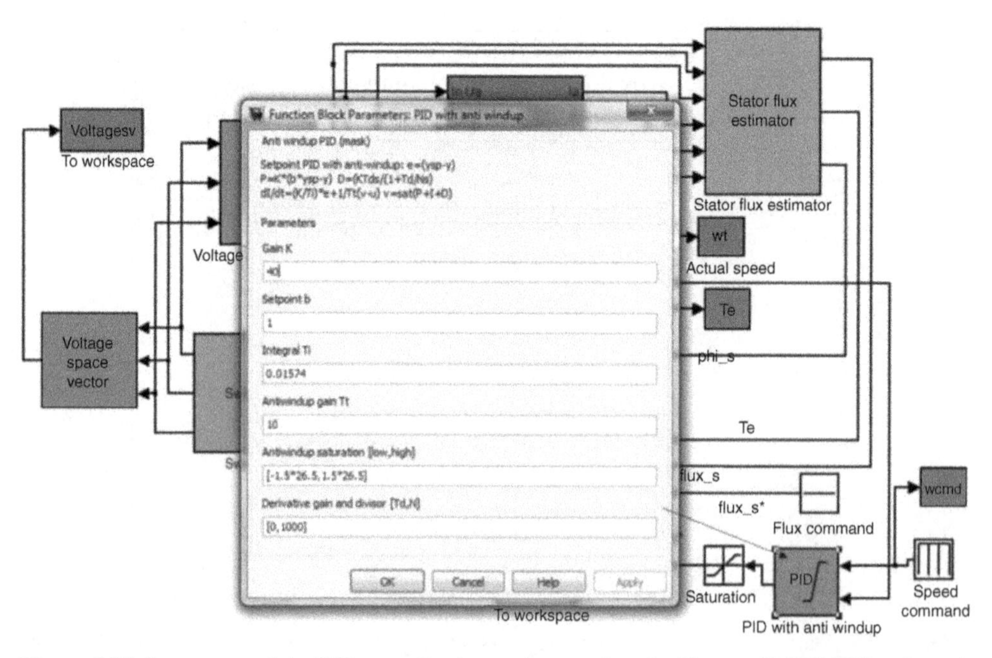

Figure 5.87 Parameters of the PID controller in speed control mode (*Source:* SIMULINK software)

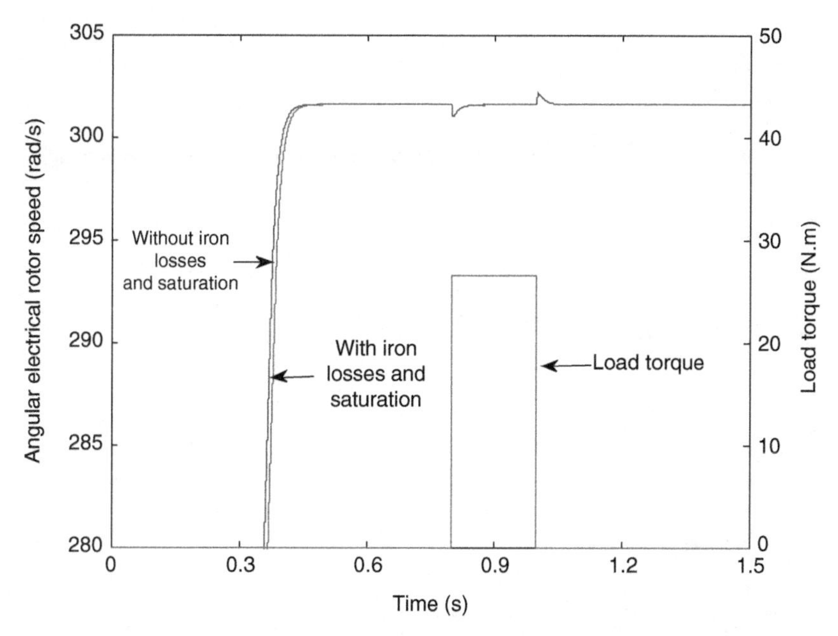

Figure 5.88 Speed responses when iron losses and saturation is included

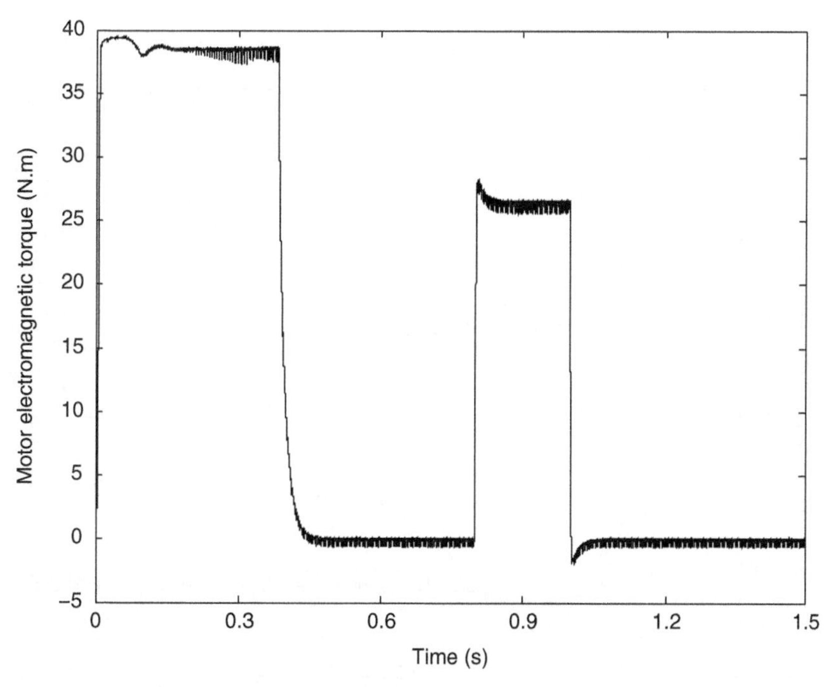

Figure 5.89 Torque responses during acceleration, loading, and unloading

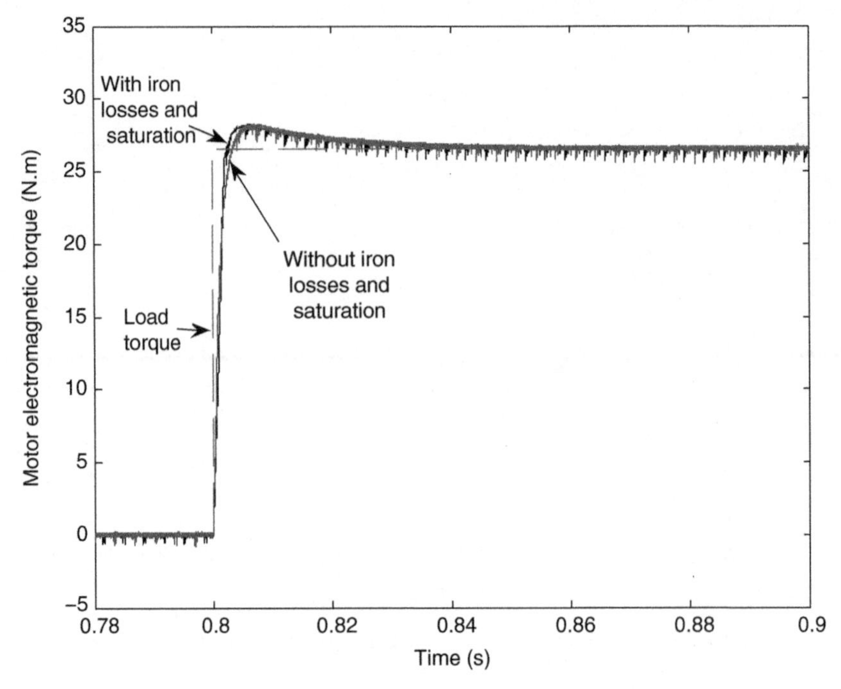

Figure 5.90 Motor torque response during the application of rated load torque

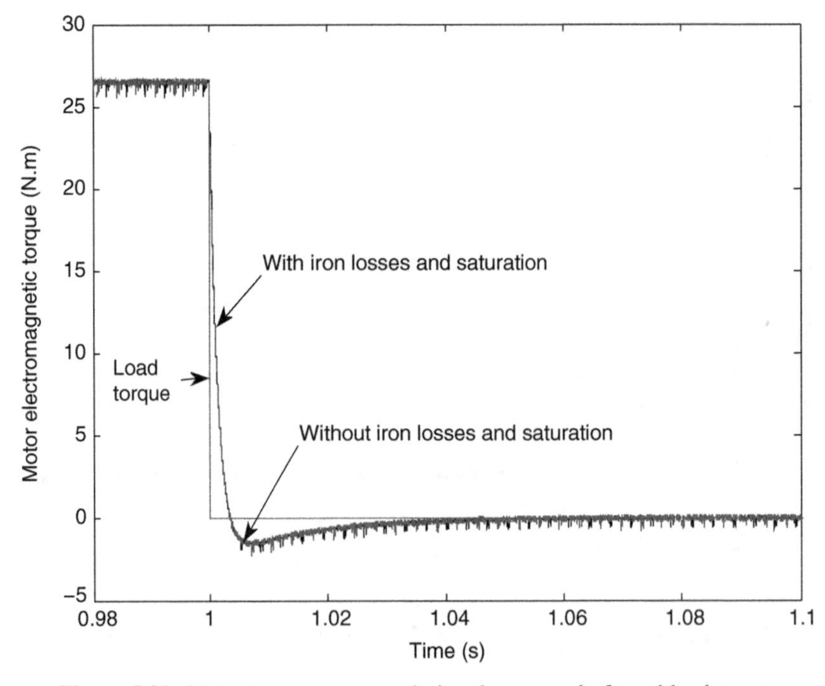

Figure 5.91 Motor torque response during the removal of rated load torque

The difference between the speed responses with and without the inclusion of iron losses and magnetic saturation is not significant, as shown in Figure 5.88. The transient response of speed is slightly bit slower with iron losses and magnetic saturation due to the power losses, which reduce the actual electromagnetic torque. The torque response is essentially the same with torque response with only iron losses consideration (Figure 5.89).

When load torque is suddenly applied, the responses of electromagnetic torque are almost identical for all the cases, as shown in Figures 5.90 and 5.91.

5.6 Modified Direct Torque Control of Induction Machine with Constant Switching Frequency

From the discussion and simulation of DTC of induction motors in previous sub-sessions, DTC has several major disadvantages such as [7]:

- problems during starting due to the absence of current controllers;
- operation in low-speed region due to high flux ripples;
- necessity of accurate estimation of stator flux and electromagnetic torque;
- changing switching frequency because of hysteresis band-based control of torque; and
- high torque ripples in high-speed operation due to the utilization of zero voltage space vectors.

Variation of switching frequency in DTC is one of the major drawbacks in many industrial applications.

There have been many modifications suggested for constant switching frequency with DTC. These modifications include DTC with space vector modulation and deadbeat control for direct calculation of switching pattern to achieve constant switching period [24], DTC with predictive control, which calculates the optimal switching instant at each cycle on the basic of RMS torque-ripple equation during one switching period obtained from the instantaneous torque variation equation [12], and Direct Self Control with fuzzy controller for choosing switching states to reduce the flux and torque errors, while still achieving fast torque response and constant switching frequency [25]. However, these modifications have resulted in the increased sophistication of the control systems, which require more computational capabilities than the traditional DTC.

The concept of DTC is also extended to introduce the new control method called Stator Flux Vector Control, which delivers fast torque response and constant switching frequency without using current controller [26]. This new concept focuses on torque control using variations of the stator flux angular velocity and stator flux control for obtaining fixed rotor flux amplitude in a feed-forward manner. The new control method also improves the sensitivities to the induction motor's parameters and stabilities, even at low-speed operation [26].

Space vector pulse width modulation (SVPWM) with two-level and three-level inverters to obtain constant switching frequency in DTC has also been proposed for the applications in electric vehicle drive [27]. Furthermore, constant frequency torque controller is introduced to achieve both constant switching frequency and reduction of torque ripples, especially when DTC is implemented with a low speed processor, which results in low sampling frequency [28].

5.7 Direct Torque Control of Sinusoidal Permanent Magnet Synchronous Motors (SPMSM)

5.7.1 Introduction

Permanent magnet synchronous machines (SPMSMs) are categorized on the basis of the waveform of the induced electromotive force; they are either sinusoidal or trapezoidal. The SPMSMs with sinusoidal wave-shape of emf are called PMSMs, and the PMSMs with trapezoidal wave-shape of emf are known as Brushless DC Machines (BLDCs) [29]. This sub-section discusses the DTC of SPMSMs, which has a sinusoidal distribution of air-gap flux.

The machines are of surface magnet structure, where the magnets are placed in the grooves of the outer periphery of the rotor lamination to provide high flux density in the air gap and also uniform cylindrical surface of the rotor for the mechanical robustness [29].

Mathematical equations for dynamic models of the SPMSMs are presented in the next sub-section.

5.7.2 Mathematical Model of Sinusoidal PMSM

The mathematical model for SPMSMs are shown in equations (5.47) to (5.52). The dynamic model is in the rotor flux-oriented reference frame. The rotating speed of rotor and rotor flux space vector is the same for the synchronous motor, and they are also equal to the rotating speed of the reference frame $\omega_a = \omega_r = \omega$:

$$\begin{aligned} v_{ds} &= R_s i_{ds} + \frac{d\psi_{ds}}{dt} - \omega_r \psi_{qs} \\ v_{qs} &= R_s i_{qs} + \frac{d\psi_{qs}}{dt} - \omega_r \psi_{ds} \end{aligned} \tag{5.47}$$

$$v_f = R_f i_f + \frac{d\psi_f}{dt} \tag{5.48}$$

$$T_e - T_L = \frac{J}{P}\frac{d\omega_r}{dt} \tag{5.49}$$

$$\begin{aligned} \psi_{ds} &= L_s i_{ds} + L_m i_f \\ \psi_{qs} &= L_s i_{qs} \end{aligned} \tag{5.50}$$

$$\psi_f = L_f i_f + L_m i_{ds} \tag{5.51}$$

$$T_e \frac{3}{2} P\left(\psi_{ds} i_{qs} - \psi_{qs} i_{ds}\right) \tag{5.52}$$

Transformation equations for stator voltage and stator current are shown in equations (5.53) to (5.55), where θ is the instantaneous angle of the rotor:

$$
\begin{aligned}
v_{ds} &= \frac{2}{3}\left[v_a \cos\theta + v_b \cos\left(\theta - \frac{2\pi}{3}\right) + v_c \cos\left(\theta - \frac{4\pi}{3}\right)\right] \\
v_{qs} &= -\frac{2}{3}\left[v_a \sin\theta + v_b \sin\left(\theta - \frac{2\pi}{3}\right) + v_c \cos\left(\theta - \frac{4\pi}{3}\right)\right]
\end{aligned}
\tag{5.53}
$$

$$
\begin{aligned}
i_a &= i_{ds}\cos\theta - i_{qs}\sin\theta \\
i_b &= i_{ds}\cos\left(\theta - \frac{2\pi}{3}\right) - i_{qs}\sin\left(\theta - \frac{2\pi}{3}\right) \\
i_c &= i_{ds}\cos\left(\theta - \frac{4\pi}{3}\right) - i_{qs}\sin\left(\theta - \frac{4\pi}{3}\right)
\end{aligned}
\tag{5.54}
$$

$$
\theta = \int_0^t \omega_r(\tau)d\tau + \theta(0) \tag{5.55}
$$

When the rotor of the synchronous motor is a permanent magnet, the magnetizing flux linkage produced by the magnet can be expressed in terms of a fictitious field current i_f as [30]

$$
\psi_m = L_m i_f \tag{5.56}
$$

The mathematical model in an arbitrary rotating reference frame can be rewritten by applying the equation (5.56) into equations (5.47) to (5.52):

$$
\begin{aligned}
v_{ds} &= R_s i_{ds} + \frac{d\psi_{ds}}{dt} - \omega_a \psi_{qs} \\
v_{qs} &= R_s i_{qs} + \frac{d\psi_{qs}}{dt} - \omega_a \psi_{ds}
\end{aligned}
\tag{5.57}
$$

$$
\begin{aligned}
\psi_{ds} &= L_s i_{ds} + \psi_m \\
\psi_{qs} &= L_s i_{qs}
\end{aligned}
\tag{5.58}
$$

$$
T_e - T_L = \frac{J}{P}\frac{d\omega}{dt} \tag{5.59}
$$

$$
T_e \frac{3}{2} P\left(\psi_{ds} i_{qs} - \psi_{qs} i_{ds}\right) = \frac{3}{2} P \psi_m i_{qs} \tag{5.60}
$$

Because of the design of SPMSMs, the machines cannot be used for main operations as with traditional synchronous motors. Due to the absence of damper winding, the SPMSMs cannot be started [30]. They are mainly used for variable speed drives, which require power electronic conversions for power supplies. Therefore, it is always necessary to measure or estimate the position of the rotor for the feedback to the control system. However, this is not a disadvantage for SPMSM-based drives because the information about rotor position is always needed in the applications of the drive systems.

The cost of BLDC machines is usually lower than the cost of SPMSMs. However, the performances of SPMSM are much better. The selection of which of the machines depends very much on the particular applications [30].

5.7.3 Direct Torque Control Scheme of PMSM

The mathematical model in equations (5.57) to (5.60) is expressed in the stationary reference frame as

$$\begin{aligned} v_{\alpha s} &= R_s i_{\alpha s} + \frac{d\psi_{\alpha s}}{dt} \\ v_{\beta s} &= R_s i_{\beta s} + \frac{d\psi_{\beta s}}{dt} \end{aligned} \tag{5.61}$$

$$\begin{aligned} \psi_{\alpha s} &= L_s i_{\beta s} + \psi_m \\ \psi_{\beta s} &= L_s i_{\beta s} \end{aligned} \tag{5.62}$$

$$T_e - T_L = \frac{J}{P}\frac{d\omega}{dt} \tag{5.63}$$

$$T_e = \frac{3}{2}P\left(\psi_{\alpha s} i_{\beta s} - \psi_{\beta s} i_{\alpha s}\right) = \frac{3}{2}P\psi_m i_{\beta s} \tag{5.64}$$

The DTC scheme for SPMSMs, which directly controls the stator flux linkage and electromagnetic torque by optimum switching table for voltage-fed inverter, is discussed in this sub-section. Equations (5.61) to (5.64) are used to estimate the stator flux linkage and electromagnetic torque [7].

Comparisons between reference and estimated values of stator flux linkage and electromagnetic torque by comparators are carried out. Stator flux comparator is two-level, and torque comparator is three-level, as expressed in equations (5.6) and (5.7), respectively.

The inputs of the optimum switching table are outputs from the comparators and the sector in the complex, which contains the estimated stator flux space vector. The complex plane is shown in Figure 5.3. The optimum switching table for DTC of SPMSMs in this sub-section is the one suggested by [1] and is shown in Table 5.2.

The next sub-section discusses the simulation of DTC of SPMSMs.

5.7.4 MATLAB/Simulink Simulation of SPMSM with DTC

5.7.4.1 Development of MATLAB/Simulink Program

The simulation program for DTC of SPMSMs is developed on the basis of the scheme presented in previous the sub-section. The structural diagram is shown in Figure 5.92.

The PID controller in Figure 5.92 is identical to the controller in Figure 5.10. The Simulink diagram for a machine model of SPMSM inside the block "SPMSM" in the above figure is shown in Figure 5.93.

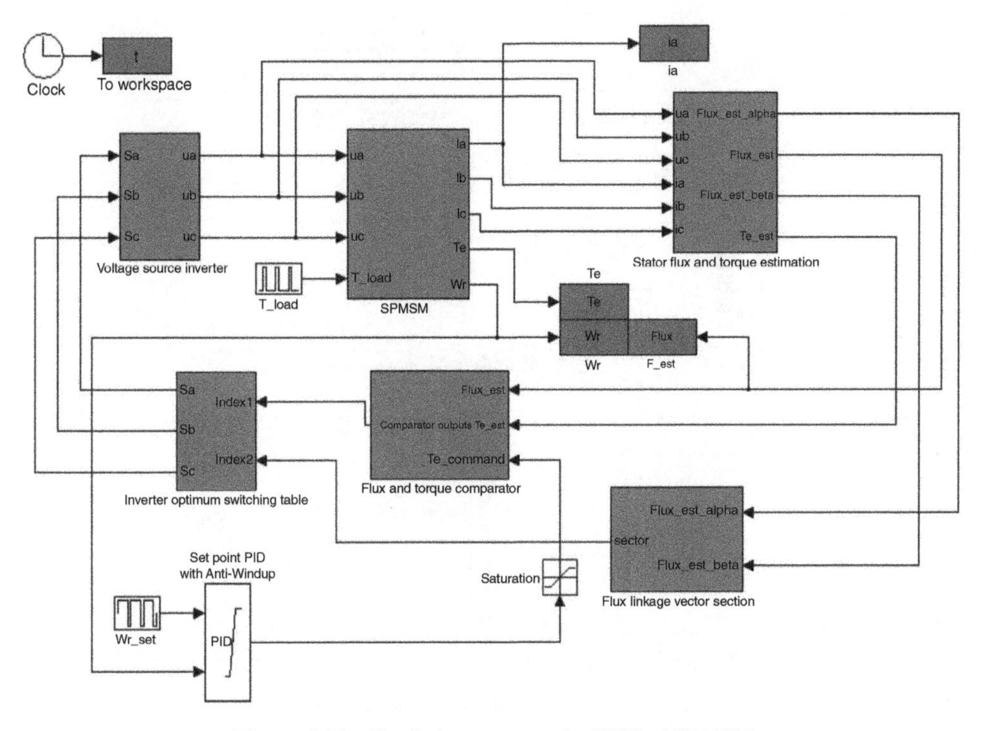

Figure 5.92 Simulation program for DTC of SPMSM

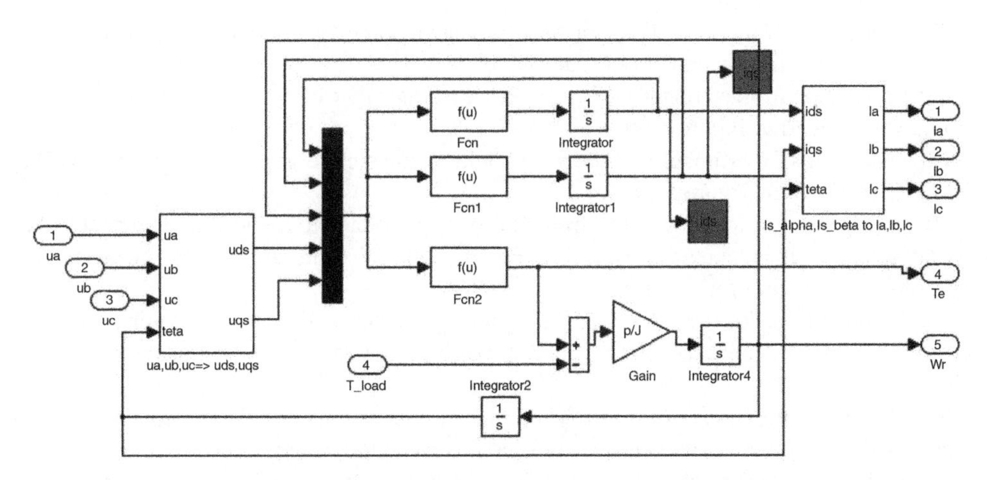

Figure 5.93 Simulink diagram of the SPMSM machine model

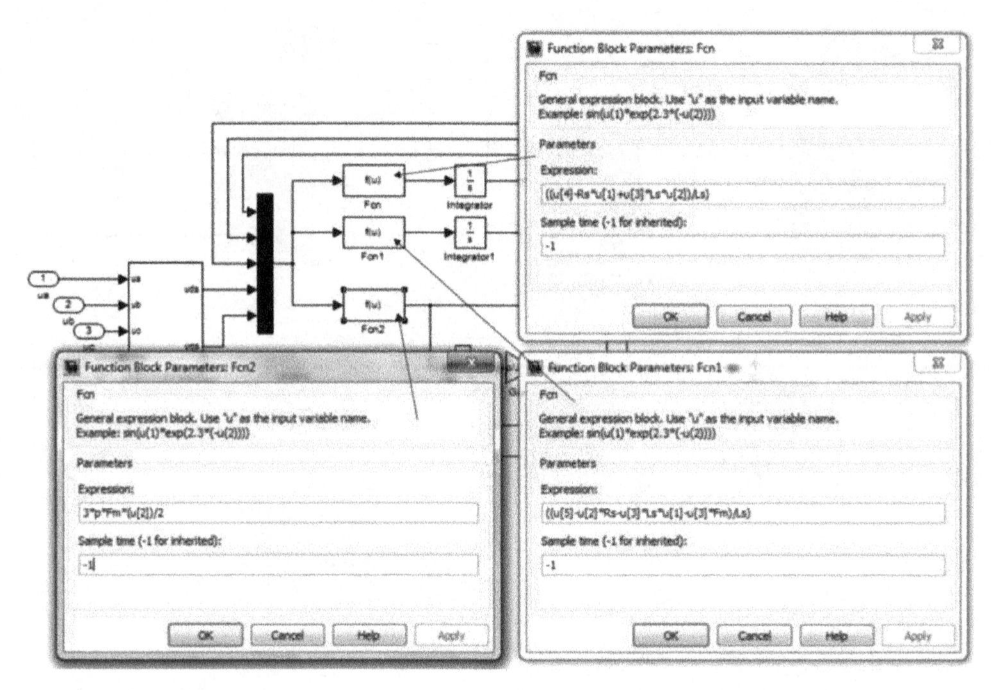

Figure 5.94 Differential equations of machine model in simulation program (*Source:* SIMULINK software)

Expressions for differential equations in equations (5.61) to (5.63) in MATLAB/Simulink are demonstrated in Figure 5.94.

Transformation from rotor flux-oriented reference frame to stationary coordinators in a machine model is demonstrated in Figure 5.95.

Transformation from stationary reference frame to rotor flux-oriented *d-q* axes in a machine model is presented in Figure 5.96.

Block diagrams for estimation of stator flux linkage and electromagnetic torque of SPMSMs for DTC is shown in Figure 5.97.

Block diagrams for transformation from phase stator voltage to α and β components of stationary reference frame are demonstrated in Figure 5.98. The same structure can be used in the case of transformation for stator current.

Block diagrams inside the sub-system "Flux linkage vector section," for determination of the section of complex plane containing instantaneous stator flux space vector, are shown in Figure 5.99.

The logical expressions for identifying the correct sector are similar to the ones in Figure 5.17 and shown again here for convenience in Figure 5.100.

Block diagrams in the sub-system "Flux and Torque Comparator," to determine the incremental or decreasing changes of the actual stator flux linkage and electromagnetic torque compared to the reference values, are illustrated in Figure 5.101.

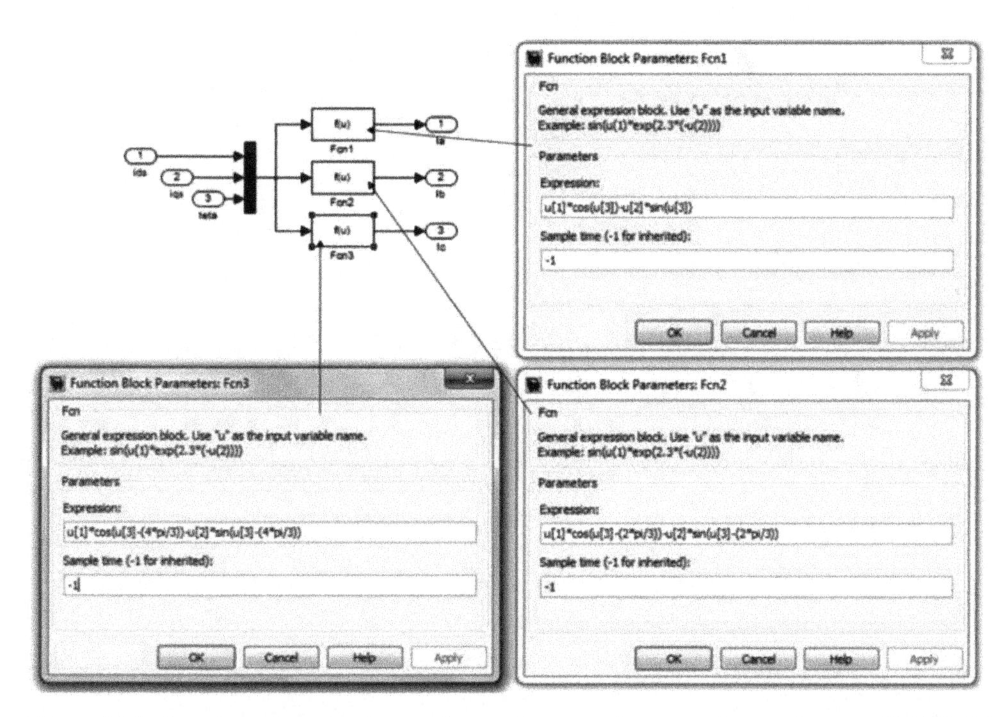

Figure 5.95 Transformation of reference frames from rotating to stationary (*Source:* SIMULINK software)

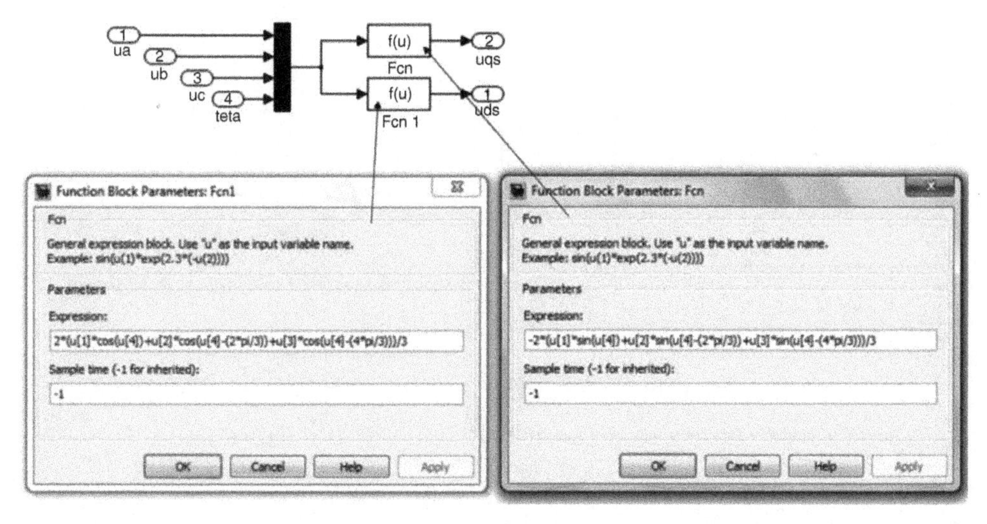

Figure 5.96 Transformation of reference frames from stationary to rotating (*Source:* SIMULINK software)

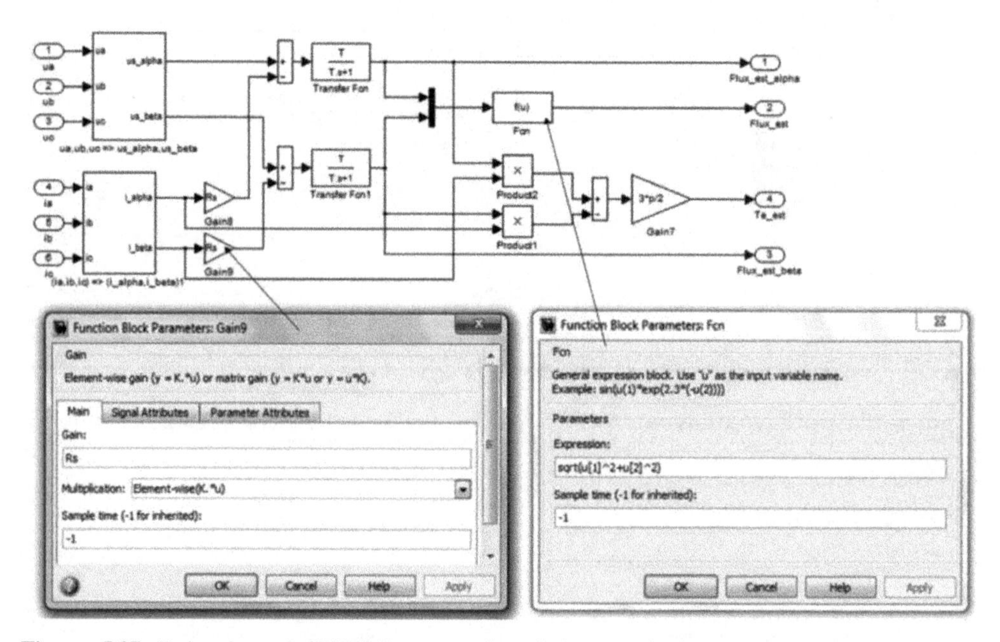

Figure 5.97 Estimation of SPMSM's stator flux linkage and electromagnetic torque (*Source:* SIMULINK software)

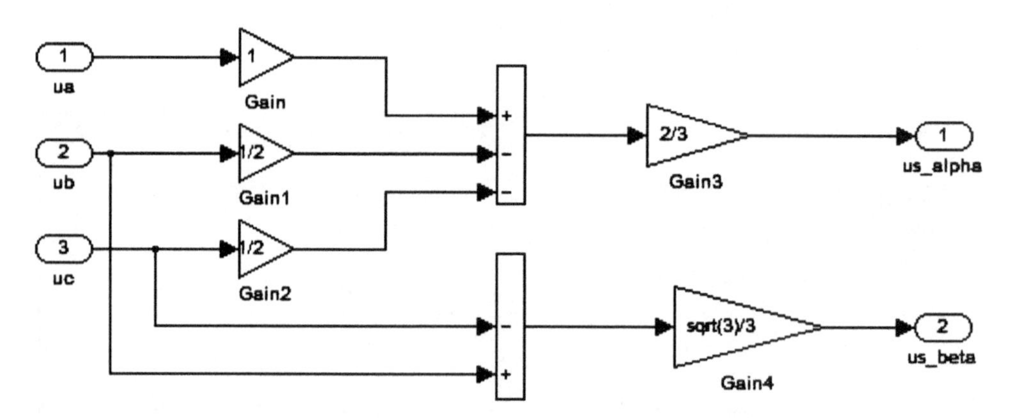

Figure 5.98 Conversion from phase voltage to α and β components in stationary reference frame

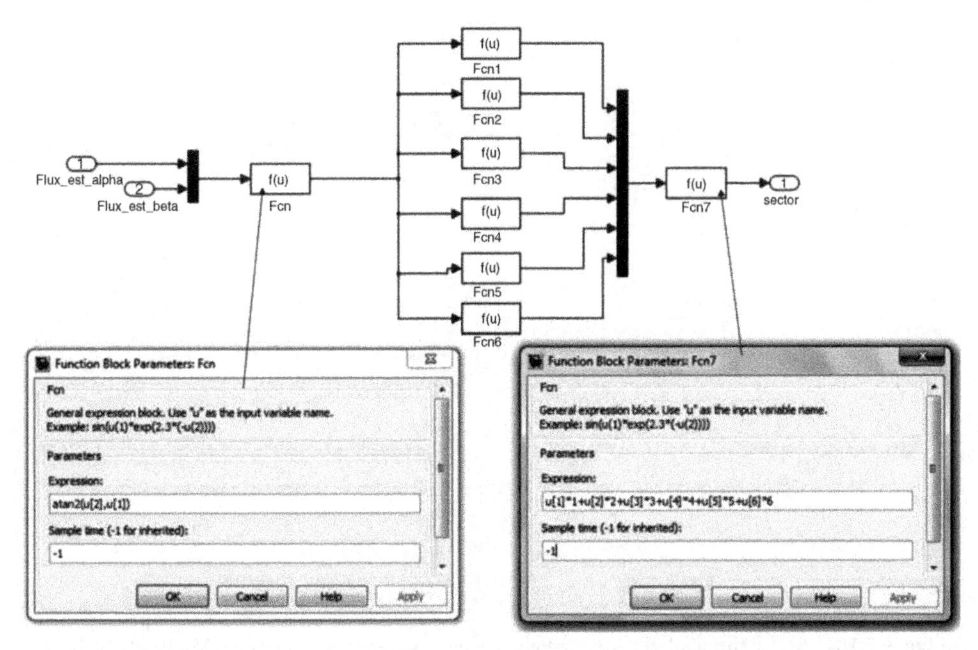

Figure 5.99 Determination of the complex plane's sector for stator flux space vector (*Source:* SIMULINK software)

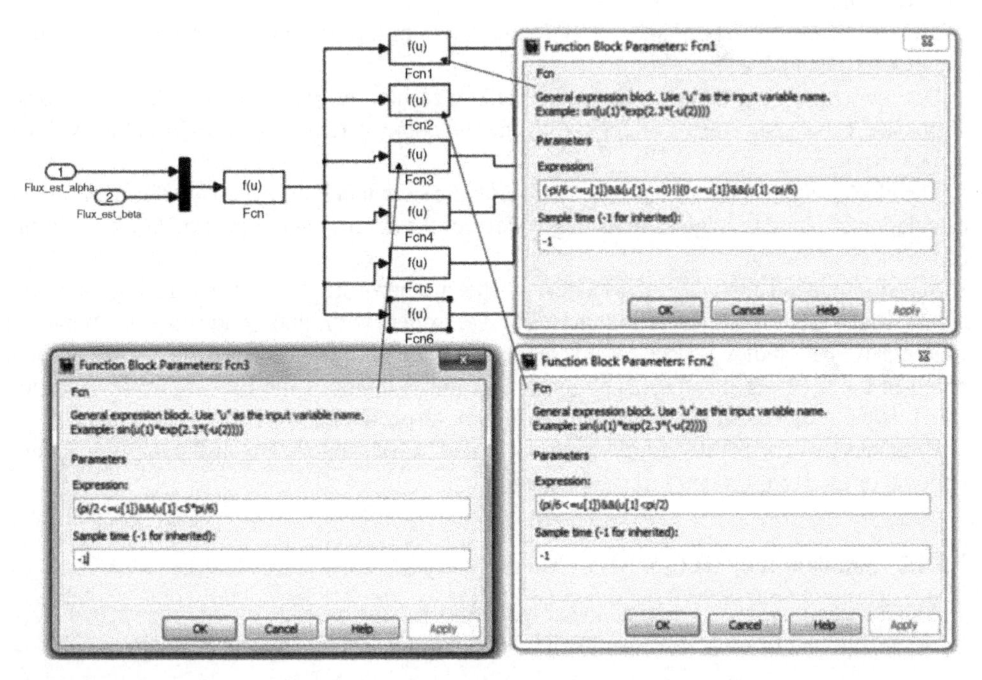

Figure 5.100 Logical expressions for identifying the sector of stator flux space vector (*Source:* SIMULINK software)

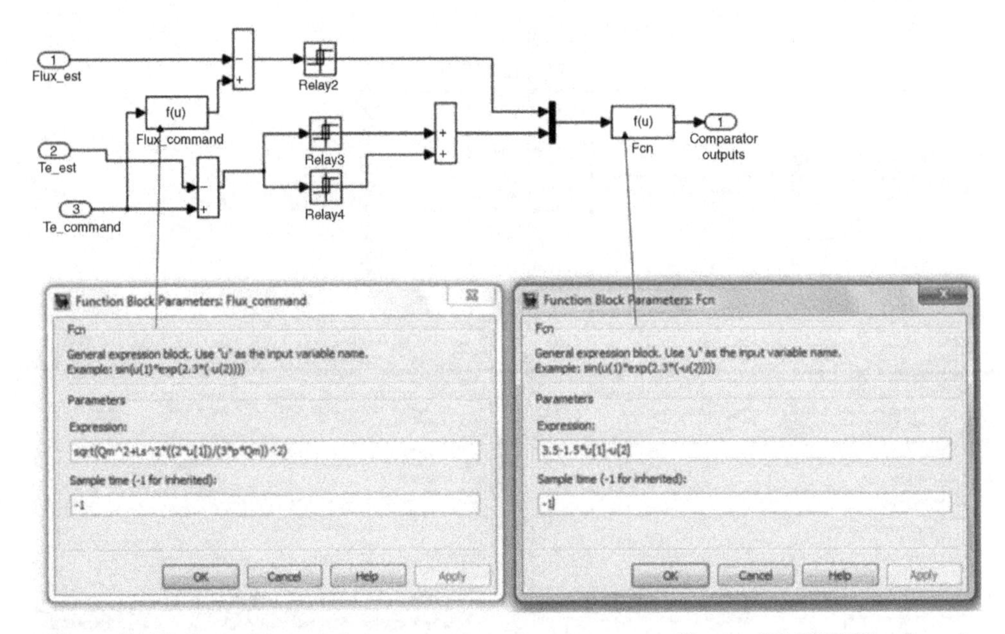

Figure 5.101 Stator flux and electromagnetic torque comparators (*Source:* SIMULINK software)

The hysteresis band for flux comparator is 1% of rated value of stator flux linkage. Similarly, the hysteresis band for torque comparator is 1% of the rated torque. These are shown in Figure 5.102.

The two levels of flux comparator and three levels of torque comparator are also shown in Figures 5.103 and 5.104.

Optimum switching table suggested by [1] is presented in Figure 5.104. The three blocks "Look-Up Table (2-D)" are used for coding the switching states of the three legs of the voltage source inverter.

The block diagrams for voltage source inverter are shown in Figure 5.105, with equations for calculation of phase voltage from the leg voltage of the inverter and Figure 5.106, with setting of the blocks "Switch."

Speed command for DTC of SPMSM in speed control mode is shown in Figure 5.107, together with the parameters of the PID controller with anti-wind up, as shown in Figure 5.10. The motor is accelerated forward to rated speed and then reversed to the same speed in the opposite direction. The upper and lower limits of the PID controller's output, which is the torque command, is 20 and −20 Nm, respectively.

Parameters of the SPMSM are shown in Figure 5.108, together with the application and removal of the load torque during the operation.

5.7.4.2 Simulation of DTC of SPMSM

The simulation program developed in sub-section 5.7.4.2 is run with a fixed step size of 10 μs for solver option of Runge–Kutta in MATLAB/Simulink. The SPMSM is accelerated and then

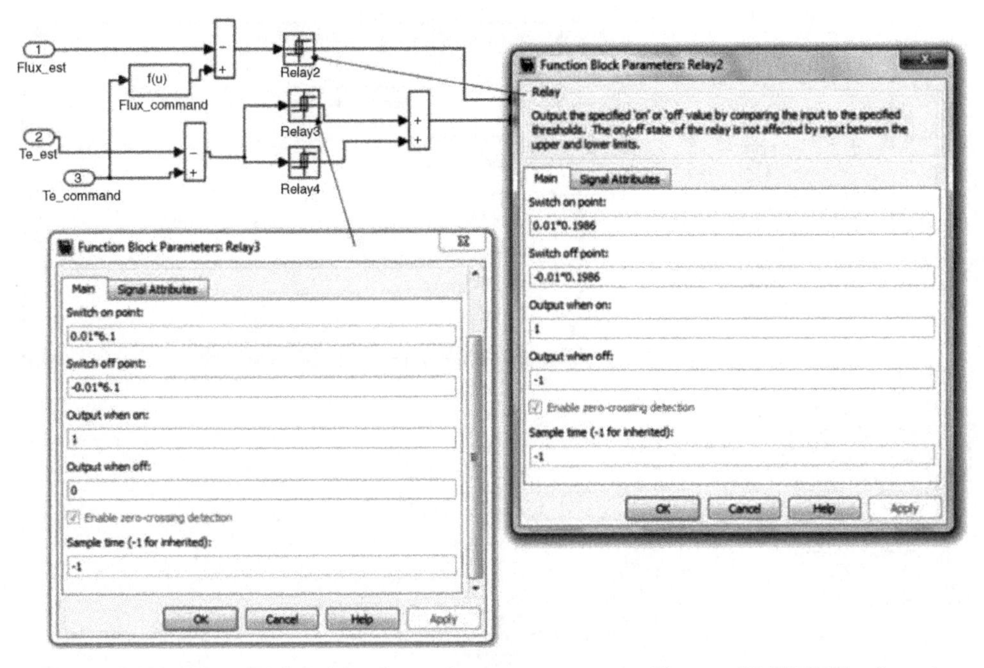

Figure 5.102 Hysteresis band for flux and torque comparators (*Source:* SIMULINK software)

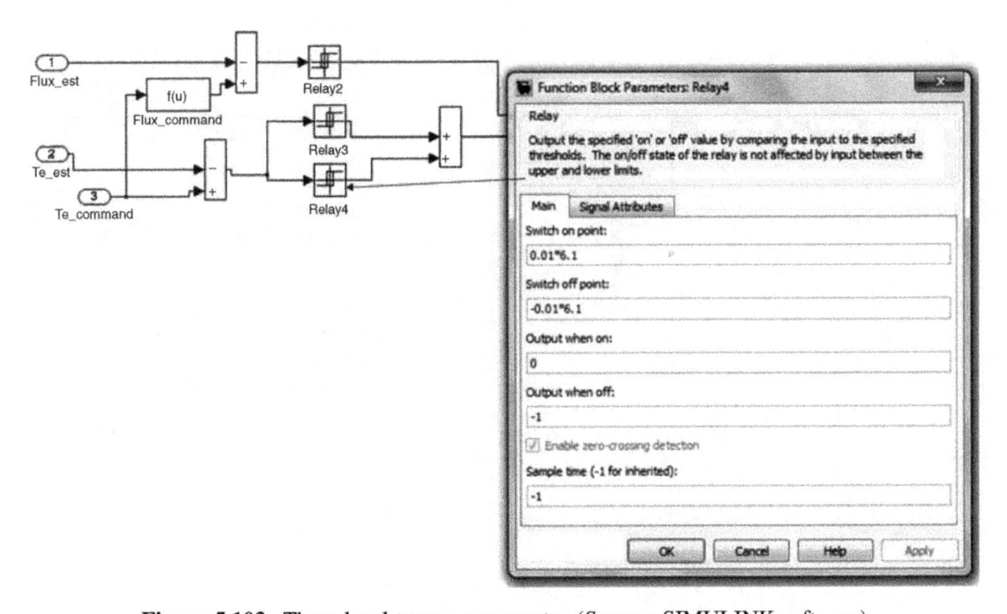

Figure 5.103 Three-level torque comparator (*Source:* SIMULINK software)

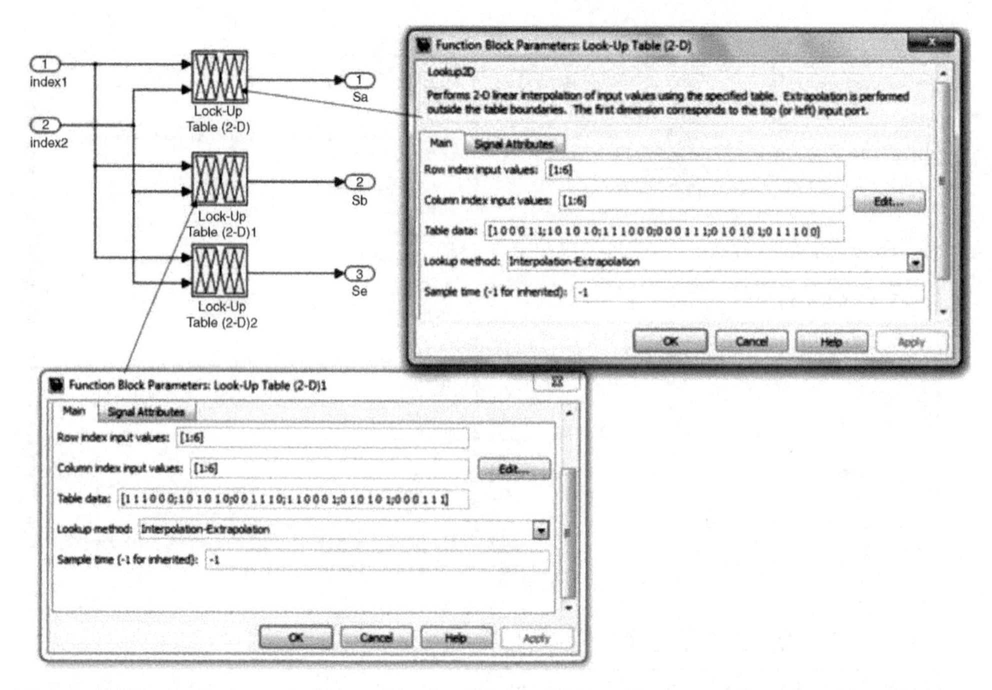

Figure 5.104 Optimum switching table for DTC of SPMSMs in MATLAB/Simulink (*Source:* SIMULINK software)

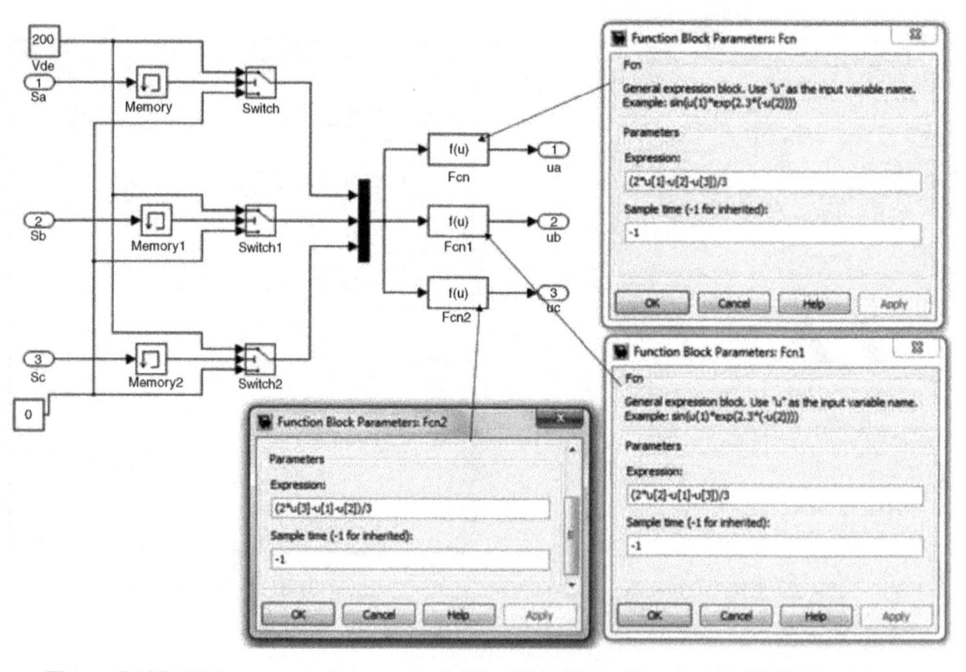

Figure 5.105 Voltage source inverter for DTC of SPMSMs (*Source:* SIMULINK software)

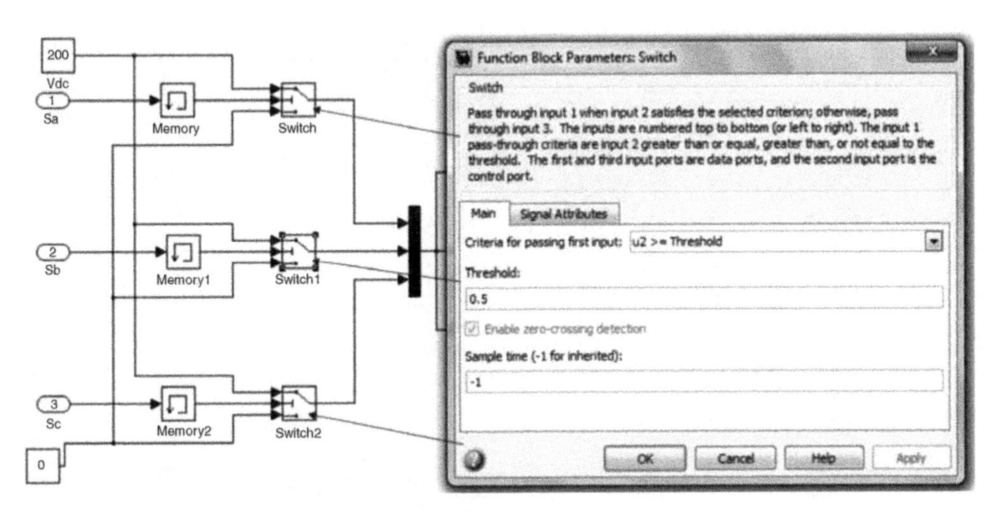

Figure 5.106 Setting of the blocks "Switch" in sub-system "Voltage source inverter" (*Source:* SIMULINK software)

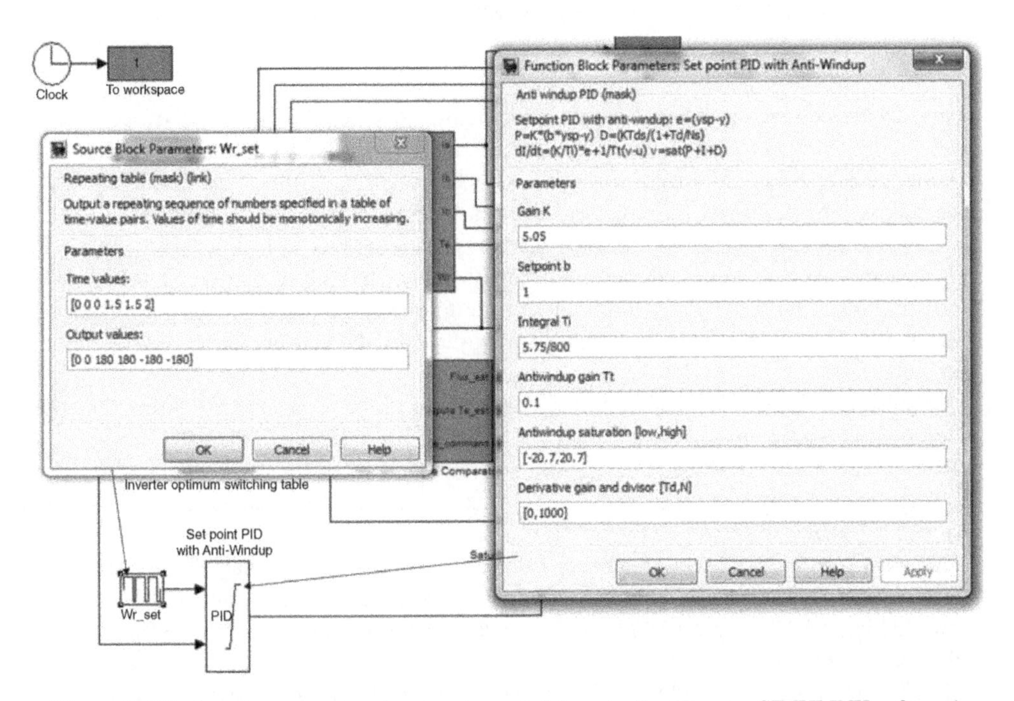

Figure 5.107 Speed command and parameters of PID controller (*Source:* SIMULINK software)

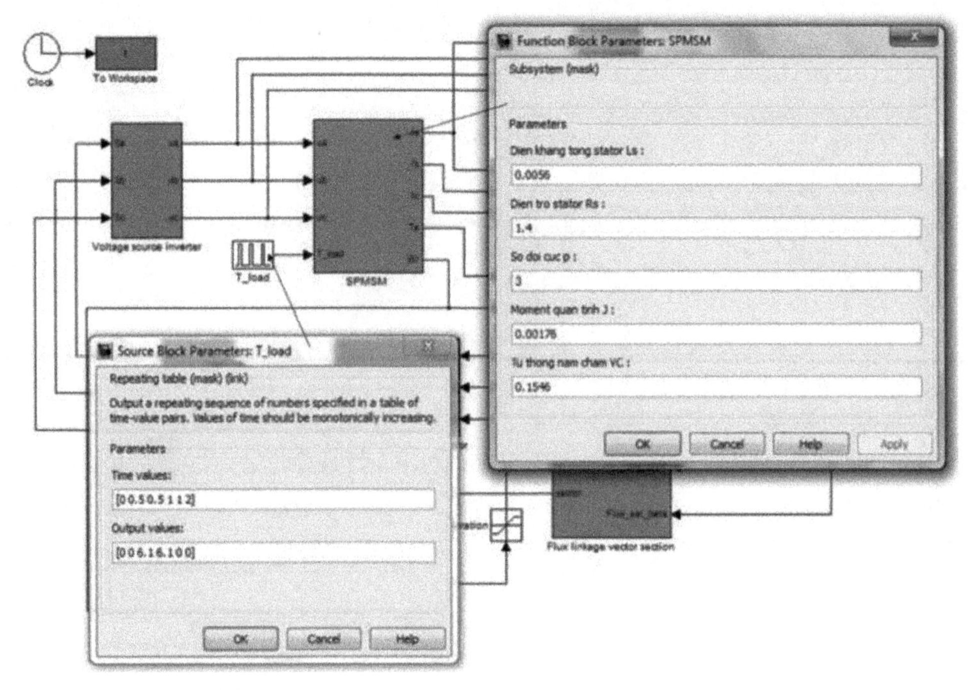

Figure 5.108 Parameters of the SPMSM (*Source:* SIMULINK software)

reversed with the speed command shown in Figure 5.107. When the motor has settled after the acceleration, the load torque is suddenly applied and removed in a step-wise approach.

Figures 5.109–5.112 show rotor speed, stator flux linkage, electromagnetic torque, and stator phase current, respectively, during start-up and acceleration. The actual rotor speed quickly meets the commanded speed after 0.05 seconds. The stator flux is fluctuated during the start-up then steady at the rated value. Torque overshoot is high during the start-up and up to the maximum limit set up by the limiter of the torque command at the output of the PID controller in Figure 5.92. Due to the high starting torque, the starting current's overshoot is also high.

Responses of actual rotor speed, motor's torque, and stator phase current during the application of load torque are shown in Figures 5.112–5.114.

Rotor speed of the SPMSM is not significantly affected when the rated load torque is suddenly applied in a step-wise approach. The speed drops slightly and then quickly returns to the commanded value, as shown in Figure 5.112.

Transient response of electromagnetic torque is also fast. Electromagnetic torque of the SPMSM quickly stabilized within the hysteresis bands around the commanded torque, as shown in Figure 5.113. Figure 5.114 also shows fast response of stator current during the application of load torque.

Stator flux linkage is slightly affected when the load torque is present, as shown in Figure 5.115.

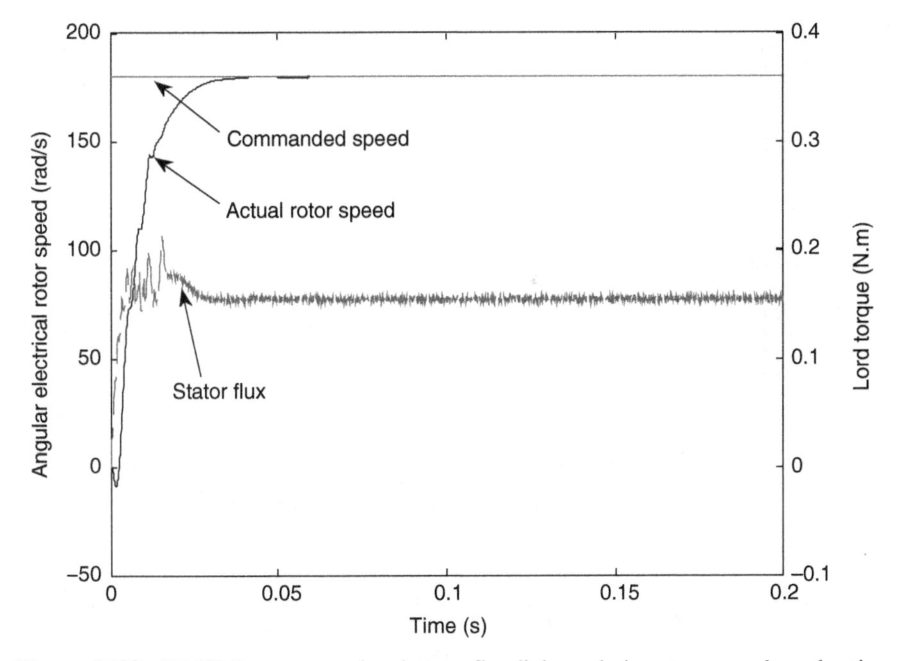

Figure 5.109 SPMSM's rotor speed and stator flux linkage during start-up and acceleration

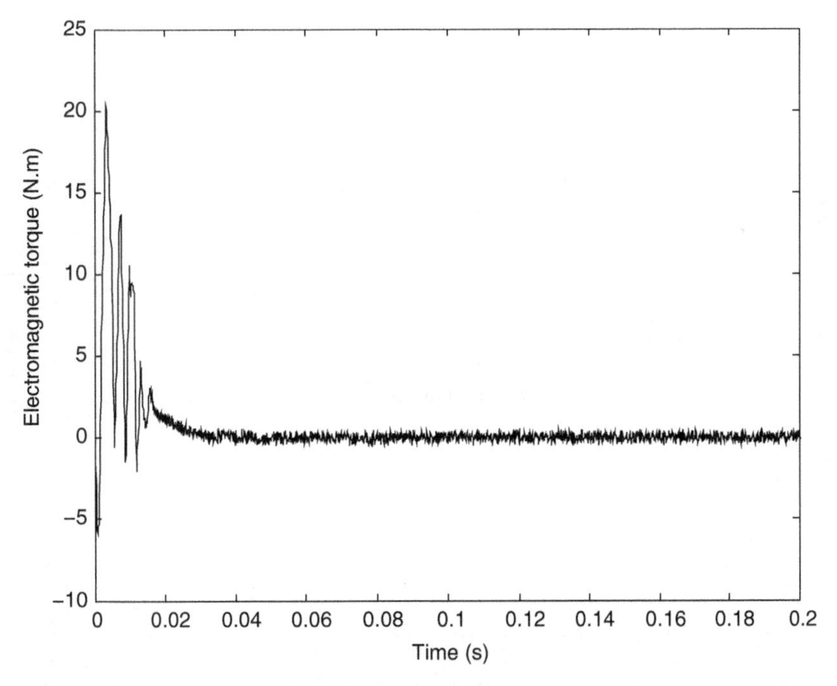

Figure 5.110 Electromagnetic torque of SPMSM during start-up and acceleration

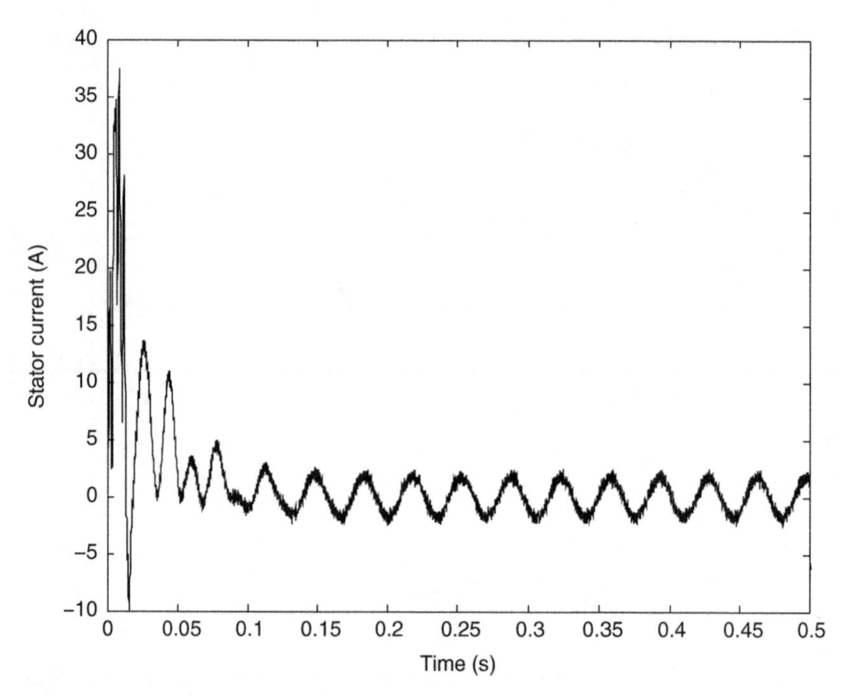

Figure 5.111 Stator phase current of the SPMSM during start-up and acceleration

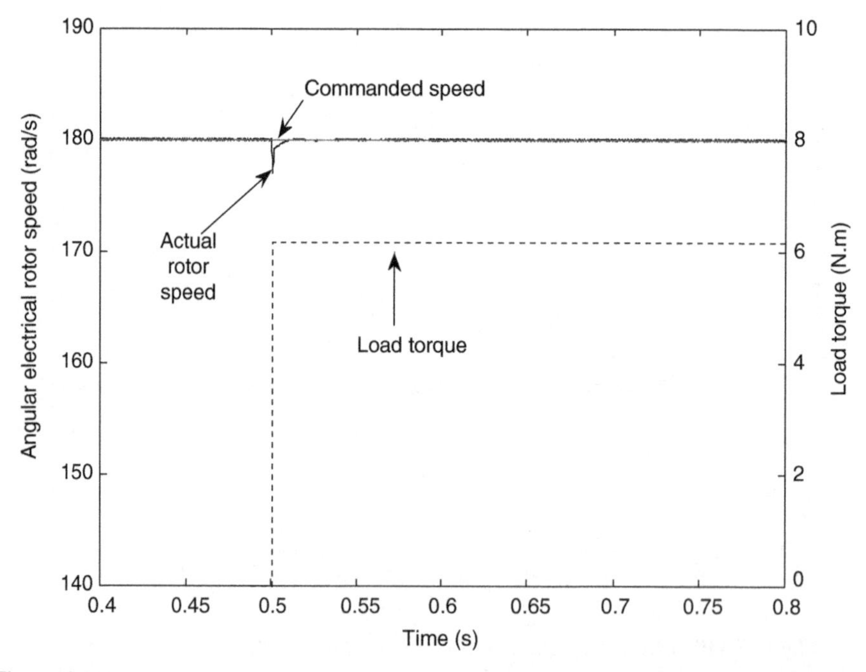

Figure 5.112 Commanded speed and actual rotor speed during the application of load torque

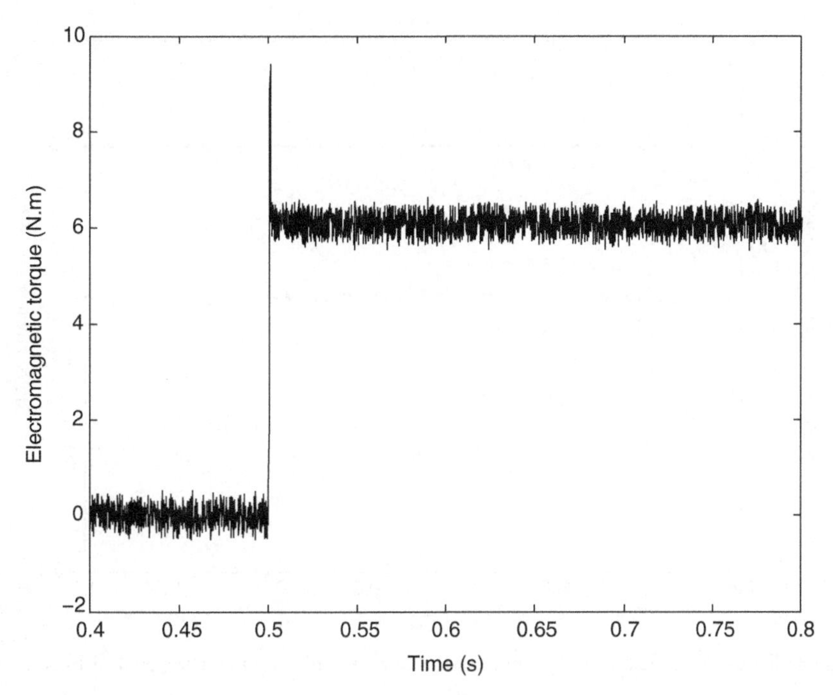

Figure 5.113 Response of SPMSM's electromagnetic torque during load torque application

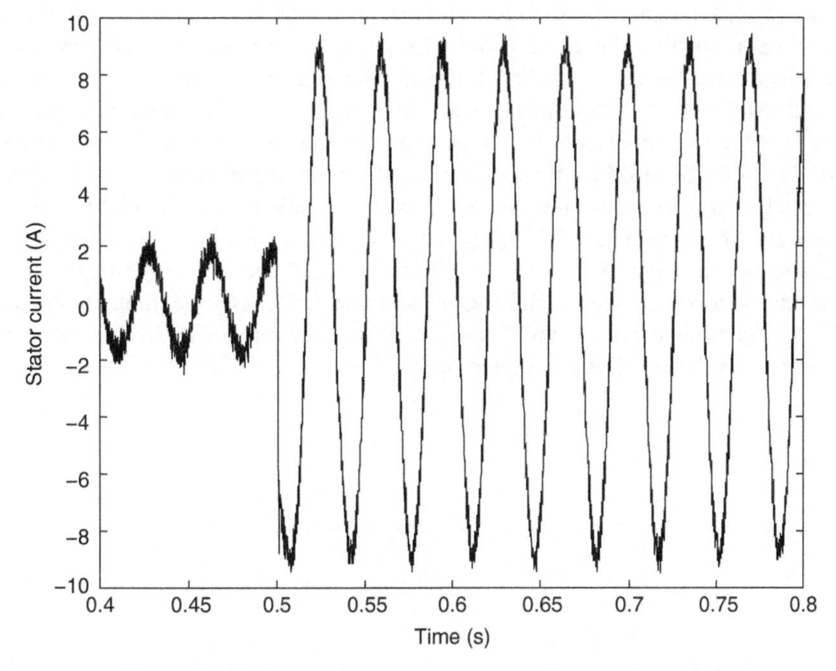

Figure 5.114 Stator phase current when load torque is applied

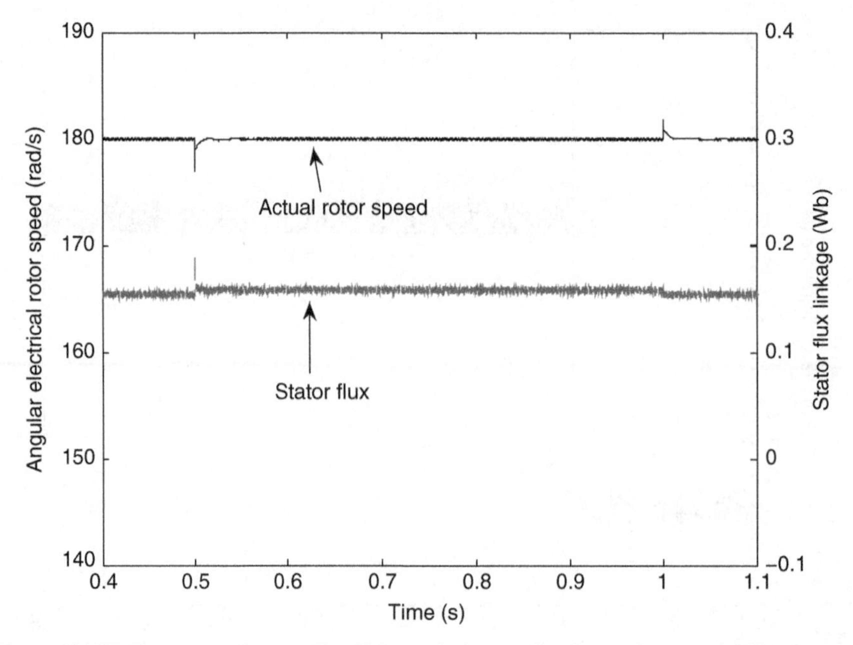

Figure 5.115 Response of stator flux linkage during application and removal of load torque

Responses of actual rotor speed, motor's torque, and stator phase current during the removal of load torque are shown in Figures 5.116–5.118. The response of speed is also slightly affected when the load torque is removed. Fast transient responses are observed for electromagnetic torque and stator current. Rotor speed, stator flux, electromagnetic torque, and stator current of the SPMSM during reversal of speed are shown in Figures 5.119–5.121, respectively.

Stator flux of the SPMSM is affected significantly during deceleration in the forward direction of speed, as well as acceleration in the backward direction. This observation is consistent with the results shown in Figure 5.109. The actual rotor speed quickly follows the commanded values, as shown in Figure 5.119. The steady state error of speed is insignificant.

Electromagnetic torque does slightly cross the lower limit set by the limiter at the output of the PID speed controller, as shown in Figure 5.120. Stator current has high overshoot during the speed reversal, as demonstrated in Figure 5.121.

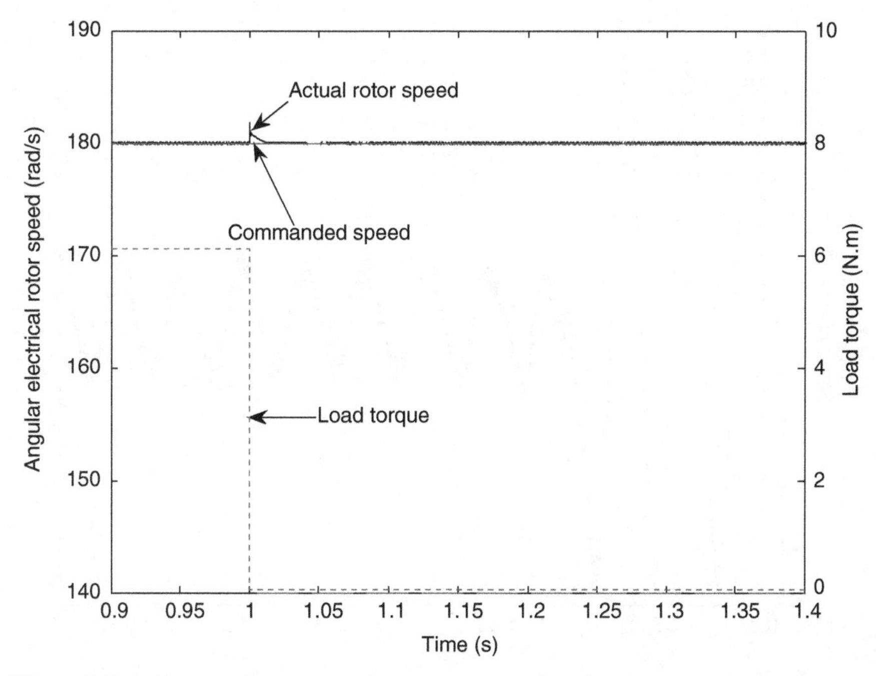

Figure 5.116 Commanded speed and actual rotor speed during the removal of load torque

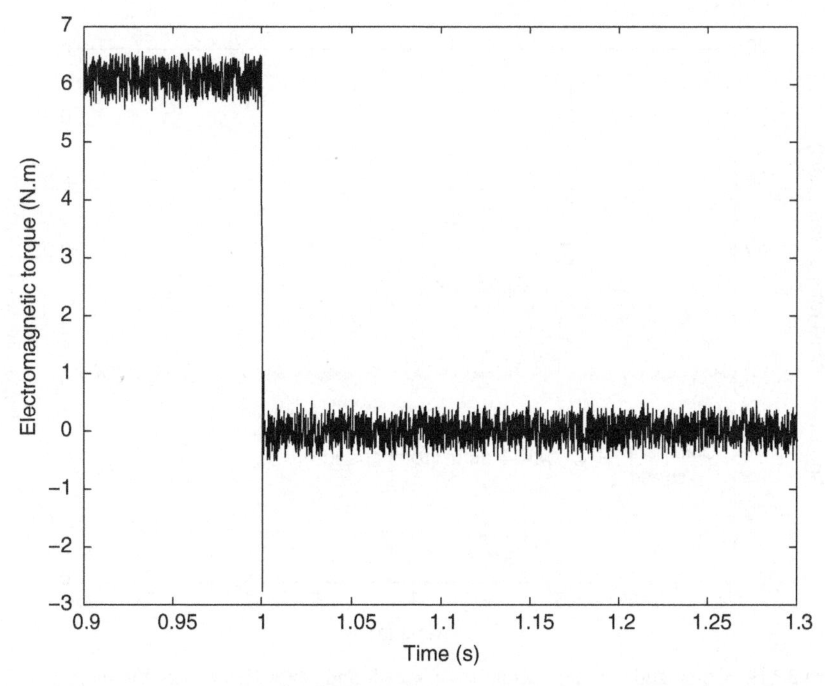

Figure 5.117 Response of the SPMSM's electromagnetic torque during load torque removal

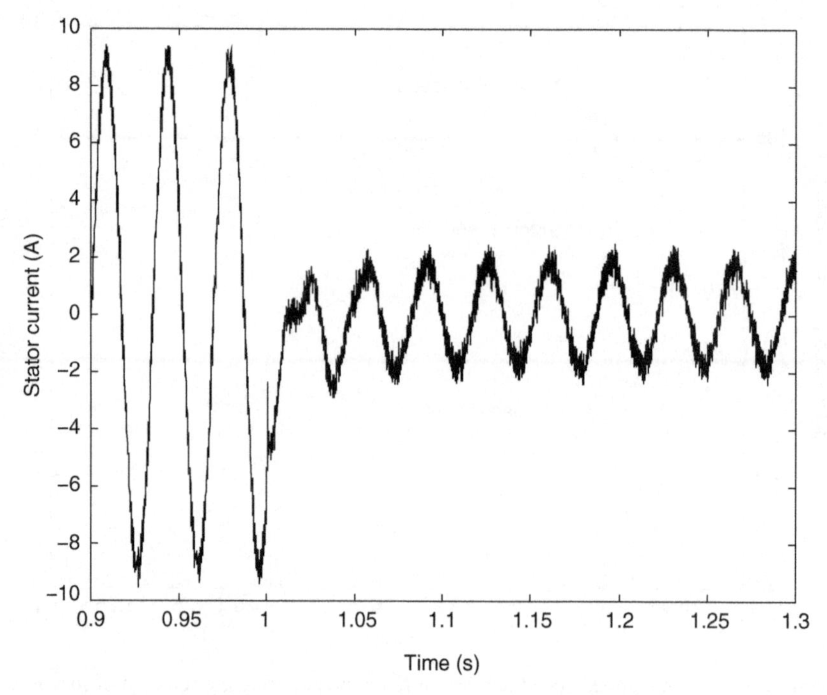

Figure 5.118 Stator current during the removal of load torque

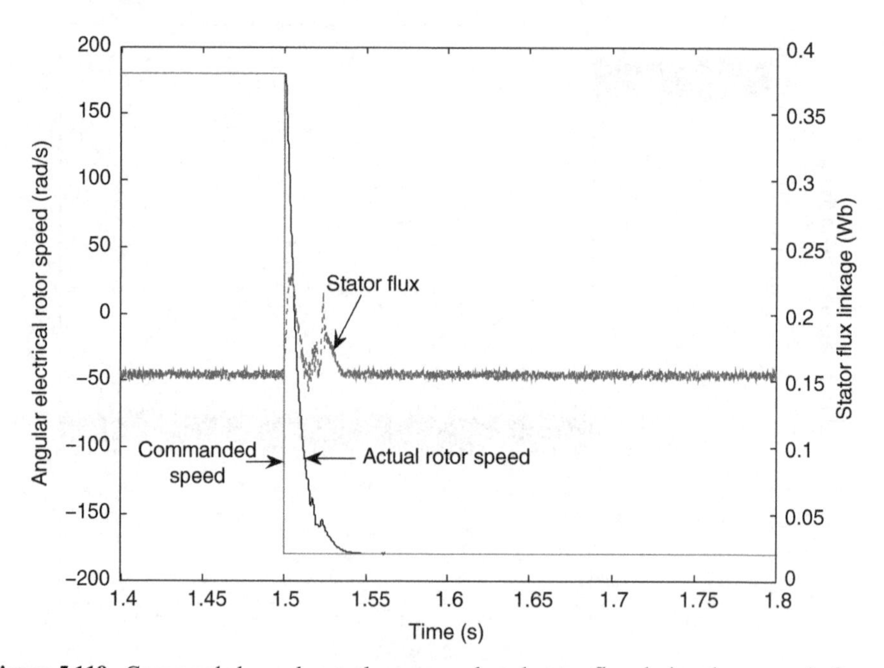

Figure 5.119 Commanded speed, actual rotor speed, and stator flux during the reversal of speed

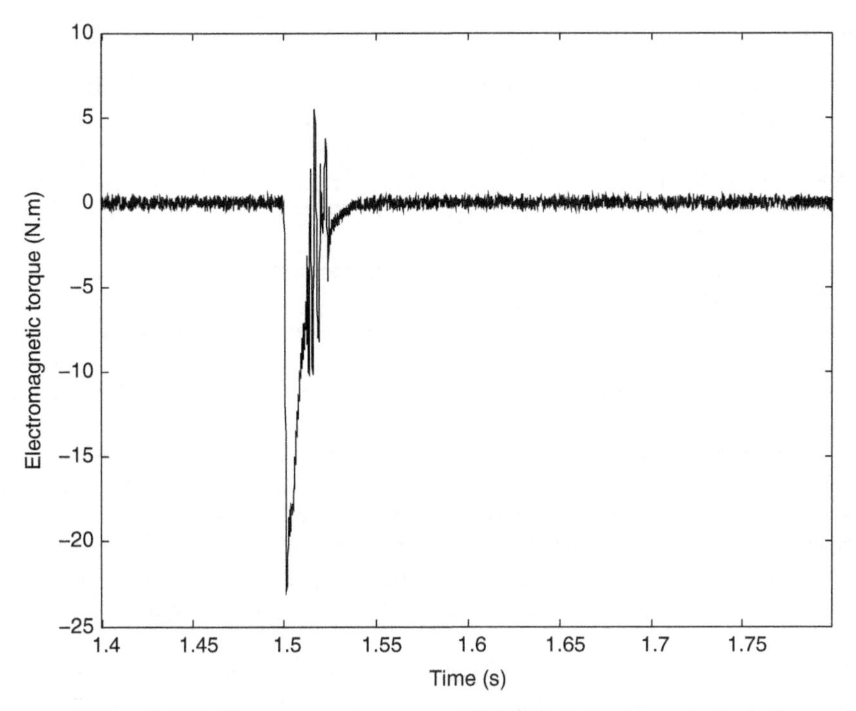

Figure 5.120 Electromagnetic torque of SPMSM during reversal of speed

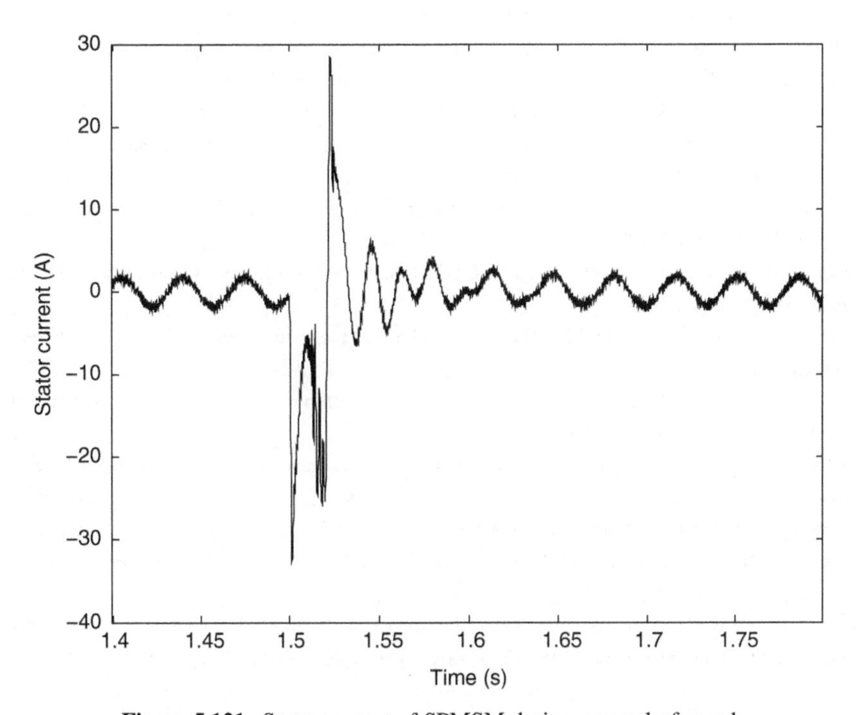

Figure 5.121 Stator current of SPMSM during reversal of speed

References

1. Takahashi, I. and Noguchi, T. (1986) A new quick-response and high-efficiency control strategy of an induction motor. *IEEE Trans. Ind. Appl.*, IA-22(5), 820–827.
2. Depenbrock, M. (1985) Direkte Selbstregelung (DSR) fur hochdynamische Drehfeldantriebe mit Stromrichterschaltung. *Etz Archiv.*, 7(7), 211–218.
3. Depenbrock, M. (1988) Direct self-control (DSC) of inverter-fed induction motors. *IEEE Trans. Power Electr.*, 3(4), 420–429.
4. Schofield, J. R. G. (1998) Variable speed drives using induction motors and direct torque control. *Dig. IEE Colloq. 'Vector Control Revisited'*, London, pp. 5/1–5/7.
5. Tiitinen, P., Pohjalainen, P., and Lalu, J. (1995) The next generation motor control method: direct torque control (DTC). *Europ. Power Electr. J.*, 5(1), 14–18.
6. Kazmierkowski, M. P. and Tunia, H. (1994) *Automatic Control of Converter-Fed Drives*. Elsevier, New York.
7. Vas, P. (1998) *Sensorless Vector and Direct Torque Control*. Oxford University Press, New York.
8. Casadei, D., Grandi, G., Serra, G., and Tani, A. (1994) Switching strategies in direct torque control of induction machines. *Int. Conf. Elect. Mach. ICEM'94*. Paris, pp. 204–209.
9. Casadei, D., Grandi, G., Serra, G., and Tani, A. (1998a) The use of matrix converters in direct torque control of induction machines. *IEEE Ann. Mtg. Ind. Elect. Soc., IECON'98*. Aachen, Germany, pp. 744–749.
10. Vas, P. (1992) *Electrical Machines and Drives: A Space-Vector Theory Approach*. Oxford, Clarendon Press.
11. Chen, J. and Li, Y. (1999) Virtual vectors based predictive control of torque and flux of induction motor and speed sensorless drives. *IEEE Ind. App. Soc. Ann. Mtg., IAS'99*, Phoenix, AZ, CD-ROM, paper No. 59_3.
12. Kang, J. K. and Sul, S. K. (1999) New direct torque control of induction motor for minimum torque ripple and constant switching frequency. *IEEE Trans. Ind. Appl.*, 35(5), 1076–1082.
13. Alfonso, D., Gianluca, G., Ignazio, M., and Aldo, P. (1999) An improved look-up table for zero speed control in DTC drives. *Europ. Conf. Power Elect. Appl. EPE'99*, Lausanne, Switzerland, CD-ROM.
14. Boldea, I. and Nasar, S. A. (1992) *Vector Control of AC Drives*. Boca Raton, FL CRC Press.
15. Levi, E. (1994) Impact of iron loss on behaviour of vector controlled induction machines. *IEEE Ind. App. Soc. Ann. Mtg., IAS'94*, Denver, CO, pp. 74–80.
16. Levi, E. (1995) Impact of iron loss on behaviour of vector controlled induction machines. *IEEE Trans. Ind. Appl.* 31(6), 1287–1296.
17. Levi, E. (1996) Rotor flux oriented control of induction machines considering the core loss. *Electr. Mach. Power Syst.*, 24(1), 37–50.
18. Levi, E. and Pham-Dihn, T. (2002) DTC of induction machines considering the iron loss. *Electr. Power Compon. Syst.*, 30(5), 557–579.
19. Pham-Dihn, T. (2003) Direct torque control of induction machines considering the iron losses, PhD thesis, Liverpool John Moores University, UK.
20. Dittrich, A. (1998) Model based identification of the iron loss resistance of an induction machine. *Int. Conf. Power Elect. Var. Speed Drives, PEVD'98*, London, IEE Conference Publication No. 456, pp. 500–503.
21. Noguchi, T., Nakmahachalasint, P., and Watanakul, N. (1997) Precise torque control of induction motor with on-line parameter identification in consideration of core loss. *Proc. Power Conv. Conf., PCC'97*, Nagaoka, Japan, pp. 113–118.
22. Wieser, R. S. (1998) Some clarifications on the impact of AC machine iron properties on space phasor models and field oriented control. *Int. Conf. Elect. Mach., ICEM'98*, Istanbul, pp. 1510–1515.

23. Sokola, M and Levi, E. (2000) A novel induction machine model and its application in the development of an advanced vector control scheme. *Int. J. Electr. Eng. Educ.*, 37(3), 233–248.
24. Habetler, T. G., Profumo, F., Pastorelli, M., and Tolbert, L. M. (1992) Direct torque control of induction machines using space vector modulation. *IEEE Trans. Ind. Appl.*, 28(5), 1045–1053.
25. Mir, S. A., Elbuluk, M. E., and Zinger, D. S. (1994) Fuzzy implementation of direct self control of induction machines. *IEEE Trans. Ind. Appl.*, 30(3), 729–735.
26. Stojic, M. D. and Vukosavic, N. S. (2005) A new induction motor drive based on the flux vector acceleration method. *IEEE Trans. Energy Convers.*, 20(1), 173–180.
27. Swarupa, L. M., Das, T. R. J., and Gopal, R. V. P. (2009) Simulaiton and analysis of SVPWM based 2-level and 3-level inverters for direct torque of induction motor. *Int. J. Elect. Eng. Res.*, 1(3), 169–184.
28. Jidin, A., Idris, N. R. N, Jatim, N. H. A, Sutikno, T., and Elbuluk, E. M. (2011) Extending switching frequency of torque ripple reduction utilizing a constant frequency torque controller in DTC of induction motors. *J. Power Electr.*, 11(2), 148–155.
29. Krishnan, R. (2010) *Permanent Magnet Synchronous and Brushless DC Drives*. CRC Press, Taylor & Francis Group.
30. Levi, E. (2001) *High Performance Drives*. Course note for the Subject ENGNG 3028, School of Engineering, Liverpool John Moores University, UK.

6

Nonlinear Control of Electrical Machines Using Nonlinear Feedback

Zbigniew Krzeminski[1] and Haitham Abu-Rub[2]
[1] *Gdansk University of Technology, Gdańsk, Poland*
[2] *Texas A&M University at Qatar, Al Rayyan, Qatar*

6.1 Introduction

A recent trend in advanced drive control has enabled an AC motor to behave like a separately excited DC motor. This goal has been almost achieved since introducing the vector control principle. The main disadvantage of vector control methods is the presence of nonlinearity in the mechanical part of the equation while changing the rotor flux linkage.

Although good results from field-oriented control (FOC) have been observed, attempts to obtain new control methods are being made. Nonlinear control of induction motors (IMs) was first presented in [1]. It introduced a novel mathematical model for IMs, which made it possible to avoid using sin/cos transformation of state variables. The model consists of two completely decoupled subsystems, mechanical and electromagnetic. It has been shown that in such a situation, it is possible to have nonlinear control and decoupling between electromagnetic torque and the rotor linkage flux. Decoupling between torque and the square of rotor flux has been obtained, after assuming some simplification [1–4]. Works that discuss nonlinear control of dynamic systems include [1–28].

When a motor is fed by voltage source inverters, and when the rotor flux linkage magnitude is kept constant, the nonlinear control system control is equivalent to the vector control method. In many other situations, this new idea gives simplicity of the structure and good response [2–4].

The main disadvantage of vector control methods is the presence of nonlinearity in the mechanical part of the equation while changing the rotor flux linkage. Direct use of vector methods to control an induction machine fed by a current inverter, or to control a

High Performance Control of AC Drives With MATLAB®/Simulink, Second Edition.
Haitham Abu-Rub, Atif Iqbal, and Jaroslaw Guzinski.

Companion website: www.wiley.com/go/aburubcontrol2e

double-fed machine, provides high complexity to the machine model, which is necessary to obtain precise control systems.

The use of variable transformation to obtain nonlinear model variables makes the control strategy easy to perform because only four state variables have been obtained with relatively simple nonlinearity form. This means it possible to use this method in the case of flux vector changing and to obtain simple system structures. In such systems, it is possible to change the rotor flux linkage with operating point without affecting the dynamics of the system. The relations occurring between the new variables make it possible to obtain novel control structures that guarantee a good response from the drive system. This is convenient for the economical operation of drive systems in which this flux is reduced if the load is decreased.

It is possible to decouple the drive system into two parts using FOC, followed by an introduction to a new variable. However, this operation provides a more complicated nonlinear feedback form than those used in the nonlinear control approach. In addition, the use of new variable further complicates the control system.

6.2 Dynamic System Linearization Using Nonlinear Feedback

The linearity of the control system using nonlinear feedback requires the accessibility to all state variables. It is assumed that all necessary variables are measured and directly calculated or estimated using a model or observer system.

The linearization process of nonlinear systems is widely discussed in [2, 4, 5, 10, 27]. To use nonlinear feedback, it is desirable to obtain a mathematical description of the nonlinear dynamic system as a system of differential equations. It is assumed that the dynamic system is described by equations [2, 5, 10, 21]:

$$\dot{x} = f(x) + \sum g_i(x)\, u_i(x) \tag{6.1}$$

$$y = Cx$$

Where:

$\mathbf{u} = [u_1, u_2, \ldots u_m]^T$ is the vector control $u \in \mathbf{R}^m$, x is a vector of state variables $\mathbf{x} \in \mathbf{R}^n$ where $\mathbf{m} \leq \mathbf{n}$, $\mathbf{f}(.)$, $\mathbf{g_i}(.)$ are vector nonlinear functions of $\mathbf{R}^n$.

Methods of differential geometry [10] make it possible to convert the above-described system by using a nonlinear change of variables and nonlinear feedback to the next form:

$$\dot{z} = Az + Bv \tag{6.2}$$

where **A** and **B** are constant **matrices**, so that a pair (**A**, **B**) is controllable, and **v** is the new control variable (input signal).

Linearization of the system using variable transformation and nonlinear feedback may be realized, as in [10]. The first step of linearization is to choose new variables:

$$z = h(x) \tag{6.3}$$

Such that the differential equations based on Lie derivatives take the form:

$$\dot{\mathbf{z}}_1 = \mathbf{A}_1\mathbf{z} \tag{6.4}$$

$$\dot{z}_k = A_k z_k + f_2(x) + g_2(x) \,.\, u \tag{6.5}$$

where $\mathbf{A_1}$ and $\mathbf{A_{2k}}$ are constant matrices and $\mathbf{f_2}(.)$ and $\mathbf{g_2}(.)$ are nonlinear vector functions; for the application for electrical machines, generally the vector $\mathbf{z}_1$ is the speed and the vector $\mathbf{z_k}$ presents the remaining chosen variables $\mathbf{z_k} = [z_2, z_3, \ldots z_n]^T$.

In the next step of linearization, it is assumed that when using nonlinear feedback, the nonlinearity in equation (6.5) will be compensated. Based on equations (6.2) and (6.5), we can write:

$$Bv = f_2(z) + g_2(z)u \tag{6.6}$$

where $\mathbf{B}$ is the adopted constant matrix and v is the new control signal.

If there is an inverse function to $\mathbf{g_2(z)}$, then the control signal used for linearization is calculated from [2, 5, 10, 27]

$$u = [g_2(z)]^{-1} \,.\, [B \,.\, v - f_2(z)] \tag{6.7}$$

After the suggestion of the nonlinear control described above, the dynamic system may be obtained as

$$\dot{\mathbf{z}}_1 = \mathbf{A}_1\mathbf{z} \tag{6.8}$$

$$\dot{z}_k = A_2 z_k + Bv \tag{6.9}$$

Linearization of a drive system with an IM using nonlinear feedback requires, in general, access to all state variables.

Figure 6.1 summarizes the idea of system linearization using nonlinear feedback [4]. Combining nonlinear feedback with the nonlinear model, and then making variable transformations, converts the highly nonlinear dynamic system, such as an IM, into a linear object. This will be shown step-by-step in this chapter.

After obtaining a linear structure of the dynamic system, it will be possible to use a simple cascaded structure of PI controllers, as shown in the next section.

6.3 Nonlinear Control of Separately Excited DC Motors

Because the separately excited DC motor has the best mechanical characteristics, and it is and one of the most widely used DC motors, we will limit our explanation of the control of this motor to our introduction to the topic. Although the DC motor is not being used for advanced drives at this time, presenting this type of motor is beneficial from an educational point of view. The mathematical model of the separated DC motor is already presented in Section (2.2.1) of Chapter 2.

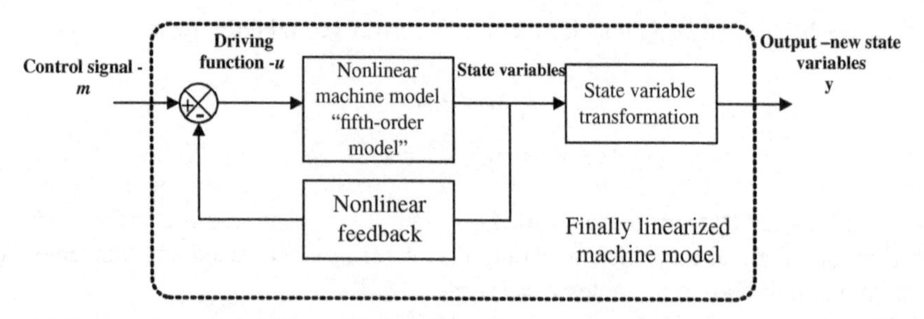

Figure 6.1 Linearization of dynamic system using nonlinear feedback: m – driving function of the linearized system; y – state variable of the linearized system; u – driving function of the nonlinear system; and x – state variable of the nonlinear system (*Source:* Based on Bogdan et al. [4])

For realizing a linearization of the nonlinearity in the mathematical model, a new variable is introduced, proportional to the motor torque [22]:

$$m = i_a \cdot i_f \tag{6.10}$$

After taking the derivative of the new variable and making some mathematical arrangements, the following model is derived [20]:

$$\frac{dm}{dt} = -\left(\frac{1}{T_f} + \frac{1}{T_a}\right) \cdot m + \frac{1}{T_a} \cdot v_1 \tag{6.11}$$

$$\frac{di_f}{dt} = -\frac{i_f}{T_f} + \frac{1}{T_f} \cdot v_2 \tag{6.12}$$

$$J\frac{d\omega_r}{dt} = \frac{m}{T_m} - \frac{m_o}{T_m} \tag{6.13}$$

The new model allows us to use linear and cascaded controllers for motor control. The output signals of the controllers are the new variables v_1 and v_2, which are described from the model as

$$v_1 = K_1 \cdot u_a \cdot i_f + \frac{T_a \cdot K_2 \cdot u_f \cdot i_a}{T_f} - K_3 \cdot i_f^2 \cdot \omega_r \tag{6.14}$$

$$v_2 = K_2 \cdot u_f \tag{6.15}$$

$$K_1 = \frac{1}{R_a} K_2 = \frac{1}{R_f} K_3 = \frac{K_\omega}{R_a}$$

The control scheme of the motor using the above procedure is shown in Figure 6.2. The speed controller commands the variable m that is proportional to the motor torque. This value is limited by the product of the actual value of the field current and the maximum value of the armature current. The torque controller commands the variable v_2, while the controller of the field current commands the control signal v_1. These two variables, along with other signals, are used in the decoupling block to compute the command values of the armature and field voltages.

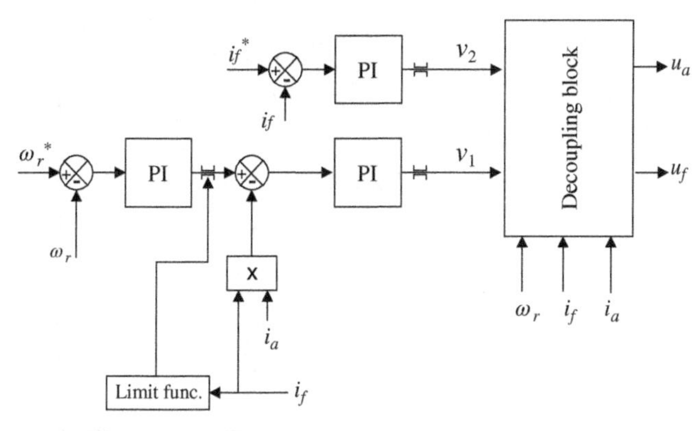

Figure 6.2 Separately excited DC motor control scheme

6.3.1 MATLAB/Simulink Nonlinear Control Model

The separately excited DC motor control model shown in Figure 6.3 consists of two blocks; a control system model and a separately excited DC motor model are shown in Figure 6.4. All the models are in per unit unless otherwise specified. The Simulink software file of separately excited DC motor model is [DCmotor_nonlinear]. The motor parameters are located in [DCmotor_nonlinear_init], which should be first initialized.

6.3.2 Nonlinear Control Systems

The motor model is linearized by a nonlinear input transformation with new input variables v_1 and v_2 as mentioned in equations (6.14) and (6.15). The natural inputs u_a and u_f can be found as follows:

$$u_a = \frac{v_1}{K_1 . i_f} - \frac{T_a . v_2 . i_a}{K_1 . i_f . T_f} - \frac{K_3}{K_1} . i_f . \omega_r \tag{6.16}$$

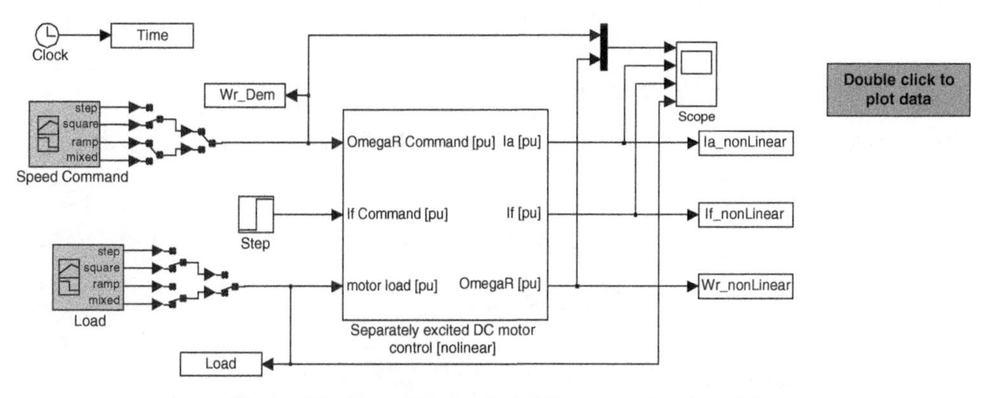

Figure 6.3 Separately excited DC motor control model

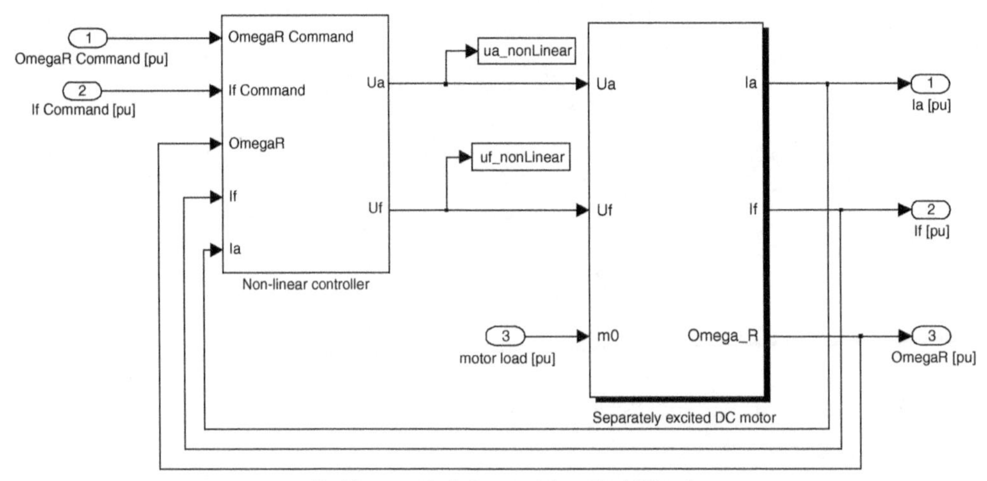

Figure 6.4 Components of separately excited DC motor control model

$$u_f = \frac{v_2}{K_2} \tag{6.17}$$

These equations are modeled in the transformation blocks shown in Figures 6.5 and 6.6.

6.3.3 Speed Controller

The speed controller is a PI controller, which calculated the demand for variable *m*, as shown in Figure 6.7. The armature current demand is calculated by solving the following equations:

$$m_{command} = U_{int} + K_{p_1}(\omega_{r_{command}} - \omega_r) \tag{6.18}$$

$$\frac{dU_{int}}{dt} = +K_{i_1}(\omega_{r_{command}} - \omega_r) \tag{6.19}$$

where $|U_{int}| \leq I_f I_{a_{max}}$ and $|m_{command}| \leq I_f I_{a_{max}}$

6.3.4 Controller for Variable m

The variable *m* controller is a PI, as shown in Figure 6.8. The input v_1 is calculated by solving the following equations:

$$v_1 = m_{int} + K_{p_2}(m_{command} - m) \tag{6.20}$$

$$\frac{dm_{int}}{dt} = K_{i_2}(m_{command} - m) \tag{6.21}$$

where $|m_{int}| \leq 3$ and $|v_1| \leq 3$

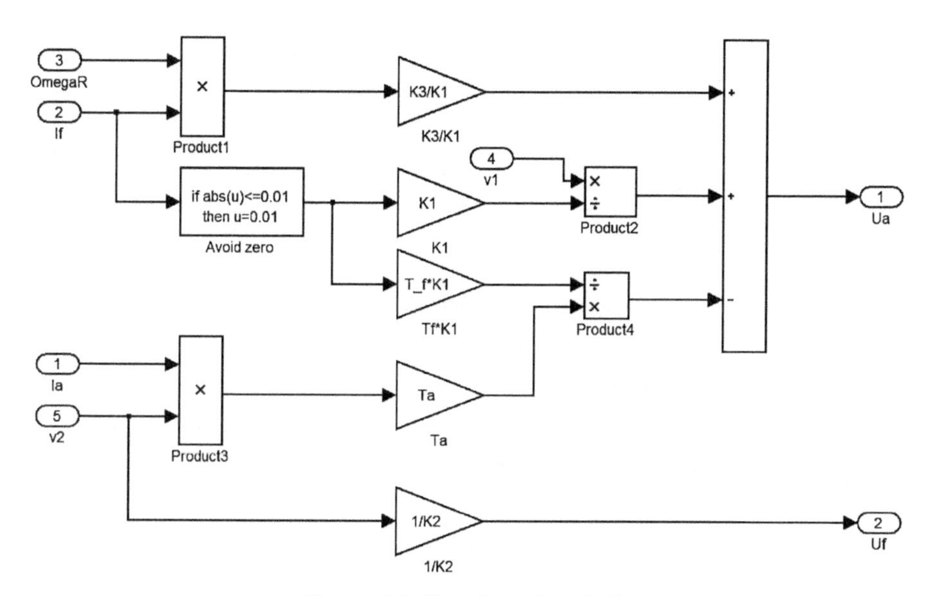

Figure 6.5 Transformation block

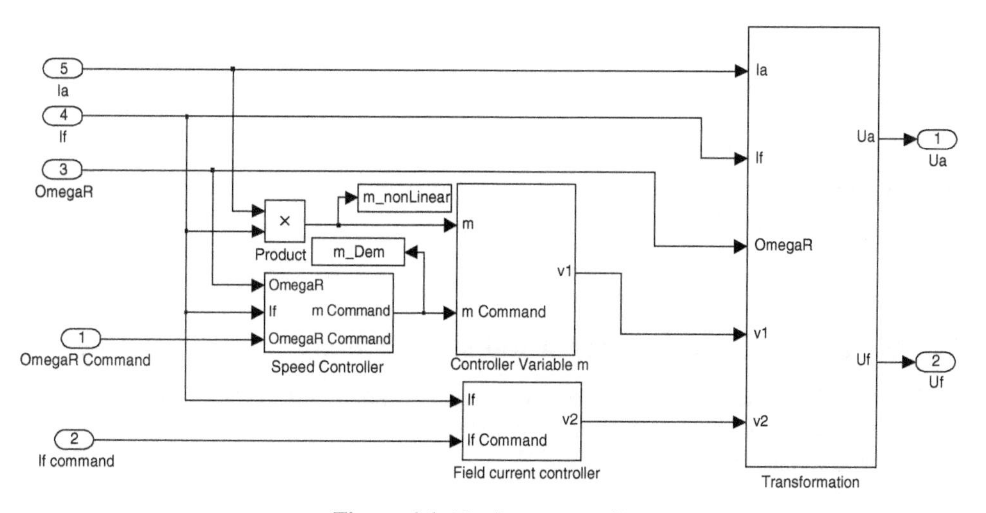

Figure 6.6 Nonlinear controller

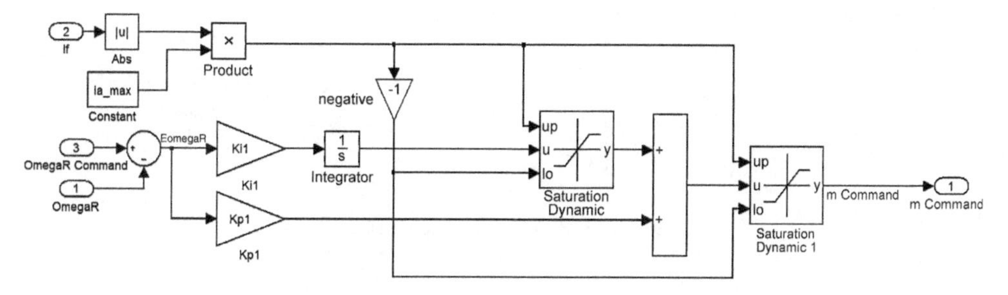

Figure 6.7 Speed controller

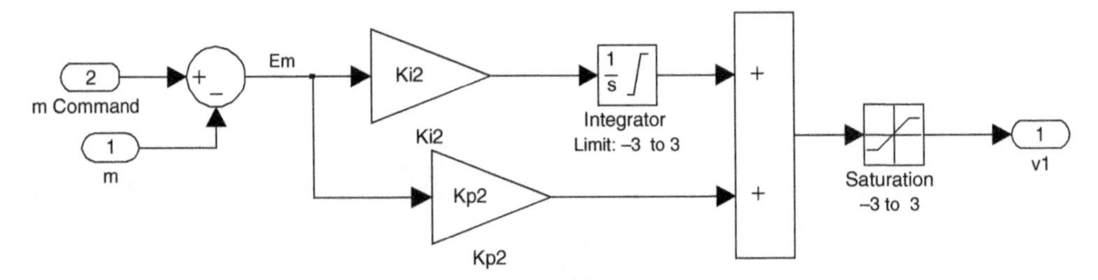

Figure 6.8 Variable m controller

6.3.5 Field Current Controller

The field controller is also a PI type. The control variable v_2 is calculated by solving the following equations (Figure 6.9):

$$v_2 = U_{f_{int}} + K_{p_3}\left(I_{f_{command}} - I_f\right) \tag{6.22}$$

$$\frac{dU_{f_{int}}}{dt} = K_{i_3}\left(I_{f_{command}} - I_f\right) \tag{6.23}$$

where, $|U_f| \leq 1$ and $|v_2| \leq 1$

6.3.6 Simulation Results

The simulation results for various steps in the speed demand with varying loads are shown in Figure 6.10. The controller tracks the demand well.

6.4 Multiscalar Model (MM) of Induction Motor

An IM is a higher-order, nonlinear object, as shown in Chapter 2. An internal coupling appears between mechanical and electromagnetic variables. Therefore, it is especially beneficial to use the idea of nonlinear control for IM control.

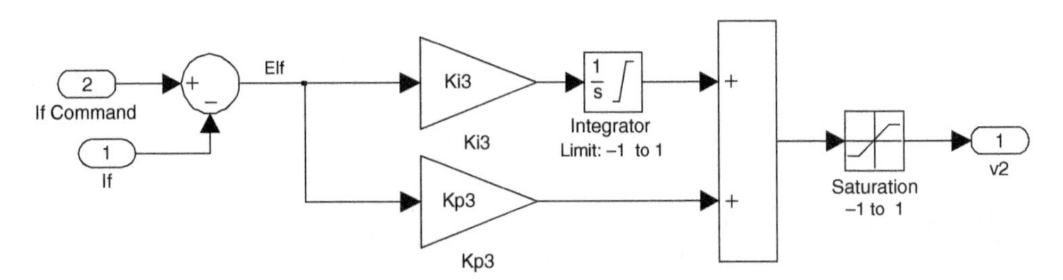

Figure 6.9 Variable v_2 controller

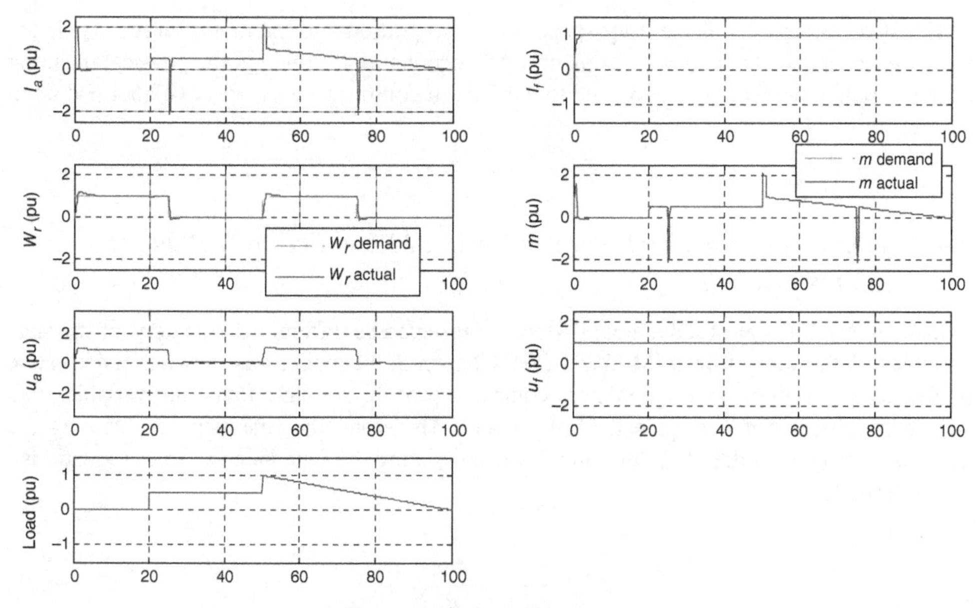

Figure 6.10 Simulation results

6.4.1 Multiscalar Variables

The nonlinear control idea using nonlinear feedback was first time presented in [1]. It creates a new model of the machine with fewer state variables, which may be linked with the rotor, stator, main flux, etc. If the variables are a function of the rotor flux, then a terminology of nonlinear rotor-oriented control systems is presented (i_s, ψ_r). Other options, such as (i_s, ψ_s), (i_s, ψ_m), may also be used. In this book, we will focus on the first option because of its popularity and less complicated decoupling system; however, for extra work, other options could be derived and examined.

The new state variables for the (i_s, ψ_r) option may be interpreted as rotor angular speed, scalar, and vector products of the stator current and rotor flux vectors, and the square of the rotor linkage flux is presented in an arbitrary reference frame (i.e. *XY*) as [1, 2, 4]

$$x_{11} = \omega_r \tag{6.24}$$

$$x_{12} = \psi_{rx} i_{sy} - \psi_{ry} i_{sx} \tag{6.25}$$

$$x_{21} = \psi_{rx}^2 + \psi_{ry}^2 \tag{6.26}$$

$$x_{22} = \psi_{rx} i_{sx} + \psi_{ry} i_{sy} \tag{6.27}$$

In these equations, ψ_{rx}, ψ_{ry}, i_{sx}, and i_{sy} are the rotor flux and stator current vectors in the coordinate system *XY*, rotating with arbitrary speed, and ω_r is the angular speed of the rotor shaft. Subscript x denotes the real frame and y the imaginary frame; therefore, the variables could be presented in a stationary $\alpha\beta$ frame or any other rotating co-ordinate system.

The state variable x_{11} is the rotor speed; x_{12} is proportional to the motor torque; x_{21} is the square of the rotor flux; and x_{22} is somehow proportional to the energy; the physical meaning of this variable, however, is not related to our idea of control, so we have not elaborated upon it here.

6.4.2 Nonlinear Linearization of Induction Motor Fed by Voltage Controlled VSI

A mathematical model of a dynamic system is presented as differential equations of the state variables. In the conventional IM model from Chapter 2, the model was shown as a derivative of five state variables. In the nonlinear control approach, we have four state variables that should be differentiated to gain the new model. Therefore, the first step of obtaining the new mathematical model of the machine is differentiating the four state variables (equations), which leads to [1, 2, 4].

$$\frac{dx_{11}}{dt} = \frac{d\omega_r}{dt} \tag{6.28}$$

$$\frac{dx_{12}}{dt} = \psi_{rx}\frac{di_{sy}}{dt} + i_{sy}\frac{d\psi_{rx}}{dt} - \psi_{ry}\frac{di_{sx}}{dt} - i_{sx}\frac{d\psi_{ry}}{dt} \tag{6.29}$$

$$\frac{dx_{21}}{dt} = 2\psi_{rx}\frac{d\psi_{rx}}{dt} + 2\psi_{ry}\frac{d\psi_{ry}}{dt} \tag{6.30}$$

$$\frac{dx_{22}}{dt} = \psi_{rx}\frac{di_{sx}}{dt} + i_{sx}\frac{d\psi_{rx}}{dt} + \psi_{ry}\frac{di_{sy}}{dt} + i_{sx}\frac{d\psi_{ry}}{dt} \tag{6.31}$$

The next step is to substitute derivatives for the currents and fluxes from the motor model (equations) in the derivatives of new variables (in equations). After using such substitutions and taking into account the relationships of the new state variables (equations), we obtain the next model of the system [1, 2, 4]:

$$\frac{dx_{11}}{d\tau} = \frac{x_{12}L_m}{JL_r} - \frac{m_o}{J} \tag{6.32}$$

$$\frac{dx_{12}}{d\tau} = -\frac{1}{T_v}x_{12} - x_{11}\left(x_{22} + \frac{L_m}{w_\sigma}x_{21}\right) + \frac{L_r}{w_\sigma}u_1 \tag{6.33}$$

$$\frac{dx_{21}}{d\tau} = -2\frac{R_r}{L_r}x_{21} + 2R_r\frac{L_m}{L_r}x_{22} \tag{6.34}$$

$$\frac{dx_{22}}{d\tau} = -\frac{x_{22}}{T_v} + x_{11}x_{12} + \frac{R_rL_m}{L_rw_\sigma}x_{21} + \frac{R_rL_m}{L_r}\frac{x_{12}^2 + x_{22}^2}{x_{21}} + \frac{L_r}{w_\sigma}u_2 \tag{6.35}$$

when T_v is the motor electromagnetic time constant:

$$T_v = \frac{w_\sigma L_r}{R_r w_\sigma + R_s L_r^2 + R_r L_m^2} \tag{6.36}$$

$$w_\sigma = L_r L_s - L_m^2 \tag{6.37}$$

$$u_1 = \psi_{rx} u_{sy} - \psi_{ry} u_{sx} \tag{6.38}$$

$$u_2 = \psi_{rx} u_{sx} + \psi_{ry} u_{sy} \tag{6.39}$$

The above model has significantly less nonlinearities than the conventional model presented in Chapter 2. Equations (6.32) and (6.34) are linear, though nonlinearities still exist in equations (6.33) and (6.35). To linearize these two equations, a new signal (m) is used in the feedback of the system according to equations (6.33) and (6.35). The two signals are defined to replace the nonlinearities in the above equations, and they are presented as follows [1, 2, 4]:

$$m_1 = -x_{11}\left(x_{22} + \frac{L_m}{w_\sigma} x_{21}\right) + \frac{L_r}{w_\sigma} u_1 \tag{6.40}$$

$$m_2 = x_{11} x_{12} + \frac{R_r L_m}{L_r w_\sigma} x_{21} + \frac{R_r L_m}{L_r} \frac{x_{12}^2 + x_{22}^2}{x_{21}} + \frac{L_r}{w_\sigma} u_2 \tag{6.41}$$

Therefore, the new mathematical model of IMs is described as [1, 2, 4] Mechanical subsystem:

$$\frac{dx_{11}}{d\tau} = \frac{x_{12} L_m}{J L_r} - \frac{m_o}{J} \tag{6.42}$$

$$\frac{dx_{12}}{d\tau} = -\frac{1}{T_v} . x_{12} + m_1 \tag{6.43}$$

Electromagnetic subsystem:

$$\frac{dx_{21}}{d\tau} = -2\frac{R_r}{L_r} x_{21} + 2R_r \frac{L_m}{L_r} x_{22} \tag{6.44}$$

$$\frac{dx_{22}}{d\tau} = -\frac{x_{22}}{T_v} + m_2 \tag{6.45}$$

The above model is linear and fully decoupled, which makes it possible to use a linear cascaded controller, as shown in Figure 6.11.

In the control system, the control signals m_1 and m_2 are generated by the PI controller of the state variables x_{12} and x_{21}, respectively. Having generated these two signals, the signals u_1 and u_2 can be then computed as [1, 2, 4]

$$u_1 = \frac{w_\sigma}{L_r}\left[x_{11}\left(x_{22} + \frac{L_m}{w_\sigma} x_{21}\right) + m_1\right] \tag{6.46}$$

$$u_2 = \frac{\omega_\sigma}{L_r}\left(-x_{11} x_{12} - \frac{R_r L_m}{L_r \omega_\sigma} x_{21} - \frac{R_r L_m}{L_r} \frac{x_{12}^2 + x_{22}^2}{x_{21}} + m_2\right) \tag{6.47}$$

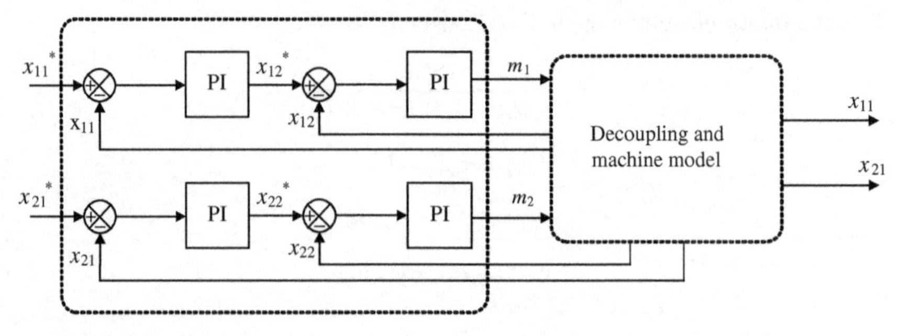

Figure 6.11 Cascaded structure for the multiscalar model control method (*Source:* Based on Krzeminski [2])

The voltage components ($u_{s\alpha}$ and $u_{s\beta}$) to be sent for the PWM algorithm are defined as [1, 2, 4]

$$u_{s\alpha} = \frac{\psi_{r\alpha}u_2 - \psi_{r\beta}u_1}{\psi_r^2} \tag{6.48}$$

$$u_{s\beta} = \frac{\psi_{r\beta}u_2 - \psi_{r\alpha}u_1}{\psi_r^2} \tag{6.49}$$

where ψ_r is the module of rotor flux linkage.

In the new space variables, the current is not directly controlled; nevertheless, it should be limited for motor protection. The current magnitude may be computed as

$$\frac{x_{12}^2 + x_{22}^2}{x_{21}} = I_s^2 \tag{6.50}$$

Therefore, the output signals of the controllers should be limited to assure that the motor current does not exceed the specific maximum current, $I_{s\ max}$. The torque, then, is limited according to

$$I_s^2 < \frac{x_{12}^2 + x_{22}^2}{x_{21}} \tag{6.51}$$

$$x_{12}^{limit} = \sqrt{I_{s\ max}^2 x_{21} - x_{22}^2} \tag{6.52}$$

6.4.3 Design of System Control

When using an MM, the torque produced by an IM is proportional only to one state variable (x_{12}), which is present in the mechanical part of the control system. The rotor flux linkage is stabilized in the electromagnetic part of the system, and its command value may be described by assuming such criterion as the minimization of energy losses in the system, or the minimization of response time in the mechanical part of the system when commands are limited [2].

The mechanical part consists of a first-order delay connected with an integral element in series. Because of the necessity for torque limiting, it is convenient to use a cascaded control structure for this subsystem. The disturbance in the mechanical sub-system is the load torque.

The electromagnetic sub-system consists of two first-order delay elements connected in a series. To limit the square of rotor flux in this situation, it is also convenient to use a cascade control structure.

Therefore, in the control system when using MM, cascaded controllers of the PI type may be applied with constant parameters that are tuned in accordance with the well-known control theory of linear systems.

In Figure 6.12, the fully decoupled system control of induction, in the case of using voltage controlled PWM, is shown [2].

6.4.4 Nonlinear Linearization of Induction Motor Fed by Current Controlled VSI

Although we operate on a stationary reference frame of initial analysis and new model derivation, we assume that our frame is the stator current in the x-axis (here x denotes the rotating, not stationary, frame with stator current). In a coordinate system, rotating with the stator current vector in the x-axis, the imaginary component equals zero ($i_{sy} = 0$). After introducing the first-order delay with a time constant T in the stator current set value channel, the differential equations of the IM are [2, 3]

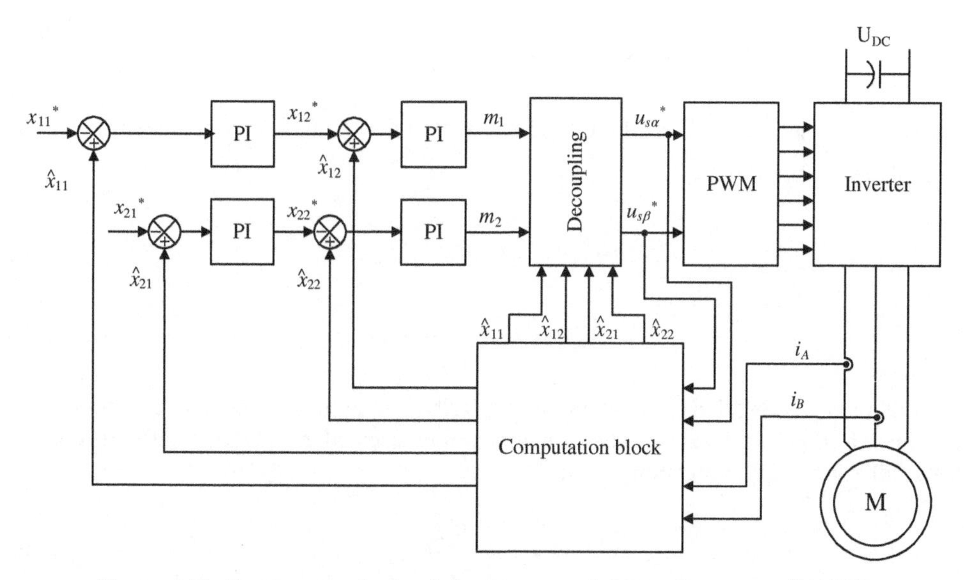

Figure 6.12 System control of an induction motor fed by voltage-controlled VSI

$$\frac{di_{sx}}{dt} = -\frac{1}{T}(i_{sx} - I_s) \tag{6.53}$$

$$\frac{d\Psi_{rx}}{dt} = -\frac{R_r}{L_r}\Psi_{rx} + (\omega_i - \omega_r)\Psi_{ry} + R_r\frac{L_m}{L_r}i_{sx} \tag{6.54}$$

$$\frac{d\Psi_{ry}}{dt} = -\frac{R_r}{L_r}\Psi_{ry} - (\omega_i - \omega_r)\Psi_{rx} \tag{6.55}$$

$$\frac{d\omega_r}{dt} = \frac{L_m}{L_r J}(-\Psi_{ry} i_{sx}) - \frac{1}{J}m_o \tag{6.56}$$

R_r, and L_r are rotor resistance and inductance, respectively, L_m is the mutual inductance, ω_r is the rotor angular speed, ω_s is the angular speed of coordinate system; J is the inertia, and m_o is the load torque.

The multiscalar model of IM is further determined from the derivations of the new MM state variables (equations), and while taking into account the above differential equations of stator current and rotor flux vectors (equations) that are similar to previous sections, we obtain the next model with less nonlinearities than a conventional one [2, 4]:

$$\frac{dx_{11}}{dt} = \frac{L_m}{JL_r}x_{12} - \frac{1}{J}m_o \tag{6.57}$$

$$\frac{dx_{12}}{d\tau} = -\frac{1}{T_v}.x_{12} + v_1 \tag{6.58}$$

$$\frac{dx_{21}}{dt} = -2\frac{R_r}{L_r}x_{21} + 2R_r\frac{L_m}{L_r}x_{22} \tag{6.59}$$

$$\frac{dx_{22}}{dt} = -\frac{1}{T_i}x_{22} - \frac{R_r L_m}{L_r}i_{sx}^2 + v_2' \tag{6.60}$$

where $\frac{1}{T_i} = \frac{1}{T} + \frac{R_r}{L_r}$, and T is a used time constant in the channel of current set value and v_1 and $v_2{}'$ are control signals:

$$v_1 = -\frac{1}{T}I_s\psi_{ry} + i_{sx}\psi_{rx}s_i \tag{6.61}$$

$$v_2' = \frac{1}{T}I_s\Psi_{rx} + i_{sx}\Psi_{ry}s_i \tag{6.62}$$

In this new model, there is a simple nonlinearity in only one differential equation. Using nonlinear feedback can compensate the nonlinearity presented in equation (6.60). This nonlinear feedback has the expression [2, 3]:

$$v_2 = v_2' - \frac{R_r L_m}{L_r}i_{sx}^2 \tag{6.63}$$

The quantity of the stator current amplitude is

$$I_s = -\frac{1}{T}\frac{\psi_{rx}v_2' - \psi_{ry}v_1}{\psi_r^2} \tag{6.64}$$

and the next relationship is

$$\frac{x_{12}^2 + x_{22}^2}{x_{21}} = I_s^2 \tag{6.65}$$

The slip frequency is

$$s_i = \frac{\psi_{ry}v_2' + \psi_{rx}v_1}{i_{sx}\psi_r^2} \tag{6.66}$$

After introducing nonlinear feedback, it is possible to obtain two linear fully decoupled subsystems, mechanical and electromagnetic. Such properties do not depend on the feeding system of the motor. The final MM has the form:

Mechanical subsystem [2, 4]:

$$\frac{dx_{11}}{dt} = \frac{L_m}{JL_r}x_{12} - \frac{1}{J}m_o \tag{6.67}$$

$$\frac{dx_{12}}{dt} = -\frac{1}{T_i}x_{12} + v_1 \tag{6.68}$$

Electromagnetic subsystem:

$$\frac{dx_{21}}{dt} = -2\frac{R_r}{L_r}x_{21} + 2\frac{R_rL_m}{L_r}x_{22} \tag{6.69}$$

$$\frac{dx_{22}}{dt} = -\frac{1}{T_i}x_{22} + v_2 \tag{6.70}$$

The multiscalar variables are not connected with the coordinate system. Therefore, it is not necessary to make the variables transform from the specified coordinate system to variables in the other coordinate system, which is selected in accordance with dynamic description or system control synthesis. This is essential for the practical realization of systems control because it gives significant simplification to the drive system.

The fully decoupled subsystems make it possible to use this method in the case of a changing flux vector and to obtain simple system structures. It is possible to use the cascade structure of PI controllers in the decoupled control subsystems, making it possible to limit the reference quantities of the variables x_{12} and x_{22}, such that the following relationship is fulfilled:

$$\frac{x_{12}^2 + x_{22}^2}{x_{21}} \le I_{s\ max}^2 \tag{6.71}$$

where $I_{s\ max}$ is the maximum allowed output current.

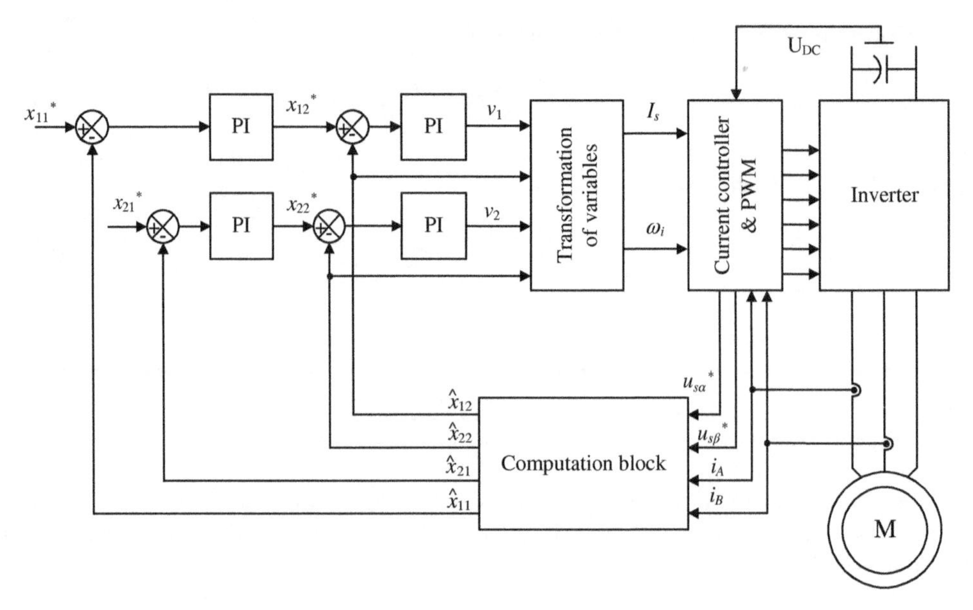

Figure 6.13 System control of an induction motor fed by current-controlled VSI

Also, it is possible, when using this method, to limit the inverter output voltage of the voltage inverter, cycloconverter, and the voltage in the DC bus of current inverter. This can be done by limiting the output quantities of controllers.

The final control system of an IM is shown in Figure 6.13. The stator current components are controlled using current controllers (e.g. hysteresis controllers). Instantaneous quantities of stator currents and voltages are used in the calculation block to compute the motor variables. To control the variables x_{12} and x_{22}, two PI controllers are used. The command value for the x_{12} controller is the output signal of the speed controller. The angular speed x_{11} is measured or calculated and used in the control feedback. The command value for the x_{22} controller is the output signal of the controller for the square of the rotor flux.

In the control system, the stator voltages are determined from the DC link voltage, with knowledge of the pulse width modulation algorithm or directly from commanded voltage in case of using voltage modulator. This method simplifies the total drive system and decreases the cost.

6.4.5 Stator-Oriented Nonlinear Control System (based on Ψ_s, i_s)

The steady state and transient electromagnetic behaviors of an IM can be described by the following equations in the stationary reference frame [3, 4]:

$$\frac{di_{sx}}{d\tau} = -\frac{L_r R_s + L_s R_r}{w_\sigma} i_{sx} + \frac{R_r}{w_\sigma}\psi_{sx} - \omega_r i_{sy} + \omega_r \frac{L_r}{w_\sigma}\psi_{sy} + \frac{L_r}{w_\sigma} u_{sx} \tag{6.72}$$

$$\frac{di_{sy}}{d\tau} = -\frac{L_rR_s + L_sR_r}{w_\sigma}i_{sy} + \frac{R_r}{w_\sigma}\psi_{sy} + \omega_r i_{sx} - \omega_r \frac{L_r}{w_\sigma}\psi_{sx} + \frac{L_r}{w_\sigma}u_{sy} \tag{6.73}$$

$$\frac{d\psi_{sx}}{d\tau} = -R_s i_{sx} + u_{sx} \tag{6.74}$$

$$\frac{d\psi_{sy}}{d\tau} = -R_s i_{sy} + u_{sy} \tag{6.75}$$

$$\frac{d\omega_r}{d\tau} = \frac{L_m}{JL_r}\left(\psi_{rx}i_{sy} - \psi_{ry}i_{sx}\right) - \frac{1}{J}m_o \tag{6.76}$$

where i_{sxy}, ψ_{sxy}, and u_{sxy} denotes vectors of stator current, stator flux, and stator voltage, respectively, and ω_r is the rotor speed. In addition, the mechanical behavior of an IM can be described by the rotor speed equation:

$$\frac{d\omega_r}{d\tau} = \frac{1}{J}\left(\psi_{sx}i_{sy} - \psi_{sy}i_{sx}\right) - \frac{1}{J}m_o \tag{6.77}$$

where the parameter w_σ is defined by:

$$w_\sigma = L_sL_r - L_m^2 \tag{6.78}$$

and L_s, L_r, and L_m are stator, rotor, and mutual inductances, respectively; R_s and R_r are stator and rotor resistances, respectively; m_o is the load torque; and J is the moment of inertia. All variables and parameters are expressed in the p.u. system. A stationary x-y reference frame is recommended for simplifications of measurements.

6.4.6 Rotor–Stator Fluxes-Based Model

The multiscalar model may be modified by selecting other motor variables other than rotor flux and stator current. Here, stator flux ψ_s and rotor flux ψ_r components could be selected to represent the motor [3, 4, 29]:

$$\frac{d\psi_{sx}}{d\tau} = -\frac{R_s}{w_\sigma}(L_r\psi_{sx} - L_m\psi_{rx}) + \omega_a\psi_{sy} + u_{sx} \tag{6.79}$$

$$\frac{d\psi_{sy}}{d\tau} = -\frac{R_s}{w_\sigma}\left(L_r\psi_{sy} - L_m\psi_{ry}\right) + \omega_a\psi_{sx} + u_{sy} \tag{6.80}$$

$$\frac{d\psi_{rx}}{d\tau} = -\frac{R_r}{\omega_\sigma}.\left(L_s\psi_{rx} - L_m\psi_{ry}\right) + (\omega_a - \omega_r).\psi_{ry} \tag{6.81}$$

$$\frac{d\psi_{ry}}{d\tau} = -\frac{R_r}{\omega_\sigma}.\left(L_s\psi_{ry} - L_m\psi_{sy}\right) - (\omega_a - \omega_r).\psi_{rx} \tag{6.82}$$

$$\frac{d\omega_r}{d\tau} = \frac{L_m}{Jw_\sigma}\left(\psi_{sx}\psi_{ry} - \psi_{sy}\psi_{rx}\right) - \frac{1}{J}m_o \tag{6.83}$$

The MM state variables could be then selected as [3, 4, 29].

$$q_{11} = \omega_r \tag{6.84}$$

$$q_{12} = \psi_{sx}\psi_{ry} - \psi_{sy}\psi_{rx} \tag{6.85}$$

$$q_{21} = \psi_{rx}^2 + \psi_{ry}^2 \tag{6.86}$$

$$q_{22} = \psi_{sx}\psi_{rx} + \psi_{sy}\psi_{ry} \tag{6.87}$$

Then, the derivatives of the above state variables, while taking into account the motor model from equations (6.79) to (6.83), yield [3, 4, 29]

$$\frac{dq_{11}}{d\tau} = \frac{L_m}{Jw_\sigma} q_{12} - \frac{m_o}{J} \tag{6.88}$$

$$\frac{dq_{12}}{d\tau} = -\frac{1}{T_v} q_{12} + q_{11}q_{22} + w_1 \tag{6.89}$$

$$\frac{dq_{21}}{d\tau} = -2\frac{R_r L_s}{w_\sigma} q_{21} + 2R_r \frac{L_m}{w_\sigma} q_{22} \tag{6.90}$$

$$\frac{dq_{22}}{d\tau} = -\frac{1}{T_v} \cdot q_{22} + \frac{R_s L_m}{\omega_\sigma} \cdot q_{22} + \frac{R_r L_m}{\omega_\sigma} \cdot \frac{q_{12}^2 + q_{22}^2}{q_{21}} + \omega_2 \tag{6.91}$$

where T_v represents machine electromagnetic time constant:

$$w_1 = u_{sx}\psi_{ry} - u_{sy}\psi_{rx} \tag{6.92}$$

$$\omega_1 = u_{sx}\psi_{ry} + u_{sy}\psi_{rx} \tag{6.93}$$

Though this model is simple, it requires an access to rotor and stator fluxes.
(Student work: design the control system and run it on MATLAB/Simulink) for this system.

6.4.7 Stator-Oriented Multiscalar Model

The first set of variables for the description of the IM dynamics was defined in [2]:

$$x_{11} = \omega_r \tag{6.94}$$

$$x_{12} = \psi_{sx} i_{sy} - \psi_{sy} i_{sx} \tag{6.95}$$

$$x_{21} = \psi_{sx}^2 + \psi_{sy}^2 \tag{6.96}$$

$$x_{22} = \psi_{sx} i_{sx} + \psi_{sy} i_{sy} \tag{6.97}$$

The differential equations for these variables form the multiscalar model of IM [3, 4, 29]:

$$\frac{dx_{11}}{d\tau} = \frac{1}{J} \cdot x_{12} - \frac{m_o}{J} \tag{6.98}$$

$$\frac{dx_{12}}{d\tau} = -\frac{1}{T_v}x_{12} + x_{11}\left(x_{22} - \frac{L_r}{w_\sigma}x_{21}\right) + \frac{L_m}{w_\sigma}u_1 \tag{6.99}$$

$$\frac{dx_{21}}{d\tau} = -2R_s x_{22} + 2u_2 \tag{6.100}$$

We need this equation to be written further as follows:

$$\frac{dx_{21}}{d\tau} = -2Rx_{22} + 2u_2 = -\frac{1}{T_v}x_{21} + m_2$$

where:

$$\frac{dx_{22}}{d\tau} = -\frac{1}{T_v}x_{22} - x_{11}x_{12} + \frac{R_r}{w_\sigma}x_{21} - \frac{x_{12}^2 + x_{22}^2}{x_{21}} + 2\frac{L_r}{w_\sigma}u_2 - \frac{L_m}{w_\sigma}u_1' \tag{6.101}$$

$$u_1 = u_{sy}\psi_{rx} - u_{sx}\psi_{ry} \tag{6.102}$$

$$u_2 = u_{sx}\psi_{sx} + u_{sy}\psi_{sx} \tag{6.103}$$

$$u_1' = u_{sx}\psi_{rx} + u_{sy}\psi_{rx} \tag{6.104}$$

$$T_v = \frac{w_\sigma L_r}{R_r w_\sigma + R_s L_r^2 + R_r L_m^2} \tag{6.105}$$

The nonlinear feedback decoupling of the form is

$$u_1 = \frac{w_\sigma}{L_r}\left[m_1 - x_{11}\left(x_{22} - \frac{L_r}{w_\sigma}x_{21}\right)\right] \tag{6.106}$$

$$u_2 = \frac{1}{2}\left(m_2 - \frac{1}{T}x_{21}\right) + R_s x_{22} \tag{6.107}$$

Then transform the system equations (6.98–6.101) to the linear systems of the following forms [3, 4, 29]:

Mechanical subsystem:

$$\frac{dx_{11}}{d\tau} = \frac{1}{J}x_{12} - \frac{m_o}{J} \tag{6.108}$$

$$\frac{dx_{12}}{d\tau} = -\frac{1}{T_v}x_{12} + m_1 \tag{6.109}$$

Electromechanical subsystem:

$$\frac{dx_{21}}{d\tau} = -\frac{1}{T_v}x_{21} + m_2 \tag{6.110}$$

where m_1 and m_2 are new inputs in the linear system and T_v time constant. The control variables u_1 and u_2, appearing in the multiscalar model of IM, are transformed on stator voltage components in the following way [3, 4, 29]:

$$u_{sx} = \frac{u_2\psi_{rx} - u_1\psi_{sy}}{\psi_{sx}\psi_{rx} + \psi_{ry}\psi_{sy}} \tag{6.111}$$

$$u_{sy} = \frac{\psi_{ry}u_2 + \psi_{rx}u_1}{\psi_{rx}\psi_{rx} + \psi_{ry}\psi_{ry}} \tag{6.112}$$

A scheme of the system with nonlinear control based on stator flux and current is presented in Figure 6.14. This system, which consists of two linear subsystems, may be controlled by means of cascaded controllers of mechanical subsystem, but only one controller of flux. The new state variables do not depend on the system coordinate. This is essential for the practical realization of control systems because it significantly simplifies the drive system. Sign ^ denotes variables estimated in the speed observer presented in [5, 6].

6.4.8 Multiscalar Control of Induction Motor

The control of IM model is shown in Figure 6.14. For control purpose, the motor model is decoupled by nonlinear state and input transformations. The model consists of motor model, various transformations, and the control system model. All the models are in per unit unless otherwise specified.

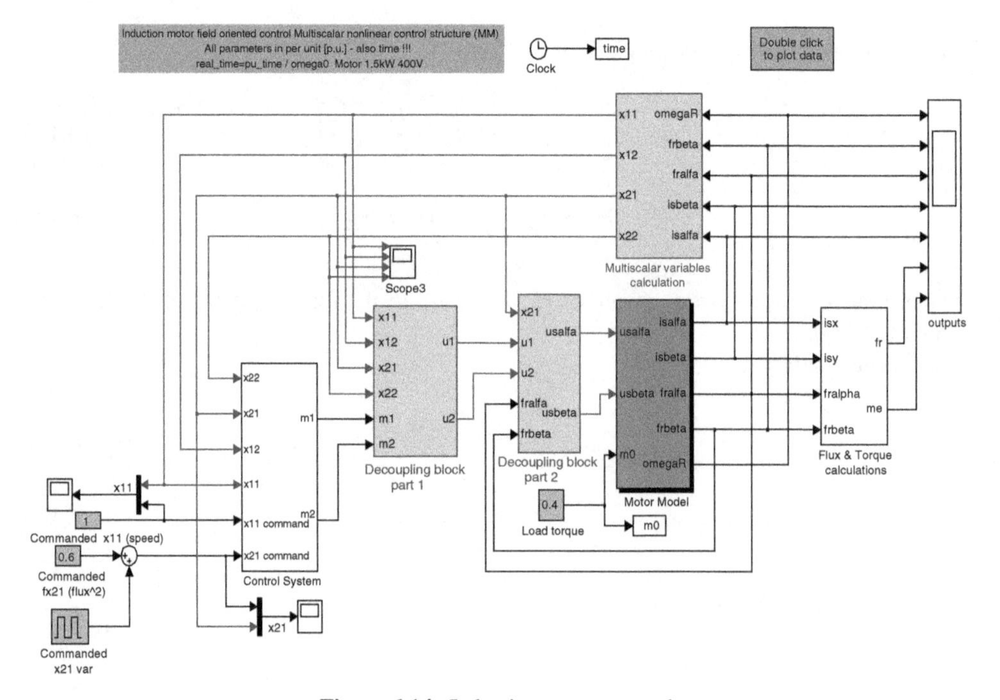

Figure 6.14 Induction motor control

The Simulink software file of multiscalar control of IM model is [MM_voltage_control]. The motor parameters are located in [IM_param], which should be first initialized.

6.4.9 Induction Motor Model

The IM model presented in per unit in a rotating frame with arbitrary speed (ω_a) is given in Chapter 2, and modeled in Figure 6.15.

The motor current and flux magnitude are calculated as

$$|I_s| = \sqrt{I_{s\alpha}^2 + I_{s\beta}^2} \tag{6.113}$$

$$|\psi_s| = \sqrt{\psi_{s\alpha}^2 + \psi_{s\beta}^2} \tag{6.114}$$

The motor torque is given by

$$M_e = \frac{L_m}{JL_r}\left(\psi_{r\alpha} i_{s\beta} - \psi_{r\beta} i_{s\alpha}\right) \tag{6.115}$$

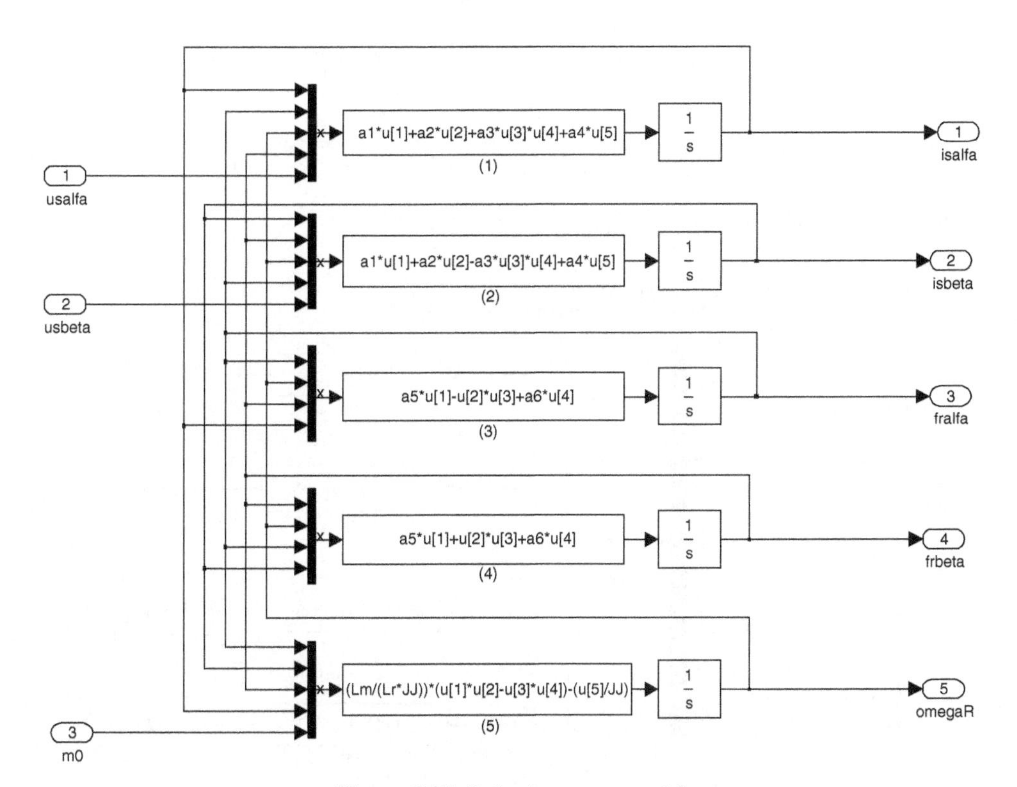

Figure 6.15 Induction motor model

6.4.10 State Transformations

The following nonlinear state transformation is adopted to convert a system into a form where nonlinearity only appears in the control channel:

$$x_{11} = \omega_r \tag{6.116}$$

$$x_{12} = \psi_{r\alpha} i_{s\beta} + \psi_{r\beta} i_{s\alpha} \tag{6.117}$$

$$x_{21} = \psi_{r\alpha}^2 + \psi_{r\beta}^2 = \psi_r^2 \tag{6.118}$$

$$x_{22} = \psi_{r\alpha} i_{s\alpha} + \psi_{r\beta} i_{s\beta} \tag{6.119}$$

The above transformation is modeled in Figure 6.16.

With the new inputs m_1 and m_2, using following input transformation, the nonlinear IM can be completely decoupled [2]:

$$u_{s\alpha} = \frac{\psi_{r\alpha} u_2 - \psi_{r\beta} u_1}{x_{21}} \tag{6.120}$$

$$u_{s\beta} = \frac{\psi_{r\alpha} u_1 - \psi_{r\beta} u_2}{x_{21}} \tag{6.121}$$

$$u_1 = \frac{w_\sigma}{L_r}\left(x_{11}\left(x_{22} + \frac{L_m}{w_\sigma} x_{21}\right) + m_1\right) \tag{6.122}$$

$$u_2 = \frac{\omega_\sigma}{L_r}\left(-x_{11} x_{12} - \frac{R_r L_m}{L_r \omega_\sigma} x_{21} - \frac{R_r L_m}{L_r} \frac{x_{12}^2 + x_{22}^2}{x_{21}} + m_2\right) \tag{6.123}$$

These equations are modeled in Figures 6.17 and 6.18.

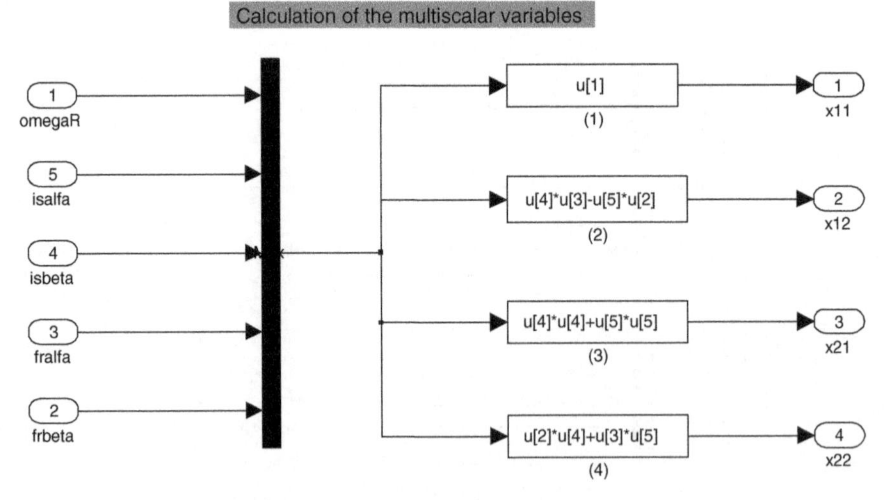

Figure 6.16 Nonlinear state transformation

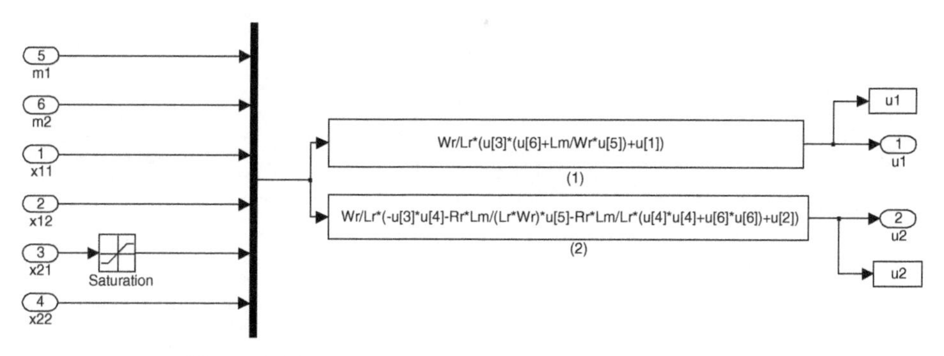

Figure 6.17 Nonlinear mapping between m_1, m_2, and u_1, u_2

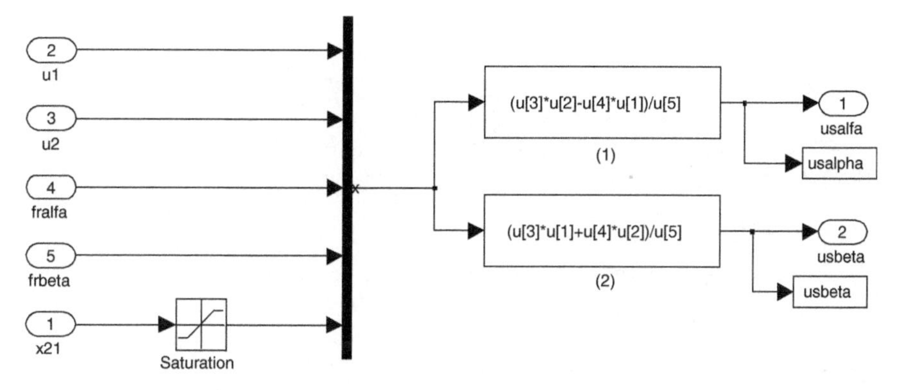

Figure 6.18 Nonlinear mapping between u_1, u_2 and usalpha, usbeta

6.4.11 Decoupled IM Model

The decoupled IM model between new multiscalar states and inputs m_1 and m_2 can be given as [2–4, 29]

$$\frac{dx_{11}}{d\tau} = \frac{L_m}{JL_r}x_{12} - \frac{1}{J}m_o \tag{6.124}$$

$$\frac{dx_{12}}{d\tau} = -\frac{1}{T_v}x_{12} + m_1 \tag{6.125}$$

$$\frac{dx_{21}}{d\tau} = -2\frac{R_r}{L_r}x_{21} + 2\frac{R_rL_m}{L_r}x_{22} \tag{6.126}$$

$$\frac{dx_{22}}{d\tau} = -\frac{1}{T_v}x_{22} + m_2 \tag{6.127}$$

where $T_v = \dfrac{w_\sigma L_r}{R_r w_\sigma + R_s L_r^2 + R_r L_m^2}$

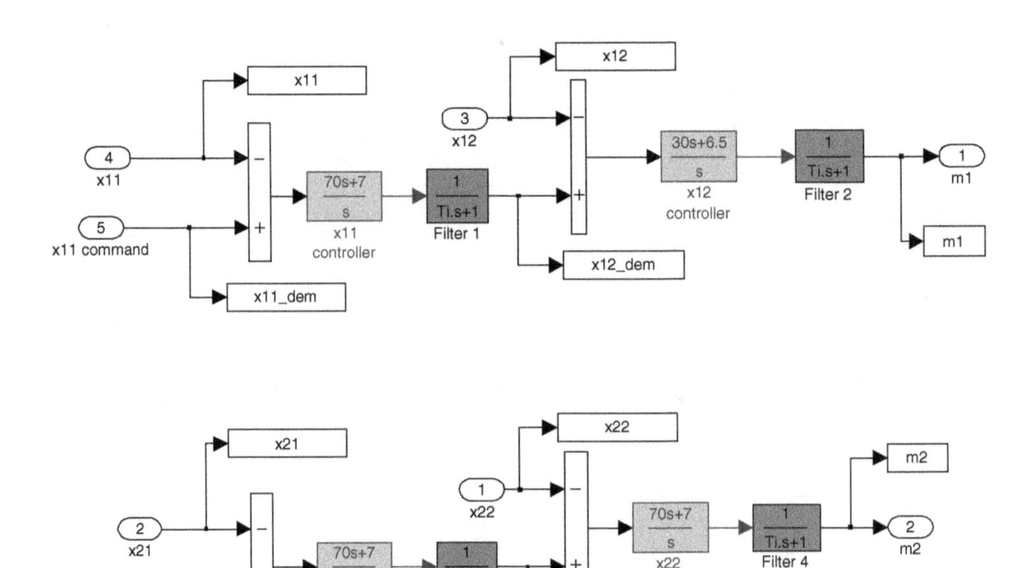

Figure 6.19 Cascaded control system for induction motor

6.4.11.1 Control System

As the system model is decoupled, the simple cascade control can be designed, which is also easy to tune. The cascade control system model is shown in Figure 6.19. The outer loop control works on system states x_{11} and x_{21} to provide demand for x_{12} and x_{22}. The inner loop controller works on x_{12} and x_{22} to achieve the required values.

6.4.11.2 Simulation Results

The simulation result for square pulse demand in speed is shown in Figures 6.20 and 6.21.

6.5 MM of Double-Fed Induction Machine (DFIM)

The mathematical model of DFIM can be derived as differential equations of five state variables. In addition to the rotor speed ω_r, stator flux ψ_s and rotor current i_r vectors are selected to represent the machine in [3, 4]

$$\frac{d\psi_{sx}}{d\tau} = -\frac{R_s}{L_s}\cdot\psi_{sx} + R_s\frac{L_m}{L_x}\cdot i_{rx} + \omega_a\psi_{sy} + u_{sx} \tag{6.128}$$

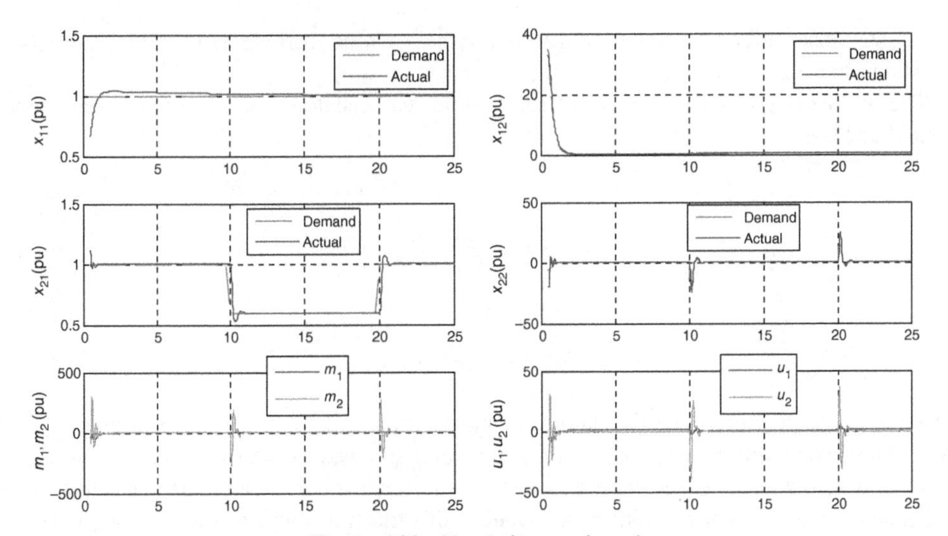

Figure 6.20 Simulation results – 1

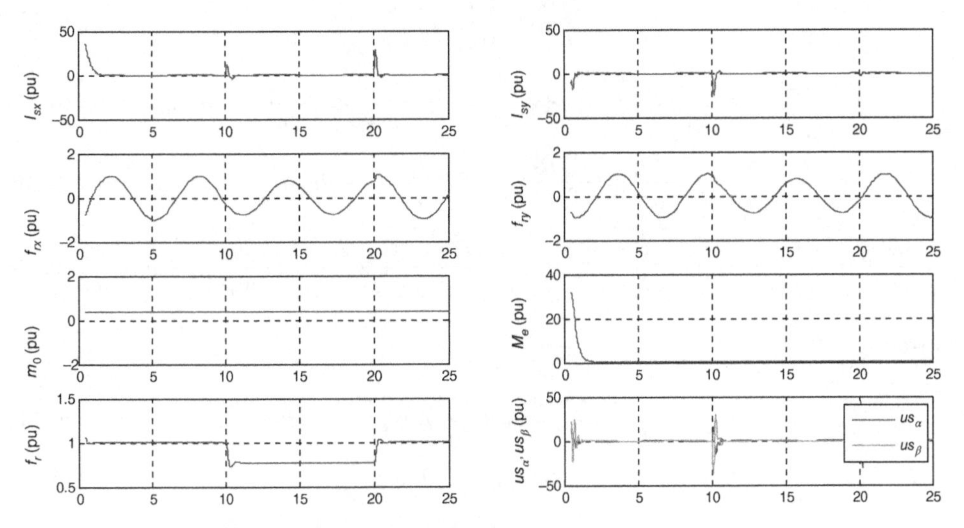

Figure 6.21 Simulation results – 2

$$\frac{d\psi_{sy}}{d\tau} = -\frac{R_s}{L_s}\cdot\psi_{sy} + R_s\frac{L_m}{L_x}\cdot i_{ry} - \omega_a\psi_{sx} + u_{sy} \tag{6.129}$$

$$\frac{di_{rx}}{d\tau} = -\frac{R_rL_r^2 + R_sL_m^2}{L_sw_\sigma}i_{rx} + \frac{R_sL_m}{L_sw_\sigma}\psi_{sx} + (\omega_a - \omega_r)i_{ry} - \omega_r\frac{L_m}{w_\sigma}\psi_{sy} + \frac{L_s}{w_\sigma}u_{rx} - \frac{L_m}{w_\sigma}u_{sx} \tag{6.130}$$

$$\frac{di_{ry}}{d\tau} = -\frac{R_rL_r^2 + R_sL_m^2}{L_sw_\sigma}i_{ry} + \frac{R_sL_m}{L_sw_\sigma}\psi_{sy} + (\omega_a - \omega_r)i_{rx} - \omega_r\frac{L_m}{w_\sigma}\psi_{sx} + \frac{L_s}{w_\sigma}u_{ry} - \frac{L_m}{w_\sigma}u_{sy} \tag{6.131}$$

$$\frac{d\omega_r}{d\tau} = \frac{L_m}{L_sJ}\left(\psi_{sx}i_{ry} - \Psi_{sy}i_{rx}\right) - \frac{1}{J}m_o \tag{6.132}$$

For stationary reference, frame ω_a is zero. For the machine working as a generator, the speed ω_r is an input to the model.

The multiscalar variables for DFIM can be selected as stator flux and rotor current, represented as [3, 4]

$$x_{11} = \omega_r \quad (6.133)$$

$$x_{12} = \psi_{sx} i_{ry} - \psi_{sy} i_{rx} \quad (6.134)$$

$$x_{21} = \psi_{sx}^2 + \psi_{sy}^2 \quad (6.135)$$

$$x_{22} = \psi_{sx} i_{rx} + \psi_{sy} i_{ry} \quad (6.136)$$

where ψ_{sx}, ψ_{sy}, i_{rx}, and i_{ry} are the stator flux and rotor current vectors in the coordinate system XY, rotating with arbitrary speed and ω_r is the angular speed of the rotor shaft.

After deriving the new state variables, and substituting the current and flux derivatives using the machine model, a new multiscalar model of this machine can be presented as [3, 4]

$$\frac{dx_{11}}{d\tau} = \frac{L_m}{JL_s} x_{12} - \frac{m_o}{J} \quad (6.137)$$

$$\frac{dx_{12}}{d\tau} = -\frac{1}{T_v} x_{12} + x_{11} x_{22} + \frac{L_m}{w_\sigma} x_{11} x_{21} + \frac{L_s}{w_\sigma} u_{r1} - \frac{L_m}{w_\sigma} u_{sf1} + u_{si1} \quad (6.138)$$

$$\frac{dx_{21}}{d\tau} = -2\frac{R_s}{L_s} x_{21} + 2R_s \frac{L_m}{L_s} x_{22} + 2u_{sf2} \quad (6.139)$$

$$\frac{dx_{22}}{d\tau} = -\frac{x_{22}}{T_v} - x_{11} x_{12} + \frac{R_s L_m}{L_s w_\sigma} x_{21} + \frac{R_s L_m}{L_s} \frac{x_{12}^2 + x_{22}^2}{x_{21}} - \frac{L_s}{w_\sigma} u_{r2} - \frac{L_m}{w_\sigma} u_{sf2} + u_{si2} \quad (6.140)$$

where

$$u_{r1} = u_{ry} \psi_{sx} - u_{rx} \psi_{sy} \quad (6.141)$$

$$u_{r2} = u_{ry} \psi_{sx} + u_{ry} \psi_{sy} \quad (6.142)$$

$$u_{sf1} = u_{sy} \psi_{sx} - u_{sx} \psi_{sy} \quad (6.143)$$

$$u_{sf2} = u_{sx} \psi_{sx} + u_{sy} \psi_{sy} \quad (6.144)$$

$$u_{si1} = u_{sx} i_{ry} - u_{sy} i_{rx} \quad (6.145)$$

$$u_{si2} = u_{sx} i_{rx} + u_{sy} i_{ry} \quad (6.146)$$

In the nonlinear control method, as well as in the vector control method, the angle between the stator and the rotor is needed either by measurement or estimation.

After using nonlinear feedback, the new linearized machine model is [3, 4]

$$\frac{dx_{11}}{d\tau} = \frac{L_m}{JL_s}x_{12} - \frac{1}{J}m_o \tag{6.147}$$

$$\frac{dx_{12}}{d\tau} = -\frac{1}{T_v}.x_{12} + m_1 \tag{6.148}$$

$$\frac{dx_{21}}{d\tau} = -2\frac{R_s}{L_s}x_{21} + 2\frac{R_s L_m}{L_s}x_{22} + 2u_{sf2} \tag{6.149}$$

$$\frac{dx_{22}}{d\tau} = -\frac{1}{T_v}.x_{12} + m_2 \tag{6.150}$$

From equations (6.148) and (6.152) and by using the definitions given in equations (6.141) and (6.142), the following equations are obtained:

$$u_{r1} = \frac{\omega_\sigma}{L_s}.\left(-x_{11}\left(x_{22} + \frac{L_m}{\omega_\sigma}.x_{21}\right) + \frac{L_m}{\omega_\sigma}.u_{sf1} - u_{si1} + m_1\right) \tag{6.151}$$

$$u_{r2} = \frac{\omega_\sigma}{L_s}.\left(-\frac{R_s L_m}{L_s \omega_\sigma}.x_{21} - \frac{R_r L_m}{L_s}.i_r^2 + x_{11}x_{12} + \frac{L_m}{\omega_\sigma}.u_{sf2} - u_{si2} + m_2\right) \tag{6.152}$$

Where i_r^2 is the rotor current square, it is expressed as follows:

$i_r^2 = i_r^2 + i_r^2 = \frac{x_{12}^2 + x_{22}^2}{x_{21}}$

The rotor voltage components to be sent for the PWM algorithm can be deduced from equations (6.141) and (6.142).

$$u_{rx} = \frac{\psi_{sx}u_{r2} - \psi_{sy}u_{r1}}{x_{21}} \tag{6.153}$$

$$u_{ry} = \frac{\psi_{sx}u_{r1} + \psi_{sy}u_{r2}}{x_{21}} \tag{6.154}$$

6.6 Nonlinear Control of Permanent Magnet Synchronous Machine

The mathematical model of Permanent Magnet Synchronous Machine (PMSM) in stationary reference frame appears as [19]

$$\frac{di_\alpha}{dt} = -\frac{R_s}{L_s}i_\alpha + \frac{1}{L_s}e_\alpha + \frac{1}{L_s}u_{s\alpha} \tag{6.155}$$

$$\frac{di_\beta}{dt} = -\frac{R_s}{L_s}i_\beta + \frac{1}{L_s}e_\beta + \frac{1}{L_s}u_{s\beta} \tag{6.156}$$

$$\frac{d\omega_r}{dt} = \frac{1}{J}\left(\psi_{s\alpha}i_\beta - \psi_{s\beta}i_\alpha\right) - \frac{1}{J}m_o \tag{6.157}$$

and

$$\psi_{f\alpha} = \psi_f \cos\theta \tag{6.158}$$

$$\psi_{f\beta} = \psi_f \sin\theta \tag{6.159}$$

$$e_\alpha = \frac{d\psi_{f\alpha}}{dt} = -\psi_f \omega_r \sin\theta = -\omega_r \psi_{f\beta} \tag{6.160}$$

$$e_\beta = \frac{d\psi_{f\beta}}{dt} = +\psi_f \omega_r \cos\theta = \omega_r \psi_{f\alpha} \tag{6.161}$$

For $\psi_s = \psi_f$:

$$\frac{d\omega_r}{dt} = \frac{1}{J}\left(\psi_{f\alpha} i_\beta - \psi_{f\beta} i_\alpha\right) - \frac{1}{J} m_o \tag{6.162}$$

The multiscalar motor variables are [19]

$$x_{11} = \omega_r \tag{6.163}$$

$$x_{12} = i_\beta \psi_{f\alpha} - i_\alpha \psi_{f\beta} \tag{6.164}$$

$$x_{21} = \psi_{f\alpha}^2 + \psi_{f\beta}^2 \tag{6.165}$$

$$x_{22} = i_\alpha \psi_{f\alpha} - i_\beta \psi_{f\beta} \tag{6.166}$$

Taking the derivatives of those two state variables, and substituting the derivatives of motor currents and fluxes, the next model is [19]

$$\frac{dx_{12}}{dt} = -\frac{R_s}{L_s}\left(i_\beta \psi_{f\alpha} - i_\alpha \psi_{f\beta}\right) + e_\alpha i_\beta + \frac{1}{L_s} e_\beta \psi_{f\alpha} - e_\beta i_\alpha - \frac{1}{L_s} e_\alpha \psi_{f\beta} + \frac{1}{L_s} u_\beta \psi_{f\alpha} - \frac{1}{L_s} u_\alpha \psi_{f\beta} \tag{6.167}$$

$$\begin{aligned}\frac{dx_{12}}{dt} = &-\frac{R_s}{L_s}\left(i_\beta \psi_{f\alpha} - i_\alpha \psi_{f\beta}\right) - \omega_r \psi_{f\beta} i_\beta - \omega_r \psi_{f\alpha} i_\alpha + \frac{1}{L_s}\omega_r \psi_{f\alpha}^2 - \frac{1}{L_s}\omega_r \psi_{f\beta}^2 \\ &+ \frac{1}{L_s} u_\beta \psi_{f\alpha} - \frac{1}{L_s} u_\alpha \psi_{f\beta}\end{aligned} \tag{6.168}$$

$$\frac{dx_{12}}{dt} = -\frac{R_s}{L_s} x_{12} - x_{11} x_{22} + \frac{1}{L_s} x_{11} x_{21} + V_1 \tag{6.169}$$

$$\begin{aligned}\frac{dx_{22}}{dt} = &-\frac{R_s}{L_s}\left(i_\alpha \psi_{f\alpha} - i_\beta \psi_{f\beta}\right) + \omega_r\left(\psi_{f\alpha} i_\beta - \psi_{f\beta} i_\alpha\right) - \frac{1}{L_s}\omega_r \psi_{f\alpha}\psi_{f\beta} + \frac{1}{L_s}\omega_r \psi_{f\alpha}\psi_{f\beta} \\ &+ \frac{1}{L_s} u_\beta \psi_{f\alpha} + \frac{1}{L_s} u_\alpha \psi_{f\beta}\end{aligned} \tag{6.170}$$

$$\frac{dx_{22}}{dt} = -\frac{R_s}{L_s} x_{22} + x_{11} x_{22} + V_2 \tag{6.171}$$

This is equivalent to [19]

$$\frac{dx_{12}}{dt} = -\frac{R_s}{L_s}x_{12} - x_{11}x_{22} + \frac{1}{L_s}x_{11}x_{21} + V_1 \tag{6.172}$$

$$\frac{dx_{22}}{dt} = -\frac{R_s}{L_s}x_{22} + x_{11}x_{22} + V_2 \tag{6.173}$$

$$V_1 = \frac{1}{L_s}u_\beta \psi_{f\alpha} - \frac{1}{L_s}u_\alpha \psi_{f\beta} \tag{6.174}$$

$$V_2 = \frac{1}{L_s}u_\beta \psi_{f\beta} + \frac{1}{L_s}u_\alpha \psi_{f\alpha} \tag{6.175}$$

The final decoupled model of the machine is [19]

$$\frac{dx_{12}}{d\tau} = -\frac{1}{T_{is}}.x_{12} + m_1 \tag{6.176}$$

$$\frac{dx_{22}}{d\tau} = -\frac{1}{T_i}.x_{22} + m_2 \tag{6.177}$$

where

$$V_1 = x_{11}x_{22} - \frac{1}{L_s}x_{11}x_{21} + \frac{1}{T_i}m_1 \tag{6.178}$$

$$V_2 = -x_{11}x_{12} + \frac{1}{T_i}m_2 \tag{6.179}$$

$$T_i = \frac{L_s}{R_s}$$

From this, the control signals are obtained as [19]

$$u_\alpha = L_s\frac{V_2\psi_{f\alpha} - V_1\psi_{f\beta}}{x_{21}} \tag{6.180}$$

$$u_\beta = L_s\frac{V_2\psi_{f\beta} + V_1\psi_{f\alpha}}{x_{21}} \tag{6.181}$$

The block diagram of the control part is shown in Figure 6.22.

6.6.1 *Nonlinear Control of PMSM for a dq Motor Model*

For a better description of the nonlinear control, we again present the motor model [19]:

$$\frac{di_d}{dt} = -\frac{R_s}{L_d}i_d + \frac{L_q}{L_d}\omega_r i_q + \frac{1}{L_d}u_d \tag{6.182}$$

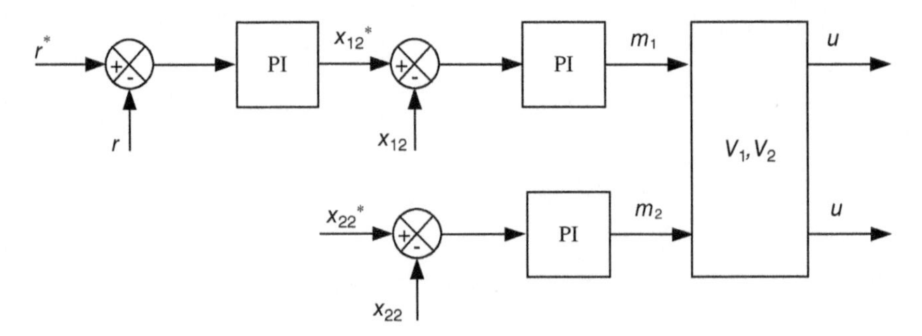

Figure 6.22 Block diagram of the PMSM with nonlinear decoupling

$$\frac{di_q}{dt} = -\frac{R_s}{L_q} i_q - \frac{L_d}{L_q} \omega_r i_d - \frac{1}{L_q} \omega_r \psi_f + \frac{1}{L_q} u_q \tag{6.183}$$

$$\frac{d\omega_r}{d\tau} = \frac{1}{J} \left(\psi_f i_q \left(L_d - L_q \right) i_d i_q - \frac{m_o}{J} \right) \tag{6.184}$$

$$\frac{d\theta}{dt} = \omega_r \tag{6.185}$$

The new driving functions are [19]

$$v_1 = \frac{L_q}{L_d} \omega_r i_q + \frac{1}{L_d} u_d \tag{6.186}$$

$$v_2 = -\frac{L_d}{L_q} \omega_r i_q - \frac{1}{L_q} \omega_r \psi_f + \frac{1}{L_q} u_q \tag{6.187}$$

Then, the motor model is [19]

$$\frac{di_d}{dt} = -\frac{R_s}{L_d} i_d + v_1 \tag{6.188}$$

$$\frac{di_q}{dt} = -\frac{R_s}{L_q} i_q + v_2 \tag{6.189}$$

The command voltage components could be used from the decoupling block by using v_1 and v_2 in the following way [19]:

$$u_d = L_d v_1 - L_q \omega_r i_q \tag{6.190}$$

$$u_q = L_q v_2 + \omega_r \left(L_d i_d + \psi_f \right) \tag{6.191}$$

This will obtain a nonlinear control scheme (decoupled) for the *dq* reference frame.

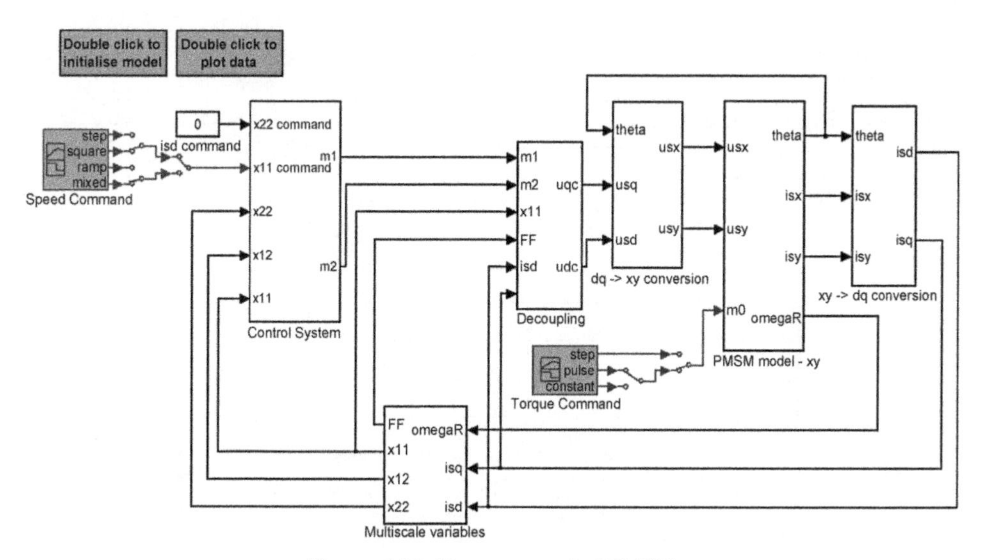

Figure 6.23 Vector control of PMSM

6.6.2 *Nonlinear Vector Control of PMSM in α–β Axis*

The control of PMSM model in the α–β (x–y) axis is shown in Figure 6.23. It consists of the PMSM model and the control system model. All the models are in per unit unless otherwise specified.

The Simulink software file of PMSM control model is [PMSM_FOC_in_alpha_beta_nonlinear]. The motor parameters are located in [PMSM_param], which should be first initialized.

6.6.3 *PMSM Model in α–β (x–y) Axis*

The PMSM model presented in per unit α–β reference frame is given in Chapter 2 and modeled in Figure 6.24.

6.6.4 *Transformations*

The currents in the x–y axis are converted into the d–q frame of reference by using the following transformation:

$$i_{sd} = i_{sx}\cos\theta + i_{sy}\sin\theta \tag{6.192}$$

$$i_{sq} = -i_{sx}\sin\theta + i_{sy}\cos\theta \tag{6.193}$$

These equations are modeled in Figure 6.25. The currents in the d–q axis are used for control purposes. The controller provides control input in the d–q axis.

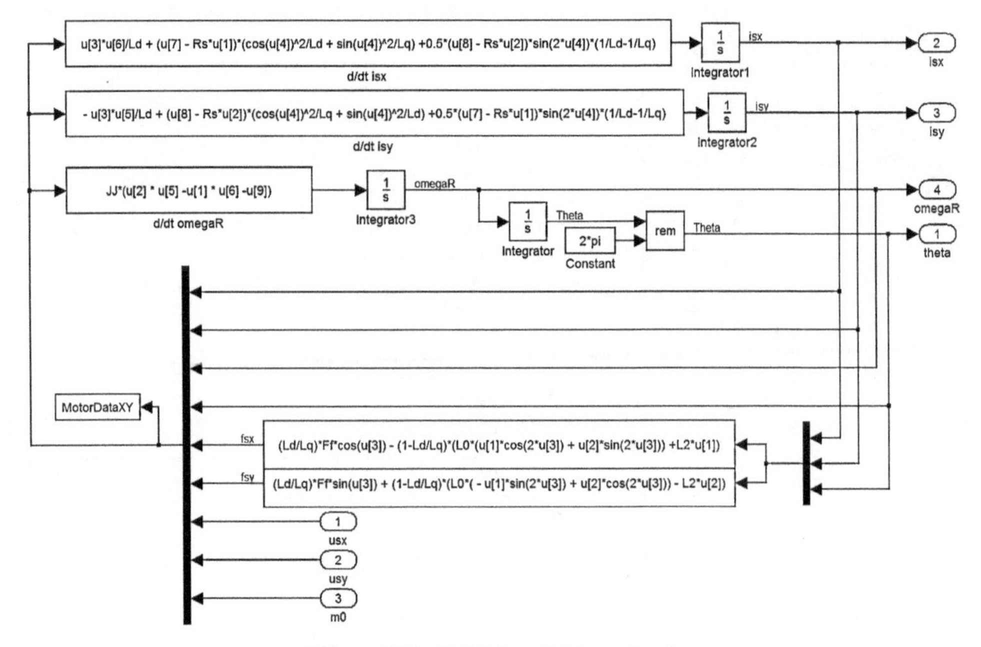

Figure 6.24 PMSM model in α–β axis

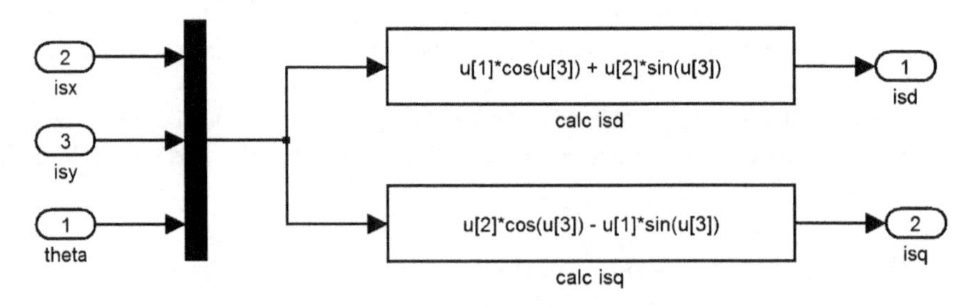

Figure 6.25 x–y to d–q transformation

The control voltage in d–q axis is then converted into the x–y axis by using the following transformation and then applying it to the machine in the x–y reference frame:

$$u_{sx} = u_{sd} \cos\theta - u_{sq} \sin\theta \tag{6.194}$$

$$i_{sy} = u_{sd} \sin\theta + u_{sq} \cos\theta \tag{6.195}$$

These above equations are modeled in Figure 6.26.

6.6.4.1 State and Input Transformations

The PMSM model is linearized by nonlinear state and input transformations. The model is linear between multiscale inputs m_1, m_2 and states x_{11}, x_{12}, and x_{22}. The transformations are given as the following:

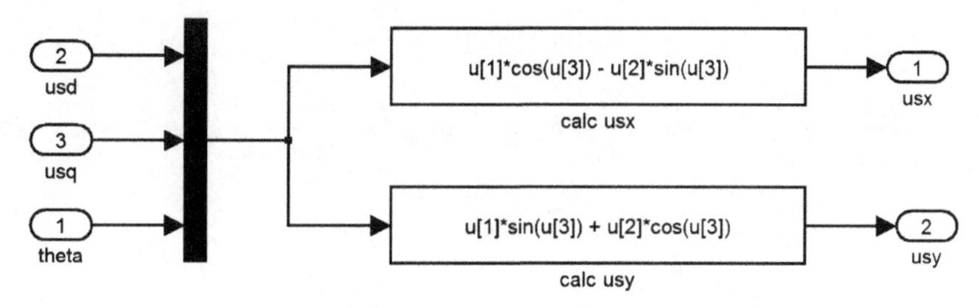

Figure 6.26 $d-q$ to $x-y$ transformation

6.6.4.2 State Transformation

The nonlinear multiscale states are calculated as

$$x_{11} = \omega_r \tag{6.196}$$

$$x_{12} = \left(\left(L_d - L_q\right) i_{sd} + \psi_f\right) . i_{sq} \tag{6.197}$$

$$x_{22} = i_{sd} \tag{6.198}$$

6.6.4.3 Input Transformation

For simplification of expressions, we define the following function:

$$FF = \left(\left(L_d - L_q\right) i_{sd} + \psi_f\right) \tag{6.199}$$

Based on equations (6.183) and (6.197), the model of this machine can be presented as follows:

$$\frac{dx_{12}}{d\tau} = -\frac{R}{L_q}.x_{12} + L_q\left(\frac{1}{L_d} - \frac{1}{L_q}\right) i_{sq}.i_{sd} - x_{11}\left(\frac{1}{L_d} FF^2 + FF.i_{sd}\right) - x_{11}\left(\frac{1}{L_d} - \frac{1}{L_{qd}}\right) L_q^2.i_{sq}^2 + u_1 \tag{6.200}$$

$$\frac{dx_{22}}{d\tau} = -\frac{R}{L_d}.x_{22} + \frac{L_q}{L_d} x_{11} i_{sq} + \frac{1}{L_d} + u_{sd} \tag{6.201}$$

where:

$$u_1 = \left(\left(\frac{L_d - L_q}{L_d}\right) i_{sq}. + \frac{FF}{L_q}\right) u_{sq} \tag{6.202}$$

The linearity of the system can be obtained as:

$$\frac{dx_{12}}{d\tau} = -\frac{R}{L_q}.x_{12} + m_1 \tag{6.203}$$

$$\frac{dx_{22}}{d\tau} = -\frac{R}{L_d}.x_{22} + m_2 \tag{6.204}$$

where:

$$m_1 = L_q\left(\frac{1}{L_d} - \frac{1}{L_q}\right)i_{sq}.i_{sd} - x_{11}\left(\frac{1}{L_d}FF^2 + FF.i_{sd}\right) - x_{11}\left(\frac{1}{L_d} - \frac{1}{L_{qd}}\right)L_q^2.i_{sq}^2 + u_1 \tag{6.205}$$

$$m_2 = \frac{L_q}{L_d}x_{11}i_{sq} + \frac{1}{L_d} + u_{sd} \tag{6.206}$$

Finally, the stator voltage can be calculated as:

$$u_{sd} = m_2 - \frac{L_q}{L_d}x_{11}i_{sq} - \frac{1}{L_d} \tag{6.207}$$

$$u_{sq} = \frac{L_q}{\left(\frac{L_q}{L_d}(L_d - L_q)i_{sq} + FF\right)}\left(m_1 - L_q\left(\frac{1}{L_d} - \frac{1}{L_q}\right)i_{sq}.i_{sd} + x_{11}\left(\frac{1}{L_d}FF^2 + FF.i_{sd}\right) + x_{11}\left(\frac{1}{L_d} - \frac{1}{L_q}\right)L_q^2.i_{sq}^2\right) \tag{6.208}$$

$$U_1 = \left[-L_q\left(\frac{1}{L_d} - \frac{1}{L_q}\right)R_s i_{sd} i_{sd} + x_{11}\left(\frac{1}{L_q}FF^2 + i_{sd}FF\right) + x_{11}\left(\frac{1}{L_d} - \frac{1}{L_q}\right)L_q^2 i_{sq}^2\right] + \frac{R_s}{L_q}m_1 \tag{6.209}$$

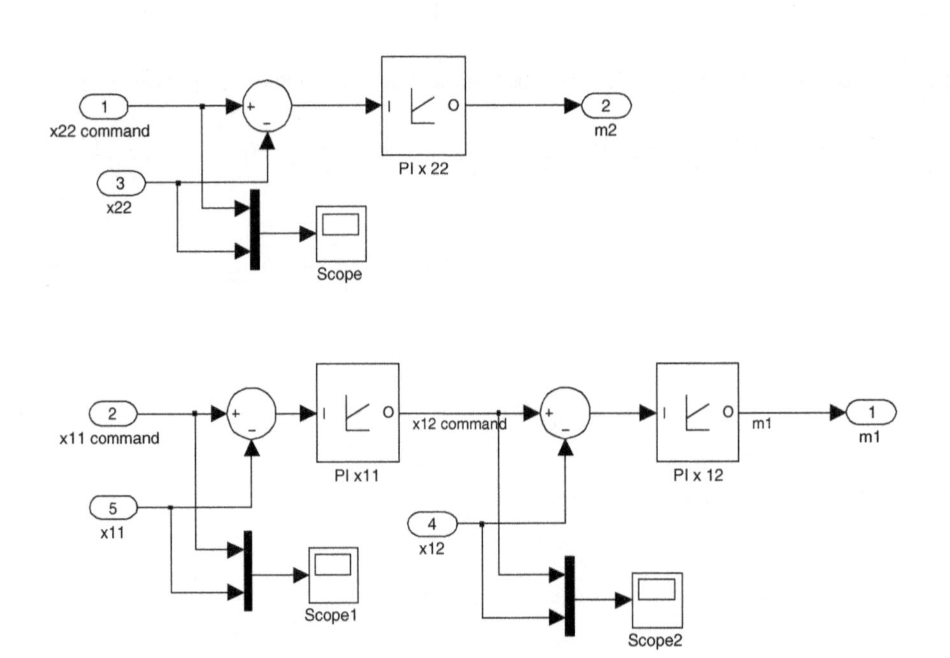

Figure 6.27 Control system model

$$U_2 = -\frac{1}{L_d} x_{11} L_q i_{sq} + \frac{R_s}{L_d} m_2 \tag{6.210}$$

$$u_{sq} = \frac{L_q\left(U_1 - \left(L_d - L_q\right) i_{sq} U_2\right)}{FF} \tag{6.211}$$

$$u_{sd} = L_d U_2 \tag{6.212}$$

6.6.5 Control System

PI control for the x_{22} channel provides the control input m_2 and a cascaded PI control for the x_{11} channel is shown in Figure 6.27. The x_{11} loop controller calculates the demand for the x_{12} loop, which in turn provides the control input m_1.

The PI controller structure is shown in Figure 6.28.

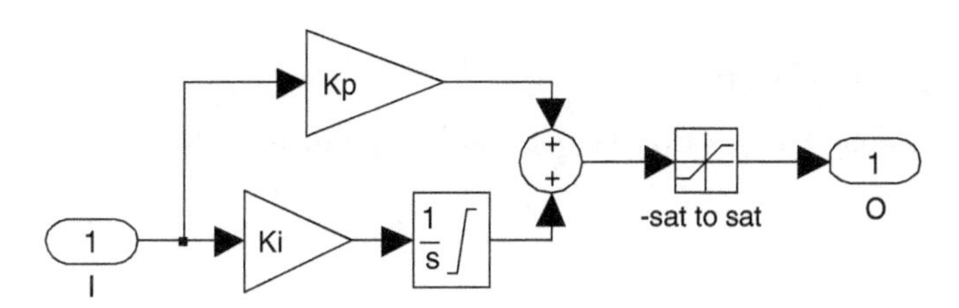

Figure 6.28 Generic PI controller

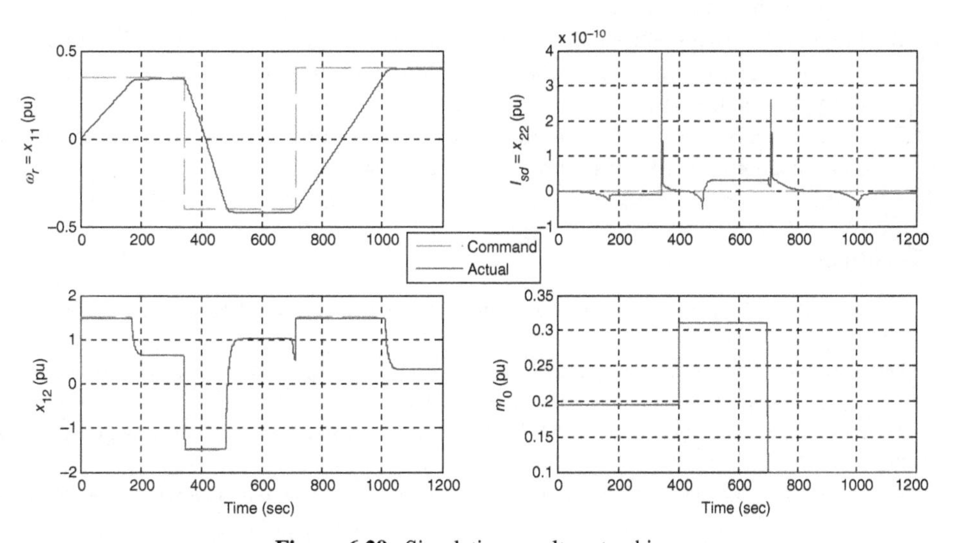

Figure 6.29 Simulation results – tracking

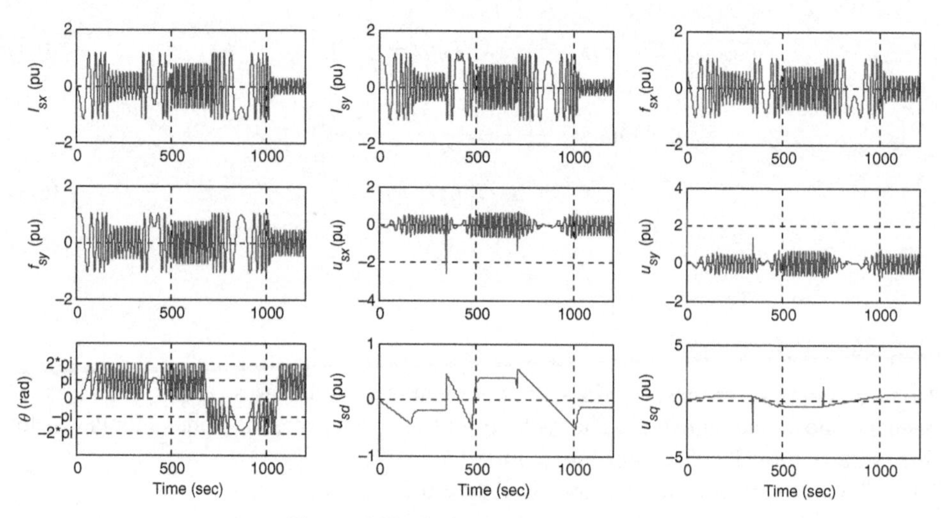

Figure 6.30 Other machine variables

6.6.6 Simulation Results

The simulation results for speed tracking are shown in Figures 6.29 and 6.30.

6.7 Problems

6.1: Select the parameters of the PI controllers.
Realize the simulation by commanding different voltages and filed currents. Investigate the motor response.

6.2: Program the two control schemes of separately excited DC motors, with linear and with nonlinear control approaches in one Simulink file. This will give you the possibility for comparison.

6.3: Program in Simulink in the multiscalar current control scheme of induction motor (hint: use the block from Figure 6.13 as an example).

References

1. Krzeminski, Z. (1987) Non-linear control of induction motor. *IFAC 10th World Congr. Auto. Contr.* Munich, pp. 349–354.
2. Krzeminski, Z. (1991) *Structures of Non-Linear Systems Control of Asynchronous Machine*. Technical University of Czestochowa, Book No 23, Poland (in Polish).
3. Krzemiński, Z. (2001) *Digital Control of AC Machines*. Gdańsk University of Technology Publishers, Gdańsk (in Polish).
4. Bogdan, M., Wilamowski, J., and Irwin, D. (2011) *The Industrial Electronics Handbook: Power Electronics and Motor Drives* (Chapters 22-I and 27-I). Taylor and Francis Group, LLC.
5. Isidori, A. (1995) *Non-linear Control Systems*, 3rd edn. Springer Verlag.

6. Bellini, A. and Figali, G. (1993) An adaptive control for induction motor drives based on a fully linearized model. *EPE 5th Europ. Conf. Power Electr. Appl.* Brighton, UK, pp. 196–201.
7. Chen, B., Ruan, Y., Xu, Y., and Liang, Q. (1990) Non-linear decoupling control of inverterfed induction motor system with feedback linearization. *IFAC 11th World Congress*. Tallin, pp. 191–196.
8. De Luca, A. and Ulivi, G. (1989) Design of an exact non-linear controller for induction motor. *IEEE Trans. Autom. Control*, 34(12), 1304–1307.
9. Georgiu, G. (1989) Adaptive feedback linearization and tracking for induction motors. *Proc. IFAC Congr.* Tbilisi, Russia, pp. 255–260.
10. Jakubczyk, B. and Respondek, W. (1980) On linearization of control systems. *Bull. Acad. Polon. Sci. Ser. Ind. Appl. Math. Astron. Phys.*, 28(9–10), 517–522.
11. Kim, G., Ha, I., Ko, M., and Park, J. (1989) Speed and efficiency control of induction motors via asymptotic decoupling. *20th Ann. IEEE Power Elect. Spe. Conf.* Milwaukee, WI, pp. 931–938.
12. Abu-Rub, H., Guzinski, J., Krzeminski, K., and Toliyat, H. (2003) Speed observer system for advanced sensorless control of induction motor. *IEEE Trans. Energy Convers.*, 18(2), 219–224.
13. Abu-Rub, H., Guzinski, J., Ahmed, S., and Toliyat, H. (2001) Sensorless torque control of induction motors. *IECON'01*. Denver, CO.
14. Krzemiński, Z. (2000) Sensorless control of the induction motor based on new observer. *PCIM*. Nürenberg.
15. Marino, R., Peresada, S., and Valigi, P. (1993) Adaptive input-output linearizing control of induction motors. *IEEE Trans. Autom. Control*, 38, 208–221.
16. Morici, R., Rossi, C., and Tonielli, A. (1993) Discrete-time non-linear controller for induction motor. *IECON'93*. Hawaje.
17. Pires, A. J. and Esteves, J. (1994) Non-linear control methodology application on variable speed AC drives. *PEMC*. Warszawa, pp. 652–657.
18. Yan, W., Dianguo, X., Yongjun, D., and Jinggi, L. (1988) Non-linear feedback decoupling control of induction motor servosystem. *2nd Int. Conf. Elect. Drives*. Poiana Brasov (Rumania), pp. C.211–217.
19. Morawiec, M., Krzemiński, Z., and Lewicki, A. (2009) The control system of PMSM with the Speer Observer, Przegl˛ad Elektrotechniczny (Electrical Review) (in Polish), pp. 48–51.
20. Baran, J. and Krzemiński, Z. (1987) Simulation of non-linear controlled DC drive. *SPD-4*. Polana Chocholowska (in Polish).
21. Krzemiński, Z. and Guziński, J. (1995) Non-linear control of DC drives. *II Polish Nat. Conf. Power Electr. Electr. Drives Cont.* Łódź (in Polish).
22. Krause, P. C. and Wasynczuk, O. (1989) *Electromechanical Motion Devices*. McGraw-Hill.
23. Buxbaum, A., Straughen, A., and Schierau, K. (1990) *Design of Control Systems for DC Drives*. Springer-Verlag.
24. Baran, J. and Szewczyk, K. (1994) Synthesis of a non-linear control system of a chopper. *Conf. Elect. Drives Power Electron., EDPE'94*. The High Tatras, Slovakia.
25. Alexandridis, A. and Tand Iracleeous, D. P. (1994) Non-linear controllers for series-connected DC motor drive. *PEMC'94*. Warsaw.
26. Vukosavić, S. N. (2007) *Digital Control of Electrical Drives*. Springer Science + Business Media, LLC.
27. Khalil, H. K. (2002) *Non-linear Systems*, 3rd edn. Prentice Hall, Upper Saddle River, NJ.
28. Abu-Rub, H., Krzeminski, Z., and Guzinski, J. (2000) Non-linear control of induction motor: idea and application. *EPE–PEMC, 9th Inter. Power Elect. Motion Cont. Conf.* Kosice, Slovac Republic, vol. 6, pp. 213–218.
29. Krzemiński, Z., Lewicki, A., and Włas, M. (2006) Properties of control systems based on non-linear models of the induction motor. *COMPEL Int. J. Comp. Math. Elect. Electr. Eng.*, 25(1), 0332–1649.

7

Five-Phase Induction Motor Drive System

7.1 Preliminary Remarks

The term "multiphase" in this chapter refers to a phase number greater than three. Three-phase electrical power is readily available as the power is generated, transmitted, and distributed in three phases. This is the most optimal number of phases for generation and transmission, as the trade-off exists between the complexity and power-handling capability of the three-phase system. The variable speed electric drive is also developed for three-phase AC machines. However, a power electronic converter, most commonly a voltage source or current source inverter, is invariably used to supply such three-phase drives. The power electronic converters do not pose any limit on their number of legs. The number of output phases in an inverter is the same as their respective number of legs. Hence, adding an additional leg to an inverter increases the number of output phases. This degree of freedom leads to an interest in developing variable speed electric drives with more than three phases. The first proposal of a variable speed five-phase induction motor drive is believed to have been made in 1969 [1]. The five-phase inverter used initially operated in the square-wave mode; however, later the pulse-width modulation (PWM) mode of operation was used. Six-phase drives have attracted much attention in the literature, after their advantages were revealed in 1983 [2]. The research on high-phase-order motor drives remained steady until the end of the 20th century. The multiphase drive attracted much attention from researchers after the advent of cheap and reliable power switching devices, as well as powerful digital signal processors. Since this chapter is dedicated to the five-phase system, the discussion is focused on five-phase drives only.

High Performance Control of AC Drives With MATLAB®/Simulink, Second Edition.
Haitham Abu-Rub, Atif Iqbal, and Jaroslaw Guzinski.

Companion website: www.wiley.com/go/aburubcontrol2e

The PWM techniques of multiphase inverters are developed as a part of the control of advanced multiphase drive systems. Sinusoidal PWM controls, applied to three-phase inverters, are also extended to five-phase inverters [2]. A selective harmonic elimination technique is employed, in addition to sinusoidal PWM. Later, space vector PWM (SVPWM) techniques were developed and implemented in multiphase voltage source inverters [3–13]. The initial approach was to use the simple extension of the SVPWM of the three-phase voltage source inverter (VSI) for the five-phase system. Then, it was realized that more elaborate schemes were needed to generate sinusoidal output and hence several schemes were developed and reported in the literature. The research on the five-phase drive system is active with a large number of publications on this subject [5–15].

High-performance electrical drive system requires fast dynamic response and precise control. Scalar and vector control are two broad techniques which are used for variable speed multiphase induction motor drives similar to three-phase induction motor drives. Though the implementation of scalar (volt/hertz = constant) control is easier, however, the dynamic response is sluggish while steady-state response is satisfactory. Flux and torque of the motor should be controlled independently for the faster dynamic response, similar to DC separately excited machine. Direct torque control (DTC) and field-oriented control are two approaches that are commonly used for high-performance control of induction motor drives [16–20]. The basic idea of high-performance control of multiphase machines is to control the electromagnetic torque and flux linkage directly and independently. The vector control is a high-performance and high-dynamic control technique, which transforms the dynamics of the induction motor to become similar to the DC motor by decoupling the torque and flux components [19, 20]. The classical DTC technique is more robust to motor parameter variations as this technique only requires the stator resistance information. It gives a faster response and easier implementation compared to field-oriented control as it does not require any trigonometric transformation matrix. DTC technique has been widely used in three-phase induction motor control and also applied to five-phase induction motor (IM) [16]. In [16, 17], DTC of five-phase IM is implemented using a two-level voltage source inverter. DTC for five-phase IM is also implemented using multilevel inverter as elaborated in [20].

This chapter describes the fundamental concept of a five-phase induction motor drive system. A typical five-phase drive system consists of a five-phase power converter, a five-phase motor, and their associated control, which is nowadays implemented in advanced digital signal processors interfaced with the PC. First, the model of the five-phase inverter is elaborated, followed by the simple control in square-wave mode known as a 10-step operation. The theoretical concept is supported by their simulation and experimental validation. More advanced control techniques, such as SVPWM, are next described. The simulation model is given subsequently for each control algorithm. Readers are encouraged to simulate the given drive system and implement their own algorithms. For further details on the five-phase drive systems and other higher-order systems, a large literature is available. A comprehensive review of the development in this area is encompassed in [21–25]. Field-oriented control and DTC of a five-phase induction motor are also discussed.

7.2 Advantages and Applications of Multiphase Drives

Multiphase drives offer some distinct advantages over their three-phase drive counterparts. The major advantages of using a multiphase machine instead of a three-phase machine are described in [5, 21–45]:

- higher torque density [5, 26–29];
- reduced torque pulsations [1, 30];
- greater fault tolerance [30–41];
- reduction in the required rating per inverter leg (and therefore simpler and more reliable power conditioning equipment) [42, 43]; and
- better noise characteristics, higher phase, number yield smoother torque due to the simultaneous increase in the frequency of the torque pulsation, and reduction of the torque ripple magnitude [44, 45].

Multiphase drives are still unsuitable for general purpose drives application and thus their application areas are restricted to some critical domains such as ship propulsion, more electric aircraft, hybrid electric vehicles, electric traction, and battery-powered electric vehicles [46–51]. The reasons behind this are primarily twofold. In high-power applications (i.e. ship propulsion), the use of multiphase drives enables reduction of the required power rating per inverter leg (phase). This is a major advantage over the converter end side, as lower switch ratings can handle reduced per-phase power. In high-power applications, series and/or parallel combinations of power electronics switches are required to process a large amount of power. The series/parallel combination of switches poses the problem of static and dynamic voltage and current sharing. Thus, reduced per-phase power is very attractive in this application. In safety-critical applications (i.e. more electric aircraft), the use of multiphase drives enables greater fault tolerance, which is of paramount importance. It has been shown that the loss of one or two phases has little impact on the behavior of the drive. Finally, in electric vehicles and hybrid electric vehicle applications, use of multiphase drives for propulsion enables reduction of the required semiconductor switch current rating; although these drives are not characterized by high power, the low voltage availability in vehicles makes the current high.

7.3 Modeling and Simulation of a Five-Phase Induction Motor Drive

A five-phase drive system consists of a five-phase AC machine, a five-phase power converter, and a controller based on microcontroller/digital signal processors/field programmable gate arrays that are controlled using a PC. The following section describes the modeling procedure of these components. At first, the model of a five-phase induction motor is presented, followed by the model development of a five-phase voltage source inverter. The simulation using MATLAB/Simulink of motor and inverter is also discussed.

7.3.1 Five-Phase Induction Motor Model

The model of a five-phase induction motor described is developed initially in phase variable form. In order to simplify the model by removing the time variation of inductance terms, a transformation is applied and a so-called *d-q-x-y-0* model of the machine is constructed. It is assumed that the spatial distribution of all the magneto-motive forces (fields) in the machine is sinusoidal, since only torque production due to the first harmonic of the field is considered. However, in a five-phase machine with concentrated type of winding, the third-harmonic component of the current is also used together with the fundamental to enhance the torque production [5]. This special feature of a five-phase machine is not considered in this book. All the other standard assumptions of the general theory of electrical machines apply (Chapter 4). A more detailed discussion of the modeling procedure is available in [52].

7.3.1.1 Phase Variable Model

A five-phase induction machine is constructed using 10 phase belts, each of 36 degrees, along the circumference of the stator. The spatial displacement between phases is therefore 72 degrees. The rotor winding is treated as an equivalent five-phase winding, with the same properties as the stator winding. It is assumed that the rotor winding has already been referred to as stator winding, using the winding transformation ratio. A five-phase induction machine can then be described with the following voltage equilibrium and flux linkage equations in matrix form (underlined symbols):

$$\begin{aligned} \underline{v}^s_{abcde} &= \underline{R}_s \underline{i}^s_{abcde} + \frac{d\underline{\psi}^s_{abcde}}{dt} \\ \underline{\psi}^s_{abcde} &= \underline{L}_s \underline{i}^s_{abcde} + \underline{L}_{sr} \underline{i}^r_{abcde} \end{aligned} \tag{7.1}$$

$$\begin{aligned} \underline{v}^r_{abcde} &= \underline{R}_r \underline{i}^r_{abcde} + \frac{d\underline{\psi}^r_{abcde}}{dt} \\ \underline{\psi}^r_{abcde} &= \underline{L}_r \underline{i}^r_{abcde} + \underline{L}_{rs} \underline{i}^s_{abcde} \end{aligned} \tag{7.2}$$

The following definition of phase voltages, currents, and flux linkages applies to equations (7.1) and (7.2):

$$\begin{aligned} \underline{v}^s_{abcde} &= [v_{as}\ v_{bs}\ v_{cs}\ v_{ds}\ v_{es}]^T \\ \underline{i}^s_{abcde} &= [i_{as}\ \ i_{bs}\ \ i_{cs}\ \ i_{ds}\ \ i_{es}]^T \\ \underline{\psi}^s_{abcde} &= [\psi_{as}\ \ \psi_{bs}\ \ \psi_{cs}\ \ \psi_{ds}\ \ \psi_{es}]^T \\ \underline{v}^r_{abcde} &= [v_{ar}\ \ v_{br}\ \ v_{cr}\ \ v_{dr}\ \ v_{er}]^T \\ \underline{i}^r_{abcde} &= [i_{ar}\ \ i_{br}\ \ i_{cr}\ \ i_{dr}\ \ i_{er}]^T \\ \underline{\psi}^r_{abcde} &= [\psi_{ar}\ \ \psi_{br}\ \ \psi_{cr}\ \ \psi_{dr}\ \ \psi_{er}]^T \end{aligned} \tag{7.3}$$

The matrices of stator and rotor inductances are given with ($\alpha = 2\pi/5$):

$$\underline{L}_s = \begin{bmatrix} L_{aas} & L_{abs} & L_{acs} & L_{ads} & L_{aes} \\ L_{abs} & L_{bbs} & L_{bcs} & L_{bds} & L_{bes} \\ L_{acs} & L_{bcs} & L_{ccs} & L_{cds} & L_{ces} \\ L_{ads} & L_{dbs} & L_{ccs} & L_{dds} & L_{des} \\ L_{aes} & L_{bes} & L_{ces} & L_{des} & L_{ees} \end{bmatrix}$$

$$\underline{L}_s = \begin{bmatrix} L_{ls}+M & M\cos\alpha & M\cos 2\alpha & M\cos 2\alpha & M\cos\alpha \\ M\cos\alpha & L_{ls}+M & M\cos\alpha & M\cos 2\alpha & M\cos 2\alpha \\ M\cos 2\alpha & M\cos\alpha & L_{ls}+M & M\cos\alpha & M\cos 2\alpha \\ M\cos 2\alpha & M\cos 2\alpha & M\cos\alpha & L_{ls}+M & M\cos\alpha \\ M\cos\alpha & M\cos 2\alpha & M\cos 2\alpha & M\cos\alpha & L_{ls}+M \end{bmatrix} \tag{7.4}$$

$$\underline{L}_r = \begin{bmatrix} L_{aar} & L_{abr} & L_{acr} & L_{adr} & L_{aer} \\ L_{abr} & L_{bbr} & L_{bcr} & L_{bdr} & L_{ber} \\ L_{acr} & L_{bcr} & L_{ccr} & L_{cdr} & L_{cer} \\ L_{adr} & L_{bbr} & L_{ccr} & L_{ddr} & L_{der} \\ L_{aer} & L_{ber} & L_{cer} & L_{der} & L_{eer} \end{bmatrix}$$

$$\underline{L}_r = \begin{bmatrix} L_{lr}+M & M\cos\alpha & M\cos 2\alpha & M\cos 2\alpha & M\cos\alpha \\ M\cos\alpha & L_{lr}+M & M\cos\alpha & M\cos 2\alpha & M\cos 2\alpha \\ M\cos 2\alpha & M\cos\alpha & L_{lr}+M & M\cos\alpha & M\cos 2\alpha \\ M\cos 2\alpha & M\cos 2\alpha & M\cos\alpha & L_{lr}+M & M\cos\alpha \\ M\cos\alpha & M\cos 2\alpha & M\cos 2\alpha & M\cos\alpha & L_{lr}+M \end{bmatrix} \tag{7.5}$$

Mutual inductances between stator and rotor windings are given with:

$$\underline{L}_{sr} = M\begin{bmatrix} \cos\theta & \cos(\theta+\alpha) & \cos(\theta+2\alpha) & \cos(\theta-2\alpha) & \cos(\theta-\alpha) \\ \cos(\theta-\alpha) & \cos\theta & \cos(\theta+\alpha) & \cos(\theta+2\alpha) & \cos(\theta-2\alpha) \\ \cos(\theta-2\alpha) & \cos(\theta-\alpha) & \cos\theta & \cos(\theta+\alpha) & \cos(\theta+2\alpha) \\ \cos(\theta+2\alpha) & \cos(\theta-2\alpha) & \cos(\theta-\alpha) & \cos\theta & \cos(\theta+\alpha) \\ \cos(\theta+\alpha) & \cos(\theta+2\alpha) & \cos(\theta-2\alpha) & \cos(\theta-\alpha) & \cos\theta \end{bmatrix}$$

$$\underline{L}_{rs} = \underline{L}_{sr}^T \tag{7.6}$$

The angle θ denotes the instantaneous position of the magnetic axis of the rotor phase "a" with respect to the stationary stator phase "a" magnetic axis (i.e. the instantaneous position of the rotor with respect to stator). Stator and rotor resistance matrices are 5×5 diagonal matrices:

$$\begin{aligned} \underline{R}_s &= diag(R_s \quad R_s \quad R_s \quad R_s \quad R_s) \\ \underline{R}_r &= diag(R_r \quad R_r \quad R_r \quad R_r \quad R_r) \end{aligned} \tag{7.7}$$

Motor torque can be expressed in terms of phase variables as:

$$T_e = \frac{P}{2}\underline{i}^T \frac{d\underline{L}_{abcde}}{d\theta}\underline{i} = \frac{P}{2}\left[\underline{i}^{sT}_{abcde} \quad \underline{i}^{rT}_{abcde}\right]\frac{d\underline{L}_{abcde}}{d\theta}\begin{bmatrix}\underline{i}^s_{abcde} \\ \underline{i}^r_{abcde}\end{bmatrix} \tag{7.8a}$$

$$T_e = P\underline{i}^{sT}_{abcde}\frac{d\underline{L}_{sr}}{d\theta}\underline{i}^r_{abcde} \tag{7.8b}$$

Substitution of stator and rotor currents from equations (7.2) to (7.3) and equation (7.6) into equation (7.8b) yields the torque equation in developed form:

$$T_e = -PM \left\{ \begin{array}{l} (i_{as}i_{ar} + i_{bs}i_{br} + i_{cs}i_{cr} + i_{ds}i_{dr} + i_{es}i_{er}) \sin\theta + (i_{es}i_{ar} + i_{as}i_{br} + i_{bs}i_{cr} + i_{cs}i_{dr} + i_{ds}i_{er}) \sin(\theta + \alpha) + \\ (i_{ds}i_{ar} + i_{es}i_{br} + i_{as}i_{cr} + i_{bs}i_{dr} + i_{cs}i_{er}) \sin(\theta + 2\alpha) + (i_{cs}i_{ar} + i_{ds}i_{br} + i_{es}i_{cr} + i_{as}i_{dr} + i_{bs}i_{er}) \\ \sin(\theta - 2\alpha) + (i_{bs}i_{ar} + i_{cs}i_{br} + i_{ds}i_{cr} + i_{es}i_{dr} + i_{as}i_{er}) \sin \mid (\theta - \alpha) \end{array} \right\} \tag{7.9}$$

7.3.1.2 Model Transformation

In order to simplify the model, it is necessary to apply a coordinate transformation, which will remove the time-varying inductances. The coordinate transformation is used in the power-invariant form. The following transformation matrix is therefore applied to the stator five-phase winding:

$$\underline{A}_s = \sqrt{\frac{2}{5}} \begin{bmatrix} \cos\theta_s & \cos(\theta_s - \alpha) & \cos(\theta_s - 2\alpha) & \cos(\theta_s + 2\alpha) & \cos(\theta_s + \alpha) \\ -\sin\theta_s & -\sin(\theta_s - \alpha) & -\sin(\theta_s - 2\alpha) & -\sin(\theta_s + 2\alpha) & -\sin(\theta_s + \alpha) \\ 1 & \cos(2\alpha) & \cos(4\alpha) & \cos(4\alpha) & \cos(2\alpha) \\ 0 & \sin(2\alpha) & \sin(4\alpha) & -\sin(4\alpha) & -\sin(2\alpha) \\ \frac{1}{\sqrt{2}} & \frac{1}{\sqrt{2}} & \frac{1}{\sqrt{2}} & \frac{1}{\sqrt{2}} & \frac{1}{\sqrt{2}} \end{bmatrix} \tag{7.10}$$

Transformation of the rotor variables is performed using the same transformation expression, except that θ_s is replaced by β, where $\beta = \theta_s - \theta$. Here, θ_s is the instantaneous angular position of the d-axis of the common reference frame with respect to the phase "a" magnetic axis of the stator, while β is the instantaneous angular position of the d-axis of the common reference frame with respect to the phase "a" magnetic axis of the rotor. Hence, the transformation matrix for rotor is

$$\underline{A}_r = \sqrt{\frac{2}{5}} \begin{bmatrix} \cos\beta & \cos(\beta - \alpha) & \cos(\beta - 2\alpha) & \cos(\beta + 2\alpha) & \cos(\beta + \alpha) \\ -\sin\beta & -\sin(\beta - \alpha) & -\sin(\beta - 2\alpha) & -\sin(\beta + 2\alpha) & -\sin(\beta + \alpha) \\ 1 & \cos(2\alpha) & \cos(4\alpha) & \cos(4\alpha) & \cos(2\alpha) \\ 0 & \sin(2\alpha) & \sin(4\alpha) & -\sin(4\alpha) & -\sin(2\alpha) \\ \frac{1}{\sqrt{2}} & \frac{1}{\sqrt{2}} & \frac{1}{\sqrt{2}} & \frac{1}{\sqrt{2}} & \frac{1}{\sqrt{2}} \end{bmatrix} \tag{7.11}$$

The angles of transformation for stator quantities and for rotor quantities are related to the arbitrary speed of the selected common reference frame through:

$$\begin{aligned} \theta_s &= \int \omega_a dt \\ \beta = \theta_s - \theta &= \int (\omega_a - \omega) dt \end{aligned} \tag{7.12}$$

where ω is the instantaneous electrical angular speed of rotation of the rotor.

7.3.1.3 Machine Model in an Arbitrary Common Reference Frame

Correlation between original phase variables and new variables in the transformed domain is governed by the following transformation expressions:

$$\begin{aligned} \underline{v}_{dq}^{s} &= \underline{A}_{s}\underline{v}_{abcde}^{s} \quad \underline{i}_{dq}^{s} = \underline{A}_{s}\underline{i}_{abcde}^{s} \quad \underline{\psi}_{dq}^{s} = \underline{A}_{s}\underline{\psi}_{abcde}^{s} \\ \underline{v}_{dq}^{r} &= \underline{A}_{r}\underline{v}_{abcde}^{r} \quad \underline{i}_{dq}^{r} = \underline{A}_{r}\underline{i}_{abcde}^{r} \quad \underline{\psi}_{dq}^{r} = \underline{A}_{r}\underline{\psi}_{abcde}^{r} \end{aligned} \tag{7.13}$$

Substituting equations (7.1) and (7.2) into equation (7.13) and the application of equations (7.10) and (7.11) yield the machine's voltage equations in the common reference frame where $p = d/dt$:

$$\begin{aligned} v_{ds} &= R_s i_{ds} - \omega_a \psi_{qs} + p\psi_{ds} & v_{dr} &= R_r i_{dr} - (\omega_a - \omega)\psi_{qr} + p\psi_{dr} \\ v_{qs} &= R_s i_{qs} + \omega_a \psi_{ds} + p\psi_{ds} & v_{qr} &= R_r i_{qr} + (\omega_a - \omega)\psi_{dr} + p\psi_{qr} \\ v_{xs} &= R_s i_{xs} + p\psi_{xs} & v_{xr} &= R_r i_{xr} + p\psi_{xr} \\ v_{ys} &= R_s i_{ys} + p\psi_{ys} & v_{yr} &= R_r i_{yr} + p\psi_{yr} \\ v_{0s} &= R_s i_{0s} + p\psi_{0s} & v_{0r} &= R_r i_{0r} + p\psi_{0r} \end{aligned} \tag{7.14}$$

Transformation of flux linkage equations (7.1) and (7.2) results in:

$$\begin{aligned} \psi_{ds} &= (L_{ls} + 2.5M)i_{ds} + 2.5Mi_{dr} & \psi_{dr} &= (L_{lr} + 2.5M)i_{dr} + 2.5Mi_{ds} \\ \psi_{qs} &= (L_{ls} + 2.5M)i_{qs} + 2.5Mi_{qr} & \psi_{qr} &= (L_{lr} + 2.5M)i_{qr} + 2.5Mi_{qs} \\ \psi_{xs} &= L_{ls}i_{xs} & \psi_{xr} &= L_{lr}i_{xr} \\ \psi_{ys} &= L_{ls}i_{ys} & \psi_{yr} &= L_{lr}i_{yr} \\ \psi_{0s} &= L_{ls}i_{0s} & \psi_{0r} &= L_{lr}i_{0r} \end{aligned} \tag{7.15}$$

Introduction of the magnetizing inductance $L_m = 2.5M$ enables equation (7.15) to be written in the following form:

$$\begin{aligned} \psi_{ds} &= (L_{ls} + L_m)i_{ds} + L_m i_{dr} & \psi_{dr} &= (L_{lr} + L_m)i_{dr} + L_m i_{ds} \\ \psi_{qs} &= (L_{ls} + L_m)i_{qs} + L_m i_{qr} & \psi_{qr} &= (L_{lr} + L_m)i_{qr} + L_m i_{qs} \\ \psi_{xs} &= L_{ls}i_{xs} & \psi_{xr} &= L_{lr}i_{xr} \\ \psi_{ys} &= L_{ls}i_{ys} & \psi_{yr} &= L_{lr}i_{yr} \\ \psi_{0s} &= L_{ls}i_{0s} & \psi_{0r} &= L_{lr}i_{0r} \end{aligned} \tag{7.16}$$

Finally, the transformation of the original torque equation (7.8b) yields

$$\begin{aligned} T_e &= \frac{5P}{2}M\left[i_{dr}i_{qs} - i_{ds}i_{qr}\right] \\ T_e &= PL_m\left[i_{dr}i_{qs} - i_{ds}i_{qr}\right] \\ T_e &= \frac{5P}{2}\left[\psi_{ds}i_{qs} - \psi_{qs}i_{ds}\right] \end{aligned} \tag{7.17}$$

The mechanical equation of rotor motion is invariant under the transformation and is

$$T_e - T_L = \frac{J}{P}\frac{d\omega}{dt} \tag{7.18}$$

where J is the inertia and P is the number of poles in machine. The difference between a three-phase machine model and a five-phase machine model lies in the extra x-y set of components that exist only in a five-phase machine. However, this extra set is non-flux- and non-torque-producing components. They simply add to the extra losses in the machine. The steady-state equivalent circuit of a five-phase induction machine can be obtained by replacing d/dt term by $j\omega_e$, where ω_e is the fundamental operating frequency of the machine, as shown in Figure 7.1.

In a five-phase machine, the x-y components are decoupled from d-q components; also there is no coupling of x-y with the rotor circuit. This is true for an n-phase AC machine with sinusoidal distributed magnetomotive force (MMF) and hence only one pair of components, i.e. d-q, produces torque and the remaining component simply causes losses in the machine. Different harmonics of the stator voltage/current map into either the d-q or x-y planes of the stationary reference frame, depending on the harmonic order, as shown in Table 7.1. This is a general property of multiphase systems, which leads to distinctions with respect to control of a multiphase machine.

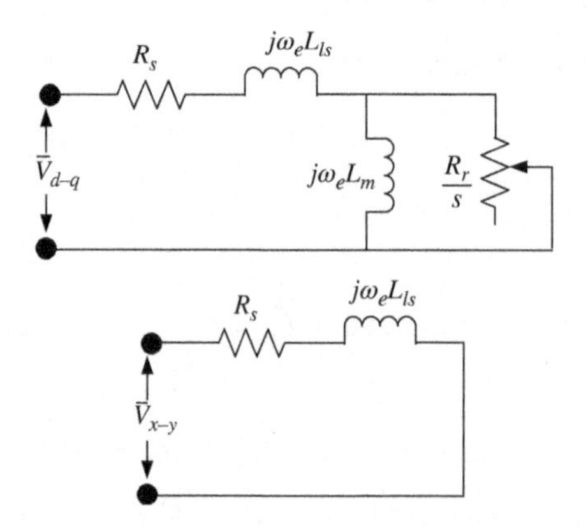

Figure 7.1 Steady-state equivalent circuit of a five-phase induction machine

Table 7.1 Harmonic mapping for five-phase machine

Component	Five-phase system
d-q	$10j \pm 1$ ($j = 0, 1, 2, \ldots$)
x-y	$10j \pm 3$ ($j = 0, 1, 2, \ldots$)
Zero sequence	$10j + 5$ ($j = 0, 1, 2, \ldots$)

Thus, the fundamental, 9th, 11th, … are produced due to *d-q* components, while 3rd, 7th, 13th, … are generated due to *x-y* components and a multiple of the fifth harmonic is produced due to zero-sequence components.

7.3.1.4 MATLAB/Simulink Model of Main Fed Five-Phase Induction Motor Drive

The MATLAB/Simulink model of a five-phase induction machine fed by a pure five-phase sine wave is illustrated in Figure 7.2. The five-phase sine wave is generated using a "sine wave" from the source library of Simulink. All the five-phases are generated from one block, by providing the appropriate phase shift of 72 degrees $\left[0 \quad \frac{2\pi}{5} \quad \frac{4\pi}{5} \quad -\frac{4\pi}{5} \quad -\frac{2\pi}{5}\right]$. The model can be used either in the phase variable form of equations (7.1) to (7.9) or in the transformed domain of equations (7.14) to ;(7.18); here the latter approach is given. The input source voltages are transformed to *d-q-x-y* and given to the machine model. The outputs of the machine models are taken as currents, speed, torque, and rotor flux. The subblocks are presented in Figure 7.2b, c.

A five-phase sinusoidal supply is given to the five-phase induction motor under no-load and loaded conditions. The resulting waveforms are presented in Figures 7.3–7.11.

7.3.1.5 No-Load Condition

The response of a five-phase motor under no-load conditions is presented in Figures 7.3–7.7. Rated voltage of value 220·sqrt (2) and frequency of 50 Hz is applied to the stator of a five-phase motor and is allowed to accelerate to its rated speed of 1500 rpm. The supply voltage for two cycles is shown in Figure 7.3. The response of the motor is similar to that obtained with a three-phase induction motor.

7.3.1.6 Rated Load Condition

To observe the behavior of the machine under loaded conditions, a rated load of 8.33 Nm is applied to the motor shaft at a time instant of 1 second. The load is applied once the motor reaches a steady-state condition after excitation and acceleration transients. The resulting waveforms are shown in Figures 7.8–7.11. It is evident from Figure 7.8 that the motor follows the load torque well. The speed response presented in Figure 7.9 clearly shows a drop in the speed of the motor, as the motor settles to a new speed of nearly 1428 rpm. Since no corrective action is incorporated into the drive, the motor continues to rotate at the reduced speed.

Normally, a PI controller is used for closed-loop operation to correct the speed error, but this is not taken up here. The rotor flux for the complete duration of simulation is presented in Figure 7.11. It shows a drop in the flux due to application of load. Thus, conclusively, it can be said that a five-phase motor shows similar behavior as that of a three-phase induction motor.

7.3.2 Five-Phase Two-Level Voltage Source Inverter Model

This section describes the modeling procedure of a five-phase voltage source inverter. A five-phase inverter has a similar front-side converter structure to that of a three-phase voltage source inverter. The fixed voltage and fixed frequency grid supply voltage is converted to DC using

(a)

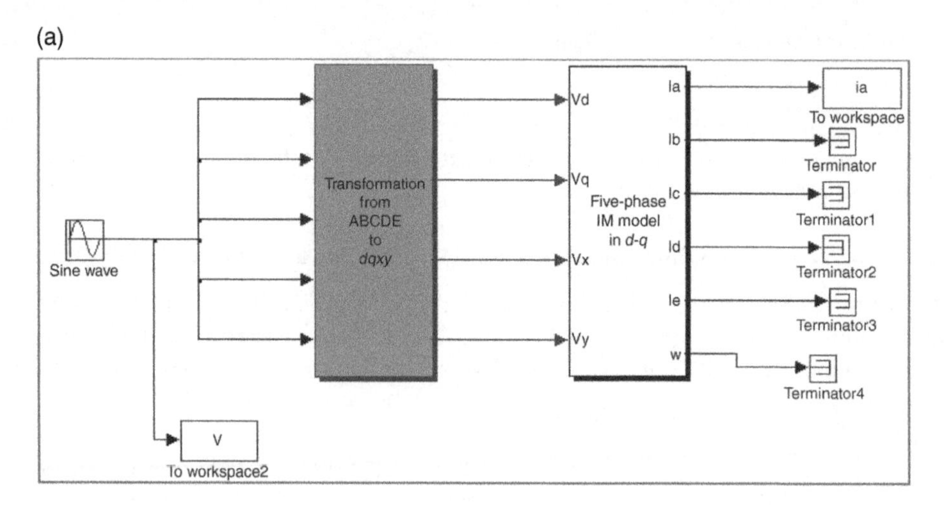

(b)

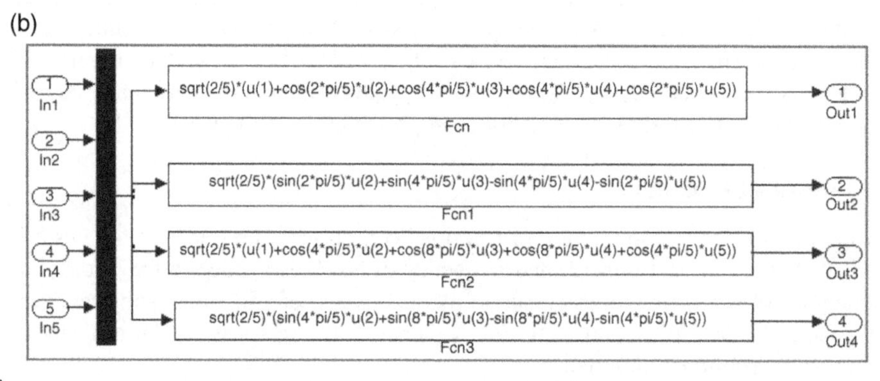

(c)

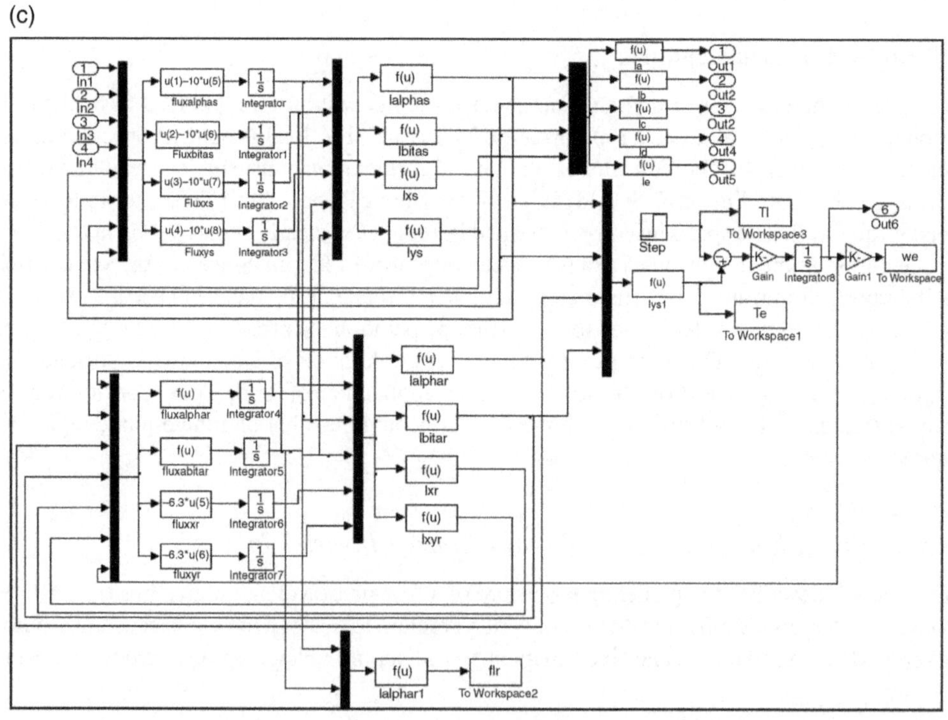

Figure 7.2 (a) Simulink block of five-phase motor fed using ideal five-phase supply (*File Name Ideal_5_Motor.mdl*); (b) subblocks of transformation from five-phase to four orthogonal phases; and (c) five-phase induction motor simulation model

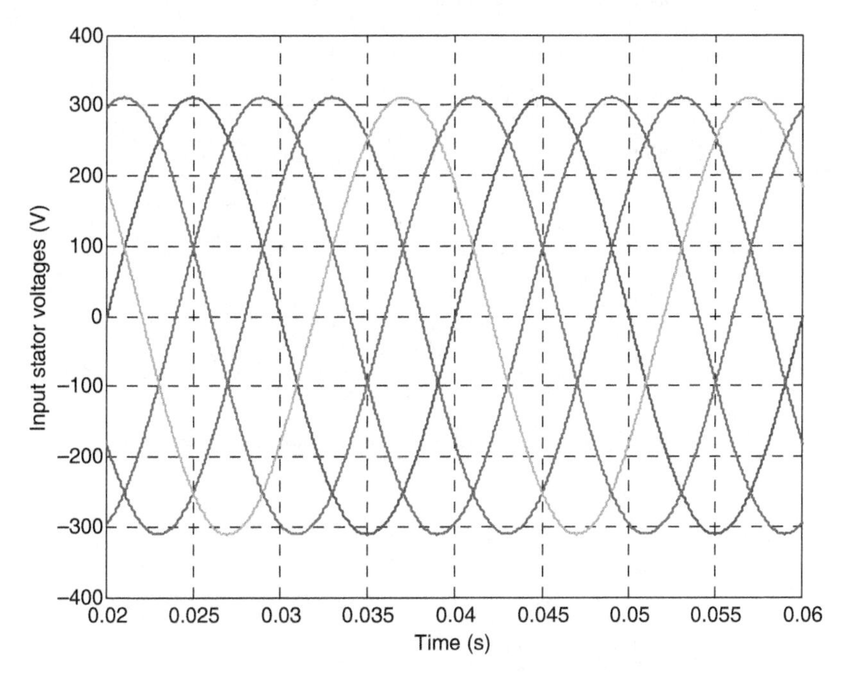

Figure 7.3 Input five-phase voltage to the stator of machine

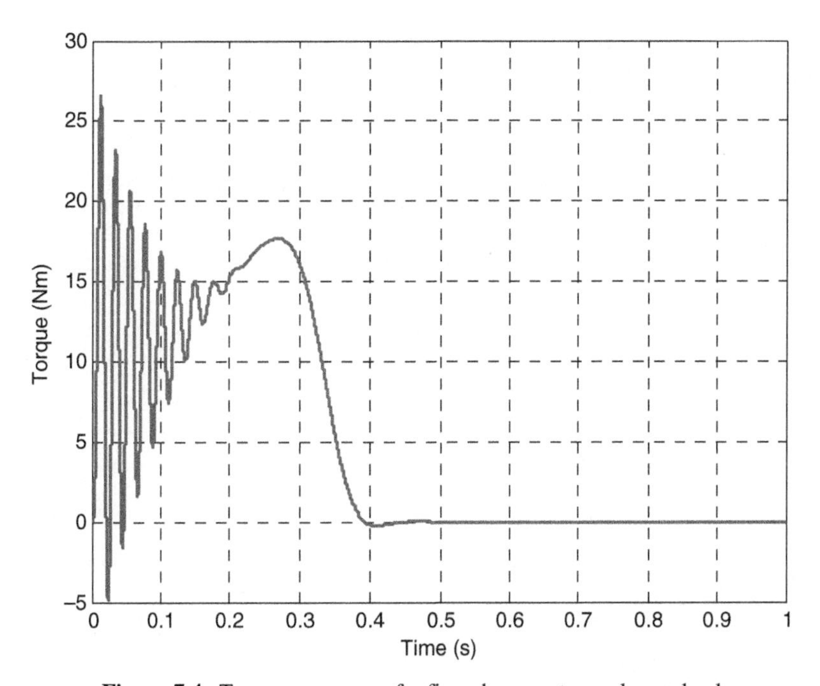

Figure 7.4 Torque response of a five-phase motor under no-load

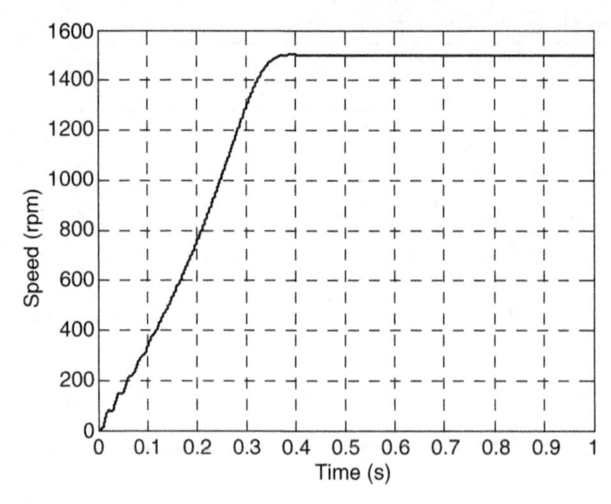

Figure 7.5 Speed response of five-phase motor under no load

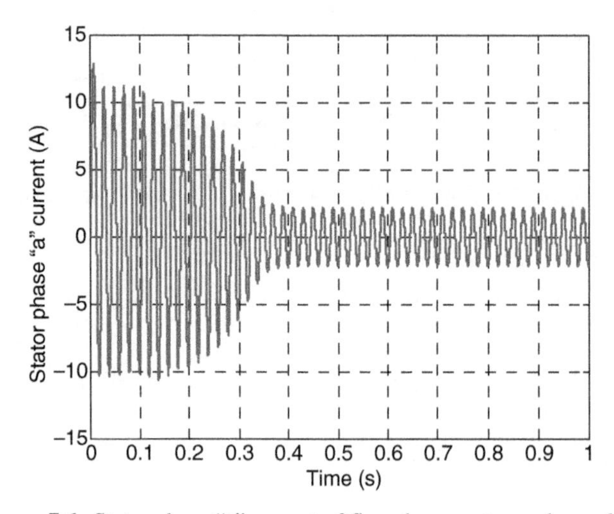

Figure 7.6 Stator phase "a" current of five-phase motor under no load

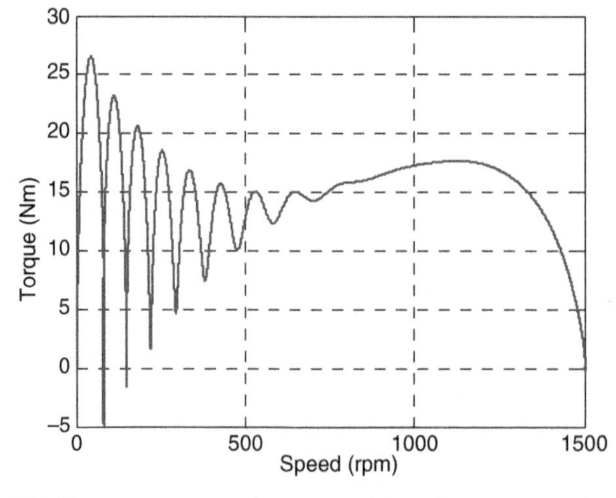

Figure 7.7 Torque versus speed response of five-phase motor under no load

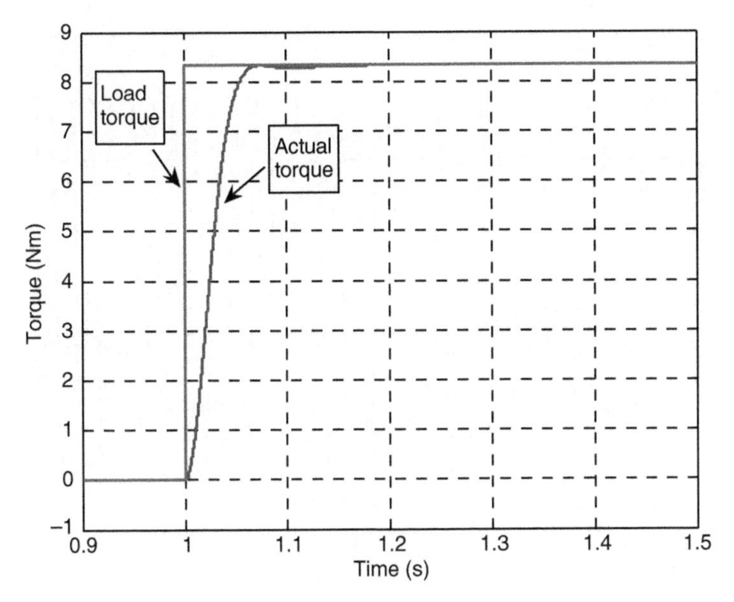

Figure 7.8 Torque response of a five-phase motor under rated load condition

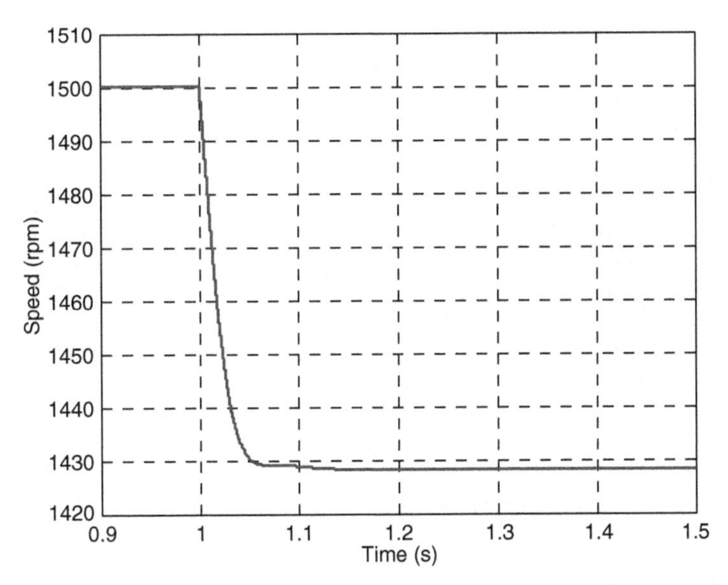

Figure 7.9 Speed response of five-phase motor under rated load condition

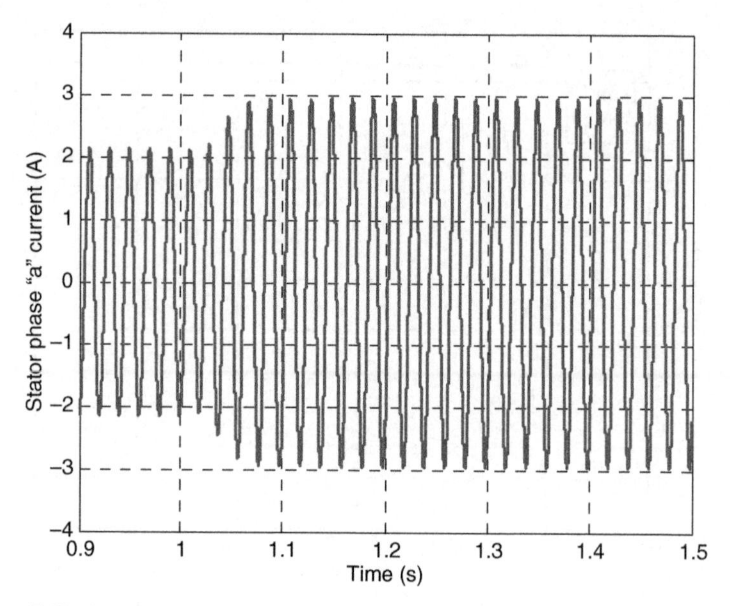

Figure 7.10 Stator phase "a" current in a five-phase motor under rated load condition

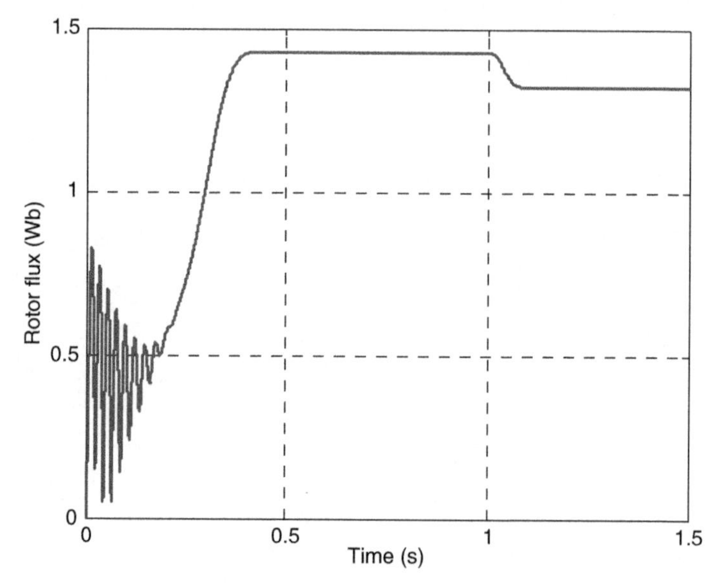

Figure 7.11 Rotor flux in a five-phase motor

either controlled (thyristor-based or power transistor-based) or uncontrolled rectifier (diode-based). The output of the rectifier (AC–DC converter) is filtered to remove the ripple in the output voltage signal. The rectified and filtered DC voltage is fed to the inverter (DC–AC) block. The inverter block outputs five-phase variable voltage and variable frequency supply

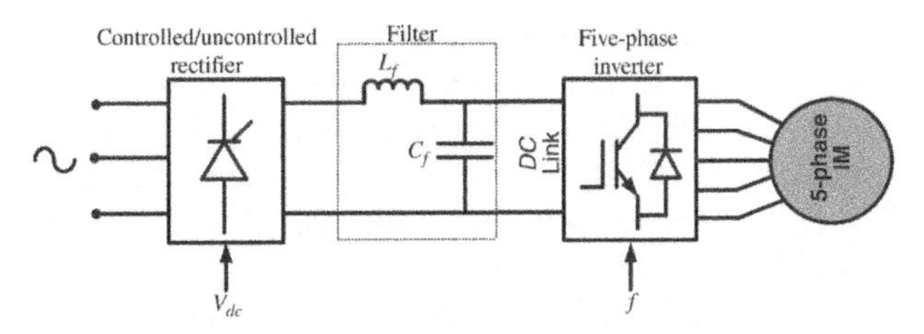

Figure 7.12 Block diagram of five-phase induction motor drive

to feed motor drives or other applications as desired. The five-phase inverter has two additional legs when compared to a three-phase inverter. The block diagram of this arrangement is shown in Figure 7.12, where the inverter is feeding a five-phase induction motor. The rectifier, filter, and five-phase inverter constitute the complete three-phase fixed voltage and fixed frequency to five-phase variable voltage and variable frequency supply system. The DC-link side or the front side will be omitted from the further discussion. The five-phase inverter block will be discussed in detail in the subsequent section.

The power circuit topology of a five-phase voltage source inverter is shown in Figure 7.2; this was first proposed by [1]. Each switch in the circuit consists of two power semiconductor devices connected in antiparallel. One of these is a fully controllable semiconductor, such as a bipolar transistor, metal oxide silicon field effect transistor (MOSFET), or insulated gate bipolar transistor (IGBT), while the second is a diode. The input of the inverter is a DC voltage, which is regarded as constant. The inverter outputs are denoted in Figure 7.13 with lower-case letters (*a, b, c, d,* and *e*), while the points of connection of the outputs to the inverter legs have symbols in capital letters (*A, B, C, D,* and *E*). The voltage of this point of connection is called the "pole voltage" or "leg voltage." The voltage between the terminal of the output of the inverter and the neutral of the load is called the "phase voltage." The voltage between the neutral of the load and the neutral of the DC link is called "common mode voltage." The voltage across the two output terminals of the load is called the "line voltages." However, in a five-phase system, there exist two different line voltages, called nonadjacent and adjacent line voltages, contrary to a three-phase system where only one line voltage is defined. This is further discussed in the next section.

The basic operating principles of the five-phase VSI are developed as follows, assuming ideal commutation and zero forward voltage drop. The upper and lower power switches of the same leg are complementary in operation, i.e. if the upper switch is "ON," the lower must

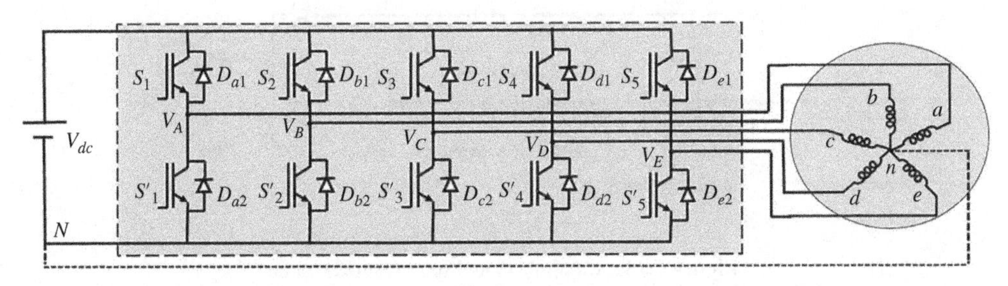

Figure 7.13 Power circuit topology of a five-phase voltage source inverter

be "OFF," and vice versa. This is done to avoid shorting the DC supply. Complementary operation of the switches is obtained by providing a 180-degree phase-shifted gate drive signal to the two upper and lower switches. However, it is important to provide a time delay between turning "ON" and turning "OFF" of the two complementary switches. This time delay is called "dead time," as both the power switches remain "OFF" simultaneously for a small duration of time. Thus, the "ON" time of each power switch is smaller compared to their corresponding "OFF" time, which is illustrated in Figure 7.14. The upper trace shows the gate drive signal for upper switch S_1 and the lower trace shows the gate drive signal for the lower switch S'_1. When switch S_1 goes "OFF," the switch S'_1 turns off after a delay of τ_d (dead time). The power modules from Semikron and other manufacturers have built-in hardware for providing this dead time. In software realization, there are also options in digital signal processors to incorporate and change the dead time. In practice, the delay time can be set between 2 and 5 μs. The dead time is set by the user and depends upon the power switching device characteristics (ON and OFF characteristics), the switching frequency of the inverter control, and the features of the gate driver circuits. The larger the dead time, the waveforms get higher distortion.

To simulate the dead time for inverter leg switching in MATLAB/Simulink, the block of Figure 7.15 can be used. The input to the block is the gate driving signal, where one signal is passed directly and the second signal is processed through "Discrete Edge Detector" (from "Discrete Control Blocks" of "Extra Library" of "SimPowerSystem" block-sets), then

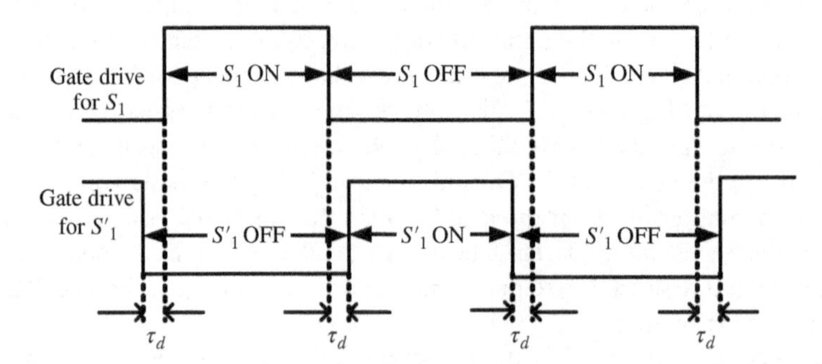

Figure 7.14 Illustration for dead time

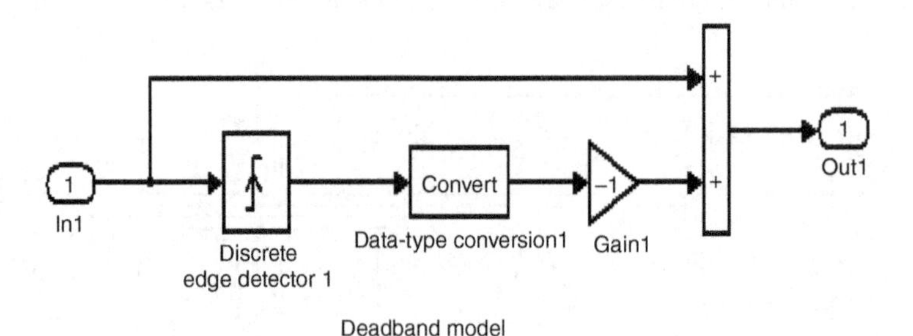

Figure 7.15 Simulink model to create dead time

multiplied by −1 (to make a complementary signal) using a "Gain" block. The outputs of the block of Figure 7.15 are two complementary gate drive signals with the desired dead time. The "Discrete Edge Detector" block is used to provide the desired dead time. This block is used when the inverter is modeled using "SimPowerSystems block-sets," using actual components.

The inclusion of dead time in the simulation model affects the output waveform and will be illustrated using a simulation example in the next section.

The relationship between pole voltage and switching signals is given as:

$$V_k = S_k V_{dc}; \quad k \in A, B, C, D, E \tag{7.19}$$

where $S_k = 1$ when the upper power switch is "ON" and $S_k = 0$ when the lower switch is "ON." If the load is assumed to be a star-connected five-phase, then the relation between the phase-to-neutral load voltage and the pole voltages can be written as:

$$\begin{aligned}
V_A(t) &= v_{an}(t) + v_{nN}(t) \\
V_B(t) &= v_{bn}(t) + v_{nN}(t) \\
V_C(t) &= v_{cn}(t) + v_{nN}(t) \\
V_D(t) &= v_{dn}(t) + v_{nN}(t) \\
V_E(t) &= v_{en}(t) + v_{nN}(t)
\end{aligned} \tag{7.20}$$

where v_{nN} is the voltage difference between the star point n of the load and the negative rail of the DC bus N, called the "common mode voltage." This common mode voltage or neutral voltage is responsible for leakage bearing currents and their subsequent failure [53–56]. The PWM techniques should take into account the minimization or elimination of common mode voltage.

By adding each term of the equation (7.20) and putting the sum of phase-to-neutral voltage zero (assuming a balanced five-phase voltage whose instantaneous sum is always zero), we obtain

$$v_{nN}(t) = (1/5)(V_A(t) + V_B(t) + V_C(t) + V_D(t) + V_E(t)) \tag{7.21}$$

Substituting equation (7.21) back into equation (7.20), the following expressions for the phase-to-neutral voltage are obtained:

$$\begin{aligned}
v_{an}(t) &= (4/5)V_A(t) - (1/5)(V_B(t) + V_C(t) + V_D(t) + V_E(t)) \\
v_{bn}(t) &= (4/5)V_B(t) - (1/5)(V_A(t) + V_C(t) + V_D(t) + V_E(t)) \\
v_{cn}(t) &= (4/5)V_C(t) - (1/5)(V_B(t) + V_A(t) + V_D(t) + V_E(t)) \\
v_{dn}(t) &= (4/5)V_D(t) - (1/5)(V_B(t) + V_C(t) + V_A(t) + V_E(t)) \\
v_{en}(t) &= (4/5)V_E(t) - (1/5)(V_B(t) + V_C(t) + V_D(t) + V_A(t))
\end{aligned} \tag{7.22}$$

Equation (7.22) can also be written using the switching function definition of equation (7.19):

$$
\begin{aligned}
v_{an}(t) &= \left(\frac{V_{dc}}{5}\right)[4S_A - S_B - S_C - S_D - S_E] \\
v_{bn}(t) &= \left(\frac{V_{dc}}{5}\right)[4S_B - S_A - S_C - S_D - S_E] \\
v_{cn}(t) &= \left(\frac{V_{dc}}{5}\right)[4S_C - S_B - S_A - S_D - S_E] \\
v_{dn}(t) &= \left(\frac{V_{dc}}{5}\right)[4S_D - S_B - S_C - S_A - S_E] \\
v_{en}(t) &= \left(\frac{V_{dc}}{5}\right)[4S_E - S_B - S_C - S_D - S_A]
\end{aligned} \tag{7.23}
$$

7.3.2.1 Ten-Step Mode of Operation

This mode of operation is an extension of the six-step operation of a three-phase voltage source inverter. The output phase voltage assumes 10 different values, so it is called the "10-step mode." The switching frequency of the power switches in this mode is equal to the fundamental output voltage frequency. Each power switch operates for half of the fundamental cycle and so it is called the 180-degree conduction mode, as well. Thus, each power switch turns "ON" and turns "OFF" only once in the whole fundamental cycle. The output is maximum in this mode and switching losses are minimal. However, the output waveform in this mode contains a considerable amount of low-order harmonics, which decreases the performance of the load. The upper power switch is "ON" when the load current is positive (current flowing from inverter to the load) and the pole voltage is positive, and the antiparallel diode across the upper switch is turned "ON" when the load current is negative while the pole voltage is positive. The lower power switch is turned "ON" when the load current is positive and the pole voltage is zero or negative (depending upon the choice of DC link, i.e. $+V_{dc}$ and 0 or $+0.5\ V_{dc}$ and $-0.5\ V_{dc}$) and the lower antiparallel diode is turned on for negative load current. This is illustrated in Figure 7.16. The possible pole voltages during the step operation of the five-phase VSI and the corresponding switches that are "ON" are listed in Table 7.2.

The switching signal for the 10-step mode of operation is given in Figure 7.17. The delay in switching between the two consecutive phases are 360 degree/5 = 72 degree or $\pi/5$. The complementary gate drive signals are also shown. The "dead time" is not considered here.

To determine the phase-to-neutral voltages for a 10-step mode, the leg voltages from Table 7.2 is substituted into equation (7.3), and the corresponding values are listed in Table 7.3 for a star-connected five-phase load. The corresponding waveform is given in Figure 7.18. It is evident that the phase-to-neutral voltage takes on four different values/levels and ten steps in one fundamental cycle.

Line-to-line voltages are now discussed for a star-connected five-phase load. In a five-phase system, there exist two different systems of line voltages, termed "adjacent line-to-line voltage" and "nonadjacent line-to-line voltage," as illustrated in Figure 7.19. The phase voltages are

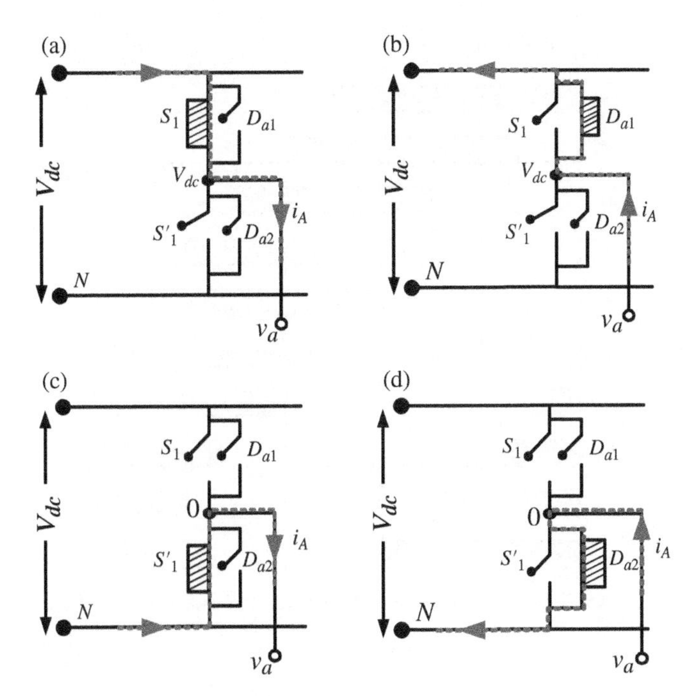

Figure 7.16 Switching state for leg A in a five-phase VSI: (a and c) $i_A > 0$; (b and d) $i_A < 0$

Table 7.2 Pole voltages of the five-phase VSI during step mode of operation

Switching mode	Switches ON	Pole voltage V_A	Pole voltage V_B	Pole voltage V_C	Pole voltage V_D	Pole voltage V_E
1	S_1, S_2, S'_3, S'_4, S_5	V_{dc}	V_{dc}	0	0	V_{dc}
2	S_1, S_2, S'_3, S'_4, S'_5	V_{dc}	V_{dc}	0	0	0
3	S_1, S_2, S_3, S'_4, S'_5	V_{dc}	V_{dc}	V_{dc}	0	0
4	S'_1, S_2, S_3, S'_4, S'_5	0	V_{dc}	V_{dc}	0	0
5	S'_1, S_2, S_3, S_4, S'_5	0	V_{dc}	V_{dc}	V_{dc}	0
6	S'_1, S'_2, S_3, S_4, S'_5	0	0	V_{dc}	V_{dc}	0
7	S'_1, S'_2, S_3, S_4, S_5	0	0	V_{dc}	V_{dc}	V_{dc}
8	S'_1 S'_2, S'_3, S_4, S_5	0	0	0	V_{dc}	V_{dc}
9	S_1 S'_2, S'_3, S_4, S_5	V_{dc}	0	0	V_{dc}	V_{dc}
10	S_1 S'_2, S'_3, S'_4, S_5	V_{dc}	0	0	0	V_{dc}

represented as (v_a, v_b, v_c, v_d, v_e), the adjacent line voltages are denoted as (v_{ab}, v_{bc}, v_{cd}, v_{de}, v_{ea}), and the nonadjacent line voltages are given as (v_{ac}, v_{bd}, v_{ce}, v_{da}, v_{eb}). The relationship between the adjacent, nonadjacent, and phase voltages are elaborated upon using a numerical example.

The adjacent line voltages are calculated from the values of the phase voltages from Table 7.3, and the resulting values are listed in Table 7.4 for the 10-step mode of operation. Similarly, nonadjacent line voltages are obtained and listed in Table 7.5.

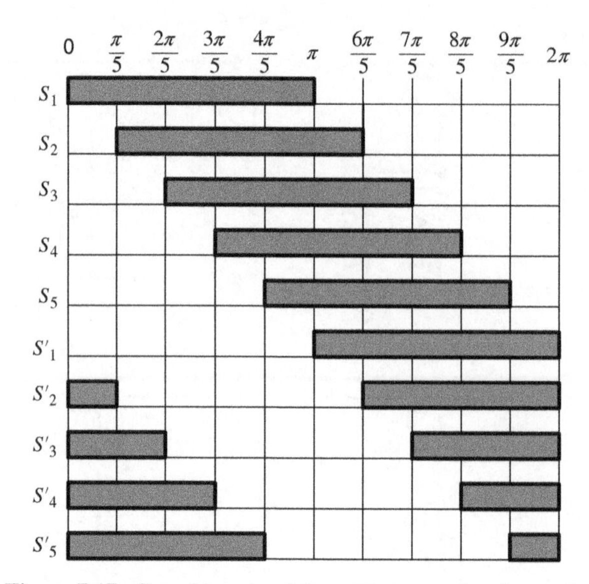

Figure 7.17 Gate drive signal for a 10-step mode of operation

Table 7.3 Phase-to-neutral voltages of a star-connected load supplied from a five-phase VSI

Mode	Switches ON	v_{an}	v_{bn}	v_{cn}	v_{dn}	v_e
1	$S_1, S_2, S'_3, S'_4, S_5$ (11001)	2/5 V_{dc}	2/5 V_{dc}	−3/5 V_{dc}	−3/5 V_{dc}	2/5 V_{dc}
2	$S_1, S_2, S'_3, S'_4, S'_5$ (11000)	3/5 V_{dc}	3/5 V_{dc}	−2/5 V_{dc}	−2/5 V_{dc}	−2/5 V_{dc}
3	$S_1, S_2, S_3, S'_4, S'_5$ (11101)	2/5 V_{dc}	2/5 V_{dc}	2/5 V_{dc}	−3/5 V_{dc}	−3/5 V_{dc}
4	$S'_1, S_2, S_3, S'_4, S'_5$ (01100)	−2/5 V_{dc}	3/5 V_{dc}	3/5 V_{dc}	−2/5 V_{dc}	−2/5 V_{dc}
5	$S'_1, S_2, S_3, S_4, S'_5$ (01110)	−3/5 V_{dc}	2/5 V_{dc}	2/5 V_{dc}	2/5 V_{dc}	−3/5 V_{dc}
6	$S'_1, S'_2, S_3, S_4, S'_5$ (00110)	−2/5 V_{dc}	−2/5 V_{dc}	3/5 V_{dc}	3/5 V_{dc}	−2/5 V_{dc}
7	$S'_1, S'_2, S_3, S_4, S_5$ (00111)	−3/5 V_{dc}	−3/5 V_{dc}	2/5 V_{dc}	2/5 V_{dc}	2/5 V_{dc}
8	$S'_1, S'_2, S'_3, S_4, S_5$ (00011)	−2/5 V_{dc}	−2/5 V_{dc}	−2/5 V_{dc}	3/5 V_{dc}	3/5 V_{dc}
9	$S_1, S'_2, S'_3, S_4, S_5$ (10011)	2/5 V_{dc}	−3/5 V_{dc}	−3/5 V_{dc}	2/5 V_{dc}	2/5 V_{dc}
10	$S_1, S'_2, S'_3, S'_4, S_5$ (10001)	3/5 V_{dc}	−2/5 V_{dc}	−2/5 V_{dc}	−2/5 V_{dc}	3/5 V_{dc}

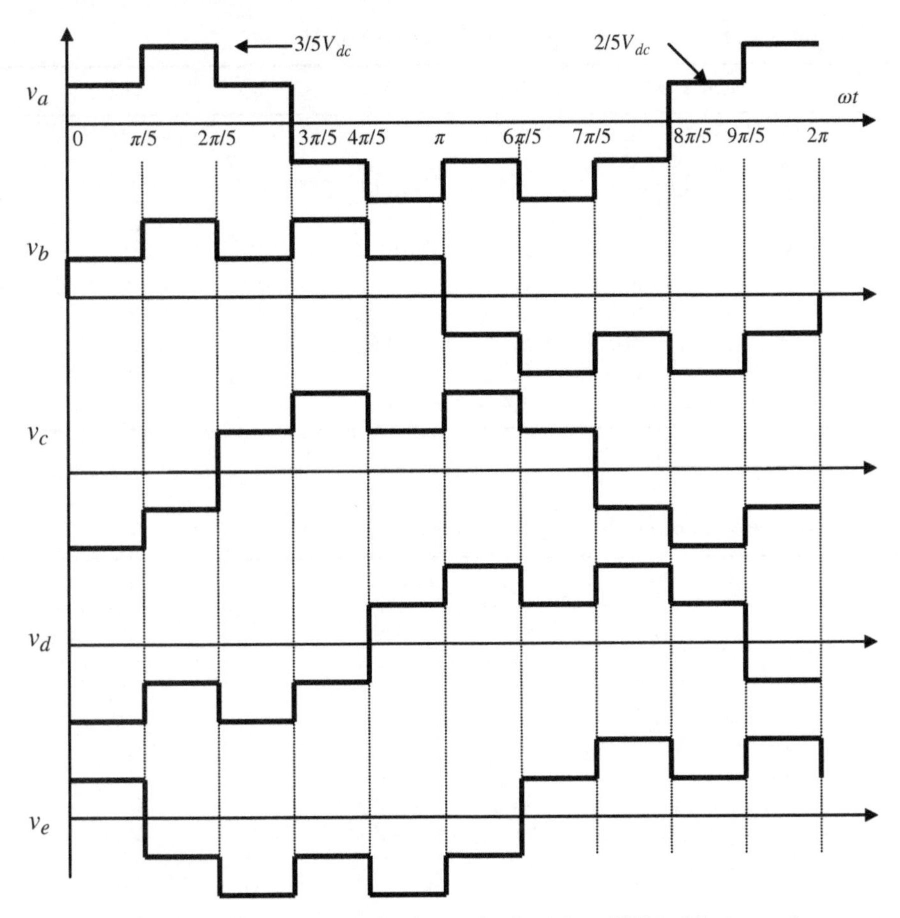

Figure 7.18 Phase-to-neutral voltage of a five-phase VSI in 10-step mode

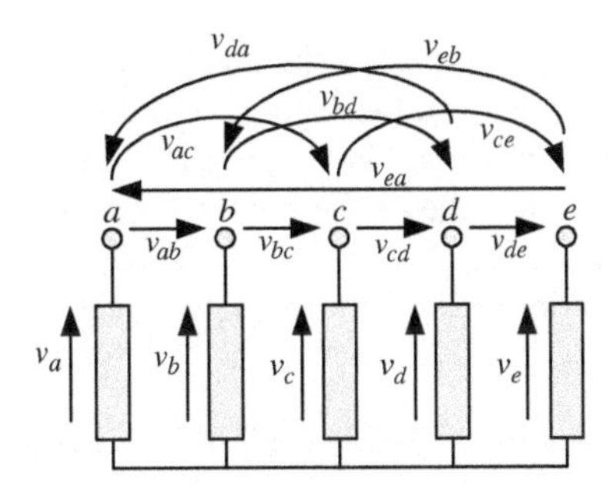

Figure 7.19 Phase and line voltages in a five-phase system

Table 7.4 Adjacent line-to-line voltages of the five-phase VSI

Mode	Switches ON	v_{ab}	v_{bc}	v_{cd}	v_{de}	v_{ea}
1	$S_1, S_2, S'_3, S'_4, S_5$	0	V_{dc}	0	$-V_{dc}$	0
2	$S_1, S_2, S'_3, S'_4, S'_5$	0	V_{dc}	0	0	$-V_{dc}$
3	$S_1, S_2, S_3, S'_4, S'_5$	0	0	V_{dc}	0	$-V_{dc}$
4	$S'_1, S_2, S_3, S'_4, S'_5$	$-V_{dc}$	0	V_{dc}	0	0
5	$S'_1, S_2, S_3, S_4, S'_5$	$-V_{dc}$	0	0	V_{dc}	0
6	$S'_1, S'_2, S_3, S_4, S'_5$	0	$-V_{dc}$	0	V_{dc}	0
7	$S'_1, S'_2, S_3, S_4, S_5$	0	$-V_{dc}$	0	0	V_{dc}
8	$S'_1, S'_2, S'_3, S_4, S_5$	0	0	$-V_{dc}$	0	V_{dc}
9	$S_1, S'_2, S'_3, S_4, S_5$	V_{dc}	0	$-V_{dc}$	0	0
10	$S_1, S'_2, S'_3, S'_4, S_5$	V_{dc}	0	0	$-V_{dc}$	0

Table 7.5 Nonadjacent line-to-line voltages of the five-phase VSI

Mode	Switches ON	v_{ac}	v_{bd}	v_{ce}	v_{da}	v_{eb}
1	$S_1, S_2, S'_3, S'_4, S_5$	V_{dc}	V_{dc}	$-V_{dc}$	$-V_{dc}$	0
2	$S_1, S_2, S'_3, S'_4, S'_5$	V_{dc}	V_{dc}	0	$-V_{dc}$	$-V_{dc}$
3	$S_1, S_2, S_3, S'_4, S'_5$	0	V_{dc}	V_{dc}	$-V_{dc}$	$-V_{dc}$
4	$S'_1, S_2, S_3, S'_4, S'_5$	$-V_{dc}$	V_{dc}	V_{dc}	0	$-V_{dc}$
5	$S'_1, S_2, S_3, S_4, S'_5$	$-V_{dc}$	0	V_{dc}	V_{dc}	$-V_{dc}$
6	$S'_1, S'_2, S_3, S_4, S'_5$	$-V_{dc}$	$-V_{dc}$	V_{dc}	V_{dc}	0
7	$S'_1, S'_2, S_3, S_4, S_5$	$-V_{dc}$	$-V_{dc}$	0	V_{dc}	V_{dc}
8	$S'_1, S'_2, S'_3, S_4, S_5$	0	$-V_{dc}$	$-V_{dc}$	V_{dc}	V_{dc}
9	$S_1, S'_2, S'_3, S_4, S_5$	V_{dc}	$-V_{dc}$	$-V_{dc}$	0	V_{dc}
10	$S_1, S'_2, S'_3, S'_4, S_5$	V_{dc}	0	$-V_{dc}$	$-V_{dc}$	V_{dc}

The pictorial representation of the adjacent and nonadjacent line voltages are given in Figures 7.20 and 7.21, respectively.

7.3.2.2 Fourier Analysis of the Five-Phase Inverter Output Voltages

In order to relate the input DC-link voltage of the inverter with the output phase-to-neutral and line-to-line voltages, Fourier analysis of the voltage waveforms is undertaken. Out of the two sets of the line-to-line voltages, discussed in the preceding subsection, only nonadjacent line-to-line voltages are analyzed, since they have higher fundamental harmonic values than the adjacent line-to-line voltages.

Using definition of the Fourier series for a periodic waveform:

$$v(t) = V_0 + \sum_{k=1}^{\infty}(A_k \cos k\omega t + B_k \sin k\omega t) \tag{7.24}$$

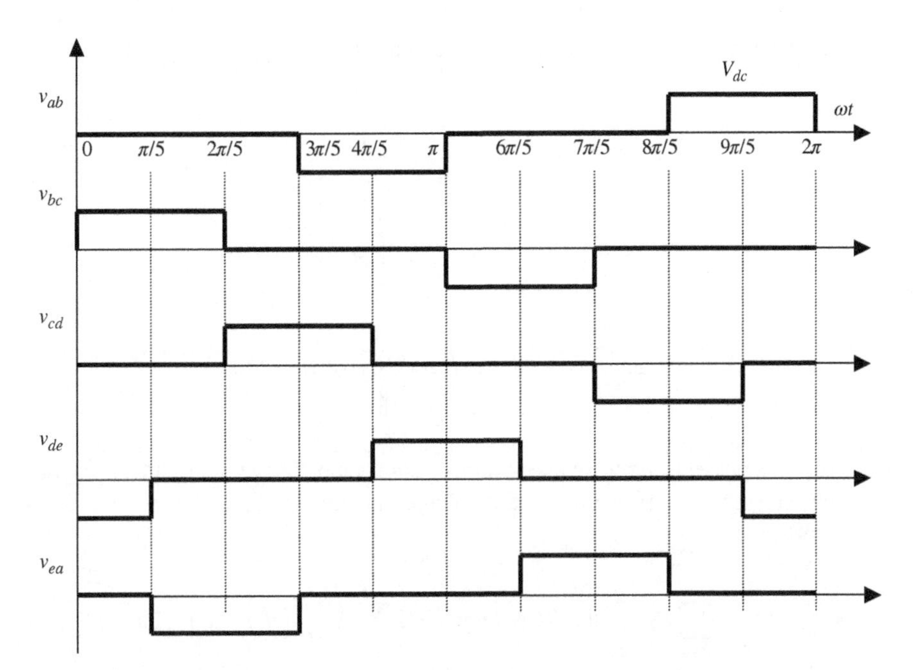

Figure 7.20 Adjacent line-to-line voltages of the five-phase VSI

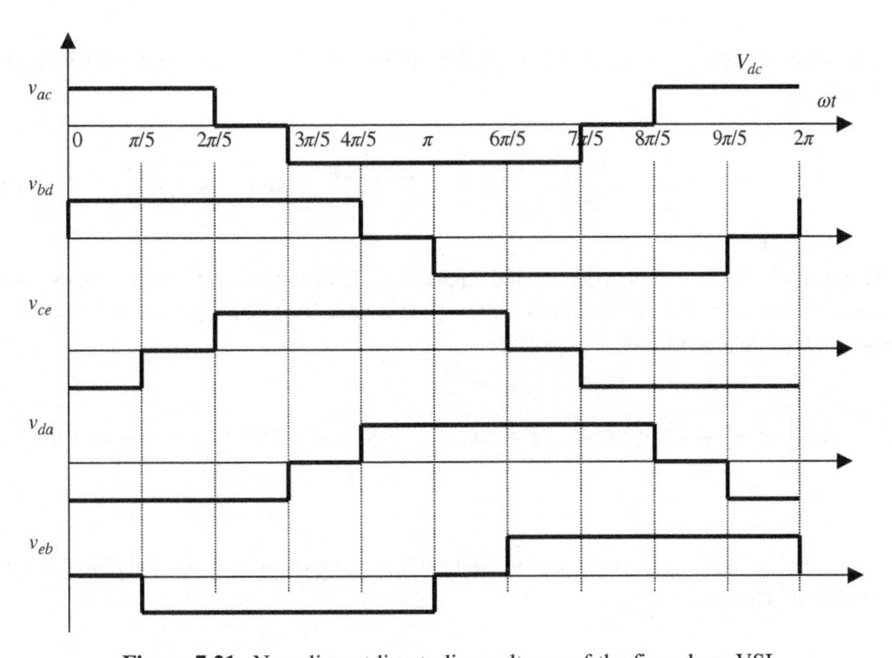

Figure 7.21 Nonadjacent line-to-line voltages of the five-phase VSI

where the coefficients of the Fourier series are given by:

$$
\begin{aligned}
V_0 &= \frac{1}{T}\int_0^T v(t)dt = \frac{1}{2\pi}\int_0^{2\pi} v(\theta)d\theta \\
A_k &= \frac{2}{T}\int_0^T v(t)\cos k\omega t dt = \frac{1}{\pi}\int_0^{2\pi} v(\theta)\cos k\theta d\theta \\
B_k &= \frac{2}{T}\int_0^T v(t)\sin\, k\omega t dt = \frac{1}{\pi}\int_0^{2\pi} v(\theta)\sin k\theta d\theta
\end{aligned}
\tag{7.25}
$$

The waveforms possess quarter-wave symmetry and can be conveniently taken as odd functions, and one can represent phase-to-neutral voltages and line-to-line voltages with the following expressions:

$$
\begin{aligned}
v(t) &= \sum_{k=0}^{\infty} B_{2k+1}\sin(2k+1)\omega t = \sqrt{2}\sum_{k=0}^{\infty} V_{2k+1}\sin(2k+1)\omega t \\
B_{2k+1} &= \sqrt{2}V_{2k+1} = \frac{1}{\pi}4\int_0^{\pi/2} v(\theta)\sin(2k+1)\theta d\theta
\end{aligned}
\tag{7.26}
$$

In the case of the phase-to-neutral voltage v_b, shown in Figure 7.18, the coefficients of the Fourier series are

$$
B_{2k+1} = \frac{1}{\pi}\frac{4}{5}V_{dc}\frac{1}{2k+1}\left[2 + \cos(2k+1)\frac{\pi}{5} - \cos(2k+1)\frac{2\pi}{5}\right]
\tag{7.27}
$$

The expression in brackets in the second equation (7.27) equals zero for all harmonics whose order is divisible by 5. For all the other harmonics, it equals 2.5. Hence, we can write the Fourier series of the phase-to-neutral voltage as:

$$
v(t) = \frac{2}{\pi}V_{dc}\left[\sin\omega t + \frac{1}{3}\sin 3\omega t + \frac{1}{7}\sin 7\omega t + \frac{1}{9}\sin 9\omega t + \frac{1}{11}\sin 11\omega t + \frac{1}{13}\sin 13\omega t + \cdots\right]
\tag{7.28}
$$

From equation (7.28), it follows that the fundamental component of the output phase-to-neutral voltage has an rms value of:

$$
V_1 = \frac{\sqrt{2}}{\pi}V_{dc} = 0.45V_{dc}
\tag{7.29}
$$

It is observed that the fundamental output of a five-phase voltage source inverter is the same as that of a three-phase voltage source inverter.

Fourier analysis of the nonadjacent line-to-line voltages is performed in the same manner. Fourier series remains to be given by equation (7.25). Taking the second voltage in Figure 7.20 and shifting the zero time instant by $\pi/10$ degrees to the left, we have the following Fourier series coefficients:

$$B_{2k+1} = \frac{1}{\pi}4\int_{\pi/10}^{\pi/2} V_{dc}\sin(2k+1)\theta d\theta = \frac{1}{\pi}4V_{dc}\frac{1}{2k+1}\cos(2k+1)\frac{\pi}{10} \tag{7.30}$$

Hence, the Fourier series of the nonadjacent line-to-line voltage is

$$v(t) = \frac{4}{\pi}V_{dc}\left[0.95\sin\omega t + \frac{0.59}{3}\sin 3\omega t - \frac{0.59}{7}\sin 7\omega t - \frac{0.95}{9}\sin 9\omega t - \frac{0.95}{11}\sin 11\omega t - \cdots\right] \tag{7.31}$$

and the fundamental harmonic rms value of the nonadjacent line-to-line voltage is

$$V_{1L} = \frac{2\sqrt{2}}{\pi}V_{dc}\cos\frac{\pi}{10} = 0.856V_{dc} = 1.902V_1 \tag{7.32}$$

7.3.2.3 MATLAB/Simulink Modeling for 10-Step Mode

The MATLAB/Simulink model is shown in Figure 7.22. The model consists of a gate drive generator and the power circuit part of the five-phase inverter. The gate drive signal generator model, shown in Figure 7.22a, is developed using a "repeating" block of Simulink, illustrated in Figure 7.22b. The inverter power circuit, shown in Figure 7.22c, is obtained using equation (7.5). The power circuit can also be modeled using an actual IGBT switch model from "SimPowerSystem" blocksets of Simulink. The repeating block is generating a square-wave gate drive signal of 50-Hz fundamental. This can be changed by changing the time axis values. The inverter equations are realized using the "function block" of Simulink. The inverter power circuit, shown in Figure 7.22c, needs the value of the DC-link voltage, which can be given from the main MATLAB command window or from the model initialization. The RL load is modeled using a first-order low-pass filter, as shown in Figure 7.22d. The complete MATLAB/ Simulink model is available in the CD-ROM provided with this book.

The simulation results for 50-Hz fundamental frequency and unit DC-link voltage are given in Figure 7.23. The phase voltage, adjacent voltage, and nonadjacent voltage and resulting current obtained using simulation are shown. The output phase voltages exhibit 10 steps and the remainder of the curves correspond to the theoretical waveforms. This is further verified by implementing this control in a prototype five-phase inverter experimentally and is shown in the next section.

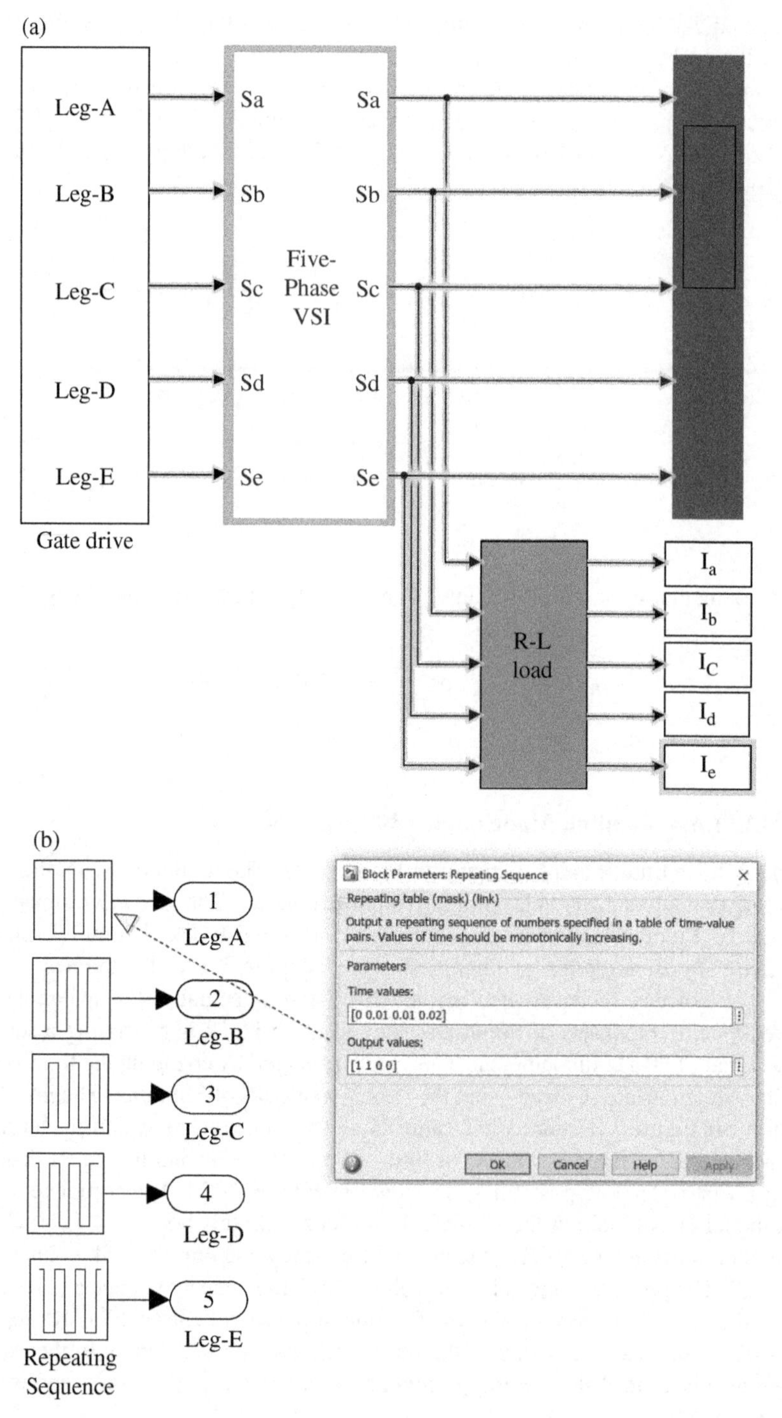

Figure 7.22 (a) MATLAB/Simulink model of a five-phase voltage source inverter for 10-step operation; (b) Gate drive signal generation for 10-step operation; (c) Simulink model of a five-phase VSI; and (d) RL load model

(c)

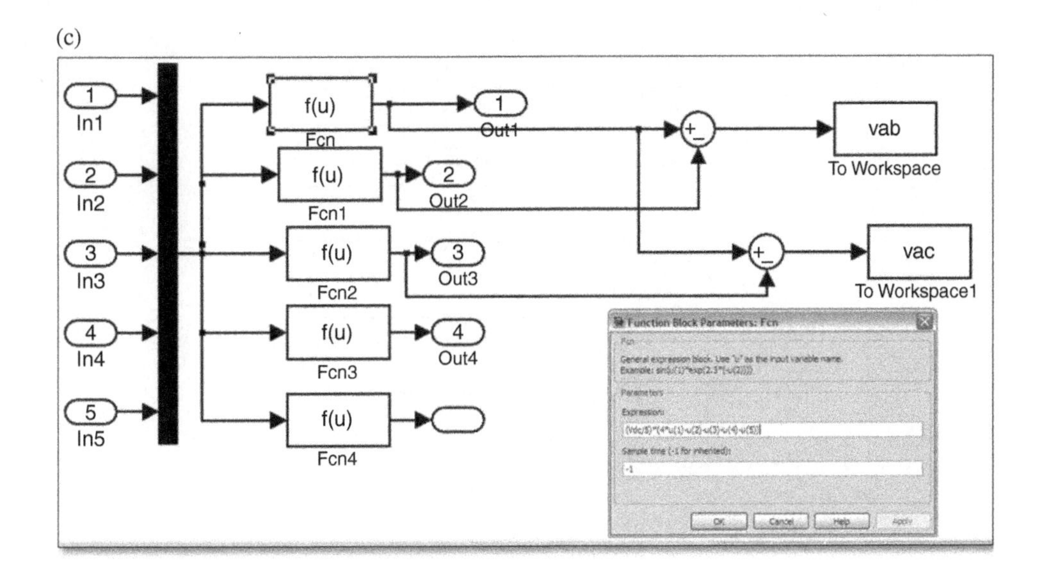

(d)

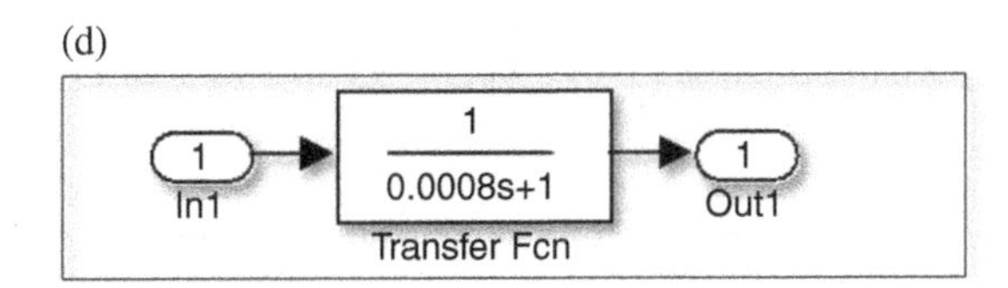

Figure 7.22 (Continued)

7.3.2.4 Prototype of a Five-Phase VSI for 10-Step Operation

The inverters control logic can be implemented using an analog circuit and its complete block diagram is shown in Figure 7.24. The other method is by using advanced microprocessors, microcontrollers, and digital signal processor to implement the control schemes.

In Figure 7.24, the supply is taken from a single-phase power supply and converted to 9-0-9 V using a small transformer. This is fed to the phase-shifting circuit, shown in Figure 7.25, to provide an appropriate phase shift for operation at various conduction angles. The phase-shifted signal is then fed to the inverting/non-inverting Schmitt trigger circuit and wave-shaping circuit (Figures 7.26 and 7.27). The processed signal is then fed to the isolation and driver circuit shown in Figure 7.28, which is then finally given to the gate of IGBTs. There are two separate circuits for upper and lower legs of the inverter.

The power circuit can be made up of IGBT, with a snubber circuit consisting of a series combination of a resistance and a capacitor with a diode in parallel with the resistance.

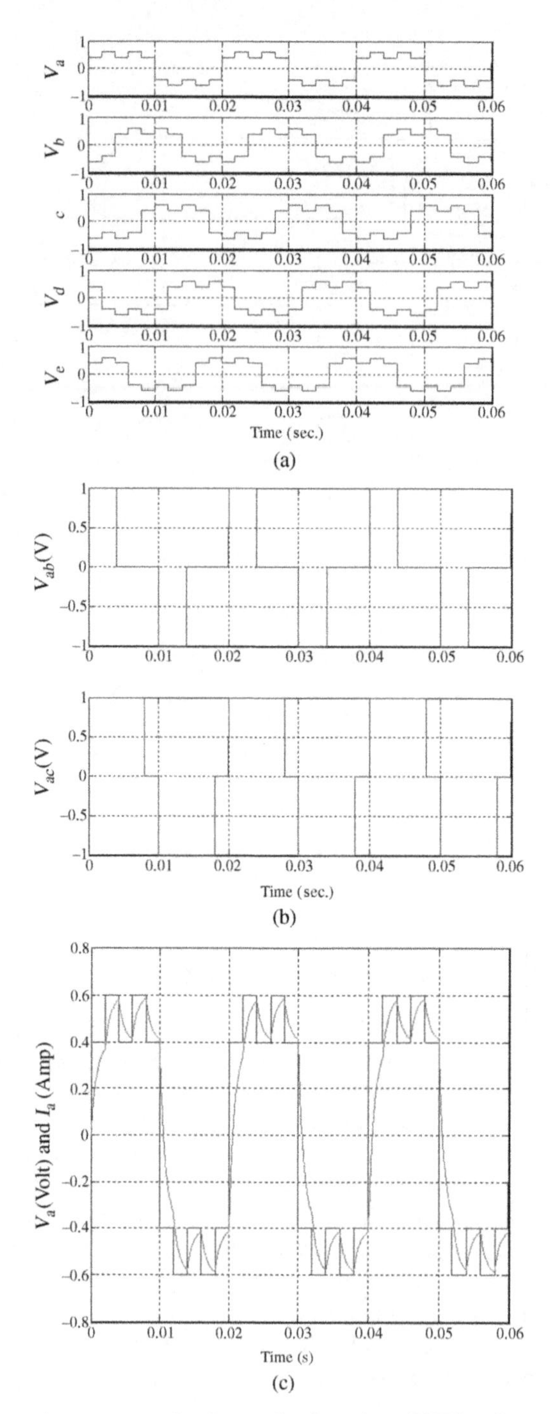

Figure 7.23 (a) Output phase-to-neutral voltage of a five-phase VSI in 10-step mode; (b) Output line voltages of a five-phase VSI in 10-step mode; and (c) output phase "a" voltage and phase "a" current of a five-phase VSI in 10-step mode for RL load

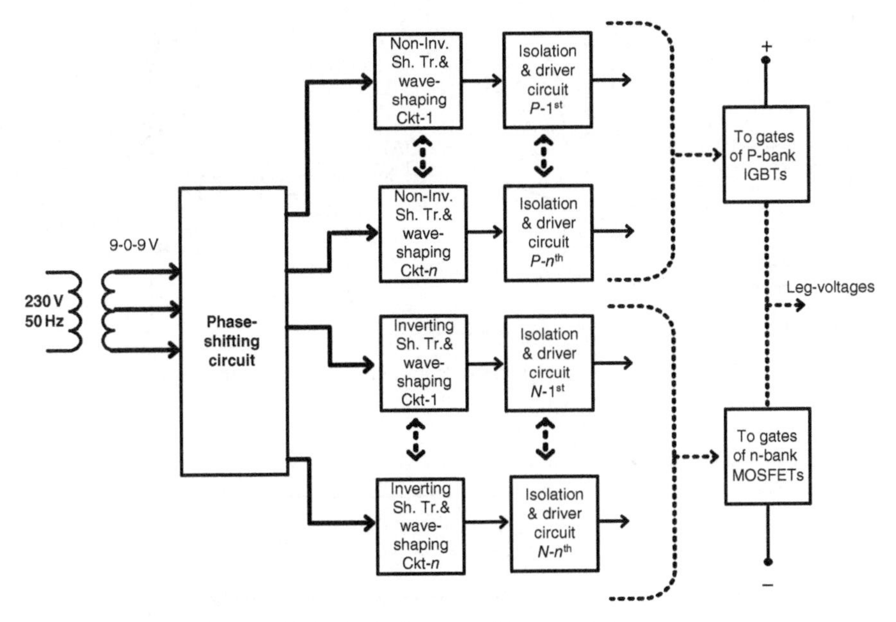

Figure 7.24 Block diagram of the complete inverter

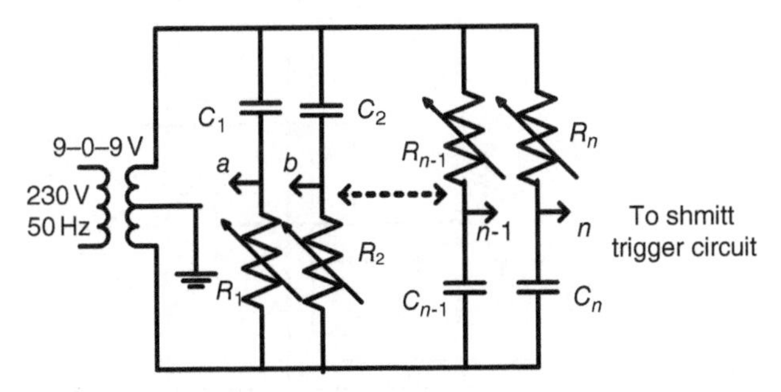

Figure 7.25 Phase-shifting network

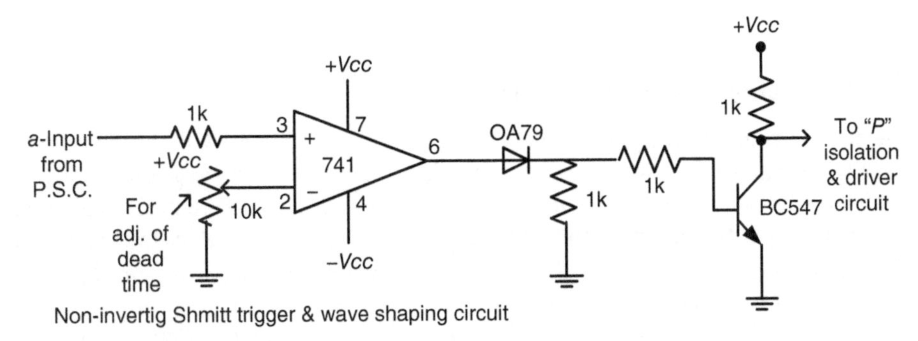

Figure 7.26 Non-inverting Schmitt trigger and wave-shaping circuit

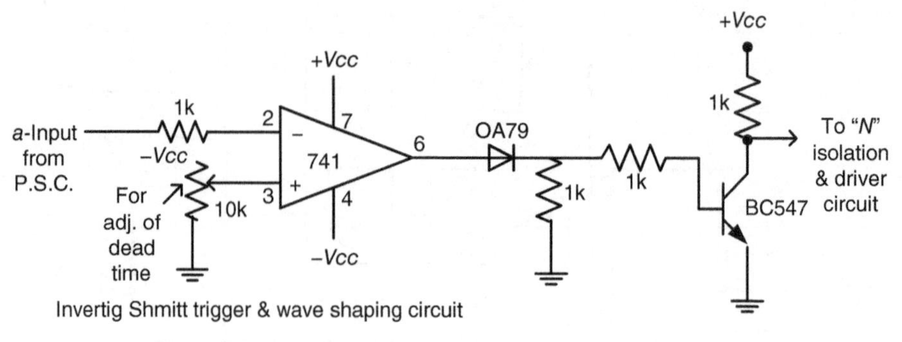

Figure 7.27 Inverting Schmitt trigger and wave-shaping circuit

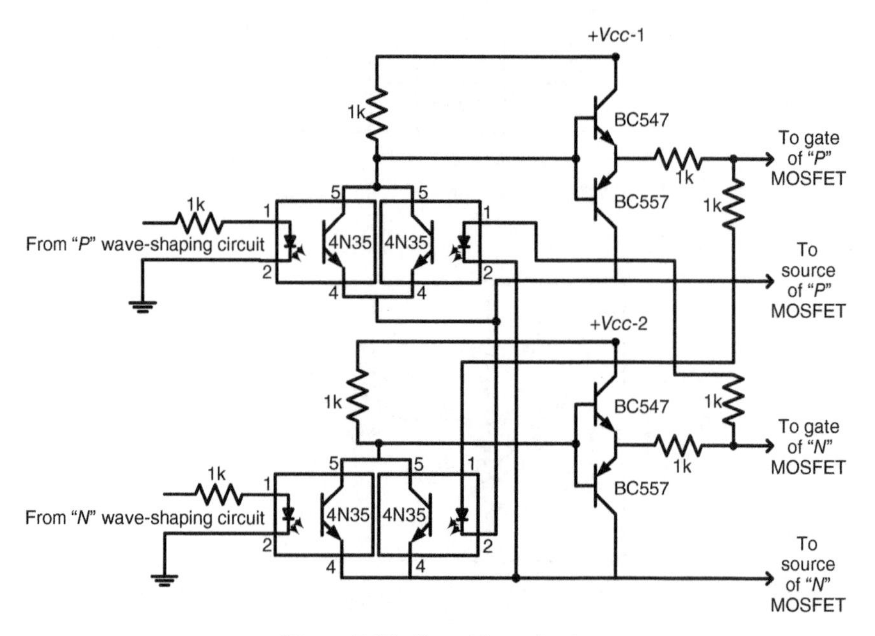

Figure 7.28 Gate driver circuit

7.3.2.5 Experimental Results for 10-Step Mode

This section gives the experimental results obtained from the five-phase voltage source inverter for stepped operation of an inverter, with a 180-degree conduction mode leading to a 10-step output with a star-connected load. A single-phase supply is given to the control circuit through the phase-shifting network. The output of the phase-shifting circuit provides the required five-phase output voltage by appropriately tuning it. These five-phase signals are then further processed to generate the gate drive circuit.

The gate drive signal is given to the IGBTs and the corresponding phase-to-neutral voltages thus generated, as shown in Figure 7.29, keeping the DC-link voltage at 60 V. It is seen that the

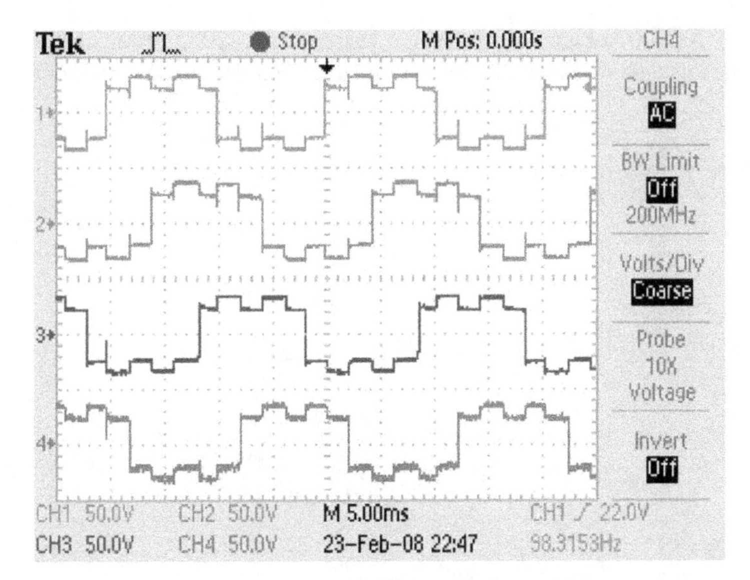

Figure 7.29 Output phase *"a-d"* voltages for 180-degree conduction mode

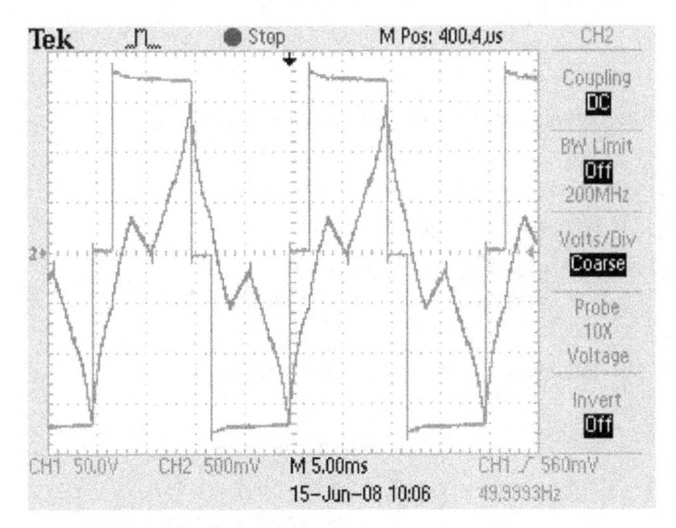

Figure 7.30 Nonadjacent line voltage and stator current

phase-to-neutral voltages have 10 steps with output voltage levels of ±0.2 V_{dc} and ±0.4 V_{dc}. Another experimental waveform is illustrated in Figure 7.30, for nonadjacent voltage and stator line current of a five-phase induction motor. The current shows a typical response of an induction machine.

Problem 7.1 Assume a balanced five-phase system with 120 V rms as the phase voltages. Determine the adjacent and nonadjacent line voltages and derive their relationship.

Solution.
The five-phase voltage is given as:

$$\begin{aligned} v_a &= 120\sqrt{2}\sin(\omega t) \\ v_b &= 120\sqrt{2}\sin\left(\omega t - 2\frac{\pi}{5}\right) \\ v_c &= 120\sqrt{2}\sin\left(\omega t - 4\frac{\pi}{5}\right) \\ v_d &= 120\sqrt{2}\sin\left(\omega t + 4\frac{\pi}{5}\right) \\ v_e &= 120\sqrt{2}\sin\left(\omega t + 2\frac{\pi}{5}\right) \end{aligned} \tag{7.33}$$

The adjacent line voltage is given as:

$$v_{ab} = v_a - v_b = 120\sqrt{2}\sin(\omega t) - 120\sqrt{2}\sin\left(\omega t - 2\frac{\pi}{5}\right) \tag{7.34}$$

The adjacent line voltage can be written in polar form as:

$$\begin{aligned} v_{ab} &= 120\sqrt{2} < 0 - 120\sqrt{2} < -72 = 120\sqrt{2} + j0.0 - 120\sqrt{2}\{\cos(-72) + j\sin(-72)\} \\ v_{ab} &= 120\sqrt{2} + j0.0 - 120\sqrt{2}\{0.309 = j0.9511\} \\ v_{ab} &= 82.92\sqrt{2} + j114.132\sqrt{2} = 199.5086 < 54 \end{aligned}$$

Similarly, we can determine other adjacent line voltages as:

$$\begin{aligned} v_{bc} &= v_b - v_c = 199.5086 < -18 \\ v_{cd} &= v_c - v_d = 199.5086 < -90 \\ v_{de} &= v_d - v_e = 199.5086 < -162 \\ v_{ea} &= v_e - v_a = 199.5086 < -126 \end{aligned}$$

This calculation can also be shown using the graphical technique, as in Figures 7.31 and 7.32. The relationship between the phase voltage and adjacent line voltage can be written as (assuming balanced voltages):

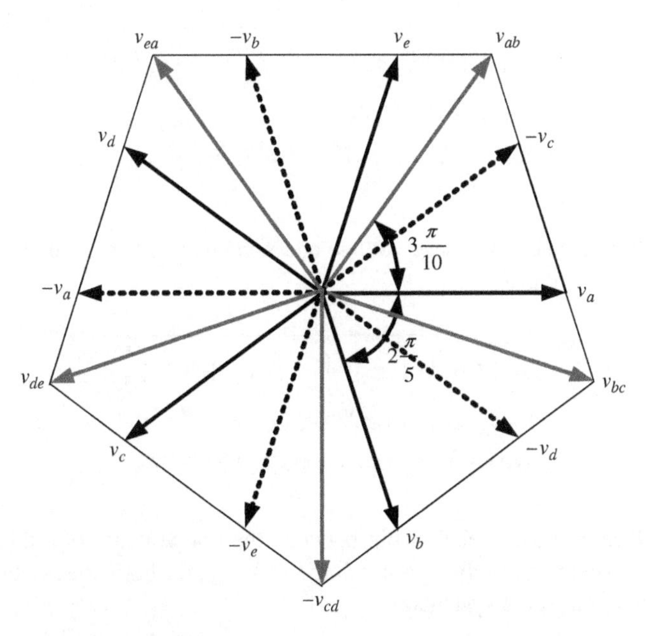

Figure 7.31 Phase and adjacent line voltage concept

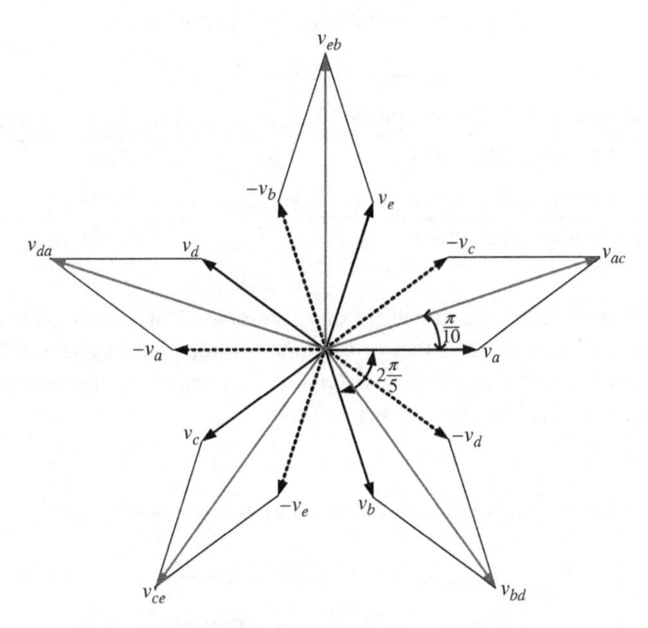

Figure 7.32 Phase and adjacent line voltage concept

$$\begin{aligned} |v_{ab}| &= \sqrt{|v_a|^2 + |v_b|^2 + |v_a| \cdot |v_b| \cos\left(3\frac{\pi}{5}\right)} \\ V_{adj-L} &= 1.1756 V_{phase} \\ |v_{ab}| &= V_{adj-L}; \quad |v_a| = |v_b| = V_{phase} \end{aligned} \tag{7.35}$$

The relationship between the phase voltage and nonadjacent line voltage can be written as (assuming balanced voltages):

$$\begin{aligned} |v_{ab}| &= \sqrt{|v_a|^2 + |v_b|^2 + |v_a| \cdot |v_b| \cos\left(\frac{\pi}{5}\right)} \\ V_{non-adj-L} &= 1.9025 V_{phase} \\ |v_{ab}| &= V_{non-adj-L}; \quad |v_a| = |v_b| = V_{phase} \end{aligned} \tag{7.36}$$

Problem 7.2 Determine the relationship between the line voltages of a three-phase system and a five-phase system, given the phase voltage as V_{phase} for both three- and five-phase systems. Assume star-connected systems.

Solution.
The line-to-line voltage in a three-phase system is given as:

$$V_{line} = \sqrt{3} V_{phase} = 1.7321\ V_{phase} \tag{7.37}$$

$$V_{adj\text{–}L} = \sqrt{1.382} V_{phase} = 1.1756 V_{phase} \tag{7.38}$$

$$V_{non\text{–}adj\text{–}L} = 1.9025 V_{phase} \tag{7.39}$$

$$\frac{\text{Adjacent line voltage five-phase}}{\text{Line voltage three-phase}} = \frac{V_{adj\text{–}L}}{V_{line}} = \frac{1.1756 V_{phase}}{1.7321 V_{phase}} = 0.6787$$

$$\frac{\text{Non-adjacent line voltage five-phase}}{\text{Line voltage three-phase}} = \frac{V_{non\text{–}adj\text{–}L}}{V_{line}} = \frac{1.9025 V_{phase}}{1.7321 V_{phase}} = 1.0984$$

Problem 7.3 Determine the phase voltage of a five-phase star-connected load for obtaining same highest line-to-line voltage as that of a standard three-phase system. (*Hint: A standard three-phase system has 400 V line-to-line voltage.*) Also determine the ratio of phase voltage of the three-phase and the five-phase system

Solution.
Given the highest line-to-line voltage of a five-phase system as 400 V, i.e. $V_{non\text{-}adj\text{-}L} = 400$ V

$$V_{non\text{–}adj\text{–}L} = 1.9025 V_{phase} = 400$$

$$V_{phase\text{–}5} = \frac{400}{1.9025} = 210.25\ \text{V}$$

Thus, the phase voltage of the five-phase load will be 210.25 V

The phase voltage of three-phase system is

$$V_{phase-3} = \frac{400}{\sqrt{3}} = 230.94\text{V}$$

$$\frac{V_{phase-3}}{V_{phase-5}} = \frac{230.94}{210.25} = 1.0984$$

Problem 7.4 Draw the waveform of common mode voltage generated in 10-step model of operation of a five-phase voltage source inverter. Assume $V_{dc} = 1$ V.

Solution.

Common mode voltage is given as:

$$v_{nN}(t) = (1/5)(V_A(t) + V_B(t) + V_C(t) + V_D(t) + V_E(t)) \tag{7.40}$$

The leg voltage waveforms for the 10-step mode of operation are given in Figure 7.33. The magnitude of the voltage is V_{dc}. To determine the common mode voltage, the instantaneous values are added for all the five legs and the resulting waveform is given in Figure 7.33.

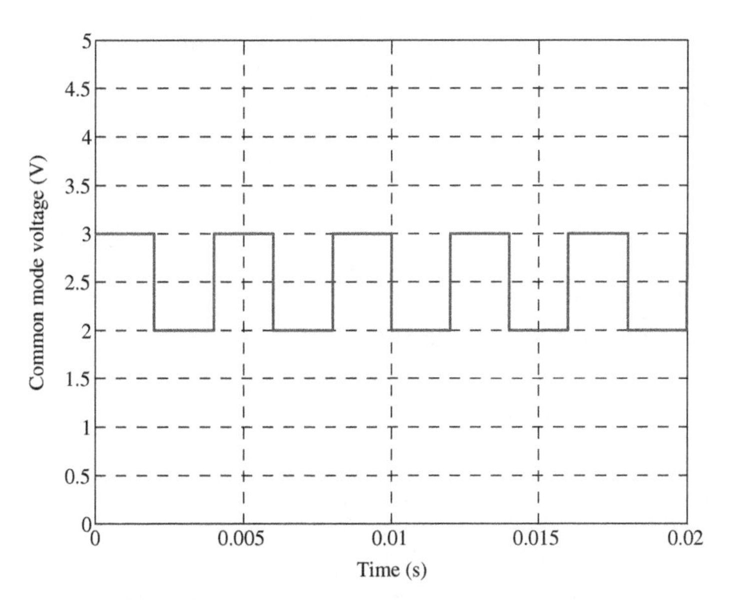

Figure 7.33 Common mode voltage of Problem 7.4

Problem 7.5 A five-phase induction motor has the following per-phase parameters referred to stator, stator resistance: 0.3 Ω, rotor resistance:0.45 Ω, stator and rotor leakage reactance: 2.1 Ω each, magnetizing reactance: 30 Ω.

Find out the parameters of an equivalent two-phase induction motor if its per-phase turns are

(a) same as that of the five-phase induction motor
(b) 5/2 times of that of the five-phase induction motor
(c) $\sqrt{5/2}$ times of that of the five-phase induction motor

Solution.
The five-phase currents of the five-phase machine are

$$i_a = I_m \cos(\omega t - \phi)$$

$$i_b = I_m \cos(\omega t - \phi - 72^\circ)$$

$$i_c = I_m \cos(\omega t - \phi - 144^\circ)$$

$$i_d = I_m \cos(\omega t - \phi - 216^\circ)$$

$$i_e = I_m \cos(\omega t - \phi - 288^\circ)$$

Case (a)

If the number of turns per phase in five-phase machine and two-phase equivalent machine are equal, then for maintaining the same mmf in both machines the currents in two-phase machine have to be increased by 5/2 times as compared with five-phase machines, i.e.

$$Ni_\alpha = N[i_a + \cos 72^\circ \cdot i_b + \cos 144^\circ \cdot i_c + \cos 216^\circ \cdot i_d + \cos 288^\circ \cdot i_e]$$

$$= [i_a + (i_b + i_e) \cdot \cos 72^\circ + (i_c + i_d) \cdot \cos 144^\circ]$$

$$= I_m \begin{bmatrix} \cos(\omega t - \phi) + (\cos(\omega t - \phi - 72^\circ) + \cos(\omega t - \phi - 288^\circ)) \cdot \cos 72^\circ \\ + ((\cos(\omega t - \phi - 144^\circ) + \cos(\omega t - \phi - 216^\circ)) \cdot \cos 144^\circ \end{bmatrix}$$

$$= I_m \begin{bmatrix} \cos(\omega t - \phi) - (2 \cdot \cos(\omega t - \phi) \cdot \cos 108^\circ) \cdot \cos 72^\circ \\ -(2 \cdot \cos(\omega t - \phi) \cdot \cos 36^\circ) \cdot \cos 144^\circ \end{bmatrix}$$

$$= I_m \left[\cos(wt - \phi)(1 - [2 \cdot (\cos 108^\circ \cdot \cos 72^\circ) + (\cos 36^\circ \cdot \cos 144^\circ)] \right]$$

$$i_\alpha = \frac{5}{2} \cdot I_m \cos(\omega t - \phi)$$

From the mmf equation,

$$Ni_\beta = N[0 + \sin 72^\circ \cdot i_b + \sin 144^\circ \cdot i_c + \sin 216^\circ \cdot i_d + \sin 288^\circ \cdot i_e]$$

$$i_\beta = [(i_b - i_e) \cdot \sin 72^\circ + (i_c - i_d) \cdot \sin 144^\circ]$$

$$= I_m \begin{bmatrix} (\cos(\omega t - \phi - 72^\circ) - \cos(\omega t - \phi - 288^\circ)) \cdot \sin 72^\circ \\ + ((\cos(\omega t - \phi - 144^\circ) - \cos(\omega t - \phi - 216^\circ)) \cdot \sin 144^\circ \end{bmatrix}$$

$$= I_m \begin{bmatrix} (2 \cdot \sin(\omega t - \phi) \cdot \sin 108^\circ) \cdot \sin 72^\circ \\ + (2 \cdot \sin(\omega t - \phi) \cdot \sin 36^\circ) \cdot \sin 144^\circ \end{bmatrix}$$

$$= I_m\left[\left(2 \cdot \sin(\omega t - \phi)[(\sin 108^\circ \cdot \sin 72^\circ) + (\sin 36^\circ \cdot \sin 144^\circ)]\right]$$

$$i_\beta = \frac{5}{2} I_m \cdot \sin(\omega t - \phi)$$

Similarly, the second set of d-axis and q-axis parameters are

$$Ni_{\alpha 1} = N[i_a + \cos 144^\circ \cdot i_b + \cos 288^\circ \cdot i_c + \cos 432^\circ \cdot i_d + \cos 576^\circ \cdot i_e]$$

$$= [i_a + (i_b + i_e) \cdot \cos 144^\circ + (i_c + i_d) \cdot \cos 288^\circ]$$

$$= I_m \left[\begin{matrix} \cos(\omega t - \phi) + (\cos(\omega t - \phi - 72^\circ) + \cos(\omega t - \phi - 288^\circ)) \cdot \cos 144^\circ \\ + ((\cos(\omega t - \phi - 144^\circ) + \cos(\omega t - \phi - 216^\circ)) \cdot \cos 288^\circ \end{matrix} \right]$$

$$= I_m \begin{bmatrix} \cos(\omega t - \phi) - (2 \cdot \cos(\omega t - \phi) \cdot \cos 108^\circ) \cdot \cos 144^\circ \\ - (2 \cdot \cos(\omega t - \phi) \cdot \cos 36^\circ) \cdot \cos 288^\circ \end{bmatrix}$$

$$= I_m[\cos(\omega t - \phi)(1 - [2 \cdot (\cos 108^\circ \cdot \cos 144^\circ) + (\cos 36^\circ \cdot \cos 288^\circ)]]$$

$$i_x = 0$$

Similarly,

$$i_y = 0$$

$$i_0 = 0$$

The five-phase voltages of the five-phase induction machine are

$$V_a = V_m \cos(\omega t)$$
$$V_b = V_m \cos(\omega t - 72^\circ)$$
$$V_c = V_m \cos(\omega t - 144^\circ)$$
$$V_d = V_m \cos(\omega t - 216^\circ)$$
$$V_e = V_m \cos(\omega t - 288^\circ)$$

If the number of turns per phase, as well as mmf in both five-phase machine and two-phase equivalent machine, are equal, then the voltages of the two-phase machine will be same as five-phase machine. The two-phase machine voltages are

$$V_a = V_\alpha = V_m \cos(\omega t - \phi)$$
$$V_\beta = V_m \cos(\omega t - \phi - 90^\circ) = V_m \sin(\omega t - \phi)$$

Remaining all components are equals to zero:

$$V_x = V_y = V_0 = 0$$

From the above analysis, it is observed that:
For five-phase machine, voltage/phase = V and current/phase = I;
For two-phase machine, voltage/phase = V and current/phase = $5I/2$;
So the change in parameters $= \dfrac{\text{volatge/current for 2} - \text{phase machine}}{\text{volatge/current for 5} - \text{phase machine}} = \dfrac{V/\left(\frac{5}{2}I\right)}{V/I} = \dfrac{2}{5}$;
So the equivalent two-phase induction per-phase parameters are
Stator resistance: 0.3·2/5 = 0.12 Ω;
Rotor resistance:0.45·2/5 = 0.18 Ω;
Stator and rotor leakage reactance: 2.1·2/5 = 0.84 Ω;
Magnetizing reactance: 30·2/5 = 12 Ω.

Case (b)
If the turns per phase of the two-phase machine are 5/2 times of five-phase machine then:
For five-phase machine, voltage/phase = V and current/phase = I;
And for two-phase machine, voltage/phase = $5V/2$ and current/phase = I;
So the change in parameters $= \dfrac{\text{volatge/current for 2} - \text{phase machine}}{\text{volatge/current for 5} - \text{phase machine}} = \dfrac{(5V/2)/I}{V/I} = \dfrac{5}{2}$;
So the equivalent two-phase induction per-phase parameters are

Stator resistance: 0.3·5/2 = 0.75 Ω;
Rotor resistance: 0.45·5/2 = 1.125 Ω;
Stator and rotor leakage reactance: 2.1·5/2 = 5.25 Ω;
Magnetizing reactance: 30·5/2 = 75 Ω.

Case (c)
If the turns per phase of the two-phase machine are $\sqrt{5/2}$ times of three-phase machine then,
For five-phase machine, voltage/phase = V and current/phase = I;
And for two-phase machine, voltage/phase = $\sqrt{5/2}V$ and current/phase = $\sqrt{5/2}I$;
So the change in parameters $= \dfrac{\text{volatge/current for 2} - \text{phase machine}}{\text{volatge/current for 5} - \text{phase machine}} = 1$;

So the equivalent two-phase induction per-phase parameters are the same as a five-phase induction motor.

Problem 7.6 A five-phase two-pole induction machine has armature currents are as follows:

$$i_a = I_m \cos(\omega t)$$
$$i_b = I_m \cos(\omega t - 72^\circ)$$
$$i_c = I_m \cos(\omega t - 144^\circ)$$
$$i_d = I_m \cos(\omega t - 216^\circ)$$
$$i_e = I_m \cos(\omega t - 288^\circ)$$

At time $t = 0$, the rotor axis is aligned with d-axis. Find the two-phase equivalent d-axis and q-axis currents.

Solution.
The transformation matrix for the five-phase machine is

$$\begin{bmatrix} i_\alpha \\ i_\beta \\ i_x \\ i_y \\ i_0 \end{bmatrix} = \sqrt{\frac{2}{5}} \begin{bmatrix} \cos\theta & \cos\left(\theta - \frac{2\pi}{5}\right) & \cos\left(\theta - \frac{4\pi}{5}\right) & \cos\left(\theta - \frac{6\pi}{5}\right) & \cos\left(\theta - \frac{8\pi}{5}\right) \\ \sin\theta & \sin\left(\theta - \frac{2\pi}{5}\right) & \sin\left(\theta - \frac{4\pi}{5}\right) & \sin\left(\theta - \frac{6\pi}{5}\right) & \sin\left(\theta - \frac{8\pi}{5}\right) \\ \cos 2\theta & \cos 2\left(\theta - \frac{2\pi}{5}\right) & \cos 2\left(\theta - \frac{4\pi}{5}\right) & \cos 2\left(\theta - \frac{6\pi}{5}\right) & \cos 2\left(\theta - \frac{8\pi}{5}\right) \\ \sin 2\theta & \sin 2\left(\theta - \frac{2\pi}{5}\right) & \sin 2\left(\theta - \frac{4\pi}{5}\right) & \sin 2\left(\theta - \frac{6\pi}{5}\right) & \sin 2\left(\theta - \frac{8\pi}{5}\right) \\ \frac{1}{\sqrt{2}} & \frac{1}{\sqrt{2}} & \frac{1}{\sqrt{2}} & \frac{1}{\sqrt{2}} & \frac{1}{\sqrt{2}} \end{bmatrix} \begin{bmatrix} i_a \\ i_b \\ i_c \\ i_d \\ i_e \end{bmatrix}$$

substitute the current values in the transformation matrix, where $\theta = \omega t = 0$.
The d-axis and q-axis components are calculated from the transformation matrix as:

$$i_\alpha = \sqrt{\frac{2}{5}}[i_a + \cos 72^\circ \cdot i_b + \cos 144^\circ \cdot i_c + \cos 216^\circ \cdot i_d + \cos 288^\circ \cdot i_e]$$
$$= [i_a + (i_b + i_e) \cdot \cos 72^\circ + (i_c + i_d) \cdot \cos 144^\circ]$$
$$= I_m \begin{bmatrix} \cos(\omega t) + (\cos(\omega t - 72^\circ) + \cos(\omega t - 288^\circ)) \cdot \cos 72^\circ \\ + ((\cos(\omega t - 144^\circ) + \cos(\omega t - 216^\circ)) \cdot \cos 144^\circ \end{bmatrix}$$

$$= I_m \begin{bmatrix} \cos(\omega t) - (2 \cdot \cos(\omega t) \cdot \cos 108^\circ) \cdot \cos 72^\circ \\ -(2 \cdot \cos(\omega t) \cdot \cos 36^\circ) \cdot \cos 144^\circ \end{bmatrix}$$

$$= I_m \left[\cos(\omega t)(1 - [2 \cdot (\cos 108^\circ \cdot \cos 72^\circ) + (\cos 36^\circ \cdot \cos 144^\circ)] \right]$$

$$i_\alpha = \frac{5}{2} \cdot I_m \cos(\omega t) = \sqrt{\frac{5}{2}} \cdot I_m \quad (\because \omega t = 0)$$

Similarly,

$$
\begin{aligned}
i_\beta &= \sqrt{\frac{2}{5}}[0 - \sin 72^\circ \cdot i_b - \sin 144^\circ \cdot i_c - \sin 216^\circ \cdot i_d - \sin 288^\circ \cdot i_e] \\
&= \sqrt{\frac{2}{5}}[-(i_b - i_e) \cdot \sin 72^\circ - (i_c - i_d) \cdot \sin 144^\circ] \\
&= 0
\end{aligned}
$$

$$
\begin{aligned}
i_x &= \sqrt{\frac{2}{5}}[i_a + \cos 144^\circ \cdot i_b + \cos 288^\circ \cdot i_c + \cos 432^\circ \cdot i_d + \cos 576^\circ \cdot i_e] \\
&= \sqrt{\frac{2}{5}}[i_a + (i_b + i_e) \cdot \cos 144^\circ + (i_c + i_d) \cdot \cos 288^\circ] \\
&= 0
\end{aligned}
$$

$$
\begin{aligned}
i_y &= \sqrt{\frac{2}{5}}[0 - \sin 144^\circ \cdot i_b - \sin 288^\circ \cdot i_c - \sin 432^\circ \cdot i_d - \sin 576^\circ \cdot i_e] \\
&= \sqrt{\frac{2}{5}}[-(i_b - i_e) \cdot \sin 144^\circ - (i_c - i_d) \cdot \sin 288^\circ] \\
&= 0
\end{aligned}
$$

$$
i_0 = \sqrt{\frac{2}{5}} \cdot \sqrt{\frac{1}{2}}[i_a + i_b + i_c + i_d + i_e] = 0
$$

Problem 7.7 A 5-hp., 400-V, 50-Hz, five-phase two-pole star-connected induction machine is supplied with the balanced five-phase supply. Find the *d-q* axis and *x-y*-axis voltages under steady-state condition.

Solution.
The voltage equations of the five-phase are as follows:

$$
\begin{aligned}
&V_a = V_m \cos(\omega t) \\
&\text{Where } V_m = \frac{400}{\sqrt{5}} \cdot \sqrt{2} = 252.9822 \\
&V_a = 252.9822 \cdot \cos(\omega t) \\
&V_b = 252.9822 \cdot \cos(\omega t - 72^\circ) \\
&V_c = 252.9822 \cdot \cos(\omega t - 144^\circ) \\
&V_d = 252.9822 \cdot \cos(\omega t - 216^\circ) \\
&V_e = 252.9822 \cdot \cos(\omega t - 288^\circ)
\end{aligned}
$$

The transformation matrix for the five-phase machine is

$$\begin{bmatrix} V_\alpha \\ V_\beta \\ V_{\alpha 1} \\ V_{\beta 1} \\ V_0 \end{bmatrix} = \sqrt{\frac{2}{5}} \begin{bmatrix} 1 & 0.3090 & -0.8090 & -0.8090 & 0.3090 \\ 0 & 0.9511 & 0.5878 & -0.5878 & -0.9511 \\ 1 & -0.8090 & 0.3090 & 0.3090 & -0.8090 \\ 0 & 0.5878 & -0.9511 & 0.9511 & -0.5878 \\ 0.7071 & 0.7071 & 0.7071 & 0.7071 & 0.7071 \end{bmatrix} \begin{bmatrix} V_a \\ V_b \\ V_c \\ V_d \\ V_e \end{bmatrix}$$

The d-axis and q-axis components are calculated from the transformation matrix as:

$$V_\alpha = \sqrt{\frac{2}{5}}[V_a + 0.3090 \cdot V_b - 0.8090 \cdot V_c - 0.8090 \cdot V_d + 0.3090 \cdot V_e]$$

substitute all voltage equations in the above equations

$$\text{Finally } V_\alpha = \frac{5}{2} \cdot V_m \cos(\omega t) = \sqrt{\frac{5}{2}} \cdot V_m = \sqrt{\frac{5}{2}} \cdot 252.98 = 400V \qquad (\because \omega t = 0)$$

Similarly,

$$V_\beta = \sqrt{\frac{2}{5}}[0 \cdot V_a + 0.9511 \cdot V_b + 0.5878 \cdot V_c - 0.5878 \cdot V_d - 0.9511 \cdot V_e]$$

$$\text{Finally } V_\beta = 0$$

Remaining all components are equals to zero:

$$V_{\alpha 1} = V_{\beta 1} = V_0 = 0$$

7.3.2.6 PWM Mode of Operation of Five-Phase VSI

If a five-phase VSI is operated in PWM mode, apart from the already described 10 states, there will be an additional 22 switching states. This is because there are five inverter legs and each of them can assume two switching states, since the discussion is limited to two-level inverters. The number of states increases in the case of multilevel inverters. The number of possible switching states is in general equal to 2^n, where n is the number of inverter legs (i.e. output phases). This correlation is valid for any two-level VSI. For a multilevel inverter, with m level and n output phases, the number of possible switching states is m^n.

These 22 possible switching states encompass three possible situations: all the states when four switches from the upper (or lower) half and one from the lower (or upper) half of the inverter are "ON" (states 11–20); two states when either all the five upper (or lower) switches are "ON" (states 31 and 32); and the remaining states with three switches from the upper (lower) half and two switches from the lower (upper) half are in conduction mode (states 21–30). When all the upper or lower switches are "ON," they lead to zero output voltage and are termed as "zero states," similar to a three-phase VSI (Chapter 3). The remaining combination of switching states is termed "active states," as they lead to positive or negative output voltages. Phase-to-neutral voltages in these 22 switching states are listed in Table 7.6.

In order to introduce space vector representation of the five-phase inverter output voltages, an ideal sinusoidal five-phase supply source is considered. Let the phase voltages of a five-phase pure balanced sinusoidal supply be given with:

Table 7.6 Phase-to-neutral voltage for PWM operation of five-phase VSI

State	Switches ON	v_a	v_b	v_c	v_d	v_e
11	$S_1, S'_2, S'_3, S'_4, S'_5$ (11000)	$4/5\ V_{dc}$	$-1/5\ V_{dc}$	$-1/5\ V_{dc}$	$-1/5\ V_{dc}$	$-1/5\ V_{dc}$
12	S_1, S_2, S_3, S'_4, S_5 (11101)	$1/5\ V_{dc}$	$1/5\ V_{dc}$	$1/5\ V_{dc}$	$-4/5\ V_{dc}$	$1/5\ V_{dc}$
13	$S'_1, S_2, S'_3, S'_4, S'_5$ (01000)	$-1/5\ V_{dc}$	$4/5\ V_{dc}$	$-1/5\ V_{dc}$	$-1/5\ V_{dc}$	$-1/5\ V_{dc}$
14	S_1, S_2, S_3, S_4, S'_5 (11110)	$1/5\ V_{dc}$	$1/5\ V_{dc}$	$1/5\ V_{dc}$	$1/5\ V_{dc}$	$-4/5\ V_{dc}$
15	$S'_1, S'_2, S_3, S'_4, S'_5$ (00100)	$-1/5\ V_{dc}$	$-1/5\ V_{dc}$	$4/5\ V_{dc}$	$-1/5\ V_{dc}$	$-1/5\ V_{dc}$
16	S'_1, S_2, S_3, S_4, S_5 (01111)	$-4/5\ V_{dc}$	$1/5\ V_{dc}$	$1/5\ V_{dc}$	$1/5\ V_{dc}$	$1/5\ V_{dc}$
17	$S'_1, S'_2, S'_3, S_4, S'_5$ (00010)	$-1/5\ V_{dc}$	$-1/5\ V_{dc}$	$-1/5\ V_{dc}$	$4/5\ V_{dc}$	$-1/5\ V_{dc}$
18	S_1, S'_2, S_3, S_4, S_5 (10111)	$1/5\ V_{dc}$	$-4/5\ V_{dc}$	$1/5\ V_{dc}$	$1/5\ V_{dc}$	$1/5\ V_{dc}$
19	$S'_1, S'_2, S'_3, S'_4, S_5$ (00001)	$-1/5\ V_{dc}$	$-1/5\ V_{dc}$	$-1/5\ V_{dc}$	$-1/5\ V_{dc}$	$4/5\ V_{dc}$
20	S_1, S_2, S'_3, S_4, S_5 (11101)	$1/5\ V_{dc}$	$1/5\ V_{dc}$	$-4/5\ V_{dc}$	$1/5\ V_{dc}$	$1/5\ V_{dc}$
21	$S'_1, S_2, S'_3, S'_4, S_5$ (01001)	$-2/5\ V_{dc}$	$3/5\ V_{dc}$	$-2/5\ V_{dc}$	$-2/5\ V_{dc}$	$3/5\ V_{dc}$
22	$S_1, S_2, S'_3, S_4, S'_5$ (11010)	$2/5\ V_{dc}$	$2/5\ V_{dc}$	$-3/5\ V_{dc}$	$2/5\ V_{dc}$	$-3/5\ V_{dc}$
23	$S_1, S'_2, S_3, S'_4, S'_5$ (10100)	$3/5\ V_{dc}$	$-2/5\ V_{dc}$	$3/5\ V_{dc}$	$-2/5\ V_{dc}$	$-2/5\ V_{dc}$
24	$S'_1, S_2, S_3, S'_4, S_5$ (01101)	$-3/5\ V_{dc}$	$2/5\ V_{dc}$	$2/5\ V_{dc}$	$-3/5\ V_{dc}$	$2/5\ V_{dc}$
25	$S'_1, S_2, S'_3, S_4, S'_5$ (01010)	$-2/5\ V_{dc}$	$3/5\ V_{dc}$	$-2/5\ V_{dc}$	$3/5\ V_{dc}$	$-2/5\ V_{dc}$
26	$S_1, S'_2, S_3, S_4, S'_5$ (10110)	$3/5\ V_{dc}$	$-2/5\ V_{dc}$	$3/5\ V_{dc}$	$3/5\ V_{dc}$	$-2/5\ V_{dc}$
27	$S'_1, S'_2, S_3, S'_4, S_5$ (00101)	$-2/5\ V_{dc}$	$-2/5\ V_{dc}$	$3/5\ V_{dc}$	$-2/5\ V_{dc}$	$3/5\ V_{dc}$
28	$S'_1, S_2, S'_3, S_4, S_5$ (01011)	$-3/5\ V_{dc}$	$2/5\ V_{dc}$	$-3/5\ V_{dc}$	$2/5\ V_{dc}$	$2/5\ V_{dc}$
29	$S_1, S'_2, S'_3, S_4, S'_5$ (10010)	$3/5\ V_{dc}$	$-2/5\ V_{dc}$	$-2/5\ V_{dc}$	$3/5\ V_{dc}$	$-2/5\ V_{dc}$
30	$S_1, S'_2, S'_3, S_4, S'_5$ (10101)	$2/5\ V_{dc}$	$-3/5\ V_{dc}$	$2/5\ V_{dc}$	$-3/5\ V_{dc}$	$2/5\ V_{dc}$

$$\begin{aligned}
v_a &= \sqrt{2}V\cos(\omega t)\\
v_b &= \sqrt{2}V\cos(\omega t - 2\pi/5)\\
v_c &= \sqrt{2}V\cos(\omega t - 4\pi/5)\\
v_d &= \sqrt{2}V\cos(\omega t + 2\pi/5)\\
v_e &= \sqrt{2}V\cos(\omega t + 2\pi/5)
\end{aligned} \tag{7.41}$$

Since a five-phase system is under consideration, the vector needs to be analyzed in a five-dimensional space. Decoupling transformation leads to two orthogonal planes, namely *d-q* and *x-y*, and a zero-sequence component. The space vectors in the two orthogonal planes are defined as equation (7.42).

$$\begin{aligned}
\underline{v}_{\alpha-\beta} &= \frac{2}{5}\left(v_a + \underline{a}v_b + \underline{a}^2 v_b + \underline{a}^4 v_d + \underline{a}^6 v_e\right)\\
\underline{v}_{x-y} &= \frac{2}{5}\left(v_a + \underline{a}^2 v_b + \underline{a}^4 v_b + \underline{a}^8 v_d + \underline{a}^{12} v_e\right)
\end{aligned} \tag{7.42}$$

where $\underline{a}^n = \exp(j2n\pi/5)$.

The space vector is a complex quantity, which represents the five-phase balanced supply with a single complex variable. Substituting equation (7.42) into equation (7.41) yields an ideal sinusoidal source, the space vector:

$$\underline{v} = V\exp(j\omega t) \tag{7.43}$$

The space vector model of a five-phase voltage source inverter can be obtained by substituting the phase voltages in equation (7.42) and determining the corresponding space vectors in the α-β and *x-y* planes, as given in Table 7.7.

Thus, it can be seen that the total of 32 space vectors, available in the PWM operation, fall into four distinct categories (three active lengths and one zero length) regarding the magnitude of the available output phase voltage. The phase voltage space vectors are summarized in Table 7.7. The ratio of phase voltage space vector magnitudes is $1:1.618:(1.618)^2$, from the smallest to the largest, respectively. All these vectors form 10 sectors of 36 degrees each and one decagon. The mapping of the α-β plane vectors in the *x-y* plane is such that the largest length vectors of α-β become the smallest length vectors of the *x-y* plane and vice versa. The medium vectors remain the same in both planes. Moreover, the *x-y* plane vectors are the third harmonic of the fundamental, as shown in Figures 7.34 and 7.35. These *x-y* vectors produce losses in the system if remaining unattenuated. Thus, while developing PWM techniques, it is necessary to reduce or eliminate completely the *x-y* comments to yield sinusoidal output voltages.

Table 7.7 Space vector table of phase voltages for a five-phase VSI

S. No.	Switching states	Space vectors in α-β plane	Space vector in x-y plane
0	0 0 0 0 0	0	0
1	0 0 0 0 1	$2/5V_{dc}\exp(j8\pi/5)$	$2/5V_{dc}\exp(j6\pi/5)$
2	0 0 0 1 0	$2/5V_{dc}\exp(j6\pi/5)$	$2/5V_{dc}\exp(j2\pi/5)$
3	0 0 0 1 1	$2/5V_{dc}2\cos(\pi/5)\exp(j7\pi/5)$	$2/5V_{dc}2\cos(2\pi/5)\exp(j4\pi/5)$
4	0 0 1 0 0	$2/5V_{dc}\exp(j4\pi/5)$	$2/5V_{dc}\exp(j8\pi/5)$
5	0 0 1 0 1	$2/5V_{dc}2\cos(2\pi/5)\exp(j6\pi/5)$	$2/5V_{dc}2\cos(\pi/5)\exp(j7\pi/5)$
6	0 0 1 1 0	$2/5V_{dc}2\cos(\pi/5)\exp(j\pi)$	$2/5V_{dc}2\cos(2\pi/5)\exp(0)$
7	0 0 1 1 1	$2/5V_{dc}2\cos(\pi/5)\exp(j6\pi/5)$	$2/5V_{dc}2\cos(2\pi/5)\exp(j7\pi/5)$
8	0 1 0 0 0	$2/5V_{dc}\exp(j2\pi/5)$	$2/5V_{dc}\exp(j4\pi/5)$
9	0 1 0 0 1	$2/5V_{dc}2\cos(2\pi/5)\exp(0)$	$2/5V_{dc}2\cos(\pi/5)\exp(j\pi)$
10	0 1 0 1 0	$2/5V_{dc}2\cos(2\pi/5)\exp(j4\pi/5)$	$2/5V_{dc}2\cos(\pi/5)\exp(j3\pi/5)$
11	0 1 0 1 1	$2/5V_{dc}2\cos(2\pi/5)\exp(j7\pi/5)$	$2/5V_{dc}2\cos(\pi/5)\exp(j4\pi/5)$
12	0 1 1 0 0	$2/5V_{dc}2\cos(\pi/5)\exp(j3\pi/5)$	$2/5V_{dc}2\cos(2\pi/5)\exp(j6\pi/5)$
13	0 1 1 0 1	$2/5V_{dc}2\cos(2\pi/5)\exp(j3\pi/5)$	$2/5V_{dc}2\cos(\pi/5)\exp(j6\pi/5)$
14	0 1 1 1 0	$2/5V_{dc}2\cos(\pi/5)\exp(j4\pi/5)$	$2/5V_{dc}2\cos(2\pi/5)\exp(j3\pi/5)$
15	0 1 1 1 1	$2/5V_{dc}\exp(j\pi)$	$2/5V_{dc}\exp(j\pi)$
16	1 0 0 0 0	$2/5V_{dc}\exp(j0)$	$2/5V_{dc}\exp(j0)$
17	1 0 0 0 1	$2/5V_{dc}2\cos(\pi/5)\exp(j9\pi/5)$	$2/5V_{dc}2\cos(2\pi/5)\exp(j8\pi/5)$
18	1 0 0 1 0	$2/5V_{dc}2\cos(2\pi/5)\exp(j8\pi/5)$	$2/5V_{dc}2\cos(\pi/5)\exp(j\pi/5)$
19	1 0 0 1 1	$2/5V_{dc}2\cos(\pi/5)\exp(j7\pi/5)$	$2/5V_{dc}2\cos(2\pi/5)\exp(j\pi/5)$
20	1 0 1 0 0	$2/5V_{dc}2\cos(2\pi/5)\exp(j2\pi/5)$	$2/5V_{dc}2\cos(\pi/5)\exp(j9\pi/5)$
21	1 0 1 0 1	$2/5V_{dc}2\cos(2\pi/5)\exp(j9\pi/5)$	$2/5V_{dc}2\cos(\pi/5)\exp(j7\pi/5)$
22	1 0 1 1 0	$2/5V_{dc}2\cos(2\pi/5)\exp(j\pi)$	$2/5V_{dc}2\cos(\pi/5)\exp(j0)$
23	1 0 1 1 1	$2/5V_{dc}\exp(j7\pi/5)$	$2/5V_{dc}\exp(j9\pi/5)$
24	1 1 0 0 0	$2/5V_{dc}2\cos(\pi/5)\exp(j\pi/5)$	$2/5V_{dc}2\cos(2\pi/5)\exp(j2\pi/5)$
25	1 1 0 0 1	$2/5V_{dc}2\cos(\pi/5)\exp(j0)$	$2/5V_{dc}2\cos(2\pi/5)\exp(j\pi)$
26	1 1 0 1 0	$2/5V_{dc}2\cos(2\pi/5)\exp(j\pi/5)$	$2/5V_{dc}2\cos(\pi/5)\exp(j2\pi/5)$
27	1 1 0 1 1	$2/5V_{dc}\exp(j9\pi/5)$	$2/5V_{dc}\exp(j3\pi/5)$
28	1 1 1 0 0	$2/5V_{dc}2\cos(\pi/5)\exp(j2\pi/5)$	$2/5V_{dc}2\cos(2\pi/5)\exp(j9\pi/5)$
29	1 1 1 0 1	$2/5V_{dc}\exp(j\pi/5)$	$2/5V_{dc}\exp(j7\pi/5)$
30	1 1 1 1 0	$2/5V_{dc}\exp(j3\pi/5)$	$2/5V_{dc}\exp(j\pi/5)$
31	1 1 1 1 1	0	0

7.3.3 PWM Schemes of a Five-Phase VSI

This section describes PWM schemes for a five-phase voltage source inverter. PWM techniques are the most basic method of energy processing in a power converter. The purpose is to obtain variable voltage and variable frequency voltages/currents at the output of the inverter. The basic idea is to modulate the pulse widths in order to alter the mean value of the voltages/currents. Several PWM schemes are developed in the literature for a five-phase VSI [6–15]; however, this chapter focuses only on the simple approaches that are extension of three-phase PWM. The popular and simple PWM schemes are elaborated on in the next subsection, along with the MATLAB/Simulink models.

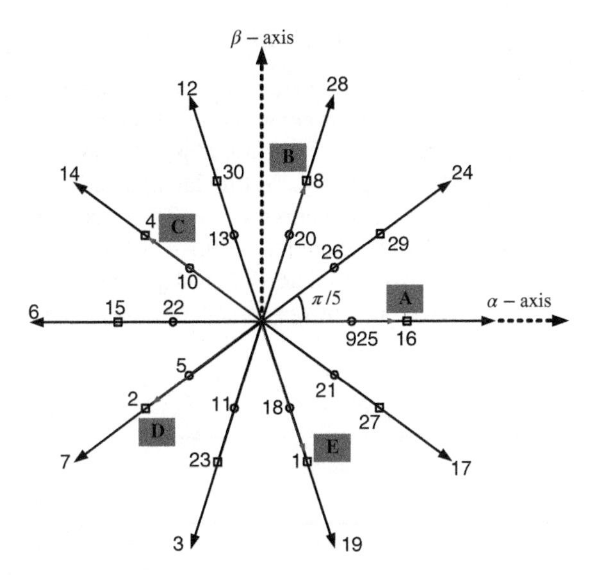

Figure 7.34 Phase voltage space vector in α-β plane

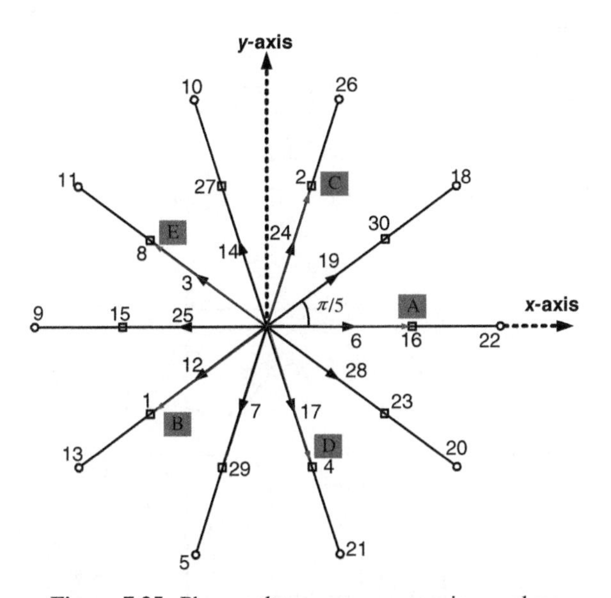

Figure 7.35 Phase voltage space vectors in x-y plane

7.3.3.1 Carrier-Based Sinusoidal PWM Scheme

Carrier-based sinusoidal PWM is the most popular and widely used PWM technique, because of its simple implementation in both analog and digital realization [57, 58]. The principle of carrier-based PWM true for a three-phase VSI is also applicable to a multiphase VSI.

The PWM signal is generated by comparing a sinusoidal modulating signal with a triangular (double-edge) or a saw-tooth (single-edge) carrier signal. The frequency of the carrier is normally kept much higher compared to the modulating signal (10–12 times). The operation of a carrier-based PWM modulator is shown in Figure 7.36 and generation of a PWM waveform is illustrated in Figure 7.37. The reference voltage signals or modulating signals are compared

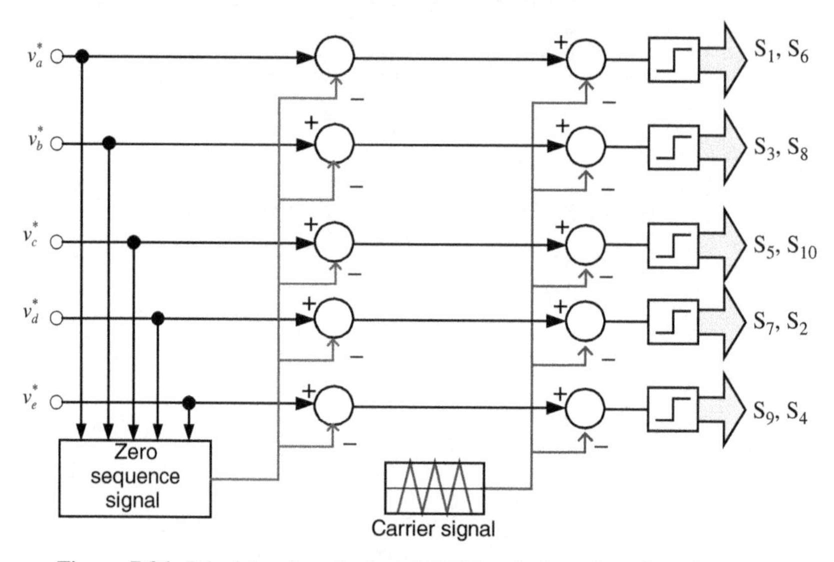

Figure 7.36 Principle of carrier-based PWM technique for a five-phase VSI

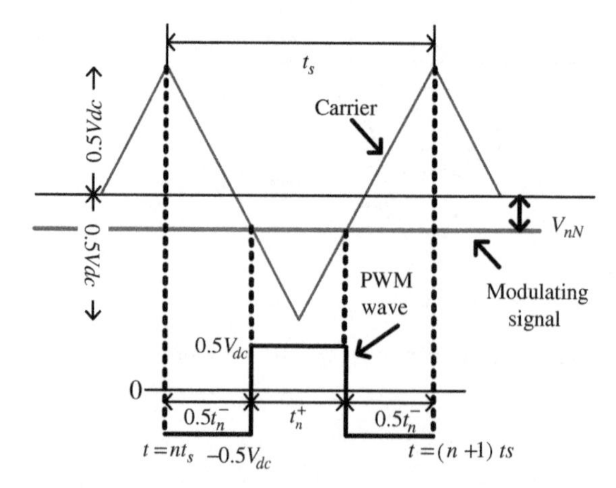

Figure 7.37 PWM waveform generation in carrier-based sinusoidal method

with the high-frequency carrier and the pulses are formed at the intersection of the modulation signals, which are five-phase fundamental sinusoidal signals (displaced in time by $\alpha = 2\pi/5$). These modulation signals are compared with a high-frequency carrier signal (saw-tooth or triangular shape) and all five switching functions for inverter legs are obtained directly. In general, modulation signal can be expressed as:

$$v_i(t) = v_i^*(t) + v_{nN}(t) \tag{7.44}$$

where $i = a, b, c, d, e$, and v_{nN} represents zero-sequence signal and v_i^* are fundamental sinusoidal signals. Zero-sequence signal represents a degree of freedom that exists in the structure of a carrier-based modulator and is used to modify modulation signal waveforms and thus to obtain different modulation schemes. Continuous PWM schemes result as long as the peak value of the modulation signal does not exceed the carrier magnitude. If the modulating signals peak exceeds the height of the carrier signals, it is termed a carrier dropping mode and leads to overmodulation.

The following relationship holds true in Figure 7.37:

$$t_n^+ - t_n^- = v_n t_s \tag{7.45a}$$

where

$$t_n^+ = \left(\frac{1}{2} + v_n\right) t_s \tag{7.45b}$$

$$t_n^- = \left(\frac{1}{2} - v_n\right) t_s \tag{7.45c}$$

where t_n^+ and t_n^- are the positive and negative pulse widths in the nth sampling interval, respectively, and v_n is the normalized amplitude of modulation signal. The normalization is done with respect to V_{dc}. Equation (7.45) is referred as the equal volt-second principle, as applied to a three-phase inverter [57, 58]. The normalized peak value of the triangular carrier wave is ±0.5 in linear region of operation. Modulator gain has the unity value while operating in the linear region and the peak value of inverter output fundamental voltage is equal to the peak value of the fundamental sinusoidal signal. Thus, the maximum output phase voltages from a five-phase VSI are limited to 0.5 p.u. Thus, the output phase voltage from a three-phase and a five-phase VSI are the same when using carrier-based PWM [59].

7.3.3.2 MATLAB/Simulink Simulation of Carrier-Based Sinusoidal PWM

Developing the Simulink model, assuming an ideal DC bus and ideal inverter model, is very simple. It follows exactly Figures 7.36 and 7.37. The inverter can either be modeled using the actual IGBT switches from the "sim-power system" blocksets or by using equation (7.23). The complete Simulink model is shown in Figure 7.38, where equation (7.23) is used to model the inverter.

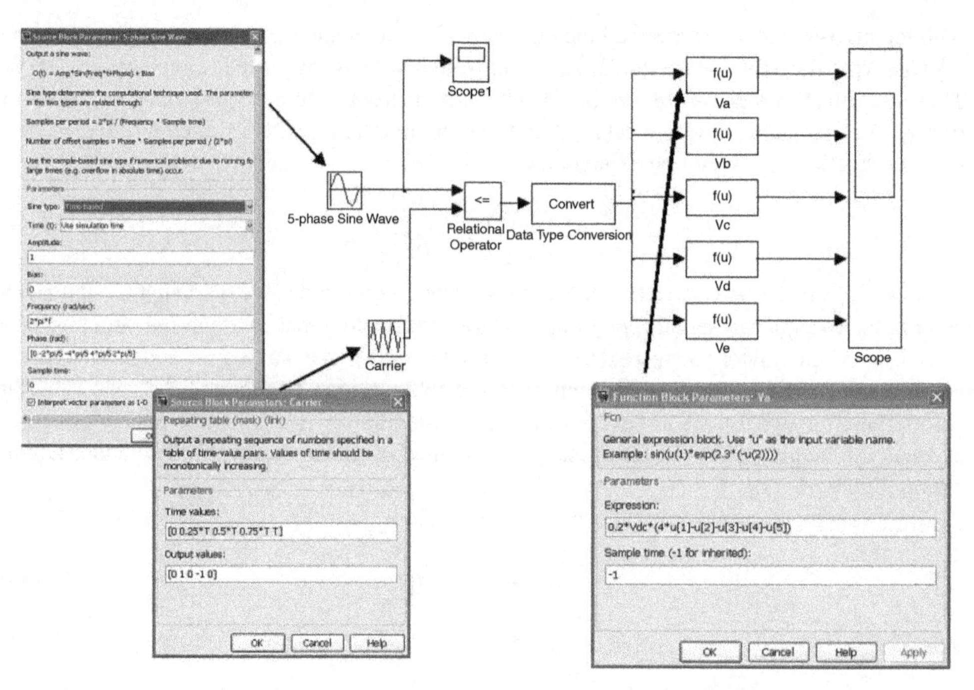

Figure 7.38 Simulink for implementing carrier-based PWM (*Source*: SIMULINK software)

The five-phase sinusoidal source is generated from the "sin wave" block from the "source" sublibrary, and the phase shifts are created by substituting appropriate phase differences in the "phase (rad)" dialog box. The triangular carrier signal is generated using the "repetitive" block from the "source" sublibrary. The switching frequency is kept at 5 kHz, the DC-link voltage is assumed as 1, and the fundamental output frequency is assumed as 50 Hz. The resulting waveforms are presented in Figures 7.36 and 7.39. The output voltage magnitude is limited to 0.5 p.u. (typically, it is the same value as that of three-phase VSI), as evident from Figure 7.39 (spectrum). The filtered output voltages are also depicted to show their sinusoidal nature.

7.3.3.3 Fifth-Harmonic Injection-Based Pulse-Width Modulation Scheme

The effect of addition of a harmonic with reverse polarity in any signal is to reduce the peak of the signal. The aim here is to bring the amplitude of the reference or modulating signal as low as possible, so that the reference can then be pushed to make it equal to the carrier, resulting in the higher output voltage and better DC bus utilization. Using this principle, third-harmonic injection PWM scheme is used in a three-phase VSI, which results in an increase in the fundamental output voltage to 0.575 V_{dc} (without harmonic injection the output is 0.5 V_{dc}) as discussed in Chapter 3. Third-harmonic voltages do not appear in the output phase voltages and are restricted to the leg voltages. Following the same principle, fifth-harmonic injection PWM scheme is used to increase the modulation index of a five-phase VSI [60] without affecting the quality of the output voltage waveform.

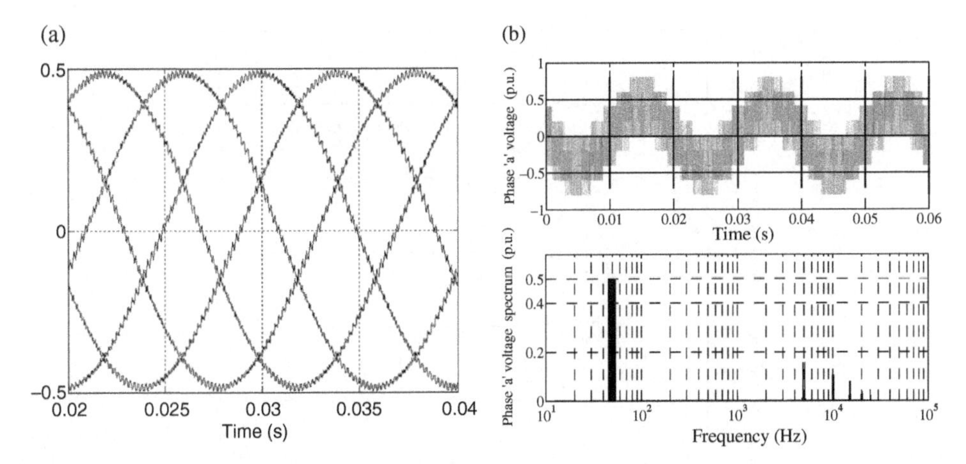

Figure 7.39 Simulation results of the carrier-based PWM scheme. (a) Filtered output voltage. (b) Harmonic spectrum phase "a" voltage

The reference leg voltages or modulating signals are given as:

$$\begin{aligned}
v_{ao}^{*} &= 0.5M_1V_{dc}\cos(\omega t) + 0.5M_5V_{dc}\cos(5\omega t)\\
v_{bo}^{*} &= 0.5M_1V_{dc}\cos(\omega t - 2\pi/5) + 0.5M_5V_{dc}\cos(5\omega t)\\
v_{co}^{*} &= 0.5M_1V_{dc}\cos(\omega t - 4\pi/5) + 0.5M_5V_{dc}\cos(5\omega t)\\
v_{do}^{*} &= 0.5M_1V_{dc}\cos(\omega t + 4\pi/5) + 0.5M_5V_{dc}\cos(5\omega t)\\
v_{eo}^{*} &= 0.5M_1V_{dc}\cos(\omega t + 2\pi/5) + 0.5M_5V_{dc}\cos(5\omega t)
\end{aligned} \tag{7.46}$$

Here, M_1 and M_5 are the peaks of the fundamental and fifth harmonic, respectively. It is to be noted that the fifth harmonic has no effect on the value of the reference waveform when $\omega t = (2k+1)\pi/10$, since $\cos(5(2k+1)\pi/10) = 0$ for all odd k. Thus, M_5 is chosen to make the peak magnitude of the reference of equation (7.46) occur where the fifth harmonic is zero. This ensures the maximum possible value of the fundamental component. The reference voltage reaches a maximum when

$$\frac{dV_{ao}^{*}}{dt} = -0.5M_1V_{dc}\sin\omega t - 0.5\cdot 5M_5V_{dc}\sin 5\omega t = 0 \tag{7.47}$$

This yields

$$M_5 = -M_1\frac{\sin(\pi/10)}{5}; \quad \text{for} \quad \omega t = \pi/10 \tag{7.48}$$

Thus, the maximum modulation index can be determined from

$$|v_{ao}^{*}| = \left|0.5M_1V_{dc}\cos(\omega t) - 0.5\frac{\sin(\pi/10)}{5}M_1V_{dc}\cos(3\omega t)\right| = 0.5V_{dc} \tag{7.49}$$

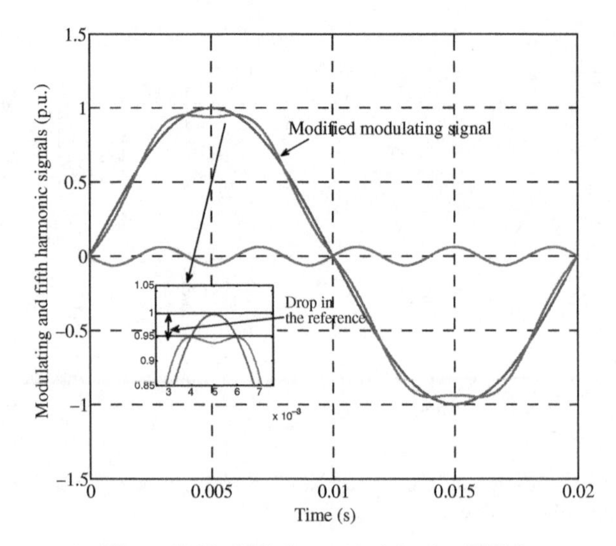

Figure 7.40 Fifth-harmonic injection PWM

The above equation gives

$$M_1 = \frac{1}{\cos(\pi/10)}; \quad \text{for} \quad \omega t = \pi/10 \tag{7.50}$$

Thus, the output fundamental voltage is increased to 5.15% higher than the value obtained using simple carrier-based PWM by injecting 6.18% fifth harmonic into the fundamental voltage. The output now reaches 0.5257 V_{dc}. The fifth harmonic is in opposite phase to that of the fundamental voltage. The effect of adding fifth harmonic in the sinusoidal reference is illustrated with the help of Figure 7.40. When fifth-harmonic = 0.0618* fundamental voltage is injected into the sinusoidal reference, the modified signal's peak reduces and its shape changes (Figure 7.40), providing more room for enhancing the reference and subsequently increasing the output of the inverter.

7.3.3.4 MATLAB/Simulink Simulation of Fifth-Harmonic Injection PWM

The Simulink model for this PWM scheme is similar to the carrier-based PWM, except the generation of the modulating signal. In the five-phase reference signals, a fifth harmonic of 0.0618 p.u. is subtracted forming the resultant modulating or reference signal. The modified modulating signals are then compared with the high-frequency carrier signal to generate the gating pulses, which are then given to the inverter model. The complete Simulink model is given in Figure 7.41. For the same simulation conditions as that of carrier-based PWM, the output PWM voltage waveforms look similar to those in Figure 7.42 and so are not shown here. The filtered inverter output voltages and the spectrum of phase "a" are presented in Figure 7.43 for the reference of 1.0515 p.u. The output voltage now increases to 0.5257 V_{dc} without going into pulse dropping mode.

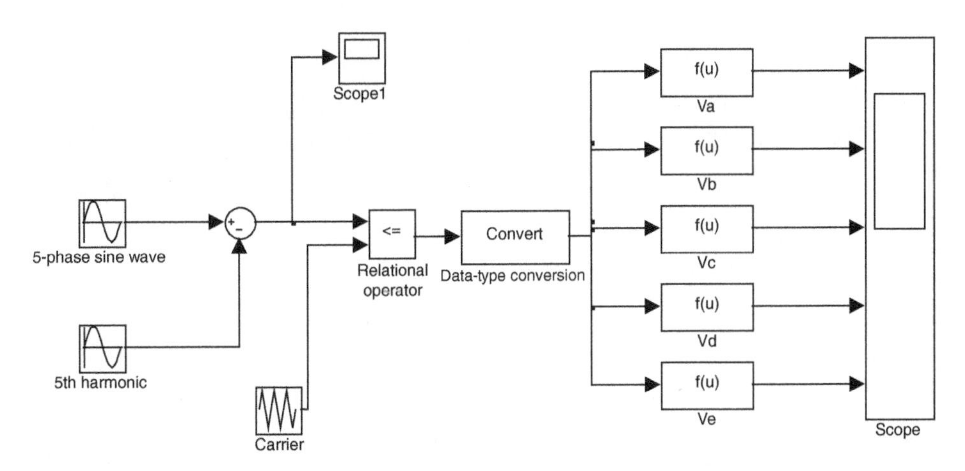

Figure 7.41 Simulink model for fifth-harmonic injection PWM scheme. (a) Filtered output voltage. (b) Harmonic spectrum phase "a" voltage

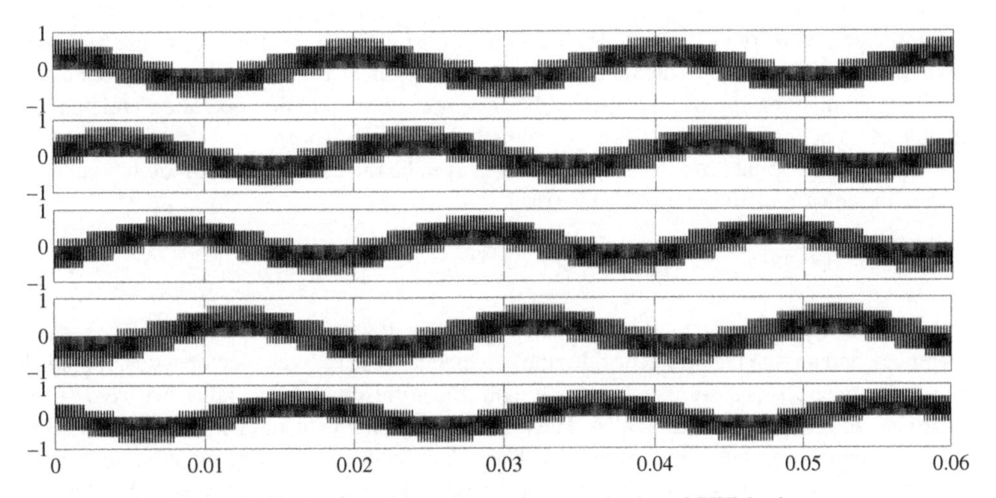

Figure 7.42 Output phase voltages from carrier-based PWM scheme

7.3.3.5 Offset Addition-Based Pulse-Width Modulation Scheme

Another way of increasing the modulation indices is to add an offset voltage to the references. The offset voltage addition is effectively adding $3n$ ($n = 1, 2, \ldots$) harmonic in a three-phase case and $5n$ ($n = 1, 2, \ldots$) in a five-phase case. This will effectively perform the same function as above. The offset voltage is given as:

$$V_{offset} = -\frac{V_{max} + V_{min}}{2} \tag{7.51}$$

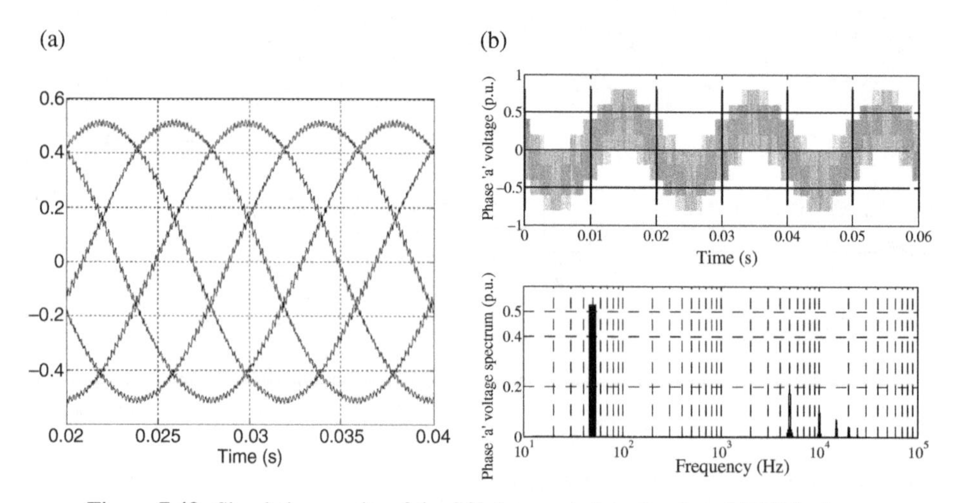

Figure 7.43 Simulation results of the fifth-harmonic injection-based PWM scheme

where $V_{max} = max(v_a, v_b, v_c, v_d, v_e)$ and $V_{min} = min(v_a, v_b, v_c, v_d, v_e)$. Note that this is the same as for a three-phase inverter. In a five-phase VSI, the offset is found as a fifth-harmonic triangular wave of magnitude 9.55% of the fundamental input reference. This peak value has been established by simulations approach. Offset addition requires only addition operation and hence is suitable for practical implementation in a digital signal processor.

A generalized formula of offset voltage, which is to be injected along with the fundamental voltage in the case of five-phase VSI, is [61]

$$V_{no} = -0.5528(V_{max} - V_{min}) + 3/5(1-2\mu)V_{DC}/2 - 3/5(1-2\mu)(V_{max} - V_{min}) \quad (7.52)$$

where V_{max} is the maximum of the five-phase references, V_{min} is the minimum of the five-phase references, and μ is the factor that decides the placement of the two zero vector states. If it is 0.5, then the two zero states are placed equally and this corresponds to symmetrical zero vector placement and conventional SVPWM. The modulating signal with and without offset addition and offset signals are shown in Figure 7.44. The reduction in the modified modulating signal is evident. The modulating signal can now be increased further to yield higher output voltage by the inverter.

The Simulink model is shown in Figure 7.45. The simulation results are not shown for the offset addition method, as nature remains the same as that of the fifth-harmonic injection method, except for improved switching harmonics.

7.3.3.6 Space Vector Pulse-Width Modulation Scheme (SVPWM)

SVPWM has become one of the most popular PWM techniques because of its easier digital implementation and better DC bus utilization, when compared to the carrier-based sinusoidal PWM method. The principle of SVPWM lies in the switching of the inverter in a special way, so as to apply a set of space vectors for a specific time. As seen in the previous section, a five-phase VSI yields 32 space vectors spanning over 360 degrees, forming a decagon with

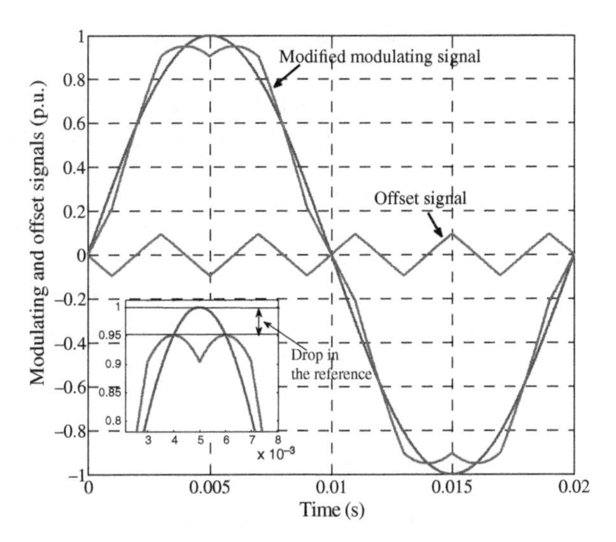

Figure 7.44 Offset addition-based PWM

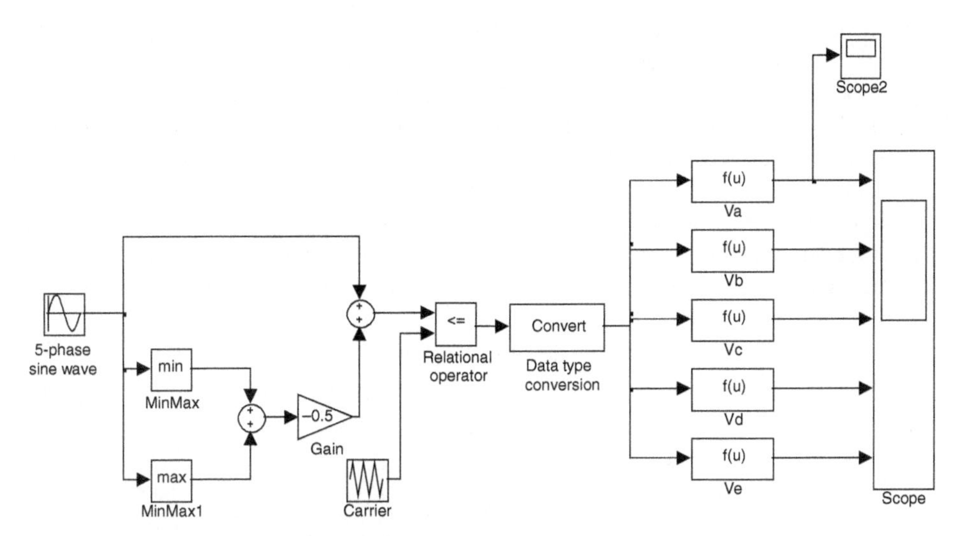

Figure 7.45 Simulink model of offset addition-based PWM

10 sectors of 36 degrees each. The reference voltage is synthesized by switching between the neighboring vectors, such that the volt-second balance is maintained. As seen in Chapter 3, two neighboring active vectors are employed to implement SVPWM, thus as an extension in a five-phase VSI, also two neighboring active space vector can be used. However, the next section shows that simple extension of three-phase SVPWM to five-phase leads to distortion in the output voltage. Hence, it is realized that instead of two, four neighboring vectors when used to implement SVPWM will lead to sinusoidal output voltages. As a rule of thumb,

$n-1$ (n phase number) numbers of active space vectors are needed to generate sinusoidal output in multiphase voltage source inverters.

Thus, there exists more than one method of implementing SVPWM in a multiphase voltage source inverter. Nevertheless, an ideal SVPWM of a five-phase inverter should satisfy a number of requirements. First, in order to keep the switching frequency constant, each switch can change state only twice in the switching period (once "ON" to "OFF" and once "OFF" to "ON," or vice versa). Second, the rms value of the fundamental phase voltage of the output must equal the rms of the reference space vector. Third, the scheme must provide full utilization of the available DC bus voltage. Finally, since the inverter is aimed at supplying the load with sinusoidal voltages, the low-order harmonic content needs to be minimized (this especially applies to the third and seventh harmonics). These criteria are used in assessing the merits and demerits of various SVPWMs. Two methods are elaborated here, one with two active space vectors and one with four active vectors.

In one case, two neighboring active space vectors and two zero-space vectors are used in one switching period to synthesize the input reference voltage. There are five legs in a five-phase inverter, each with two power switches whose operations are complementary. In one switching period, each power switch will change its state twice (from "OFF" to "ON" and then from "ON" to "OFF"); hence in total, 10 switchings take place in one switching period. The switching patterns are preformulated and stored in a lookup table. The switching is done in such a way that in the first switching half-period, the first zero vector is applied, followed by two active state vectors and then by the second zero state vector. The second switching half-period is the mirror image of the first. The symmetrical SVPWM is achieved in this way. This method is the simplest extension of space vector modulation of three-phase VSIs. The time of application of each active and zero-space vectors are obtained using the simple trigonometric relation considered in Figure 7.46.

Assuming the time of application of the right-hand side vector as t_a and the time of application of the left-hand side vector as t_b, and the time of application of zero-space vector is t_o where the total switching period is t_s. From Figure 7.46, the following relation holds good:

$$\underline{v}_s^* t_s = \underline{v}_a t_a + \underline{v}_b t_b \tag{7.53}$$

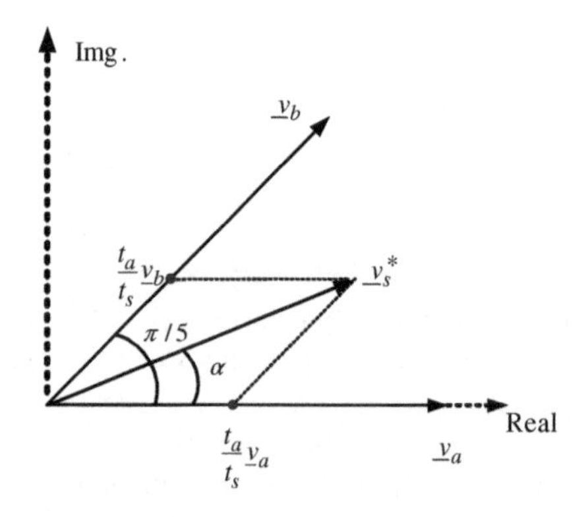

Figure 7.46 Principle of SVPWM for a five-phase VSI

In Cartesian form, this equation can be written as:

$$|\underline{v}_s^*|(\cos(\alpha) + j\sin(\alpha))t_s = |\underline{v}_a|(\cos(0) + j\sin(0))t_a + |\underline{v}_b|\left(\cos\left(\frac{\pi}{5}\right)\right) + j\sin\left(\frac{\pi}{5}\right)t_b \quad (7.54)$$

Now equating the real and imaginary parts, the following relations are obtained:

$$t_a = \frac{|\underline{v}_s^*|\sin(k\pi/5-\alpha)}{|\underline{v}_l|\sin(\pi/5)}t_s \quad (7.55a)$$

$$t_b = \frac{|\underline{v}_s^*|\sin(\alpha-(k-1)\pi/5)}{|\underline{v}_l|\sin(\pi/5)}t_s \quad (7.55b)$$

$$t_o = t_s - t_a - t_b \quad (7.55c)$$

Here, k is the sector number (k = 1–10) and large vector length is $|\underline{v}_{al}| = |\underline{v}_{bl}| = |\underline{v}_l| = \frac{2}{5}V_{dc}\,2\cos(\pi/5)$. The corresponding medium vector length, which will be needed in the subsequent expression, is $|\underline{v}_{am}| = |\underline{v}_{bm}| = V_m = (2/5)V_{dc}$. Symbol $\underline{v}_s^*$ denotes the reference space vector, while $|\underline{x}|$ is the modulus of a complex number $\underline{x}$. The largest possible fundamental peak voltage magnitude that may be achieved using this scheme corresponds to the radius of the largest circle that can be inscribed within the decagon. The circle is tangential to the midpoint of the lines connecting the ends of the active space vectors, as shown in Figure 7.47. The trajectory of the output will follow the outermost circle if operating in the overmodulation region (not discussed here), and the trajectory will follow the inner circle when operating in maximum linear modulation. The trajectory will be along the decagon for the 10-step operation of the inverter. Thus, the maximum fundamental peak output voltage V_{max} is $V_{max} = |\underline{v}_l|\cos(\pi/10)$, and $V_{max} = (2/5)\,2\cos(\pi/5)\cos(\pi/10)V_{dc} = 0.61554V_{dc}$. The maximum peak fundamental output in the 10-step

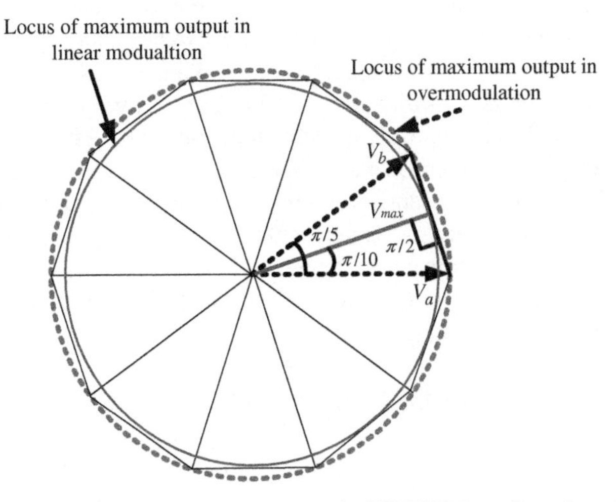

Figure 7.47 Maximum possible output in SVPWM for a five-phase VSI

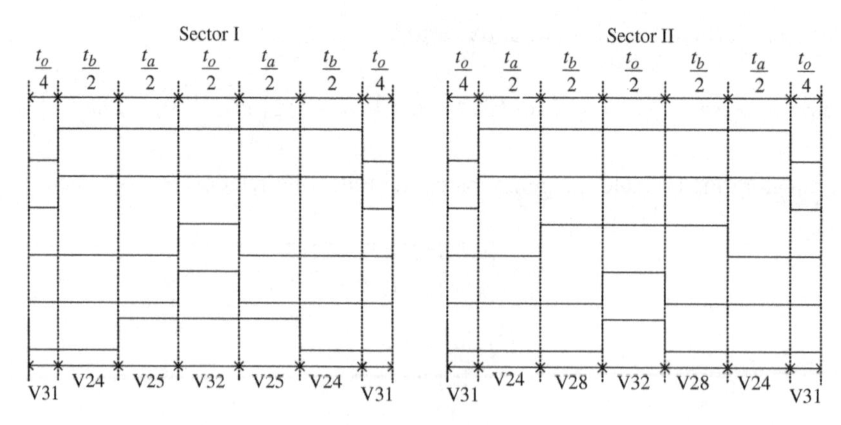

Figure 7.48 Switching pattern for SVPWM using only two active vectors

mode is, from equation (7.29), given with $V_{max,10step} = \frac{2}{\pi}V_{dc}$. Thus, the ratio of the maximum possible fundamental output voltage with SVPWM and in the 10-step mode is $V_{max}/V_{max,10step} = 0.96689$.

The sequence of vectors applied in sectors I and II and corresponding switching patterns are shown in Figure 7.48, where states of five inverter legs take values of –1/2 and 1/2 (referencing to the midpoint of the DC supply is applied) and the five traces illustrate, from top to bottom, legs *A*, *B*, *C*, *D*, and *E*, respectively. In odd sector numbers, the left-hand side (with respect to the reference) vector is applied first, followed by the right-hand side vector; while in even sectors, the right-hand side vector is applied first, followed by the left-hand side vector. For the implementation of this scheme, these switching patterns are stored in a lookup table.

The above method leads to unwanted low-order harmonics in the output phase voltages of the inverter, as shown in the next section. The reason is the free flow of *x-y* components. Application of two adjacent medium active space vectors, together with two large active space vectors in each switching period, makes it possible to maintain zero average values in the second plane (*x-y*) and consequently providing sinusoidal output. When using two large length and two medium length active vectors, their mapping in the *x-y* plane is such that they cancel out each other, as illustrated in Figure 7.49 for sector I and it follows for the rest of the sectors. Since vector numbers 16 and 25 are opposite to each other, and so are 24 and 29, their lengths are different (ratio of lengths of larger to smaller is 1.618). Thus, if the time of application of the smaller vector is increased in the same proportion, they will have equal volt-second and will cancel each other, eliminating the *x-y* components and generating sinusoidal output.

Use of four active space vectors per switching period requires the calculation of four application times, labeled here as t_{al}, t_{bl}, t_{am}, and t_{bm}. The expressions used for calculation of dwell times of various space vectors are [9]

$$\underline{v}_s^* t_s = \underline{v}_{al} t_{al} + \underline{v}_{bl} t_{bl} + \underline{v}_{am} t_{am} + \underline{v}_{bm} t_{bm}, \tag{7.56}$$

where

$$|\underline{v}_{al}| = |\underline{v}_{bl}| = |\underline{v}_l| = \frac{2}{5} V_{dc} 2\cos(\pi/5) \tag{7.57a}$$

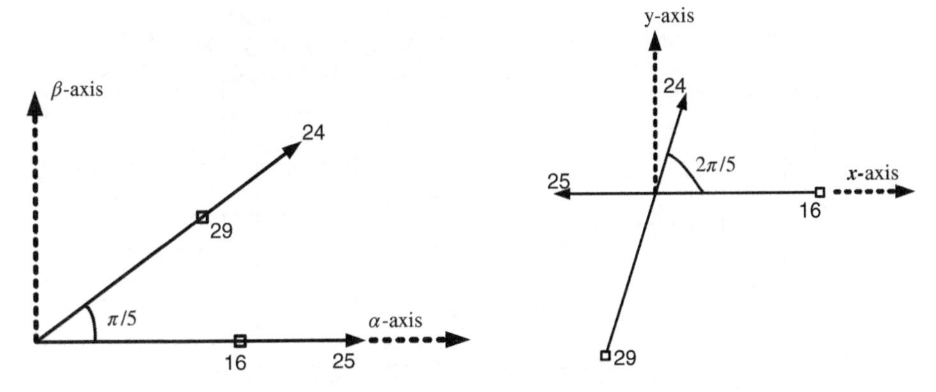

Figure 7.49 Principle of SVPWM with four active vectors

$$|\underline{v}_{am}| = |\underline{v}_{bm}| = |\underline{v}_{m}| = V_m = \frac{2}{5}V_{dc} \tag{7.57b}$$

and

$$\frac{t_{al}}{t_{am}} = \frac{t_{bl}}{t_{bm}} = \frac{|\underline{v}_l|}{|\underline{v}_m|} = \tau = 1.618 \tag{7.57c}$$

Separating the real and imaginary parts of equation (7.56) and substituting equation (7.57a), the following equations result

$$t_{al} = \frac{|v_s^*|}{V_m \sin(\pi/5)}\left(\frac{\tau}{1+\tau^2}\right)t_s \sin\left(\frac{\pi}{5}k - \alpha\right) \tag{7.58a}$$

$$t_{bl} = \frac{|v_s^*|}{V_m \sin(\pi/5)}\left(\frac{\tau}{1+\tau^2}\right)t_s \sin\left(\alpha - (k-1)\frac{\pi}{5}\right) \tag{7.58b}$$

$$t_{am} = \frac{|v_s^*|}{V_m \sin(\pi/5)}\left(\frac{\tau}{1+\tau^2}\right)t_s \sin\left(\frac{\pi}{5}k - \alpha\right) \tag{7.58c}$$

$$t_{bm} = \frac{|v_s^*|}{V_m \sin(\pi/5)}\left(\frac{\tau}{1+\tau^2}\right)t_s \sin\left(\alpha - (k-1)\frac{\pi}{5}\right) \tag{7.58d}$$

$$t_0 = t_s - t_{al} - t_{bl} - t_{am} - t_{bm} \tag{7.58e}$$

where $t_a = t_{al} + t_{am}$; $t_b = t_{bl} + t_{bm}$. This allocates 61.8% more dwell time to a large space vector compared to a medium space vector, thus satisfying the constraints of producing zero average voltage in the x-y plane. The constraint imposed on the time of application of each vector is that they cannot be less than zero, and also the sum of time of application of active and zero vectors cannot exceed the switching time. With these constraints, the maximum possible output with this approach is 0.5257 V_{dc}, which is almost 16% less than with the previous method (method

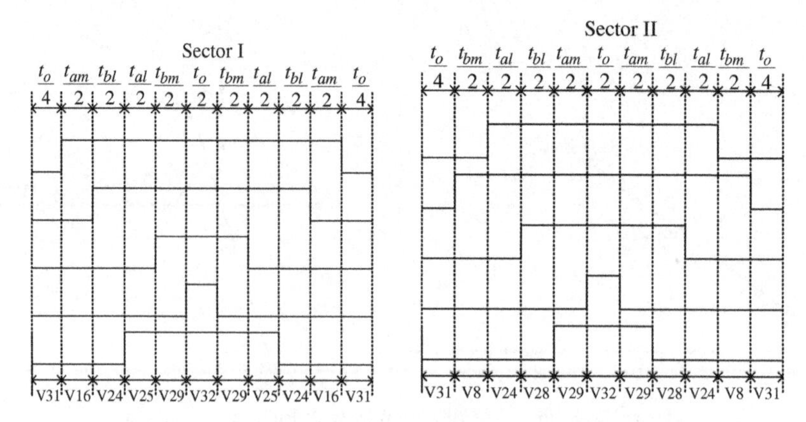

Figure 7.50 Switching pattern for SVPWM using four active vectors

using only two neighboring active space vectors). The switching patterns for the two sectors are shown in Figure 7.50. The switching pattern is a symmetrical PWM with two commutations per each inverter leg. The space vectors are applied in odd sectors using the sequence [$\underline{v}_{31}$, $\underline{v}_{ah}$, $\underline{v}_{bm}$, $\underline{v}_{am}$, $\underline{v}_{bl}$, $\underline{v}_{32}$, $\underline{v}_{bl}$, $\underline{v}_{am}$, $\underline{v}_{bm}$, $\underline{v}_{al}$, $\underline{v}_{31}$], while the sequence is [$\underline{v}_{31}$, $\underline{v}_{bl}$, $\underline{v}_{am}$, $\underline{v}_{bm}$, $\underline{v}_{ah}$ $\underline{v}_{32}$, $\underline{v}_{al}$, $\underline{v}_{bm}$, $\underline{v}_{am}$, $\underline{v}_{bl}$, $\underline{v}_{31}$] in even sectors.

7.3.3.7 MATLAB/Simulink Model of SVPWM

The SVPWM can be developed in MATLAB/Simulink using different approaches. The model presented here uses the Simulink and MATLAB function block. The MATLAB/Simulink model is presented in Figure 7.51. The five-phase voltage reference is generated using the "sine wave" generator and then transformed into α-β and x-y components. The reference space vector magnitude and angle are thus generated from the "reference voltage generator" block. The magnitude is constant, while the angle changes from 0 to pi and then −pi to 0 as a saw-tooth waveform. The angle is held for one sample time, so as to keep its value constant during the calculation of times of application of vectors. Another repetitive signal (time [0 Ts], amplitude [0 Ts]) is used to compare with the angle to determine the location (sector) of the reference

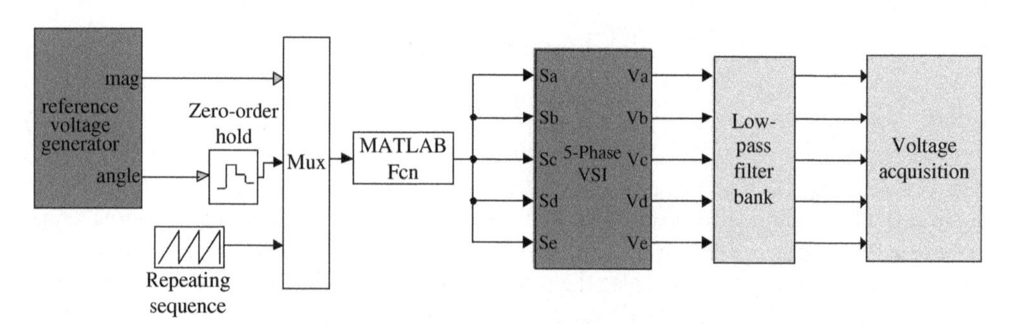

Figure 7.51 MATLAB/Simulink model for implementing SVPWM

signal. The MATLAB function block contains the switching table and sector determination algorithm (given in the soft copy). The MATLAB function block outputs the switching functions and is given to the five-phase inverter model. The algorithm inside the MATLAB function block can be changed to implement two-vector or four-vector SVPWM. The inverter produces a space vector-modulated voltage waveform that can be given to the five-phase load.

Simulation results are shown for two SVPWM techniques, using two active vectors (Figure 7.52) and using four active vectors (Figure 7.53). The fundamental frequency is kept at 50 Hz, the switching frequency is chosen as 5 kHz, the output is set at maximum value (0.6115 V_{dc} for two vectors case and 0.5257 V_{dc} for four vectors case), and the DC-link voltage V_{dc} is set to 1 p.u. The output phase voltages are distorted if only two active vectors are

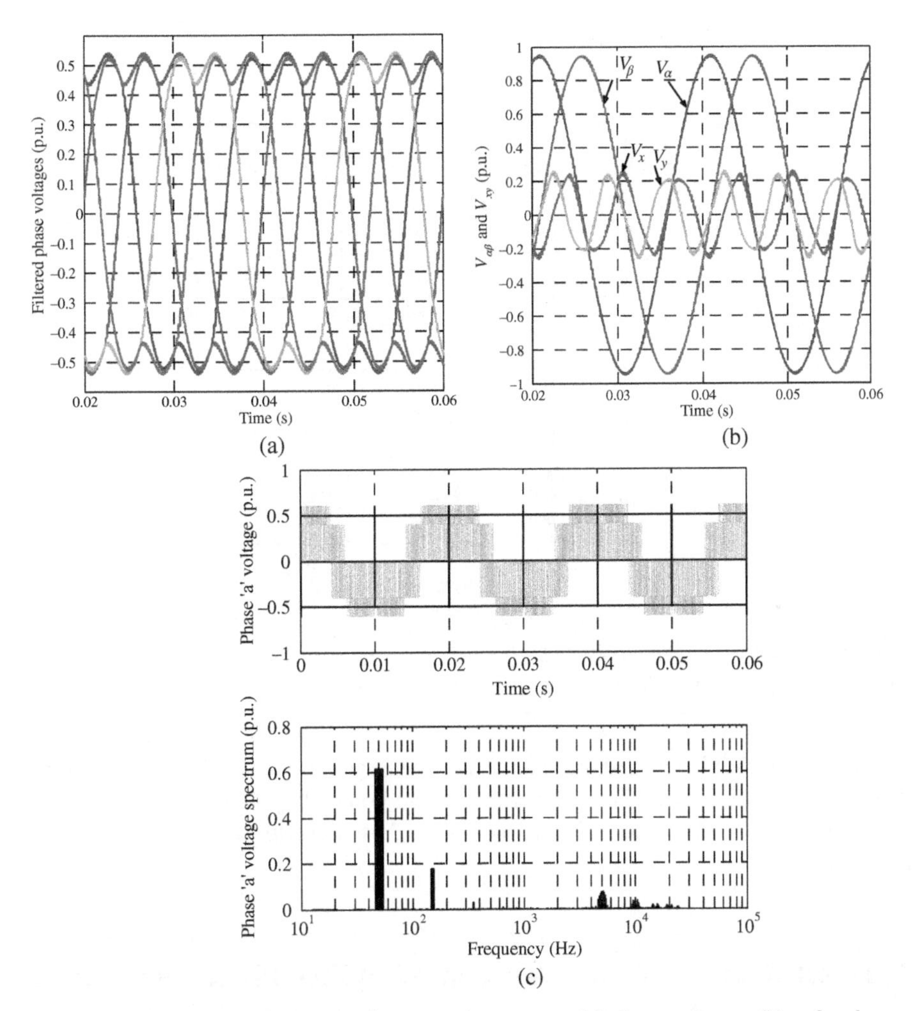

Figure 7.52 SVPWM simulation results for two active vectors: (a) phase voltages; (b) α-β and x-y axes voltages; and (c) phase “a” voltage and its spectrum

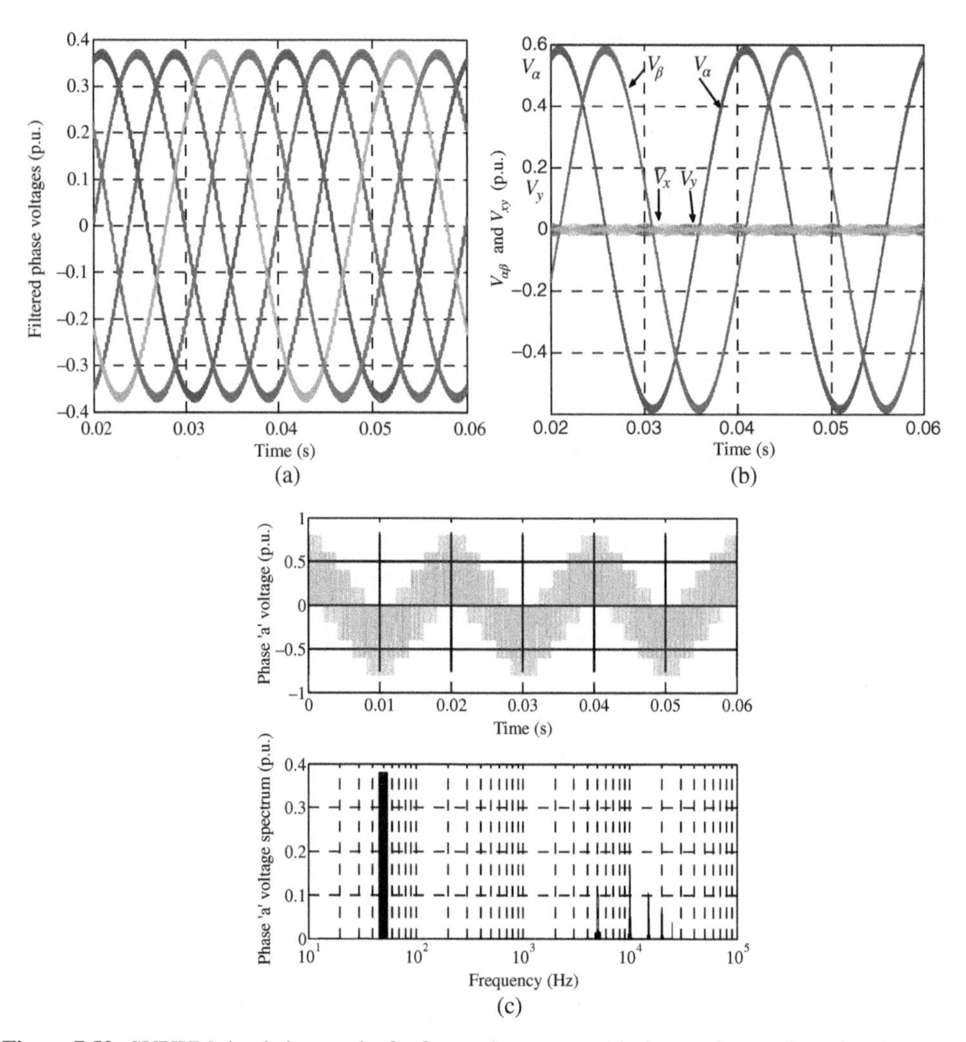

Figure 7.53 SVPWM simulation results for four active vectors: (a) phase voltages; (b) α-β and x-y axes voltages; and (c) phase "a" voltage and its spectrum

employed for SVPWM. The distortion appears due to the presence of the x-y plane vectors, as evident from Figure 7.52b. The phase voltages are completely sinusoidal when using four active vectors for the implementation of SVPWM. This is due to the fact that the x-y components are completely eliminated.

7.4 Direct Rotor Field-Oriented Control of Five-Phase Induction Motor

Field-oriented control refers to the control technique in which an induction machine emulates a separately excited DC machine. The torque and flux producing currents are decoupled and controlled independently, giving rise to fast dynamics of machine. This control technique is also called vector control, since the stator current space vector is controlled in this method. The detail description of field orientation and the vector control principle is discussed in Chapter 4. The purpose here is to distinguish between the control of a three-phase and a five-phase machine. Essentially, only two current components are required to produce torque in an induction machine, regardless of the phase number. In general, the *d*-axis current control of the flux in the machine is maintained constant for speed control in the base speed region and is reduced in appropriate proportion in the field-weakening region (speed above the nominal or rated value). The *q*-axis current component controls the torque in the machine, since the *d*-axis current is kept constant (see equation (7.17)).

Since only two current components are needed for vector control, its principle remains the same for a multiphase (more than three) machine as that of a three-phase machine. The only difference lies in the transformation of the current from two (α-β) to three (i_a, i_b, and i_c) in the case of a three-phase machine and two (α-β) to five (i_a, i_b, i_c, i_d, and i_e) in a five-phase machine, as illustrated in Figure 7.54 for speed control in the base speed region. The commanded speed and the actual speed (obtained from the speed sensor or speed observer in sensorless mode) are compared to determine the speed error, which is processed in a PI controller to determine the commanded torque. The output of the PI controller (commanded torque) is limited to rated torque or sometimes twice the rated torque (for fast acceleration). The commanded torque when multiplied by constant K_1. gives the *q*-axis current component. This current component is multiplied by constant K_2 to obtain slip speed. The expression for constants K_1 and K_2 are given as:

$$K_1 = \frac{1}{P}\frac{L_r}{L_m}\frac{1}{\psi_r^*} = \frac{1}{P}\frac{L_r}{L_m^2}\frac{1}{i_{ds}^*} = 0.431 \quad K_2 = \omega_{sl}^*/i_{qs}^* = \frac{L_m}{T_r\psi_r^*} = \frac{1}{T_r i_{ds}^*} = 4.527 \tag{7.59}$$

Slip speed is integrated to obtain the position and then added to the rotor position, this gives the rotor flux position ϕ_r. The letter "*P*" in Figure 7.54 represents the number of pole pairs of the motor and θ gives the mechanical rotor position. This is multiplied by the number of pole

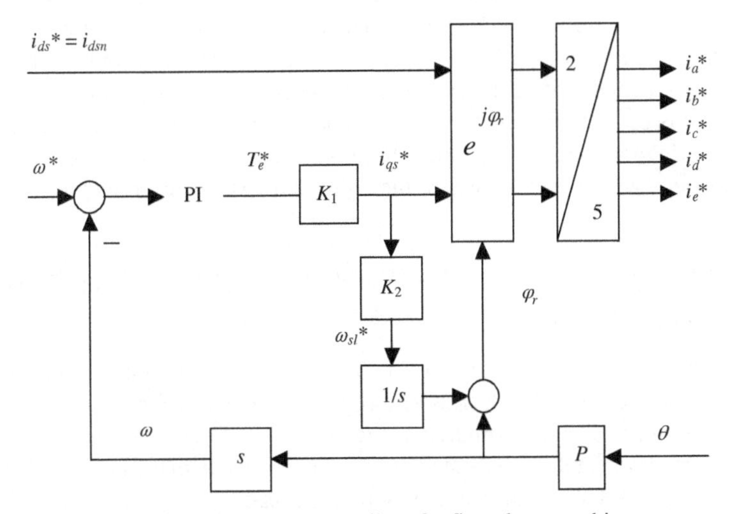

Figure 7.54 Vector controller of a five-phase machine

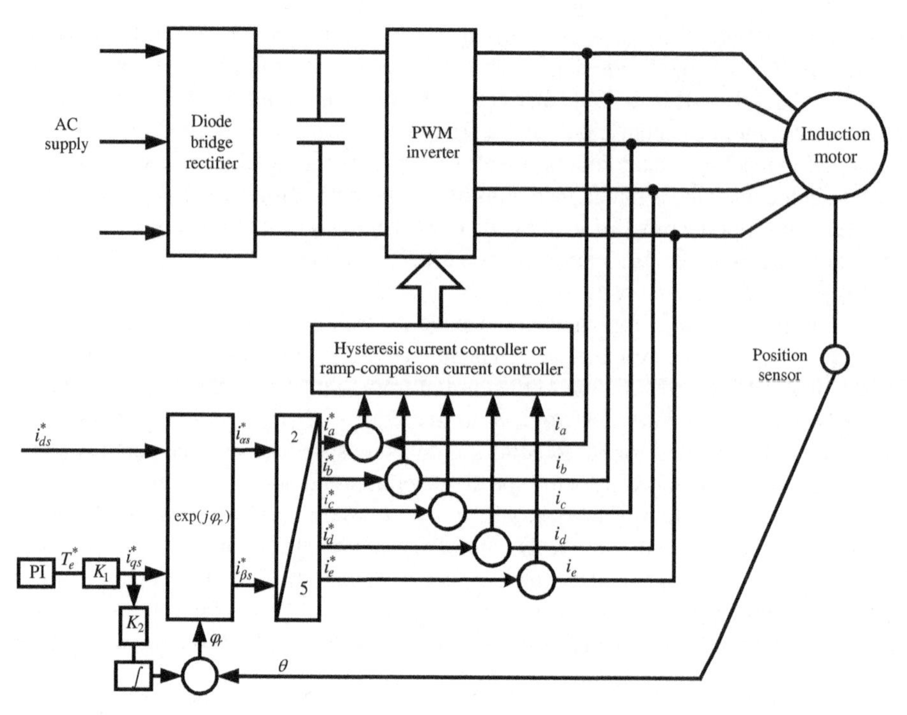

Figure 7.55 Block schematic of indirect rotor field-oriented control of a five-phase induction machine

pairs of the machine to obtain an electrical rotor position. The vector rotator block j^{ϕ_r} transforms the d-q-axis current to the stationary α-β-axis current, which is finally transformed into five-phase currents. Thus, generated are the reference five-phase currents and are reconstructed by a current controlled five-phase PWM voltage source inverter. The current control can be performed using a simple hysteresis current controller or using a ramp comparison current controller. A block schematic of an indirect rotor field oriented control (FOC) of a five-phase induction motor is given in Figure 7.55. In the block diagram, the torque reference is applied if the machine is in torque-controlled mode. In case of speed control mode, the reference torque is generated from the PI controller where the input is the speed error signal.

7.4.1 MATLAB/Simulink Model of Field-Oriented Control of Five-Phase Induction Machine

MATLAB/Simulink model of a five-phase induction machine in indirect field-oriented controlled mode for speed control in base speed region is shown in Figures 7.56 and 7.57. The vector controller block takes speed reference, the rotor flux reference, and actual speed as input; the rotor flux is kept constant in base speed region and is reduced in the field-weakening region. The block shows only the actual speed as input, but inside the block, reference speed and

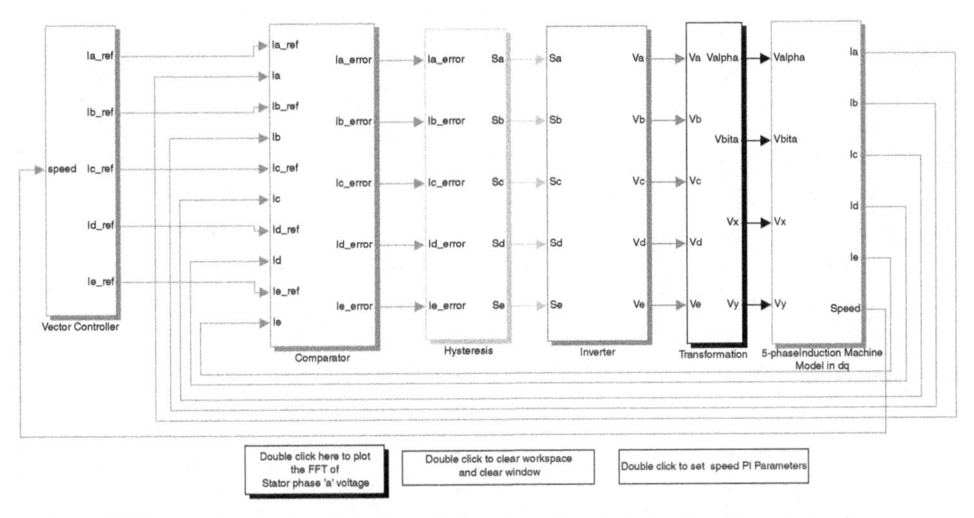

Figure 7.56 Simulink model for indirect field-oriented control of a five-phase induction motor.

reference rotor flux are provided. The vector controller generates the stator current references, which are compared with the actual stator current in the comparator block. The current errors thus generated are given to the hysteresis block that produces the gating signals for five-phase voltage source inverter. The phase voltages produced by the inverter is transformed using equation (7.10) (with $\theta_s = 0$, since the motor model is in the stationary reference frame) into α-β and x-y components and is given as the input to the five-phase induction motor model.

The resulting waveforms for the five-phase vector controlled drive system are shown in Figure 7.58. Per-phase equivalent circuit parameters of the 50-Hz, five-phase induction machine are $R_s = 10\,\Omega$, $R_r = 6.3\,\Omega$. $L_{ls} = L_{lr} = 0.04$ H, and $L_m = 0.42$ H. Inertia and the number of pole pairs are equal to 0.03 kg/m^2 and 2, respectively. Rated phase current, phase-to-neutral voltage, and per-phase torque are 2.1 A, 220 V, and 1.67 Nm, respectively. Rated (rms) rotor flux is 0.5683 Wb. Parameters of the speed PI controller are obtained using standard tuning procedure. The hysteresis band is set to ±2.5% of the rated peak phase current (i.e. ±0.07425 A). The torque limit is at all times equal to twice the rated motor torque (i.e. 16.67 Nm). DC-link voltage equals 587 V ($=\sqrt{2}\times 415$ V) and provides approximately 10% voltage reserve at rated frequency. The constants K_1 and K_2 of the vector controller are 0.431 and 4.527, respectively. Rotor flux reference (i.e. stator d-axis current reference) is ramped from $t = 0$ to $t = 0.01$ seconds to twice the rated value. In order to obtain a forced excitation, it is kept at twice the rated value from $t = 0.01$ to $t = 0.05$ seconds. Next, it is reduced from twice the rated value to the rated value in a linear fashion from $t = 0.05$ to $t = 0.06$ seconds and it is kept constant for the rest of the simulation period. Speed command of 1200 rpm is applied at $t = 0.3$ seconds in a ramp-wise manner from $t = 0.3$ to $t = 0.35$ seconds and is further kept unchanged. Operation takes place under no-load conditions initially, followed by loading and then reversing. Reference and actual rotor flux, reference and actual torque, speed and stator phase "a" reference, and actual current are shown in Figure 7.57. After the initial transient

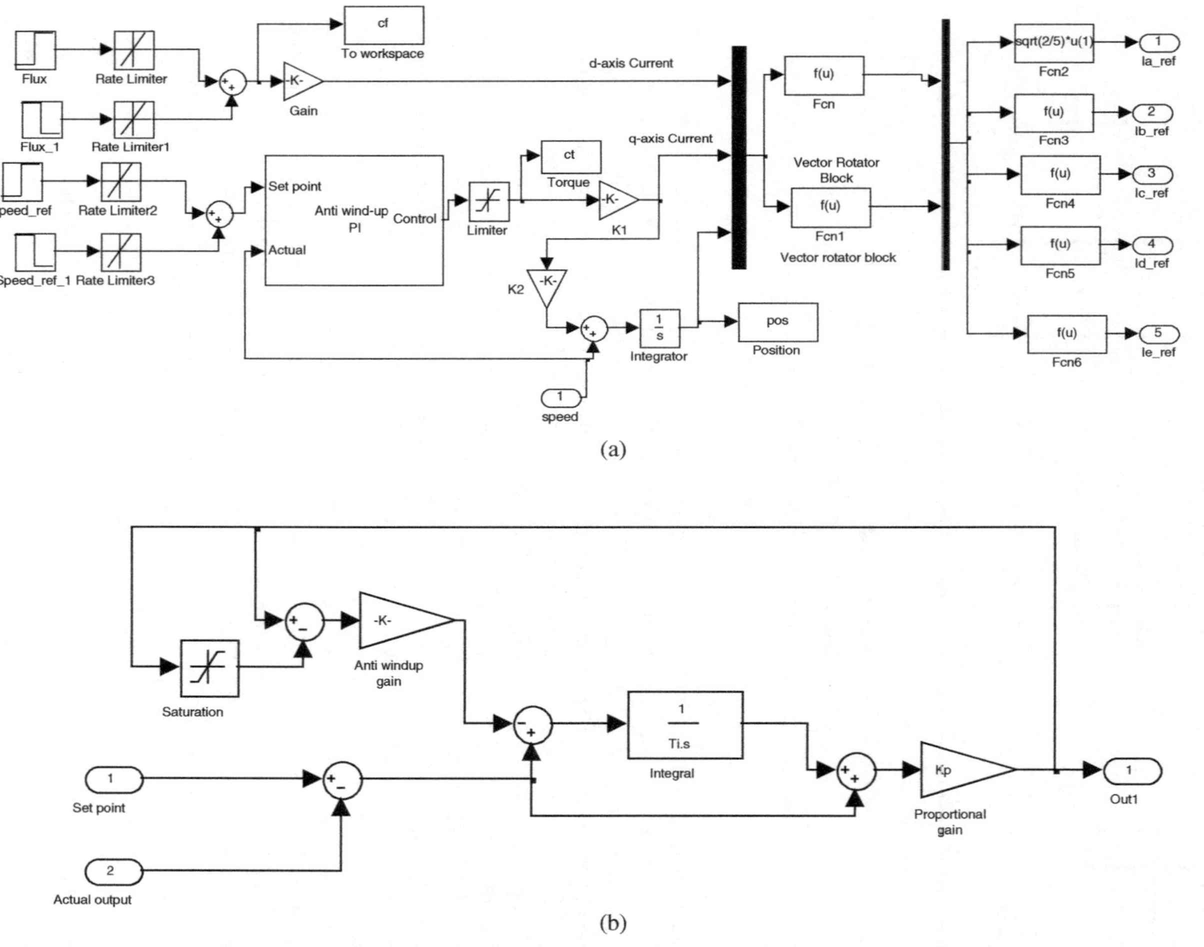

Figure 7.57 (a) Simulink of vector controller to generate five-phase reference current; and (b) anti-windup speed PI controller

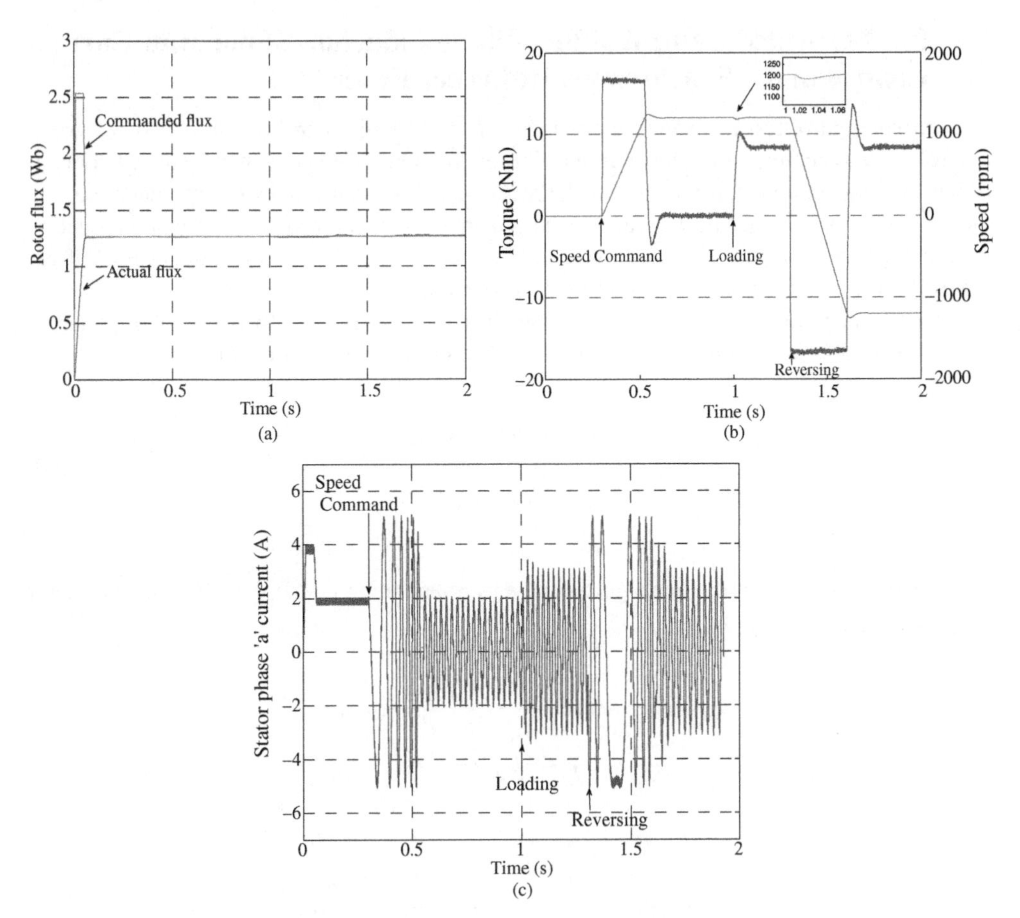

Figure 7.58 Indirect rotor field-oriented control of five-phase induction motor: (a) rotor flux; (b) speed and torque; and (c) stator phase "a" current

rotor flux settles to the reference value, it does not change anymore, showing decoupled torque and flux control (vector control). Acceleration takes place with the maximum allowed value of the motor torque. The motor phase current tracks the reference well and consequently, torque response closely follows torque reference. Application of the load torque at $t = 1$ seconds causes an inevitable dip in speed, of the order of 30 rpm in this case. Motor torque quickly follows the torque reference and enables rapid compensation of the speed dip (<100 ms). The motor torque settles at the value equal to the load torque after around 100 ms and the motor current becomes rated at the end of the transient. Reverse speed command is given at $t = 1.3$ seconds and actual torque closely follows the reference, leading to the speed reversal, with torque in the limit, in the shortest possible time interval (~350 ms). Thus, decoupled torque and flux control are achieved in a five-phase induction motor drive system.

7.5 Field-Oriented Control of Five-Phase Induction Motor with Current Control in the Synchronous Reference Frame

The current control can be exercised either for phase current control (Section 7.4) or current control in the synchronous reference frame. Current references are generated by the vector controller for phase current control, while voltage references are generated when current control in the synchronous reference frame is employed. The mathematical model of five-phase induction machine, obtained in Section 7.2, is modified to incorporate the vector control using current control in the synchronous reference frame as follows.

Under the condition of rotor flux orientation, the stator flux is given by (the x-y component is not considered as it is supposed to be zero for the sinusoidal supply condition):

$$
\begin{aligned}
\psi_{ds} &= \frac{L_m}{L_r}\psi_r + \sigma L_s i_{ds} \\
\psi_{qs} &= \frac{L_m}{L_r}\psi_r + \sigma L_s i_{qs}
\end{aligned}
\tag{7.60}
$$

where $\sigma = 1 - \frac{L_m^2}{L_s L_r}$ is the total leakage coefficient. Substituting equation (7.60) into the stator voltage equation (7.14) yields

$$
\begin{aligned}
v_{ds} &= R_s i_{ds} + \sigma L_s \frac{di_{ds}}{dt} + \frac{L_m}{L_r}\frac{d\psi_r}{dt} - \omega_r \sigma L_s i_{qs} \\
v_{qs} &= R_s i_{ds} + \sigma L_s \frac{di_{qs}}{dt} + \frac{L_m}{L_r}\frac{d\psi_r}{dt} - \omega_r \sigma L_s i_{ds}
\end{aligned}
\tag{7.61}
$$

$$
\begin{aligned}
i_{ds} + \frac{\sigma L_s}{R_s}\frac{di_{ds}}{dt} &= \frac{1}{R_s}v_{ds} - \frac{L_m}{L_r R_s}\frac{d\psi_r}{dt} + \frac{\omega_r \sigma L_s}{R_s} i_{qs} \\
i_{ds} + \frac{\sigma L_s}{R_s}\frac{di_{qs}}{dt} &= \frac{1}{R_s}v_{qs} + \frac{L_m}{L_r R_s}\frac{d\psi_r}{dt} + \frac{\omega_r \sigma L_s}{R_s} i_{ds}
\end{aligned}
\tag{7.62}
$$

As seen from equation (7.62), the two stator current components are not decoupled. Hence, decoupling circuit needs to be introduced for true decoupled d- and q-axis currents. If the output variables of current controllers are defined as:

$$
\begin{aligned}
v_{ds}^1 &= R_s\left(i_{ds} + \frac{\sigma L_s}{R_s}\frac{di_{ds}}{dt}\right) \\
v_{qs}^1 &= R_s\left(i_{qs} + \frac{\sigma L_s}{R_s}\frac{di_{qs}}{dt}\right)
\end{aligned}
\tag{7.63}
$$

then the required reference values of axis voltages v_{ds}^* and v_{qs}^*:

$$
\begin{aligned}
v_{ds}^* &= v_{ds}^1 + e_d \\
v_{qs}^* &= v_{qs}^1 + e_q
\end{aligned}
\tag{7.64}
$$

where auxiliary variables are defined as:

$$e_d = R_s\left(\frac{L_m}{L_r R_s}\frac{d\psi_r}{dt} - \omega_r\frac{\sigma L_s}{R_s}i_{qs}\right)$$
$$e_q = R_s\left(\frac{L_m}{L_r R_s}\frac{d\psi_r}{dt} - \omega_r\frac{\sigma L_s}{R_s}i_{ds}\right) \tag{7.65}$$

Equation (7.65) describes the decoupling circuit which is used in conjunction with the voltage fed rotor flux-oriented controlled five-phase induction motor drive. Usually, the derivative of the rotor flux term is neglected for simplicity without compromising the dynamics of the machine. The structure of the vector controller is shown in Figure 7.59.

The constant in Figure 7.59a is given as $\omega_{sl}^* = K_1 i_{qs}^* \Rightarrow K_1 = \omega_{sl}^*/i_{qs}^* = \frac{L_m}{T_r\psi_r^*}\frac{1}{T_r i_{ds}^*}$. Since the speed control in the base speed region is considered, the d-axis current is held constant equal to the rated value. The reference q-axis current is generated using the PI controller, which processes the speed error. The d-axis current error and q-axis current error are further processed in the PI controller and the auxiliary variables are subtracted from the outputs to generate the required reference voltages in the rotational reference frame. These voltages are then transformed to the stationary reference frame and then transformed into five-phase voltage references. These five-phase voltage references are given to the PWM block. This block then generates the appropriate switching signals for the PWM voltage source inverter. The inverter finally produces the required voltages that are impressed at the stator of the five-phase induction machine. The PWM that can be used for generating the voltages can be carrier-based sinusoidal method or space vector method.

The simulation result for the indirect rotor FOC with current control in the synchronously rotating reference frame is given in [62]. It is evident from the results shown in [62] that the x-y components of the stator current are still present, despite the use of the PWM method, which claims to completely eliminate the x-y component. The presence of x-y components of current is the consequence of the dead time in the switching signals of the IGBTs of the inverter.

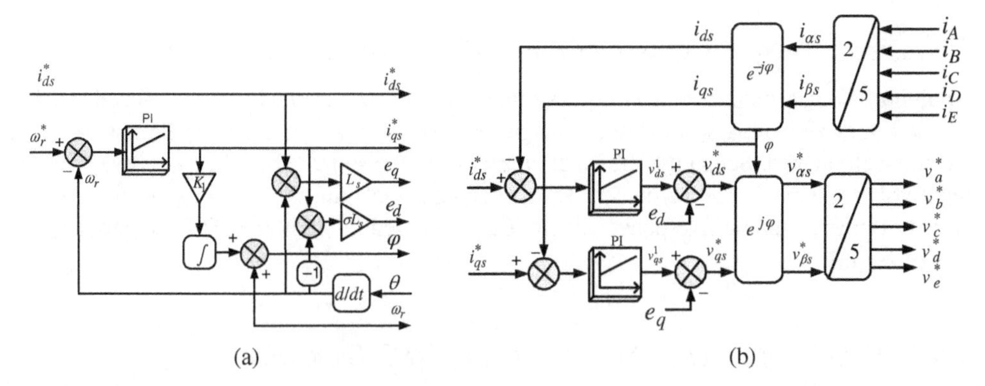

Figure 7.59 Vector control of five-phase induction machine: (a) reference current generation and (b) reference voltage generation

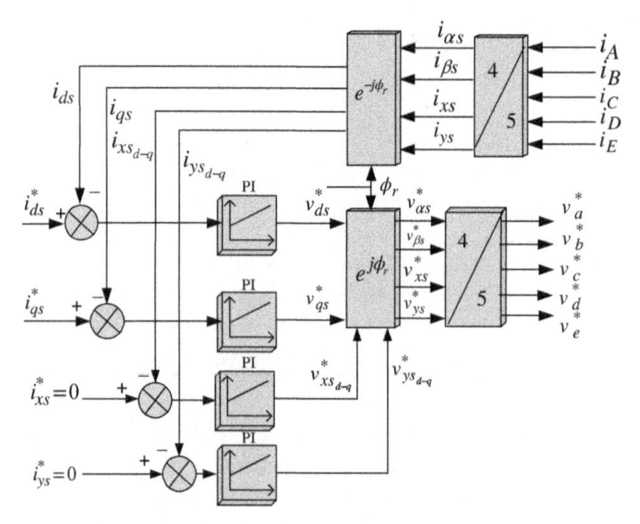

Figure 7.60 Vector control scheme of a five-phase induction machine with four current controllers in synchronous reference frame (*Source*: Based on Jones et al. [62])

To eliminate completely the *x-y* components of the current and consequently the distortion in the stator current, two additional current controllers need to be added in the vector controller of Figure 7.59, resulting in the modified vector control scheme. In the modified current control scheme, two extra pairs of current controllers are introduced that operate in the same reference frame as for the *d-q* components and have the same gains. The reference currents for *x-y* are set to zero. The *d-q* current parameters are the same as in the previous case (Figure 7.59) and the rotor flux position calculation also remains the same. The simulation result of the modified current control scheme is also depicted in [62]. The current waveform now shows complete elimination of the *x-y* components of current and the distortion in the stator current is minimized (Figure 7.60).

7.6 Direct Torque Control of a Five-Phase Induction Motor

Applying DTC technique to a five-phase induction motor supplied by a two-level voltage source inverter will have following challenges. Firstly, the *x-y* component of the stator current has to be eliminated. Secondly, the input current from the source should be in phase with the input voltage to maximize the output. The switching signals for the inverter switches are generated assuming that the five-phase inverter output is used to apply the DTC technique to the five-phase IM.

7.6.1 Control of Inverter Switches Using DTC Technique

Application of Direct Torque Control (DTC) technique for a five-phase IM is available in [16, 17]. In DTC realization, only one space vector or one switching vector is used at one instant of time to generate the voltage applied to the motor terminal. In case of a three-phase inverter, the space vectors diagram has only two vectors in one sector. Each space vector is distinct (has only one possible switching states). However, in a five-phase space vector diagram, there are six space vectors in each sector, e.g. in sector I, vectors available are 9, 16, 25 and 26, 29, 24. Three vectors lies along the same line (9, 16, 25) and (26, 29, 24). The short vectors 9 and 26 are not used for DTC realization since they produce high distortion in the stator current due to the fact that they map into the largest length vectors in the x-y plane. The vectors to be used are 16, 25 and 29, 24. For DTC realization, only one vector is needed along one direction. Hence, instead of two vectors only one virtual vector is generated by combining them together. This is called virtual vectors because they are not real. The following section describe the work of Payami et al. [17] for realization of DTC applied to a five-phase IM. The concept of virtual vector is explained below.

7.6.2 Virtual Vector for Five-Phase Two-Level Inverter

A five-phase two-level inverter generates 2^5, i.e. 32 switching states. Each switching state has corresponding space vector in α-β and x-y planes which is shown in Figure 7.61. The voltage vectors generated in α-β and x-y planes have a magnitude of $0.642V_{dc}$, $0.40V_{dc}$, and $0.247V_{dc}$. The volt-second balance technique is used for the synthesis of a virtual vector from two switching

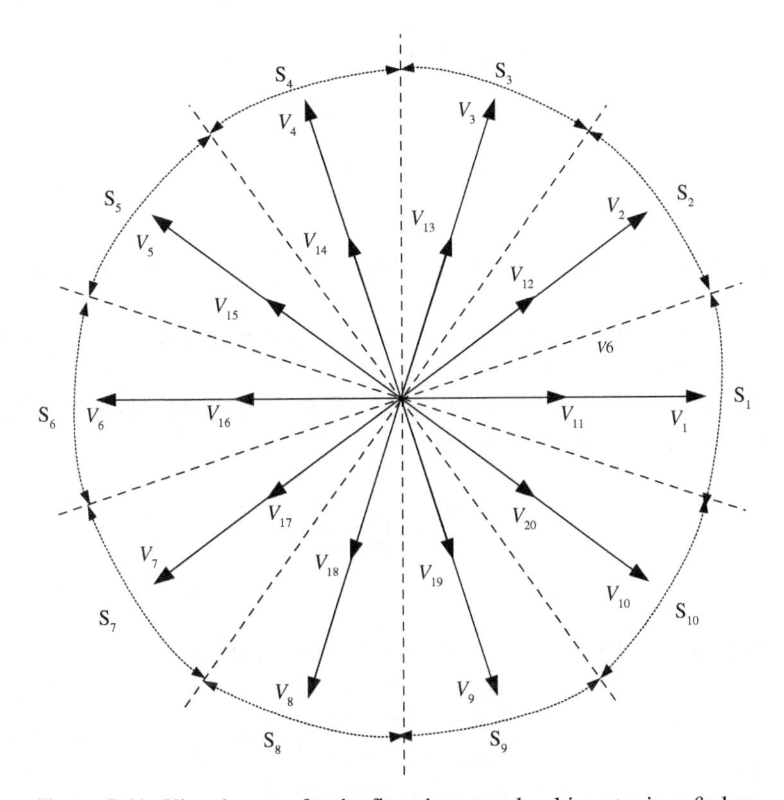

Figure 7.61 Virtual vector for the five-phase two-level inverter in α-β plane

states. For this, two switching states in the α-β plane are selected in one sample time. For example, the voltage vectors 16 and 25 in the α-β plane are in the opposite direction with different magnitudes in the x-y plane, as shown in Figures 7.34 and 7.35, respectively. The magnitude of vectors 16 and 25 is $0.4V_{dc}$ and $0.642V_{dc}$ in the α-β plane and $0.4V_{dc}$ and $0.242V_{dc}$ in the x-y-plane, respectively. The dwell time for voltage vector 25 is calculated as $0.618T_s$ [17].

For voltage vector 16, it is $0.382T_s$, where T_s is the sample time in seconds. The magnitude of the resultant virtual vector using voltage vectors 16 and 25 in the α-β plane is $0.553V_{dc}$. Similarly, using vectors 9 and 16 in the α-β plane, the virtual vector of magnitude $0.342V_{dc}$ is produced. Figure 7.61 shows the virtual vectors synthesized from actual voltage vectors in the α-β plane. The large virtual vectors from V_1 to V_{10} are obtained using medium and large voltage vectors. The small virtual vectors from V_{11} to V_{20} are obtained from medium and small voltage vectors in the α-β plane. The direction of virtual vectors is same as that of the synthesizing vectors in the α-β plane.

Problem 7.8 Show the computation of length of virtual vector in sector I of Figure 7.61.

Solution.
The lengths of the virtual vector length is computed using the volt-second principle in both $\alpha\beta$ and xy planes. The following equations can be written:

$$T_s\overline{V}_{\alpha\beta}^{ref} = T_1\overline{V}_{\alpha\beta 1} + T_2\overline{V}_{\alpha\beta 2} \tag{7.66}$$

$$T_s\overline{V}_{xy}^{ref} = T_1\overline{V}_{xy1} + T_2\overline{V}_{xy2} \tag{7.67}$$

$$T_s = T_1 + T_2 \tag{7.68}$$

where

$\overline{V}_{\alpha\beta}^{ref}$ is the resultant voltage vector in $\alpha\beta$ plane and $\overline{V}_{xy}^{ref}$ is the resultant voltage vector in xy plane.

T_1 and T_2 are the dwell time or time of application for vectors V_1 and V_2, respectively, and T_s is the switching period or sampling time.

In equation (7.67), the reference voltage magnitude in xy plane should be zero, $\overline{V}_{xy}^{ref} = 0$ and the ratio of dwell times T_1 and T_2 is given as:

$$\frac{T_1}{T_2} = \frac{|V_{xy2}|}{|V_{xy1}|} \tag{7.69}$$

Vectors in sector I is 16 and 25 with magnitudes and angles as:

$$\begin{aligned} \overline{V}_{16\alpha\beta} &= 0.4V_{dc}\angle 0 \\ \overline{V}_{16xy} &= 0.4V_{dc}\angle 0 \\ \overline{V}_{25\alpha\beta} &= 0.642V_{dc}\angle 0 \\ \overline{V}_{25xy} &= 0.242V_{dc}\angle 180 \end{aligned} \tag{7.70}$$

Put the values of $\overline{V}_{16xy}$ and $\overline{V}_{25xy}$ in equation (7.67):

$$\overline{V}_{xy}^{ref} = 0 = T_1(0.4V_{dc}\angle 0) + T_2(0.242V_{dc}\angle 180)$$
$$T_1 0.4V_{dc} - T_1 0.242V_{dc} = 0$$
$$\frac{T_1}{T_2} = \frac{0.242}{0.4}$$
$$\frac{T_1 + T_2}{T_2} = \frac{0.242 + 0.4}{0.4}$$
$$\frac{T_2}{T_s} = \frac{0.4}{0.642}; \frac{T_1}{T_s} = \frac{0.242}{0.642}$$

From equation (7.66),

$$\overline{V}_{\alpha\beta}^{ref} = \frac{T_1}{T_s}(0.4V_{dc}\angle 0) + \frac{T_2}{T_s}(0.642V_{dc}\angle 0)$$
$$\overline{V}_{\alpha\beta}^{ref} = \frac{0.242}{0.642}(0.4V_{dc}\angle 0) + \frac{0.4}{0.642}(0.642V_{dc}\angle 0) = 0.553V_{dc}\angle 0$$

7.6.2.1 Direct Torque Control

Torque equation for a five-phase induction machine is given in equation (7.17) as:

$$T_e = \frac{5P}{2}\left[\psi_{ds} i_{qs} - \psi_{qs} i_{ds}\right] \tag{7.71}$$

Torque is to be represented in terms of fluxes:

$$\begin{aligned}
\psi_{ds} &= L_s i_{ds} + L_m i_{dr} \\
\psi_{dr} &= L_r i_{dr} + L_m i_{ds}; i_{dr} = \frac{\psi_{dr} - L_m i_{ds}}{L_r} \\
\psi_{ds} &= L_s i_{ds} + L_m\left(\frac{\psi_{dr} - L_m i_{ds}}{L_r}\right) = L_s\left(1 - \frac{L_m^2}{L_s L_r}\right) i_{ds} + \frac{L_m}{L_r}\psi_{dr} \\
\psi_{ds} &= \sigma L_s i_{ds} + \frac{L_m}{L_r}\psi_{dr}
\end{aligned} \tag{7.72}$$

where T_e is electromagnetic torque, P is the number of stator poles, $\psi_{ds}, \psi_{qs}, \psi_{dr}$, and ψ_{qr} are the direct and quadrature axis stator and rotor flux components, respectively. i_{ds}, i_{qs}, i_{dr}, and i_{qr} are the direct and quadrature axis stator and rotor current components, respectively. L_m is magnetizing inductance of equivalent circuit of IM in d-q plane, L_r is leakage inductance of rotor winding, and L_s is leakage inductance of stator winding.

The currents in terms of fluxes are obtained as:

$$i_{ds} = \frac{1}{\sigma L_s}\left(\psi_{ds} - \frac{L_m}{L_r}\psi_{dr}\right)$$
$$i_{qs} = \frac{1}{\sigma L_s}\left(\psi_{qs} - \frac{L_m}{L_r}\psi_{qr}\right) \tag{7.73}$$

Substituting the currents from equation (7.73) into equation (7.71), the following is obtained:

$$T_e = \frac{5P}{2}\frac{L_m}{\sigma L_s L_r}\left[\psi_{qs}\psi_{dr} - \psi_{ds}\psi_{qr}\right] \tag{7.74}$$

The position of stator and rotor fluxes (θ_{fs}, θ_{fr}, respectively) are shown in Figure 7.62, and the two fluxes rotate at synchronous speed (ω_e) such that a fixed angle γ_{sr} exist between them.

The d-q axes components of the stator and rotor fluxes can be written as:

$$\psi_{ds} = \psi_s \cos\left(\theta_{fs}\right); \psi_{qs} = \psi_s \sin\left(\theta_{fs}\right)$$
$$\psi_{dr} = \psi_s \cos\left(\theta_{fr}\right); \psi_{qr} = \psi_s \sin\left(\theta_{fr}\right) \tag{7.75}$$

Substituting equation (7.75) into equation (7.74), the following is obtained:

$$T_e = \frac{5P}{2}\frac{L_m}{\sigma L_s L_r}\psi_s\psi_r\left[\sin\left(\theta_{fs}\right)\cos\left(\theta_{fr}\right) - \cos\left(\theta_{fs}\right)\sin\left(\theta_{fr}\right)\right]$$
$$T_e = K\psi_s\psi_r \sin\left(\theta_{fs} - \theta_{fr}\right)$$
$$T_e = K\psi_s\psi_r \sin\left(\gamma_{sr}\right) \tag{7.76}$$

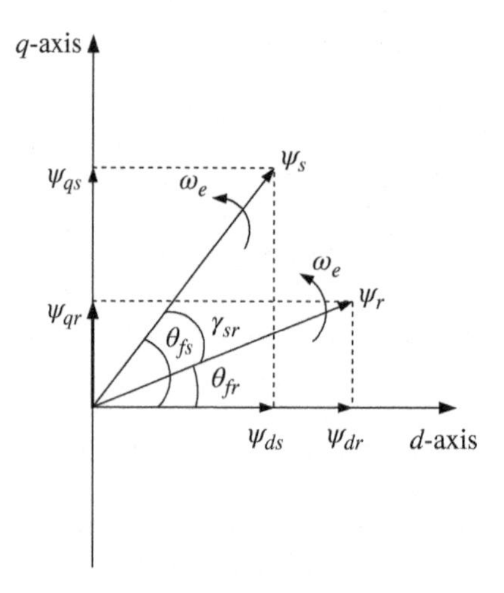

Figure 7.62 Stator and rotor flux positions and their components along d-q axes

where γ_{sr} is the angle between stator flux (ψ_s) and rotor flux (ψ_r). As per equation (7.76), torque can be controlled by either controlling the magnitude of the stator and rotor fluxes, called direct flux control or by controlling the torque angle γ_{sr}, and this is called DTC.

The basic block diagram for two-level inverter-fed DTC of a five-phase IM is shown in Figure 7.63. Flux controller and torque controller are given in Figures 7.64 and 7.65,

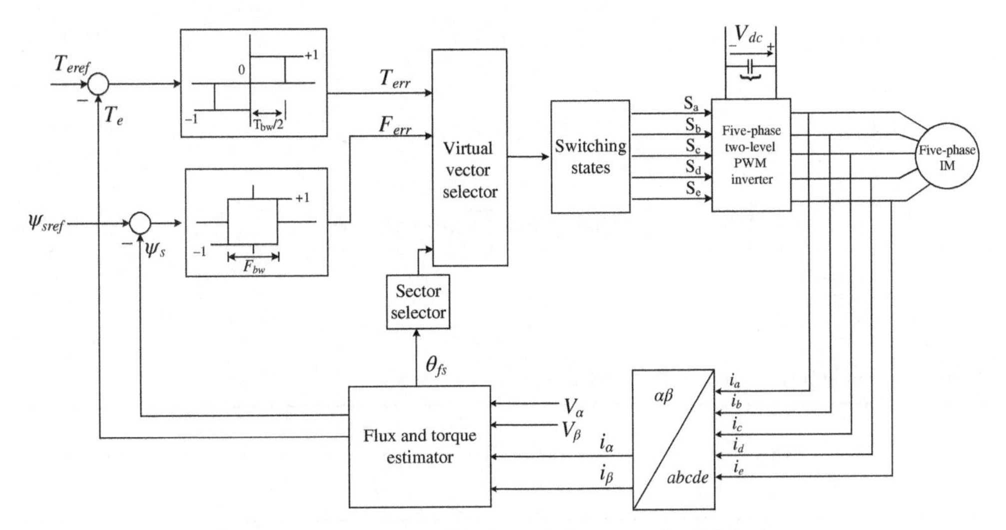

Figure 7.63 Basic DTC scheme using a two-level inverter

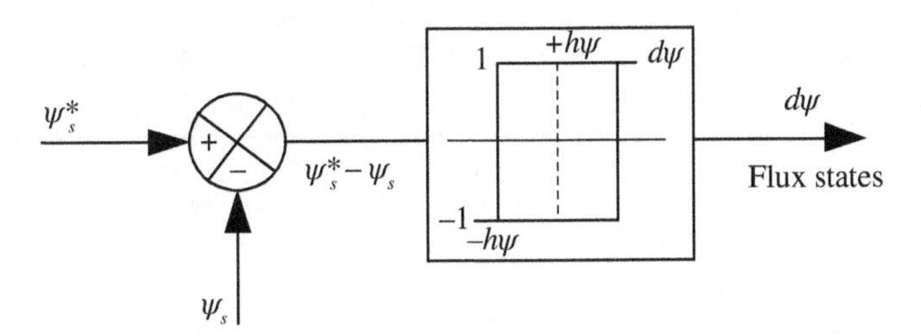

Figure 7.64 Two-level hysteresis flux controller

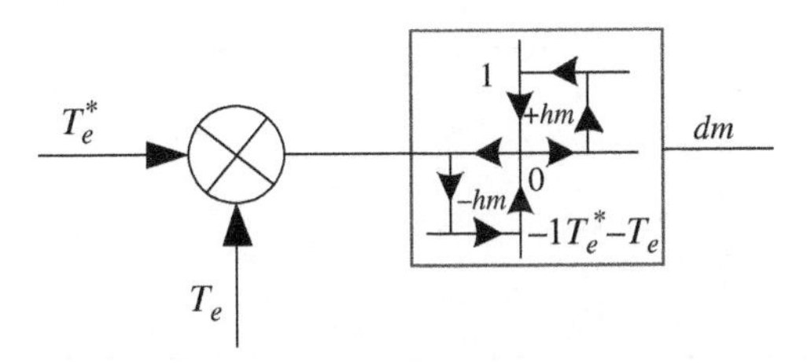

Figure 7.65 Three-level torque controller

respectively. Depending upon the torque (T_{err}), flux (F_{err}) error commands and sector location of the virtual vector is selected as per the lookup tables given in Tables 7.8 and 7.9. Lookup table for speed greater than 0 is given in Table 7.8 and for less than 0 is given in Table 7.9. Corresponding to virtual vectors, respective space vectors are applied for their dwell time [17].

The stator voltage equation is given as:

$$\overline{V_s} = R_s\overline{i_s} + \frac{d\overline{\psi_s}}{dt} \tag{7.77}$$

$$\frac{d\overline{\psi_s}}{dt} = \overline{V_s} - R_s\overline{i_s} \tag{7.78}$$

$$\overline{\psi_s} = \int(\overline{V_s} - R_s\overline{i_s})dt + \psi(0) = \int\overline{V_s}dt + \psi_{t=0} = \overline{V_s}\Delta t + \psi|_{t=0} \tag{7.79}$$

Table 7.8 Lookup table for speed greater than 0

		Sector number									
F_{err}	T_{err}	1	2	3	4	5	6	7	8	9	10
1	1	V_3	V_4	V_5	V_6	V_7	V_8	V_9	V_{10}	V_1	V_2
	0	V_0	V_0	V_0	V_0	V_0	V_0	V_0	V_0	V_0	V_0
	−1	V_{10}	V_1	V_2	V_3	V_4	V_5	V_6	V_7	V_8	V_9
−1	1	V_4	V_5	V_6	V_7	V_8	V_9	V_{10}	V_1	V_2	V_3
	0	V_0	V_0	V_0	V_0	V_0	V_0	V_0	V_0	V_0	V_0
	−1	V_8	V_9	V_{10}	V_1	V_2	V_3	V_4	V_5	V_6	V_7

Table 7.9 Lookup table for speed less than 0

		Sector number									
F_{err}	T_{err}	1	2	3	4	5	6	7	8	9	10
1	1	V_2	V_3	V_4	V_5	V_6	V_7	V_8	V_9	V_{10}	V_1
	0	V_0	V_0	V_0	V_0	V_0	V_0	V_0	V_0	V_0	V_0
	−1	V_9	V_{10}	V_1	V_2	V_3	V_4	V_5	V_6	V_7	V_8
−1	1	V_4	V_5	V_6	V_7	V_8	V_9	V_{10}	V_1	V_2	V_3
	0	V_0	V_0	V_0	V_0	V_0	V_0	V_0	V_0	V_0	V_0
	−1	V_8	V_9	V_{10}	V_1	V_2	V_3	V_4	V_5	V_6	V_7

Flux control logic is given as:

1. If $\psi_s^* - \psi_s \geq h\psi$ then $d\psi = 1$
2. If $-h\psi \leq \psi_s^* - \psi_s \leq h\psi$ and $d\psi = 1$
3. If $-h\psi \leq \psi_s^* - \psi_s \leq h\psi$ and $d\psi = -1$
4. $\psi_s^* - \psi_s \leq -h\psi$ then $d\psi = -1$

Torque control logic is gives as:

1. If $T_e^* - T_e \geq hm$ then $dm = +1$
2. If $0 \leq T_e^* - T_e \leq hm$ and $dm = +1$ then $dm = +1$
3. If $T_e^* - T_e \leq -hm$ then $dm = -1$
4. $-hm \leq T_e^* - T_e \leq 0$ and $dm = -1$ then $dm = -1$
5. $0 \leq T_e^* - T_e \leq hm$ and $dm = -1$ then $dm = 0$
6. If $-hm \leq T_e^* - T_e \leq 0$ and $dm = +1$ then $dm = 0$

The three level torque comparator is shown in Figure 7.65.

Figure 7.66 shows the principle of DTC for a two-level five-phase voltage source inverter voltage vector. The *d-q* axis plane for the stator flux ψ_s is divided into 10 sectors S1–S10. For

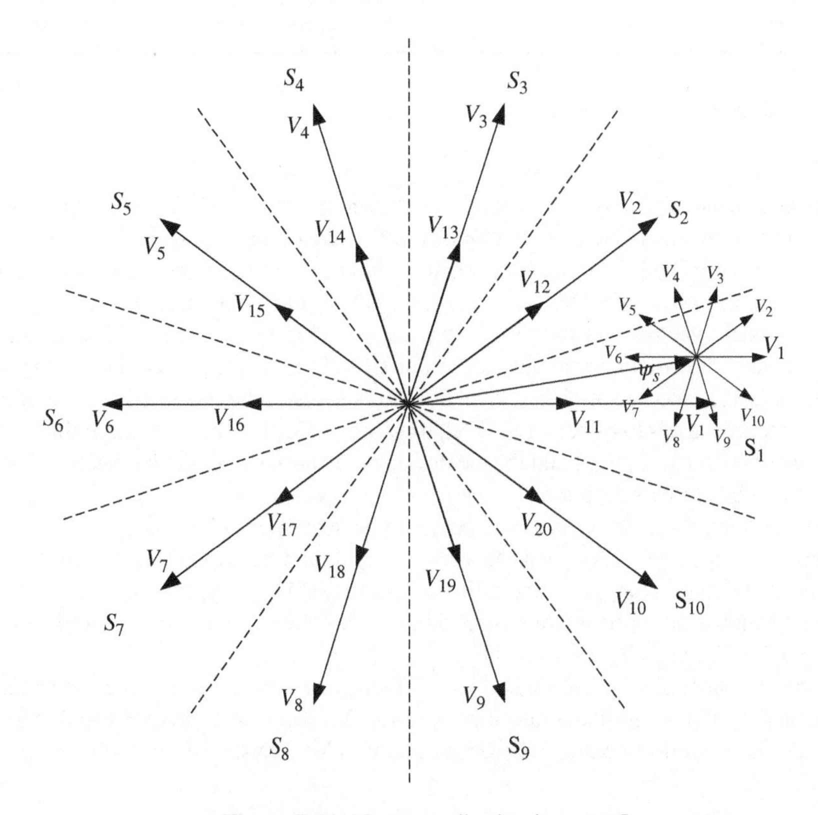

Figure 7.66 Vector application in sector I

instance, the stator flux ψ_s shown in the figure falls into sector S1 at an angle referenced to the d-axis of the stationary reference frame. The rotor flux vector ψ_r lags ψ_s by an angle γ_{sr} is shown in Figure 7.62. Let's now examine the impact of the stationary voltage vectors on stator flux and angle between stator and rotor flux. For the initial shown position at $t = 0$, when voltage vector V_2 is selected, the stator flux vector will become $\overline{\psi_s} = \overline{V_2}\Delta t + \psi_s|_{t=0}$ after a short time interval Δt, leading to an increase in flux magnitude and torque angle. If voltage vector V_1 or V_{10} is selected, new stator flux is lower than initial one and torque angle is also lower than initial value. It will cause a decrease in the torque and flux magnitude. Similarly, the selection of other vectors, e.g. V_3 and V_1, can make one variable increase and the other decrease. Therefore, stator flux and torque can be controlled separately by proper selection of the inverter voltage vectors.

DTC scheme can be implemented using the lookup tables. Both sensored and sensorless schemes can be implemented following the concept of three-phase induction motor DTC.

Problem 7.9 Implement DTC of a five-phase Induction motor with the per-phase parameters:

Rated voltage	$U_n = 220$ v	Stator resistance	$R_s = 10\ \Omega$
Rated current	$I_n = 2.1$ A	Rotor resistance	$R_r = 6.3\ \Omega$
Stator inductance	$L_s = 0.46$ H	Mutual inductance	$L_m = 0.4$ H
Rotor inductance	$L_r = 0.46$ H	Number of poles	$p = 4$
Moment of inertia	$J = 0.03$ kg.m^2	Friction coefficient	$f_r = 0.008$ N.m.s/rd

Using 10 large space vectors in sensored mode.

Solution.

The space vectors are divided into 10 sectors, as shown in Figure 7.67. Only one space vector is needed from two available adjacent vectors from the most outer decagon.

The sector is identified from the flux position. In each sector, the next switching vector is assumed based on the combination between flux error and torque error. For example, if the stator flux linkage vector is in the "i" sector, Figure 7.67, and the flux has to be increased (FI) and the electromagnetic torque has to be positive (TP), then the switching voltage vector to be chosen is V_{i+1}. However, if the stator flux linkage has to be increased (FI) and the electromagnetic torque needs to be negative (TN), thus vector V_{i-1} is to be selected. If the stator flux linkage has to be decreased (FD) and the electromagnetic torque needs to be positive (TP), then the vector V_{i+4} has to be selected.

In the same way, if the stator flux linkage has to be decreased (FD) and the electromagnetic torque has to be negative (TN), then the vector V_{i-4} should be applied. If the torque error is within the error limit, i.e. $dT_e = 0$, the null voltage vector V_0 or V_{31} is selected.

Table 7.10 shows the space vector lookup table. In this table, $d\Phi = 1$ stands for FI, $d\Phi = 0$ for FD, $dT_e = 1$ for TP, and $dT_e = -1$ for TN.

The control system block diagram shown in Figure 7.68 calculates the electromagnetic torque generated by five-phase induction motor, stator flux, and rotor angular speed. The three-level hysteresis controllers compare the torque and two-level hysteresis controller compare the

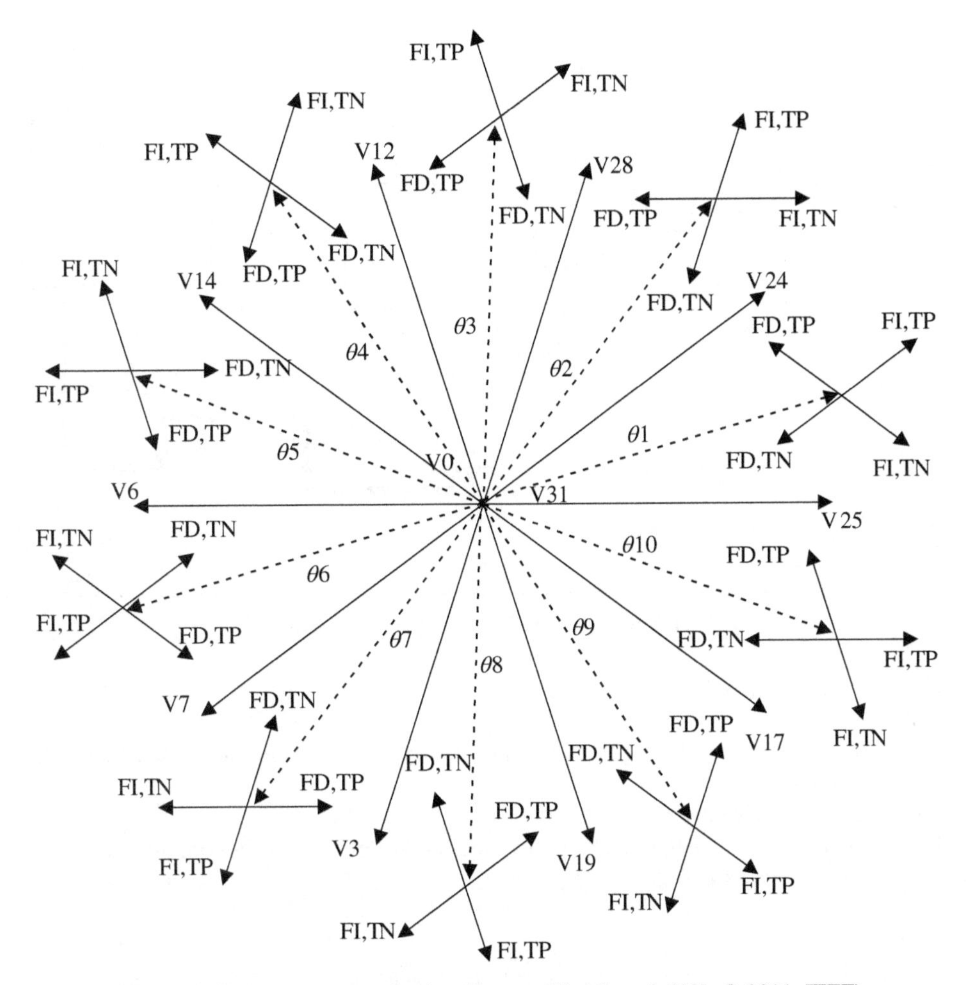

Figure 7.67 Space vector selection (*Source*: Khaldi et al. [18]. © 2011, IEEE)

Table 7.10 Lookup table for vector selection

$\Delta\Phi$	ΔT_e	Sector 1	Sector 2	Sector 3	Sector 4	Sector 5	Sector 6	Sector 7	Sector 8	Sector 9	Sector 10
1	1	V_{24}	V_{28}	V_{12}	V_{14}	V_6	V_7	V_3	V_{19}	V_{17}	V_{25}
1	0	V_0	V_{31}	V_0	V_{31}	V_0	V_{31}	V_0	V_{31}	V_0	V_{31}
1	−1	V_{17}	V_{25}	V_{24}	V_{28}	V_{12}	V_{14}	V_6	V_7	V_3	V_{19}
0	1	V_{14}	V_6	V_7	V_3	V_{19}	V_{17}	V_{25}	V_{24}	V_{28}	V_{12}
0	0	V_{31}	V_0	V_{31}	V_0	V_{31}	V_0	V_{31}	V_0	V_{31}	V_0
0	−1	V_7	V_3	V_{19}	V_{17}	V_{25}	V_{24}	V_{28}	V_{12}	V_{14}	V_6

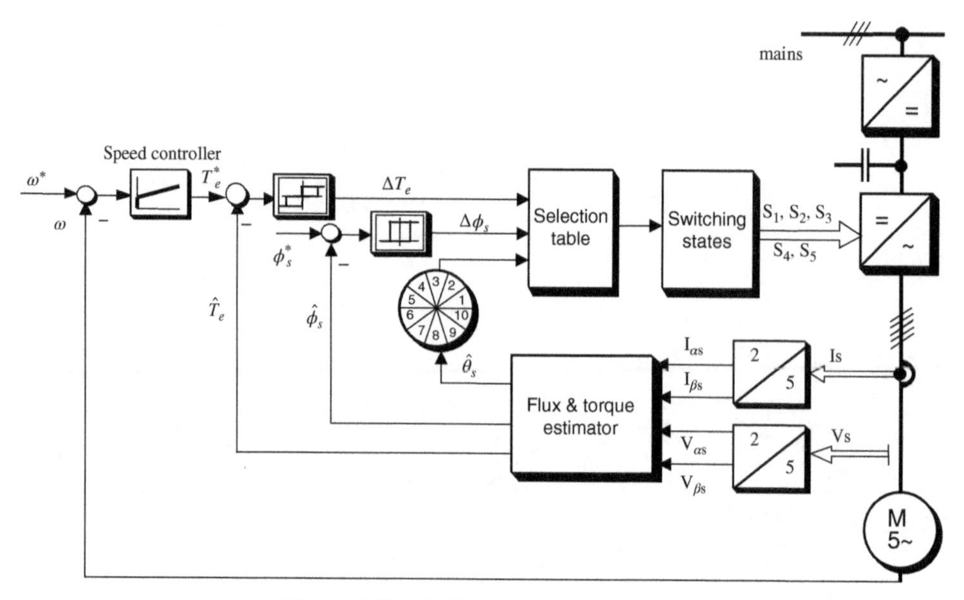

Figure 7.68 DTC control block diagram

flux reference values with the actual quantities and generate an optimal switching signal to drive the inverter.

Flux and Torque Estimator

Flux estimator computes the stator flux linkages and torque estimator computes the electromagnetic torque using the following equations:

$$\psi_{\alpha s} = \int (V_{\alpha s} - R_s i_{\alpha s}) dt$$
$$\psi_{\beta s} = \int (V_{\beta s} - R_s i_{\beta s}) dt \tag{7.80}$$

$$T_e = \frac{5P}{2}\left[\psi_{\alpha s} i_{\beta s} - \psi_{\beta s} i_{\alpha s}\right] \tag{7.81}$$

Simulation Model

Simulink model is shown in Figure 7.69. Reference flux and reference speed are given as user input. The responses obtained are shown in Figure 7.70. Each block is further shown in Figure 7.69a–d.

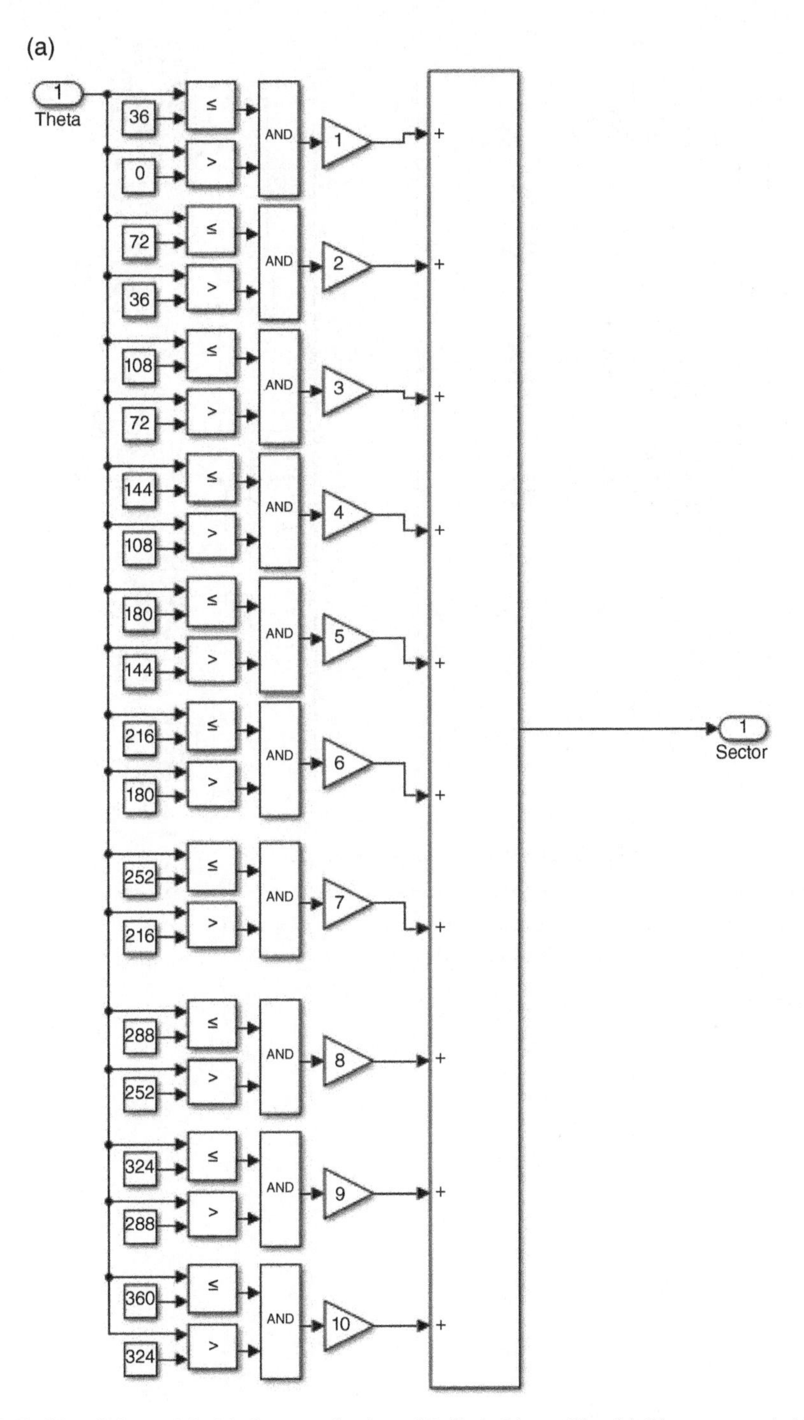

Figure 7.69 Simulink model. (a) Sector selection. (b) Switching table. (c) Vector to switching state conversion. (d) Flux and torque estimator

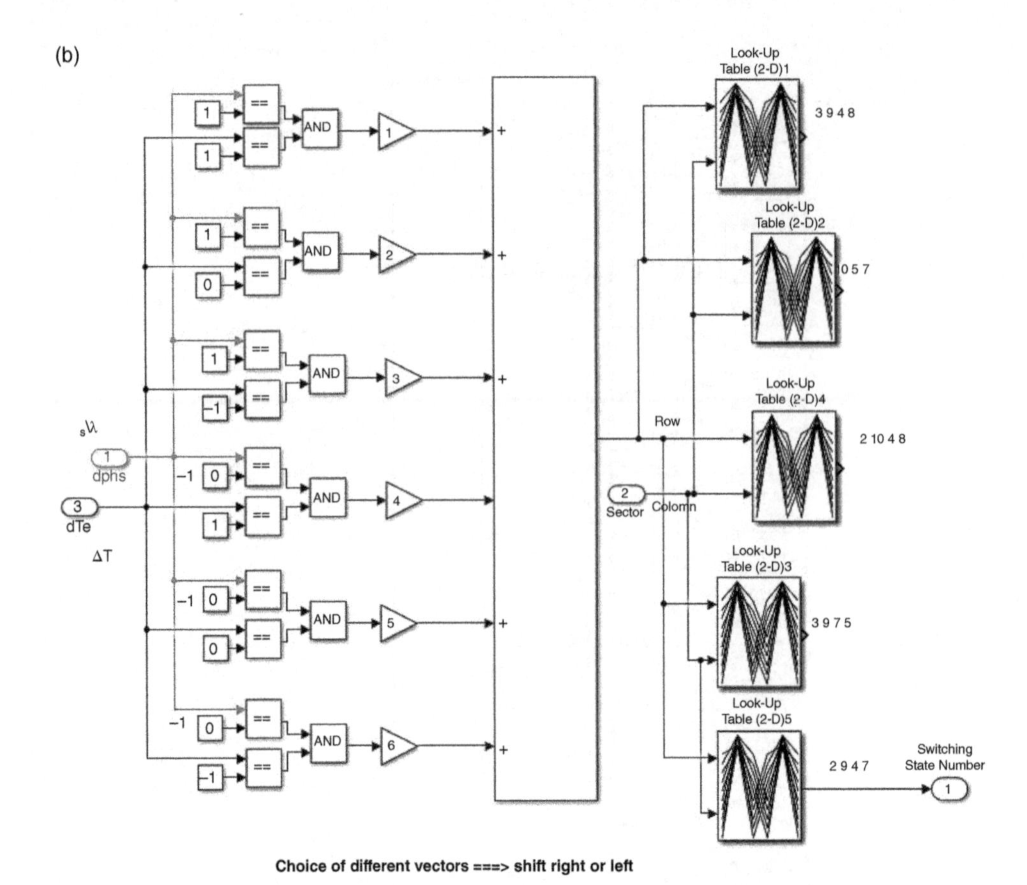

[3 4 5 6 7 8 9 10 12;11 12 11 12 11 12 11 12 11 12; 9 10 1 2 3 4 5 6 7 8 ; 5 6 7 8 9 10 1 2 3 4; 11 12 11 12 11 12 11 12 11 12; 7 8 9 10 1 2 3 4 5 6]

[2 3 4 5 6 7 8 9 10 1;11 12 11 12 11 12 11 12 11 12;10 1 2 3 4 5 6 7 8 9; 4 5 6 7 8 9 10 1 2 3;11 12 11 12 11 12 11 12 11 12; 8 9 10 1 2 3 4 5 6 7]

[2 3 4 5 6 7 8 9 10 1;11 12 11 12 11 12 11 12 11 12;10 1 2 3 4 5 6 7 8 9; 5 6 7 8 9 10 1 2 3;11 12 11 12 11 12 11 12 11 12; 7 8 9 10 1 2 3 4 5 6]

Figure 7.69 (Continued)

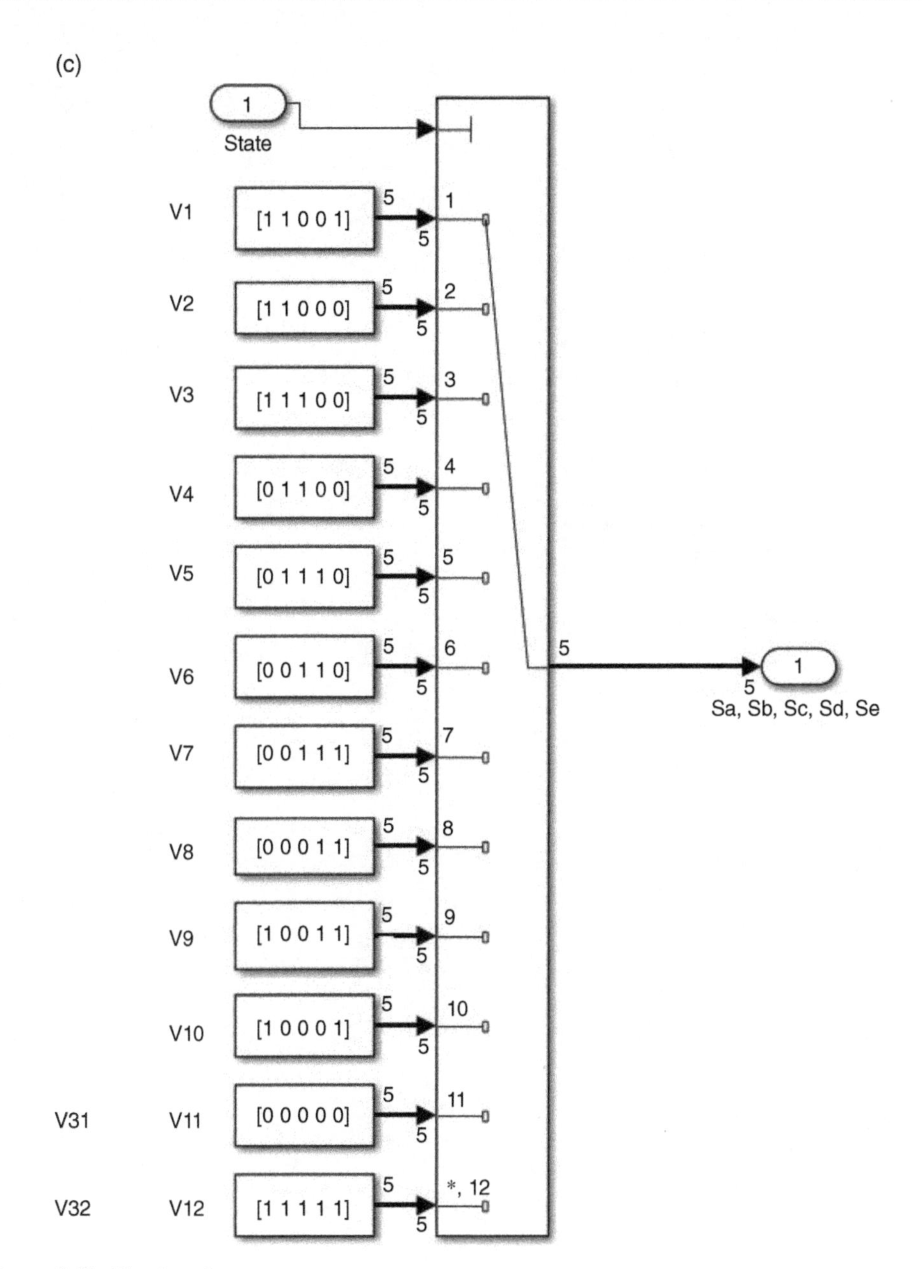

Figure 7.69 (Continued)

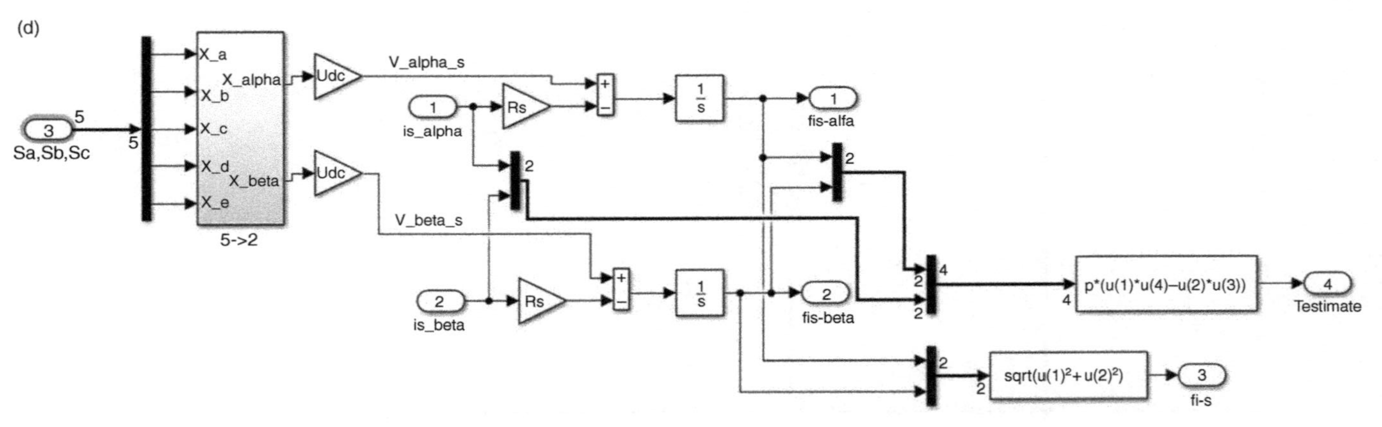

Figure 7.69 (Continued)

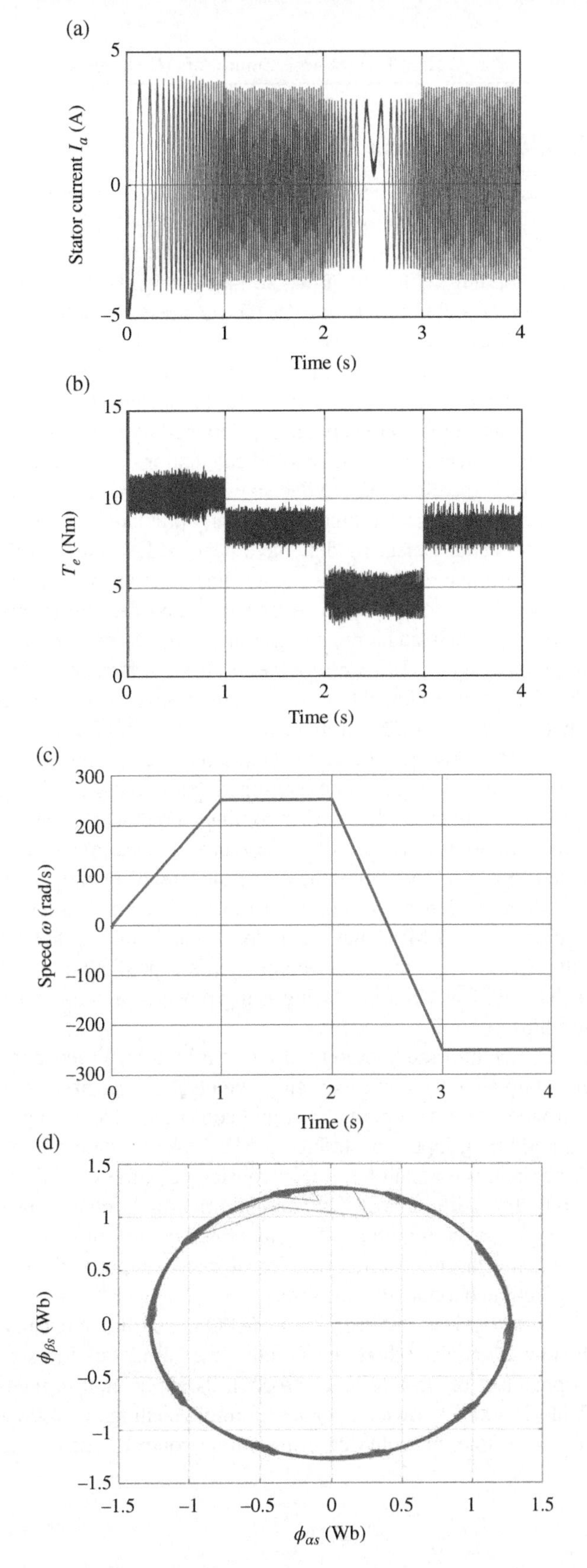

Figure 7.70 Five-phase IM DTC response (a) Stator current, phase 'a', (b) Torque, (c) Speed and (d) Stator flux locus

7.7 Model Predictive Control (MPC)

In the last 20 years, model predictive control (MPC) has achieved a significant level of acceptability and success in practical process control applications and has been mainly applied in power plants and petroleum refineries. With the development of the modern microcontroller, digital signal processors, and field programmable gate arrays, MPC applications have been found in a variety of areas including chemicals, food processing, automotive and aerospace applications, and power electronics and drives [63–68].

MPC algorithms use an explicit process model to predict the future response of a process or plant. A cost function represents the desired behavior of the system. An optimization problem is formulated, where a sequence of future actuation is obtained by minimizing the cost function. The first element of the sequence is applied and all calculations are repeated for every sample period. The process model therefore plays a decisive role in the controller. The chosen model must be able to capture the process dynamics to precisely predict future outputs and be simple to implement and understand. Linear models have been widely used, as they can be easily obtained by system identification technique, or by linearization of first-principle nonlinear models. The cost function, in consequence, is quadratic and the constraints are in the form of linear inequalities. For such quadratic programming (QP) problem, active set methods (AS), and interior point methods (IP) can provide the best performance. This formulation of linear MPC (LMPC) has been implemented successfully in several commercial predictive control products, such as Aspentech's DMC Plus, Honeywell's RMPCT, Adersa's PFC and HIECON, and ABB's 3DMPC. However, if the plant exhibits severe nonlinearities, the usefulness of predictive control based on a linear model is limited, particularly if it is used to transfer the plant from one operating point to another [69]. Adopting a nonlinear model in an MPC scheme is an obvious solution to improve the control performance. In fact, predictive control with nonlinear internal models (NLMPC) has been one of the most relevant and challenging problems. Because the optimization problem to find control action has become nonlinear, the convexity of the optimization problem in LMPC has been lost, which makes the online application extremely difficult. However, in order to obtain accurate predictions of process behavior, the application of NLMPC has been increasing using powerful microcontroller and efficient optimization algorithms [70–72].

MPC is an attractive control methodology to modern industry because it has the ability to handle both input and output constraints explicitly, which allows plant operation closer to constraints. If the controlled processes operate at output constraints, the most profitable operation can be obtained. In addition, input constraints in MPC take account of actuator limitations, which means safe operations on plant have been considered at the same time as its most profitable condition [69]. In addition, MPC handles multivariable control problems naturally, which makes it easily meet the requirement of integrated control of complex plants in profit maximization. By monitoring the future behavior of many output variables, MPC can keep a process operating both economically and safely [73].

However, a widely recognized shortcoming of MPC is that it can usually only be used in applications with slow dynamics, where the sample time is measured in seconds or minutes. Hence, the development of the fast MPC to broaden its application in industry has been an active research field. One well-known technique for implementing the fast MPC is to compute the entire control law offline, in which case the online controller can be implemented as a lookup table [74, 75]. This method works well for systems with small state and input dimensions (i.e. no more than five), few constraints, and short time horizons. On the other hand, optimization algorithms, such QP reformulations, warm-starting, and early termination in IP, have been investigated to speed up the computation of control action in MPC, and high-quality controls using these improved algorithms have been obtained [69, 75].

7.7.1 MPC Applied to a Five-Phase Two-Level VSI

Since the power electronic converters have finite switching states, MPC can be effectively used in such converter systems for their current and power control called "finite state MPC." This section elaborates on finite state MPC for a five-phase voltage source inverter for their current control.

The basic block schematic of the proposed strategy is illustrated in Figure 7.71. The discrete load model, called the "Predictive Model," is used to precalculate the behavior of the current in the next sampling interval. The precalculated current sample is then fed to the optimizer along with the reference current (obtained from the external loop). The optimizer calculates the cost function for all the possible switching combinations of the inverter. Thus, it generates the optimal switching state corresponding to the global minimum cost function in each sampling interval and passes it on to the gate drive of the inverter. This is how the optimal solution is obtained. This concept is entirely different from the conventional PWM techniques, where the symmetrical switching patterns are generated and ensures that each leg changes the state at least twice in the same switching interval, keeping constant the switching frequency spectrum. In contrast to the MPC, the two switchings of each leg are not guaranteed and hence the switching frequency is variable. However, this method is powerful owing to the fact that the concept is simple and intuitive. The controller can incorporate many desired features by simply modifying the cost function.

Owing to the fact that a five-phase inverter generates a large number of space vectors (30 active and 2 zero), many possible solutions can exist to implement MPC, as discussed in the latter section. Nevertheless, in this section, few solutions are elaborated on and there still remains many more to explore.

The choice of the cost function is the most important aspect of the MPC. A judicious choice leads to the optimal solution of the control objective. Thus, the cost function should include all the parameters to be optimized within the imposed constraints. In the current control, the most important variable is the current-tracking error. Thus, the most simple and straightforward

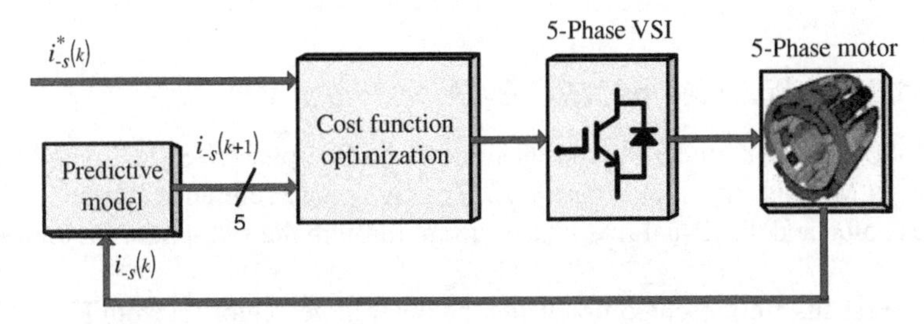

Figure 7.71 Model predictive current controller for five-phase drive

choice is the absolute value of the current error. The other choices could be the square of the current error, integral of the current error, or the rate of change of error, etc. Specifically, in a five-phase drive system, there exist two orthogonal subspaces, namely α-β and x-y. Thus, in the case of a five-phase drive system, the current errors in both planes have to be considered for devising a cost function. In general, for the square of current error, the cost function is given as:

$$\begin{aligned} \hat{g}_{\alpha\beta} &= \left|i_{\alpha}^{*}(k) - \hat{i}_{\alpha}k(k+1)\right| + \left|i_{\beta}^{*}(k) - i_{\beta}(k+1)\right| \\ \hat{g}_{xy} &= \left|i_{x}^{*}(k) - \hat{i}_{x}k(k+1)\right| + \left|i_{y}^{*}(k) - \hat{i}_{y}(k+1)\right| \end{aligned} \tag{7.82}$$

The final cost function can be expressed as:

$$J_{\alpha\beta xy} = \left\|\hat{g}_{\alpha\beta}\right\|^{2} + \gamma\left\|\hat{g}_{xy}\right\|^{2} \tag{7.83}$$

where ||.|| denotes modulus and γ is a tuning parameter that offer degrees of freedom to put emphasis on α-β or x-y subspaces. Comparative studies are made to emphasize the effect of the choice of the tuning parameter on the performance of the controller.

The load is assumed as a five-phase RLE (resistance, inductance, and back emf). The discrete time model of the load suitable for current prediction is obtained from [66]:

$$\hat{\underline{i}}(k+1) = \frac{T_s}{L}(\underline{v}(k) - \hat{\underline{e}}(k)) + \underline{i}(k)\left(1 - \frac{RT_s}{L}\right) \tag{7.84}$$

where R and L are the resistance and inductance of the load, T_s is the sampling interval, $\underline{i}$ is the load current space vector, $\underline{v}$ is the inverter voltage space vector used as a decision variable, and $\hat{\underline{e}}$ is the estimated back e mf obtained from [66]:

$$\underline{i} = [i_a \quad i_b \quad i_c \quad i_d \quad i_e]^t, \quad \underline{v} = [v_a \quad v_b \quad v_c \quad v_d \quad v_e]^t \tag{7.85}$$

$$\hat{\underline{e}}(k) = \underline{v}(k) + \frac{L}{T_s}\underline{i}(k-1) - \underline{i}(k)\left(R + \frac{L}{T_s}\right) \tag{7.86}$$

where $\hat{\underline{e}}(k)$ is the estimated value of $\underline{e}(k)$. However, for simulation purposes, the amplitude and frequency of back emf are assumed as constant.

7.7.2 MATLAB/Simulink of MPC for Five-Phase VSI

The MATLAB/Simulink model for current control of a five-phase VSI fed five-phase RL load using finite set MPC is shown in Figure 7.72. The five-phase reference current is given as the commanded value and the actual five-phase current through the five-phase RL load is recorded and compared.

Simulation results are presented first using the outer large vector set from Figure 7.73 and a zero vector, thus in total 11 vectors are used, and the cost function is such that it minimizes

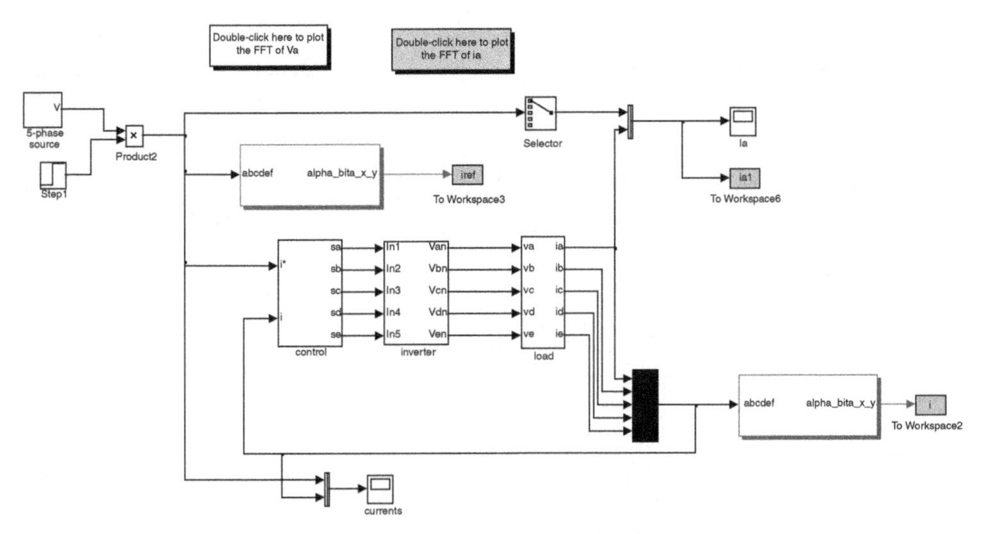

Figure 7.72 Finite set MPC of a five-phase VSI

current-tracking errors only in the α-β plane. This is followed using cost function, which takes into account the current-tracking error minimization in both the α-β and x-y planes. The sampling time T_s is kept equal to 50 μs. The fundamental frequency of the sinusoidal reference current is chosen as 50 Hz. The other parameters are $R = 10\,\Omega$, $L = 10\,\text{mH}$, and $V_{dc} = 200$ V. The five-phase reference current amplitude is at first kept at 4 A and then stepped up to 8 A in the first quarter of the second fundamental cycle. The optimization algorithm is implemented using the "s" function of MATLAB.

7.7.3 Using Eleven Vectors with $\gamma = 0$

The resulting waveforms for using only 10 outer large space vectors and 1 zero vector are shown in Figure 7.73. The current-tracking error is forced to become zero only in the α-β plane. It is evident from Figure 7.73 that the α-β components of the current are sinusoidal, while there are extra x-y components produced because of the vector of the x-y plane, which are firmly tied to the vectors of the α-β plane. Moreover, the optimizer does not take into account the tracking error in the second plane. The loci of the current in the two different planes are illustrated in Figure 7.73. The α-β plane shows a circular trajectory, while the x-y plane is irregular but close to the circle. To further explore the performance of the controller current and inverter voltage spectrum, they are depicted in Figure 7.73f. The total harmonic distortion (THD) is calculated up to 500 harmonics for both voltage and current waveforms. The current waveform shows a THD of 19.67% and the highest lower-order harmonic is third equal to 19.55% of the fundamental (7.9 A). The voltage spectrum offers a THD of 42.36% and the highest lower-order harmonic is one-third with their magnitude equal to 25.82% of the fundamental (82.7 V). The switching harmonics are distributed and the average could be estimated around 7–8 kHz.

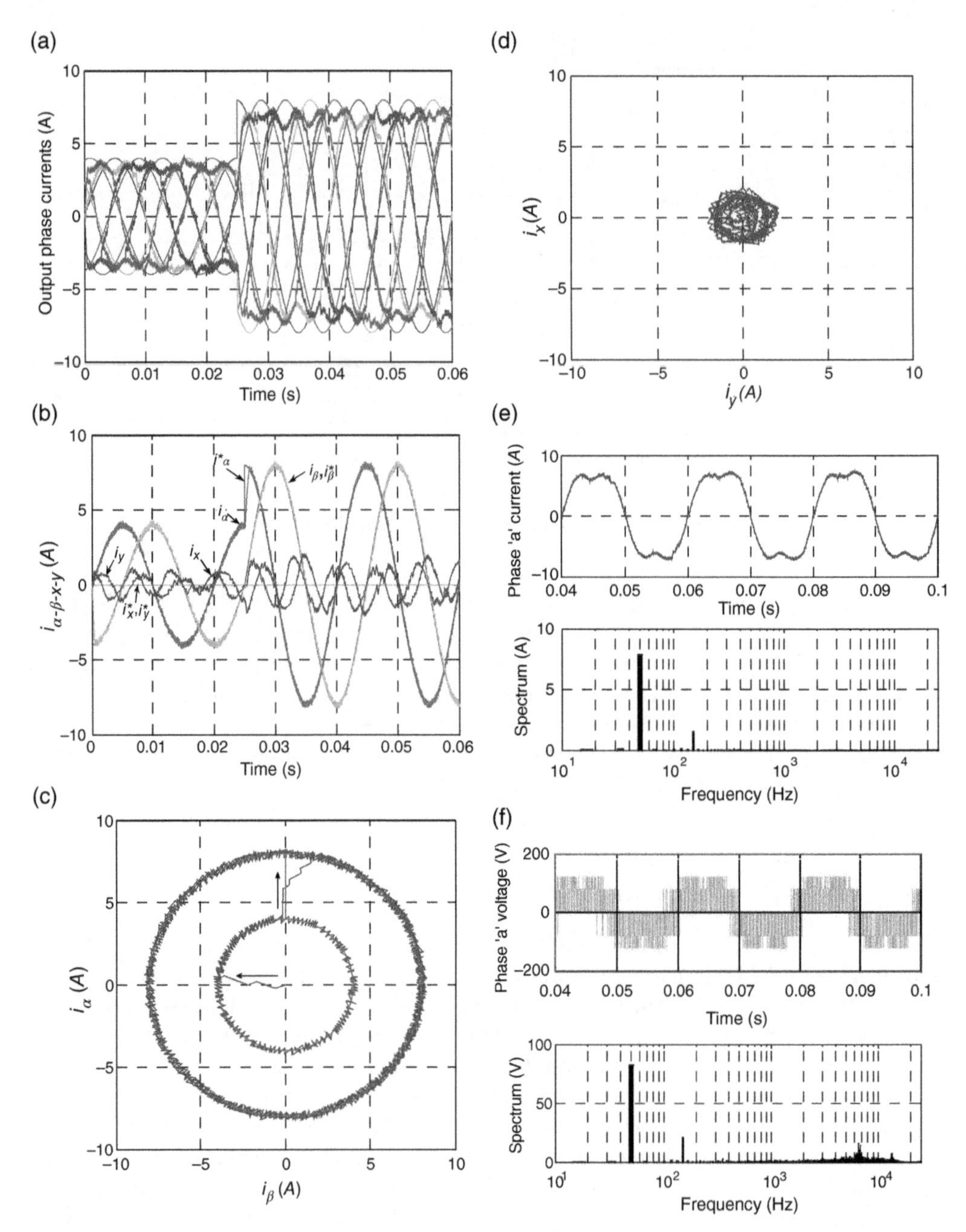

Figure 7.73 MPC performance with $\gamma = 0$: (a) actual and reference five-phase currents; (b) transformed current; (c) trajectory of α-axis current; (d) trajectory of x-axis current; (e) spectrum of current; and (f) spectrum of voltage

7.7.4 Using Eleven Vectors with γ = 1

The simulation results are further shown to eliminate the x-y components of the currents, as they are undesirable for distributed winding AC machine drives (causing extra losses). The cost function thus includes the x-y current-tracking error, and the resulting waveforms are shown in Figure 7.74. The results obtained show effective elimination of the x-y components, leading

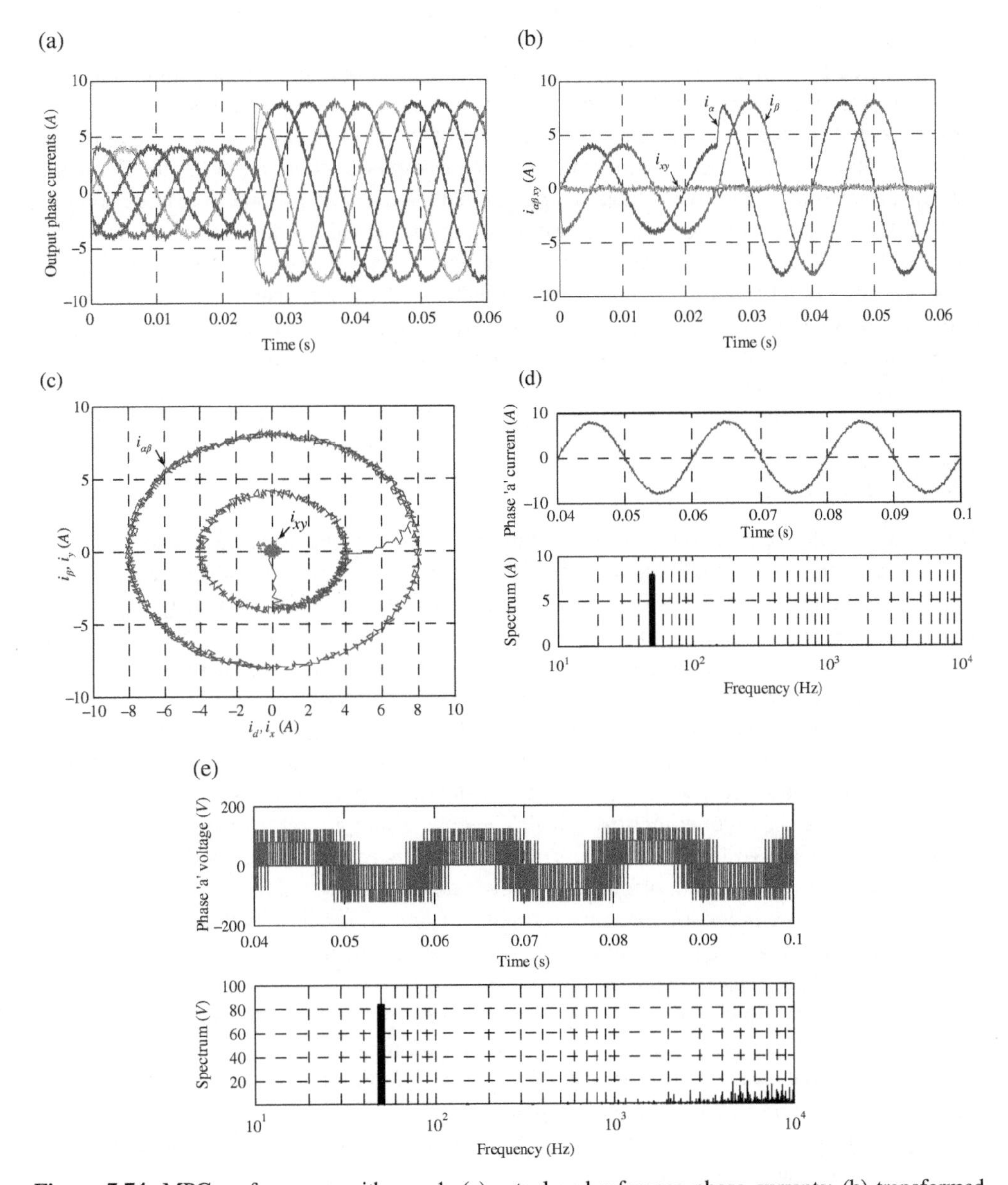

Figure 7.74 MPC performance with $\gamma = 1$: (a) actual and reference phase currents; (b) transformed current; (c) trajectory of α-β and x-y axis currents; (d) spectrum of current; and (e) spectrum of phase voltage

to sinusoidal output voltages and currents. Hence, using only 11 large outer space vector sets, sinusoidal output voltages and currents are obtained that were not possible with the SVPWM technique.

7.8 Summary

This chapter discussed the structure of a five-phase drive system and its control properties and elaborated the advantages and limitations of a five-phase induction motor drive system and identified its potential application areas. Modeling of a five-phase voltage source inverter was illustrated for step mode of operation and PWM mode of operation. Fast Fourier transform (FFT) analysis was given for the step mode of operation of five-phase VSI. Experimental results were also given for step mode of operation of the five-phase voltage source inverter. This was followed by the phase variable model of a five-phase induction machine. The model was then transformed into two orthogonal planes, namely *d-q* and *x-y* and the torque equation was obtained. The vector control principle was then presented for the five-phase induction motor drive system. The difference between the vector control principle applied to a three-phase system and a five-phase system was highlighted. The theoretical background was supported by the simulation model using MATLAB/Simulink software. The modern control technique such as field-oriented control, DTC, and MPC is discussed with reference to the application to a five-phase IM.

7.9 Problems

D 7.1: Calculate the rms voltage of the output phase voltage and adjacent line and nonadjacent line voltage for 10-step mode of operation of a five-phase voltage source inverter operating at 100-Hz fundamental frequency in terms of DC-link voltage V_{dc}.

D 7.2: A five-phase voltage source inverter is supplying a five-phase load and operating in 10-step mode of operation. The fundamental output frequency is assumed to be 25 Hz. Calculate the conduction period of each power switch of the inverter.

D 7.3: A five-phase induction machine is supplied from a five-phase voltage source inverter. The five-phase VSI is controlled using SVPWM technique and sinusoidal supply is to be supplied to the motor. The inverter switching frequency is 5 kHz and the DC bus voltage at the inverter input is 300 V. The inverter output frequency is 50 Hz. The reference phase voltage rms is 100 V. Identify the space vectors that will be applied and their time of application for the time instants: 1, 5, and 10.5 ms.

D 7.4: A five-phase induction machine is supplied from a five-phase voltage source inverter and controlled in square-wave mode with conduction angle as 144 degrees. The DC-link voltage is assumed as 1 p.u.,

(a) Write down the switching state and corresponding leg voltage values and leg voltage space vectors in the stationary reference frame.

(b) Calculate the phase-to-neutral voltages and determine the corresponding space vectors in both *d-q* and *x-y* planes.

(c) Determine the possible line-to-line voltages and corresponding space vectors in *d-q* and *x-y* planes.

D 7.5: Determine the effect of dead time on the output phase-to-neutral voltage of a five-phase voltage source inverter operating in 10-step mode. Consider the dead time of 20 and 50 μs. (*Hint: Develop the Simulink model to solve this problem*)

D 7.6: A five-phase induction machine rated at 200 V (adjacent line voltage) is supplied from a five-phase PWM voltage source inverter operating at 5-kHz switching frequency and employing a simple carrier-based sinusoidal PWM scheme. The motor is assumed to be operating with *v/f = constant* control law and the rated voltage is applied at 50 Hz. To enhance the output of the inverter, fifth-harmonic injection PWM scheme is adopted. Determine the amplitude and frequency of the fifth-harmonic components to be injected for motor operation at 50, 25, and 10 Hz.

D 7.7: A five-phase induction machine is supplied from a five-phase voltage source inverter and the inverter is modulated using SVPWM with sinusoidal output. The operation of the five-phase motor is assumed in *v/f = constant* mode. The inverter switching frequency is 2 kHz and the DC bus voltage at the inverter input is 600 V.

(a) Determine the maximum value of the inverter output phase-to-neutral fundamental rms voltage for this PWM mode. This is treated as the rated motor's voltage and the rated frequency of the motor is assumed as 60 Hz.

(b) Find the set of the space vectors that will be applied and the time of their applications for the output frequency of 25 Hz, for the time instants of: 1.5, 4, and 8 ms

(c) The five-phase induction motor is now supplied from 120 Hz. Calculate the new reference space vector and determine which of the space vectors will be applied and for how long, for the time instants of 2, 10, and 15 ms.

D 7.8: Develop a complete MATLAB/Simulink model of a four-pole, 220-V rms, five-phase induction motor drive system for open-loop *v/f = constant* control mode. The five-phase voltage source inverter may be simulated using IGBT model from "Simpower System" blocksets. The inverter dead time may be given as 50 ms and the switching frequency of the inverter is assumed as 5 kHz. The same model should be implemented in real time with TMS320F2812 DSP and the voltage signal to the DSP, corresponding to rated speed is 3.3 V. The speed reference should be given as voltage signal to the control block. Per-phase equivalent circuit parameters of the 50 Hz, five-phase induction machine are $R_s = 10\ \Omega$, $R_r = 6.3\ \Omega$, $L_{ls} = L_{lr} = 0.04$ H, and $L_m = 0.42$ H. Inertia and the number of pole pairs are equal to 0.03 kg/m^2 and 2, respectively. Rated phase current, phase-to-neutral voltage and per-phase torque are 2.1 A, 220 V, and 1.67 Nm, respectively.

D 7.9: Determine the voltages in the α-β and x-y plane for the following phase voltages in the stationary reference frame. Use the transformation matrix (assume star-connected load and $\alpha = 72$ degrees). You can use MATLAB/Simulink simulation and plot the voltages in the two orthogonal planes. (Assume $V_3 = 0.5\ V_1$ and $V_5 = 0.25\ V_1$)

References

1. Ward, E. E. and Harer, H. (1969) Preliminary investigation of an inverter-fed 5-phase induction motor. *Proc. Int. Electr. Eng.*, 116(6), 980–984.
2. Klingshirn, E. A. (1983) High phase order induction motors (part I and II). *IEEE Trans. Power App. Syst.*, PAS_102(1), 47–59.

3. Pavithran, K. N., Parimelalagan, R., and Krishnamurthy, M. R. (1988) Studies on inverter-fed five phase induction motor drive. *IEEE Trans. Power Electron.*, 3(2), 224–233.
4. Xu, H., Toliyat, H. A., and Petersen, L. J. (2009) Five-phase induction motor drives with DSP-based control system. *Proc. IEEE Int. Electr. Mach. And Drives Conf. IEMDC2001*. Cambridge, MA, pp. 304–309.
5. Xu, H., Toliyat, H. A., and Petersen, L. J. (2001) Rotor field oriented control of a five-phase induction motor with the combined fundamental and third harmonic injection. *Proc. IEEE Applied Power Elec. Conf. APEC*. Anaheim, CA, pp. 608–614.
6. Iqbal, A. and Levi, E. (2005) Space vector modulation scheme for a five-phase voltage source inverter. *Proc. Europ. Power Electr. Appl. Conf., EPE*. Dresden, CD-ROM, paper 0006.
7. Ojo, O., Dong, G., and Wu, Z. (2006) Pulse width modulation for five-phase converters based on device turn-on times. *Proc. IEEE Ind. Appl. Soc. Ann. Mtg IAS*. Tampa, FL, CD-ROM, paper IAS15, p. 7.
8. Ryu, H. M., Kim, J. H., and Sul, S. K. (2005) Analysis of multi-phase space vector pulse width modulation based on multiple d-q space concept. *IEEE Trans. Power Electron.*, 20(6), 1364–1371.
9. De Silva, P. S. N., Fletcher, J. E., and Williams, B.W. (2004) Development of space vector modulation strategies for five-phase voltage source inverters. *Proc. IEE Power Elect., Mach. Drives Conf., PEMD*. Edinburgh, pp. 650–655.
10. Dujic, D., Jones, M., and Levi, E. (2009) Generalised space vector PWM for sinusoidal output voltage generation with multiphase voltage source inverters. *Int. J. Ind. Electron. Drives*, 1(1), 1–13.
11. Iqbal, A. and Levi, E. (2006) Space vector PWM techniques for sinusoidal output voltage generation with a five-phase voltage source inverter. *Elect. Power Comp. Syst.*
12. Duran, M. J., Toral, S., Barrero, F., and Levi, E. (2007) Real time implementation of multidimensional five-phase space vector pulse width modulation. *Electron. Lett.* 43(17), 949–950.
13. Ryu, H. M., Kim, J. H., and Sul, S. K. (2005) Analysis of multiphase space vector pulse-width modulation based on multiple *d-q* spaces concept. *IEEE Trans. Power Electron.*, **20**(6), 1364–1371.
14. Zheng, L., Fletcher, J. E., Williams, B. W., and He, X. (2011) A novel direct torque control scheme for a sensorless five-phase induction motor drive. *IEEE Trans. Ind. Electron.*, 58(2), 503–513.
15. Liliang, G. and Fletcher, J. E. (2010) A space vector switching strategy for a 3-level five-phase inverter drives. *IEEE Trans. Ind. Electron.*, 57(7), 2332–2343.
16. Gao, L., Fletcher, J. E., and Zheng, L. (2011) Low-speed control improvements for a two level five-phase inverter-fed induction machine using classic direct torque control. *IEEE Trans. Ind. Electron.*, 58(7), 2744–2754.
17. Payami, S. and Behera, R. K. (2017) An improved DTC technique for low-speed operation of a five-phase induction motor. *IEEE Trans. Ind. Electron.*, 64(5), 3513–3523.
18. Khaldi, B. S., Abu-Rub, H., Iqbal, A., Kennel, R., Mahmoudi, M. O., and Boukhetala, D. (2011) Sensorless direct torque control of five-phase induction motor drives. *IECON 2011 37th Ann. Conf. IEEE Ind. Electr. Soc.* Melbourne, VIC, pp. 3501–3506.
19. Tatte, Y. N., Aware, M. V., Pandit, J. K., and Nemade, R. (2018) Performance improvement of three-level five-phase inverter-fed DTC-controlled five-phase induction motor during low-speed operation. *IEEE Trans. Ind. Appl.*, 54(3), 2349–2357.
20. Payami, S., Behera, R. K., Iqbal, A., and Al-Ammari, R. (2015) Common-mode voltage and vibration mitigation of a five-phase three-level NPC inverter-fed induction motor drive system. *IEEE J. Emerg. Sel. Topics Power Electron.*, 3(2), 349–361.
21. Singh, G. K. (2002) Multi-phase induction machine drive research: A survey. *Electr. Pow. Syst. Res.*, 61, 139–147.

22. Jones, M. and Levi, E. (2002) A literature survey of state-of-the-art in multiphase AC drives. *Proc. 37th Int. Univ. Power Eng. Conf. UPEC*. Stafford, UK, pp. 505–510.
23. Levi, E., Bojoi, R., Profumo, F., Toliyat, H. A., and Williamson, S. (2007) Multi-phase induction motor drives: A technology status review. *IET Electr. Power Appl.*, 1(4), 489–516.
24. Levi, E. (2008) Guest editorial. *IEEE Trans. Ind. Electron.*, 55(5), 1891–1892.
25. Levi, E. (2008) Multi-phase machines for variable speed applications. *IEEE Trans. Ind. Electron.*, 55(5), 1893–1909.
26. Toliyat, H. A., Rahimian, M. M., and Lipo, T. A. (1991) dq Modeling of five-phase synchronous reluctance machines including third harmonic air-gap MMF. *Proc. IEEE Ind. Appl. Soc. Ann. Mtg*, pp. 231–237.
27. Toliyat, H. A., Rahmania, M. M., and Lipo, T. A. (1992) A five phase reluctance motor with high specific torque. *IEEE Trans. Ind. Appl.*, 28(3), 659–667.
28. Duran, M. J., Salas, F., and Arahal, M. R. (2008) Bifurcation analysis of five-phase induction motor drives with third-harmonic injection. *IEEE Trans. Ind. Electron.*, 55(5), 2006–2014.
29. Shi, R., Toliyat, H. A., and El-Antably, A. (2001) Field oriented control of five-phase synchronous reluctance motor drive with flexible 3rd harmonic current injection for high specific torque. *Proc. 36th IEEE Ind. Appl. Conf.* pp. 2097–2103.
30. Williamson, S. and Smith, S. (2003) Pulsating torques and losses in multiphase induction machines. *IEEE Trans. Ind. Appl.* 39(4), 986–993.
31. Green, S., Atkinson, D. J., Jack, A. G., Mecrow, B. C., and King, A. (2003) Sensorless operation of a fault tolerant PM drive. *IEE Proc. Electr. Power Appl.*, 150(2), 117–125.
32. Wang, J. B., Atallah, K., and Howe, D. (2003) Optimal torque control of fault-tolerant permanent magnet brushless machines. *IEEE Trans. Magn.*, 39(5), 2962–2964.
33. Fu, J. R. and Lipo, T. A. (1994) Disturbance-free operation of a multiphase current-regulated motor drive with an opened phase. *IEEE Trans. Ind. Appl.*, 30(5), 1267–1274.
34. Suman, D. and Parsa, L. (2008) Optimal current waveforms for five-phase permanent magnet motor drives under open-circuit fault. *Proc. IEEE Power Ener. Soc. Gen. Mtg: Conv. Del. Elect. Ener. 21st Cent.*, pp. 1–5.
35. Parsa, L. and Toliyat, H. A. (2004) Sensorless direct torque control of five-phase interior permanent magnet motor drives. *Proc. IEEE Ind. Appl. Soc. Mtg, IAS*, 2, 992–999.
36. Chiricozzi, E. and Villani, M. (2008) Analysis of fault-tolerant five-phase IPM synchronous motor. *Proc. IEEE Int. Symp. Ind. Elect.*, pp. 759–763.
37. Toliyat, H. A. (1998) Analysis and simulation of five-phase variable speed induction motor drives under asymmetrical connections. *IEEE Trans. Power Electron.*, 13(4), 748–756.
38. Husain, T., Ahmed, S. K. M., Iqbal, A., and Khan, M. R. (2008) Five-phase induction motor behaviour under faulted conditions. *Proc. INDICON*, pp. 509–513.
39. Casadei, D., Mengoni, M., Serra, G., Tani, A., and Zarri, L. (2008) Optimal fault-tolerant control strategy for multi-phase motor drives under an open circuit phase fault condition. *Proc. 18th Int. Conf. Elect. Mach., ICEM*, pp. 1–6.
40. Bianchi, N., Bologani, S., and Pre, M. D. (2007) Strategies for fault-to-lerant current control of a fivephase permanent magnet motor. *IEEE Trans. Ind. Appl.*, 43(4), 960–970.
41. Jacobina, C. B., Freitas, I. S., Oloveira, T. M., da Silva, E. R. C., and Lima, A. M. N. (2004) Fault tolerant control of five-phase AC motor drive. *35th IEEE Power Elect. Spec. Conf.*, pp. 3486–3492.
42. Heising, C., Oettmeier, M., Bartlet, R., Fang, J., Staudt, V., and Steimel, A. (2009) Simulation of asymmetric faults of a five-phase induction machine used in naval applications. *Proc. 35th IEEE Ind. Elect. Conf. IECON*, pp. 1298–1303.

43. Parsa, L. and Toliyat, H. A. (2005) Five-phase permanent magnet motor drives for ship propulsion applications. *Proc. IEEE Electric Ship Technol. Symp.*, pp. 371–378.
44. Golubev, A. N. and Ignatenko, S. V. (2000) Influence of number of stator-winding phases on the noise characteristics of an asynchronous motor. *Russ. Electr. Eng.*, 71(6), 41–46.
45. McCleer, P. J., Lawler, J. S., and Banerjee, B. (1993) Torque notch minimization for five-phase quasi square wave back EMF PMSM with voltage source drives. *Proc. IEEE Int. Symp. Ind. Elect.*, pp. 25–32.
46. Zhang, X., Zhnag, C., Qiao, M., and Yu, F. (2008) Analysis and experiment of multi-phase induction motor drives for electrical propulsion. *Proc. Int. Conf. Elect. Mach., ICEM*, pp. 1251–1254.
47. Sadehgi, S. and Parsa, L. (2010) Design and dynamic simulation of five-phase IPM machine for series hybrid electric vehicles. *Proc. Green Tech. Conf.*, pp. 1–6.
48. Chan, C. C., Jiang, J. Z., Chen, G. H., Wang, X. Y., and Chau, K. T. (1994) A novel polyphase multipole square wave PM motor drive for electric vehicles. *IEEE Trans. Ind. Appl.*, 30(5), 1258–1266.
49. Jayasundara, J. and Munindradasa, D. (2006) Design of multi-phase in-wheel axial flux PM motor for electric vehicles. *Proc. Int. Conf. Ind. Info. Syst.*, pp. 510–512.
50. Abolhassani, M. T. (2005) A novel multiphase fault tolarent high torque density PM motor drive for traction application. *Proc. IEEE Int. Conf. Elect. Mach.*, pp. 728–734.
51. Scuiller, F., Charpentier, J., Semail, E., and Clenet, S. (2007) Comparision of two 5-phase PM machine winding configurations, application on naval propulsion specification. *Proc. Elect. Mach. Drives Conf. IEMDC*, pp. 34–39.
52. Jones, M. (2005) A novel concept of series-connected multi-phase multi-motor drive systems. PhD. Thesis, Liverpool John Moores University, Liverpool, UK.
53. Chen, S., Lipo, T. A., and Fitzgerald, D. (1996) Sources of induction motor gearing currents caused by PWM inverters. *IEEE Trans. Ener. Convers.*, 11, 25–32.
54. Wang, F. (2000) Motor shaft voltages and bearing currents and their reduction in multi-level medium voltage PWM voltage source inverter drive applications. *IEEE Trans. Ind. Appl.*, 36, 1645–1653.
55. Rodriguez, J., Moran, L., Pontt, J., Osorio, R., and Kouro, S. (2003) Modeling and analysis of common-mode voltages generated in medium voltage PWM-CSI drives. *IEEE Trans. Power Electron.*, 18(3), 873–879.
56. Leonhard, W. (1985) *Control of Electrical Drives*. Springer-Verlag.
57. Grahame, D. and Lipo, T. A. (2003) *Pulse Width Modulation for Power Converters*. IEEE Power Engineering Series by Wiley Inter-Science.
58. Kazmierkowski, M. P., Krishnan, R., and Blaabjerg, F. (2002) *Control in Power Electronics-Selected Problems*. Academic Press, New York.
59. Iqbal, A. and Moinuddin, S. (2009) Comprehensive relationship between carrier-based PWM and space vector PWM in a five-phase VSI. *IEEE Trans. Power Electron.*, 24(10), 2379–2390.
60. Iqbal, A., Levi, E., Jones, M., and Vukosavic, S. N. (2006) Generalised Sinusoidal PWM with harmonic injection for multi-phase VSIs. *IEEE 37th Power Elect. Spec. Conf. (PESC)*, Jeju, Korea, 18–22 June, CD_ROM, paper no. ThB2–3, pp. 2871–2877.
61. Ojo, O. and Dong, G. (2005) Generalized discontinuous carrier-based PWM modulation scheme for multi-phase converter machine systems. *Proc. IEEE Ind. Appl. Soc. Ann. Mtg, IAS*, Hong Kong, CD-ROM, paper IAS 38, p. 3.
62. Jones, M., Dujic, D., Levi, E., and Vukosavic, S. N. (2009) Dead-time effects in voltage source inverter fed multi-phase AC motor drives and their compensation. *Proc. 13th Int. Power Elect. Conf. EPE'09*, Barcelona, 5–8 September, pp. 1–10.

63. Rchalet, J. (1993) Industrial applications of model based predictive control. *Automatica*, 29(5), 1251–1274.
64. Qin, S. J. and Badgwell, T. A. (2003) A survey of industrial model predictive control technology. *Control Eng. Pract.*, 11, 733–764.
65. Linder, A. and Kennel, R. (2005) Model predictive control for electric drives. *Proc. 36th Ann. IEEE PESC*. Recife, Brazil, pp. 1793–1799.
66. Kouro, S., Cortes, P.,Vargas, R., Ammann, U., and Rodriguez, J. (2009) Model predictive control:A simple and powerful method to control power converters. *IEEE Trans. Ind. Electron.*, 56(6), 1826–1839.
67. Cortes, P., Kazmierkowski, P., Kennel, R. M., Quevedo, D. E., and Rodriguez, J. (2008) Predictive control in power electronics and drives. *IEEE Trans. Ind. Electron.*, 55(12), 4312–4321.
68. Barrero, F., Arahal, M. R., Gregor, R., Toral, S., and Duran, M. J. (2009) A proof of concept study of predictive current control for VSI driven asymmetrical dual three-phase AC machines. *IEEE Trans. Ind. Electron.*, 56(6), 1937–1954.
69. Arahal, M. R., Barrero, F., Toral, S., Duran, M., and Gregor, R. (2009) Multi-phase current control using finite state model predictive control. *Control Eng. Pract.*, 17, 579–587.
70. Maciejowski, J. M. (2002) *'Predictive Control with Constraints*, Prentice Hall.
71. Mayne, D. Q., Rawlings, J. B., Rao, C. V., and Scokaert, P. O. M. (2000) Constrained model predictive control: Stability and optimality. *AUTOMATICA*, 36(6), 789–814.
72. Lazar, M., Heemels, W., Bemporad, A., and Weiland, S. (2007) Discrete-time non-smooth NMPC: Stability and robustness. *Assess. Fut. Direc. NMPC*, LNCIS. Springer, Heidelberg, 358, pp. 93–103.
73. Magni, L. and Scattolini, R. (2007) Robustness and robust design of MPC for non-linear discrete time systems. In: Findeisen, R., Allgöwer, F., and Biegler, L. T. (eds) *Assessment and Future Directions of Nonlinear Model Predictive Control*. Lecture Notes in Control and Information Sciences, vol 358. Springer, Berlin, Heidelberg. https://doi.org/10.1007/978-3-540-72699-9_19.
74. Franklin, G. F., Powell, J. D., and Emami-Naeini, A. (1994) *Feedback Control of Dynamics Systems*, 3rd edn. Addison-Wesley.
75. Bemporad, A., Morari, M., Dua, V., and Pistikopoulos, E. N. (2002) The explicit linear quadratic regulator for constrained systems. *Automatica*, 38(1), 3–20.

8

Sensorless Speed Control of AC Machines

8.1 Preliminary Remarks

Sensorless speed control of induction machine (IM) drives has, in the past decade, become a mature technology for a wide speed range [1–3].

The elimination of the rotor speed sensor, without affecting performance, is a major trend in advanced drives control systems [1]. The advantages of speed sensorless AC drives are reduced hardware complexity, lower costs, elimination of the sensor cable, better noise immunity, increased reliability, access to both sides of the shaft, less maintenance requirements, and higher robustness. An encoder is expensive and a problematic factor. The special motor shaft extension increases the drive's price. The use of encoders affects the reliability, particularly in hostile environments. In general, the operation in explosive, corrosive, or chemically aggressive environments requires a motor without a speed sensor.

A variety of different solutions for sensorless AC drives have been proposed, mostly in the past two decades. Their advantages and limits are reviewed in many survey papers [1–60].

Many methods are generally accepted as better solutions for high sensorless performance, for example, model reference adaptive system (MRAS), Kalman Filters, adaptive nonlinear flux observer, sliding mode observers, and other improvements [7–14].

8.2 Sensorless Control of Induction Motor

Two basic approaches are used for sensorless control. The first includes methods that model the induction motor by its state equations [1]. A sinusoidal magnetic field in the air gap is assumed.

High Performance Control of AC Drives With MATLAB®/Simulink, Second Edition.
Haitham Abu-Rub, Atif Iqbal, and Jaroslaw Guzinski.

Companion website: www.wiley.com/go/aburubcontrol2e

The models are either implemented as open-loop structures, like the stator model [1, 11], or as closed-loop models, like adaptive observers [5, 17]. The adaptive flux observers are now receiving considerable attention and many achieving new solutions because of their high precision and relative robustness against machine parameter deviation [1, 10].

Open-loop models and even adaptive observers have stability limits at very low stator frequencies. The rotor-induced voltage at such operating points is then zero or close to zero, which renders the induction motor an unobservable system.

The basic limitation for sensorless operation is the DC offset components in the stator current and voltage acquisition channels at very low speeds [1]. At lower speeds, voltage distortions caused by the PWM inverter become significant.

The second approach used for low speed sensorless operation is the signal injection technique [1]. Carrier injection methods for sensorless control are sophisticated, and the design must match the properties of the motor [1, 4, 18, 41]. This makes the method unfeasible for practical application, so will not be discussed in this book.

Eliminating the speed sensor from the drive system requires estimation of the state variables, for example, motor speed and machine flux using stator variables. In this chapter, we will discuss a few simple solutions for sensorless AC motor drives.

8.2.1 *Speed Estimation Using Open-Loop Model and Slip Computation*

The rotor speed in the vector control scheme can be computed from a difference between the synchronous speed and the slip speed, using the following known equation:

$$\hat{\omega} = \frac{\psi_{s\alpha}\psi_{s\beta(k)} - \psi_{s\beta}\psi_{s\alpha}}{\psi_s^2} - \hat{\omega}_{2r} \tag{8.1}$$

where

$$\omega_{si} = \frac{R_r\left(\psi_{s\alpha}i_{s\beta} - \psi_{s\beta}i_{s\alpha}\right)}{\psi_s^2} \tag{8.2}$$

The stator flux components (or rotor flux in the rotor flux-oriented system) can be computed using open-loop models, as explained in earlier chapters, or by using closed-loop observer systems. These observers can be used for speed computation in addition to the flux components, as will be discussed in the following sections.

8.2.2 *Closed-Loop Observers*

The accuracy of the open-loop models decreases as the mechanical speed decreases [1]. The performance depends on how the machine parameters can be identified. Those parameters have the largest influence on system operation at lower speeds. The robustness against parameter deviation can be significantly improved by using closed-loop observers.

8.2.2.1 Observer 1

The Observer 1 is based on the voltage $\boldsymbol{\psi}_s$, $\boldsymbol{\psi}_r$ model of the induction motor, with combination of the rotor and stator fluxes and stator current relationships [61]:

$$\tau_s' \frac{d\boldsymbol{\psi}_s}{d\tau} + \boldsymbol{\psi}_s = k_r\boldsymbol{\psi}_r + \mathbf{u}_s \tag{8.3}$$

To prevent problems of voltage drift and to offset errors, instead of pure integrators, low pass filters are used. The limitation of the estimated stator flux is tuned to the stator flux nominal value. In addition, the extra compensation part is added, as presented in [62].

Based on equations (8.9)–(8.11), the rotor Observer 1 equations are [61]

$$\frac{d\hat{\boldsymbol{\psi}}_{s\alpha}}{d\tau} = \frac{-\hat{\boldsymbol{\psi}}_{s\alpha} + k_r\hat{\boldsymbol{\psi}}_{r\alpha} + \hat{\mathbf{u}}_{s\alpha}}{\tau_s'} - k_{ab}\left(\mathbf{i}_1 - \hat{\mathbf{i}}_1\right) \tag{8.4}$$

$$\hat{\boldsymbol{\psi}}_r = \frac{1}{k_r}\left(\hat{\boldsymbol{\psi}}_s - \sigma L_s \mathbf{i}_s\right) \tag{8.5}$$

where k is observer gain.

The estimated value of the $\mathbf{i}_s$ current appearing in equation (8.5) is

$$\hat{\mathbf{i}}_s = \frac{\hat{\boldsymbol{\psi}}_s - k_r\hat{\boldsymbol{\psi}}_r}{\sigma L_s} \tag{8.6}$$

The rotor flux Observer I structure is presented in Figure 8.1.

The rotor flux calculator receives the observed stator flux components and the measured stator current components, both in the stationary $\alpha-\beta$ axis frame. These input signals require high precision, which is achieved by using the correcting error signal in equation (2.4) derived from the difference between the estimated and the measured stator current vectors. The stator current

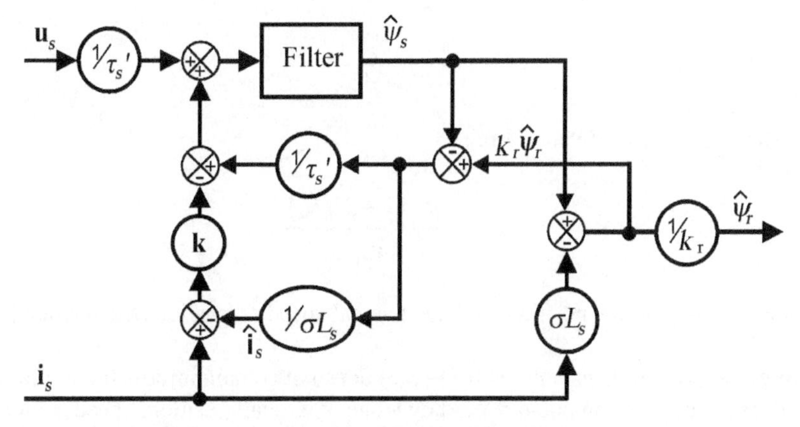

Figure 8.1 Structure of the proposed flux Observer I system

calculator in equation (2.6) receives the estimated stator flux and rotor flux. Any existing error may be further eliminated by providing the current error as a negative feedback signal in the stator flux observer system. The stator current error signal $(\mathbf{i}_s - \hat{\mathbf{i}}_s)$ in the negative feedback of the flux observer is multiplied by a gain k to increase its robustness and its stability behavior, as well as to help eliminate additional disturbances related to the DC drift and computational and measurement errors.

Generally, the observer matrix gain k can be tuned experimentally and set at a constant real value. The tuning should be done for a wide range of operating speeds and varying load conditions. Finally, a behaviorally chosen optimum constant value of k should be selected. As was shown in the results sections, in spite of that simple tuning process, the described observer works properly, even for the low motor small voltage frequency supply.

The next calculation is of a rotor flux vector angle needed for transforming from the stationary to the rotating frame, and vice versa for a vector controller. The flux position calculator receives stator flux components in the α-axis and β-axis stationary frame. The derivative of this angle gives the rotor flux speed $\omega_{\psi r}$.

Based on the estimated rotor flux components, the flux magnitude and angle position are

$$|\hat{\boldsymbol{\Psi}}_r| = \sqrt{\hat{\psi}_{r\alpha}^2 + \hat{\psi}_{r\beta}^2} \tag{8.7}$$

$$\hat{\rho}_{\psi r} arc\, tg \frac{\hat{\psi}_{r\beta}}{\hat{\psi}_{r\alpha}} \tag{8.8}$$

If the main goal of the drive controller is to control the torque, then the drive system does not require a rotor speed signal. However, rotor velocity is required to perform closed-loop speed control. In the proposed estimation method, the motor mechanical speed signal is obtained by subtracting the slip from the rotor flux speed:

$$\hat{\omega}_r = \hat{\omega}_{\psi r} - \hat{\omega}_2 \tag{8.9}$$

where the rotor flux pulsation is

$$\hat{\omega}_{\psi r} = \frac{d\hat{\rho}_{\psi r}}{d\tau} \tag{8.10}$$

and rotor slip pulsation is obtained from

$$\hat{\omega}_2 = \frac{\left(\hat{\psi}_{r\alpha} i_{s\beta} - \hat{\psi}_{r\beta} i_{s\alpha}\right)}{|\hat{\boldsymbol{\Psi}}_r|^2} \tag{8.11}$$

The rotor flux magnitude, position, and mechanical speed estimation structure are presented in Figure 8.2.

The Observer I system in equations (8.4)–(8.6) does not contain information on rotor speed, therefore does not require computation or knowledge of its shaft position. Speed computation is performed in a separate block, as shown in equations (8.9)–(8.11), which does not affect the

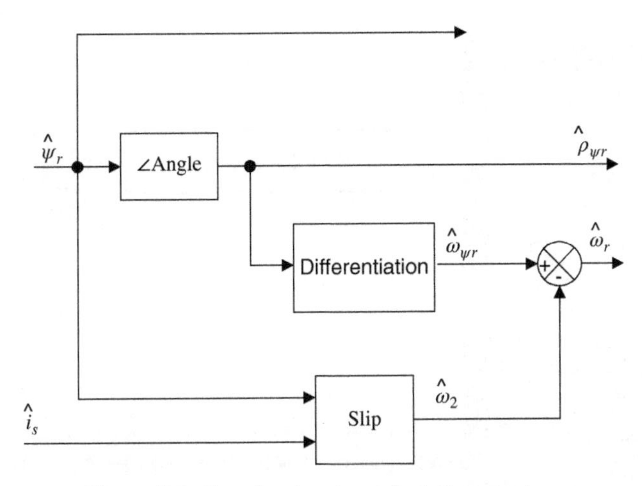

Figure 8.2 Speed estimation calculation structure

precision of the proposed observer. This eliminates any additional error associated with computing or even measuring such signals, particularly at extremely low frequencies. It is robust in that it does not require additional techniques for parameters tuning or DC drift elimination.

8.2.2.1.1 Simulink Model of Observer 1

The observer 1 for the estimation of stator and rotor fluxes is depicted by

$$\frac{d\hat{\psi}_{sx}}{d\tau} = \frac{1}{\tau_s'}\left(-\hat{\psi}_{sx} + k_r\hat{\psi}_{rx} + u_{sx}\right) - k\left(i_{sx} - \hat{i}_{sx}\right) \tag{8.12}$$

$$\frac{d\hat{\psi}_{sy}}{d\tau} = \frac{1}{\tau_s'}\left(-\hat{\psi}_{sy} + k_r\hat{\psi}_{ry} + u_{sy}\right) - k\left(i_{sy} - \hat{i}_{sy}\right) \tag{8.13}$$

$$\hat{\psi}_{rx} = \frac{\hat{\psi}_{sx} - \sigma L_s\hat{i}_{sx}}{k_r} \tag{8.14}$$

$$\hat{\psi}_{ry} = \frac{\hat{\psi}_{sy} - \sigma L_s\hat{i}_{sy}}{k_r} \tag{8.15}$$

The above equations are modeled in Figure 8.3.

8.2.2.1.2 Stator Currents

The stator current components can be computed from

$$\hat{i}_{sx} = \frac{\hat{\psi}_{sx} - k_r\hat{\psi}_{rx}}{\sigma L_s} \tag{8.16}$$

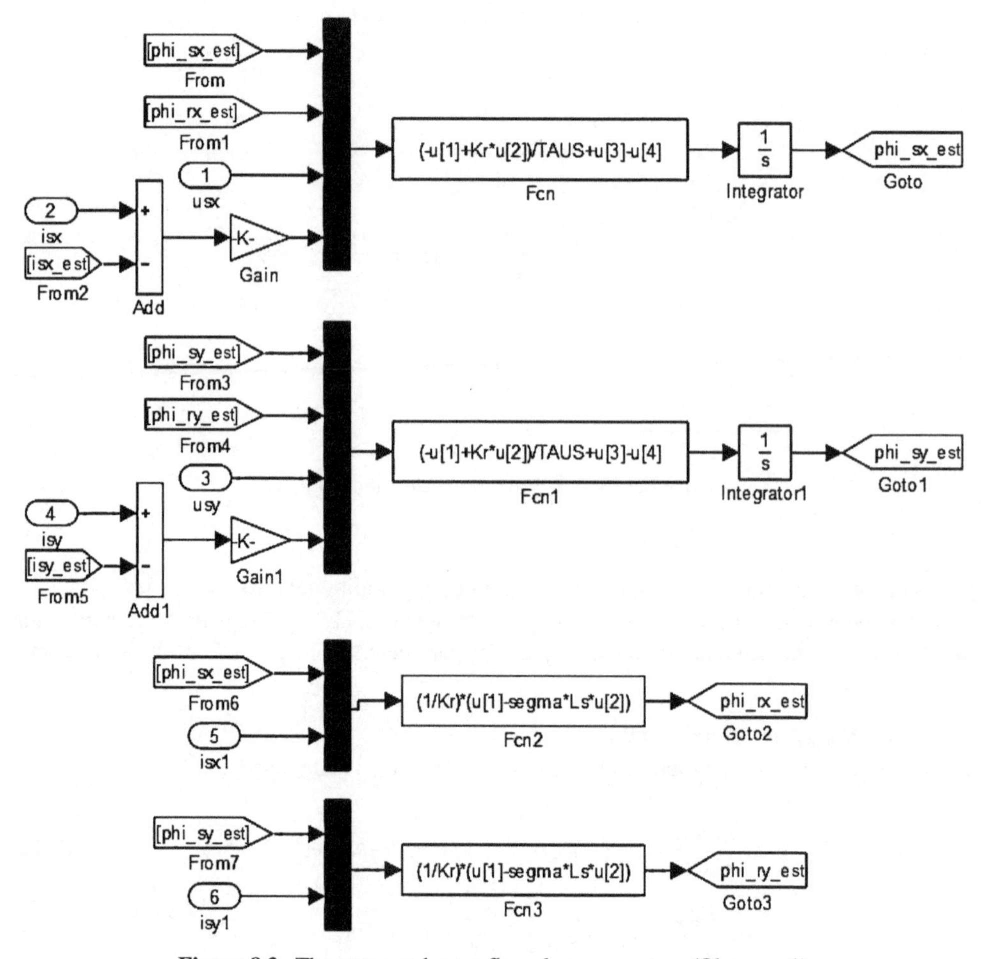

Figure 8.3 The stator and rotor flux observer system (Observer 1)

$$\hat{i}_{sy} = \frac{\hat{\psi}_{sy} - k_r \hat{\psi}_{ry}}{\sigma L_s} \tag{8.17}$$

The above equations are modeled in Figure 8.4.

8.2.2.2 Observer 2

The second flux observer system (Figure 8.5) was recently described in [20]; however, it was combined with an additional observer for speed computation. It is important to note that such a solution, as in [20], did not give satisfactory performance at low-speed ranges. The approach in this paper is to compute the rotor speed, not from an additional speed observer, as in [20], but using slip computation equations (8.1) and (8.2). This approach of speed computation extended the operating point of this observer significantly, more than that presented in [20]. In [20], it

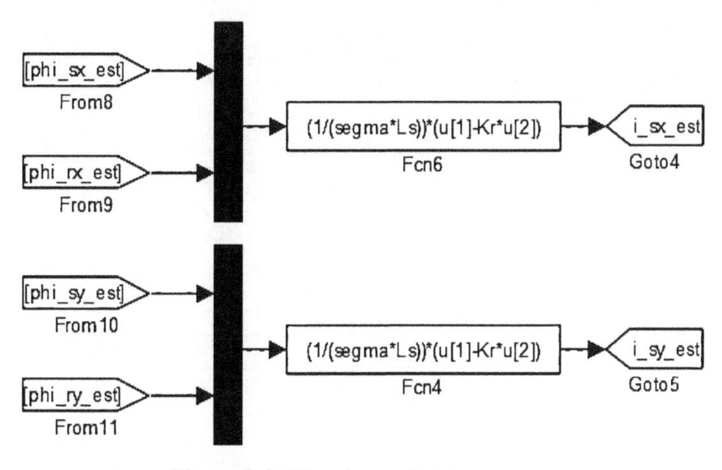

Figure 8.4 The observed stator currents

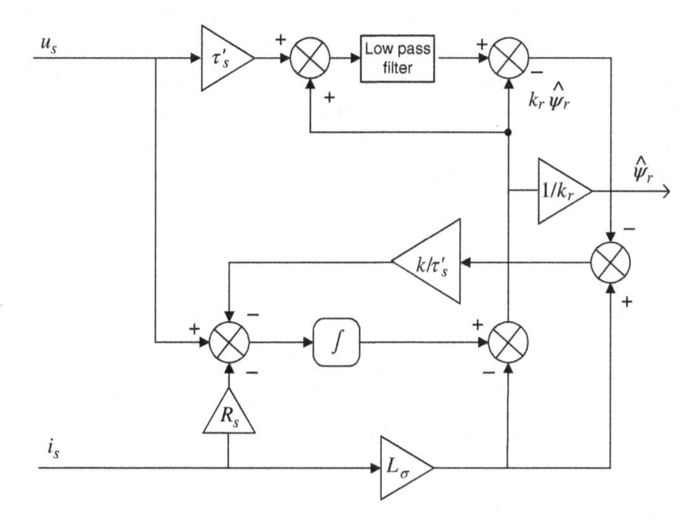

Figure 8.5 Version 2 of observer system (*Source*: Modified from Abu-Rub and Oikonomou [20])

was possible to operate the motor down to 3% of rated speed, while in this procedure all observers operated correctly, even around zero speed. The mathematical description of the third observer is given as [20].

$$\tau'_s \frac{d\hat{\psi}_s}{dt} + \hat{\psi}_s = k_r\hat{\psi}_r + u_s \tag{8.18}$$

$$\tau'_s \frac{d\hat{\psi}'_s}{dt} + \hat{\psi}'_s = k_r\hat{\psi}'_r + u_s \tag{8.19}$$

$$\frac{d\hat{\psi}_s''}{dt} = u_s - R_s i_s - k\Delta u_s \tag{8.20}$$

$$\hat{\psi}_r = \frac{1}{k_r}\left(\hat{\psi}_s'' - L_\sigma i_s\right) \tag{8.21}$$

$$\hat{\psi}_\sigma'' = -L_\sigma i_s \tag{8.22}$$

$$\hat{\psi}_\sigma' = \hat{\psi}_s' - k_r\hat{\psi}_r \tag{8.23}$$

$$\Delta\psi_\sigma = \hat{\psi}_\sigma'' - \hat{\psi}_\sigma' \tag{8.24}$$

$$\Delta u_s = \frac{1}{\tau_s}\Delta\hat{\psi}_\sigma \tag{8.25}$$

which is

$$\Delta u_s = R_s \Delta i_s \tag{8.26}$$

Rotor speed and flux position are computed using the same procedure as earlier described systems equations (8.1–8.8):

The observers' gains were selected experimentally and are shown in the following sections. Although a constant gain was used in our verifications, it was possible to run the controller around zero frequency with sufficient robustness against parameter variation. The influence of the selected gain on the accuracy and dynamics of the observer will be presented in the experimental section of this paper.

8.2.2.3 Observer 3

This kind of observer [63, 64] is a speed observer and based on the Luenberger observer system (Figure 8.6). The observer may be used for rotor flux and speed estimation. It has been developed and discussed for almost 20 years. The differential equations of the speed observer are [63, 64]

$$\frac{d\hat{i}_{sx}}{dt} = a_1\hat{i}_{sx} + a_2\hat{\psi}_{rx} + a_3\omega_r\hat{\psi}_{ry} + a_4u_{sx} + k_i\left(i_{sx} - \hat{i}_{sx}\right) \tag{8.27}$$

$$\frac{d\hat{i}_{sy}}{dt} = a_1\hat{i}_{sy} + a_2\hat{\psi}_{ry} - a_3\omega_r\hat{\psi}_{rx} + a_4u_{sy} + k_i\left(i_{sy} - \hat{i}_{sy}\right) \tag{8.28}$$

$$\frac{d\hat{\psi}_{rx}}{d\tau} = a_5\hat{i}_{sx} + a_6\hat{\psi}_{rx} - \zeta_y - k_2\left(\hat{\omega}_r\hat{\psi}_{ry} - \zeta_y\right) \tag{8.29}$$

$$\frac{d\hat{\psi}_{ry}}{d\tau} = a_5\hat{i}_{sy} + a_6\hat{\psi}_{ry} + \zeta_x + k_2\left(\hat{\omega}_r\hat{\psi}_{rx} - \zeta_x\right) \tag{8.30}$$

$$\frac{d\zeta_x}{d\tau} = k_1\left(i_{sy} - \hat{i}_{sy}\right) \tag{8.31}$$

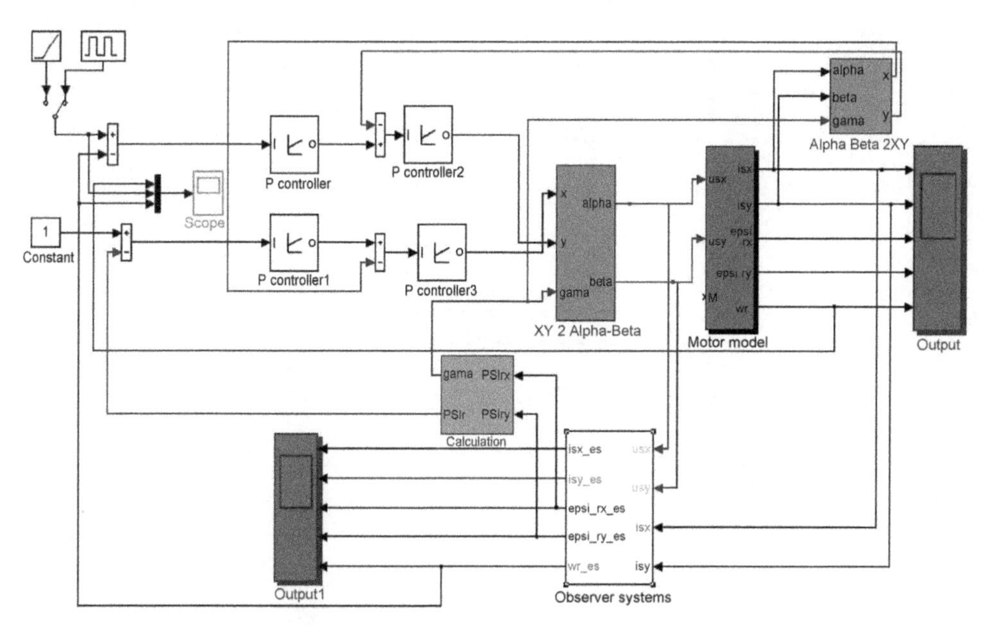

Figure 8.6 Control of induction motor model with the speed observer system [*Observer_three*]

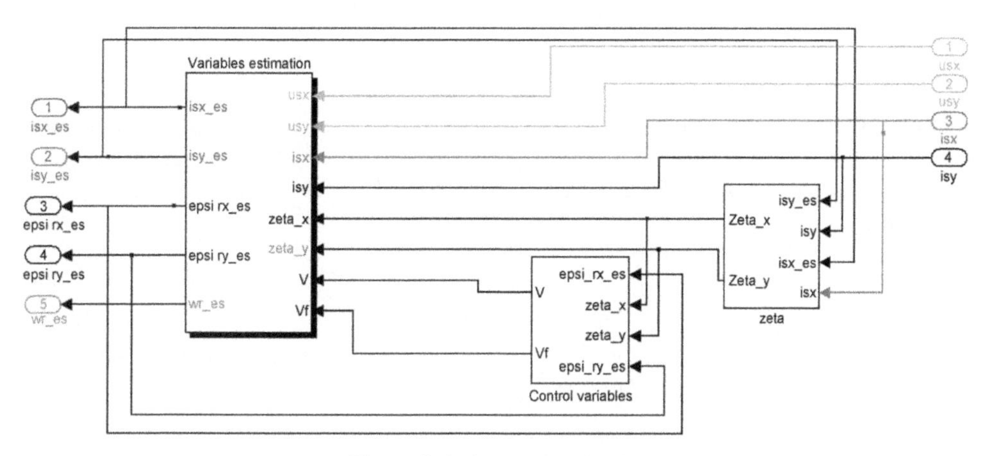

Figure 8.7 Observer 3 details

$$\frac{d\zeta_y}{d\tau} = k_1\left(i_{sx} - \hat{i}_{sx}\right) \tag{8.32}$$

$$\hat{\omega}_r = S\left(\sqrt{\frac{\zeta_x^2 + \zeta_y^2}{\hat{\psi}_{rx}^2 + \hat{\psi}_{ry}^2}} + k_4\left(V - V_f\right)\right) \tag{8.33}$$

where ^ denotes estimated variables; k_1; k_2; k_3 are the observer gains; and S is the sign of speed (Figures 8.7 and 8.8). The values ζ_x, ζ_y are the components of disturbance vector; V is the

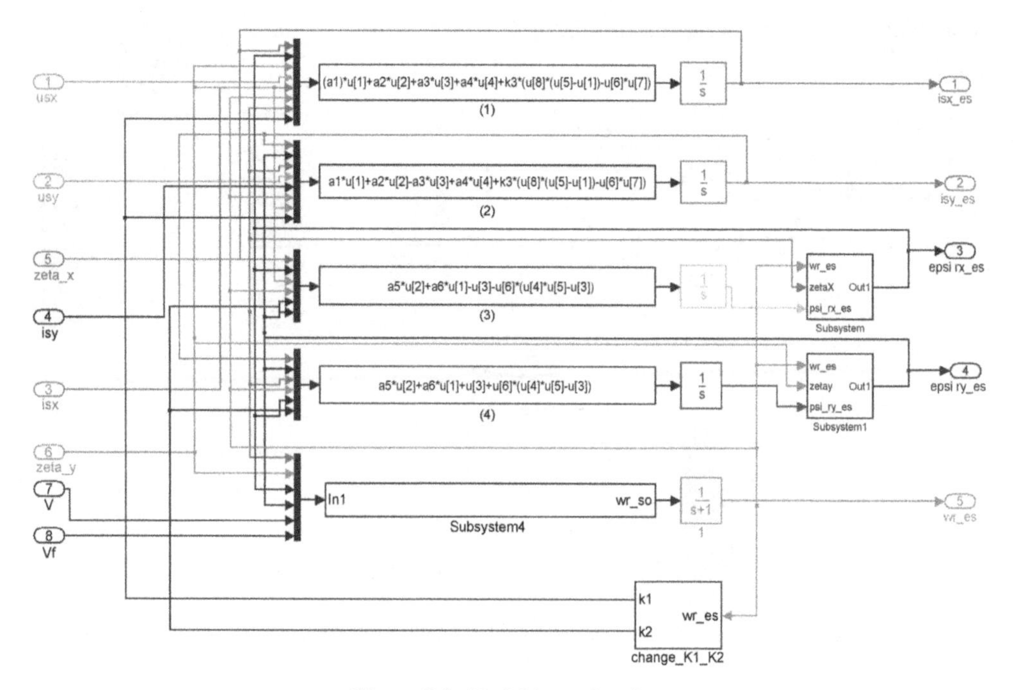

Figure 8.8 Variables estimation

control signal obtained through experiments; and V_f is the filtered signal V. The observer coefficients k_1 to k_4 have small values and are selected experimentally. The coefficients depend on the operating point, for example, torque, speed, etc. The coefficient k_2 depends on the rotor speed:

$$k_2 = a + b \cdot \left|\hat{\omega}_{rf}\right| \tag{8.34}$$

where a and b are constants and $\hat{\omega}_{rf}$ is the absolute value of the estimated and filtered rotor speed.The control signals are defined as

$$V = \hat{\psi}_{rx}\zeta_y - \hat{\psi}_{ry}\zeta_x \tag{8.35}$$

$$\frac{dV_f}{d\tau} = \frac{1}{T_1}\left(V_1 - V_f\right) \tag{8.36}$$

This observer is appropriate for very low speed operations.

The observer is simulated in Simulink in [65]. The Simulink software file of control of the induction motor model with the speed observer system is [*Observer_three*]. The motor parameters are located in [*PAR_AC*], which should be first executed.

8.2.3 MRAS (Closed-Loop) Speed Estimator

Although several schemes are available for sensorless operation of a vector-controlled drive, the MRAS is popular because of its simplicity [1, 2, 25–36]. The model reference approach takes advantage of using two independent machine models for estimating the same state variable [36–45, 47, 49]. The estimator that does not contain the speed to be computed is considered a reference model. The other block, which contains the estimated variable, is regarded as an adjustable model (Figure 8.9). The estimation error between the outputs of the two computational blocks is used to generate a proper mechanism for adapting the speed. The block diagram of the MRAS-based speed estimation is shown in Figure 8.9.

In this scheme, the outputs of the reference model and the adjustable model denoted in Figure 8.9 by $\Psi_r^{(1)}$ and $\Psi_r^{(2)}$ are two estimates of the rotor flux vectors.

The difference between the two estimated vectors is used to feed a PI controller (could be the only gain). The output of the controller is used to tune the adjustable model. The tuning signal actuates the rotor speed, which makes the error signal zero. The adaptation mechanism of MRAS is a simple gain, or in this example, a PI controller algorithm:

$$\omega^{est} = K_p\varepsilon + K_i\int \varepsilon dt \tag{8.37}$$

where the input of the PI controller is

$$\varepsilon = \psi_{\beta r}^{(1)}\psi_{\alpha r}^{(2)} - \psi_{\alpha r}^{(1)}\psi_{\beta r}^{(2)} \tag{8.38}$$

and K_p and K_i are the parameters of the PI controller.

8.2.3.1 Simulink MRAS Model

The MRAS consists of two models, the reference one and the tuned one. Both models may be formed using the following equations.

By letting the reference frame speed $\omega_a = 0$, the motor basic equations are

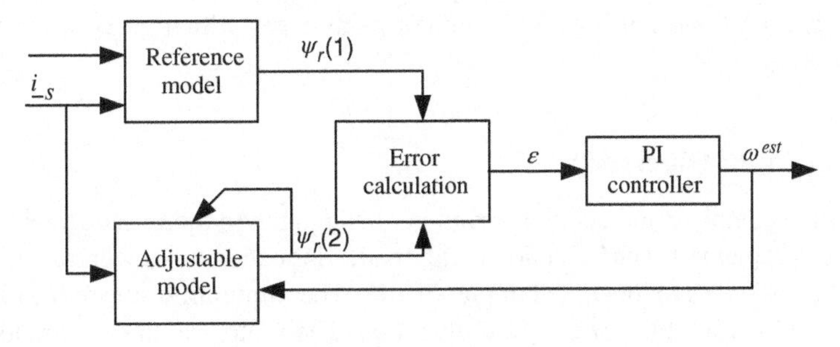

Figure 8.9 Block diagram of MRAS

$$\underline{v}_s = R_s \underline{i}_s + \frac{d\underline{\psi}_s}{dt} \tag{8.39}$$

$$0 = R_r \underline{i}_r + \frac{d\underline{\psi}_r}{dt} - j\omega_r \underline{\psi}_r \tag{8.40}$$

The extraction of the stator flux vector and rotor current vector enables the computation of the rotor flux vector using the following known equation:

$$\begin{aligned} \frac{d\underline{\psi}_r}{dt} &= \frac{L_r}{L_m}\left[\underline{v}_s - R_s \underline{i}_s - \sigma L_s \frac{d\underline{i}_s}{dt}\right] \\ \frac{d\underline{\psi}_r}{dt} &= \left[-\frac{1}{T_r} + j\omega_r\right]\underline{\psi}_r + \frac{L_m}{T_r}\underline{i}_s \end{aligned} \tag{8.41}$$

The first equation (8.41) can be used to calculate the rotor flux vector on the basis of the measured stator current and commanded voltage (or computed using PWM vectors). This first equation is independent of rotor speed; therefore, it may represent the reference model of Figure 8.9. On the other hand, the second equation (8.41) requires stator currents only and is dependent on the rotor speed. Therefore, this equation may represent the adjustable (adaptive) model of Figure 8.9.

In matrix form those equations (models) appear as

$$\begin{aligned} p\begin{bmatrix} \psi_{\alpha r} \\ \psi_{\beta r} \end{bmatrix} &= \frac{L_r}{L_m}\left[\begin{bmatrix} v_{\alpha s} \\ v_{\beta s} \end{bmatrix} - \begin{bmatrix} (R_s + \sigma L_s p) & 0 \\ 0 & (R_s + \sigma L_s p) \end{bmatrix}\begin{bmatrix} i_{\alpha s} \\ i_{\beta s} \end{bmatrix}\right] \\ p\begin{bmatrix} \psi_{\alpha r} \\ \psi_{\beta r} \end{bmatrix} &= \begin{bmatrix} -1/T_r & -\omega_r \\ \omega_r & -1/T_r \end{bmatrix}\begin{bmatrix} \psi_{\alpha r} \\ \psi_{\beta r} \end{bmatrix} + \frac{L_m}{T_r}\begin{bmatrix} i_{\alpha s} \\ i_{\beta s} \end{bmatrix} \end{aligned} \tag{8.42}$$

where $\sigma = 1 - \left(L_m^2/L_s L_r\right)$ and $p = \frac{d}{dt}$

Figure 8.10 describes the Simulink diagram of MRAS with both references (upper) and adaptive (lower) models used for motor speed estimation. The currents (i_α and i_β) and voltages (V_α and V_β) are taken from motor terminals for the speed estimation.

Indeed, it is possible to use different models either as a reference or adaptive system. Below is an additional example of the reference system in which Observer one is used as reference model. This is robust and not sensitive to motor parameters, which gives better results, even for very low frequencies.

8.2.3.2 MRAS with Observer 1

In this section, a complete and detail description of an induction motor model with MRAS and Observer one is explained. The induction motor model, induction motor controller, MRAS estimator, and the observer are illustrated in Figure 8.11. The Simulink software file of vector control of IM model is [*IM_VC_sensor_less_observer*]. The motor parameters are located in the file model properties within the initial function. Parameters should run automatically.

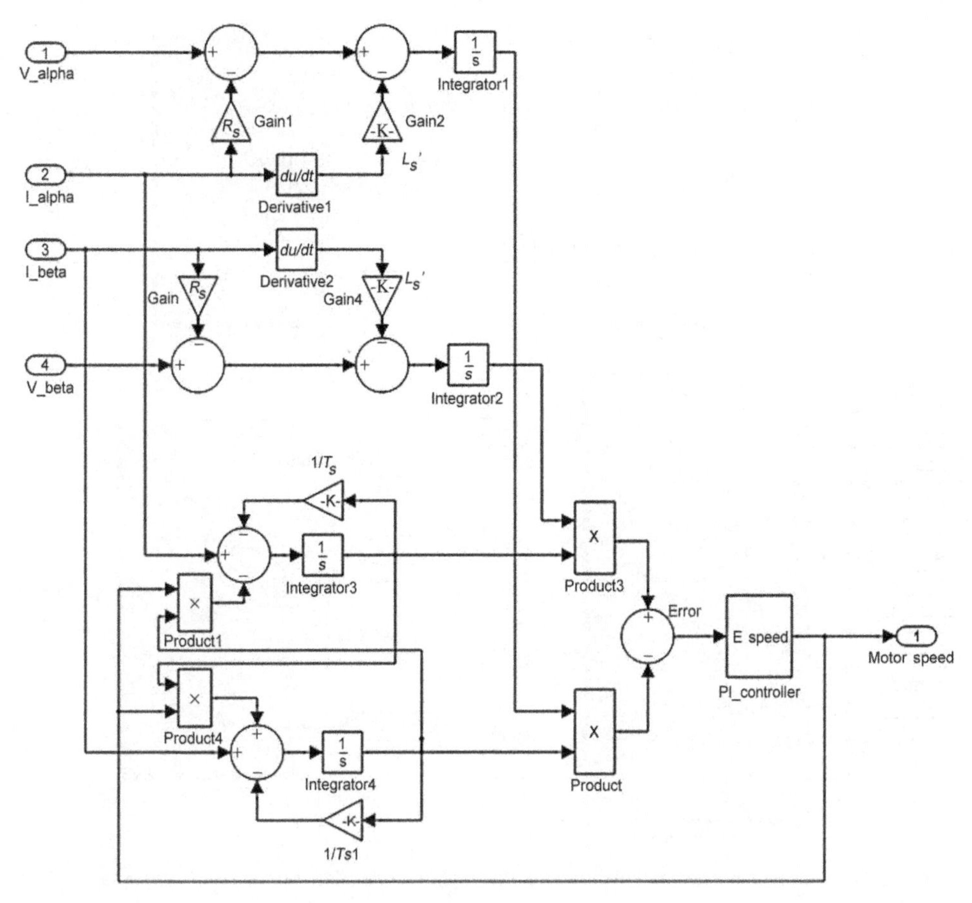

Figure 8.10 Simulink diagram of the MRAS estimator (both reference and adaptive models)

Description of induction motor with vector controller is given in an earlier chapter; therefore, it will not be repeated.

The Simulink block of the observer is shown in Figure 8.12. The sensorless flux observer is used for the estimation of stator and rotor fluxes. The observer can also be used for other variable estimations.

As shown in Figure 8.12, the adaptive system is connected with inputs (fluxes) obtained from the sensorless observer, which is used as the reference model. The reference model is used for adjusting the speed in an adaptive model. By this method, a robust scheme is possible to be designed because the sensorless flux observer is insensitive to motor parameter deviation.

8.2.3.3 Simulation Results

The motor parameters for simulation are stator and rotor resistances $R_s = R_r = 0.045$ p.u., stator and rotor inductances $L_s = L_r = 1.927$ p.u., motor mutual inductance $L_m = 1.85$ p.u., motor load moment of inertia $J = 5.9$ p.u., motor load torque, and $m_o = 0.7$ p.u. The simulation result for a

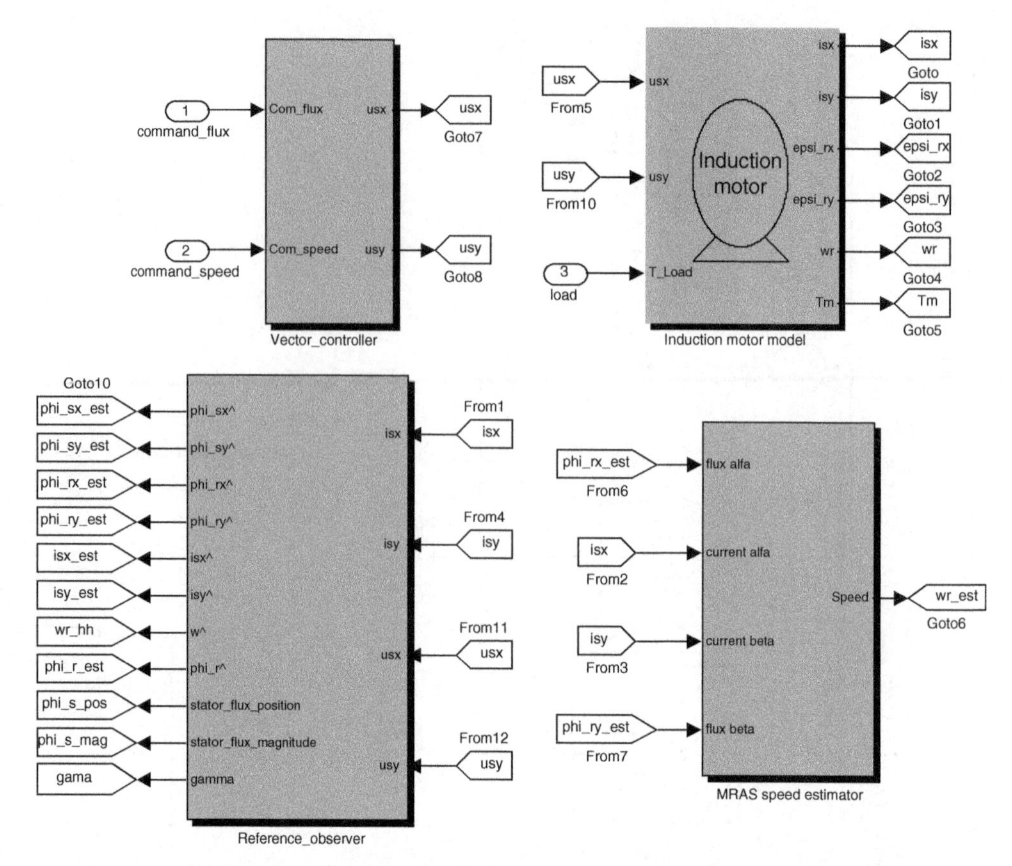

Figure 8.11 Simulink blocks of induction motor, vector controller, reference observer, and MRAS observer

square pulse demand in speed is shown in Figure 8.13. The simulation time is set at 10 seconds. A step speed command of 0.9 p.u. magnitude is given at 0 seconds and then becomes zero at five seconds. A step load torque command is applied at two seconds and then makes t zero at six seconds The rotor flux command is set at 1 p.u. The corresponding simulation results of the motor speed ω_r, currents I_d and I_q, rotor flux, load torque, and applied voltage in the p.u. system are shown in Figure 8.13. Proper limiting of the control variables is not provided. This should be taken into account for right control.

8.2.4 The Use of Power Measurements

The use of instantaneous reactive power and instantaneous active power provides a simplification of the speed estimation process [21, 22].

Reference [61] provides new definitions of instantaneous powers in three-phase (***p*** and ***q***) circuits, based on the instantaneous voltages and currents. The power definitions are

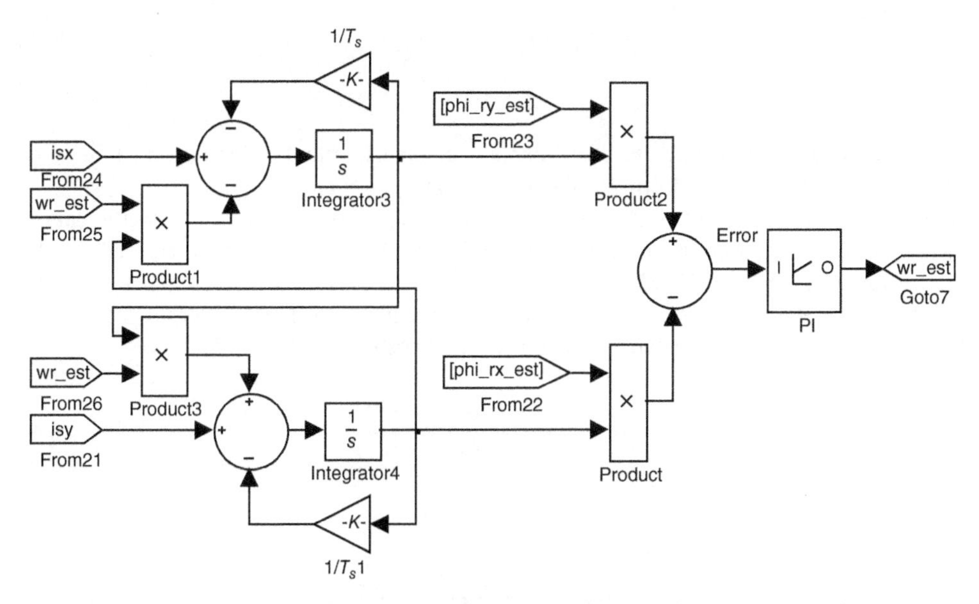

Figure 8.12 Simulink diagram of adaptive system with input from sensorless observer

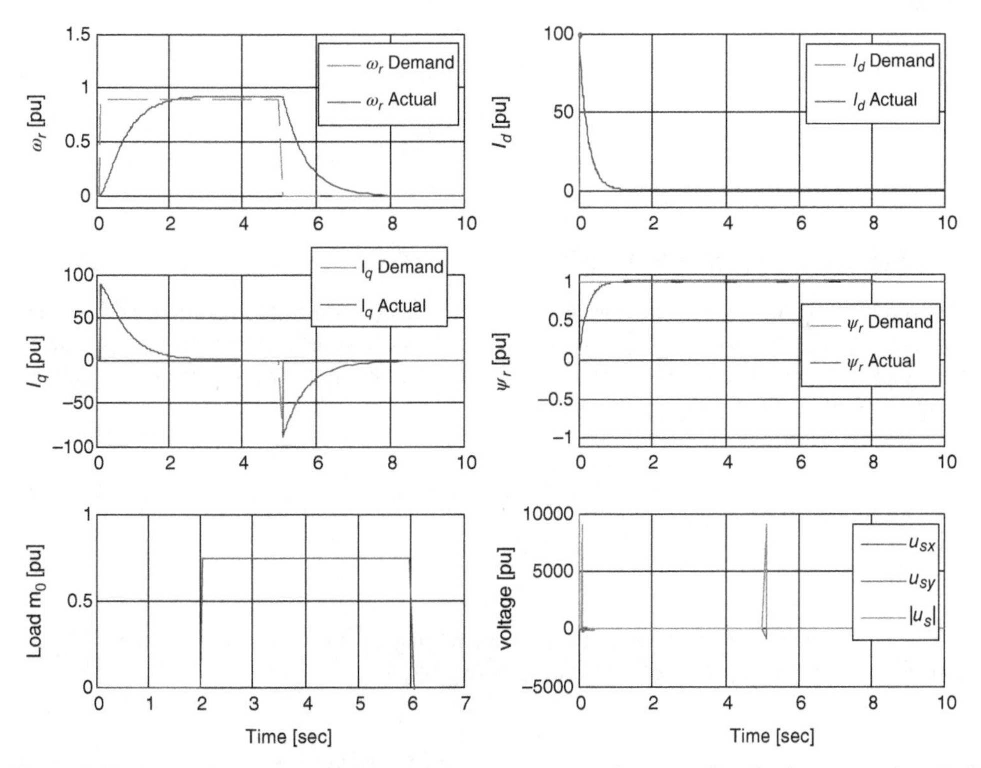

Figure 8.13 Simulation results of motor speed w_r, currents I_d and I_q, rotor flux, load torque, and applied voltage in the p.u. system

$$\boldsymbol{p} = u_{sx}i_{sx} + u_{sy}i_{sy} \tag{8.43}$$

$$\boldsymbol{q} = u_{sy}i_{sx} - u_{sx}i_{sy} \tag{8.44}$$

The used power $\boldsymbol{P}$ and $\boldsymbol{Q}$ for speed estimation are obtained after filtration (using a first-order filter) of $\boldsymbol{p}$ and $\boldsymbol{q}$ calculated from the above equations.

Rotor angular speed may be determined by using combinations between power equations, multiscalar variables, and the machine model (differential equations of stator current and rotor flux vector components) [21, 22].

For simplification, the left-hand side of the machine model is equal to zero in steady state. Accordingly, the rotor speed can be computed in different ways [21, 22]:

$$\omega_r = \frac{a_1 I_s^2 + a_2 z_3 + a_a P}{a_3 z_2} \tag{8.45}$$

or

$$\omega_r = \frac{-a_2 z_2 - \omega_{si} I_s^2 + a_4 Q}{I_s^2 + a_3 z_3} \tag{8.46}$$

where

$$z_2 = \psi_{rx} i_{sy} - \psi_{ry} i_{sx} \tag{8.47}$$

$$z_3 = \psi_{rx} i_{sx} + \psi_{ry} i_{sy} \tag{8.48}$$

8.3 Sensorless Control of PMSM

The rotor speed and produced torque can be computed using the Luenbergera back emf observer. Assuming that the dynamics of the mechanical system are much slower than the dynamics of the electric variables, we may assume that the motor back emf does not change during an observer sampling period, so that

$$\frac{de_\alpha}{dt} = 0 \tag{8.49}$$

$$\frac{de_\beta}{dt} = 0 \tag{8.50}$$

The used Luenberger observer for the permanent magnet synchronous motor (PMSM) back emf in the stationary $\alpha\beta$ frame is described as [66–68]

$$\frac{d\hat{i}_\alpha}{dt} = \frac{1}{L_s}u_\alpha - \frac{R_s}{L_s}\hat{i}_\alpha + \frac{1}{L_s}\hat{e}_\alpha + K_{i\alpha}\left(i_\alpha - \hat{i}_\alpha\right) \tag{8.51}$$

$$\frac{d\hat{i}_\beta}{dt} = \frac{1}{L_s}u_\beta - \frac{R_s}{L_s}\hat{i}_\beta + \frac{1}{L_s}\hat{e}_\beta + K_{i\beta}\left(i_\beta - \hat{i}_\beta\right) \tag{8.52}$$

$$\frac{d\hat{e}_\alpha}{dt} = K_{e\alpha}\left(i_\alpha - \hat{i}_\alpha\right) \tag{8.53}$$

$$\frac{d\hat{e}_\beta}{dt} = K_{e\beta}\left(i_\beta - \hat{i}_\beta\right) \tag{8.54}$$

where ^ is the estimated value; Ls is the stator inductance, assuming that $L_s = L_d$; and $K_{i\alpha}$, $K_{i\beta}$, $K_{e\alpha}$, $K_{e\beta}$, are observer gains (when using symmetrical motor $K_{i\alpha} = K_{i\beta} = K_i$, $K_{e\alpha} = K_{e\beta} = K_e$).

The state variables in the observer are stator currents i_α, i_β; however, the back emf signals e_α, e_β are disturbances that should be identified in the observer. The block showing such an operation is illustrated in Figure 8.14.

The rotor angle can be computed from the following:

$$\theta = -arc\ tg\frac{\hat{e}_\alpha}{\hat{e}_\beta} \tag{8.55}$$

where

$$\sin\hat{\theta} = -\frac{\hat{e}_\alpha}{|\hat{e}|} \tag{8.56}$$

$$\cos\hat{\theta} = -\frac{\hat{e}_\beta}{|\hat{e}|} \tag{8.57}$$

$$|\hat{e}| = \sqrt{(\hat{e}_\alpha)^2 + (\hat{e}_\beta)^2} \tag{8.58}$$

Rotor speed is identified by differentiating the rotor angle:

$$\hat{\omega}_r = \frac{d\hat{\theta}}{dt} \tag{8.59}$$

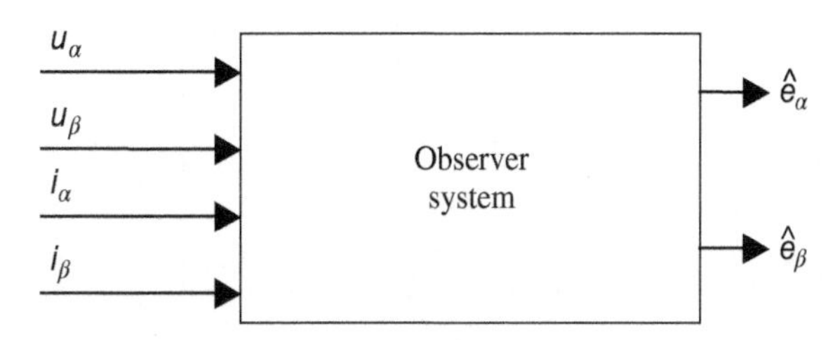

Figure 8.14 Emf observer

$$|\hat{\omega}_r| = \frac{|\hat{e}|}{k_e} \tag{8.60}$$

In this example, the sign of the speed needs to be additionally identified, for example, by verifying the sign of angle θ derivative.

Motor electromagnetic torque can be computed by

$$\hat{t}_e = \frac{\hat{i}_\alpha \hat{e}_\alpha + \hat{i}_\beta \hat{e}_\beta}{\hat{\omega}_r} \tag{8.61}$$

The schematic of the back emf observer system is shown in Figure 8.15 [68].

8.3.1 Control System of PMSM

The schematic of the sensorless drive system with PMSM is shown in Figure 8.16.

The machine torque is obtained by

$$\hat{t}_e = \psi_f \hat{i}_q + (L_d - L_q)\hat{i}_d \hat{i}_q \tag{8.62}$$

8.3.2 Adaptive Backstepping Observer

The machine model in the rotating d–q frame is [68, 69]

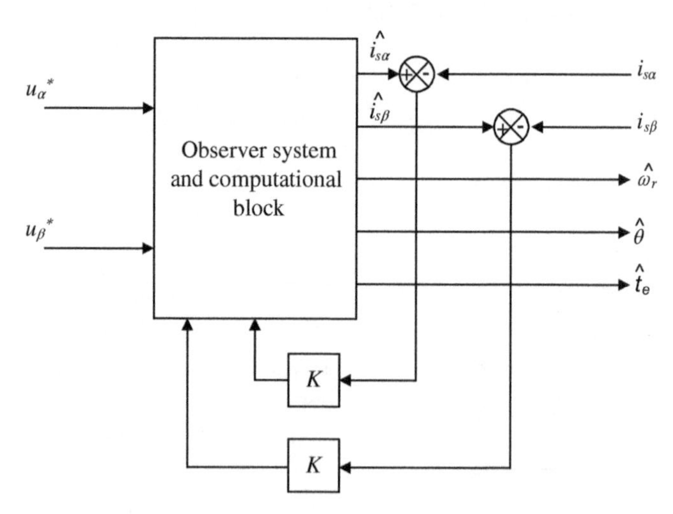

Figure 8.15 Schematic of the back emf observer system (*Source*: Based on Zawirski [68])

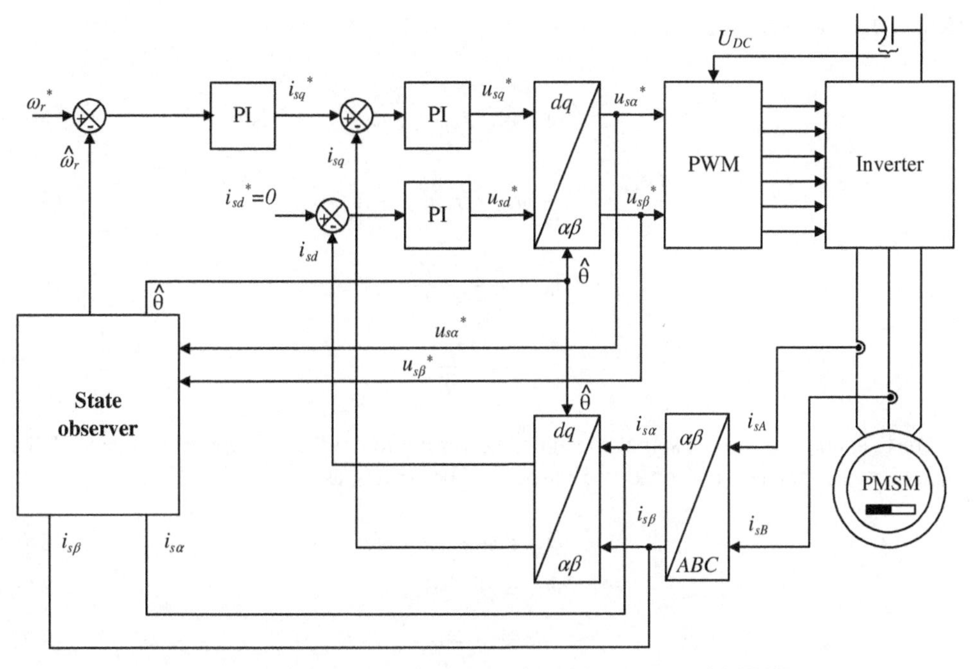

Figure 8.16 Schematic of the sensorless drive system with PMSM

$$u_d = R_s i_d + \frac{d\psi_{sd}}{dt} - \omega_{\psi r}\psi_{sq} \tag{8.63}$$

$$u_q = R_s i_q + \frac{d\psi_{sq}}{dt} + \omega_{\psi r}\psi_{sd} \tag{8.64}$$

$$\psi_{sd} = L_d i_d + \psi_f \tag{8.65}$$

$$\psi_{sq} = L_q i_q \tag{8.66}$$

It is possible to obtain [68, 69]

$$\frac{d\psi_{sd}}{dt} = u_d - R_s i_d + \omega_{\psi r}\psi_{sq} = u_d - R_s\left(\frac{\psi_{sd} - \psi_f}{L_d}\right) + \omega_{\psi r}\psi_{sq} = -\frac{R_s}{L_d}\psi_{sd} + \omega_{\psi r}\psi_{sq} + \frac{R_s}{L_d}\psi_f + u_d \tag{8.67}$$

$$\frac{d\psi_{sq}}{dt} = u_q - R_s i_q - \omega_{\psi r}\psi_{sd} = u_q - R_s\frac{\psi_{sq}}{L_q} - \omega_{\psi r}\psi_{sd} = -\frac{R_s}{L_q}\psi_{sq} - \omega_{\psi r}\psi_{sd} + u_q \tag{8.68}$$

$$i_d = \frac{\psi_{sd}}{L_d} + \frac{\psi_f}{L_d} \tag{8.69}$$

$$i_q = \frac{\psi_{sq}}{L_q} \tag{8.70}$$

Hence, the observer equations may appear as [68, 69]

$$\frac{d\hat{\psi}_{sd}}{dt} = -\frac{\hat{R}_s}{L_d}\hat{\psi}_{sd} + \hat{\omega}_{\psi r}\hat{\psi}_{sq} + \frac{\hat{R}_s}{L_d}\psi_f + u_d + k_d \tag{8.71}$$

$$\frac{d\hat{\psi}_{sq}}{dt} = -\frac{\hat{R}_s}{L_q}\hat{\psi}_{sq} - \hat{\omega}_{\psi r}\hat{\psi}_{sd} + u_q + k_q \tag{8.72}$$

$$i_d = \frac{\hat{\psi}_{sd}}{L_d} + \frac{\psi_f}{L_d} \tag{8.73}$$

$$i_q = \frac{\hat{\psi}_{sd}}{L_d} \tag{8.74}$$

where k_d and k_q are control signals designed according to the 'backstepping' theory [69]. The errors between the real and observed values can be written as

$$\widetilde{\psi}_{sd} = \psi_{sd} - \widetilde{\psi}_{sd} \tag{8.75}$$

$$\widetilde{\psi}_{sq} = \psi_{sq} - \widetilde{\psi}_{sq} \tag{8.76}$$

$$\widetilde{\omega}_r = \omega_r - \hat{\omega}_r \tag{8.77}$$

$$\widetilde{R}_s = R_s - \hat{R}_s \tag{8.78}$$

Then the response of the error can be represented by [68, 69]

$$\frac{d\widetilde{\psi}_{sd}}{dt} = -\frac{\widetilde{R}_s}{L_d}\widetilde{\psi}_{sd} + \widetilde{\omega}_{\psi r}\widetilde{\psi}_{sq} + \frac{\widetilde{R}_s}{L_d}\psi_f + u_d + k_d \tag{8.79}$$

$$\frac{d\widetilde{\psi}_{sq}}{dt} = -\frac{\widetilde{R}_s}{L_q}\widetilde{\psi}_{sq} - \widetilde{\omega}_{\psi r}\widetilde{\psi}_{sd} + u_q + k_q \tag{8.80}$$

$$i_d = \frac{\widetilde{\psi}_{sd}}{L_d} + \frac{\psi_f}{L_d} \tag{8.81}$$

$$i_q = \frac{\widetilde{\psi}_{sq}}{L_q} \tag{8.82}$$

8.3.3 Model Reference Adaptive System for PMSM

Recently, the sensorless control of the PMSM has attracted wide attention due to their wide use in the industry. Several techniques have evolved in order to estimate the motor speed. Again, open-loop speed estimators, closed-loop estimators, back emf-based speed estimators, the high frequency signal injection method, and many others are used. Although several schemes are available for sensorless operation of vector-controlled PMSM drives, one of the most popular

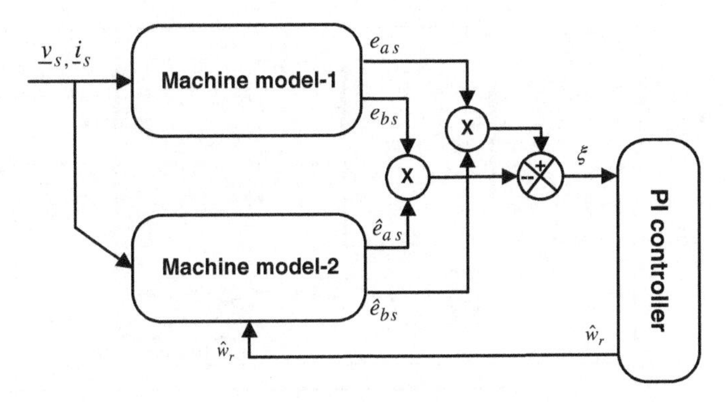

Figure 8.17 Block diagram of MRAS estimator for PMSM

schemes is the MRAS because of the ease of implementation [37–49]. The model reference approach makes use of two independent machine models of different structures to estimate the same state variable (back emf, rotor flux, reactive power, etc.) on the basis of different sets of input variables. The error between the outputs of the two estimators with the described earlier strategy is used to generate an adaptive mechanism for identifying rotor speed in the adjustable model. The schematic block of an MRAS for PMSM is illustrated in Figure 8.17.

The PMSM stator voltage equation in the stationary frame can be used as the reference model:

$$\underline{v}_{\alpha s} = R_s \underline{i}_{\alpha s} + L_s \frac{d\underline{i}_{\alpha s}}{dt} + e_{\alpha s} \tag{8.83}$$

$$\underline{v}_{\beta s} = R_s \underline{i}_{\beta s} + L_s \frac{d\underline{i}_{\beta s}}{dt} + e_{\beta s} \tag{8.84}$$

where R_s and L_s are stator winding resistance and inductance per phase, respectively, and e_s is the motor back emf.

The adaptive system contains a motor speed that is supposed to be identified in adaptive way according to Figure 8.12. In this scheme, the outputs of the motor model-1 and the motor model-2 are two estimates of the back emf's.

The speed tuning signal actuates the rotor speed, which makes the error signal converge to zero. The adaptation mechanism of the MRAS-based speed estimation method is a PI controller algorithm:

$$\hat{\omega}_r = K_p \xi + K_i \int \xi dt \tag{8.85}$$

where the input of the PI controller is

$$\xi = e_{\alpha s} \hat{e}_{\beta s} - e_{\beta s} \hat{e}_{\alpha s} \tag{8.86}$$

and K_p and K_i are controller gains and ξ is the error signal.

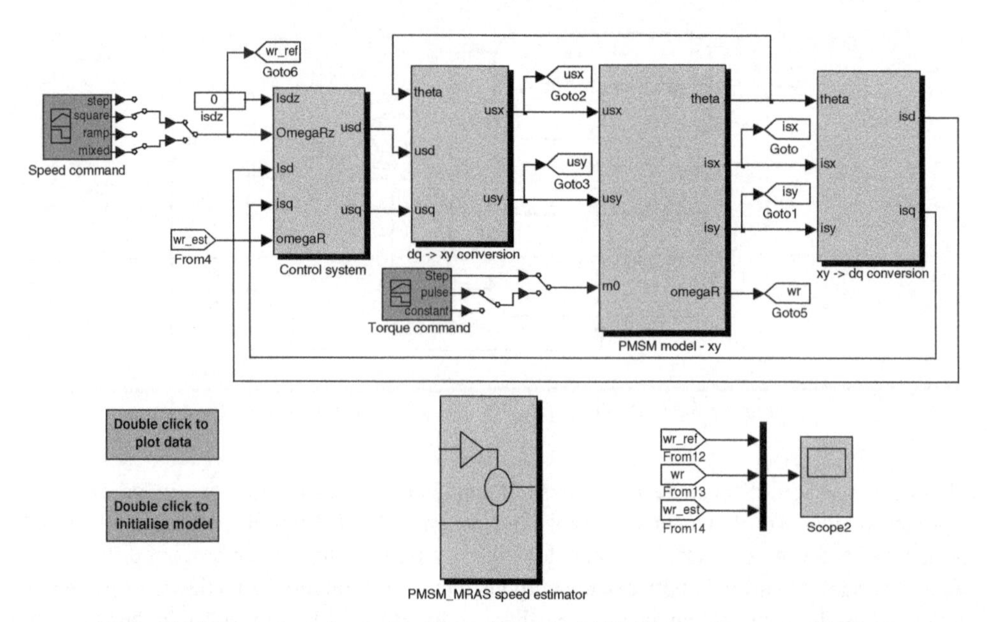

Figure 8.18 Complete simulation block diagram of MRAS-based control of PMSM

The complete Simulink diagram of the MRAS block for the control of the PMSM is shown in Figure 8.18. The MRAS block takes currents and voltages from the motor terminals. The description of the MRAS block is explained in Figure 8.19. The Simulink software file of vector control of the PMSM model is [*PMSM_FOC_alfa_beta*]. The motor parameters are located in [*PMSM_param*], which should be first executed.

8.3.4 Simulation Results

Selected simulation results for an MRAS sensorless drive are shown in Figures 8.20 and 8.21.

8.4 MRAS-Based Sensorless Control of Five-Phase Induction Motor Drive

Sensorless operation of a vector-controlled three-phase drive is widely discussed in the literature; however, this is not true for multiphase AC machines, where a limited number of publications have appeared in the literature [25–27, 59]. One of the most popular solutions is the MRAS, which is similar to three-phase drives [29–36].

As was shown in Chapter 6, multiphase machine models can be transformed into an orthogonal frame of decoupled equations. The d–q reference frame contributes to torque and flux production. Hence, in rotor field-oriented control (FOC), the rotor flux linkage is kept in the d-axis so that the q component of the rotor flux is zero. Thus, the produced torque and the rotor flux

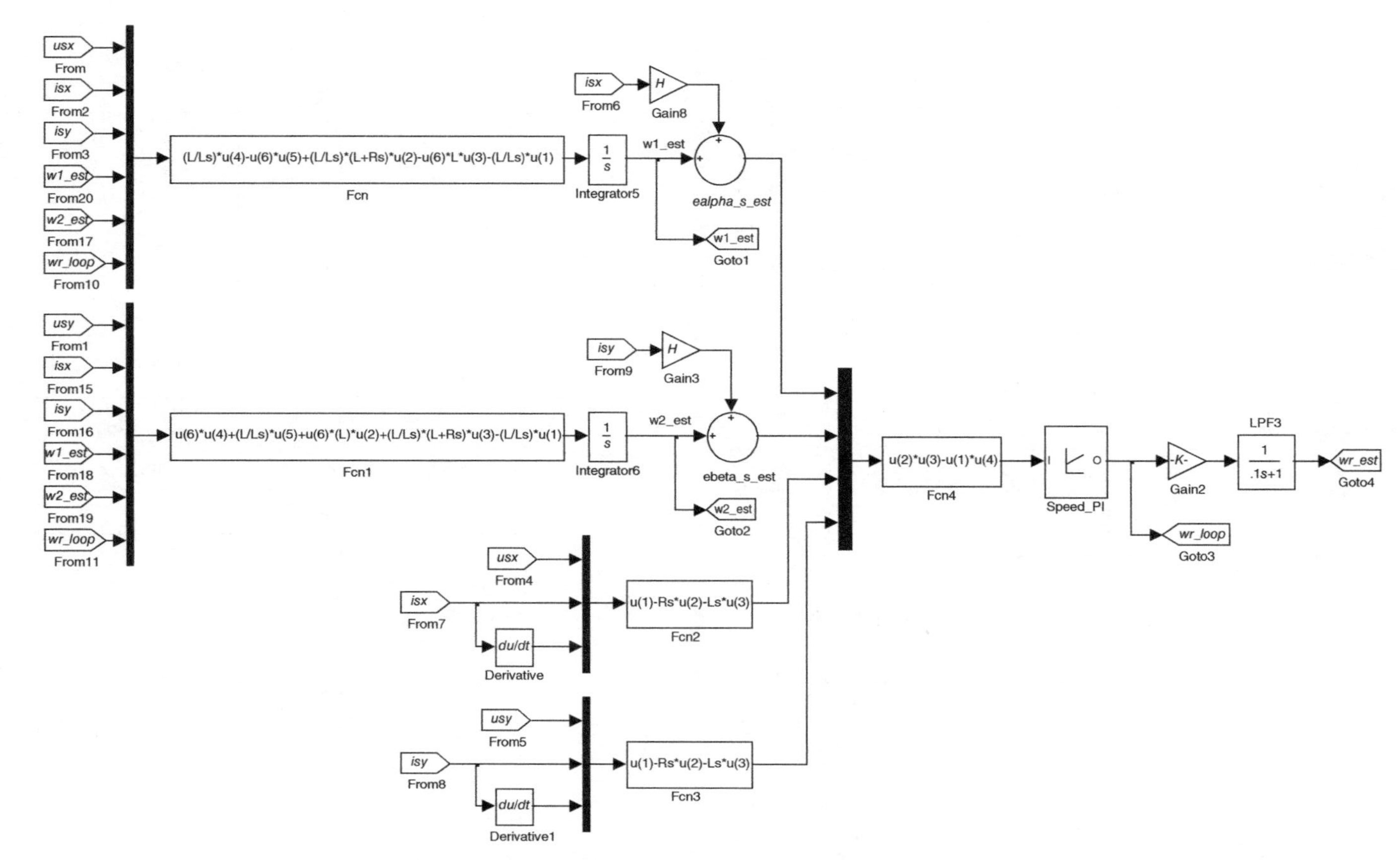

Figure 8.19 Simulink diagram of MRAS estimator for PMSM

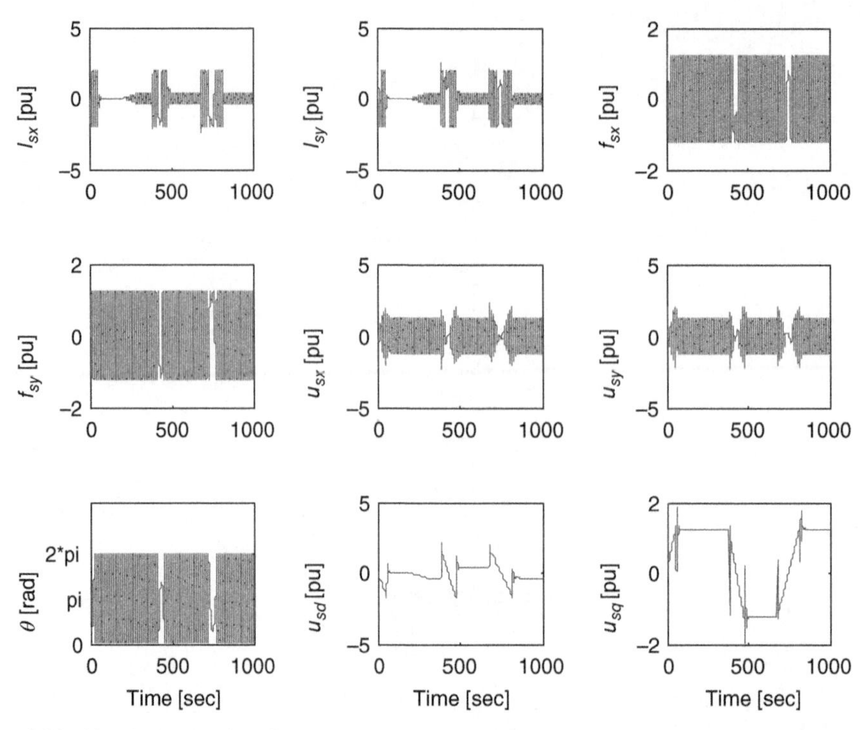

Figure 8.20 Simulation results of currents (I_{sx}, I_{sy},), field fluxes (f_{sx}, f_{sy}), and rotor angle and voltages (u_{sx}, u_{sy}, u_{sd}, u_{sq})

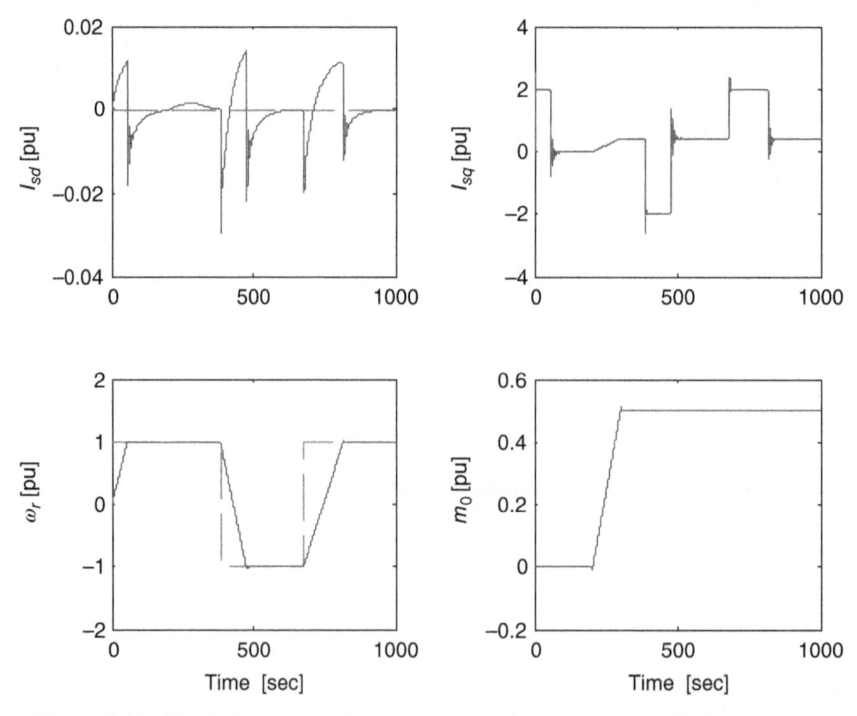

Figure 8.21 Simulation results of currents (I_{sd} and I_{sq}), rotor speed, and load torque

can be controlled independently by the d and q components of the stator current. The decoupled control of the torque and flux using the rotor flux-oriented control for a five-phase IM is illustrated in Chapter 6.

This section focuses on the MRAS-based sensorless control of a five-phase IM, with current control in the stationary reference frame. API controller is used to compensate the rotor speed in the adaptive model. Phase currents are controlled using the hysteresis current control method.

The difference between the five-phase machine model, given in Chapter 6, and the three-phase machine model is the presence of x–y component. Motor x–y components are fully decoupled from d–q components, in addition to the one decoupling from the other. Since the rotor winding of a squirrel cage motor is short-circuited, the x–y components do not appear in the rotor winding. Since the stator x–y components are fully decoupled from the d–q components and one from the other, the equations for x–y components do not need to be further considered. This means that the model of the five-phase IM in an arbitrary reference frame becomes identical to the model of a three-phase IM. Hence, the same principles of RFOC can be used.

The mathematical model of a five-phase induction motor makes it possible that the MRAS used for a three-phase machine can be easily extended to a multiphase machine. The MRAS for the five-phase induction motor drive in the stationary reference frame is shown in Figure 8.22.

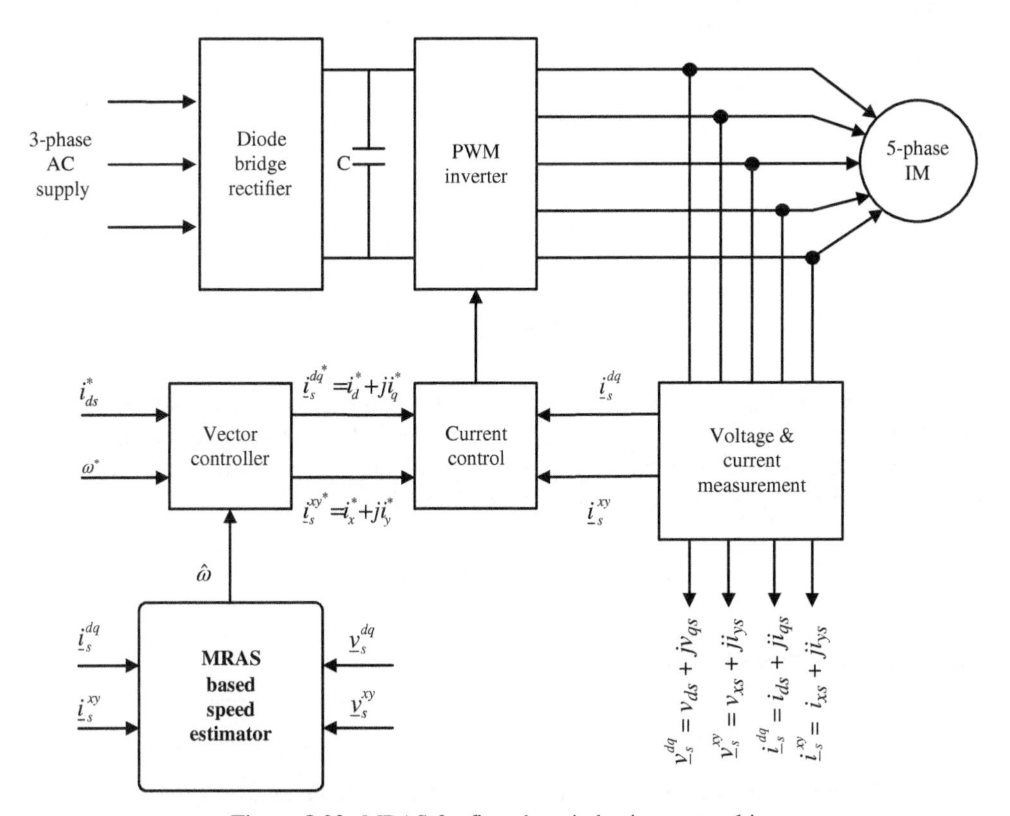

Figure 8.22 MRAS for five-phase induction motor drive

8.4.1 MRAS-Based Speed Estimator

The schematic block of an MRAS-based speed estimator is the same as that for IM or PMSM. In such a scheme, the outputs of the reference (motor model) and the adjustable models denoted by ψ_r and $\hat{\psi}_r$ are two estimates of the rotor flux space vector, which are obtained from the machine model in the stationary reference frame. The rotor flux components in the stationary reference frame are obtained as

$$\frac{d\psi_{dr}}{dt} = \frac{L_r}{L_m}\left(v_{ds} - R_s i_{ds} - \sigma L_s \frac{di_{ds}}{dt}\right) \tag{8.87}$$

$$\frac{d\psi_{qr}}{dt} = \frac{L_r}{L_m}\left(v_{qs} - R_s i_{qs} - \sigma L_s \frac{di_{qs}}{dt}\right) \tag{8.88}$$

$$\frac{di_{xs}}{dt} = \frac{1}{L_{ls}}\left(v_{xs} - R_s i_{xs}\right) \tag{8.89}$$

$$\frac{di_{ys}}{dt} = \frac{1}{L_{ls}}\left(v_{ys} - R_s i_{ys}\right) \tag{8.90}$$

The adaptive model is based on the well-known rotor equations (current model):

$$T_r \frac{d\psi_{dr}}{dt} = L_m i_{ds} - \psi_{dr} - \omega_r T_r \psi_{qr} \tag{8.91}$$

$$T_r \frac{d\psi_{qr}}{dt} = L_m i_{qs} - \psi_{qr} + \omega_r T_r \psi_{dr} \tag{8.92}$$

$$\frac{d\psi_{xr}}{dt} = -\frac{R_{xr}}{L_{lr}} \cdot \psi_{xr} \tag{8.93}$$

$$\frac{d\psi_{yr}}{dt} = -\frac{R_{yr}}{L_{lr}} \cdot \psi_{yr} \tag{8.94}$$

The MRAS model is using the reference model of equations (8.87–8.90) and the adaptive model of equations (8.91–8.94) as two models for the rotor flux estimation. The angular difference between the two rotor fluxes is used as the speed tuning signal (error signal). The adaptation mechanism of MRAS is based on using a simple PI controller:

$$\hat{\omega}_r = K_p \xi + K_i \int \xi dt \tag{8.95}$$

where the input of the PI controller is

$$\xi = \psi_{ds} \hat{\psi}_{qs} - \psi_{qs} \hat{\psi}_{ds} \tag{8.96}$$

and K_p and K_i are arbitrary positive constants (parameters of the PI controller) and j is the error signal. The complete Simulink diagram of the MRAS block for the control of the five-phase induction motor is shown in Figures 8.23 and 8.24. The MRAS block takes currents and

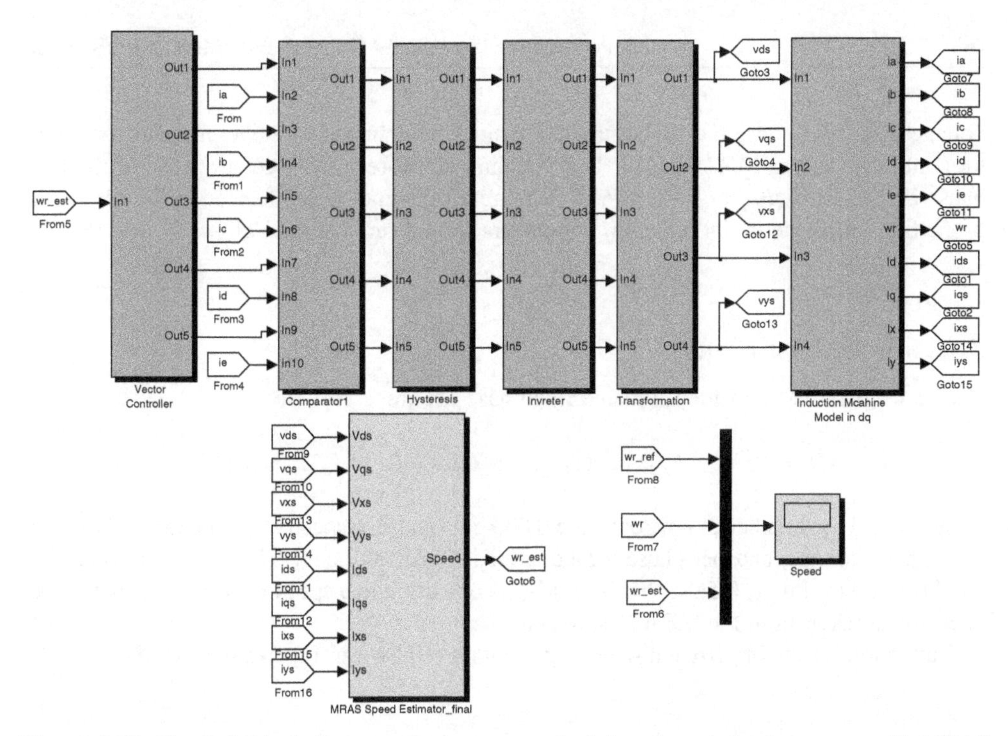

Figure 8.23 Simulink block diagram of vector control of five-phase induction motor with MRAS estimator

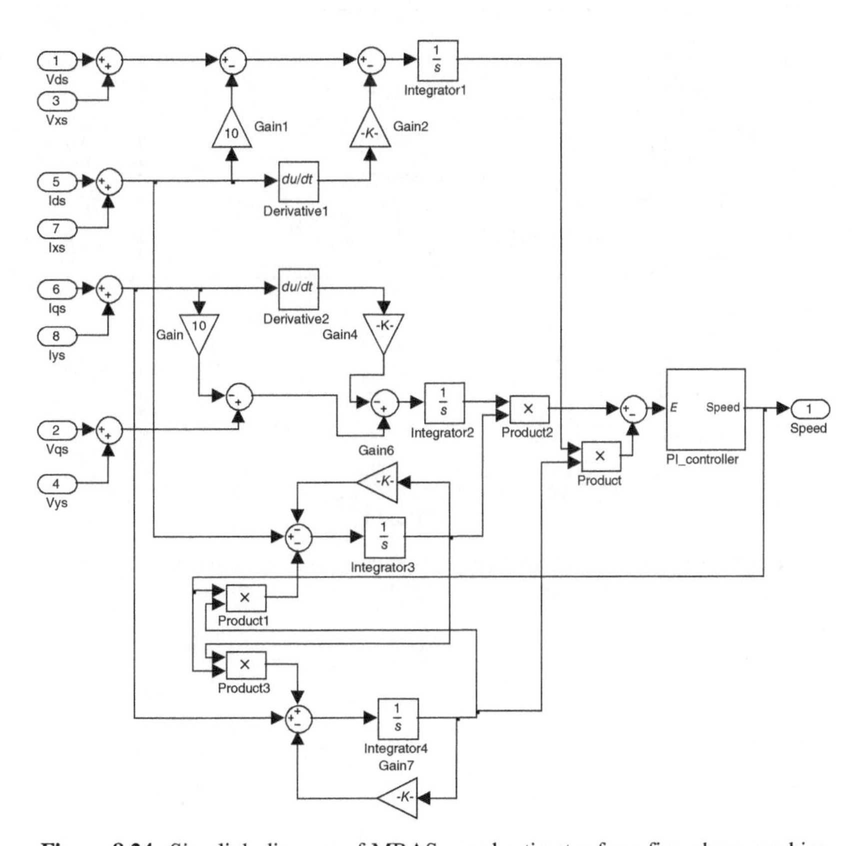

Figure 8.24 Simulink diagram of MRAS speed estimator for a five-phase machine

voltages from motor terminals. The simulation block diagram shows a five-phase motor, a vector controller, and an MRAS block. The Simulink software file of vector control of the five-phase IM model is [*Five_phase_MRAS*]. The motor parameters are located in the file model properties within the initial function. Parameters should run automatically.

8.4.2 Simulation Results

The motor parameters of the simulated five-phase IM are

$$R_s = 10\,\Omega, \quad R_r = 6.3\,\Omega, \quad L_{ls} = L_{lr} = 0.04\,\text{H}, \quad L_m = 0.42\,\text{H}.$$

Inertia and the number of pole pairs are 0.03 kgm^2 and 2, respectively. The rated phase current, phase voltage, and per-phase torque are 2.1 A, 220 V, and 1.67 Nm, respectively. The rated (rms) rotor flux is 0.5683 Wb. The drive is operating in a speed control mode with feedback signal taken from the MRAS estimator.

Simulation results of different operating points are shown in Figures 8.25–8.28.

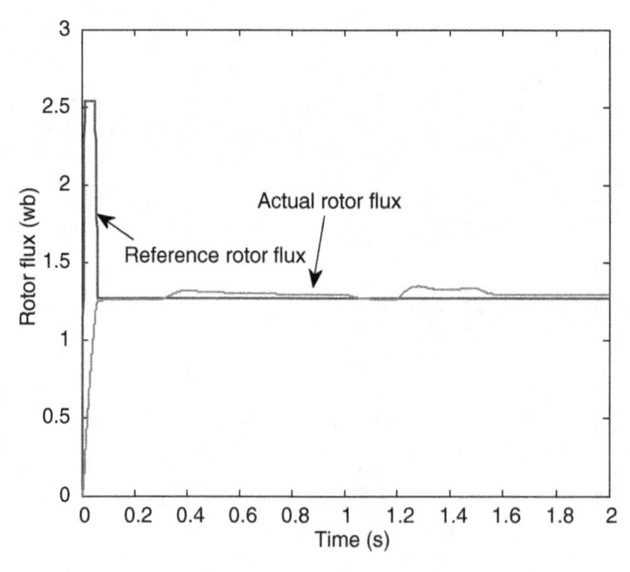

Figure 8.25 Actual and reference rotor flux of five-phase induction motor

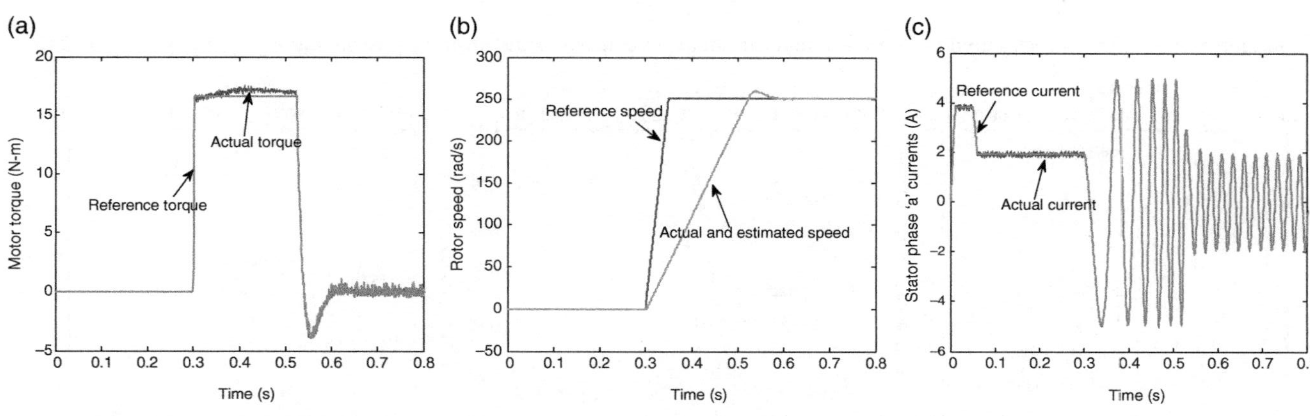

Figure 8.26 IRFOC of a five-phase IM: excitation and acceleration transient: (a) motor torque; (b) motor speed; and (c) phase current

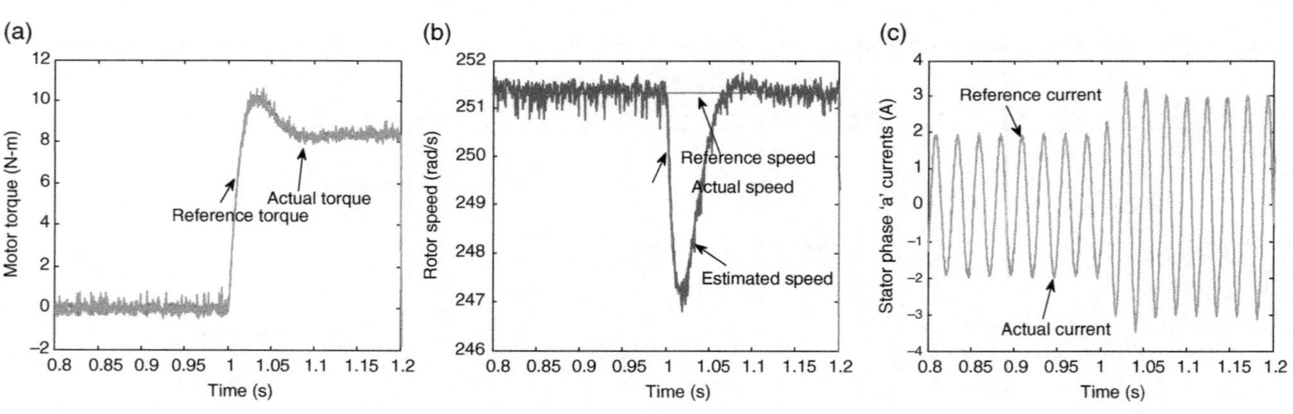

Figure 8.27 IRFOC of a five-phase IM: disturbance rejection behavior: (a) motor torque; (b) motor speed; and (c) phase current

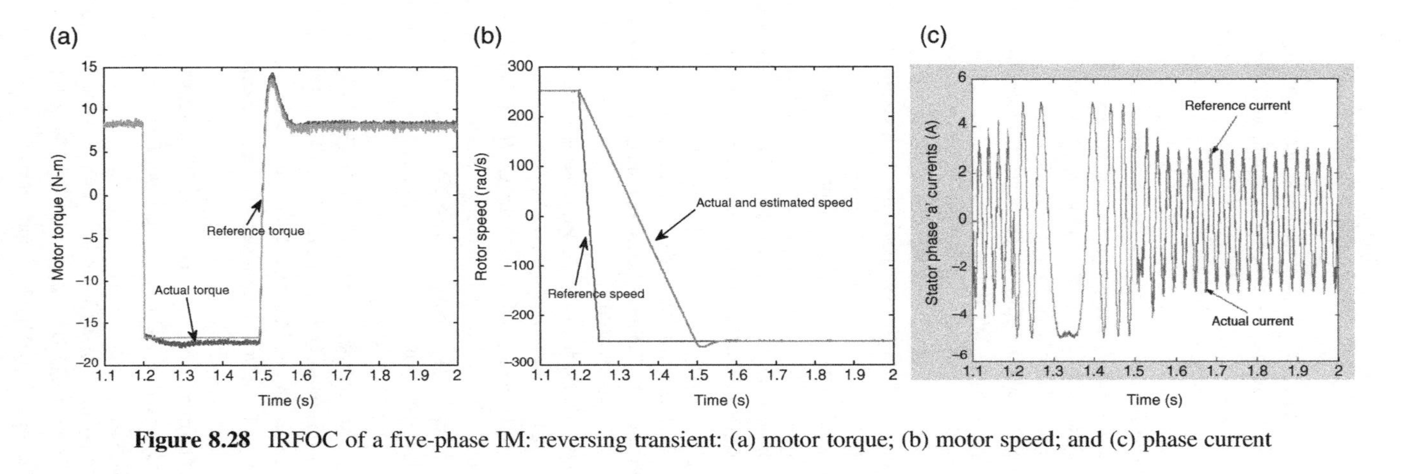

Figure 8.28 IRFOC of a five-phase IM: reversing transient: (a) motor torque; (b) motor speed; and (c) phase current

References

1. Holtz, J. (2006) Sensorless control of induction machines: With or without signal injection? *IEEE Trans. Ind. Elect.*, 53(1), 7–30.
2. Holtz, J. (2002) Sensorless control of induction motor drives. *Proc. IEEE*, 90(8), 1359–1394.
3. Holtz, J. and Quan, J. (2002) Sensorless vector control of induction motors at very low speed using a non-linear inverter model and parameter identification. *IEEE Trans. Ind. Appl.*, 38(4), 1087–1095.
4. Lascu, C., Boldea, I., and Blaabjerg, F. (2005) Very-low speed variable-structure control of sensorless induction machine drives without signal injection. *IEEE Trans. Ind. Appl.*, 41(2), 591–598.
5. Kubota, H. (2003) Closure to discussion of regenerating-mode low-speed operation of sensorless induction motor drive with adaptive observer. *IEEE Trans. Ind. Appl.*, 39(1), 20.
6. Depenbrock, M., Foerth, C., and Koch, S. (1999) Speed sensorless control of induction motors at very low stator frequencies. *8th Europ. Power Elect. Conf. (EPE)*, Lausanne.
7. Schauder, C. (1992) Adaptive speed identification for vector control of induction motors without rotational transducers. *IEEE Trans. Ind. Appl.*, 28, 1054–1061.
8. Abu-Rub, H., Guzinski, J., Krzeminski, Z., and Toliyat, H. (2004) Advanced control of induction motor based on load angle estimation. *IEEE Trans. Ind. Elect.*, 51(1), 5–14.
9. Yan, Z., Jin, C., and Utkin, V. I. (2000) Sensorless sliding-mode control of induction motors. *IEEE Trans. Ind. Elect.*, 47(6), 1286–1297.
10. Ohyama, K., Asher, G. M., and Sumner, M. (2006) Comparative analysis of experimental performance and stability of sensorless induction motor drives. *IEEE Trans. Ind. Elect.*, 53(1), 178–186.
11. Armstrong, G. J. and Atkinson, D. J. (1997) A comparison of model reference adaptive system and extended Kalman filter estimators for sensorless vector drives. *Proc. Int. Conf. Power Elect. Appl. (EPE)*, Singapore, Singapore; 26–29 May 1997, pp. 1.424–1.429.
12. Salvatore, N., Cascella, G. L., Stasi, S., and Cascella, D. (2007) Stator flux oriented sliding mode control of sensorless induction motor drives by Kalman filter. *33rd Ann. Conf. IEEE Ind. Elect. Soc. (IECON)*, Taipei, Taiwan, 5–8 November 2007, pp. 956–961.
13. Kim, Y-R., Sul, S. K., and Park, M. H. (1994) Speed sensorless vector control of induction motor using extended Kalman filter. *IEEE Trans. Ind. Appl.*, 30(5), 1225–1233.
14. Boldea, I. (2008) Control issues in adjustable-speed drives. *IEEE Ind. Elect. Mag.*, 8, 32–50.
15. Hinkkanen, M. and Luomi, J. (2004) Stabilization of regenerating-mode operation in sensorless induction motor drives by full-order flux observer design. *IEEE Trans. Ind. Elect.*, 51(6), 1318–1328.
16. Guzinski, J., Abu-Rub, H., and Toliyat, H. (2003) An advanced low-cost sensorless induction motor drive. *IEEE Trans. Ind. Appl.*, 39(6), 1757–1764.
17. Abu-Rub, H., Guzinski, J., Krzeminski, K., and Toliyat, H. (2003) Speed observer system for advanced sensorless control of induction motor. *IEEE Trans. Ener. Conv.*, 18(2), 219–224.
18. Briz, F., Degner, M. W., Diez, A., and Lorenz, R. D. (2001) Measuring, modeling, and decoupling of saturation-induced saliencies in carrier-signal injection-based sensorless AC drives. *IEEE Trans. Ind. Appl.*, 37, 1356–1364.
19. Briz, F., Degner, M. W., Guerrero, J. M., and Diez, A. (2001) Improving the dynamic performance of carrier signal injection based sensorless AC drives. *Proc. 9th Eur. Conf. Power Elect. Appl. (EPE-2001)*, 27–29 August, Graz, CD-ROM.
20. Abu-Rub, H. and Oikonomou, N. (2008) Sensorless observer system for induction motor control. *39th IEEE Power Elect. Spec. Conf. (PESC08)*, Rhodes, 15–19 June 2008.
21. Abu-Rub, H. and Hashlamoun, W. (2001) A comprehensive analysis and comparative study of several sensorless control system of asynchronous motor. *Europ. Trans. Elect. Power*, 11(3), 203–210.

22. Abu-Rub, H. and Titi, J. (2005) Comparative analysis of fuzzy and Pi sensorless control systems of induction motor using power measurement. *4th Int. Workshop CPE 2005 Compatibility in Power Electronics*, Gdynia, Poland (in English).
23. Cao-Minh Ta, Uchida, T., and Hori, Y. (2001). MRAS-based speed sensorless control for induction motor drives using instantaneous reactive power. *IECON'01. 27th Annual Conference of the IEEE Industrial Electronics Society (Cat. No.37243)*, Denver, CO, 2001, pp. 1417–1422 vol. 2. doi: https://doi.org/10.1109/IECON.2001.975989.
24. Vas, P. (1998) *Sensorless Vector and Direct Torque Control.* Oxford University Press, Oxford.
25. Rajashekara, K., Kawamura, A., and Matsuse, K. (1996) *Sensorless Control of AC Motors.* IEEE Press, Piscataway, NJ.
26. Elbulk, M. E., Tong, L., and Husain, I. (2002) Neural network based model reference adaptive systems for high performance motor drives and motion controls. *IEEE Trans. Ind. Appl.*, 38, 879–886.
27. Abu-Rub, H., Khan, M. R., Iqbal, A., and Ahmed, S. M. (2010) MRAS-based sensorless control of a five-phase induction motor drive with a predictive adaptive model. *Ind. Elect. (ISIE), 2010 IEEE Int. Symp. Ind. Elect.*, Bari, Italy, 4–7 July 2010, pp. 3089–3094.
28. Khan, M. R., Iqbal, I., and Mukhtar, A. (2008) MRAS-based sensorless control of a vector controlled five-phase induction motor drive. *Elect. Power Syst. Res.*, 78(8), 1311–1321.
29. Khan, M. R. and Iqbal, I. (2008) MRAS-based sensorless control of series-connected five-phase two-motor drive system. *Korian J. Elect. Eng. Tec.*, 3(2), 224–234.
30. Luo, Y-C. and Lin, C-C. (2010) Fuzzy MRAS based speed estimation for sensorless stator field oriented controlled induction motor drive. *Int. Symp. Comp. Comm. Cont. Auto. (3CA)*, vol. **2**, Tainan, Taiwan; 5–7 May 2010, pp. 152–155.
31. Li, Z., Cheng, S., and Cai, K. (2008) The simulation study of sensorless control for induction motor drives based on MRAS. *Asia Simul. Conf.: 7th Inter. Conf. Syst. Simul. Sci. Comp., ICSC 2008*, Beijing, China; 10–12 October 2008, pp. 235–239.
32. Rashed, M., Stronach, F., and Vas, P. (2003) A new stable MRAS-based speed and stator resistance estimators for sensorless vector control induction motor drive at low speeds. *Ind. Appl. Conf., 38th IAS Ann. Mtg*, vol. **2**, Salt Lake City, UT; 12–16 October 2003, pp. 1181–1188.
33. Sayouti, Y., Abbou, A., Akherraz, M., and Mahmoudi, H. (2009) MRAS-ANN based sensorless speed control for direct torque controlled induction motor drive. *Int. Conf. Power Eng., Ener. Elect. Drives. POWERENG'09*, Lisbon, Portugal; 18–20 March 2009, pp. 623–628.
34. Ta, C-M., Uchida, T., and Hori, Y. (2001) MRAS-based speed sensorless control for induction motor drives using instantaneous reactive power. *Ind. Elect. Soc., IECON'01. 27th Ann. Conf. IEEE*, vol. **2**, pp. 1417–1422.
35. Yushui, H. and Dan, L. (2008) Realization of sensorless vector control system based on MRAS with DSP. *Cont. Conf., CCC 2008*, United States; 29 November 2001 through 2 December 2001, pp. 691–694.
36. Gao, L., Guan, B., Zhou, Y., and Xu, L. (2010) Model reference adaptive system observer based sensorless control of doubly-fed induction machine. *Int. Conf. Elect. Mach. Syst. (ICEMS)*, Incheon, South Korea; 10–13 October 2010, pp. 931–936.
37. Xiao, J., Li, B., Gong, X., Sheng, Y., and Chai, J. (2010) Improved performance of motor drive using RBFNN-based hybrid reactive power MRAS speed estimator. *IEEE Inter. Conf. Info. Auto. (ICIA)*, Harbin, Heilongjiang; China; 20–23 June 2010, pp. 588–593.
38. Kim, Y. S., Kim, S. K., and Kwon, Y. A. (2003) MRAS based sensorless control of Permanent magnet synchronous motor. *SICE Ann. Conf. Fukui*, Fukui University, Japan, 4–6 August, pp. 1632–1637.

39. Shahgholian, G., Rezaei, M. H., Etesami, A., and Yousefi, M. R. (2010) Simulation of speed sensor less control of PMSM based on DTC method with MRAS. *IPEC, Conf. Proc.*, Singapore, Singapore; 27–29 October 2010, pp. 40–45.
40. Xu, H. and Xie, J. (2009) A vector-control system based on the improved MRAS for PMSM. *Inter. Works. Intell. Syst. Appl., ISA 2009*, Wuhan; China; 23–24 May 2009, pp. 1–5.
41. Bingyi, Z., Xiangjun, C., Guanggui, S., and Guihong, F. (2005) A Position sensorless vector-control system based on MRAS for low speed and high torque PMSM drive. *Proc. 8th Int.l Conf. Elect. Mach. Syst. ICEMS*, vol. **2**, Nanjing, China, 27–29 September 2005.
42. Zhao, G., Liu, J., and Feng, J. (2010) The research of speed sensorless control of PMSM based on MRAS and high frequency signal injection method. *Int. Conf. Comp. Mech., Cont. Elect. Eng. (CMCE)*, vol. **4**, Changchun, China; 24–26 August 2010, pp. 175–178.
43. Xingming, Z., Xuhui, W., Feng, Z., Xinhua, G., and Peng, Z. (2010) Wide-speed-range sensorless control of Interior PMSM based on MRAS. *Int. Conf. Elect. Mach. Syst. (ICEMS)*, Incheon, South Korea; 10–13 October 2010, pp. 804–808.
44. Liu, Y., Wan, J., Li, G., Yuan, C., and Shen, H. (2009) MRAS speed identification for PMSM based on fuzzy PI control. *4th IEEE Conf. Ind. Elect. Appl., ICIEA*, Xi'an; China; 25–27 May 2009, pp. 1995–1998.
45. Yan, W., Lin, H., Li, H., and Lu, J. (2009) A MRAS based speed identification scheme for a PM synchronous motor drive using the sliding mode technique. *Int. Conf. Mech. Auto. ICMA*, Changchun, China; 9–12 August 2009, pp. 3656–3661.
46. Zhou, F., Yang, J., and Li, B. (2008) a novel speed observer based on parameter-optimized MRAS for PMSMs. *IEEE Int. Conf. Network., Sens. Cont., ICNSC* 2008, Sanya; China; 6–8 April 2008.
47. Yaojing, F. and Kai, Y. (2010) Research of sensorless control for permanent magnet synchronous motor systems. *Int. Conf. Elect. Mach. Syst. (ICEMS)*, Incheon, South Korea; 10–13 October 2010, pp. 1282–1286.
48. Kojabadi, H. M. and Ghribi, M. (2006) MRAS-based adaptive speed estimator in PMSM drives. *9th IEEE Int. Works. Adv. Mot. Cont.*, Istanbul, Turkey; 27–29 March 2006, pp. 569–572.
49. Baohua, L., Jianhua, Y., and Weiguo, L. (2009) Study on speed sensorless SVM-DTC system of PMSM. *9th Int. Conf. Elect. Meas. Instr., ICEMI'09*, Beijing, China; 16–19 August 2009, pp. 2914–2919.
50. Kang, J., Zeng, X., Wu, Y., and Hu, D. (2009) Study of position sensorless control of PMSM based on MRAS. *IEEE Int, Conf. Ind. Tech., ICIT 2009*, Churchill, VIC, Australia; 10–13 February 2009, pp. 1–4.
51. Finch, J. W. and Giaouris, D. (2008) Controlled AC electrical drives. *IEEE Ind. Elect.*, 55(2), 481–491.
52. Vas, P. (1998) *Sensorless Vector and Direct Torque Control.* Oxford University Press, Oxford.
53. Acarnley, P. P. and Watson, J. F. (2006) Review of position-sensorless operation of brushless permanent-magnet machines. *IEEE Trans. Ind. Elect.*, 53(2), 352–362.
54. Tae-Hyung, K., Hyung-Woo, L., and Mehrdad, E. (2004) Advanced sensorless drive technique for multi-phase BLDC motors. *Proc. IECON 2004*, 2–6 November, Busan, South Korea, CD-ROM.
55. Hwang, T. S. and Seok, J-K. (2007) Observer-based ripple force compensation for linear hybrid stepping motor drives. *IEEE Trans. Ind. Elect.*, 54(5), 2417–2424.
56. White, D. C. and Woodson, H. H. (1959) *Electromechanical Energy Conversion.* John Wiley & Sons, New York.
57. Nahid-Mobarakeh, B., Meibody-Tabar, F., and Sargos, F-M. (2007) Back emf estimation-based sensorless control of PMSM: Robustness with respect to measurement errors and inverter irregularities. *IEEE Trans. Ind. Appl.*, 43(2), 485–494.
58. García, P., Briz, F., Raca, D., and Lorenz, R. D. (2007) Saliency-tracking-based sensorless control of AC machines using structured neural networks. *IEEE Trans. Ind. Appl.*, 43(1), 77–86.

59. Elloumi, M., Ben-Brahim, L., and Al-Hamadi, M. (1998) Survey of speed sensorless controls for IM drives. *Proc. IEEE IECON'98*, vol. **2**, Aachen, Ger; 31 August 1998 through 4 September 1998, pp. 1018–1023.
60. Ben-Brahim, L., Tadakuma, S., and Akdag, A. (1999) Speed control of induction motor without rotational transducers. *IEEE Trans. Ind. Appl.*, 35(4), 844–850.
61. Abu-Rub, H., Guziński, J., Rodriguez, J., Kennel, R., and Cortés P. (2010) Predictive current controller for sensorless induction motor drive. *Proc. IEEE-ICIT 2010 Int. Conf. Ind. Tech.*, 14–17 March, Viña del Mar, Chile.
62. Parsa, L. and Toliyat, H. (2004) Sensorless direct torque control of five-phase interior permanent magnet motor drives. *IEEE Ind. Appl. Conf., IEEE-IAS 2004, Ann. Mtg*, Seattle, WA, United States; 3–7 October 2004.
63. Krzemiński, Z. (1999) A new speed observer for control system of induction motor. *Proc. IEEE Int. Conf. Power Elect. Drive Syst., PESC'99*, Poznan, Poland, 1–3 September 2008, pp. 555–560.
64. Krzeminski, Z. (2008) Observer of induction motor speed based on exact disturbance model. *Int. Conf. EPE-PEMC' 2008*, Poznan, Poland; 1–3 September 2008.
65. Batran, A., Abu-Rub, H., and Guziński, J. (2005) Wide range sensorless control of induction motor using power measurement and speed observer. *11th IEEE Int. Conf. Meth. Models Auto. Robot, MMAR '05*, Miedzyzdroje, Poland.
66. Akagi, H., Kanazawa, Y., and Nabae, A. (1983) *Generalised Theory of the Instantaneous Reactive Power in Three-Phase Circuits*. IPEC, Tokyo.
67. Parasiliti, F., Petrella, R., and Tursini, M. (2002) *Speed Sensorless Control of an Interior PM Synchronous Motor*. IEEE, Pittsburgh, PA.
68. Zawirski, K. (2005) *Control of Permanent Magnet Synchronous Motors (in Polish)*. Poznan, Poland.
69. Pajchrowski, T. and Zawirski, K. (2005) Robust speed control of servodrive based on ANN. *IEEE ISIE*, 20–23 June, Dubrovnik, Croatia.
70. Foo, G. and Rahman, M. F. (2008) A novel speed sensorless direct torque and flux controlled interior permanent magnet synchronous motor drive. *39th IEEE Power Elect. Spec. Conf. (PESC 2008)*, 15–19 June, Rhodes.

9

Selected Problems of Induction Motor Drives with Voltage Inverter and Inverter Output Filters

9.1 Drives and Filters – Overview

Nowadays drives with induction motors and voltage-type inverters are commonly used as adjustable speed drives (ASD) in industrial systems [1].

The inverters are built with insulated gate bipolar transistors (IGBT), whose dynamic parameters are very high, i.e. the on- and off-switch times are extremely short. Fast switching of power devices causes high *dv/dt* at the rising and falling edges of the inverter output waveform.

High *dv/dt* in modern inverters is the source of numerous adverse effects in the drives systems [2, 3]. The main negative effects are faster motor bearings degradation, over-voltages on motor terminals, failure or degradation of motor winding insulation due to partial discharges, increase of motor losses, and higher electromagnetic interference levels. Other negative effects are increased by the long cables used to connect the inverter and motor.

The prevention or limitation of the negative effects of *dv/dt* is possible if proper passive or active filers are installed in the drive. In particular, passive filters are preferable for industrial applications (Figure 9.1).

Passive filters used in induction motor drivers are called inerter output filters or motor filters. Depending on filter structures and their parameters, the following filters are specified [4, 5]:

- differential mode filters;
- common mode (CM) filter; and
- *dv/dt* filters.

High Performance Control of AC Drives With MATLAB®/Simulink, Second Edition.
Haitham Abu-Rub, Atif Iqbal, and Jaroslaw Guzinski.

Companion website: www.wiley.com/go/aburubcontrol2e

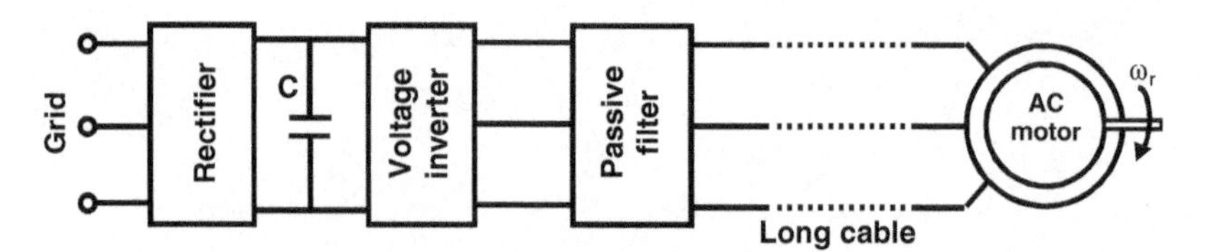

Figure 9.1 Induction motor drives with the inverter output filter

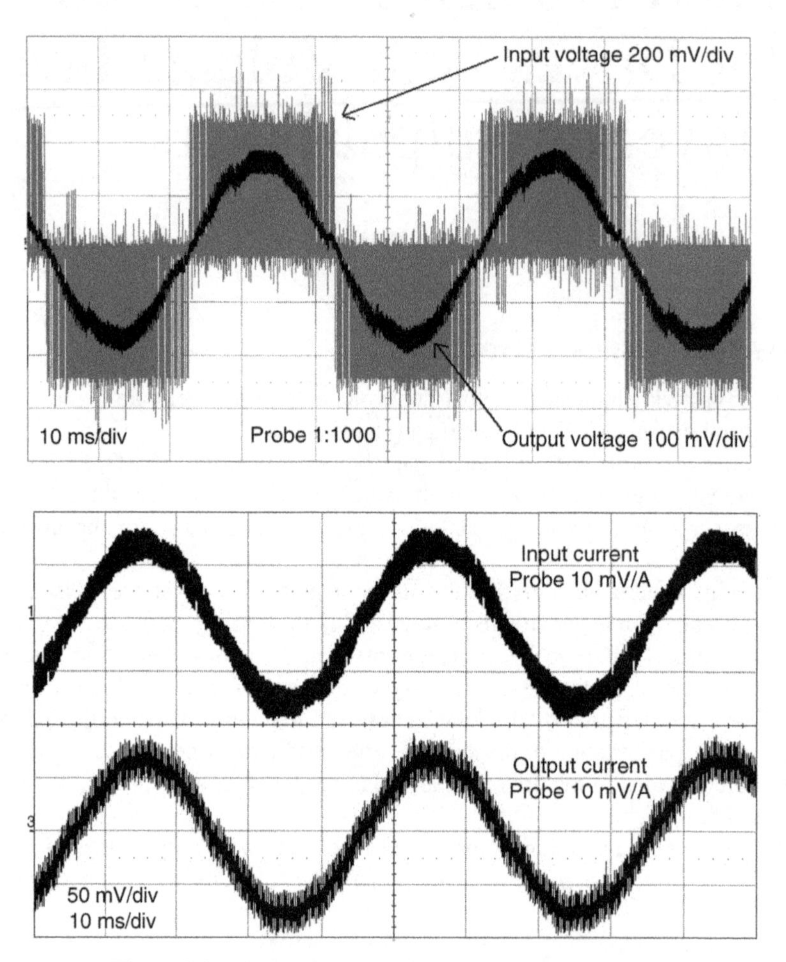

Figure 9.2 The results of differential mode filter operation

With differential mode filters, the motor supply voltage is smoothed to almost a sinusoidal shape contrary to the inverter output voltage (filter's input), which is composed of a series of short rectangular pulses (Figure 9.2). The differential mode filters are also known as sinusoidal filters or LC filters.

The CM filters are used mainly to limit the motor leakage currents, which flow through motor parasitic capacitances. The major part of leakage current flow is through the motor bearing to the motor case and to the ground.

The role of *dv/dt* filters is to eliminate the wave reflection effect in long cables, in order to avoid over-voltages on the motor terminal, as well as to prevent failure of the motor windings insulation [6]. The wave reflections are due to incompatibility between the cable and motor wave impedances. In extreme conditions, the peak of the motor terminal over-voltages can reach twice the value of the inverter supply voltage.

This chapter presents the basic problems associated with electric drives and inverter output filters. The basic mathematical transformations, CM voltage descriptions, and filter structures are first presented. Next, the filter models and design procedures are shown. An essential part of this chapter is dedicated to modifications in the original estimation and control algorithms required in the use of differential mode filters in the drive system. Also, the diagnostic solutions for selected faults in the drive with filters are presented.

This chapter is supported by simulation and experimental validation. Selected simulation models are discussed for better understanding and analysis. More simulation examples could be found in [7].

9.2 Three-Phase to Two-Phase Transformations

For a deeper analysis of electric drives with filters, knowledge of filter models is required.

The natural frame of coordinates for a three-phase system is a three-axis co-ordinate with a 120-degree shift, but for a better analysis of the system, the use of the orthogonal co-ordinates is more useful, as well as easier to understand. Therefore, for the next analysis, proper coordinate transformations are presented. Two kinds of transformations are used, one for constant vectors magnitudes and another for constant power of the three-phase and two-phase frames of reference.

The transformation matrix for the conversion from a three-phase ABC co-ordinate to a two-phase $\alpha\beta 0$ co-ordinate for constant vector magnitude is described as

$$\mathbf{A}_W = \begin{bmatrix} \frac{1}{3} & \frac{1}{3} & \frac{1}{3} \\ \frac{2}{3} & -\frac{1}{3} & -\frac{1}{3} \\ 0 & \frac{1}{\sqrt{3}} & -\frac{1}{\sqrt{3}} \end{bmatrix} \tag{9.1}$$

and the transformation matrix for constant power of the system is

$$\mathbf{A}_P = \begin{bmatrix} \frac{1}{\sqrt{3}} & \frac{1}{\sqrt{3}} & \frac{1}{\sqrt{3}} \\ \frac{\sqrt{2}}{\sqrt{3}} & -\frac{1}{\sqrt{6}} & -\frac{1}{\sqrt{6}} \\ 0 & \frac{1}{\sqrt{2}} & -\frac{1}{\sqrt{2}} \end{bmatrix} \tag{9.2}$$

Using equations (9.1) or (9.2), each of the model variables x can be transformed from ABC to two-phase $\alpha\beta 0$ co-ordinates according to

$$\begin{bmatrix} x_0 \\ x_\alpha \\ x_\beta \end{bmatrix} = \mathbf{A}_W \begin{bmatrix} x_A \\ x_B \\ x_C \end{bmatrix} \tag{9.3}$$

where the variables retain the same magnitude in each co-ordinate.

The next relation is used when the constant power of both systems is fulfilled:

$$\begin{bmatrix} x_0 \\ x_\alpha \\ x_\beta \end{bmatrix} = \mathbf{A}_P \begin{bmatrix} x_A \\ x_B \\ x_C \end{bmatrix} \tag{9.4}$$

The parameter transformations are required for system modeling, so the resistances matrix is

$$\mathbf{R}_{ABC}^{W} = \mathbf{R}_{ABC}^{P} = \begin{bmatrix} R & 0 & 0 \\ 0 & R & 0 \\ 0 & 0 & R \end{bmatrix} \tag{9.5}$$

The three-phase choke with symmetrical coils on the toroidal core is

$$\mathbf{M}_{ABC}^{W} = \mathbf{M}_{ABC}^{P} = \begin{bmatrix} 3M & 0 & 0 \\ 0 & 0 & 0 \\ 0 & 0 & 0 \end{bmatrix} \tag{9.6}$$

and for a three-phase choke with an E-shape, the core is

$$\mathbf{M}_{ABC}^{W} = \mathbf{M}_{ABC}^{P} = \begin{bmatrix} 0 & 0 & 0 \\ 0 & \frac{3}{2}M & 0 \\ 0 & 0 & \frac{3}{2}M \end{bmatrix} \tag{9.7}$$

In equation (9.6), it is evident that the toroidal three-phase symmetrical choke has parameters only for the CM, whereas in equation (9.7), the E-core choke has parameters only for the differential mode. Such a conclusion is essential for choosing the correct choke core selection in the filters design process.

9.3 Voltage and Current Common Mode Component

As mentioned at the beginning of this chapter, unwanted bearing currents appear in the drives with voltage inverters. This occurs because the voltage inverter with classical pulse width modulation (PWM) is a source of CM voltage, which forces the bearing current to flow.

To clarify the CM voltage effect, the voltage inverter structure presented in Figure 9.3 is analyzed.

In voltage inverters, only one of eight transistor switching combinations is possible. Due to the voltage notation presented in Figure 9.3, the inverter output voltages for each of the eight states are presented in Table 9.1.

It is noted that the zero-component voltage changes according to changes in the transistors switching combination. The peak value of the u_0 voltage is very high and equal to the inverter DC link voltage U_d. The u_0 frequency is equal to the inverter PWM switching frequency. A typical u_0 voltage waveform is presented in Figure 9.4.

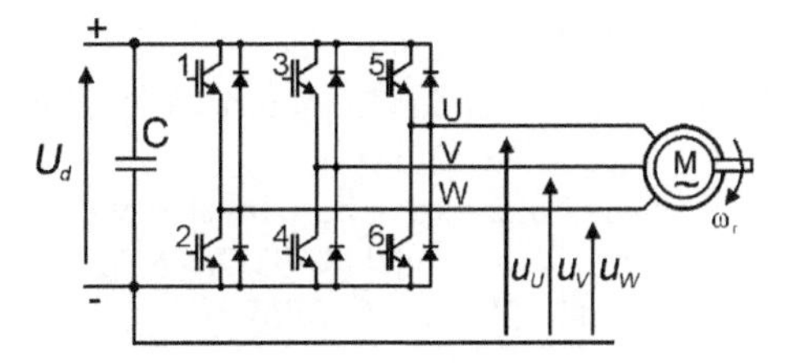

Figure 9.3 Structure of the voltage inverter with output voltages notations (*Source*: Modified from Muetze and Binder [10])

Table 9.1 Voltage inverter output voltages for possible switching combination

	Binary notation for transistor switching combination							
	100	110	010	011	001	101	000	111
u_U	U_d	U_d	0	0	0	U_d	0	U_d
u_V	0	U_d	U_d	U_d	0	0	0	U_d
u_W	0	0	0	U_d	U_d	U_d	0	U_d
u_0	$\frac{U_d}{\sqrt{3}}$	$\frac{2U_d}{\sqrt{3}}$	$\frac{U_d}{\sqrt{3}}$	$\frac{\sqrt{2}U_d}{\sqrt{3}}$	$\frac{U_d}{\sqrt{3}}$	$\frac{2U_d}{\sqrt{3}}$	0	$\sqrt{3}U_d$
u_α	$\frac{\sqrt{2}U_d}{\sqrt{3}}$	$\frac{U_d}{\sqrt{6}}$	$-\frac{U_d}{\sqrt{6}}$	$-\frac{\sqrt{2}U_d}{\sqrt{3}}$	$-\frac{U_d}{\sqrt{6}}$	$-\frac{U_d}{\sqrt{6}}$	0	0
u_β	0	$\frac{U_d}{\sqrt{2}}$	$\frac{U_d}{\sqrt{2}}$	0	$-\frac{U_d}{\sqrt{2}}$	$-\frac{U_d}{\sqrt{2}}$	0	0

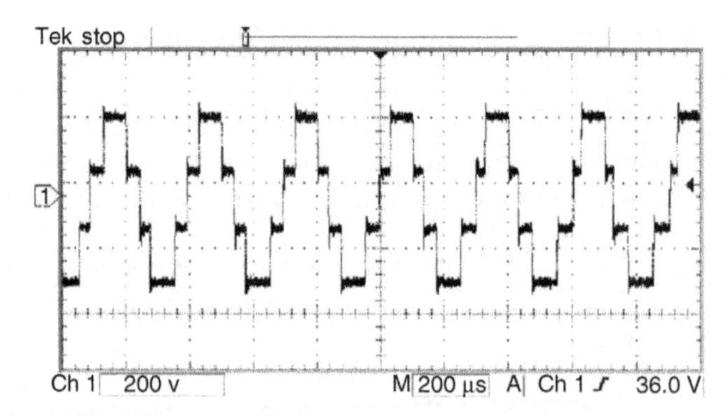

Figure 9.4 The CM voltage waveform in a voltage inverter

9.3.1 MATLAB/Simulink Model of Induction Motor Drive with PWM Inverter and Common Mode Voltage

The MATLAB/Simulink model of induction motor drive *Symulator.mdl* is presented in Figure 9.5.

The *Symulator.mdl* model consists of a three-phase inverter with a space vector PWM algorithm (*PWM_SFUN*) and induction motor model (*MOTOR_SFUN*). Both blocks are Simulink S-functions from C code. The C code files are as follows:

- *declarations.h* – list of predefined values;
- *pwm.h* – declaration of global variables for PWM inverter model;
- *pwm.c* – source file of PWM inverter model;
- *motor.h* – declaration of global variables for induction motor model;
- *motor.c* – source file of induction motor model.

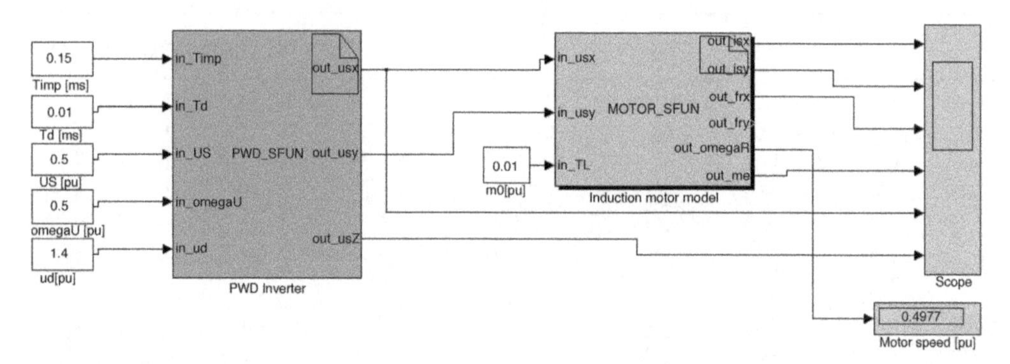

Figure 9.5 MATLAB/Simulink model of induction motor drive with PWM inverter and common mode voltage (*Symulator.mdl*)

The commanded values are:

- *Timp* – inverter switching period in milliseconds;
- *Td* – inverter dead time in milliseconds;
- *US* – module of the commanded inverter output voltage;
- *omegaU* – inverter output voltage pulsation;
- *ud* – inverter DC link supply voltage;
- *m*0 – motor load torque.

All variables and parameters are in per unit (p.u.) system, where the fundamental base values are presented in Table 9.2.

The simulated drive is working according to the V/f = const principle. No ramp is added for stator voltage pulsation, so when the simulation starts the motor direct start-up is performed.

The model of four poles, $P_N = 1.5$ kW, $U_N = 380$ V induction motor is used. The motor parameters are given in the *pwm.h* file. Examples of simulation results for motor commanded frequency $\omega_u = 0.5$ p.u. (i.e. 25 Hz) and at DC link voltage $u_d = 1.4$ p.u. (i.e. 540 V) are given in Figures 9.6 and 9.7. In Figure 9.6, the process of the motor start-up is presented for the complete simulation time. In Figure 9.7, the inverter output voltage of α and β components, as well as the CM voltage, are presented in wider zoom. The CM voltage is related to the negative terminal of the inverter DC link supply.

The complete MATLAB/Simulink model is available in the CD-ROM provided with this book.

Table 9.2 Base values for p.u. system used in the MATLAB/Simulink model

Definition	Description
$U_d = \sqrt{3}U_n$	Base voltage
$I_b = \sqrt{3}I_n$	Base current
$Z_b = U_b/I_b$	Base impedance
$m_b = (U_b I_b p)/\omega_0$	Base torque
$\Psi_b = U_b/\omega_0$	Base flux
$\omega_b = \omega_0/p$	Base mechanical speed
$L_b = \Psi_b/I_b$	Base inductance
$C_b = 1/(\omega_b Z_b)$	Base capacitance
$J_b = m_b/(\omega_b \omega_0)$	Base inertia
$\tau = \omega_0 t$	Relative time

Where $\omega_0 = 2\pi f_N$ is nominal grid pulsation, e.g. $f_N = 50$ Hz or $f_N = 60$ Hz.

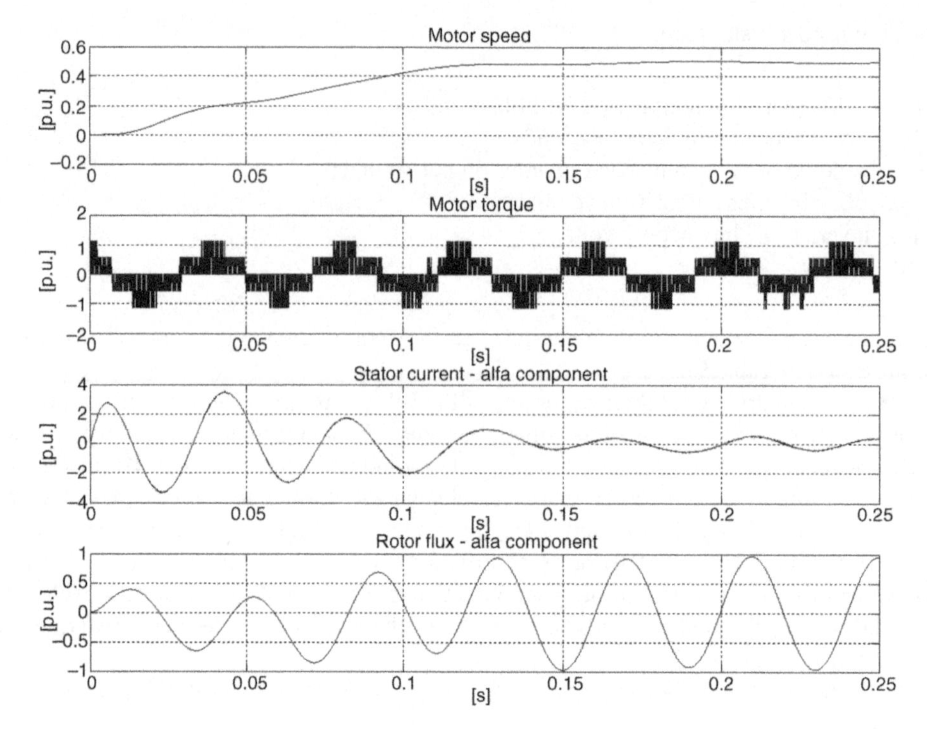

Figure 9.6 Simulation results for motor start-up

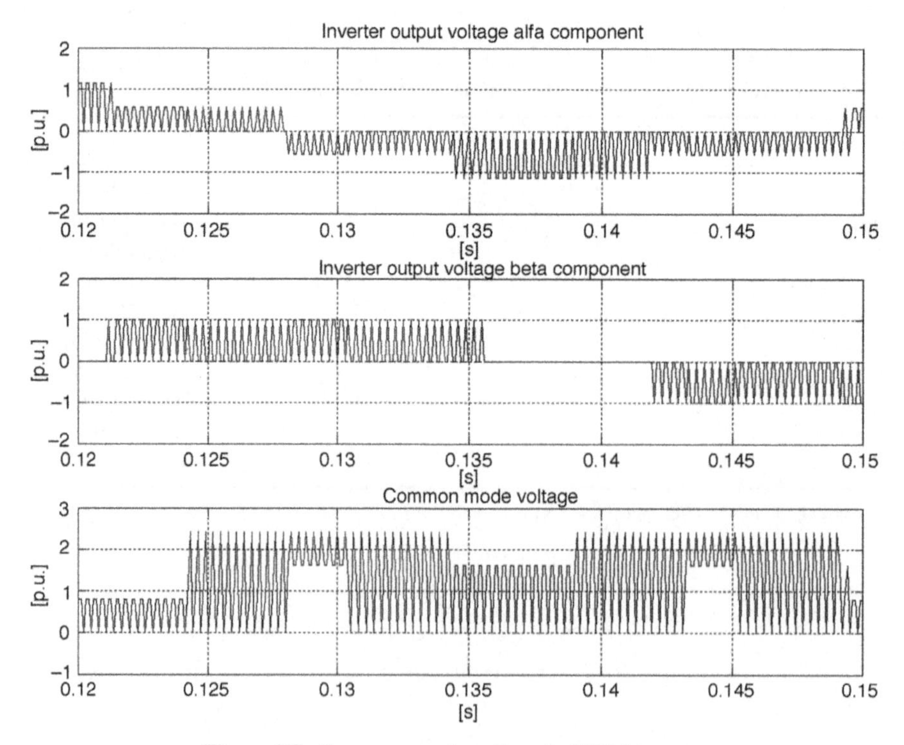

Figure 9.7 Common mode voltage in PWM inverter

9.4 Induction Motor Common Mode Circuit

The motor has some parasitic capacitances with small values, in the order of picofarads. The CM voltage on the stator windings creates a shaft voltage by capacitive coupling through the motor air gap. Therefore, electrostatic discharges are generated through the bearing lubricating film. Such parasitic currents cause a reduction in the lifetime of the bearings. The motor's main parasitic capacitances are presented in Figure 9.8.

In Figure 9.8, the following motor parasitic capacitances are labeled as:

- C_{wf} – parasitic capacitance between stator windings and stator frame;
- C_{wr} – parasitic capacitance between stator winding and rotor winding;
- C_{rf} – parasitic capacitance between the rotor and frame;
- C_b – equivalent capacitance of motor bearings;
- C_{ph} – parasitic capacitance between stator windings.

All the parasitic capacitances, except C_{ph}, are significant for the CM. So in the next explanation, the phase-to-phase capacitance C_{ph} is omitted.

The C_{wf} capacitance is dependent on the motor's mechanical size, according to empirical relation [10]:

$$C_{wf} = 0.00024 \cdot H^2 - 0.039 \cdot H + 2.2 \tag{9.8}$$

where C_{wf} is in [nF] and motor mechanical size H is in [mm].

The bearing capacitance is dependent on the bearing dimensions and mechanical properties. The value of C_b is of the order of picofarads and close to the C_{wr} value. In practice, it is assumed that $C_b \approx C_{wr}$. The motor parasitic capacitance between the stator windings and the rotor C_{wr} is dependent on motor size, winding type, and motor magnetic circuit. The C_{wr} calculation is complicated and requires a lot of data, which can be obtained only from the motor manufactures. The same goes for the C_{rf} calculations. More in-depth explanations can be found in [10].

In spite of the small values of the capacitances, the parasitic current can reach large values, because the switching frequency in modern inverters is high. Part of the parasitic current circulates inside the motor, while some flows to the motor case and to the ground. The current flow path to the ground is presented in Figure 9.9.

An equivalent circuit of the motor and feeder cable for CM current is presented in Figure 9.10.

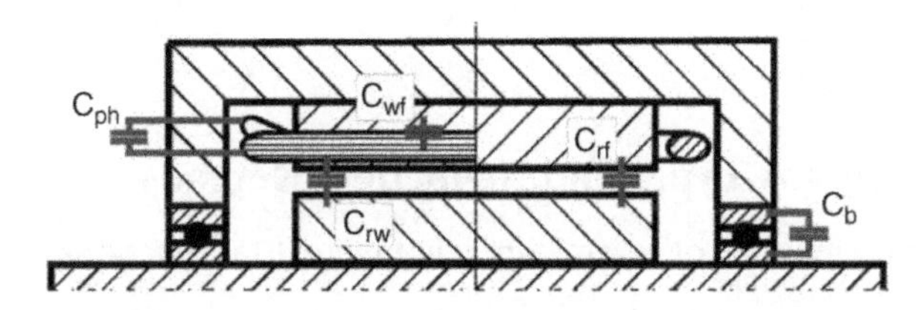

Figure 9.8 The capacitances of the common mode current circuit of an induction motor (*Sources*: Binder and Muetze [8]; Muetze [9])

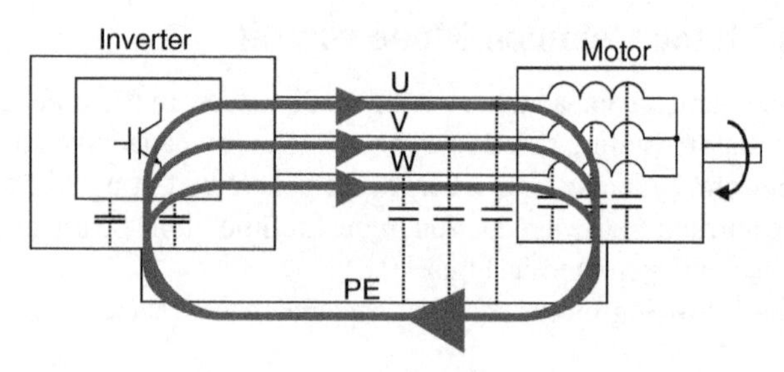

Figure 9.9 Common mode current path flow in electric drives with voltage inverter

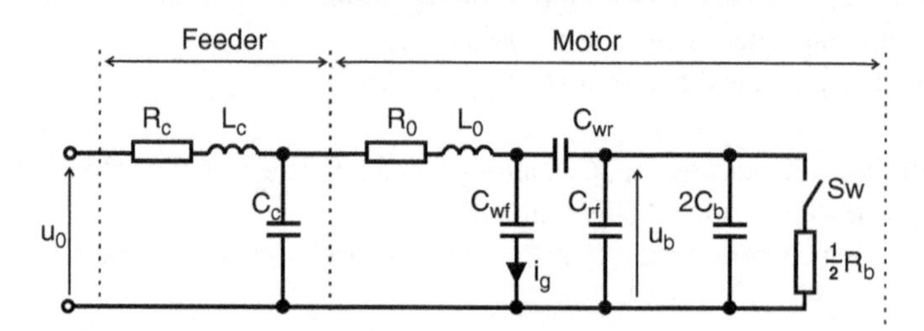

Figure 9.10 An equivalent circuit of the motor and feeder cable for CM current (*Sources*: Binder and Muetze [8]; Muetze [9])

In Figure 9.10, the motor Common Mode circuit elements are:

- ***Feeder cable parameters:***
  - R_c – cable resistance;
  - L_c – cable inductance;
  - C_c – cable capacitance.
- ***Motor parameters:***
  - R – stator winding cupper resistance;
  - L – stator winding inductance;
  - R_b – equivalent resistance of motor bearings;
  - Sw_1, Sw_2, – switches, when closed indicates the breakdown of the bearing lubricating film in appropriate bearing.

9.5 Bearing Current Types and Reduction Methods

The motor bearing currents are of several types. In Figure 9.11, four classes of bearing current are presented.

The capacitance bearing current has a small value close to 510 mA at temperature $T_b \approx 25°C$ and motor speed $n \approx 100$ rpm. For higher temperatures and motor speeds, the current increases

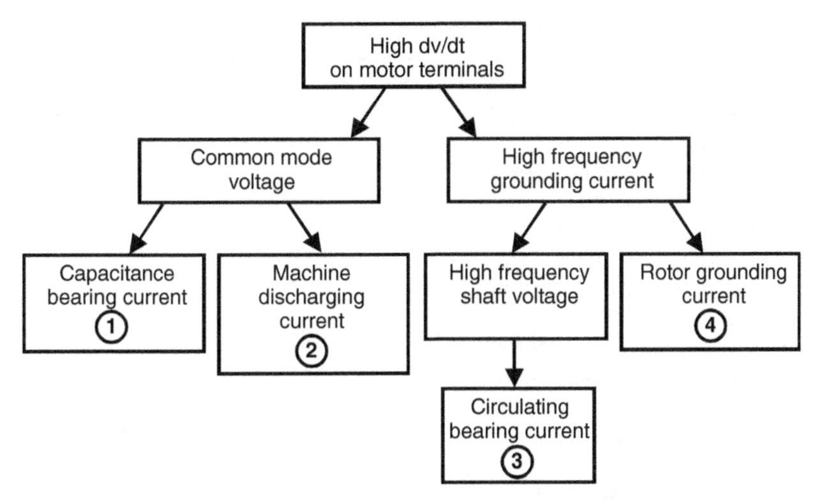

Figure 9.11 Bearing current classes (*Source*: Muetze and Binder [10])

but will not exceed 200 mA. Contrary to other bearing current values, the current is negligible for bearing degradation analysis.

The machine discharging current appears when the bearing oil film breaks down. According to the literature, the value of that current peak is up to 3 A [10].

The bearing circulating current i_{bcir} is induced by a motor shaft voltage u_{sh}. The u_{sh} appears due to parasitic motor flux ψ_{cir}. In [10], the practical condition for i_{bcir} calculation is given by

$$i_{bcir(max)} \leq 0.4 \cdot i_{g(max)} \tag{9.9}$$

where $i_{g(max)}$ is the maximum value of the measured grounding current.

An in-depth explanation on the particular bearing currents calculation and measurements can also be found in [10].

The rotor grounding current can appear when a galvanic connection between the rotor and ground appears, i.e. caused by a load machine coupled with a motor shaft.

In [10], the types of bearing currents are correlated with motor mechanical size H, as follows:

- if $H < 100$ mm, then machine discharging current, i_{bEDM}, is dominant;
- if 100 mm $< H <$ 280 mm, then both i_{bEDM} and circulating bearing currents i_{bcir} are dominant;
- if $H > 280$ mm, then circulating bearing currents are dominant.

To prevent bearing current flow, the methods presented in Table 9.3 can be used.

In addition to the use of passive filters for CM, current active methods are also used, such as series active filter (Figure 9.12), semi-active filters [12], or by changing the PWM algorithm.

Active filters are unpopular in MV drives because they operate at a high switching frequency, which is not recommended from a loss point of view. In this book, the CM active filters as well semi-active filters are not discussed and only the PWM algorithms modifications reducing the CM current will be described in this chapter.

Table 9.3 Bearing current prevention methods [10]

Reduction of electric discharge current i_{bEDM}:
• Ceramic bearings;
• CM passive filters;
• Active compensation systems of CM voltage;
• Decreasing of inverter switching frequency;
• Motor shaft grounding by brushes use;
• Conductive grease in the bearings.
Reduction of circulating bearing current i_{bcir}:
• CM choke;
• Active compensation systems of CM voltage;
• Decreasing of inverter switching frequency;
• One or two insulated bearings use;
• One or two ceramic bearings use;
• *dv/dt* filter.
Reduction of rotor grounding current i_{gr}:
• CM choke;
• Active compensation systems of CM voltage;
• Decreasing of inverter switching frequency;
• One or two insulated bearings use;
• One or two ceramic bearings use;
• Screened cable for motor supply.

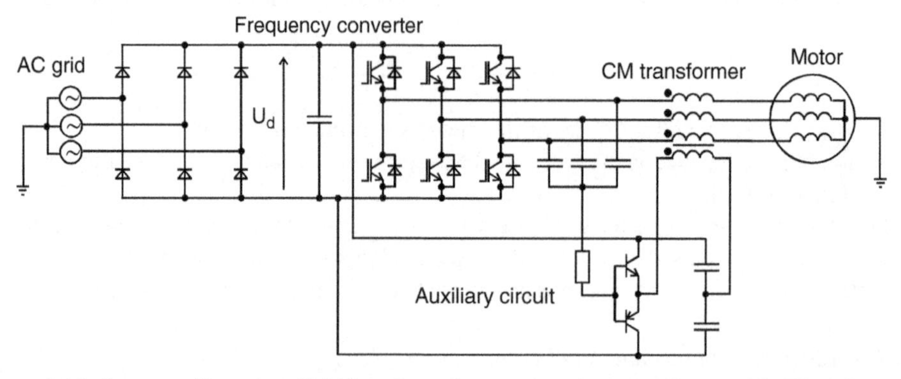

Figure 9.12 System with active CM filter for voltage source inverter (*Source*: Modified from Sun et al. [11])

9.5.1 Common Mode Choke

Among the numerous methods for bearing current reduction, the most popular is known from small electronics circuits is a three-phase CM choke (Figure 9.13).

A CM choke is built in the form of three symmetric coils wound on a toroidal-shaped core. The choke has a negligible inductance in a differential mode for the orthogonal components $\alpha\beta$; however, it has a significant inductance, L_{10}, for the path of the CM current:

$$L_{10} = 3M_1 \tag{9.10}$$

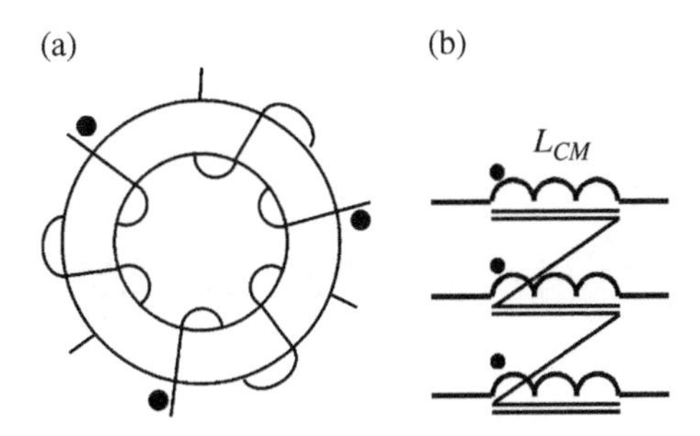

Figure 9.13 Structure (a) and schematic symbol (b) of the common mode choke for three-phase system

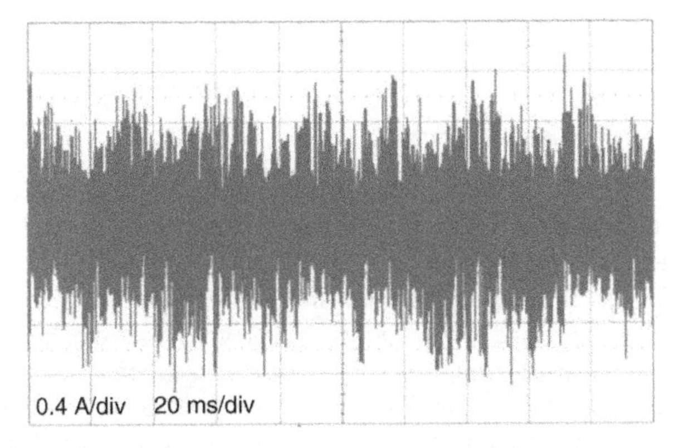

Figure 9.14 Waveform of the current measured in the motor grounding wire in 1.5 kW industrial drive **without** CM choke use

where M_1 is the inductance of each CM mode choke.

Figures 9.14 and 9.15 present the results of using CM choke in the 1.5 kW induction motor industrial drive.

In both Figures 9.14 and 9.15, the current in the motor grounding wire is measured. Previously, the CM choke used had current peaks up to 1 A (Figure 9.14). When the CM choke of inductance M_1 = 14 mH was installed, the current peaks were strongly limited (Figure 9.15).

To achieve maximum limiting of CM current, a maximum choke inductance is needed. A realization of the CM choke is based on using a maximum number of windings on the selected toroidal-shaped core. It is important for a CM choke that all windings are identical, to ensure low leakage inductances. For differential mode circuits only, the CM choke leakage inductance can appear.

For high power motors, large-sized toroids for CM chokes cannot be accessible, so CM current reduction can be achieved by a few CM chokes connected in series.

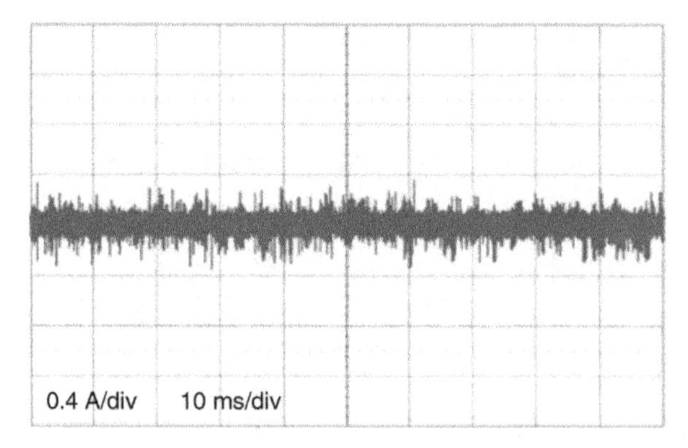

Figure 9.15 Waveform of the current measured in the motor grounding wire in 1.5 kW industrial drive **with** CM choke use

9.5.2 Common Mode Transformers

The value of the CM voltage may be reduced by using a CM transformer. The difference between the CM choke and the CM transformer is the additional winding shortened by the resistor (Figure 9.16).

In the CM transformer, an additional fourth winding is wound with the same number of turns as the choke phase windings. A CM transformer makes it possible to reduce the CM current in the motor by up to 25%, while using less core volume than when using a CM choke [13].

If a CM transformer is used, then the CM circuit consists of an additional inductance L_t and resistance R_t. The equivalent circuit of the CM transformer is shown in Figure 9.17.

An equivalent circuit of the CM transformer, inverter, feeder cable, and motor is given in Figure 9.18.

In Figures 9.17 and 9.18, R_t is operating as a dumping resistance for a resonant circuit. If the CM transformer leakage inductance is neglected, the CM current circuit is [14].

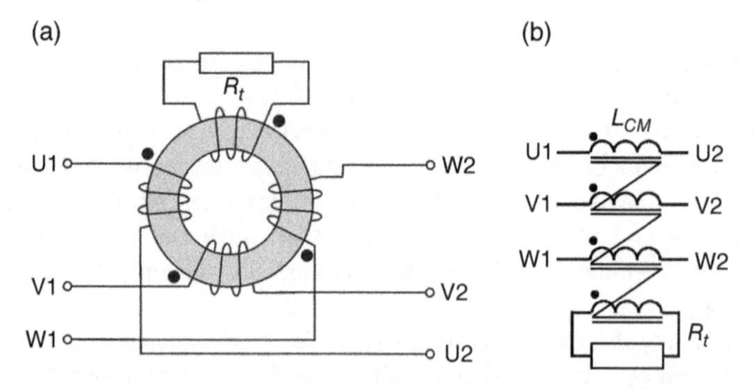

Figure 9.16 Common mode transformer: (a) structure and (b) schematic

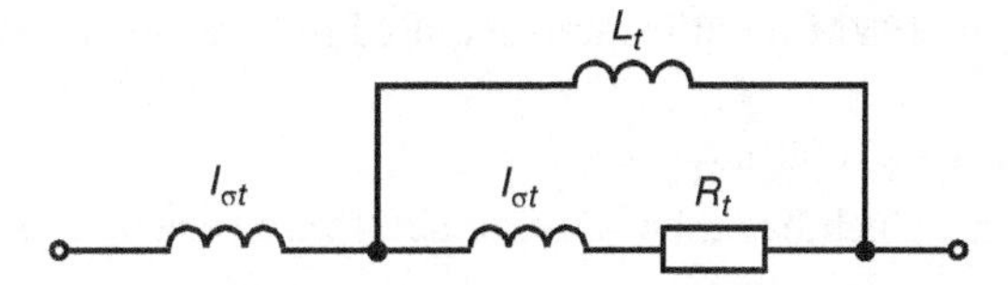

Figure 9.17 Equivalent circuit of CM transformer ($l_{\sigma t}$ – leakage inductance of the transformer)

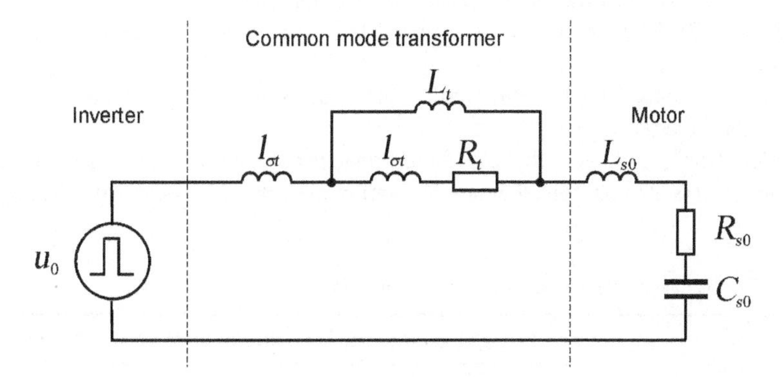

Figure 9.18 Equivalent circuit of CM transformer, inverter, and induction motor

$$I_0(s) = \frac{sL_tC_{s0} + R_tU_d}{s^3L_tL_{s0}C_{s0} + s^2(L_t + L_{s0})C_{s0}R_t + sL_t + R_t} \tag{9.11}$$

The analysis of equation (9.11) calculates the R_t value that limits both maximal and rms CM current values. According to [14], the proper condition for R_t selection is

$$2Z_{00} \leq R_t \leq \frac{1}{2}Z_{0\infty} \tag{9.12}$$

where Z_{00} and $Z_{0\infty}$ are wave impedances of the circuit presented in Figure 9.18, if $R_t = 0$ and $R_t = \infty$.

9.5.3 Common Mode Voltage Reduction by PWM Modifications

Typical methods for limiting the CM current, such as different configurations of CM chokes or active compensation of CM voltages, are mostly expensive solutions and require additional hardware.

However, limiting the CM current in the drive system is possible in a way that does not require inverter reconstruction. The CM current can be limited by only changing the PWM algorithm. Changing the PWM should be done in such a way that it eliminates or decreases the CM voltage generated by the inverter. Different methods based on PWM algorithm modifications are suggested in the literature [15].

The following are the PWM modifications required for CM reduction:

- elimination of zero voltage vectors;
- use of active vectors, which have the same value of zero sequence voltage u_0.

Such approaches come under the analysis shown in Table 9.4.

In Table 9.4, the active vectors are named as parity or non-parity vectors. Each of the parity and non-parity vectors has the same value of zero component u_0.

For classical PWM methods, an example of a voltage u_{NO} waveform is presented in Figure 9.19.

The u_{NO} is the voltage between the mid-point of the star connected load and the negative potential of the inverter supply.

It is evident from Figure 9.19 that the highest changes of the voltage u_{NO} happen during zero vector changes. It is also obvious that an elimination of such vectors can significantly limit the

Table 9.4 Zero sequence vectors of inverter output voltage

	Arrangement of switches states of three-phase voltage inverter							
Vectors type				Active			Zero	
Vectors notation	U_{W4}	U_{w6}	U_{w2}	U_{w3}	U_{w1}	U_{w5}	U_{w0}	U_{w7}
Vector binary notation	100	110	010	011	001	101	000	111
Vector decimal notation	4	6	2	3	1	5	0	7
Vectors number	1	2	3	4	5	6	0	7
U_{NO}	$\frac{1}{3}U_d$	$\frac{2}{3}U_d$	$\frac{1}{3}U_d$	$\frac{2}{3}U_d$	$\frac{1}{3}U_d$	$\frac{2}{3}U_d$	0	U_d
u_0 (coordinate system $\alpha\beta$)	$\frac{U_d}{\sqrt{3}}$	$\frac{2U_d}{\sqrt{3}}$	$\frac{U_d}{\sqrt{3}}$	$\frac{2U_d}{\sqrt{3}}$	$\frac{U_d}{\sqrt{3}}$	$\frac{2U_d}{\sqrt{3}}$	0	$\sqrt{3}U_d$
Notation(1)	NP	P	NP	p	NP	P	Z	Z

(1)Vector: NP – non-parity (odd), P – parity (even), Z – zero.

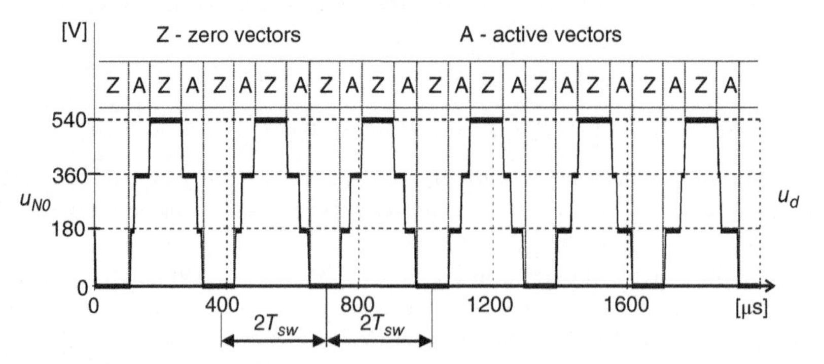

Figure 9.19 Example of u_{NO} waveform with classical space vector PWM algorithm (the inverter DC input voltage is $U_d = 540$ V)

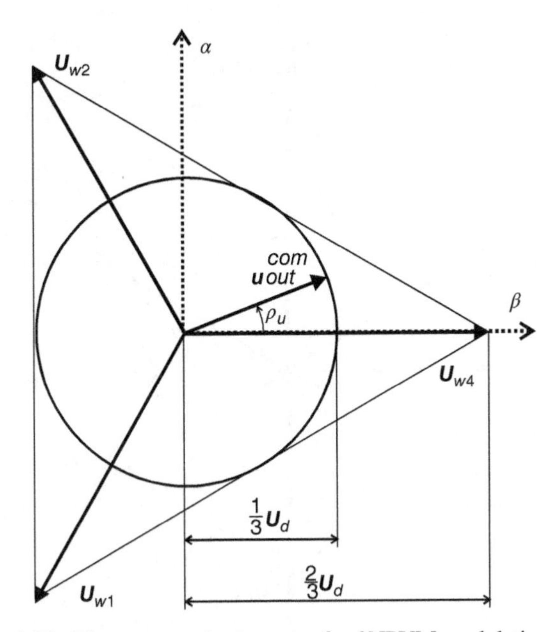

Figure 9.20 The output voltage vector for 3NPVM modulation strategy

CM current. If only active vectors that are equivalent to u_0 are used, then the CM current will decrease. For example, in [16] a PWM algorithm without using zero vectors is presented. With the algorithm only, the output voltage is built with parity and non-parity active vectors.

The disadvantage of the PWM based only on parity or odd active vectors is that without overmodulation, the inverter output voltage is decreased. This is explained in Figure 9.20 for a three non-parity active vectors (3NPAV) algorithm.

In Figure 9.21, the inverter output voltage vector $\mathbf{u}_{out}^{com}$ is generated using three non-parity vectors $\mathbf{U}_{w4}$, $\mathbf{U}_{w2}$, $\mathbf{U}_{w1}$:

$$\mathbf{u}_{out}^{com} \cdot T_{imp} = \mathbf{U}_{w4} t_4 + \mathbf{U}_{w2} t_2 + \mathbf{U}_{w1} t_1 \tag{9.13}$$

The vectors switching times t_4, t_2 and t_1 are [17]

$$t_4 = \frac{1}{3}\left(1 + \frac{2U_{out}^{com}}{U_d}\left(\cos\left(\frac{\pi}{3} + \rho_u\right) + \sin\left(\frac{\pi}{6} + \rho_u\right)\right)\right) \tag{9.14}$$

$$t_2 = \frac{1}{3}\left(1 - \frac{2U_{out}^{com}}{U_d}\cos\left(\frac{\pi}{3} + \rho_u\right)\right) \tag{9.15}$$

$$t_1 = \frac{1}{3}\left(1 - \frac{2U_{out}^{com}}{U_d}\sin\left(\frac{\pi}{6} + \rho_u\right)\right) \tag{9.16}$$

$$T_{imp} = t_4 + t_2 + t_1 \tag{9.17}$$

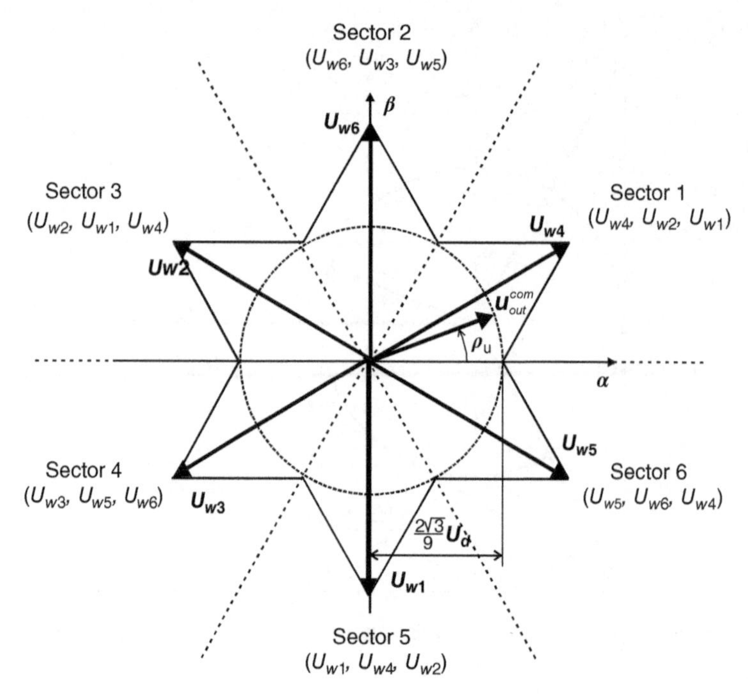

Figure 9.21 Output voltage vector for 3AVM modulation strategy

where U_{out}^{com} and ρ_u are the length and position angle of the inverter output voltage vector $\mathbf{u}_{out}^{com}$, respectively.

By using non-parity active vectors, the CM voltage is

$$u_0 = \frac{U_d}{\sqrt{3}} \tag{9.18}$$

It is obvious that the constant value of u_0 does not allow for zero sequence current flow, i_0, through the internal machine's capacitances. The disadvantage of the 3NPAV is that the maximum inverter output voltage is limited to $U_d/3$.

A higher use of the inverter output voltage with simultaneous CM current reduction is made possible by using the modulation method with only active vectors. The method is close to the classical PWM and was proposed in [17] where a 3AVM is presented. In the 3AVM method, the position of the voltage vector is divided into six sectors and displaced by 30 degrees from the original sectors in the classical space vector modulation SVPWM (Figure 9.21).

When the 3AVM algorithm is used, both the value and frequency of u_0 are decreased; therefore, the CM current is significantly limited. Simultaneously, the inverter output voltage amplitude equals $2\sqrt{3}/9{\cdot}U_d$. It is 15.5% more in comparison with the 3NPAV and 3PAV.

Another modulation method that reduces the CM current is the active zero vector control method (AZVC) presented in [17, 18]. In the AZVC algorithm, the zero voltage vectors are

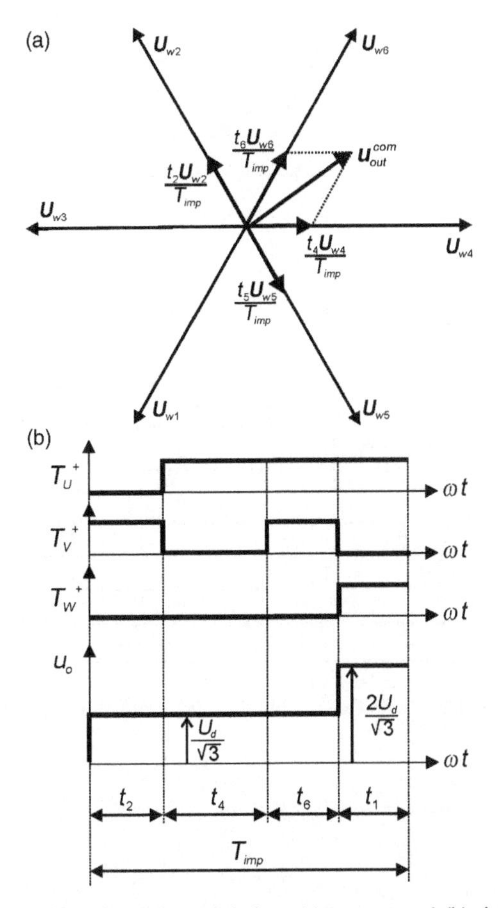

Figure 9.22 AZVC-2 modulation: (a) vectors and (b) timings

replaced by two opposite active vectors (AZVC-2) (Figure 9.22) or one active vector (AZVC-1) (Figure 9.23).

The algebraic relations for AZVC-1 and AZVC-2, for the situation presented in Figures 9.22 and 9.23, are

$$\mathbf{u}_{out}^{com} \cdot T_{imp} = \mathbf{U}_{w4}t_4 + \mathbf{U}_{w6}t_6 + \mathbf{U}_{w2}t_2 + \mathbf{U}_{w5}t_5 \quad \text{for AZVC-2} \tag{9.19}$$

$$\mathbf{u}_{out}^{com} \cdot T_{imp} = \mathbf{U}_{w4}t_4 + \mathbf{U}_{w6}t_6 + \mathbf{U}_{w3}t_3 + \mathbf{U}_{w4}t_4^* \quad \text{for AZVC-1} \tag{9.20}$$

$$t_4 = T_{imp} \cdot \frac{U_{out\alpha}^{com} \cdot U_{w6\beta} - U_{out\beta}^{com} \cdot U_{w6\alpha}}{U_d \cdot w_t} \tag{9.21}$$

$$t_6 = T_{imp} \cdot \frac{-U_{out\alpha}^{com} \cdot U_{w4\beta} + U_{out\beta}^{com} \cdot U_{w4\alpha}}{U_d \cdot w_t} \tag{9.22}$$

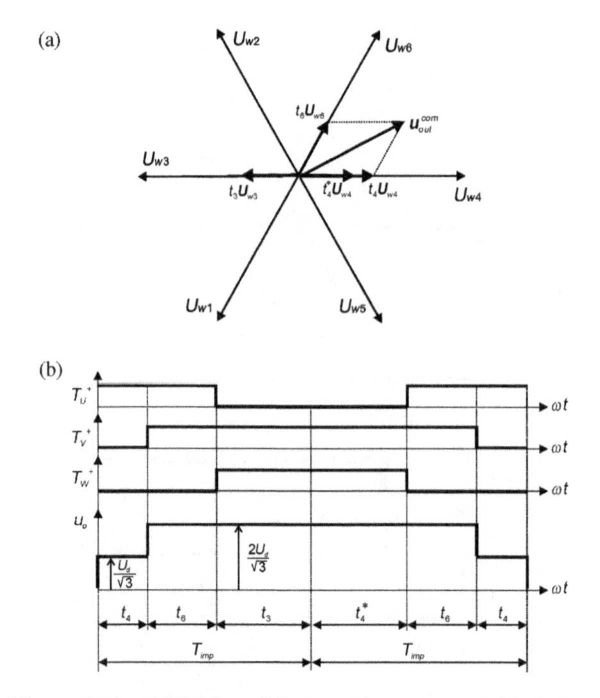

Figure 9.23 AZVC-1 modulation: (a) vectors and (b) timings

$$t_2 = t_5 = \frac{1}{2}\left(T_{imp} - t_4 - t_6\right) \quad \text{for AZVC-2} \tag{9.23}$$

$$t_3 = t_4^{*} = \frac{1}{2}\left(T_{imp} - t_4 - t_6\right) \quad \text{for AZVC-1} \tag{9.24}$$

where

$$w_t = U_{w4\alpha}U_{w6\beta} - U_{w4\beta}U_{w6\alpha} \tag{9.25}$$

In both AZVC methods, the same maximum inverter output voltage is obtained and is the same as in the classical SWPWM algorithm.

The disadvantage of the modified PWM method is with the proper inverter output currents measurement. In classical SVPWM, current measurements are synchronized with the SVPWM operation and performed under zero voltage vector generation. Current values sampled in the middle of the zero vector's duration are identical with the values of the inverter output current's first harmonic component. When using PWM modulations without zero vectors, it is essential to use additional filtering elements for inverter output currents.

9.6 Inverter Output Filters

9.6.1 Selected Structures of Inverter Output Filters

The improvement of the AC motor operation in inverter fed drives is possible, if the shape of the stator voltage becomes as close as possible to the sinusoidal.

A motor fed by such waveforms shows higher efficiency as a result of decreasing miscellaneous losses in the machine. Motors fed by non-sinusoidal waveforms have higher eddy current losses. For high switching frequency, those losses are the dominating losses in the machine, compared to copper and hysteresis losses [19–21].

In inverter fed drives, the use of the voltage inverter output filters reduces the disturbance levels in the current and voltage waveforms.

Motor side filters may be categorized into three basic types:

1. Sinusoidal output filters (LC filters, sine wave filters, differential mode filters);
2. Common mode filters (CM filters);
3. *du/dt* filters.

The three types of filters can be used separately or combined together in different combinations, for example, connecting a sine filter with a CM filter. In this chapter, different filters configurations are analyzed, as shown in Figures 9.24–9.30.

The filter shown in Figure 9.24 is a combination of a sinusoidal filter and a CM filter. Connection of such filters makes it possible to obtain sinusoidal voltage and current at the filter output and also limits the CM current. In such situations, it is possible to use a long cable between the filter and the motor. The length of the cable will be limited only by the allowed voltage drop in the cable.

Elements L_1, C_1, and R_1 represent the sinusoidal output filter, while M_1, M_2, R_0, and C_0 represent the CM filter. These filters create a closed circuit during inverter operation through

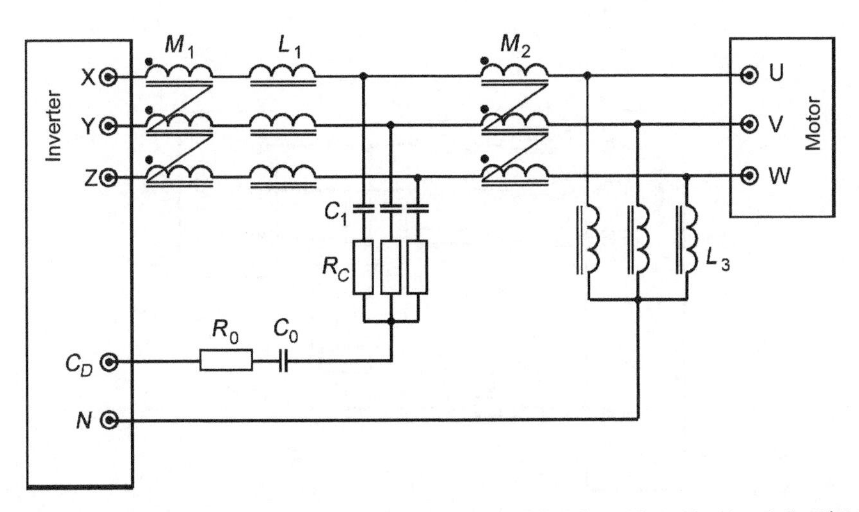

Figure 9.24 Inverter output filter – Structure 1 (*Source*: Modified from Krzeminski and Guziński [22])

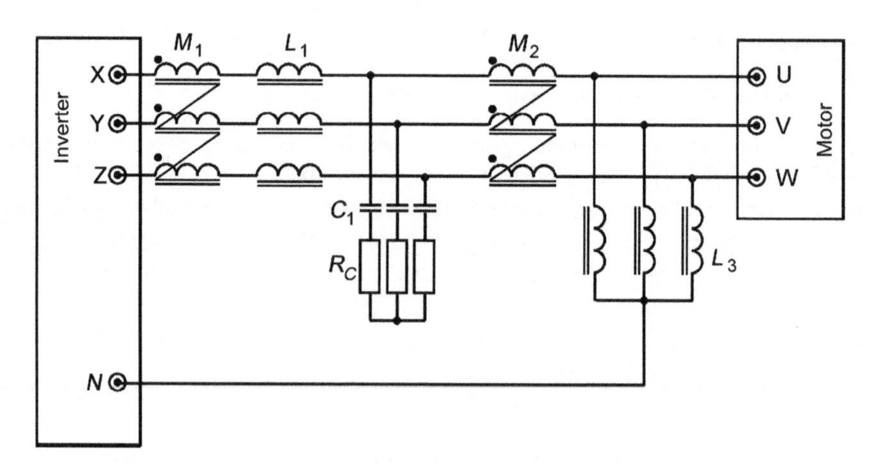

Figure 9.25 Inverter output filter – Structure 2

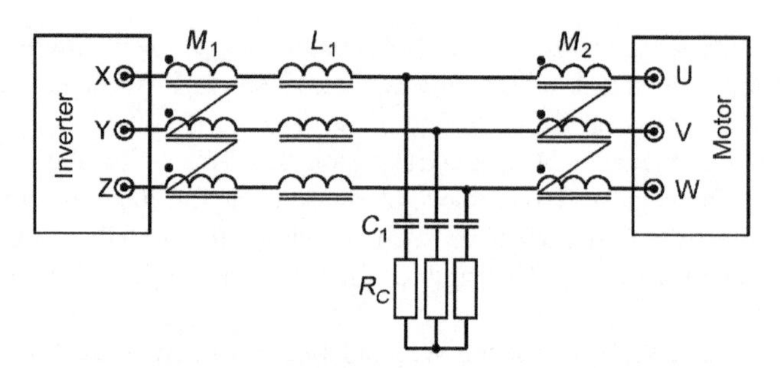

Figure 9.26 Inverter output filter – Structure 3

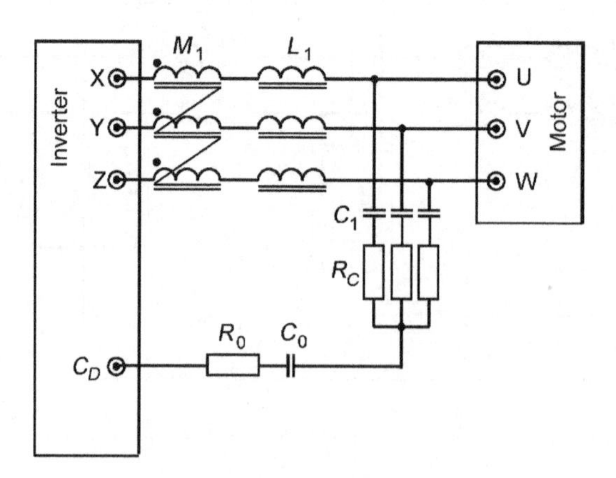

Figure 9.27 Inverter output filter – Structure 4

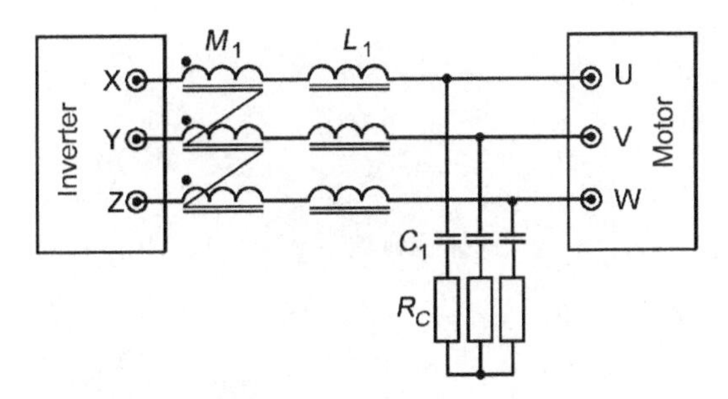

Figure 9.28 Inverter output filter – Structure 5

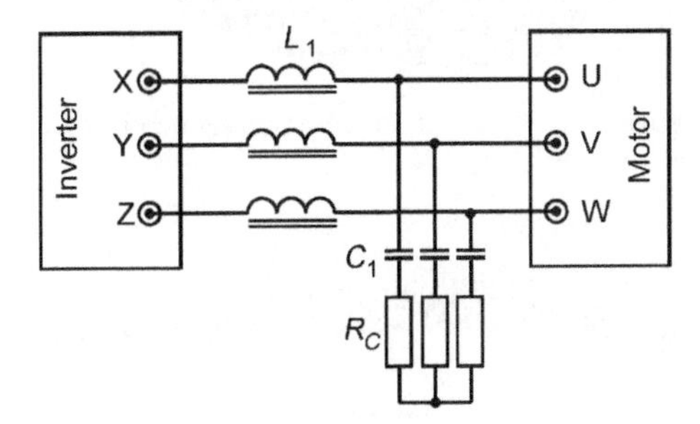

Figure 9.29 Inverter output filter – Structure 6

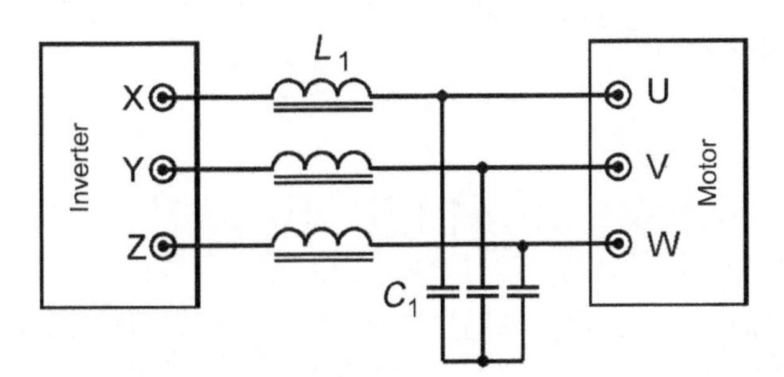

Figure 9.30 Inverter output filter – Structure 7

the capacitor in the DC bus. The coupled choke, M_2, limits the CM current in the circuit external to the filter and inverter. Chokes L_1 are single phase, while chokes M_1 and M_2 are three-phase coupled chokes made on the toroidal cores (Figure 9.31).

Resistances R_1 and R_0 are the damping elements protecting the system from oscillations. The input and output terminals of the filter are labeled XYZ and UVW, respectively. Terminal C_D should remain connected to the DC link voltage. In most cases, it is accessible for the user to use the positive polarity of the DC bus battery, which corrects the external braking resistor.

Figure 9.31 Three-phase common mode choke view

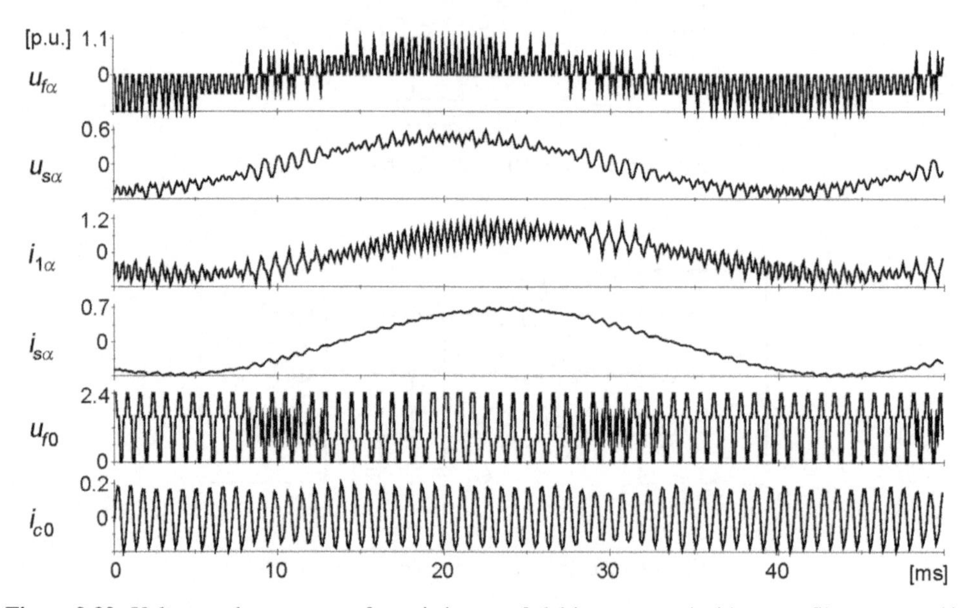

Figure 9.32 Voltage and current waveforms in inverter fed drives operated with output filter presented in Figure 9.24

Unfortunately, connection to DC link, could exclude the possibility of using residual current devices for electric shock protection in frequency converters, because the connection used forces CM current to flow in the protective conductor PE [23].

Examples of the voltage and current waveforms in the inverter fed drive, with the filter from Figure 9.24, are shown in Figure 9.32.

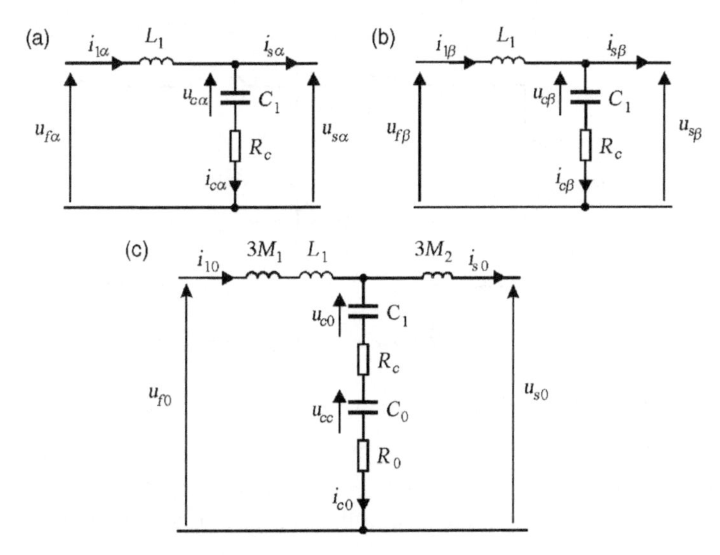

Figure 9.33 Equivalent circuits of the filter for components: (a) α; (b) β; and (c) 0

In Figure 9.32, the smoothed motor current and voltage waveforms are clearly noted. The level of smoothing of the voltage and current depends on the filter parameters and motor load. For the proper selection of filter parameters, it is important to assume rated motor loading to ensure minimum total harmonic distortion (THD) in the voltage and current and also to ensure minimum voltage drop in the filter.

In the filter from Figure 9.25, the branch containing the parameters R_0 and C_0 were dropped, which aided in increasing CM current in the motor compared with the solution from Figure 9.24. However, such a solution excludes the necessity of using a resistor R_0 with relatively high power and therefore also large dimensions. With high transistor switching frequency, a large current flows through R_0 and C_0. This forces the use of a high power resistor, R_0 and the necessity of using a special capacitor, C_0.

In filter Structure 3 (Figure 9.26), the more practical topology of the three-phase capacitors is used. Commercially accessible three-phase capacitors are mostly in delta connection.

The filter shown in Figure 9.27 is presented in [4, 5]. This is a limited version of the filter from Figure 9.24, from which the chokes L_3 and M_2 were excluded. Eliminating the elements R_0 and C_0 provides the structure shown in Figure 9.28.

Figure 9.29 presents a filter from which the CM choke, M_2, was removed as, according to experimental verifications, in the external circuit for the filter and inverter there is negligible CM current. Such a structure is only a sinusoidal filter containing damping resistor R_1. A sinusoidal filter is practically the only filter that causes complications in the control of electric drives. The use of sinusoidal filter with the damping resistor R_1 is essential only with the inaccurate generation of voltages and incomplete compensation of dead time or in systems without such compensation. In this case, there exist small voltages with frequencies close to the filter resonance frequency, while the resistor R_1 protects the system against damage. If the higher

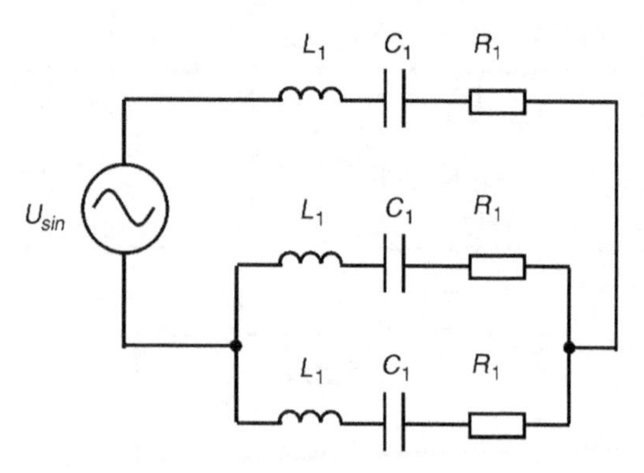

Figure 9.34 Equivalent circuit of the inverter and sinusoidal filter

harmonics are very small, then it is possible to exclude the resistor R_1 from the circuit and obtain a simpler and less expensive version of the sinusoidal filter, as shown in Figure 9.30.

9.6.2 Inverter Output Filters Design

Selection of inverter output filters is a complex task. This chapter presents a method for selecting parameters of the filter shown in Figure 9.24. This selection procedure is presented in [5, 13, 22].

The selection methods discussed in this chapter also fit the structures shown in Figures 9.25–9.30.

The combined filter structure from Figure 9.24 may complicate the selection of its parameters. Nevertheless, assuming that $M_1 \gg L_1$ and $C_0 \ll C_1$, this allows separate selection of filter elements components $\alpha\beta$ and for CM.

When selecting filter parameters, it is convenient to use a transformation from a three-phase system to a rectangular one, while maintaining the power of the system during transformation. For simplification, the leakage inductance of the coupled chokes M_1 and M_2, as well as the circuit branch that contains the choke L_3, is neglected. Note that L_3 is chosen based on the motor power.

The structure of the filter shown in Figure 9.24, for the orthogonal co-ordinate system $\alpha\beta 0$, is presented in Figure 9.33.

Resistances R_{L1}, R_{M1}, and R_{M2} are copper resistances of the chokes L_1, M_1, and M_2, respectively. Considerations on the selection of individual elements of the filter are presented in the following sections.

9.6.2.1 Selection of Differential Mode Filter Elements

The first selections are the elements of sinusoidal filter. The selection of the sinusoidal filter decides:

- assumed acceptable level of output voltage distortion;
- maximum allowed voltage drop in the filter;
- switching frequency of the power switches.

Selection of filter elements requires specific compromise between voltage total harmonic distortion (THD_U), weight, dimension, cost of the filter, and current parameters of the inverter. It is commonly assumed that THD_U does not exceed 5% under full load. The level of THD in the stator voltage has an effect mainly on the motor efficiency. Since one additional reason for using a filter in drive systems is to avoid voltage wave reflection on the motor terminals, a filter with a THD higher than 5% will also fit the set tasks while being economically acceptable; however, the efficiency of the motor will be lower. It is important to note that the transistors' switching frequency has a significant effect on the value of filter inductances and capacitances. This switching frequency is much lower in drive systems than in UPS systems.

For further analysis on the selection of filter parameters, it is assumed that the pulse frequency is the inverse of the time per duration of two adjacent passive vectors (z) and four active vectors (n) in a cycle $z1$-$n1$-$n2$-$z2$-$n2$-$n1$. This corresponds to a notation of a sampling period for a sinusoidal PWM.

As a first element of the filter, it is selected as choke L_1. One of the methods for selecting a choke for the sinusoidal filter is identifying L_1 on the basis of choke reactance X_1, which is obtained according to the assumed AC current component ΔI_s for inverter switching frequency f_{imp} [5]. A three-phase voltage inverter with sinusoidal filter and squirrel cage induction motor can be represented as f_{imp}, as in an equivalent circuit shown in Figure 9.34.

The equivalent circuit, shown in Figure 9.34, results from the situation that during a sinusoidal PWM and space vector PWM, one of the DC link polarities is connected to one output phase, while the other polarity is connected to two parallel output phases. This is a correct assumption, assuming that the selection of the filter elements is done for the whole control range; this is a maximum inverter output voltage without over-modulation. Therefore, in such inverter controls, it is possible to neglect zero vectors. In this schematic, the stator copper resistance and the motor leakage inductance are neglected because their values are too small compared with filter parameters with high pulse frequency. The schematic in Figure 9.34 is reduced, as shown in Figure 9.35.

In Figures 9.34 and 9.36, an inverter is replaced by a sinusoidal voltage source U_P. The rms value of U_P voltage under full control, maximum voltage without over-modulation, is

$$U_{P\,max} = \frac{U_d}{\sqrt{2}} \tag{9.26}$$

where U_d is the average value of the DC bus voltage.

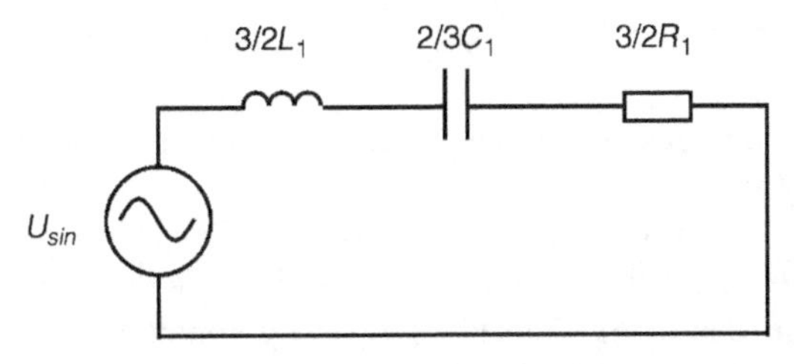

Figure 9.35 Reduced equivalent circuit of the inverter and sinusoidal filter

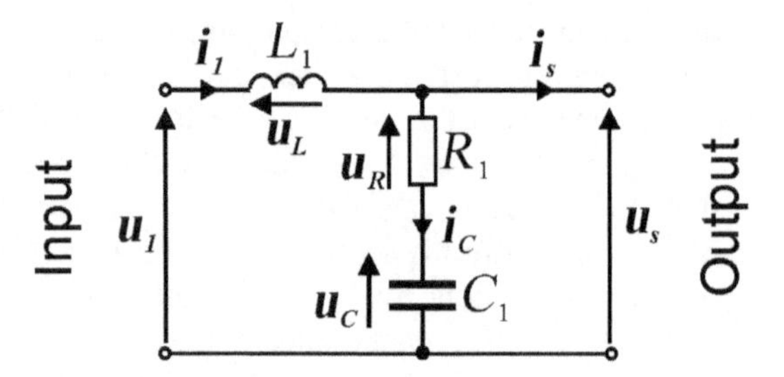

Figure 9.36 Equivalent circuit of the filter for α component

The voltage U_d, in the case of feeding an inverter from a three-phase grid through a diode rectifier, is

$$U_d = \sqrt{2}U_n\frac{6}{\pi}\sin\frac{\pi}{6} \tag{9.27}$$

where U_n is the rms value of the line voltage on the rectifier terminals. Therefore,

$$U_{P\,max} = U_n\frac{6}{\pi}\sin\frac{\pi}{6} \approx 0.95U_n \tag{9.28}$$

In Figure 9.35, the element smoothing the current is mainly the choke L_1. The reactance of this choke is much greater than the reactance of the capacitor for a given switching frequency, therefore can be written as

$$2{\cdot}\pi{\cdot}f_{imp}{\cdot}\frac{3L_1}{2}{\cdot}\Delta I_s = \frac{U_d}{\sqrt{2}} \tag{9.29}$$

where ΔI_s means switching ripple inductor current. Ripple frequency is the equal transistor's switching frequency.

It is assumed that ΔI_s should be lower than 20% of the rated current.

After conversion of equation (9.29) the relationship for an inductance of choke L_1 is

$$L_1 = 2\frac{U_d}{\sqrt{2}{\cdot}2{\cdot}\pi{\cdot}f_{imp}{\cdot}3{\cdot}\Delta I_s} \tag{9.30}$$

An inductance L_1, computed according to equation (9.5), is much higher than the leakage inductance of the motor [5].

Because of the dimension, weight, and cost, the filter inductance L_1 should be as small as possible. Small L_1 means small voltage drop on the filter and less losses in the drive.

Another method of identifying inductance L_1 is by computing its value on the basis of assumed voltage drop in the filter ΔU_1 and on the known maximum value of the first voltage component $U_{out\ 1h}$. In drives systems, it is assumed that the voltage drop in the filter for the first component f_{out1h} should be less than 5% of the motor rated voltage under full current loading I_n [13]. With such assumptions, an inductance is calculated as

$$L_1 = \frac{\Delta U_1}{2\pi f_{out1h} I_n} \tag{9.31}$$

After identifying L_1, the next parameter to be selected is C_1, which is described from the relationship of the resonance frequency of the system:

$$f_{res} = \frac{1}{2\pi\sqrt{L_1 C_1}} \tag{9.32}$$

hence:

$$C_1 = \frac{1}{4\pi^2 f_{res}^2 L_1} \tag{9.33}$$

where f_{res} is the resonance frequency of the filter.

To assure good filtering capabilities of the filter, the resonance frequency should be lower than the switching frequency of the transistors f_{imp}. At the same time, to avoid the resonance phenomenon, the capacitance C_1 should be higher than the maximum frequency of the first harmonic of the inverter output voltage f_{wy1h}. The safety range is [13]

$$10 \cdot f_{out1h} < f_{res} < \frac{1}{2} \cdot f_{imp} \tag{9.34}$$

The upper range of C_1 is described by the maximum value of the current flowing through this capacitor during rated operating conditions of the drive. The value of this current for the first harmonic cannot exceed 10% of the rated motor current.

The risk of resonance existence in the sinusoidal filter occurs during drive operation in the over-modulation region. In such a situation, the current flowing through the branch with C_1 contains multiples of the fundamental harmonic, in addition to the fundamental harmonic.

A filter damping resistor should be selected in such a way that during the existence of the resonance, the current flowing in a capacitor should not exceed 20% of the rated inverter current. Hence, for computing C_1, it is necessary to know the voltage component with a frequency close to the filter resonance frequency. Relationships are given in [24]. In practice, because of the complicated analytical analysis, the harmonic component of resonance frequency is identified in simulation.

Another method of damping resistor selection is by choosing a value of R_1 so that the total stray losses on the filter damping resistors are kept at the acceptable level of 0.1% of the converter power rating [5]. With a resistance R_1 selected in such way, a filter quality factor is identified as

$$Q = \frac{Z_0}{R_1} \tag{9.35}$$

where Z_0 is the natural frequency of the filter:

$$Z_0 = \sqrt{\frac{L_1}{C_1}} \tag{9.36}$$

To ensure sufficient attenuation, filters should have the same time minimum power losses and quality factor in the range 5–8 [5].

Knowing the parameters of the transverse branch of the sinusoidal filter C_1, R_1, and assuming the sinusoidal shape of the filter output voltage, the power of the element R_1 for the first harmonic of filter output current and voltage should be computed:

$$P_{R11har} = \frac{U_{sph}^2}{R_1} \tag{9.37}$$

where U_{sph} is phase filter output voltage.

With power selection of resistors R_1, it is important to take into consideration that the filter output voltage contains a component with a frequency equal to the transistors switching frequency. With proper design of the filter, this component is small. However, because of its high frequency, the impedance of the transverse branch of C_1 and R_1 for this frequency is small.

This can cause a flow of essential current with high frequency in this branch. The power of R_1 for the voltage component with a modulation frequency is

$$P_{R1imp} = \frac{U_{s\,imp}^2}{R_1} \tag{9.38}$$

where U_{simp} is the filter output voltage component for PWM frequency.

The value of U_{simp} corresponds to the values that result from the assumed THD of the voltage in the design process.

For the chosen elements of the sinusoidal filter, it is possible to find its frequency characteristic based on the known filter transfer function presented as a two-port network in Figure 9.36.

Elements L_f, C_f, and R_f correspond to the values $3/2L_1$, $2/3C_1$, and $3/2R_1$, respectively, for the reduced filter scheme shown in Figure 9.24. Figure 9.37 presents a sinusoidal filter that has a two-port network consisting of two impedances, Z_1 and Z_2.

Impedances Z_1 and Z_2 are described as

$$Z_1(j\omega) = j\omega L_f \tag{9.39}$$

$$Z_2(j\omega) = R_f + \frac{1}{j\omega C_f} \tag{9.40}$$

or in the form of a transfer function:

$$Z_1(s) = sL_f \tag{9.41}$$

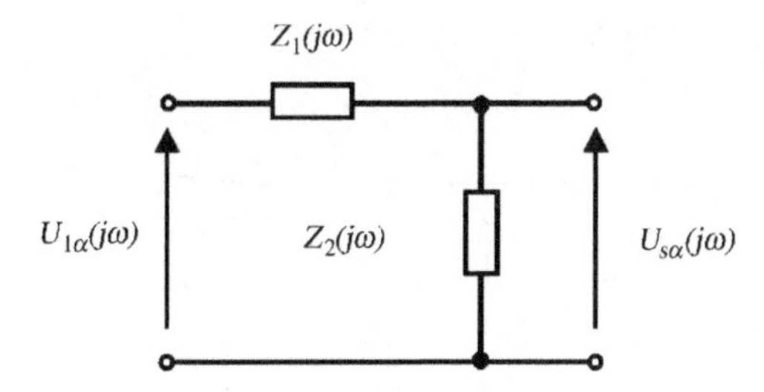

Figure 9.37 Equivalent circuit of the two-port network – sinusoidal filter for component α

$$Z_2(s) = R_f + \frac{1}{sC_f} \tag{9.42}$$

The transfer function of the filter for voltages is described as

$$G_{fu}(s) = \frac{U_{s\alpha}}{U_{1\alpha}} = \frac{Z_{wy}(s)}{Z_{we}(s)} = \frac{Z_2(s)}{Z_1(s) + Z_2(s)} = \frac{R_f + \frac{1}{sC_f}}{sL_f + R_f + \frac{1}{sC_f}} \tag{9.43}$$

which is

$$G_{fu}(s) = \frac{U_{s\alpha}}{U_{1\alpha}} = \frac{sR_fC_f + 1}{s^2L_fC_f + sR_fC_f + 1} \tag{9.44}$$

The cutoff frequency of the filter is described according to

$$f_{3\text{dB}} = \frac{1}{2\pi\sqrt{L_fC_f}} \tag{9.45}$$

while the damping coefficient is

$$\xi = \frac{R_f}{2\sqrt{L_f/C_f}} = \frac{1}{2Q} \tag{9.46}$$

and Q is the filter quality factor of the filter:

$$Q = \frac{\sqrt{L_f/C_f}}{R_f} \tag{9.47}$$

Examples of phase and magnitude characteristics of the sinusoidal filter are shown in Figure 9.38.

Knowledge of filter parameters and characteristic of parameters is essential for the correct consideration of such filters in drive systems.

An operation of a filter, presented in Figure 9.24, is shown in Figures 9.39 and 9.40 for a rated load of a 5.5 kW squirrel cage induction motor.

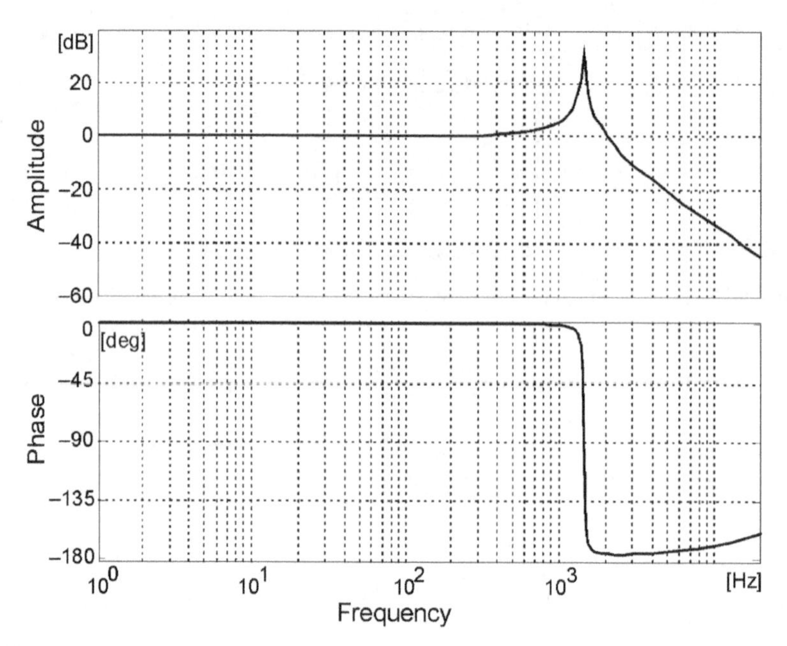

Figure 9.38 Phase and magnitude plots of a sinusoidal filter (filter chosen for a 1.5 kW motor; $L_f = 4$ mH, $C_f = 3$ μF, $R_f = 1$ Ω)

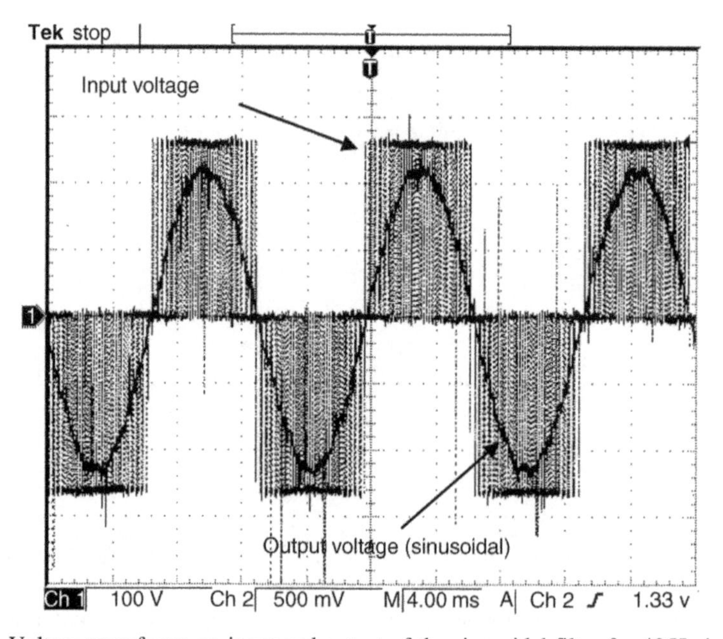

Figure 9.39 Voltage waveforms on input and output of the sinusoidal filter for 40 Hz inverter output voltage frequency

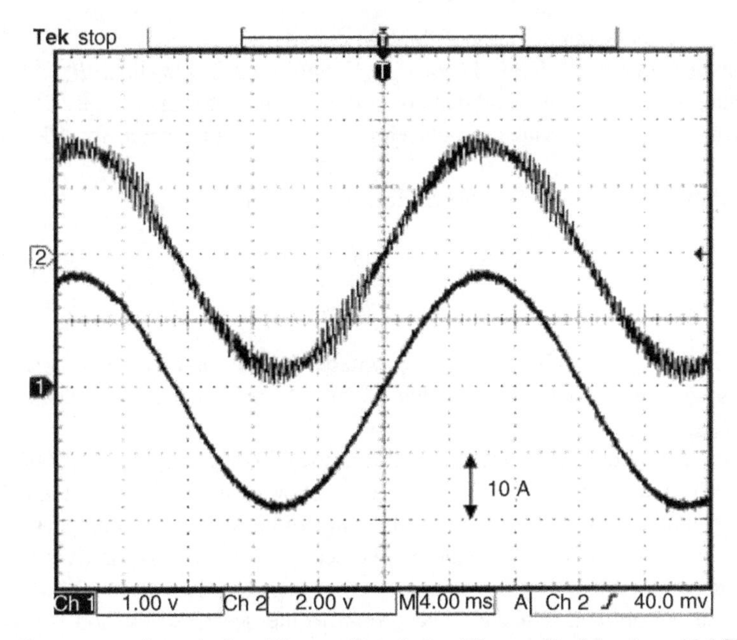

Figure 9.40 Current waveforms before (Chapter 1) and after (Chapter 2) of the sinusoidal filter for 40 Hz inverter output voltage frequency

With a fully loaded motor, voltage and current waveforms are significantly filtered with minimum voltage drop on the filter and minimum phase shifts of voltages and currents before and after the filter.

9.6.2.2 Selection of Elements for Zero Sequence (Common Mode)

Selection of the filter elements for CM starts from choosing CM chokes M_1 and M_2. In the integrated filter structure shown in Figure 9.24, it is assumed that both chokes are identical. The inductances of CM chokes should be the same. During the selection of filter elements, it is assumed that $M_1 \gg L_1$. While selecting M_1 and M_2, it is important to pay special attention to the possibility of their practical realization. Such possibilities result from the parameters of the used toroidal cores. In practical solutions, the measurement of CM voltage u_{cm} is used, which enables description of the flux in the choke core [5]:

$$\psi_{cm} = \frac{1}{N_{cm}} \int u_{cm} dt \tag{9.48}$$

where N_{cm} is the number of turns in one phase of the CM choke.

The magnetic field density in the choke is

$$B_{cm} = \frac{\psi_{cm}}{S_{cm}} = \frac{1}{S_{cm} N_{cm}} \int u_{cm} dt \tag{9.49}$$

where S_{cm} is the cross-sectional area of the CM choke.

The computed magnetic field density B_{cm} should be lower than B_{sat}, which is the saturated value of the magnetic field density in the core used for the design of the CM choke. Manufacturers of magnetic materials currently offer dedicated cores for CM chokes with $B_{sat} = 1\ldots\ 1.2\ \text{T}$.

The CM choke inductance with known dimensions is described as

$$L_{cm} = \frac{\mu S_{cm} N_{cm}^2}{l_{cm}} \tag{9.50}$$

where l_{cm} is an average path length of the magnetic field in the CM choke core.

Neglecting the small leakage inductance of the CM choke, it is possible to assume that $M_1 = M_2 = L_{cm}$.

For a constant product $S_{cm}N_{cm}$, the peak value of the CM current $I_{cm\ max}$ is proportional to the ratio l_{cm}/N_{cm}. Decreasing the dimensions of chokes by reducing the flux path length to l_{cm} and increasing the number of turns N_{cm} has the advantage of limiting the current peak value $I_{cm\ max}$. At the same time, it should be noted that the planned number of turns can fit in the window of the selected core.

Searching for an optimal CM choke according to the described method is a complicated, iterative process requiring special measuring tools that allow for the identification of u_{cn} and $I_{cm\ max}$. Therefore, to simplify the process of the filter design, manufacturers of magnetic materials frequently suggest specialized solutions of cores and complete CM chokes depending on motor power.

The capacitance C_0 is determined using the known values of inductances M_1 and M_2, and the resonance frequency f_{rez}:

$$C_0 = \frac{1}{4\pi^2 f_{rez}^2 L_{cm}} \tag{9.51}$$

while the resistance R_0:

$$R_0 = \frac{\sqrt{L_{cm}/C_{cm}}}{Q_0} \tag{9.52}$$

where Q_0 is the assumed quality factor of the CM filter in the range 5–8.

The power of the resistor R_0:

$$P_{R_0} = \frac{U_{cm}^2}{R_0} \tag{9.53}$$

where U_{cm} is rms value of the CM voltage, which is

$$U_{cm} \leq \frac{U_d}{\sqrt{2}} \tag{9.54}$$

In the case of filters, where it is essential to use chokes L_3, it is assumed that the current flowing through the choke L_1 is close to the 10% rated motor current:

$$L_3 \approx \frac{1}{2\pi f_n} \frac{U_n}{0.1 \cdot I_n} \tag{9.55}$$

where U_n, I_n, and f_n are voltage, current, and rated frequency of the inverter, respectively.

Operation of the CM filter, shown in Figure 9.24, is presented in Figures 9.41 and 9.42.

In comparison with a drive without a filter, the CM current in a shielded conductor PE for a 5.5 kW motor (Figure 5.20) is limited after using a CM filter, as shown in Figure 5.21.

9.6.3 Motor Choke

Motor chokes are used in many inverter fed drives. They represent the simplest form of inverter output filters. In three-phase systems, motor chokes are three-phase or three single-phase bulks connected between motor and inverter terminals, as shown in Figure 9.43.

The main task of the motor choke in AC drives is to limit the voltage rise (*dv*/*dt*) on the motor terminals, which decreases the danger of damaging the insulation of the motor.

Motor chokes are also used for limiting short circuit currents before running the internal short circuit protection of the inverter.

While neglecting the copper resistance, a motor choke could be represented as a linear inductance, as shown in Figure 9.44.

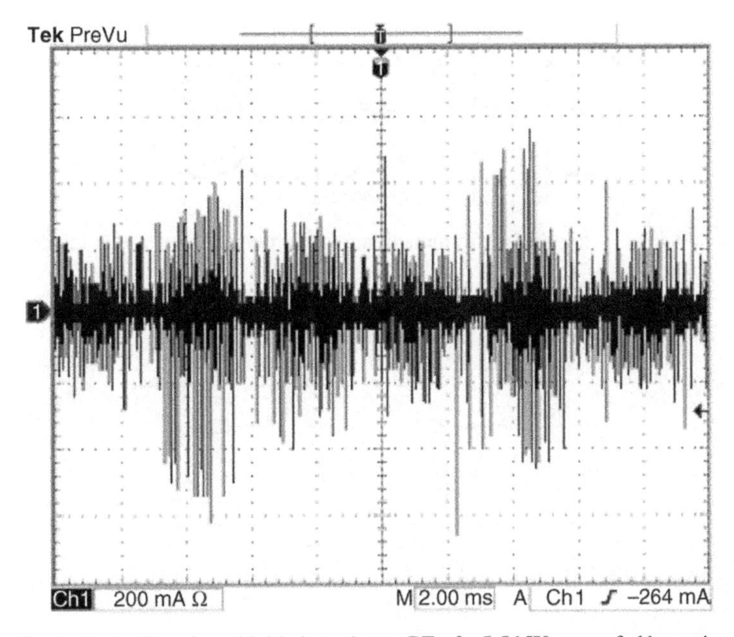

Figure 9.41 Current waveform in a shielded conductor PE of a 5.5 kW motor fed by an inverter **without** output filter

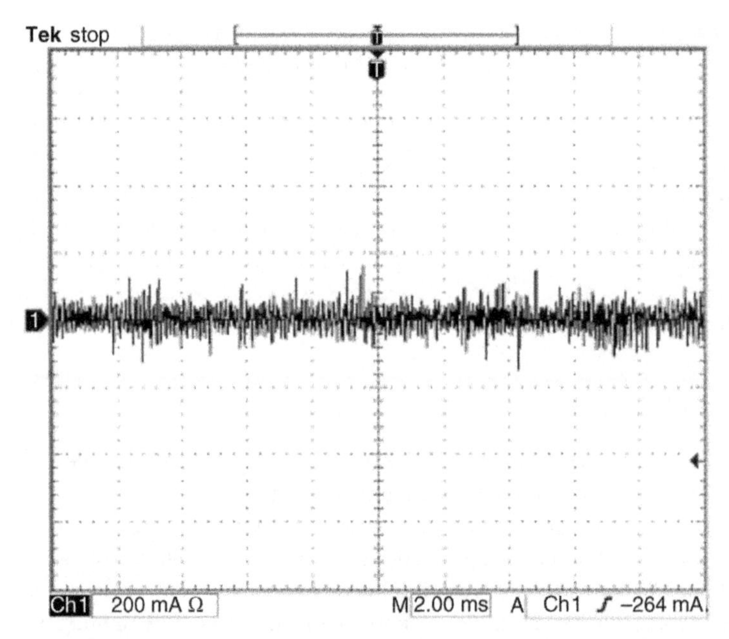

Figure 9.42 Current waveform in a shielded conductor PE of a 5.5 kW motor fed by an inverter **with** output filter

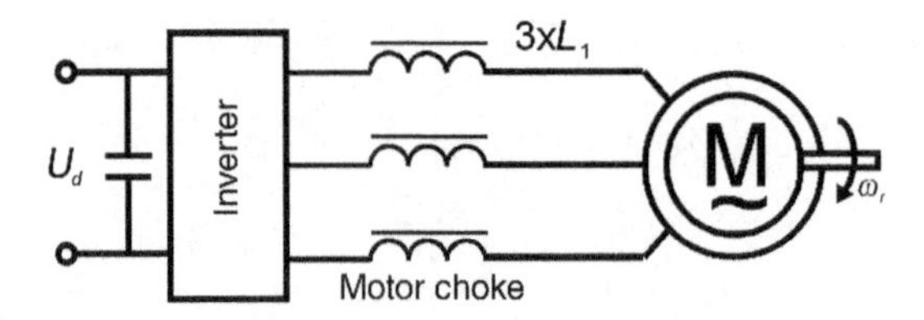

Figure 9.43 Electric drive with induction motor, voltage inverter, and motor choke

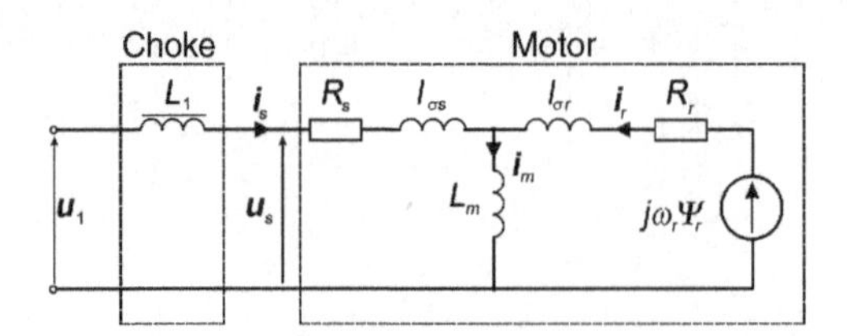

Figure 9.44 Equivalent circuit of the motor choke and induction motor

According to Figure 5.23, a motor choke may be represented as

$$\frac{d\mathbf{i}_s}{d\tau} = \frac{1}{L_1}(\mathbf{u}_1 - \mathbf{u}_s) \tag{9.56}$$

where L_1 is the motor choke inductance.

The selection of motor choke L_1 is performed by assuming that the voltage drop on the motor choke of the first component f_{out1h} is lower than 5% of the rated motor voltage under loading the filter with rated current I_n:

$$L_1 = \frac{\Delta U_1}{2\pi f_{out1h} I_n} \tag{9.57}$$

Figure 9.45 shows examples of currents and voltages waveforms from an experimental setup with a squirrel cage induction motor and motor choke.

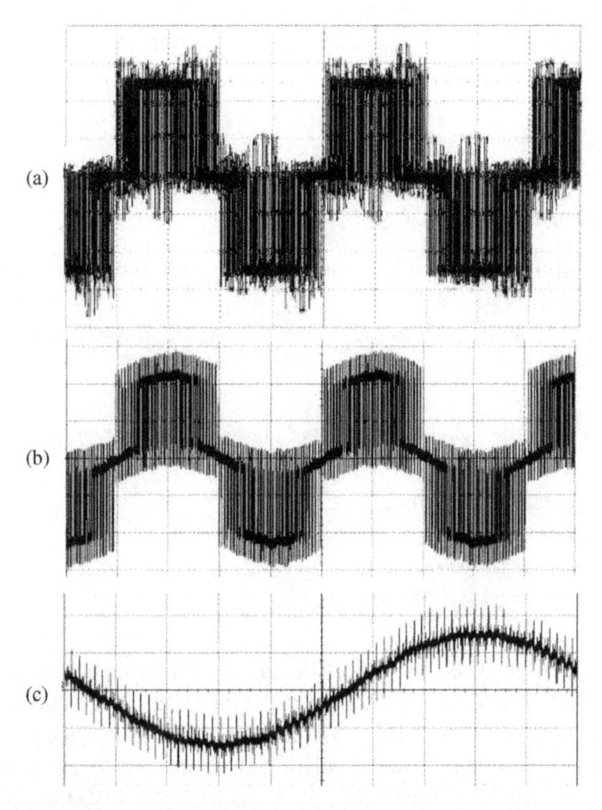

Figure 9.45 Waveforms registered from experimental setup with induction motor and motor choke: (a) inverter output voltage; (b) voltage on motor terminals; (c) motor current (a, b – 200 V/div, 5 ms/div; c – 5 A/div, 2 ms/div, motor 1.5 kW 300 V, choke L_1 = 11 mH, motor loaded with rated condition – registered waveforms are not synchronized)

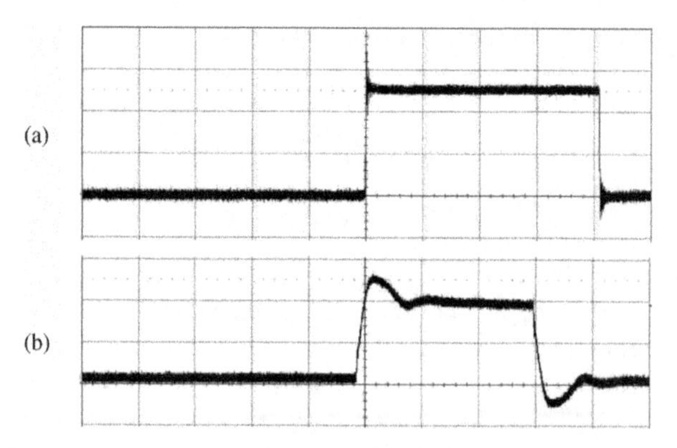

Figure 9.46 Waveforms: (a) inverter output voltage; (b) motor current for the drive with motor choke (scale 10 μs/div, 200 V/div – waveforms are not synchronized)

In the system with motor choke, the voltage feeding the motor has a pulse shape superimposed with an additional small component of the sinusoidal voltage.

The motor choke effect on *du*/*dt* is shown in Figure 9.46.

In the waveforms shown in Figure 9.46, it is possible to note the clear limit of motor voltage *dv*/*dt* after connecting a motor choke.

9.6.4 MATLAB/Simulink Model of Induction Motor Drive with PWM Inverter and Differential Mode LC Filter

The MATLAB/Simulink model of induction motor drive with inverter and LC filter *sys_LC_filter.mdl* is presented in Figure 9.47.

The simulated system is prepared as the single file *sys_LC_filter.mdl*. The model consists of the following blocks (subsystems):

- Three-phase inverter with space vector PWM algorithm (*PWM&Inverter*);
- LC filter model (*Filter model*);
- induction motor model (*Motor model*);
- frame of references transformation (*dq2xy, xy2abc*);
- flux and speed observer (*Observer*);
- PI controllers, commanded signals, and scopes.

The commanded values are speed and flux. The motor control algorithm is the classical field-oriented control (FOC). Both control systems and the observer are the classical solutions, without any changes due to LC filter use. To assure the system works correctly, the dynamics of the drive is limited, because the aim of the simulation is to present the LC filter properties only.

The induction motor model is four poles, $P_N = 1.5$ kW, $U_N = 300$ V. All the system parameters, including filter and motor parameters, are given in the preload function (*File* → *Model properties* → Callbacks).

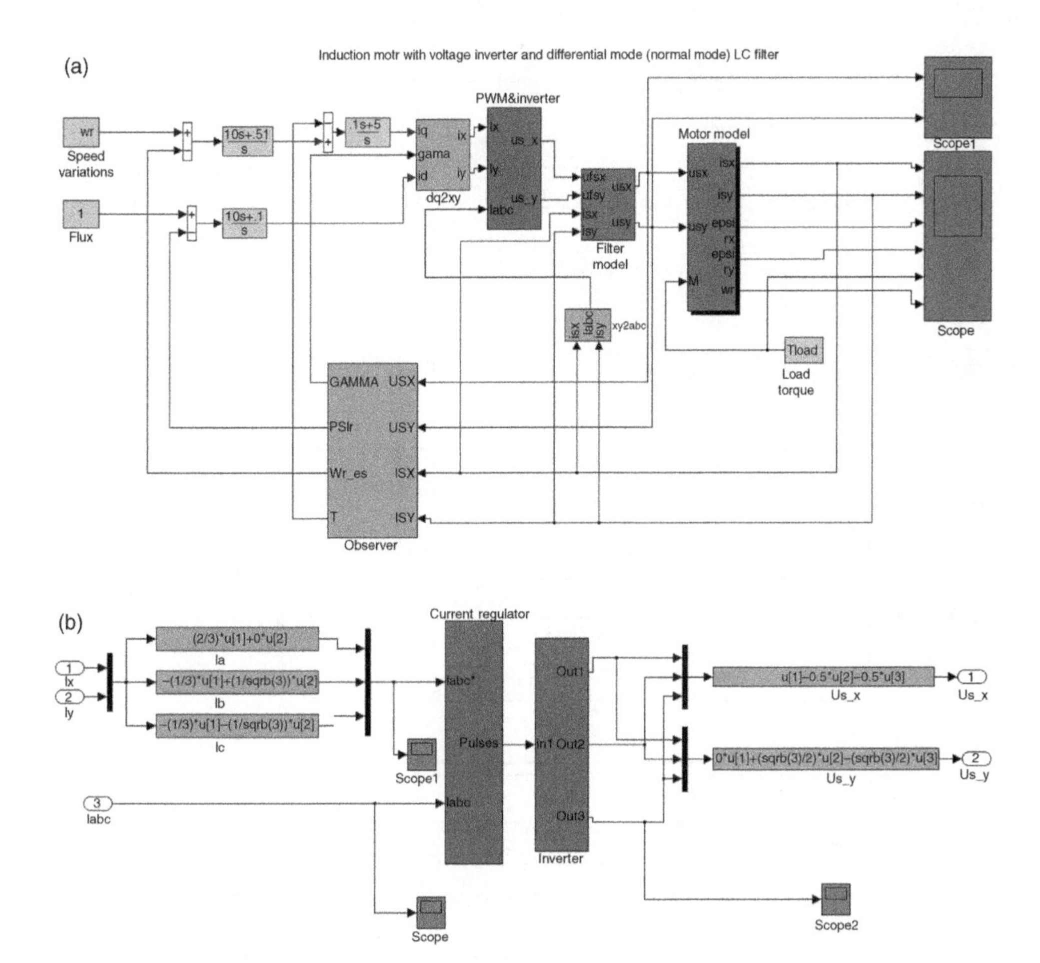

Figure 9.47 MATLAB/Simulink model of induction motor drive with PWM inverter and LC filter -*sys_LC_filter.mdl:* (a) whole structure; (b) inverter and PWM subsystem; (c) LC filter subsystem, and (d) observer subsystem

In the simulation test, the constant motor flux is retained, whereas the commanded motor speed is changed. Due to the speed variations, the motor voltage supply frequency changes. The LC filter assures that motor voltages and currents are sinusoidal in shape. In Figures 9.48 and 9.49, the example of the motor voltages and currents are presented.

The complete MATLAB/Simulink model is available in the CD-ROM provided with this book.

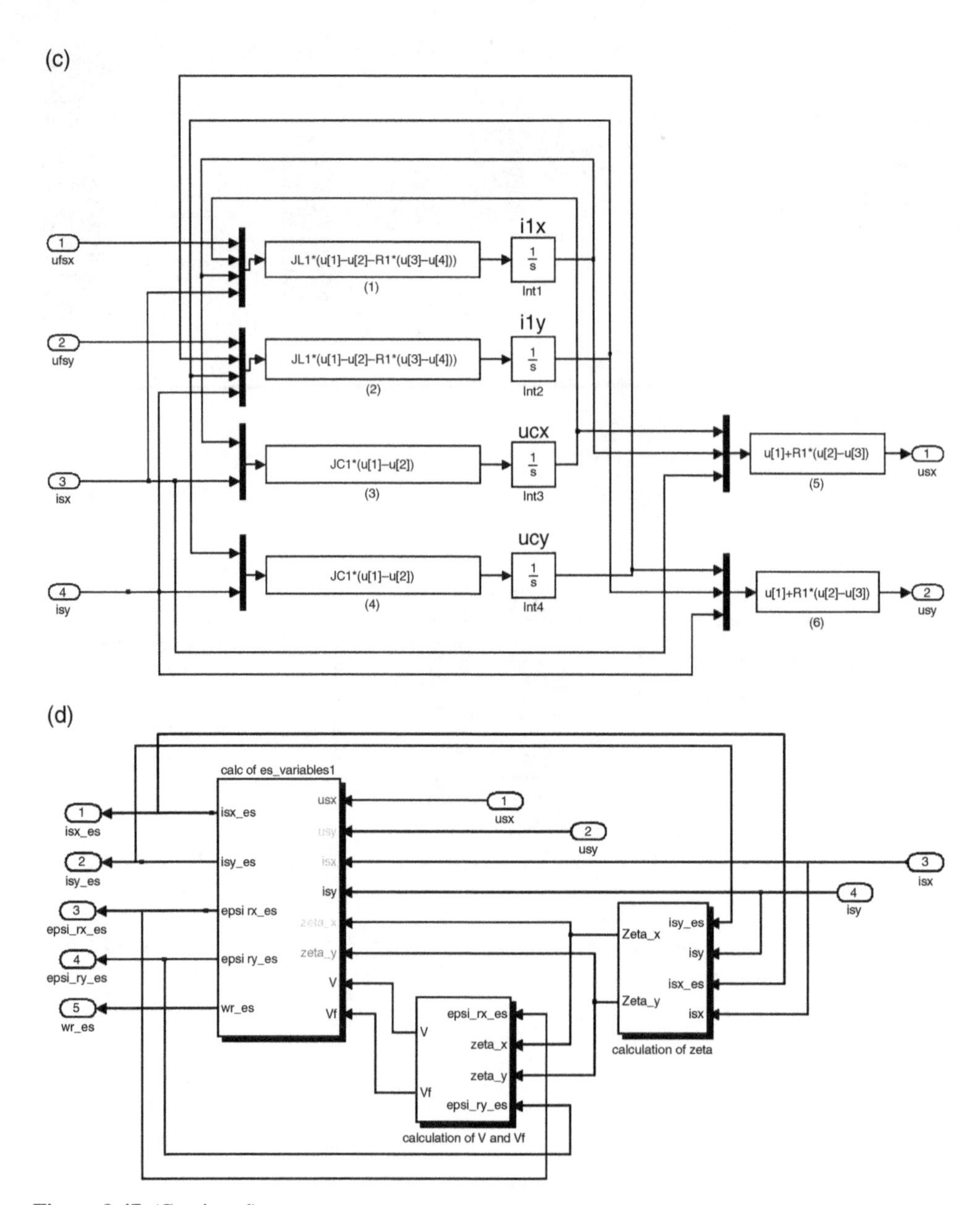

Figure 9.47 (Continued)

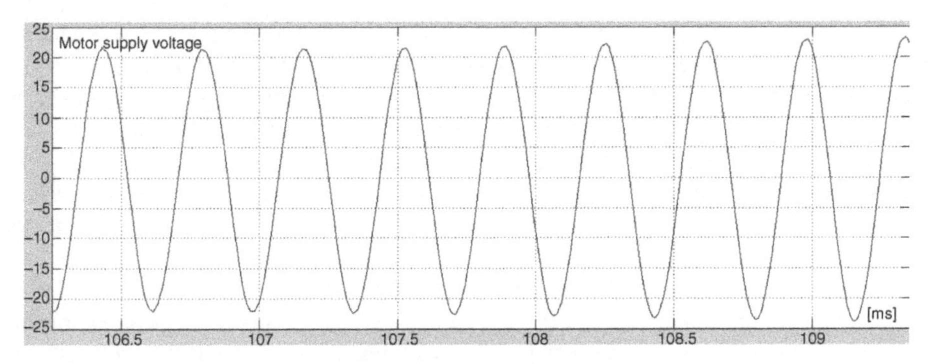

Figure 9.48 Motor supply voltage

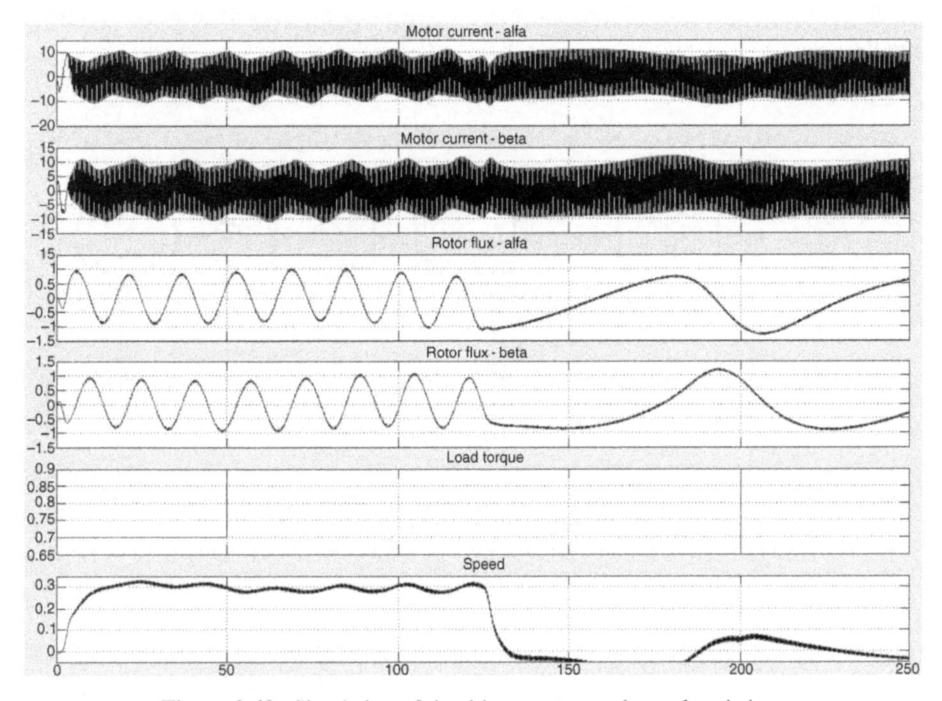

Figure 9.49 Simulation of the drive start-up and speed variations

9.7 Estimation Problems in the Drive with Filters

9.7.1 Introduction

For motor closed-loop control purposes, the actual controlled state variables have to be known. In most control systems, the controlled variables are mechanical speed, magnetic flux, and motor torque. The speed is easy to measure; however, measuring other variables is not so simple. In spite of the easy speed measurement in modern drives, the requirements for speed sensor elimination exist. To solve these problems, numerous estimation methods are used and the variables are calculated online. Then the sensors are limited to only the current and voltage sensors installed inside the converter. To prevent noise, all sensors should be installed inside the converter box. A drive with such limited sensors is known as a sensorless drive. The general structure of the sensorless AC drive is presented in Figure 9.50.

In the literature, numerous sensorless solutions are proposed. A comprehensive review can be found in [25]. But most of the solutions presented are applicable for the drives without motor filter use.

In the case of motor filter use, the estimation process is more complicated. Some of the solutions propose the installation of voltage and current sensors outside the converter for direct motor current and voltage measurement [26, 27]. Unfortunately, as was mentioned earlier, this is not an applicable solution. The more useful solution is to implement the filter model in the estimation algorithm, where the sensor structure will be the same as in drives without filter use. Then, the general structure of the sensorless drive with the LC filter is as presented in Figure 9.51.

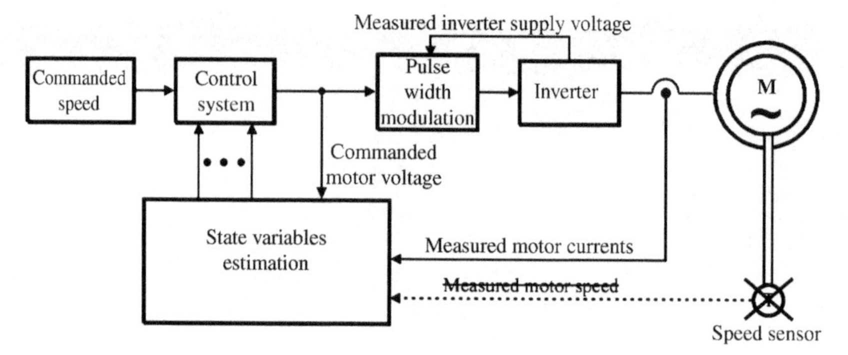

Figure 9.50 General structure of the sensorless AC drive

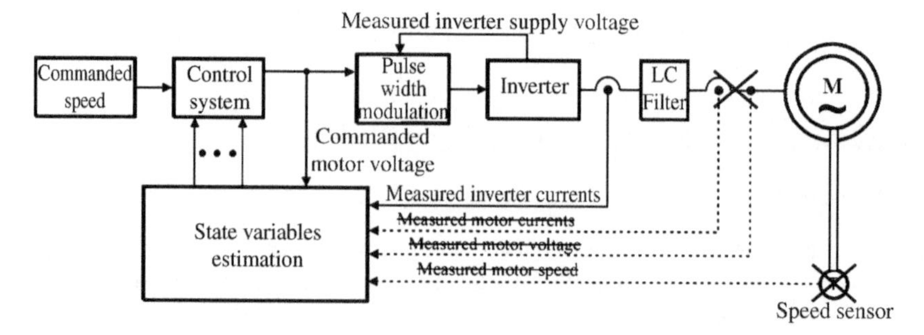

Figure 9.51 General structure of the sensorless AC drive with motor LC filter use

Only the differential model filter has influence on motor control and estimation algorithms.

Only a limited amount of literature presents estimation solutions for driver with filters, some of which are proposed in [28–30]. The most common estimation method is to add the filter model dependencies into some known observer structure and to change the observer correction parts. Some of the possible solutions are presented in the following sections.

9.7.2 Speed Observer with Disturbances Model

The operation of modern complex asynchronous motor electric drives requires application of the calculating system, which computes online all the state variables used in control systems. Nowadays, frequent solutions in state variable calculations are to use a state observer. In this chapter, the state observer presented in [31] has been converted into a system with a sine filter.

The base observer structures for the drive, without the sine filter in $\alpha\beta$ references, are

$$\frac{d\hat{i}_{s\alpha}}{d\tau} = -\frac{R_s L_r^2 + R_r L_m^2}{L_r w_\delta}\hat{i}_{s\alpha} + \frac{R_s L_m}{L_r w_\delta}\hat{\psi}_{r\alpha} + \frac{L_m}{w_\delta}\xi_\beta + \frac{L_r}{w_\delta}u_{s\alpha}^{com} + k_3\left(i_{s\alpha} - \hat{i}_{s\alpha}\right) \tag{9.58}$$

$$\frac{d\hat{i}_{s\beta}}{d\tau} = -\frac{R_s L_r^2 + R_r L_m^2}{L_r w_\delta}\hat{i}_{s\beta} + \frac{R_r L_m}{L_r w_\delta}\hat{\psi}_{r\beta} + \frac{L_m}{w_\delta}\xi_\alpha + \frac{L_r}{w_\delta}u_{s\beta}^{com} + k_3\left(i_{s\beta} - \hat{i}_{s\beta}\right) \tag{9.59}$$

$$\begin{aligned}\frac{d\hat{\psi}_{r\alpha}}{d\tau} &= -\frac{R_r}{L_r}\hat{\psi}_{r\alpha} - \xi_\beta + R_r\frac{L_m}{L_r}\hat{i}_{s\alpha} - k_2 S_b \hat{\psi}_{r\alpha} + S k_2 k_3 \hat{\psi}_{r\beta}(S_b - S_{bF}) \\ &\quad + S k_5\left((S_x - S_{xF})\hat{\psi}_{r\alpha} - (S_b - S_{bF})\hat{\psi}_{r\beta}\right)\end{aligned} \tag{9.60}$$

$$\begin{aligned}\frac{d\hat{\psi}_{r\beta}}{d\tau} &= -\frac{R_r}{L_r}\hat{\psi}_{r\beta} + \xi_\beta + R_r\frac{L_m}{L_r}\hat{i}_{s\beta} - k_2 S_b \hat{\psi}_{r\beta} - S k_2 k_3 \hat{\psi}_{r\alpha}(S_b - S_{bF}) \\ &\quad + S k_5\left(-(S_x - S_{xF})\hat{\psi}_{r\beta} - (S_b - S_{bF})\hat{\psi}_{r\alpha}\right)\end{aligned} \tag{9.61}$$

$$\frac{d\hat{\xi}_\alpha}{d\tau} = -\hat{\omega}_{\psi r}\hat{\xi}_\beta - k_1\left(i_{s\beta} - \hat{i}_{s\beta}\right) \tag{9.62}$$

$$\frac{d\hat{\xi}_\beta}{d\tau} = \hat{\omega}_{\psi r}\hat{\xi}_\alpha + k_1\left(i_{s\alpha} - \hat{i}_{s\alpha}\right) \tag{9.63}$$

$$\frac{dS_{bF}}{d\tau} = \frac{1}{T_{Sb}}(S_b - S_{bF}) \tag{9.64}$$

$$\frac{d\hat{\omega}_{rF}}{d\tau} = \frac{1}{T_{KT}}(\hat{\omega}_r - \hat{\omega}_{rF}) \tag{9.65}$$

$$\frac{dS_{xF}}{d\tau} = \frac{1}{T_{Sx}}(S_x - S_{xF}) \tag{9.66}$$

$$S = \begin{cases} 1 & \text{if} \quad \hat{\omega}_{\psi r} > 0 \\ -1 & \text{if} \quad \hat{\omega}_{\psi r} \le 0 \end{cases} \tag{9.67}$$

$$S_x = \hat{\xi}_\alpha \hat{\psi}_{r\alpha} + \hat{\xi}_\beta \hat{\psi}_{r\beta} \tag{9.68}$$

$$S_b = \hat{\xi}_\alpha \hat{\psi}_{r\beta} - \hat{\xi}_\beta \hat{\psi}_{r\alpha} \tag{9.69}$$

$$\hat{\omega}_{\psi r} = \hat{\omega}_{rF} + R_r \frac{L_m}{L_r} \left(\frac{\hat{\psi}_{r\alpha} \hat{i}_{s\beta} + \hat{\psi}_{r\beta} \hat{i}_{s\alpha}}{\hat{\psi}_{r\alpha}^2 + \hat{\psi}_{r\beta}^2} \right) \tag{9.70}$$

$$\hat{\omega}_r = \frac{\hat{\zeta}_\alpha \hat{\psi}_{r\alpha} + \hat{\zeta}_\alpha \hat{\psi}_{r\alpha}}{\hat{\psi}_{r\alpha}^2 + \hat{\psi}_{r\beta}^2} \tag{9.71}$$

where ξ_α, ξ_β are components of the motor electromotive forces; $\omega_{\psi r}$ is the angular speed of the motor flux vector; k_1, k_2, k_3, k_4, k_5 are the observer gains; S_x, S_{xF}, S_b, S_{bF} are additional variables used to stabilize the observer work; T_{Sb}, T_{KT}, and T_{Sx} are the time constants of the filters; and S is the sign of the rotor flux speed.

The state observer equations (9.58)–(9.71) need to be modified to assure the proper function of the electric drive, as presented in [32].

For the system with the sine filter, the observer equations (9.58)–(9.71) are extended with LC filter model equations.

The circuit of the LC filter, presented in Figure 9.52, can be described as

$$\frac{du_{c\alpha}}{d\tau} = \frac{i_{c\alpha}}{C_1} \tag{9.72}$$

$$\frac{du_{1\alpha}}{d\tau} = \frac{u_{1\alpha} - R_1 i_{1\alpha} - R_c i_{c\alpha} - u_{c\alpha}}{(L_1 + L_{\sigma M1})} \tag{9.73}$$

$$i_{c\alpha} = i_{1\alpha} - i_{s\alpha} \tag{9.74}$$

$$u_{s\alpha} = R_c(i_{1\alpha} - i_{s\alpha}) + u_{c\alpha} \tag{9.75}$$

$$\frac{du_{c\beta}}{d\tau} = \frac{i_{c\beta}}{C_1} \tag{9.76}$$

$$\frac{di_{1\beta}}{d\tau} = \frac{u_{1\beta} - R_1 i_{1\beta} - R_c i_{c\beta} - u_{c\beta}}{(L_1 + L_{\sigma M1})} \tag{9.77}$$

$$i_{c\beta} = i_{1\beta} - i_{s\beta} \tag{9.78}$$

$$u_{s\beta} = R_c\left(i_{1\beta} - i_{s\beta}\right) + u_{c\beta} \tag{9.79}$$

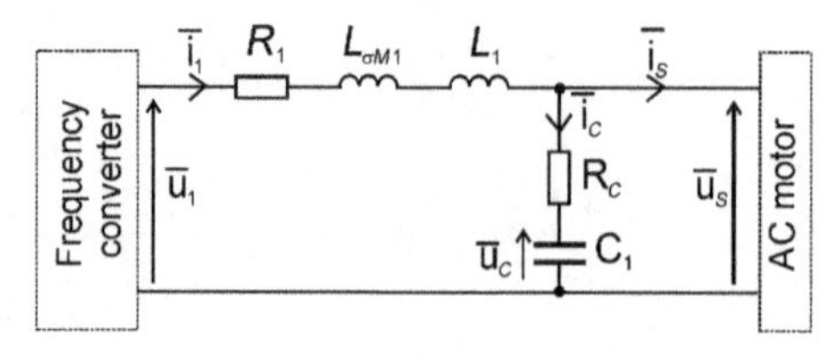

Figure 9.52 Equivalent circuit of the differential inverter output LC filter

The correction parts in equations (9.58) and (9.59) are changed to the difference between measured and calculated inverter output current. Instead of the commanded motor voltages in equations (9.58) and (9.59), the commanded inverter output voltages are used in equations (9.51) and (9.52). In equations (9.2) and (9.6) of the LC filter model, the new correction terms are added.

Equations of the modified state observer for electric drives with inverter output filter are

$$\frac{d\hat{i}_{s\alpha}}{d\tau} = -\frac{R_s L_r^2 + R_r L_m^2}{L_r w_\delta}\hat{i}_{s\alpha} + \frac{R_r L_m}{L_r w_\delta}\hat{\psi}_{r\alpha} + \frac{L_m}{w_\delta}\xi_\beta + \frac{L_r}{w_\delta}\hat{u}_{s\alpha} + k_3\left(i_{1\alpha} - \hat{i}_{1\alpha}\right) \tag{9.80}$$

$$\frac{d\hat{i}_{s\beta}}{d\tau} = -\frac{R_s L_r^2 + R_r L_m^2}{L_r w_\delta}\hat{i}_{s\beta} + \frac{R_r L_m}{L_r w_\delta}\hat{\psi}_{r\beta} - \frac{L_m}{w_\delta}\xi_\alpha + \frac{L_r}{w_\delta}\hat{u}_{s\beta} + k_3\left(i_{1\beta} - \hat{i}_{1\beta}\right) \tag{9.81}$$

$$\begin{aligned}\frac{d\hat{\psi}_{r\alpha}}{d\tau} = &-\frac{R_r}{L_r}\hat{\psi}_{r\alpha} - \xi_\beta + R_r\frac{L_m}{L_r}\hat{i}_{s\alpha} - k_2 S_b \hat{\psi}_{r\alpha} - S k_2 k_3 \hat{\psi}_{r\beta}(S_b - S_{bF}) \\ &+ S k_5\left((S_x - S_{xF})\hat{\psi}_{r\alpha} - (S_b - S_{bF})\hat{\psi}_{r\beta}\right)\end{aligned} \tag{9.82}$$

$$\begin{aligned}\frac{d\hat{\psi}_{r\beta}}{d\tau} = &-\frac{R_r}{L_r}\hat{\psi}_{r\beta} + \xi_\beta + R_r\frac{L_m}{L_r}\hat{i}_{s\beta} - k_2 S_b \hat{\psi}_{r\beta} - S k_2 k_3 \hat{\psi}_{r\alpha}(S_b - S_{bF}) \\ &+ S k_5\left(-(S_x - S_{xF})\hat{\psi}_{r\beta} - (S_b - S_{bF})\hat{\psi}_{r\alpha}\right)\end{aligned} \tag{9.83}$$

$$\frac{d\hat{\xi}_\alpha}{d\tau} = -\hat{\omega}_{\psi r}\hat{\xi}_\beta - k_1\left(i_{1\beta} - \hat{i}_{1\beta}\right) \tag{9.84}$$

$$\frac{d\hat{\xi}_\beta}{d\tau} = \hat{\omega}_{\psi r}\hat{\xi}_\alpha + k_1\left(i_{1\alpha} - \hat{i}_{1\alpha}\right) \tag{9.85}$$

$$\frac{d\hat{u}_{c\alpha}}{d\tau} = \frac{i_{1\alpha} - \hat{i}_{s\alpha}}{C_1} \tag{9.86}$$

$$\frac{d\hat{u}_{c\beta}}{d\tau} = \frac{i_{1\beta} - \hat{i}_{s\beta}}{C_1} \tag{9.87}$$

$$\frac{d\hat{i}_{1\alpha}}{d\tau} = \frac{u_{1\alpha}^{com} - \hat{u}_{s\alpha}}{L_1} + k_A\left(i_{1\alpha} - \hat{i}_{1\alpha}\right) - k_B\left(i_{1\beta} - \hat{i}_{1\beta}\right) \tag{9.88}$$

$$\frac{d\hat{i}_{1\beta}}{d\tau} = \frac{u_{1\beta}^{com} - \hat{u}_{s\beta}}{L_1} + k_A\left(i_{1\beta} - \hat{i}_{1\beta}\right) + k_B\left(i_{1\alpha} - \hat{i}_{1\alpha}\right) \tag{9.89}$$

$$\hat{u}_{s\alpha} = \hat{u}_{c\alpha} + \left(i_{1\alpha} - \hat{i}_{s\alpha}\right)R_c \tag{9.90}$$

$$\hat{u}_{s\beta} = \hat{u}_{c\beta} + \left(i_{1\beta} - \hat{i}_{s\beta}\right)R_c \tag{9.91}$$

$$\frac{dS_{bF}}{d\tau} = \frac{1}{T_{Sb}}(S_b - S_{bF}) \tag{9.92}$$

$$\frac{d\hat{\omega}_{rF}}{d\tau} = \frac{1}{T_{KT}}(\hat{\omega}_r - \hat{\omega}_{rF}) \tag{9.93}$$

$$\frac{dS_{xF}}{d\tau} = \frac{1}{T_{Sx}}(S_x - S_{xF}) \tag{9.94}$$

$$S = \begin{cases} 1 & \text{if} \quad \hat{\omega}_{\psi r} > 0 \\ -1 & \text{if} \quad \hat{\omega}_{\psi r} \leq 0 \end{cases} \tag{9.95}$$

$$S_x = \hat{\xi}_\alpha \hat{\psi}_{r\alpha} + \hat{\xi}_\beta \hat{\psi}_{r\beta} \tag{9.96}$$

$$S_b = \hat{\xi}_\alpha \hat{\psi}_{r\beta} - \hat{\xi}_\beta \hat{\psi}_{r\alpha} \tag{9.97}$$

$$\hat{\omega}_{\psi r} = \hat{\omega}_{rF} + R_r \frac{L_m}{L_r} \left(\frac{\hat{\psi}_{r\alpha} \hat{i}_{s\beta} + \hat{\psi}_{r\beta} \hat{i}_{s\alpha}}{\hat{\psi}_{r\alpha}^2 + \hat{\psi}_{r\beta}^2} \right) \tag{9.98}$$

$$\hat{\omega}_r = \frac{\hat{\zeta}_\alpha \hat{\psi}_{r\alpha} + \hat{\zeta}_\alpha \hat{\psi}_{r\alpha}}{\hat{\psi}_{r\alpha}^2 + \hat{\psi}_{r\beta}^2} \tag{9.99}$$

where k_A and k_B are the additional observer gains.

The small resistance R_1 and leakage inductance $L_{\sigma M1}$ could be omitted in the modified observer procedure.

9.7.3 Simple Observer Based on Motor Stator Models

Some estimations of algorithms are based on the motor stator circuit model. These methods are simple in implementation, but have an inseparable problem with voltage drift [33]. In the literature, some suggestions for motor stator-based estimation improvements are proposed [34, 35]. For example, in [35], the observer is based on the voltage model of the induction motor with combination of rotor and stator fluxes and stator current relationships:

$$\tau_s' \frac{d\boldsymbol{\psi}_{s\alpha}}{d\tau} + \boldsymbol{\psi}_{s\alpha} = k_r \boldsymbol{\psi}_r + \mathbf{u}_s \tag{9.100}$$

where $\hat{\boldsymbol{\psi}}_s = \left[\hat{\psi}_{s\alpha}, \hat{\psi}_{s\beta}\right]^T$ is stator flux vector $\hat{\boldsymbol{\psi}}_r = \left[\hat{\psi}_{r\alpha}, \hat{\psi}_{r\beta}\right]^T$, $\mathbf{u}_s = [U_{s\alpha}, U_{s\beta}]^T$, $\tau_s' = \sigma L_s / R_s$ is time constant, and $k_r = L_m/L_r$ is the rotor coupling factor.

To prevent problems of voltage drift and offset errors, instead of pure integrators, low pass filters (LPFs) are used. The limitation of the estimated stator flux is tuned to the stator flux nominal value. In addition, the extra compensation part is added, as presented in [35].

This section presents the observer system from [35], which is changed adequately to fit the drive with the LC filter requirements – the structure of the observer is extended, taking into account the model of the filter [35]. For the drive with the LC filter, the extra filter simulator relations are used. In the filter dynamics simulator block, the estimated values of the motor current and voltage are calculated as $\hat{\mathbf{i}}_s$ and $\hat{\mathbf{u}}_s$. These estimated variables are calculated on the basis of the inverter commanded voltage $\mathbf{u}_1^{com}$ and the measured inverter output current i_1.

The simulator calculations are performed in an open loop, according to equations (9.18)–(9.23) filter model:

$$\frac{d\hat{\mathbf{u}}_s}{d\tau} = \frac{\mathbf{i}_1 - \hat{\mathbf{i}}_s}{C_1} \tag{9.101}$$

$$\frac{d\hat{\mathbf{i}}_1}{d\tau} = \frac{\mathbf{u}_1 - \hat{\mathbf{u}}_s}{L_1} \tag{9.102}$$

where $\mathbf{i}_1 = [i_{1\alpha}, i_{1\beta}]^T$, $\hat{\mathbf{i}}_1 = \left[\hat{i}_{1\alpha}, \hat{i}_{1\beta}\right]^T$, $\hat{\mathbf{i}}_s = \left[\hat{i}_{s\alpha}, \hat{i}_{s\beta}\right]^T$, $\mathbf{u}_1 = [u_{1\alpha}, u_{1\beta}]^T$, $\hat{\mathbf{u}}_s = \left[\hat{u}_{s\alpha}, \hat{u}_{s\beta}\right]^T$.

The other rotor flux and stator flux observer equations are computed as

$$\frac{d\hat{\boldsymbol{\psi}}_{s\alpha}}{d\tau} = \frac{-\hat{\boldsymbol{\psi}}_{s\alpha} + k_r\hat{\boldsymbol{\psi}}_{r\alpha} + \hat{\mathbf{u}}_{s\alpha}}{\tau_s'} - k_{ab}\left(\mathbf{i}_1 - \hat{\mathbf{i}}_1\right) \tag{9.103}$$

$$\hat{\boldsymbol{\psi}}_r = \frac{\hat{\boldsymbol{\psi}}_s - \sigma L_s \hat{\mathbf{i}}_s}{k_r} \tag{9.104}$$

where $k_{AB} = \begin{bmatrix} k_A & -k_B \\ k_B & k_A \end{bmatrix}$ is the observer gains matrix.

The rotor flux magnitude and angle position are

$$|\hat{\boldsymbol{\psi}}_r| = \sqrt{\hat{\psi}_{r\alpha}^2 + \hat{\psi}_{r\beta}^2} \tag{9.105}$$

$$\hat{\rho}_{\psi r} = arc\,tg\frac{\hat{\psi}_{r\beta}}{\hat{\psi}_{r\beta}} \tag{9.106}$$

The estimated current $\hat{\mathbf{i}}_s$, appearing in equation (9.103), is

$$\hat{\mathbf{i}}_s = \frac{\hat{\boldsymbol{\psi}}_s - k_r\hat{\boldsymbol{\psi}}_r}{\sigma L_s} \tag{9.107}$$

The rotor flux pulsation is

$$\hat{\omega}_{\psi r} = \frac{d\hat{\rho}_{\psi r}}{d\tau} \tag{9.108}$$

The rotor slip is obtained from

$$\hat{\omega}_2 = \frac{\hat{\psi}_{r\alpha}\hat{i}_{s\beta} - \hat{\psi}_{r\beta}\hat{i}_{s\alpha}}{|\hat{\boldsymbol{\psi}}_r|^2} \tag{9.109}$$

The rotor mechanical speed is the difference between the rotor flux pulsation and slip frequencies:

$$\hat{\omega}_r = \hat{\omega}_{\psi r} - \hat{\omega}_2 \tag{9.110}$$

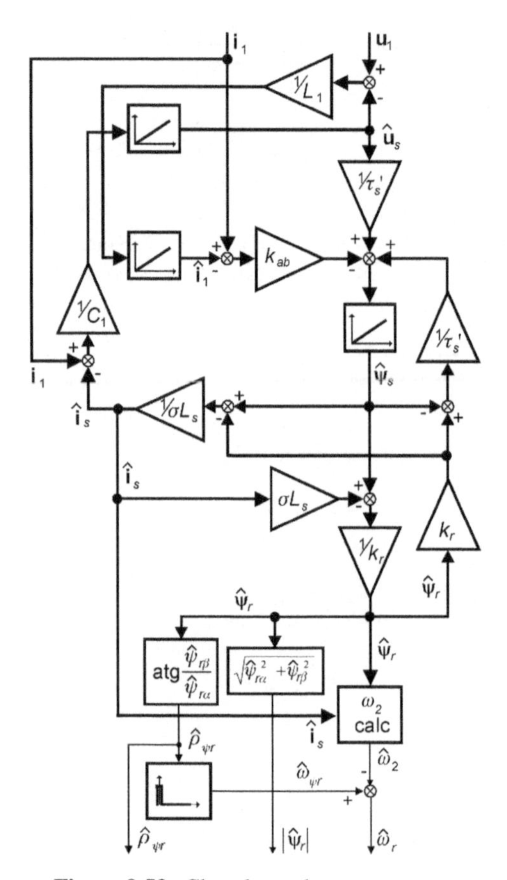

Figure 9.53 Close-loop observer structure

The presented closed-loop flux observer has high robustness of the ASD, with the observer based on the stator model, and the results can be found in [35]. The observer is not affected by stator resistance and other motor parameters mismatch. This significantly extends the stable operating region, even without precise parameter tuning (Figure 9.53).

9.8 Motor Control Problems in the Drive with Filters

9.8.1 Introduction

The differential mode motor filter has considerable influence on the control process. This is because each filter adds a voltage drop and a phase shift between the current and voltages on the filter input and output. For example, a typical LC filter for a 5.5 KW motor, under nominal load and frequency, gives ca. 5% and 5 degrees of voltage drop and phase shift, respectively. An example of real voltages on the filter input and output is presented in Figure 9.54.

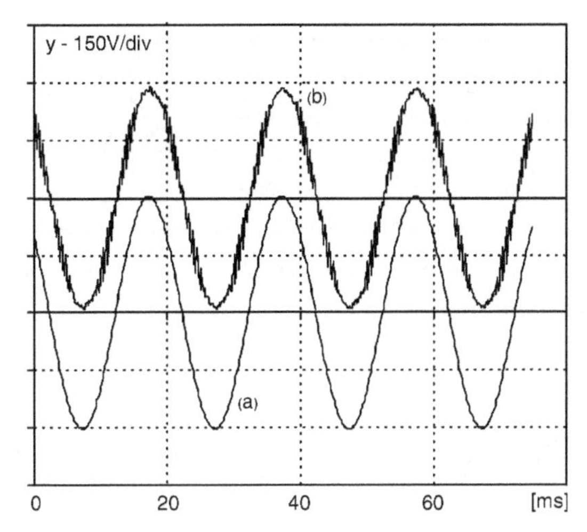

Figure 9.54 Drive with LC filter – waveforms of: (a) commanded inverter output voltage; and (b) real motor supply voltage

As a result, most of the sophisticated sensorless drives cannot work properly when the filter is installed. It is also necessary to take into account the filter's components in the control process.

The general concept is to extend the corresponding steering algorithm with its subordinated control system for some of the filter state variable controls.

A differential mode LC filter is a two-dimensional linear stationary controlled system. The controlled state variables are motor supply voltage $\mathbf{u}_s$ and inverter output current $\mathbf{i}_1$, whereas the control quantity is an inverter output voltage $\mathbf{u}_1$. Current $\mathbf{i}_1$ represents the filter's internal variables.

The basic structure of the LC filter control system is presented in Figure 9.55.

In the control structure shown in Figure 9.55, the commanded signal is a motor supply voltage $\mathbf{u}_s^{com}$. The commanded signal is compared with a real motor supply voltage $\mathbf{u}_s$ and the proper controller evaluates the desired inverter output voltage $\mathbf{u}_1$, which should be directly set to the controlled system, LC filter. For the control process, the motor current $\mathbf{i}_s$ is a disturbance. The disturbance influence on the control process is possible, if it takes into account the control algorithm. The disturbance has to be compensated for, as presented in Figure 9.56.

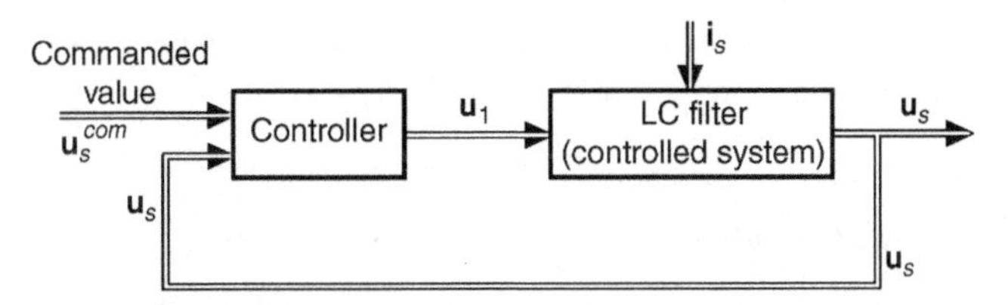

Figure 9.55 Structure of the LC filter control system

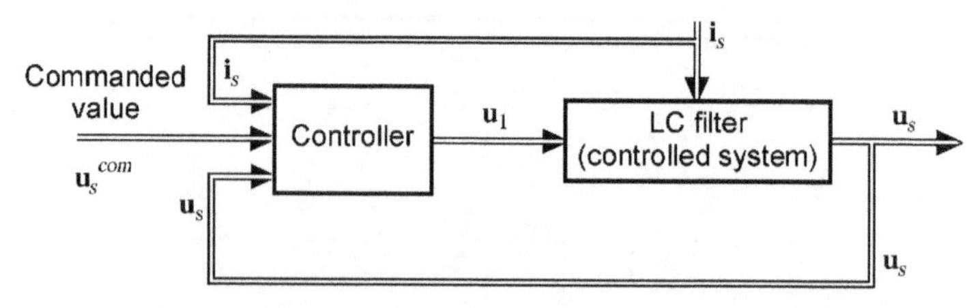

Figure 9.56 Structure of the LC filter control system

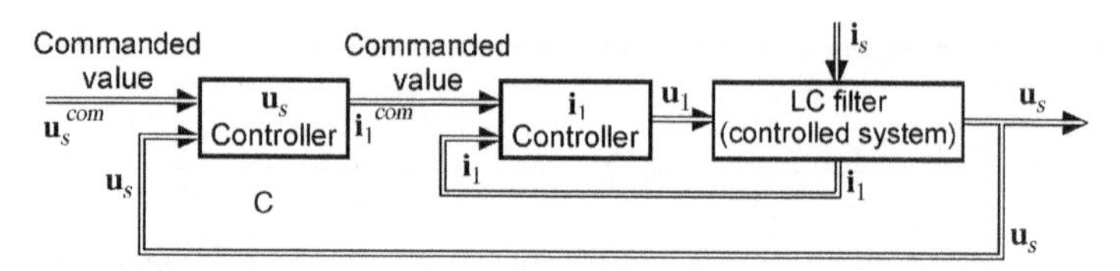

Figure 9.57 Structure of the multi-loop LC filter control system

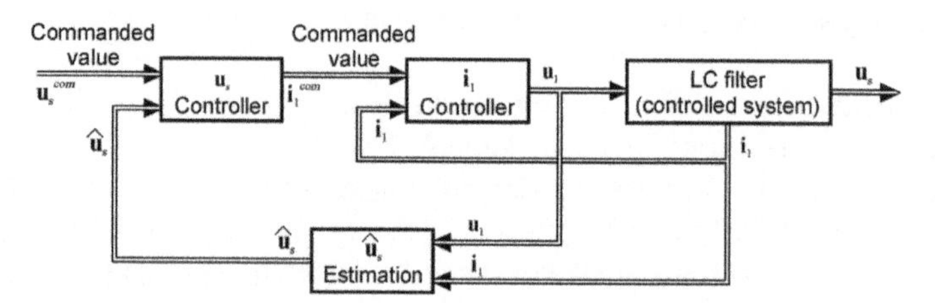

Figure 9.58 Structure of the multi-loop LC filter control system with motor voltage estimation

In the control systems presented in Figures 9.55 and 9.56, the inverter output current $\mathbf{i}_1$ is not controlled, which is unadvisable because of the lack of inverter current protection. To control the inverter current and motor voltage, the multi-loop control system proposed in [27] can be used (Figure 9.57).

The control system presented in Figure 9.57 requires information on the actual $\mathbf{i}_1$ and $\mathbf{u}_s$ signals. Current $\mathbf{i}_1$ is easily measured, but $\mathbf{u}_s$ requires sensors outside the inverter, which is not recommended. So instead of the $\mathbf{u}_s$ measurement, the estimation process for $\hat{\mathbf{u}}_s$ can be used. The corresponding system structure is presented in Figure 9.58.

The multi-loop control structure of the LC filter can be applied in numerous motor regulation algorithms. Some of them are presented in the following sections.

9.8.2 Field-Oriented Control

Among the induction motor closed-loop drives, the most popular is the FOC system. The control principle is widely presented in the literature [36], as well as in previous chapters of this book. Here the FOC is used in the drive with an LC filter and only the necessary information is given.

The base FOC structure is presented in Figure 9.59 and the vectors relations in Figure 9.60.

In Figure 9.61, the operation of the classical FOC system for the driver without an LC filter is presented. The same operations are presented in Figure 9.62, but with the LC filter installed.

In both cases, the controller setting and control structure are kept the same. The influence of the observer is eliminated, because all control variables are assumed as measured online. When Figures 9.61 and 9.62 are compared, they shows that when an LC filter is installed, the system's

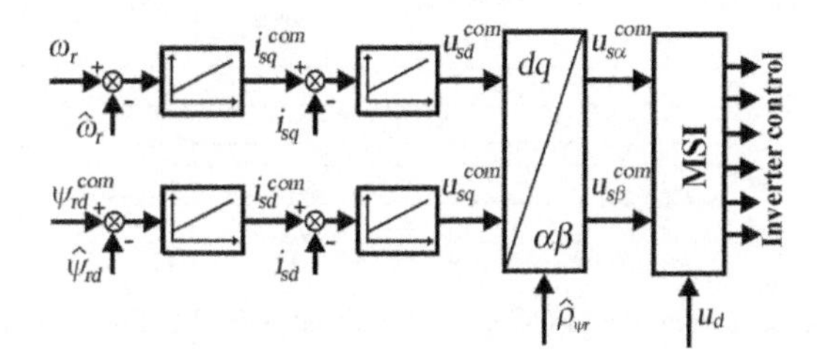

Figure 9.59 Classical field oriented control structure

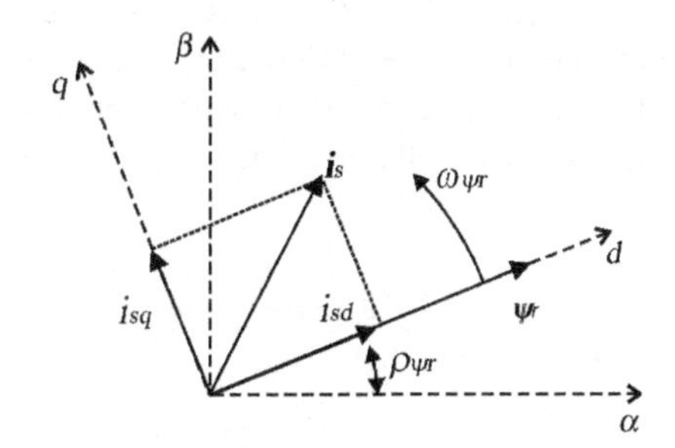

Figure 9.60 Relations between vectors in FOC system

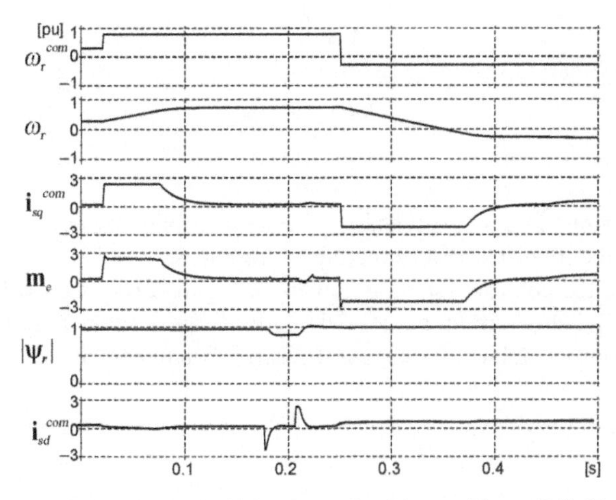

Figure 9.61 Operation of the classical FOC system in driver **without** LC filter – system without observers

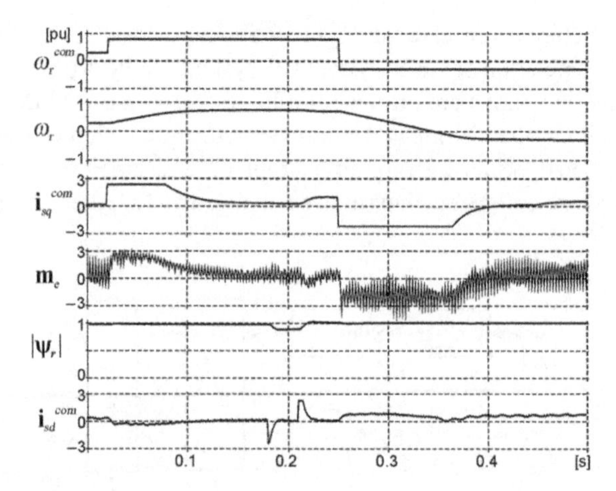

Figure 9.62 Operation of the classical FOC system in driver with LC filter – system without observers

performance degenerates. It is especially visible in the motor torque waveforms, where high frequency oscillations appear as a result of the controlled object structure change.

This disadvantage can be compensated for if the FOC structure is modified with respect to the multi-loop filter control conception. The additional control variables of capacitor voltages $u_{c\alpha}$, $u_{c\beta}$ and inverter currents $i_{1\alpha}$, $i_{1\beta}$ are controlled by additional blocks.

Assuming that $u_{s\alpha} \approx u_{c\alpha}$ and $u_{s\beta} \approx u_{c\beta}$, the LC filter model is described in the dq frame of references:

$$\frac{du_{sd}}{d\tau} = \frac{i_{cd}}{C_1} \tag{9.111}$$

$$\frac{di_{1d}}{d\tau} = \frac{u_{1d} - u_{sd}}{L_1} \tag{9.112}$$

$$\frac{du_{sq}}{d\tau} = \frac{i_{cq}}{C_1} \tag{9.113}$$

$$\frac{di_{1q}}{d\tau} = \frac{u_{1q} - u_{sq}}{L_1} \tag{9.114}$$

$$i_{cd} = i_{1d} - i_{sd} \tag{9.115}$$

$$i_{cq} = i_{1q} - i_{sq} \tag{9.116}$$

In the multi-loop, structured u_{sd}, u_{sq}, and i_{1d}, i_{1q} are controlled by correct PI elements.

In Figure 9.63, the modified FOC structure is presented.

The operation of the control system, presented in Figure 9.63, is show in Figure 9.64.

The modified FOC structure properties (Figure 9.61) are similar to those in classical FOC (Figure 9.58). The only notable difference is the lower dynamics of the drive with an LC filter,

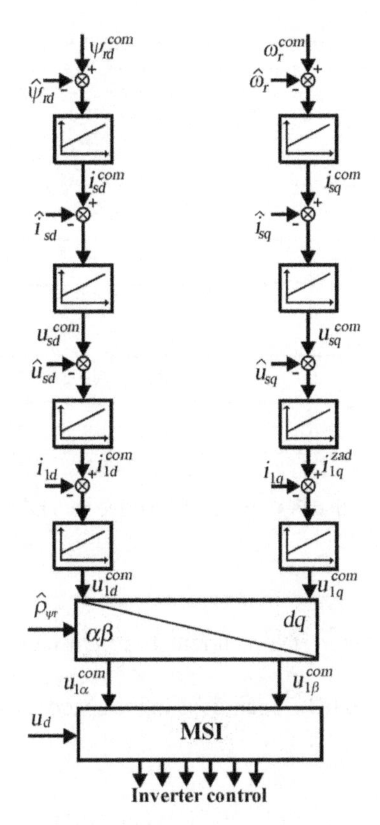

Figure 9.63 Field oriented control structure modified due to LC filter use

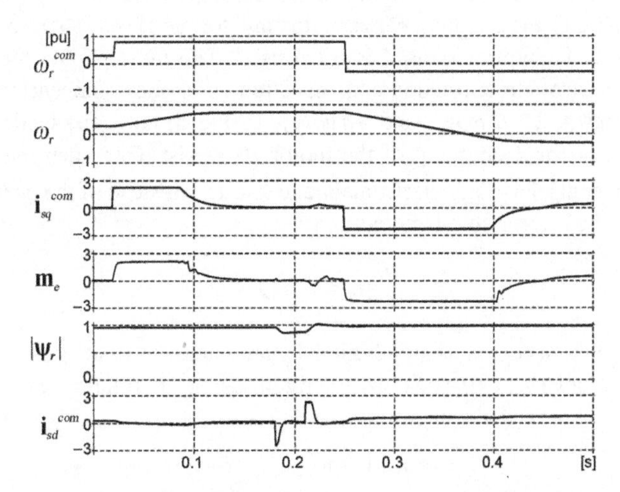

Figure 9.64 Operation of the modified FOC system for driver with LC filter – system without observers

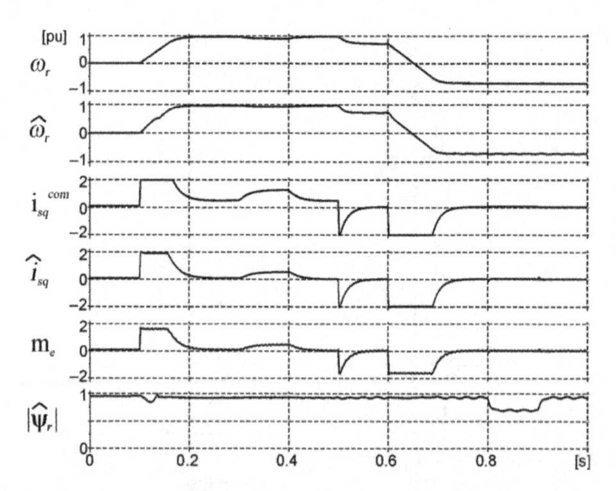

Figure 9.65 Operation of the modified FOC system for a drive with LC filter in sensorless mode – system with variables estimation

which is the result of the higher stator current time constant. However, the difference is insignificant.

The examples of the modified FOC operations in the speed sensorless mode are presented in Figures 9.65–9.67.

9.8.3 Nonlinear Field-Oriented Control

The most popular industrial induction motor (IM) control is rotor field oriented control (RFOC) [36]. In classical RFOC, the coupling between flux and torque exists. Therefore, to improve the RFOC properties, a decoupling control is often used. The most popular decoupling system relies on adding electromotive rotation compensation components appearing in motor model equations (9.1) and (9.2): $a_3\omega_r\psi_{ry}$ and $-a_3\omega_r\psi_{rx}$ to the motor commanded voltages u_{sx}^{com}, u_{sy}^{com} Other solutions for decoupling of the motor also exist. One such method is proposed in [37], in order to control a motor electromagnetic torque t_e instead of the q current component (in [37] the controlled t_e variable is noted as x):

$$t_e = i_{sq}\psi_{rd} \tag{9.117}$$

The motor torque t_e is accepted as an additional state variable. With such an assumption, the motor model equations (9.1)–(9.5) can be rewritten, using the rotating dq coordinates, as

$$\frac{di_{sd}}{d\tau} = a_1 i_{sd} + a_2\psi_{rd} + t_e\frac{\omega_{\psi r}}{\psi_{rd}} + a_4 u_{sd} \tag{9.118}$$

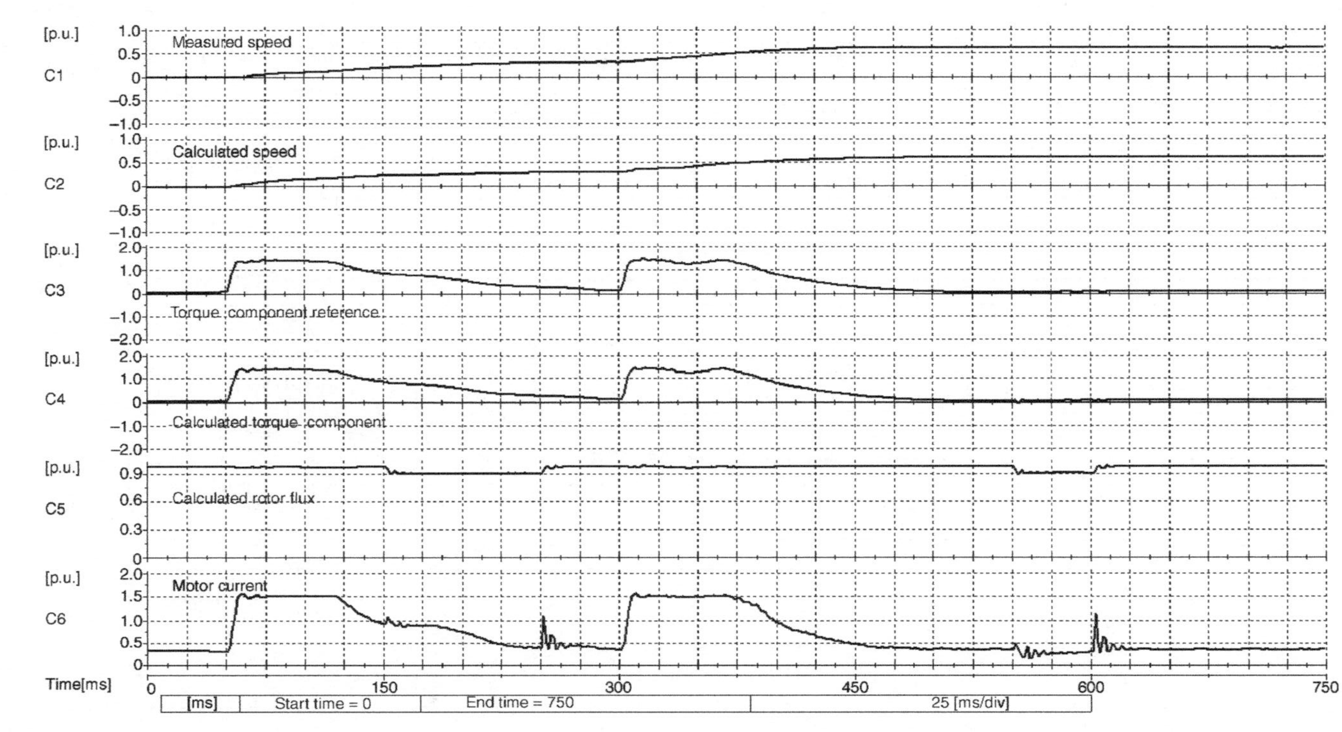

Figure 9.66 Operation of the modified FOC system for a drive with LC filter in sensorless mode – system with variables estimation – experiment 1 ($C_1 - \omega_r$, $C_2 - \hat{\omega}_r$, $C_3 - \hat{i}_{sq}^{com}$, $C_4 - \hat{i}_{sq}$, $C_5 - |\hat{\mathbf{\Psi}}_r|$, $C_6 - |i_s|$)

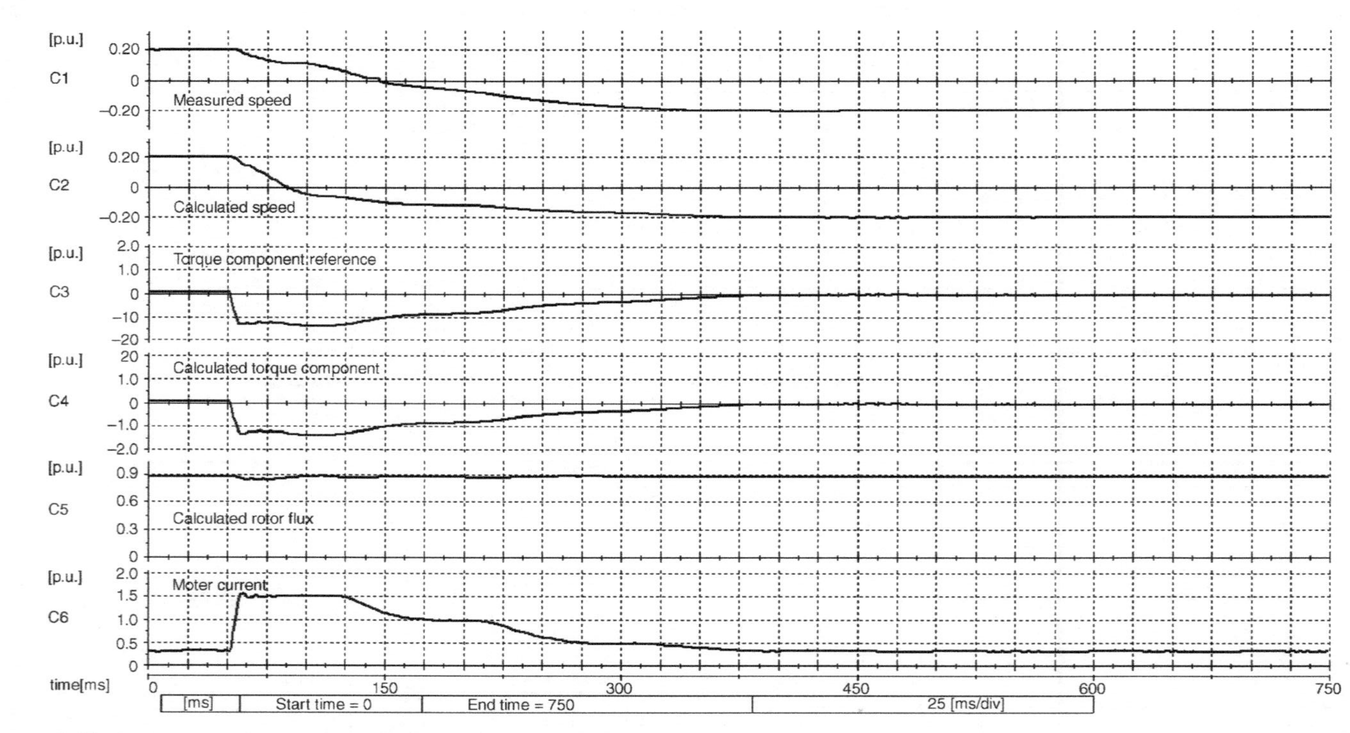

Figure 9.67 Operation of the modified FOC system for a drive with LC filter in sensorless mode – *system* with variables estimation – experiment 2 ($C_1 - \omega_r$, $C_2 - \hat{\omega}_r$, $C_3 - \hat{i}_{sq}^{com}$, $C_4 - \hat{i}_{sq}$, $C_5 - |\hat{\boldsymbol{\Psi}}_r|$, $C_6 - |i_s|$)

$$\frac{dt_e}{d\tau} = \left(a_5 - \frac{R_s L_r}{w\sigma}\right)t_e + a_6 t_e \frac{i_{sd}}{\psi_{rd}} - \omega_{\psi r}\psi_{rd}\left(i_{sd} + a_3\psi_{rd}\right) + a_4\psi_{rd}u_{sq} \tag{9.119}$$

$$\frac{d\psi_{rd}}{d\tau} = a_5\psi_{rd} + a_6 i_{sd} \tag{9.120}$$

$$\frac{d\omega_r}{d\tau} = \frac{1}{J}\left(\frac{L_m}{L_r}t_e - t_L\right) \tag{9.121}$$

where $\omega_{\psi r} = a_6 i_{sq}/\psi_{rd} + \omega_r$ and i_{sd}, i_{sq}, u_{sd}, u_{sq}, ψ_{rd} are motor stator current, voltages, and rotor flux in dq coordinates.

Equations (9.118)–(9.121) describe the model of the induction motor in dq coordinates, which is a nonlinear and coupled system. If the following, control variables v_1 and v_2 are

$$v_1 = a_6 t_e \frac{i_{sd}}{\psi_{rd}} - \omega_{\psi r}\psi_{rd}\left(i_{sd} + a_3\psi_{rd}\right) + a_4\psi_{rd}u_{sq} \tag{9.122}$$

$$v_2 = \omega_{\psi r}\frac{t_e}{\psi_{rd}} + a_2\psi_{rd} + a_4 u_{sd} \tag{9.123}$$

the motor model dq equations (9.7)–(9.10) are converted into two linear decoupled subsystems:

$$\frac{di_{sd}}{d\tau} = a_1 i_{sd} + v_2 \tag{9.124}$$

$$\frac{d\psi_{rd}}{d\tau} = a_5\psi_{rd} + a_6 i_{sd} \tag{9.125}$$

$$\frac{dt_e}{d\tau} = (a_5 - R_s a_4)t_e + v_1 \tag{9.126}$$

$$\frac{d\omega_r}{d\tau} = \frac{1}{J}\left(\frac{L_m}{L_r}t_e - t_L\right) \tag{9.127}$$

The base structure of the nonlinear filed oriented control method is presented in Figure 9.68.

In Figure 9.68, the angle $\rho_{\psi r}$ represents the position of the rotor flux vector and u_d is the voltage in the inverter DC link. A block marked as $dq/\alpha\beta$ represents the Park transformation from the rotating dq into the stationary $\alpha\beta$ frame of references. The $\alpha\beta$ coordinates are natural for PWM space vector algorithm.

9.8.3.1 Extended Control System

In order to assure control of the system with the LC filter, the motor control structure presented in previous section has been extended by using additional controllers.

The cascaded multi-loop PI controllers are used to control the motor supply voltages u_{sd}, u_{sq} and inverter output currents i_{1d}, i_{1q} [26, 27]. In the filter control subsystem, the disturbance compensation on the stator voltage controllers is used.

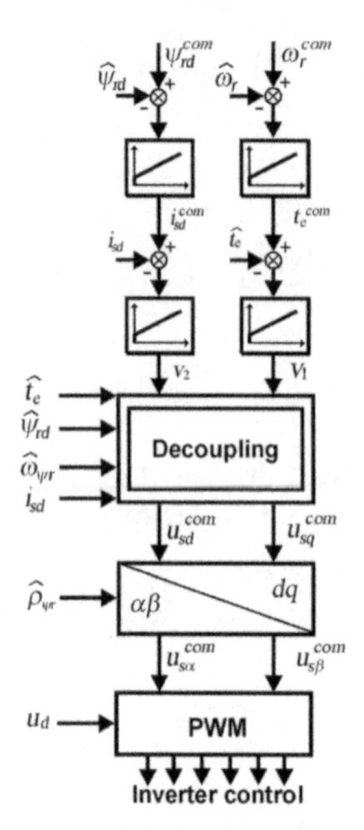

Figure 9.68 Base control system for decoupled rotor field-oriented method

In order to eliminate a phase shift in the PI units, the control of the filter state variables is done by the synchronous dq coordinates synchronized with the rotor flux vector – the same coordinates as used in the basic RFOC control system.

The disadvantage of multi-loop solutions is a necessity of the motor voltage and current knowledge. Unfortunately, the real sensors are not practical in this solution, as noted in the introduction. The variables are calculated online in the complex observer structure.

In this section, instead of the full multi-loop structure, only the PI controllers for motor stator voltage are implemented. The whole structure of the extended control system of the drive with induction motor, inverter, and LC filter is presented in Figure 9.69.

In the extended control system presented in Figure 9.66, two additional PI controllers appear. The controllers directly control the motor stator voltage. The inverter output current is not controlled explicitly. This happens because the capacitor current is small, contrary to the motor current. The motor current controllers are present in the FOC structure, so that the inverter output current is controlled indirectly.

The commanded inverter voltage components, u_{1d}^{com} and $\underline{u}_{1q}^{com}$, are transformed to the stationary $\alpha\beta$ coordinates, $u_{1\alpha}^{com}$ and $u_{1\beta}^{com}$, and are treated as the inputs of the PWM block.

9.8.4 Nonlinear Multiscalar Control

In this section, the nonlinear multiscalar control system, designed to be operated with the LC filter, is presented. For the control, only one sensor is used to measure the inverter DC supply voltage and two sensors are used to measure the two inverter output currents. The control

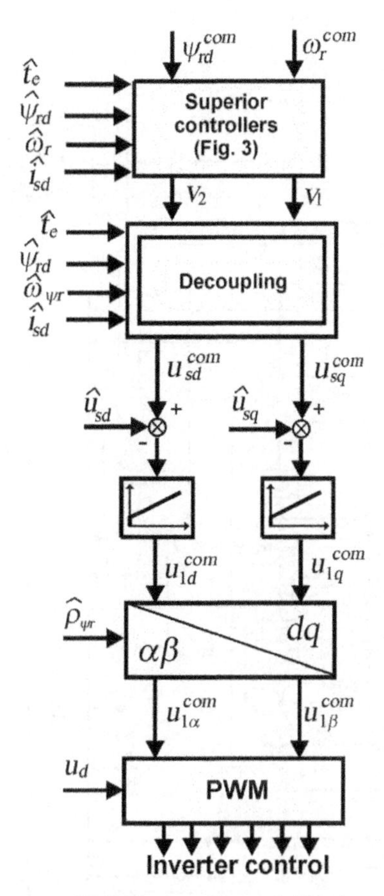

Figure 9.69 Induction motor and LC filter nonlinear field-oriented control system structure

system is divided into two subsystems, the superior motor control and the subordinate LC filter control. The control structure is part of the motor speed sensorless system presented in Figure 9.70.

9.8.4.1 Motor Control Subsystem

In the motor control subsystem, the nonlinear control using nonlinear feedback is used. That control is based on the methods of the differential geometry and was first adapted to the electric drive in [37] (MMB) and presented in other papers later, but only for the drive without LC filter. In MMB, the nonlinear and decoupled object is converted into a linear one with use of the new state variables and the nonlinear feedbacks. The four MMB state variables are

$$x_{11} = \omega_r \tag{9.128}$$

$$x_{12} = \psi_{r\alpha} i_{s\beta} - \psi_{r\beta} i_{s\alpha} \tag{9.129}$$

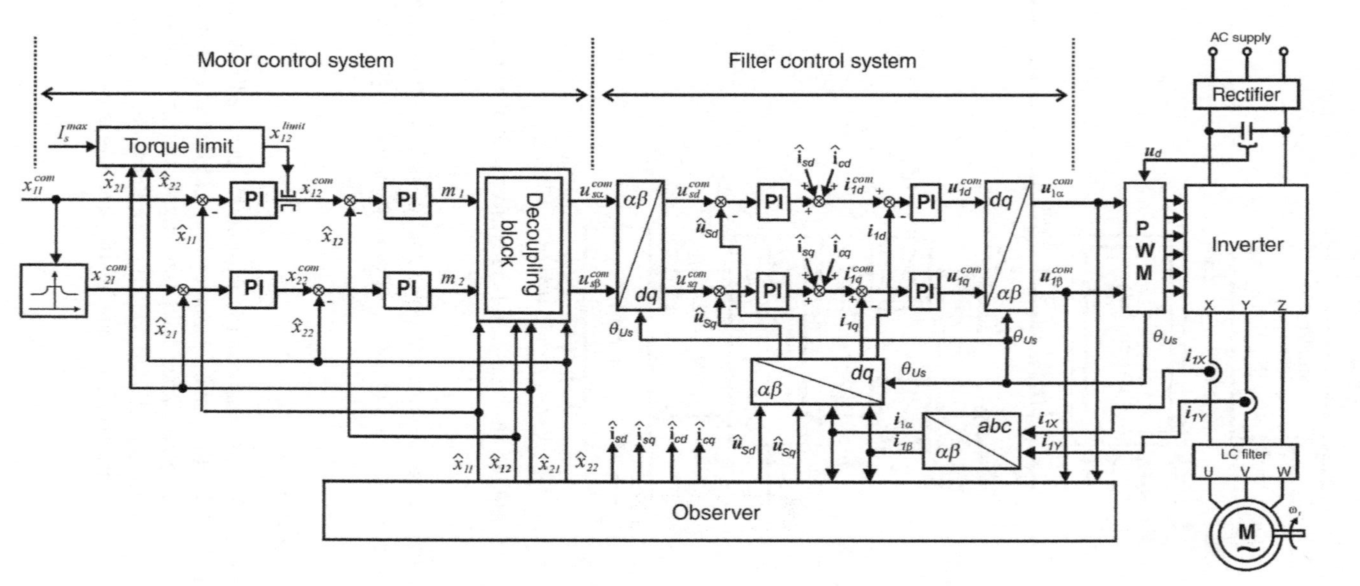

Figure 9.70 Speed sensorless multiscalar control system

$$x_{21} = \psi_{r\alpha}^2 + \psi_{r\beta}^2 \tag{9.130}$$

$$x_{22} = \psi_{r\alpha} i_{s\alpha} + \psi_{r\beta} i_{s\beta} \tag{9.131}$$

where x_{11} is rotor angular speed; x_{12} is proportional to the motor torque; x_{21} is square of rotor flux; x_{22} is scalar product of the stator current and rotor flux vectors; and i_s and ψ_r are the magnitudes of the stator current and rotor flux, respectively.

With the multiscalar variables $x_{11}, x_{12}, x_{21}, x_{22}$ equations (9.128)–(9.131), the next model of the motor is

$$\frac{dx_{11}}{d\tau} = x_{12}\frac{L_m}{JL_r} - \frac{T_L}{J} \tag{9.132}$$

$$\frac{dx_{12}}{d\tau} = -\frac{1}{T_V}x_{12} - x_{11}\left(x_{22} + x_{21}\frac{L_m}{w_\sigma}\right) + u_1\frac{L_r}{w_\sigma} \tag{9.133}$$

$$\frac{dx_{21}}{d\tau} = 2R_r\left(-\frac{1}{x_{21}} + x_{22}\frac{L_m}{L_r}\right) \tag{9.134}$$

$$\frac{dx_{22}}{d\tau} = -\frac{1}{T_V}x_{22} + x_{11}x_{22} + x_{21}R_r\frac{L_m}{L_r w_\sigma} + \frac{R_r L_m}{L_r}\frac{\left(x_{12}^2 + x_{22}^2\right)}{x_{21}} + u_2\frac{L_r}{w_\sigma} \tag{9.135}$$

where $T_v = w_\sigma L_r / \left(R_r w_\sigma + R_s L_r^2 + R_r L_m^2\right)$.

The compensation of nonlinearities in equations (9.12)–(9.15) leads to the new driving functions m_1 and m_2:

$$u_1 = \frac{w_\sigma}{L_r}\left[x_{11}\left(x_{22} + x_{21}\frac{L_m}{w_\sigma}\right) + m_1\right] \tag{9.136}$$

$$u_1 = \frac{w_\sigma}{L_r}\left(-x_{11}x_{12} - x_{21}\frac{R_r L_m}{L_r w_\sigma} - \frac{R_r L_m}{L_r}\frac{x_{12}^2 + x_{22}^2}{x_{21}} + m_2\right) \tag{9.137}$$

The components of the commanded motor stator voltage vector in the stationary frame are

$$u_{s\alpha}^{com} = \frac{\psi_{r\alpha}u_2 - \psi_{r\beta}u_1}{x_{21}} \tag{9.138}$$

$$u_{s\beta}^{com} = \frac{\psi_{r\beta}u_2 - \psi_{r\alpha}u_1}{x_{21}} \tag{9.139}$$

With equations (9.8)–(9.11) and (9.16)–(9.19), the asynchronous motor model is decoupled and converted into two separate linear subsystems, mechanical equations (9.20) and (9.21), and electromagnetic equations (9.22) and (9.23):

$$\frac{dx_{11}}{d\tau} = x_{12}\frac{L_m}{JL_r} - \frac{T_L}{J} \tag{9.140}$$

$$\frac{dx_{12}}{d\tau} = -\frac{1}{T_v}x_{12} - m_1 \tag{9.141}$$

$$\frac{dx_{21}}{d\tau} = 2R_r\left(-\frac{1}{x_{21}} + x_{22}\frac{L_m}{L_r}\right) \tag{9.142}$$

$$\frac{dx_{22}}{d\tau} = -\frac{1}{T_v}x_{22} + m_2 \tag{9.143}$$

The fully decoupled subsystems make it possible to use this method in the case of a changing flux vector and to obtain simple system structure, which is not easily achieved in the case of vector control methods. It is possible to use cascade structure of simple proportional – integral PI controllers in the decoupled, linear control subsystems.

The motor stator current does not exist in the control system distinctly. So to limit the motor current to the maximum allowed $I_{s\ max}$, the output of the x_{11} controller is variably limited according to the relation:

$$x_{12}^{\lim it} = \sqrt{I_{s\ max}^2 x_{21} - x_{22}^2} \tag{9.144}$$

9.8.4.2 LC Filter Control Subsystem

In the case of a drive without the LC filter, the MMB output variables $u_{s\alpha}^{com}$ and $u_{s\beta}^{com}$ are commanded for the PWM block. The PWM controls the inverter transistors to obtain the commanded voltage on the inverter output.

When the LC filter is used, the inverter output voltage changes the form of the motor supply voltage. So the error in the MMB control loop will appear. To solve this problem, the multiloop controller loops, presented in previous section, is used.

The filter output sensors are not used because of the proper observer use. To omit the phase shift of PI controllers, the filter variables are operated in the rotating frame of references. The frame of reference is noted as the *dq* system, where the position of the *d*-axis is connected with the commanded inverter output vector $\mathbf{u}_s$.

9.9 Predictive Current Control in the Drive System with Output Filter

In this section, the predictive current controller (PCC) is presented in the IM speed sensorless system with a FOC method and load angle regulation, to show how the system is modified due to motor choke installation.

9.9.1 Control System

Among different control algorithms, the control of the load angle FOC exists [38]. The structure of the load angle control seems simpler than the FOC because they do not need Park transformations. The load angle δ is noted as the angle between the $\boldsymbol{\psi}_r$ and $\mathbf{i}_s$ vectors. The control of the vector amplitudes and mutual position makes it possible to control the motor electromagnetic torque T_e:

$$T_e = k \cdot \mathrm{Im}\left(\boldsymbol{\psi}_r^* \mathbf{i}_s\right) = k|\boldsymbol{\psi}_r||\mathbf{i}_s| \sin \delta \tag{9.145}$$

where k is the constant of proportionality.

The FOC load angle control system base structure, based on equation (9.145), is presented in Figure 9.71.

In the control system shown in Figure 9.71, the measured variables are the inverter output currents and inverter DC link voltage. The motor currents and motor voltages are not measured. The load angle is controlled by commanding the motor slip frequency ω_2. The sum of ω_2 and the motor speed ω_r is noted as current commanded angular frequency ω_i^{com}. The signal ω_i^{com}, along with the commanded stator current magnitude i_s^{com}, is simultaneously controlled by the current controller algorithm. The current controller cooperates with the PWM procedure. The commanded values of I_s and ω_i are set by speed and flux module controllers, with the output signals I_s^{com} and δ^{com}:

$$I_s^{com} = \sqrt{\left(i_{sd}^{com}\right)^2 + \left(i_{sq}^{com}\right)^2} \tag{9.146}$$

$$\delta^{com} = arc\ tg\left(\frac{i_{sq}^{com}}{i_{sd}^{com}}\right) \tag{9.147}$$

The amplitudes of the $\boldsymbol{\psi}_r$ and $\mathbf{i}_s$ vectors are kept constant, while the motor torque is controlled by changing the angle δ. Unfortunately, in transients, some coupling and interactions appear between the controlled variables, due to inherent nonlinearities and couplings in the induction motor.

To prevent these negative features from appearing in the drive, the nonlinear control principles [39] are implemented [37].

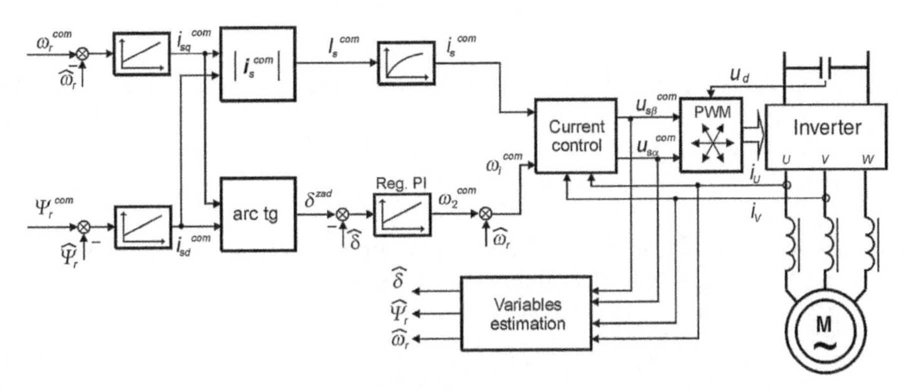

Figure 9.71 Base Samples notation for switching period (^denotes variables evaluated in estimation block)

The nonlinear control principle is implemented for the FOC system presented in Figure 9.71. The linearization and decoupling of the Figure 9.71 structure are based on the equations of the current controlled induction motor model described by the next dependencies:

$$\frac{di_s}{d\tau} = \frac{1}{T_i}\left(i_s - I_s^{com}\right) \tag{9.148}$$

$$\frac{d\delta}{d\tau} = -\frac{R_r L_m}{L_r}\frac{i_s}{\psi_r} + \omega_i - \omega_r \tag{9.149}$$

$$\frac{d\psi_r}{d\tau} = -\frac{R_r}{L_r}\psi_r + \frac{R_r L_m}{L_r} i_s \cos\delta \tag{9.150}$$

$$\frac{d\omega_r}{d\tau} = -\frac{1}{J_M}\left(\frac{L_m}{L_r}\psi_r i_s - t_L\right) \tag{9.151}$$

where i_s and u_s are stator current and voltage modules, respectively, ψ_r is rotor flux module; I_s^{com} is commanded stator current; and T_i is the time constant of the LPF related to the inertia of elements appearing in the current control loop.

Equation (9.148) exists only in the control system and is added to model the physical limitations of the stator current pulsation changes.

Based on the analysis of equations (9.149) and (9.150), the linearization and decoupling of the system are performed. To assure system stability, the relationship of cos δ in equation (9.150) should be positive; otherwise, positive feedback can appear in the control loop. To insure $\cos\delta > 0$, the angle δ is limited to the $(-\pi/2\ldots\ \pi/2)$ range and then a new control variable ψ_r^* is introduced:

$$\psi_r^* = L_m i_s \cos\delta \tag{9.152}$$

From equation (9.152), the commanded stator current module is

$$i_s^{com} = \frac{\psi_r^*}{L_m \cos\delta} \tag{9.153}$$

The LPF equation (9.148) is replaced by the LPF dependency for the rotor flux magnitude:

$$\frac{d\psi_{ri}^*}{dt} = \frac{1}{T_\psi}\left(\psi_{ri}^* - \psi_r^*\right) \tag{9.154}$$

where ψ_{ri}^* is an output signal of the LPF and T_ψ is LPF inertia.

According to equation (9.154), equation (9.153) has the form:

$$i_s^{com} = \frac{\psi_{ri}^*}{L_m \cos\delta} \tag{9.155}$$

If the control signal equation (9.152) is substituted into equation (9.149), then the dynamic of the load angle is transformed into the form:

$$\frac{d\delta}{d\tau} = -\frac{R_r}{L_r}\frac{\psi_{ri}^*}{\psi_r \cos\delta} + \omega_i - \omega_r \tag{9.156}$$

which is still nonlinear. To transform equation (9.156) into a linear form, the commanded motor current pulsation should be

$$\omega_i^{com} = \omega_r + \frac{R_r}{L_r}\frac{\psi_{ri}^*}{\psi_r \cos\delta} + \frac{1}{T_\delta}(\delta^* - \delta) \tag{9.157}$$

where δ^* and T_δ are desired load angle and time constant of the load angle dynamic system, respectively.

Regarding equations (9.152) and (9.157), the next dynamic system for the load angle and rotor flux control is

$$\frac{d\delta}{d\tau} = \frac{1}{T_\delta}(\delta^* - \delta) \tag{9.158}$$

$$\frac{d\psi_r}{d\tau} = \frac{R_r}{L_r}(\psi_{ri}^* - \psi_r) \tag{9.159}$$

It is noted that the dynamic system equations (9.158) and (9.159) are linear and decoupled. The structure of the nonlinear FOC with the load angle controller is presented in Figure 9.72.

In the system shown in Figure 9.72, the commanded value of the load angle is

$$\delta^{com} = arc\ tg\left(\frac{i_{sq}^{com}}{\hat{i}_{sd}}\right) \tag{9.160}$$

The commanded stator current amplitude is calculated in control block 1, based on equation (9.155), and the commanded stator current pulsation is calculated in control block 2, based on equation (9.157). Both blocks are indicated in Figure 9.72.

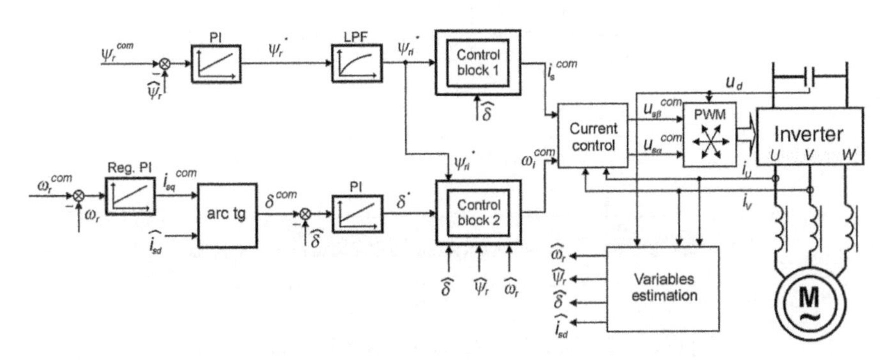

Figure 9.72 Structure of the nonlinear FOC with load angle control

9.9.2 Predictive Current Controller

The integral part of the FOC system is the IM stator current controller. Many different methods for current control are presented in the literature. A comprehensive survey is presented in [40].

Good results of current control are reported in [41], where a variety of predictive control methods (PCM) were used.

The filters significantly affect the structure of the system, so some of the predictive controllers cannot work properly. The simplest version of the inverter output filter is a motor choke, which reduces the reflected waves in long cables connected to the motor, prevents overvoltages on motors and inverter transistors and reduces the radio frequency interference level. However, in drives with motor chokes, the predictive controller's algorithm should also be modified.

In the control system presented in this section, the motor stator current is controlled. The controller uses actual values of the induction motor electromagnetic forces $\mathbf{e}$ (emf) to obtain proper current regulation. Values of the emf are calculated in the observer system described in the next section.

The notation for the presented current controller is explained in Figure 9.73.

The basic assumption is that the PWM inverter is working in such a way that the inverter output voltage $\mathbf{u}_s$ is equal to its commanded value $\mathbf{u}_s^{com}$:

$$\mathbf{u}_s \approx \mathbf{u}_s^{com} \tag{9.161}$$

The current controller principle is based on the dynamic equation that describes the model of the system. An equivalent model of the presented system contains three parts: inductance, resistance, and emf. Because a motor choke is installed in the ASD, it is acceptable to neglect an equivalent resistance of the load. With such an assumption, the system dynamic is

$$\frac{d\mathbf{i}_s}{d\tau} = \frac{1}{L_1 + L_\sigma}\left(\mathbf{u}_s^{com} - \mathbf{e}\right) \tag{9.162}$$

where $\mathbf{u}_s^{com}$ is the inverter commanded voltage vector and T_{imp} is the inverter switching frequency period.

For small T_{imp}, equation (9.162) is approximated as

$$\frac{\mathbf{i}_s(k) - \mathbf{i}_s(k-1)}{T_{imp}} = \frac{1}{L_1 + L_\sigma}\left(\mathbf{u}_s^{com}(k-1) - \mathbf{e}(k-1)\right) \tag{9.163}$$

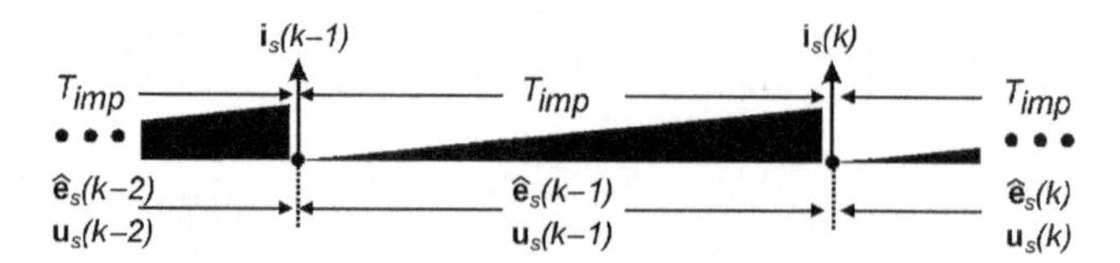

Figure 9.73 Samples notation for switching period *Timp*

If $(k-1)$, ..., (k) is the time period taken into account, the variables $\mathbf{i}_s(k)$ and $\mathbf{e}(k-1)$ are unknown and should be predicted on the basis of previous known examples.

In the presented system, emf is calculated online in the estimation block:

$$\hat{\mathbf{e}} = \omega_r \hat{\boldsymbol{\psi}}_r \tag{9.164}$$

Motor emf equation (9.74) is calculated inside an observer structure presented in Section 9.9.3.

The samples of $\hat{\mathbf{e}}(k-2)$ and $\hat{\mathbf{e}}(k-3)$, calculated in the observer, are memorized and used to predict the value of $\hat{\mathbf{e}}(k-1)$:

$$\hat{\mathbf{e}}^{pred}(k-1) = \mathbf{C}_{EMF}\hat{\mathbf{e}}(k-2) \tag{9.165}$$

where

$$\mathbf{C}_{EMF} = \begin{bmatrix} \cos(\Delta\varphi_e) & \sin(\Delta\varphi_e) \\ -\sin(\Delta\varphi_e) & \cos(\Delta\varphi_e) \end{bmatrix} \tag{9.166}$$

and $\Delta\varphi_e$ is $\mathbf{e}$ vector angular position change:

$$\Delta\varphi_e = arc\,tg\frac{\hat{e}_\alpha(k-3)\hat{e}_\beta(k-2) - \hat{e}_\alpha(k-2)\hat{e}_\beta(k-3)}{\hat{e}_\alpha(k-2)\hat{e}_\alpha(k-3) + \hat{e}_\beta(k-2)\hat{e}_\beta(k-3)} \tag{9.167}$$

For the calculated $\hat{\mathbf{e}}^{pred}$ value, the current sample at (k) is predicted as

$$\mathbf{i}_s^{pred}(k) = \mathbf{i}_s(k-1) + \frac{T_{imp}}{L_1 + L_\sigma}\left(\mathbf{u}_s^{com}(k-1) - \hat{\mathbf{e}}^{pred}(k-1)\right) \tag{9.168}$$

The current regulation errors at the instant $(k-1)$ and (k) are calculated as

$$\Delta\mathbf{i}_s(k-1) = \mathbf{i}_s^{com}(k-1) - \mathbf{i}_s^{pred}(k-1) \tag{9.169}$$

$$\Delta\mathbf{i}_s(k) = \mathbf{i}_s^{com}(k) - \mathbf{i}_s^{pred}(k) \tag{9.170}$$

In order to minimize the stator current regulation error at the $(k+1)$ instant, the proper voltage vector $\mathbf{u}_s^{com}(k)$ should be applied [35]:

$$\mathbf{u}_s^{com}(k) = \frac{L_1 + L_\sigma}{T_{imp}}\left(\mathbf{i}_s^{com}(k+1) - \mathbf{i}_s^{pred}(k) + \mathbf{D}_{Is}\right) + \hat{\mathbf{e}}^{pred}(k) \tag{9.171}$$

where $\mathbf{D}_{Is}$ is the current controller correction feedback:

$$\mathbf{D}_{Is} = W_1\mathbf{C}_{EMF}\Delta\mathbf{i}_s(k) + W_2\mathbf{C}_{2EMF}\Delta\mathbf{i}_s(k-1) \tag{9.172}$$

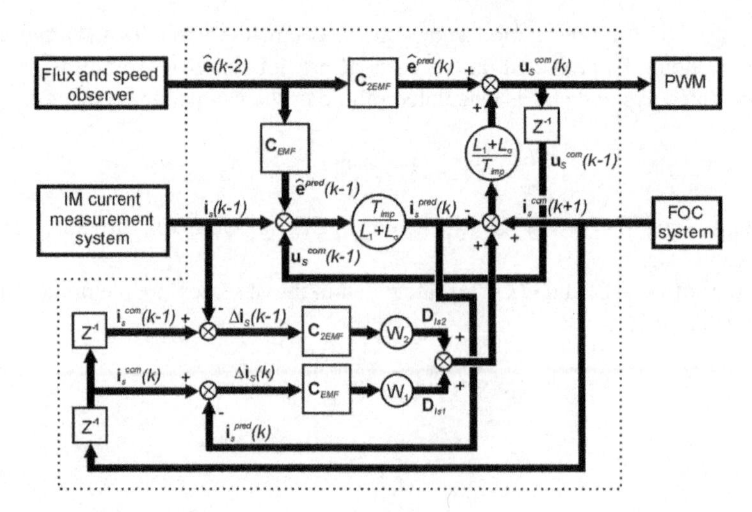

Figure 9.74 Predictive current controller structure for ASD with motor choke

and

$$\hat{\mathbf{e}}^{pred}(k) = \mathbf{C}_{2EMF}\hat{\mathbf{e}}(k-2) \tag{9.173}$$

$$\mathbf{C}_{2EMF} = \begin{bmatrix} \cos(2\Delta\varphi_e) & \sin(2\Delta\varphi_e) \\ -\sin(2\Delta\varphi_e) & \cos(2\Delta\varphi_e) \end{bmatrix} \tag{9.174}$$

and W_1 and W_2 are tuned parameters.

The PCC structure is presented in Figure 9.74.

An important parameter of the system is the inductance that appears in current controller relationships. Therefore, it is important to add the motor choke inductance L_1 to the current controller equations.

9.9.3 EMF Estimation Technique

In the case of the ASD with motor choke, the disturbance observer is modified in a similar way to the solution presented in [42].

In this section, the disturbance observer [31] is presented, taking into account the presence of the motor choke. In the observer, the motor emf is treated as a disturbance with components in the $\alpha\beta$ coordinates calculated using the exact disturbance model [43]:

$$\frac{d\boldsymbol{\xi}}{d\tau} = \frac{R_r}{L_r}\boldsymbol{\xi} + R_r\frac{L_m}{L_r}\omega_r\mathbf{i}_s + j\hat{\omega}_r\boldsymbol{\xi} \tag{9.175}$$

where

$$\xi_\alpha = \psi_{r\alpha}\omega_r \tag{9.176}$$

$$\xi_\beta = \psi_{r\beta}\omega_r \tag{9.177}$$

$$\boldsymbol{\xi} = \left[\xi_\alpha \ \xi_\beta\right]^T \tag{9.178}$$

In the case of the ASD with motor choke, the disturbance observer is modified in a similar way to the solution presented in [42]. In the case of the ASD with motor choke, the L_1 inductance is added to the motor model (Figure 9.75).The equations of the observer for the ASD with motor choke are

$$\frac{d\hat{\mathbf{i}}_s}{d\tau} = -\frac{R_s L_r^2 + R_r L_m^2}{L_r w_{\sigma 1}}\hat{\mathbf{i}}_s + \frac{R_r L_m}{L_r w_{\sigma 1}}\hat{\boldsymbol{\psi}}_r - j\frac{L_m}{w_{\sigma 1}}\hat{\boldsymbol{\xi}} + \frac{L_r}{w_{\sigma 1}}\mathbf{u}_1^{com} + k_1\left(\mathbf{i}_s - \hat{\mathbf{i}}_s\right) \tag{9.179}$$

$$\frac{d\hat{\boldsymbol{\psi}}_r}{d\tau} = \frac{R_r}{L_r}\hat{\boldsymbol{\psi}}_r + R_r\frac{L_m}{L_r}\hat{\mathbf{i}}_s + j\hat{\boldsymbol{\xi}} + \mathbf{e}_\psi \tag{9.180}$$

$$\frac{d\hat{\boldsymbol{\xi}}}{d\tau} = \frac{R_r}{L_r}\hat{\boldsymbol{\xi}} + R_r\frac{L_m}{L_r}\omega_r\hat{\mathbf{i}}_s + j\hat{\omega}_r\hat{\boldsymbol{\xi}} + jk_4\left(\mathbf{i}_s - \hat{\mathbf{i}}_s\right) \tag{9.181}$$

$$\frac{dS_{bF}}{d\tau} = k_{fo}(S_b - S_{bF}) \tag{9.182}$$

where $\mathbf{e}_\psi = \left[-k_2 S_b\hat{\psi}_{r\alpha} + k_3\hat{\psi}_{r\beta}(S_b - S_{bF}) - k_2 S_b\hat{\psi}_{r\beta} - k_3\hat{\psi}_{r\alpha}(S_b - S_{bF})\right]^T$; k_1, k_2, and k_4 are observer gains; S_b is the observer stabilizing component; S_{bF} is the S_b filtered value; and

$$w_{\sigma 1} = L_r(L_s + L_1) - L_m^2 \tag{9.183}$$

The rotor mechanical speed is estimated as

$$\hat{\omega}_r = \frac{\hat{\xi}_\alpha\hat{\psi}_{r\alpha} + \hat{\xi}_\beta\hat{\psi}_{r\beta}}{\boldsymbol{\Psi}_r^2} \tag{9.184}$$

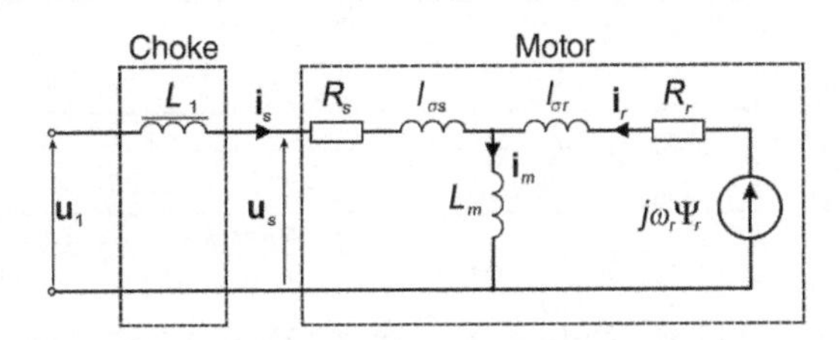

Figure 9.75 Induction motor drive with voltage inverter and motor choke

In equations (9.179)–(9.182), due to the assumption of the small step of the observer calculation, the derivative of estimated speed is neglected. For the PCC, the value of the motor emf **e** is equal to ξ, calculated in equation (9.181).

Figure 9.76 presents the operation of the drive only with PCC, without the FOC loop. The controller operates correctly with a current regulation error of less than 5%. At the 100 ms instant, the inductance of the motor choke L_1 is set to zero. It is noticeable that if the choke parameters are not taken into account, the drive is not operating correctly.

Figure 9.77 presents the operation of the full control system without nonlinear feedback. The structure of the system is based on the scheme presented in Figure 9.71. In steady state, the system correctly controls the commanded speed and flux, but in transients, the interaction between both the regulation systems appears. In the case of the speed reverse, a huge flux error appears. In the real system, the drive operating without the decoupled control will not work in a

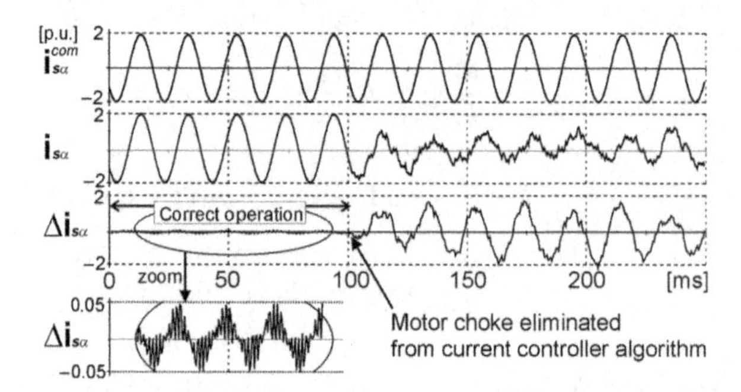

Figure 9.76 Current controller operation – at instant 100 ms the motor choke L_1 is eliminated from current controller equations

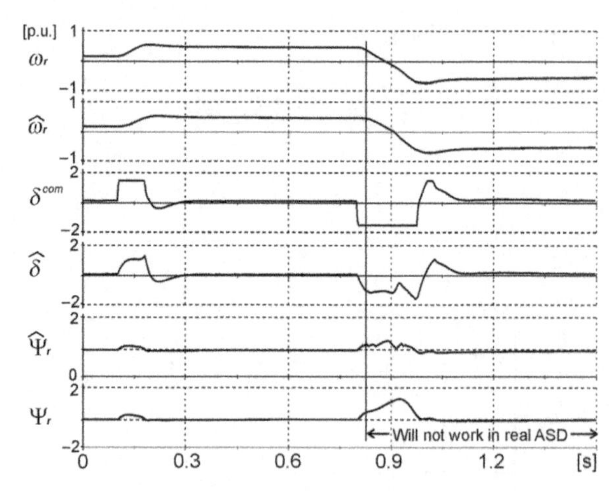

Figure 9.77 The speed sensorless adjustable speed drive (ASD) in case of the speed variation in the control system without linearization feedback – the control structure is as presented in Figure 9.71

stable manner. The stable work of this system is possible only when the dynamics of the system are limited. In the real system, the flux will be limited due to magnetic circuit saturation. This is not observed in the simulation, because a linear model of the motor is assumed.

In Figures 9.78–9.81, the operation in the speed sensorless nonlinear FOC system with load angle controller is presented.

In Figure 9.78, the speed changes are commanded. In comparison with Figure 9.77, better properties are observed. The decrease of the motor speed at 0.1 seconds has no influence on the motor flux. The estimated load angle is the same as the commanded value. Only for speed

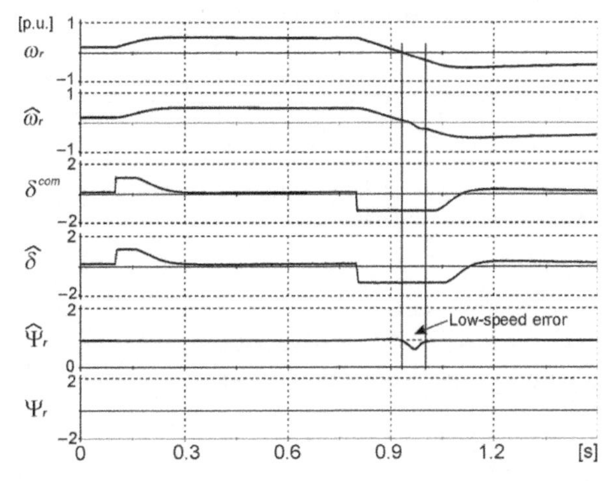

Figure 9.78 The speed sensorless induction motor control in case of the speed variation in the control system **with linearization feedback** for speed variations – the control structure is as presented in Figure 9.72

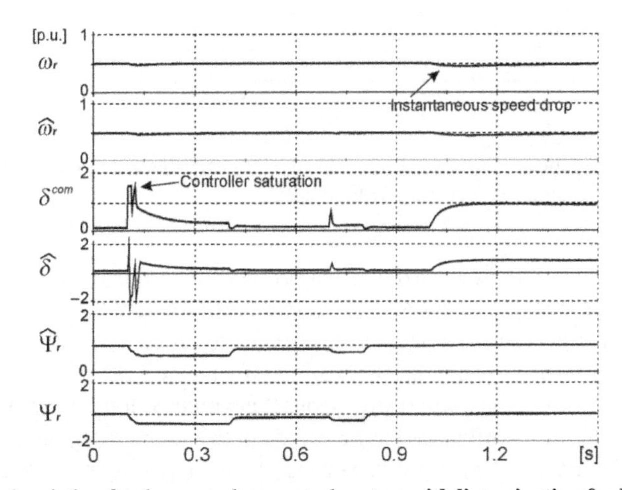

Figure 9.79 Speed variation for the sensorless control system **with linearization feedback** with flux and load torque variations – control structure as in Figure 9.72

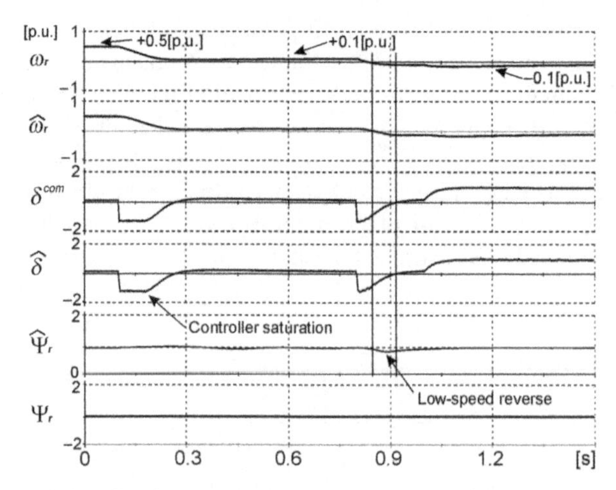

Figure 9.80 Speed variation for the sensorless control system **with linearization feedback** and low speed variations – control structure from Figure 9.72

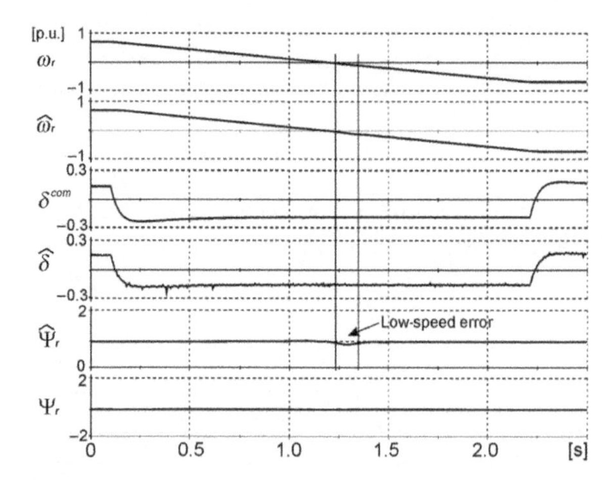

Figure 9.81 Speed variation in the sensorless control system **with linearization feedback** for slow speed reverse – the control structure is presented in Figure 9.72

reverse is the calculated flux decreased, whereas the real flux is kept constant. It is the result of passing the observer through an unstable point that typically appears in the low-speed regenerative mode. Such a phenomenon is typical for induction motor observers and is reported in [44]. Fortunately, the phenomenon has negligible influence on how the system works.

In Figure 9.79, the operation of the ASD under flux and load torque variations is presented. The flux variations have no influence on the speed control loop, except at the moments where controller saturation occurs. Under the step change of the load torque at one seconds instant, a small speed drop appears until it is compensated for by the speed controller at 1.3 seconds instant.

In Figure 9.80, the ASD operation under speed variations, including low speed ranges, is presented. The system correctly keeps the commanded speed and flux. Interactions between torque and flux regulations result from the controller's saturation, when passing the regenerative mode during speed reverse.

A slow reverse of the proposed ASD with feedback linearization is presented in Figure 9.81. In comparison with the previous fast speed reverse, the estimated flux, which appears during passing through the regenerative mode, is smaller.

9.10 Problems

Problem 9.1:

Design the differential filter (sinusoidal output filter) with L, C, and R damping resistance for the drive with a voltage inverter and induction motor. The filter topology is presented in Figure 9.29. The three-phase induction motor data are $P_n = 3$ kW, $U_n = 380$ V (Y connected), and $I_n = 6.4$ A. The inverter DC supply voltage is 540 V, transistor switching frequency is $f_{imp} = 3.3$ kHz, and inverter output maximal frequency is $f_{out\ 1\mathrm{h}} = 100$ Hz. The desired system property is switching ripple current $\Delta I_s = 20\%$.

Solution

First the inductor is selected. The motor ripple current is

$$\Delta I_s = 0.1 \cdot I_n = 0.2 \cdot 6.4 = 1.28\mathrm{A} \tag{9.185}$$

The L_1 inductance is

$$L_1 > 2\frac{U_d}{\sqrt{2} \cdot 2 \cdot \pi \cdot f_{imp} \cdot 3 \cdot \Delta I_s} = 2\frac{540}{\sqrt{2} \cdot 2 \cdot \pi \cdot 3300 \cdot 3 \cdot 1.28} = 9.5\ \mathrm{mH} \tag{9.186}$$

The voltage drop on L_1 for the motor nominal frequency f_n and nominal load is

$$\Delta U_1 = I_n 2\pi f_n L_1 = 6.4 \cdot 2\pi \cdot 50 \cdot 0.0095 = 19\mathrm{V} \tag{9.187}$$

The assumed $L_1 = 9.5$ mH satisfies the limiting current ripple and permitted voltage drop.The L_1 nominal current is equal to the motor nominal $I_{L1n} = I_n = 6.4$A.

Next the C_1 is selected. The C_1 is related to the assumed resonant frequency f_{res}. The value of the $f_{out1\mathrm{h}}$ should satisfy:

$$\begin{aligned} &10 \cdot f_{out1\ \mathrm{h}} < f_{res} < \frac{1}{2} \cdot f_{imp} \\ &10 \cdot 100 < f_{res} < \frac{1}{2} \cdot 3300 \\ &1000 < f_{res} < 1650 \end{aligned} \tag{9.188}$$

so the resonant frequency is selected:

$$f_{res} = 1333 \tag{9.189}$$

hence:

$$C_1 = \frac{1}{4\pi^2 f_{res}^2 L_1} = \frac{1}{4\cdot\pi^2\cdot 1333^2\cdot 0.0095} = 1.5\ \mu\text{F} \tag{9.190}$$

For the capacitor, the nominal voltage should be at least as much as the inverter supply voltage:

$$C_1 > U_d = 540\ \text{V} \tag{9.191}$$

Because the C_1 current consists of a high frequency ripple $f_{\text{imp}} = 3.3$ kHz, the proper capacitor type should be selected.

The values of L_1 and C_1 calculated in the above way satisfy that the filter output voltage THD factor is not greater than 5%.

For the selected L_1 and C_1, the filter characteristic impedance is

$$Z_0 = \sqrt{\frac{L_1}{C_1}} = \sqrt{\frac{9.5e-3}{1.5e-6}} = 80\ \Omega \tag{9.192}$$

if damping resistance $R_1 = 10\ \Omega$, so the filter quality factor is

$$Q = \frac{Z_0}{R_1} = \frac{80}{10} = 8 \tag{9.193}$$

The calculated quality factor is in the acceptable range of $Q = 5$–8, which gives a good filter quality and satisfies the dumping properties.

The nominal voltage of R_1 is the same as for C_1.

The total power of R_1 is the sum of two parts, related to $f_{\text{out 1har}}$ and f_{imp}. For both frequencies, the impedances of C_1 and R_1 connected in series are

$$Z_{R1\text{1har}} \approx X_{C1\text{1har}} = \frac{1}{2\pi f_{out\text{1har}} C_1} = \frac{1}{2\pi\cdot 100\cdot 1.5e-6} = 1062\ \Omega \tag{9.194}$$

$$Z_{R1imp} \approx X_{C1imp} = \frac{1}{2\pi f_{imp} C_1} = \frac{1}{2\pi\cdot 3300\cdot 1.5e-6} = 32\ \Omega \tag{9.195}$$

It is noticeable that the most important R_1 power part is related to the ripple current. So, finally the power of R_1 should be

$$P_{R1} > R_1\left[\left(\frac{U_n}{\sqrt{3}Z_{R1\text{ 1har}}}\right)^2 + \left(\frac{U_{THD}}{Z_{R1\text{ imp}}}\right)^2\right] = 10\left[\left(\frac{380}{\sqrt{3}\cdot 1062}\right)^2 + \left(\frac{0.05\cdot 540}{32}\right)^2\right] = 7.5\ \text{W} \tag{9.196}$$

and $P_{R1} = 10$ W was selected.

Problem 9.2:

Design the differential filter (normal mode filter) with L, C, and R dampings for the drive with voltage inverter and induction motor. The filter topology is presented in Figure 9.29. The three-phase induction motor data are $P_n = 5.5$ kW, $U_n = 400$ V (Y connected), and $I_n = 11$ A. The inverter DC supply voltage is 540 V, transistors switching frequency is $f_{imp} = 5$ kHz, and inverter output maximal frequency is $f_{out\ 1h} = 80$ Hz. The desired system properties are switching ripple current $\Delta I_s = 20\%$. When the filter is calculated, prepare the simulation in Simulink (Hint: use the *sys_LC_filter.mdl* example).

Solution

Because the design process is the same as in Problem 9.1, so some comments are omitted and mainly the calculations and results are presented.

The motor ripple current:

$$\Delta I_s = 0.2 \cdot I_n = 0.2 \cdot 11 = 2.2\ \text{A} \tag{9.197}$$

The L_1 inductance is

$$L_1 > 2 \frac{U_d}{\sqrt{2} \cdot 2 \cdot \pi \cdot f_{imp} \cdot 3 \cdot \Delta I_s} = 2 \frac{540}{\sqrt{2} \cdot 2 \cdot \pi \cdot 5000 \cdot 3 \cdot 2.2} = 1.8\ \text{mH} \tag{9.198}$$

The voltage drop on L_1 is

$$\Delta U_1 = I_n 2\pi f_n L_1 = 11 \cdot 2\pi \cdot 50 \cdot 0.0018 = 6.2\ \text{V} \tag{9.199}$$

The resonant frequency f_{res} should fulfill condition:

$$800 < f_{res} < 2500 \tag{9.200}$$

so f_{res} is selected:

$$f_{res} = 1650 \tag{9.201}$$

Capacitance C_1 is

$$C_1 = \frac{1}{4\pi^2 f_{res}^2 L_1} = \frac{1}{4 \cdot \pi^2 \cdot 1650^2 \cdot 0.0018} = 5.2\ \mu\text{F} \tag{9.202}$$

The filter characteristic impedance is

$$Z_0 = \sqrt{\frac{L_1}{C_1}} = \sqrt{\frac{1.8e-3}{5.2e-6}} = 18.6\ \Omega \tag{9.203}$$

For the quality factor $Q = 5$, the R_1 is

$$R_1 = \frac{Z_0}{Q} = \frac{18.6}{5} = 3.7\,\Omega \tag{9.204}$$

The impedances $Z_{R1\text{1har}}$ and Z_{R1imp} are

$$Z_{R1\text{1har}} \approx X_{C1\text{1har}} = \frac{1}{2\pi f_{out\text{1har}} C_1} = \frac{1}{2\pi \cdot 80 \cdot 5.2e-6} = 382\,\Omega \tag{9.205}$$

$$Z_{R1imp} \approx X_{C1imp} = \frac{1}{2\pi f_{imp} C_1} = \frac{1}{2\pi \cdot 5000 \cdot 5.2e-6} = 6\,\Omega \tag{9.206}$$

The power of R_1 should be no less than

$$P_{R1} > R_1\left[\left(\frac{U_n}{\sqrt{3}Z_{R1\ \text{1har}}}\right)^2 + \left(\frac{U_{THD}}{Z_{R1\ imp}}\right)^2\right] = 3.7\left[\left(\frac{400}{\sqrt{3} \cdot 382}\right)^2 + \left(\frac{0.05 \cdot 540}{6}\right)^2\right] = 76\ \text{W} \tag{9.207}$$

Power $P_{R1} = 100$ W is selected.

Questions

9.1: Why does the CM voltage appears in PWM voltage inverters?

Answer: The common mode voltage is inherent for PWM voltage inverters. It is the result of transistors switching sequence. It should be explained by the analysis of zero voltage components for each voltage vector, as presented in Table 9.2.

9.2: Explain the bearing current phenomena.

Answer: The bearing current phenomena are explained in Section 9.4.

9.3: Why is the differential mode filter used?

Answer: The main purpose of the differential mode filter use is to obtain the sinusoidal motor supply voltage. As the result of the supply voltage shape improvement, the motor has greater efficiency, is less noisy, and the disturbance level is decreased.

9.4: Which type of filter is correct for bearing current limitation?

Answer: The common mode filter limits the common mode current and bearing currents.

9.5: Specify the method for motor protection in the drive with a voltage inverter.

Answer: This is explained in Section 9.5 and Table 9.3.

9.6: Describe the design process of the differential mode filter.

Answer: The design process is presented in Section 9.6.1.

9.7: Describe the design process of the common mode filter.

Answer: The design process is presented in Section 9.6.1.

9.8: Sketch the structure of the CM choke. Where should the CM choke be installed?

Answer: The structure of the CM choke is presented in Figure 9.13, with a description given is Section 9.5.1. The CM choke is installed on the output of the voltage inverter?

9.9: If the long cable connection is used to connect the inverter and motor, where should the LC filter be installed? Explain why.

Answer: The LC filter should always be installed close to the motor. It is because the sinusoidal shaped voltage is in the whole cable, which connects the inverter/filter with the motor. When the voltage in cable is sinusoidal in shape, the risk of inducted noise is strongly limited in comparison with the rectangular high *dv/dt* shape.

References

1. Bose, B. K. (2009) Power electronics and motor drives recent progress and perspective. *IEEE Trans. Ind. Elec.*, 56(2), 250–259.
2. Erdman, J., Kerkman, R., Schlegel, D., and Skibinski, G. (1995) Effect of PWM inverters on AC motor bearing currents and shaft voltages. *IEEE APEC Conf.*, Dallas, TX.
3. Muetze, A. and Binder. A. (2003) High frequency stator ground currents of inverter-fed squirrel-cage induction motors up to 500 kW. *EPE Conf.*, Toulouse.
4. Akagi, H. (2002) Prospects and expectations of power electronics in the 21st century. *Power Conv. Conf. PCC'2002.*, Osaka, 2–5 April 2002.
5. Akagi, H., Hasegawa H., and Doumoto, T. (2004) Design and performance of a passive EMI filter for use with a voltage-source PWM inverter having sinusoidal output voltage and zero common-mode voltage. *IEEE Trans. Power Elect.*, 19(4), 1069–1076.
6. Palma, L. and Enjeti, P. (2002) An inverter output filter to mitigate dv/dt effects in PWM drive systems. *Seventeenth Annu. IEEE Appl. Power Electron. Conf. Expo., APEC'02*, 10–14 March, 2002, Dallas, USA.
7. Guzinski, J., Abu-Rub, H., and Strankowski, P. (2015) *Variable Speed AC Drives with Inverter Output Filters*. Wiley, Hoboken, NJ.
8. Binder, A. and Muetze, A. (2008) Scaling effects of inverter-induced bearing currents in AC machines. *IEEE Trans. Ind. Appl.*, 44(3), 769–776.
9. Muetze, A. (2004) Bearing currents in inverter-fed AC motors. PhD dissertation, Darmstadt University of Technology, Aachen, Germany.
10. Muetze, A. and Binder, A. (2007) Practical rules for assessment of inverter-induced bearing currents in inverter-fed AC motors up to 500 kW. *IEEE Trans. Ind. Elect.*, 54(3), 1614–1622.
11. Sun, Y., Esmaeli, A., and Sun, L. (2006) A new method to mitigate the adverse effects of PWM inverter. *1st IEEE Conf. Ind. Electron. Appli. ICIEA'06*, 24–26 May, Singapore.
12. Xu, W., Kaizheng, H., Bin, X. (2007) A new inverter output passive filter topology for PWM motor drives. *The Eighth Int. Conf. Electron. Meas. Instrum. ICEMI'2007*, 16–18 August, 2007, Xian, China.
13. Xiyou, C., Bin, Y., and Tu, G. (2002) The engineering design and optimization of inverter output RLC filter in AC motor drive system. *IECON*, USA, 5–8 November 2002.
14. Ogasawara, S. and Akagi, H. (1996) Modeling and damping of high–frequency leakage currents in PWM inverter-fed AC motor drive systems. *IEEE Trans. Ind. Appl.*, 22(5), 1105–1114.
15. Zitselsberger, J. and Hofmann, W. (2003) Reduction of bearing currents by using asymmetric, space-vector-based switching patterns. *EPE'03*, Toulouse.

16. Cacciato, M., Consoli, A., Scarcella, G., and Testa, A. (1999) Reduction of common-mode currents in PWM inverter motor drives. *IEEE Trans. Ind. Appl.*, 35(2), 469–476.
17. Hofmann, W. and Zitzelsberger, J. (2006) PWM-control methods for common mode voltage minimization: A survey. *Int. Symp. Power Elect., Elect. Dr., Auto. Mot., SPEEDAM 2006*, 23–26 May, Taormina, Italy.
18. Lai, Y. S. and Shyu, F.-S. (2004) Optimal common–mode voltage reduction PWM technique for inverter control with consideration of the dead-time effects, Part I: Basic development. *IEEE Trans. Ind. Appl.*, 40 (6), 1605–1612.
19. Ruderman, A. and Welch, R. (2005) Electrical machine PWM loss evaluation basics. *Int. Conf. Ener. Effic. Mot. Dr. Syst. (EEMODS)*, vol. **1**, Heidelberg, Germany.
20. Ruifangi, L., Mi, C. C., and Gao, D. W. (2008) Modeling of iron losses of electrical machines and transformers fed by PWM inverters. *IEEE Trans. Magn.*, 44(8).
21. Yamazaki, K. and Fukushima, N. (2009) Experimental validation of iron loss model for rotating machines based on direct eddy current analysis of electrical steel sheets. *Elect. Mach. Dr. Conf., IEMDC'09*, 3–6 May, Miami, FL.
22. Krzeminski, Z. and Guziński, J. (2005) Output filter for voltage source inverter supplying induction motor. *Int. Conf. Power Elect., Intell. Mot. Power Qual., PCIM'05*. 7–9 June, Nuremberg.
23. Czapp, S. (2008) The effect of Earth fault current harmonics on tripping of residual current devices. *Int. School Non-sinusoidal Curr. Compens.*, Łagów, Poland.
24. Franzo, G., Mazzucchelli, M., Puglisi, L., and Sciutto G. (1985) Analysis of PWM techniques using uniform sampling in variable-speed electrical drives with large speed range. *IEEE Trans. Ind. Appl.*, IA-21 (4), 966–974.
25. Rajashekara, K., Kawamura, A., and Matsuse, K. (1996) Sensorless control of AC motor drives. *IEEE Ind. Elect. Soc.*, IEEE Press, USA.
26. Kawabata, T., Miyashita T., and Yamamoto, Y. (1991) Digital control of three phase PWM inverter with LC filter. *IEEE Trans. Power Elect.*, 6(1), 62–72.
27. Seliga, R. and Koczara, W. (2001) Multiloop feedback control strategy in sine-wave voltage inverter for an adjustable speed cage induction motor drive system. *Eur. Conf. Power Elect. Appl., EPE'2001.*, 27–29 August, Graz, Austria.
28. Adamowicz, M. and Guziński, J. (2005) Control of sensorless electric drive with inverter output filter. *4th Int. Symp. Auto. Cont. AUTSYM 2005*, 22–23 September, Wismar, Germany.
29. Guzinski, J. and Abu-Rub, H. (2008) Speed sensorless control of induction motors with inverter output filter. *Int. Rev. Elect. Eng.*, 3(2), 337–343.
30. Salomaki, J. and Luomi, J. (2006) Vector control of an induction motor fed by a PWM inverter with output LC filter. *EPE J.*, 16(1), 37–43.
31. Krzemiński, Z. (2000) Sensorless control of the induction motor based on new observer. *PCIM 2000*, Nürnberg, Germany.
32. Guziński, J. (2008) Closed loop control of AC drive with LC filter. *13th Int. Power Elect. Mot. Conf. EPE–PEMC 2008*, 1–3 September, Poznań, Poland.
33. Holtz, J. (1995) The representation of AC machine dynamics by complex signal flow graph. *IEEE Trans. Ind. Elect.*, 42(3), 263–271.
34. Abu-Rub, H. and Oikonomou, N. (2008) Sensorless observer system for induction motor control. *39th IEEE Power Elect. Spec. Conf., PESCO'08*, 15–19 June, Rodos, Greece.
35. Abu-Rub, H., Guziński, J., Rodriguez, J., Kennel, R., and Cortés, P. (2010) Predictive current controller for sensorless induction motor drive. *IEEE-ICIT 2010*, Vina del Mar, Valparaiso, Chile.
36. Vas, P. (1990) *Vector Control of AC Machines*. Oxford University Press, Oxford.

37. Krzemiński, Z. (1987) Non-linear control of induction motor. *Proc. 10th IFAC World Cong. Auto. Cont.*, vol. 3, Munich.
38. Bogalecka, E. (1992) Control system of an induction machine. *EDPE1992*, High Tatras, Stara Lesna, Slovakia.
39. Isidori, A. (1995) *Non-Linear Control Systems*, 3rd edn. Springer-Verlag, London.
40. Kaźmierkowski, M. P., and Malesani, L. (1998) Current control techniques for three-phase voltage-source PWM converters: A survey. *IEEE Trans. Ind. Elect.*, 45(5), 691–703.
41. Tsuji, M., Ohta, T., Izumi, K., and Yamada, E. A. (1997) *Speed Sensorless Vector-Controlled Method for Induction Motors Using q-Axis Flux*. IPEMC, Hangzhou, China.
42. Guzinski, J. (2009) Sensorless AC drive control with LC filter. *13th Eur. Conf. Power Elect. App. EPE 2009*, 8–10 September, Barcelona.
43. Krzemiński, Z. (2008) Observer of induction motor speed based on exact disturbance model. *Proc. Int. Conf. EPE-PEMC 2008*, Poznan, Poland; 1–3 September 2008.
44. Kubota, H., Sato, I., Tamura, Y., Matsuse, K., Ohta, H., and Hori, Y. (2002) Regenerating-mode low-speed operation of sensorless induction motor drive with adaptive observer. *IEEE Trans. Ind. App.*, 38(4), 1081–1086.

10

Medium Voltage Drives – Challenges and Trends

Haitham Abu-Rub[1], Sertac Bayhan[2], Shaikh Moinoddin[3], Mariusz Malinowski[4], and Jaroslaw Guzinski[5]

[1]*Department of Electrical and Computer Engineering, Texas A&M University at Qatar, Doha, Qatar*
[2]*Qatar Environment and Energy Research Institute, Hamad Bin Khalifa University, Doha, Qatar*
[3]*Aligarh Muslim University, Aligarh, India*
[4]*Warsaw University of Technology, Warsaw, Poland*
[5]*Gdansk University of Technology, Gdansk, Poland*

10.1 Introduction

The ever-growing demand for electrical energy and continuous rise in energy prices compel to think that the energy must be used more efficiently. Modern power electronic technology with high efficiency and appropriate control approaches are needed in energy-intensive industries to decrease the immense waste of energy and to improve power quality. The electric motor drive systems being the major energy consumers still have the highest potential for improvement regarding efficient energy consumption. High-power motors, mostly operating in medium voltage (MV), are of most interest due to their enormous energy consumption [1].

MV drives have found extensive applications in several industries, such as in oil and gas, petrochemical, mining, water/waste, pulp/paper, cement, chemical, power generation, metal production and processes, traction, and marine drive sectors. To improve power quality, system response and to reduce operation cost and energy loss, the installed MV drives should be

This chapter was partially published in IEEE Power Electronics Magazine and the copyright of this paper* was purchased to publish it here as a book chapter. *H. Abu-Rub, S. Bayhan, S. Moinoddin, M. Malinowski and J. Guzinski, "Medium-Voltage Drives: Challenges and existing technology," in IEEE Power Electronics Magazine, vol. 3, no. 2, pp. 29–41, June 2016.

High Performance Control of AC Drives With MATLAB®/Simulink, Second Edition.
Haitham Abu-Rub, Atif Iqbal, and Jaroslaw Guzinski.

Companion website: www.wiley.com/go/aburubcontrol2e

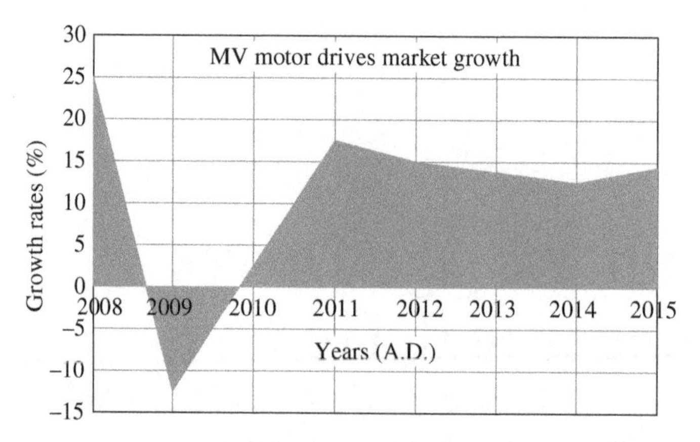

Figure 10.1 Global medium voltage drive market growth

adjustable-speed drives (ASDs). Figure 10.1 shows the global market growth of MV drives from 2008 to 2015 [2]. Due to recession during 2009 and 2010, there was a little or negative growth in the global market. Due to fracking of unconventional shale gas and oil in Americas, the sales of MV drives increased during 2012 and 2013, whereas in Europe growth was recovering [3].

Table 10.1 shows the benefits of ASDs for pumping applications. Control techniques with transformerless solutions lead to a further decrease in the payback time [4]. Nevertheless, the deployment of such drives is associated with several requirements and challenges. Significant challenges are with the power line side (e.g. power quality, resonance, and power factor), motor side (e.g. *dv/dt*, common-mode (CM) voltage, motor derating caused by generated harmonics, resonance, torsional vibration, and traveling wave reflections), and semiconductor devices (switching losses, reliability) [1, 4].

Unfortunately, various challenges related to converters, line side as well as motor side, are faced when installing MV drives. Hence, the motivation behind this chapter is to point out the

Table 10.1 Benefits of ASDs for pumping applications

Speed level	Benefits
Soft start of motor	• No network voltage dips • Reduced mechanical stress
High speed	• Maximum capacity • Best productivity
Low speed	• Best energy efficiency • Reduced operating costs
Precise and optimal speed	• Best efficiency point (BEP) of pumps • Increased lifetime of equipment
Soft stop of motor	• No water hammering • Reduced mechanical stress

challenges and problems faced when using MV drives and indicating the direction for proper use and understanding of such drives. The chapter presents the existing MV drives technologies and emphasizes the need for further development and enhancement of MV drives.

10.2 Medium Voltage Drive Topologies

MV drives are classified to cover a power range from 0.2 MW to almost 40 MW at the MV level of 2.3 kV up to 13.8 kV [4–6]. However, most of the installed MV drives in industries are in the range of 1–4 MW with voltage ratings from 3.3 to 6.6 kV [6]. A typical block diagram of MV drive is shown in Figure 10.2. Small size, lower cost, high efficiency, and reliability, fault protection, ease of installation and maintenance, high dynamic performance, and regenerative capability in some applications are the essential requirements for the MV drives. A list of some of the industrial drives is presented in Table 10.2. In this table, the power rating, devices and topology used, and control methods are presented. Furthermore, an overview of the popular converter topologies in MV drives is summarized in Table 10.3 [28, 29].

The main disadvantage of the multilevel inverters (MIs) is the complexity of power circuit and controls. However, the use of MIs in MV drives offers improved power quality, lower switching losses, high voltage capability, and lower *dv/dt* [36].

There are different types of power semiconductor switches that could be adopted for MV drives. Some of them are the injection-enhanced gate transistors (IEGTs), the integrated gate-commutated thyristors (IGCTs), and insulated-gate bipolar transistor (IGBT). The gate drive circuits of IEGTs are more reliable than IGCT drive circuits. Furthermore, IGBT gate drive circuit is simpler and has fewer components than IEGT gate drives circuit. Hence, IGBT gate drive circuits are more reliable than IEGT [37].

The modern power semiconductor switches have peak voltage blocking capability of nearly 6.5 kV, which restricts the maximum voltage ratings of the inverter and the motor in MV high-power drives. The apparent power that can be obtained is limited with available MV IGBT switches that have current conducting capability of 750 A (peak). Series or parallel combinations of semiconductor switches are used to overcome the limits of switch ratings, but with these arrangements, a balance of the current/voltage between devices is achieved using extra measurements [38].

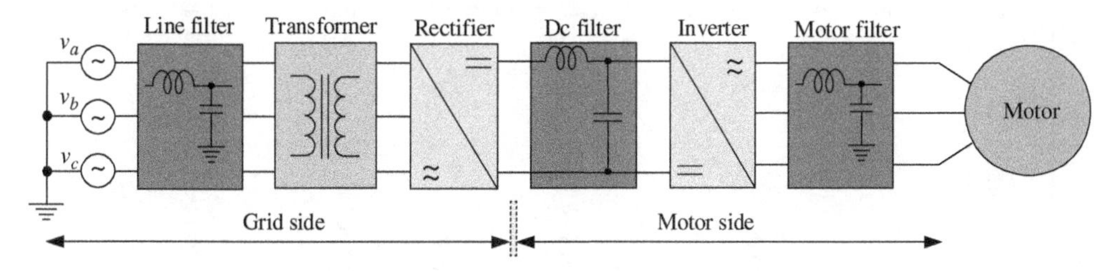

Figure 10.2 Typical MV adjustable-speed drive

Table 10.2 Market overview of industrial MV drives

Manufacturer	Power	Voltage (kV)	Topology	Semiconductor	Control method
ABB[a]	0.315–5.0 MW	2.3–4.16	3L-neutral-point-clamped-voltage	IGBT	
	0.25–72 MW	4.0–6.90	source converter (NPC-VSC)	IGCT	
	2.0–36 MW	6.0–13.8	5L-NPC-VSC		Direct torque control
	3.0–36 MW	Up to	5L-H-bridge neutral-point-clamped		(DTC)
	2.0–72 MW	3.3	(HNPC)-VSC		
		2.1–10			
SIEMENS[b]	4–70 MVA	3.3–13.8	Modular multilevel converter (MMC)-	IGBT	CLVC
	3.4–31.5 MVA	3.3–4.16	VSC	IGBT, IGCT	CLVC
	0.14–48.8 MVA	2.3–11	3L-NPC-VSC	IGBT	CLVC
	1–85 MVA	1.4–10.3	HB-VSC	Thyristor	CLVC
	6–13.7 MVA	3.3–7.2	LCI-CSI	LV/MV IGBT	CLVC
			M2L		
SIEMENS[b]	0.15–14.2 MW	2.3–4.16	MMC-VSC	IGCT	
	0.82–18 MW	3.3–7.2	3L-NPC-VSC	IGBT	Voltage/frequency (V/f)
	2.8–31.5 MW	2.3–12	5L-NPC-VSC	LV/MV IGBT	Field-oriented control
	0.8–85 MW	4–7.2			(FOC)
TMEIC[c]	4.0–120 MVA	1.25–7.2	3L-HNPC	IGBT, gate-commutated	
			5L-HNPC	thyristor (GCT)	V/f
			5L-PWM	IGCT, IEGT	
AMTECH[d]	0.25–12.5 MW	3.3–11	7L-cascaded H-bridge (CHB)-VSC	IGBT	
			(3.3 kV)		V/f
			9L-CHB-VSC (4.16 kV)		FOC
			13L-CHB-VSC (6.6 kV)		
			19L-CHB-VSC (11 kV)		
ALSTOM[e]	1.4–7.2 MVA	Up to	2L-VSC	HV IGBT	
	2.2–8.0 MVA	4.16	4L-FLC-VSC	Gate turn-off thyristor	
	7.0–9.5 MVA	Up to	3L-NPC-VSC	(GTO)	FOC
	8.3–13.5 MVA	4.16	Pulse width modulation (PWM)-CSI		

		Up to 3.3			
		Up to 10			
Schneider Electric[f]	0.5–10 MVA	2.3–6.6	3L-NPC-VSI	HV-IGBT	V/f
					FOC
Fuji Electric[g]	5.2–10.5 MVA	0.28–8.3	9L-NPC-VSC	HV-IGBT	V/f
			17L-NPC-VSC		FOC
DELTA GROUP[h]	0.28–9.52 MVA	3.3–11	7L-CHB-VSC (3.3 kV)	HV-IGBT	
			9L-CHB-VSC (4.16 kV)		
			13L-CHB-VSC (6.6 kV)		V/f
			19L-CHB-VSC (10 kV)		FOC
			21L-CHB_VSC (11 kV)		
EATON Corporation[i]	0.22–4.29 MVA	2.4–13.8	3L-NPC-VSC	IGBT	V/f
Toshiba International Corporation[j]	0.22–3.73 MVA	2.4–4.16	9L-NPC-VSC	IGBT	V/f
					FOC
WEG			EquipamentosElétricos[k]	Up to 3.36 MVA	2.3–4.16
3L-NPC-VSC	HV-IGBT	V/f			
5L-NPC-VSC					
Hitachi[l]	0.31–10 MVA	2.4–11	2~9L-CHB-VSC	HV-IGBT	FOC
INGETEAM[m]	0.8–36 MVA	2.3–6.9	3L-NPC-VSC	HV-IGBT, IGCT	V/f, FOC
Rockwell Automation[n]	1.5–25.4 MVA	2.4–6.6	3L-NPC-VSC	HV-IGBT	FOC
			5L-NPC-VSC	SGCT, SCR	
Yaskawa[o]	0.15–3.73 MVA	2.4–4.16	9L-CHB-VSI	IGBT	V/f
			17L-CHB-VSI		FOC
Benshaw[p]	0.224–7.466 MVA	1.5–4.16	M2L	1.7KV IGBT	V/f and SVC
Converteam[q]	0.4–40 MVA	1.25–6.6	3L-NPC-VSI	MV/HV IGBT,GTO,IGCT	FVC
LS Ind. Systems[r]	0.2–12.5 MVA	3–11	CHB	IGBT	V/f, SLVC

(continued)

Table 10.2 (*Continued*)

Manufacturer	Power	Voltage (kV)	Topology	Semiconductor	Control method
Rongxin Power[s]	0.2–14 MW	3–13.8	CHB	IGBT	VC
GE[t]	0.11–101 MVA	2.3–10	NPP, NNPP	IGBT	V/f, VC
Severn Drives & Energy[u]	0.2–14 MW	3–13.8	CHB		V/f, SLVC, SVC

References
[a] [7].
[b] [8].
[c] [9].
[d] [10].
[e] [11].
[f] [12].
[g] [13].
[h] [14].
[i] [15].
[j] [16].
[k] [17].
[l] [18].
[m] [19].
[n] [20].
[o] [21].
[p] [22].
[q] [23].
[r] [24].
[s] [25].
[t] [26].
[u] [27].

Table 10.3 Overview of popular converter topologies in MV drives

Topology	Description
1. Three-phase 5L-HNPC [7, 9, 28] 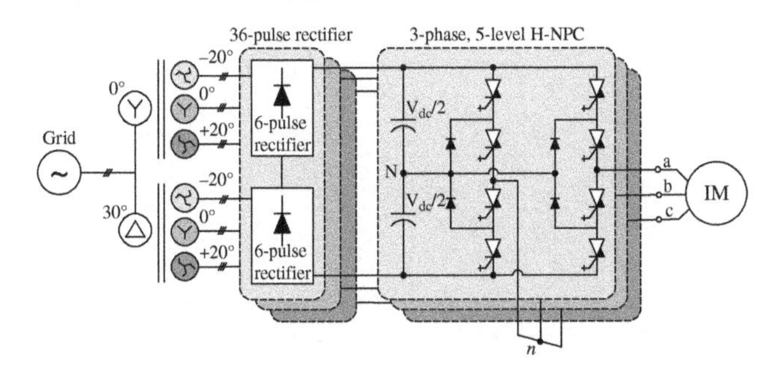	The 5L-HNPC bridge inverter is developed from the three-level NPC inverter topology. This inverter has some unique features that have promoted its application in the MV drive industry. The inverter phase voltage contains five voltage levels instead of three levels for the NPC inverter. This leads to lower *dv/dt* and total-harmonic-distortion (THD). The inverter does not have any switching devices in series, which eliminates the device dynamic and static voltage sharing problems. However, this topology requires three isolated dc supplies, which increases the complexity and cost of the dc supply system [1].
2. Three-phase 4L-FLC [11, 29] 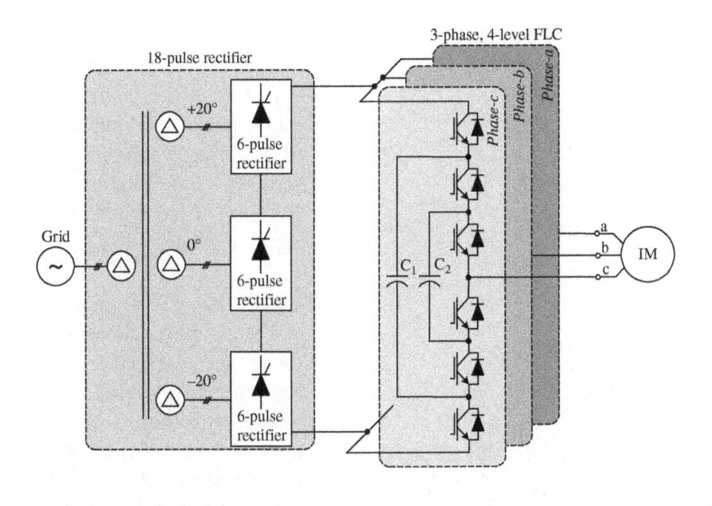	This topology has evolved from the two-level inverter by adding dc capacitors to the cascaded switches. There are three complementary switch pairs in each of the inverters. Therefore, only three independent gate signals are required for each phase. The flying-capacitor inverter can produce an inverter phase voltage with four voltage levels. In this topology, some voltage levels can be obtained by more than one switching state. The switching state redundancy is a common phenomenon in multilevel converters, which provides great flexibility for

(*continued*)

Table 10.3 (*Continued*)

Topology	Description
	the switching pattern design. However, the practical use of the flying-capacitor inverter seems limited due to the use of a large number of capacitors and complex control scheme [1].
3. Three-phase *n*L-CHB [10, 14, 18, 21] 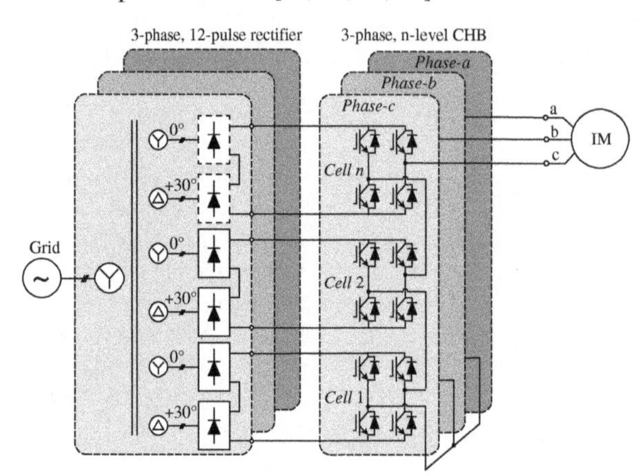	Cascaded H-bridge (CHB) multilevel inverter is one of the most popular converter topologies used in MV drives. It is composed of multiple units of single-phase H-bridge power cells. The H-bridge cells are normally connected in cascade on their ac side to achieve medium-voltage operation and low-harmonic distortion. In practice, the number of power cells in a CHB inverter is mainly determined by its operating voltage and manufacturing cost. The use of identical power cells leads to a modular structure, which is an effective means for cost reduction. However, the main disadvantage and limitation of this topology are that the need for a large number of isolated voltage sources that increase the converter cost.
4. Three-phase 5L-NPC [8, 17, 20] 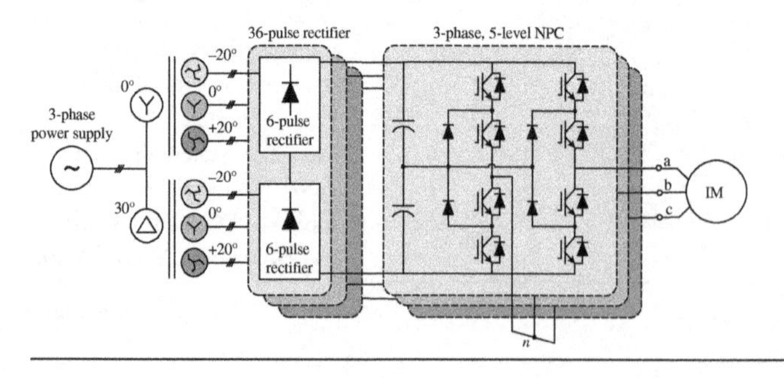	The diode-clamped multilevel inverter employs clamping diodes and cascaded dc capacitors to produce ac voltage waveforms with multiple levels. The main features of the NPC inverter include reduced *dv*/*dt* and THD in its ac output voltages. More importantly, the inverter can be used in the

Table 10.3 (*Continued*)

Topology	Description
	MV drive to reach a certain voltage level without switching devices in series. In this topology, capacitors have been used to generate an intermediate voltage level. However, the voltages on these capacitors are unequal that result in unbalancing dc-link voltage.
5. Three-phase 13L-MMC [22, 30] 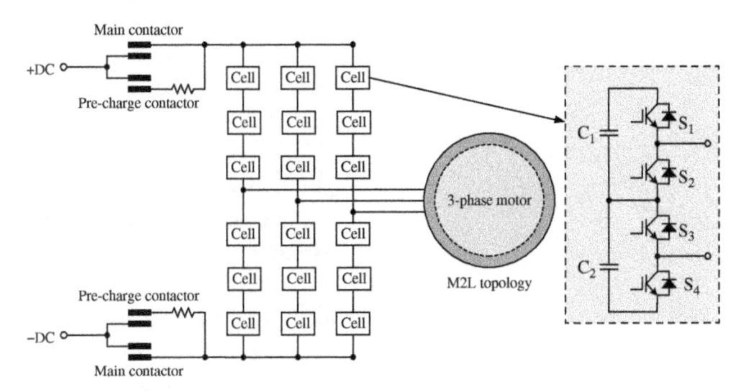	The modular multilevel converter (MMC) is one of the promising topologies in multicell converter family. It employs a cascade connection of submodules to reach the desired system voltage while producing a high-quality multilevel output voltage waveform (18 cell [4160 V], 30 cell [6900 V], 36 cell [7200 V]). The submodule is a building block of the MMC and can be configured in various forms by using IGBT devices (1700 V) and DC.
6. 3L-NPP [31] 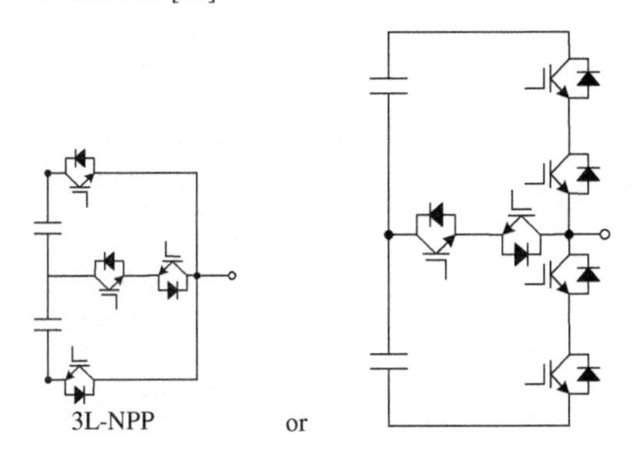	Neutral point piloted (NPP) converter (also known as T-type NPC converter) The harmonic content of voltage waveform is the lowest for *R-L* load, i.e. 30.85%. The strength of fundamental waveform is high 156.2%. The maximum voltage stress across semiconductors is $0.25Vs$. The turn-on and turn-off voltage stress is the lowest. Hence, switching frequency is very high. Used for medium- to high-speed drives at lower voltage. Switching losses are even

(*continued*)

Table 10.3 (*Continued*)

Topology	Description
	more reduced by alternate switching technique for high-frequency application. The output has even lower amount of low-order harmonics and larger amount of higher-order harmonic content. Neutral current is allowed to flow. So current is cut off at low value. This reduces EMI. Associated filter circuit cost and energy loss are the lowest. Lowest rating and hence cheaper semiconductor is used. Largest number of devices is used. Hence, installation is more costly. Gating circuitry is more intricate.
7. 5L-NNPP [31, 32] 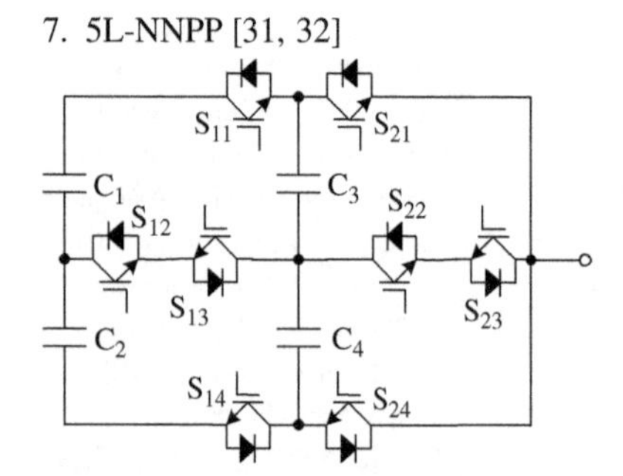	Single phase of five-level NNPP converter is nested by the two modular structures of the phase-leg of three-level NPP converter rather than stacks of two flying-capacitor converters together. If the total dc-link voltage is equal to 4E, then the dc-link capacitors C_1 and C_2 are charged to 2E, which is half of the total dc-link voltage, and the floating capacitors C_3 and C_4 are charged to E, which is one-fourth of the total dc-link voltage. Therefore, all switches have the same voltage stress, which is one-fourth of the total dc-link voltage.

Table 10.3 (*Continued*)

Topology	Description
8. 4L-NNPC [33] 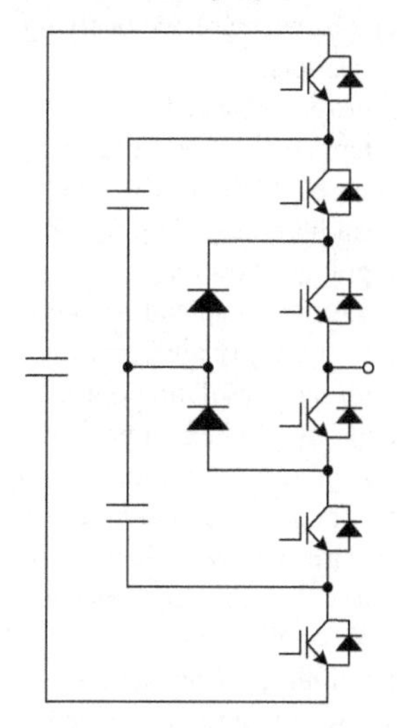	The four-level VSC for medium-voltage applications called NNPC. The proposed topology can operate over a wide range of 2.4–7.2 kV without any power semiconductor in series. The proposed converter has fewer components as compared with classic multilevel converters. Moreover, the voltage across the power semiconductors is only one-third of the dc link (and equal for all semiconductors). The space vector modulation (SVM) strategy, which benefits from the switching state redundancy, has been used to control the output voltage and stabilize voltages of the flying capacitor (FCs).
9. 7L-NANPC [31, 34] 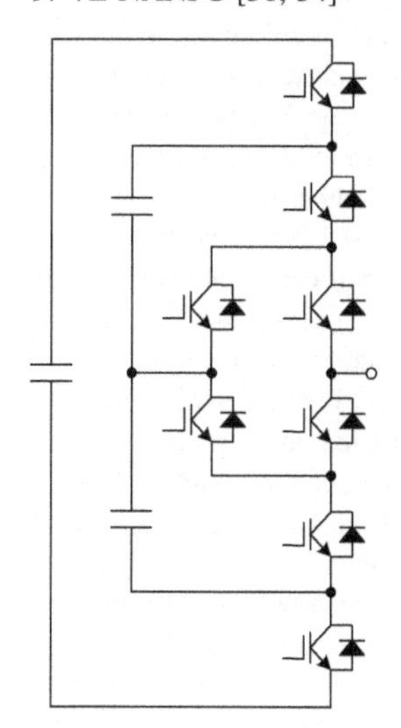	Similar to the NNPC converter, if the base cell is active neutral-point-clamped (ANPC) cell, then a nested active neutral point clamped (NANPC) converter can be derived. The proposed seven-level topology, which is a combination of flying capacitor and nested neutral point clamped (NNPC) converter, has less number of components when compared to classic multilevel topologies with the same number of level. A SVPWM technique has been developed to control output voltages and regulate voltages of flying capacitors. The developed SVPWM technique employs redundant

(*continued*)

Table 10.3 (*Continued*)

Topology	Description
	switching states to control and balance voltage of flying capacitors.
10. CCIL [35] 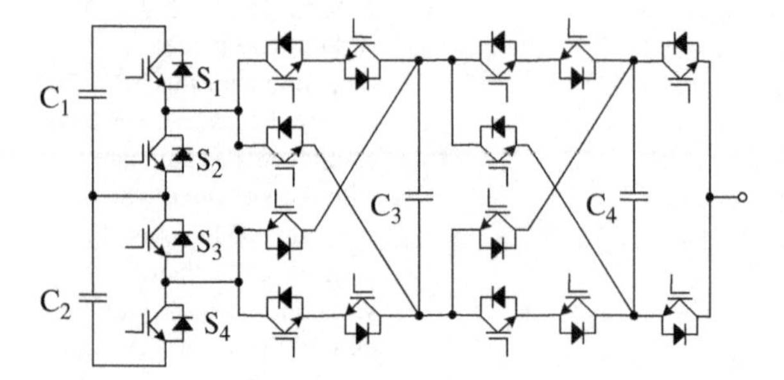	The cross-connected intermediate-level voltage source inverter (CCIL) has been introduced in [35]. This topology shows an interesting potential for high-power-density applications with high-resolution outputs and only limited number of components. Additionally, the multiple ways to tune this inverter family (redundant or not, boosting or not) enable wide application possibilities. Comparison to known topologies revealed that the complexity (amount of stages and to some extent switches) of the CCIL is reduced at the same resolution, but on the other hand, the total blocking voltage of active devices (the sum of all blocking voltages) is higher by a factor of a least 2. This drawback can be significant when considering voltage ranges where series connection would be required. Reliability could be increased on given applications, but this depends if the blocking voltage of the switches is high enough not to require a large number of series connections. For this reason, boosting capability can penalize and should be considered carefully.

10.3 Challenges and Requirements of MV Drives

10.3.1 Power Quality and LC Resonance Suppression

Harmonics in the voltage and current waveforms of utility grid is a crucial problem that should be effectively resolved. Diode-based rectifiers draw distorted current from the grid and cause notches in the voltage waveforms. This results in numerous problems in the power grid such as equipment failure, computer data loss, malfunction of communications equipment, etc. Various standards, such as IEEE 519-2014, IEC 1000-3-2, and IEC 61000-3-2, define the limit of harmonics injected into the power grid [6, 28, 29].

To reduce current harmonics or to compensate the input power factor, an *LC* line-side filter is a common solution. However, the low-damping *LC* resonances may cause undesired oscillations or overvoltages in the grid side because of the low impedance of the MV grid. This may result in a destruction of the switching devices or other components of the rectifier circuits. Solutions to this problem should assure low harmonics and low *dv/dt* using just a reactor instead of an *LC* filter or using a small filter.

10.3.2 Inverter Switching Frequency

High *dv/dt* is generated with the use of high switching frequency semiconductor devices in power electronic converters, which can produce CM voltage and currents, electromagnetic interference (EMI), shaft voltages, bearing currents, and high voltage stress that affect the reduction of insulation life of motors and transformers [39].

The harmonic distortion of the output waveforms increases with the decrease of inverter switching frequency. MIs provide voltage/current waveforms with improved harmonic spectrum and lower *dv/dt*, which limits the insulation stress on the motor windings. However, a higher number of switching devices in MIs tends to reduce their overall reliability and efficiency. On the other hand, an inverter with a lower number of output voltage levels requires a large *LC* output filter to decrease the motor winding insulation stress. The challenge is to reduce waveform distortions and total harmonic distortion when the lower switching frequency is used, to ensure high power quality, as well as to allow fast transient operations. The switching loss due to the fast transition is an important issue that should be considered in MV drives. The maximum and minimum modulation depth and power factor ranging between 1 and −1 are the critical operating points of the MIs. When applying continuous modulation methods, some switching devices reach their maximum allowable junction temperature while other switches remain much cooler. Unbalanced distributions of junction temperatures depend on the type of modulation method used. These problems can be resolved at the expense of additional effort and cost. The operating cost can be reduced with the minimization of switching losses. This also enables to reduce the cooling requirements. Hence, the cost and size of the drive are reduced. The switching losses of MV semiconductor devices contribute to the major portion of the total device losses. Hence, a reduction of switching frequency allows increasing the maximum output power. On the other hand, harmonic distortion at the line and motor side increases with the decrease of switching frequency [40].

A comparison of losses in three-level neutral point clamped (3L-NPC), a three-level flying capacitor (3L-FLC), a four-level flying capacitor (4L-FLC), and nine-level series connected H-

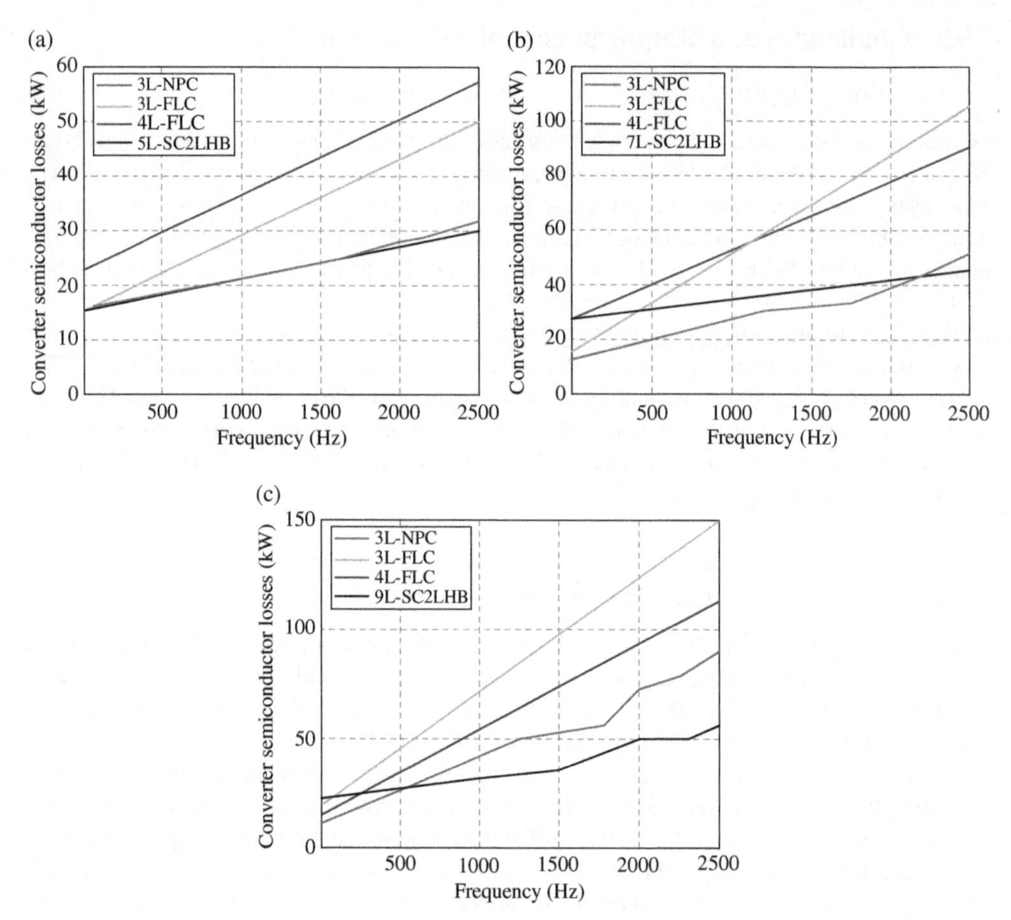

Figure 10.3 NPC, FLC, series-connected converter semiconductor losses versus switching frequency at (a) 2.3, (b) 3.3, (c) 4.16 kV (*Sources*: Kouro et al. [28]; Kouro et al. [29])

bridge (9L-SCHB) is reported in [41, 42]. Figure 10.3 shows switching losses as a function of frequency at different classes of output voltage of 2.3, 3.3, 4.16 kV of various types of MIs [42]. It can be observed that in all types of converters, the losses increase with an increase in the switching frequency and with the voltage. The smallest losses were found in 9L-SC2HB MI. At 2.3 kV, losses in 3L-NPC and SC2LHB MIs are almost the same, but at 4.16 kV, losses in 3L-NPC MI are almost double that of SC2LHB [42].

10.3.3 Motor Side Challenges

10.3.3.1 High *dv*/*dt* and Wave Reflection

The high switching frequency of power devices results in high *dv*/*dt* at the rising and falling edges of the inverter output waveform. Such high rate of change of voltages may result in failure of the motor winding insulation due to partial discharges. Furthermore, such rapid voltage transition induces rotor shaft voltages that cause current flow into the shaft bearing, which finally leads to motor bearings failure [43].

The switching pattern of the power switches affects the wave reflection value, which is produced by the mismatch between the cable and the inverter and the motor wave impedances. The motor cable works as a transmission line where the voltage pulses will travel very fast, up to 150–200 m/μs [42]. When the pulses take more than half of the rise time to move from the inverter to the motor, a full-wave reflection occurs. For that worst case, the wave reflections will double the voltage on the motor terminals at each switching transient. The critical cable length for 500 V/μs is in the 100-m range and for 10 000 V/μs in the 5-m range [1]. The wave reflection coefficient Γ is depended on the ratio between motor and cable wave inductances $\Gamma = (Z_{motor} - Z_{cable})/(Z_{motor} - Z_{cable})$. Nevertheless, cable diameter (Z_{cable}) is around 80–180 Ω that is much smaller than motor wave impedance, which is around 2–0.4 kΩ [44].

The high *dv/dt* also causes EMI on the cables between inverter and motor. The expensive shielded cables are used to avoid these effects; nevertheless, the electromagnetic emission may affect the operation of nearby installed electronic equipment. In the inverter, the *dv/dt* still depends on the switching characteristics of the power devices, and it could still be problematic if no output filter is used. To get guaranteed low THD in both motor and line ends, commonly, passive filters are employed. The high value of the inductor in the *LC* filter must be used in most high power drive systems, but that causes a higher voltage drop across the inductor.

The increase in capacitor value of the filter reduces *LC* resonant frequency, which is affected by the parallel connection of the filter capacitor and motor magnetizing inductance. This leads to instability in drive system. To overcome this issue, the active damping could be proposed and at the same time to suppress the *LC* resonance to achieve high efficiency [45]. Furthermore, the use of *LC* filter introduces a phase shift between the voltages at the output of the feeding converter and the voltage at the motor terminal [46]. This phase shift may pose a problem in the control if not taken into account. Hence, the control algorithm should be modified accordingly.

10.3.3.2 Common-Mode Voltage

The CM voltage on the motor side is produced because of the switching actions of the power converters. This phenomenon has to be taken into consideration while designing the motor drive [47]. CM voltage is mostly responsible for the ground leakage current through stray capacitances that ultimately may damage the motor bearing. Replacement of the bearing is an expensive and time-consuming process, and hence, unplanned maintenance must be avoided. Normally the bearings should be replaced/maintained during the scheduled or planned maintenance. Great effort has been put to minimize the CM voltage in MV drives to save the drive system from catastrophic failure. The most widely used approach is to modify the PWM strategy toward minimizing the CM voltages. Another approach is to employ passive filters at the output of PWM inverter. However, the weight and cost of the drive system will increase [47].

For the multilevel inverter (MLIs), the CM voltage is similar to the traditional two-level inverter, but with a lower value. Therefore, this topic is still a subject of research, and several contributions have been reported in recent years [48]. An effective solution to the issues aforementioned might be found mainly by offering lower switching frequency drive system with very low-harmonic content. Other possible solutions for such problem could be grounding the brushes on the motor shaft [43] and using *dv/dt* resistant winding insulation [49]. Furthermore, a dual-inverter-fed open-end winding (neutral of winding is removed) induction motor

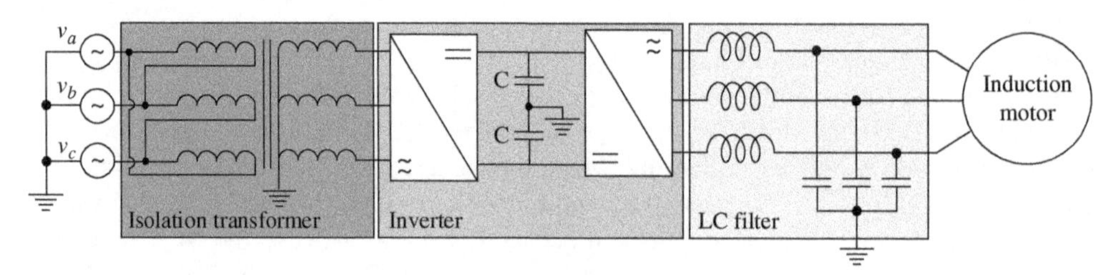

Figure 10.4 Use of isolating transformer

drive with two isolated dc power supplies for CM voltage elimination and to maintain dc-link capacitor voltage balancing is reported in [39].The neutral point of the DC link and motor or star point of the output filter capacitors has been grounded through grounding network using isolation transformers to reduce CM voltage. The usage of the isolation transformer is shown in Figure 10.4.

At low modulation index, the three-phase system looks continuous, which leads to very low dwell time resulting in an increase in CM voltage. There will also be high spikes in DC side, which affect the reliability and performance of the motor drive. To overcome this, the dwell time is modified to decrease CM voltage. The adjusted dwell time is compensated in the subsequent cycles of the switching periods [50].

10.3.3.3 Use of Inverter Output Filters

The passive and active filter-based solutions are employed to mitigate the problems arising due to PWM actions [46, 51, 52]. Presently, passive filtering is commonly used for such problems. Passive filters are hardware circuits that are installed at the output terminals of the converter structure [53]. The most common approach is using filters based on inductors and capacitors (low-pass *LC* filters) as well the CM chokes or CM transformers [53]. For reducing the overvoltages at the motor terminals (especially in the case of long cables connection), the differential mode *LC* filters are used [43]. On the other hand, the differential mode *LC* filter makes it extremely difficult to apply precise control in these motors. Control system design at low-speed conditions is complicated because of phase shift between voltages at input and output of the *LC* filter installed at the output of an inverter. Also, there is a voltage drop across the *LC* [53]. Usually, it is assumed in the drive control that the output voltage and current of an inverter are equal to the motor input voltage and current. The region of proper motor operation shall remain limited if there are any discrepancies. Hence, it is necessary that the measurement circuits or the control algorithms of the electric drive should be modified [54].

Using MLI in drive system results in reduction in motor winding insulation stress. However, the reliability and efficiency of the inverter may decrease. The practice is to make it possible to use very small *LC* filters or even exclude them by ensuring low-harmonic voltage waveforms while keeping low switching frequency. Furthermore, the control strategy is chosen that it must actively damp filter oscillations and allow fast dynamic operation with very low switching frequencies [55].

10.3.3.4 Regeneration Capability and Power Factor Correction

For all electric devices, generally, high power factor operation is desired. Therefore, rectifiers with low current harmonics and those capable of operating at almost unity power factor are required as utility interfaces for many ASDs. The requirement of high power factor is especially important for MV drives due to their high power rating. The rectifier design is a decisive factor for optimum size, cost, and efficiency of the MV drive.

The most popular solution, for high power regenerative loads, is the use of a multilevel active front-end (AFE) converter that provides the regeneration at reduced harmonics and operation at high power factor and active/reactive power combination [56]. Therefore, it is important in some applications to use an AFE multilevel converter providing harmonic mitigation in the power grid and almost unity power factor for all operating points. It is highly advantageous for MV applications, which require regeneration capability. Figure 10.5 shows a cycle of power and speed trajectories [56]. The use of multilevel AFE solutions makes it possible to improve input power factor, power quality, and total cost and volume by possibly eliminating line-side transformer. Back-to-back converters (converter with AFE) have almost double the cost of a single converter. It makes a worthy investment in achieving the solution with nearly perfect sinusoidal input currents at nearly unity power factor and regenerative capability.

10.3.3.5 Torsional Vibration

In the MV drive, torsional vibration can occur because of the large inertias of the motor and its mechanical load. They occur when the natural frequency of the mechanical system matches the frequency of torque pulsations caused by the motor current harmonics [57].

Excessive torsional vibrations at resonance can result in high twisting torque and, hence break of shafts and couplings. The undamped response is shown in Figure 10.6a [58]. The

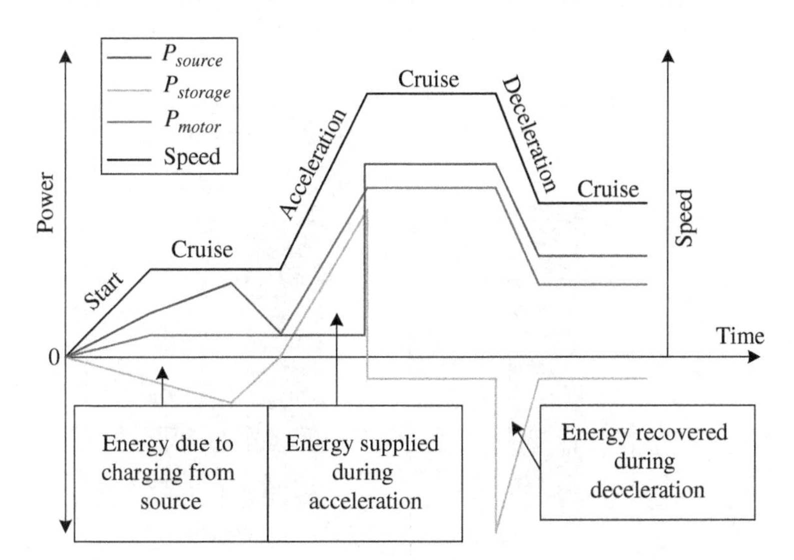

Figure 10.5 Power (source, storage, motor) and speed trajectories

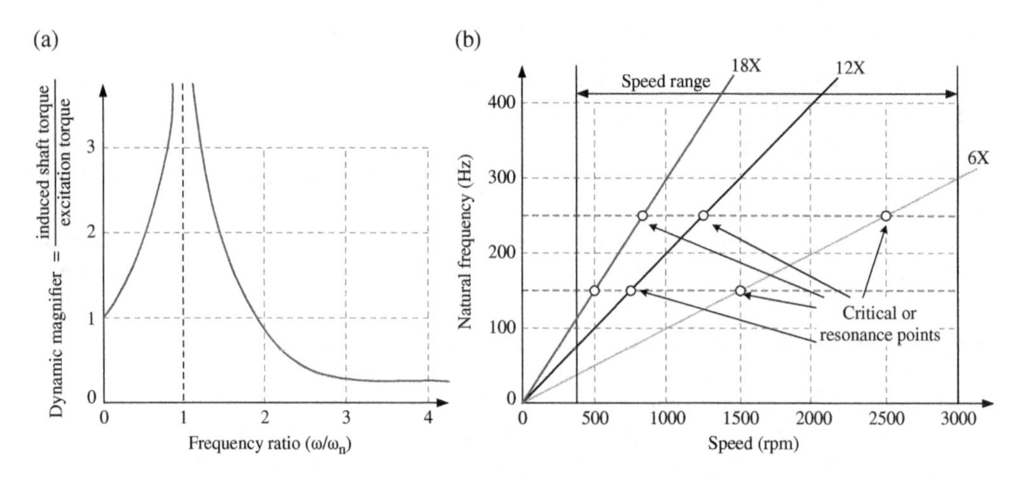

Figure 10.6 (a) Undamped response. (b) Campbell diagram

torsional vibrations can also cause damage to other mechanical components in the system. Pulse width modulation control with harmonic elimination can be helpful to overcome such problem. The operating points that excite the shaft in PWM inverter-fed motors can be predicted without torque sensors as shown in Figure 10.6b [58].

10.3.3.6 Transformerless Solution

Complex multiwinding transformers can mitigate harmonics through phase shifting in modular converters. An isolation transformer can represent 30–50% of electric drive system size and 50–70% of the system's weight. The comparison of the average drive system space and weight with and without transformer is shown in Figure 10.7. In addition to soaring raw material costs,

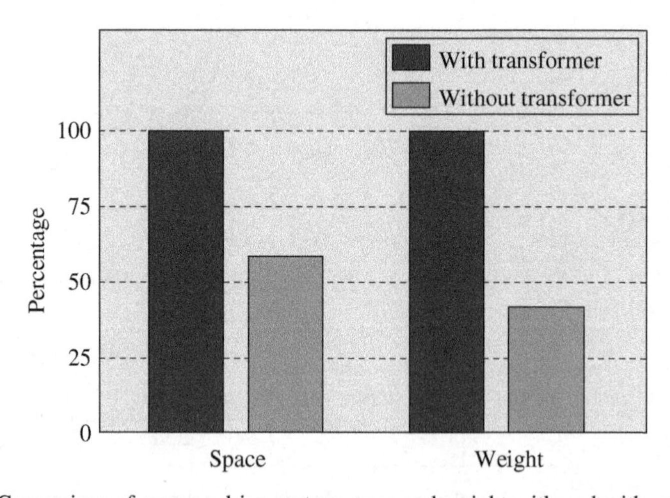

Figure 10.7 Comparison of average drive system space and weight with and without a transformer

an isolation transformer adds to total drive costs with extra cabling, air conditioning to cool the transformer, engineering time, concrete pad construction for outdoor transformers, and overall installation [59].

Issues such as cost, space, weight, and platform balance can be improved significantly with fewer transformers for an offshore platform. To control pump motors on down hole wells, each used drive requires a multiton transformer. The offshore platforms typically require 20 MV drives (sometimes up to 40 or more) [59]. A significant amount of cooling system or air conditioning is required in a tropical environment because a 1000 kVA transformer generates up to 2 kW of heat energy. The transformerless solution gives substantial energy savings [60]. This feature is also important in applications such as utility distribution systems and high-voltage vehicle drives [61]. Solutions are desired to reduce CM voltage, produce sinusoidal output waveform, and limit *dv/dt* in a transformerless MV drive.

The high-level CM voltage stress on the motor is imposed on the transformer and cable insulation even though transformers protect the motor from CM voltage. To withstand the CM voltage stress, specific transformer and cable insulation is required. In a transformerless solutions, MV variable-speed drive requires very good insulation on the motor side to sustain high CM voltage stress. Hence, the integrated CM DC choke is used to block the CM voltage and reduce motor neutral to ground voltage. The structure of the integrated choke and its connection diagram in transformerless motor drives are shown in Figure 10.8 [62].

Transformerless MV motor drive usually requires a high level of motor insulation to overcome CM voltage stress or requires the use of an additional inductor with approximately same impedance as the transformer to be replaced [63]. The use of shunt active hybrid filters solves this problem [51, 64].

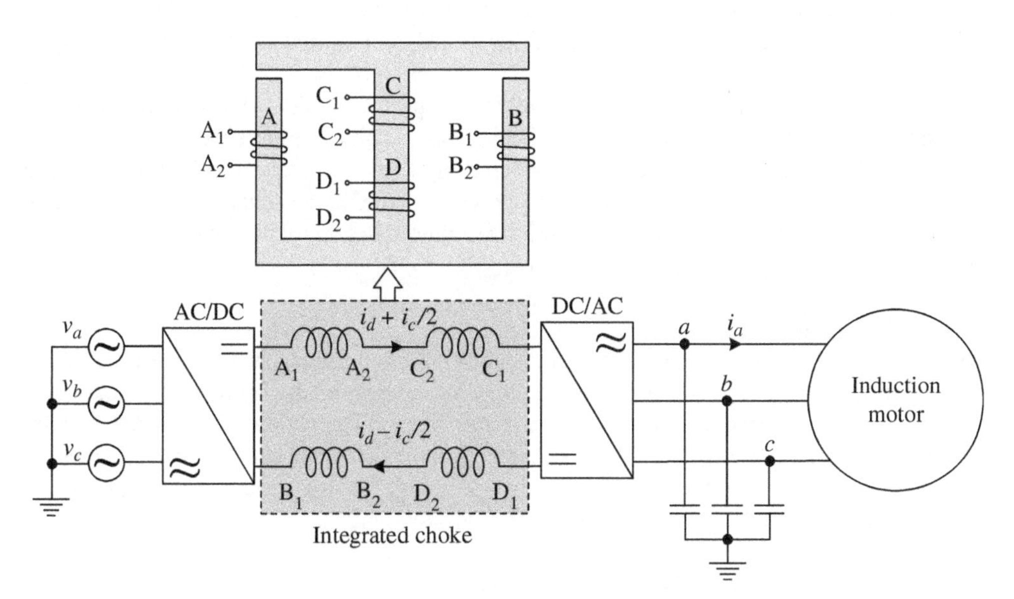

Figure 10.8 The structure of the integrated choke and its connection diagram in transformerless motor drives

10.3.3.7 Fault Detection and Condition Monitoring

Under severe operating conditions, there should be accurate techniques for fault diagnosis and condition monitoring of MV drives to avoid financial losses due to unscheduled and forced shutdown. Therefore, condition monitoring and fault diagnosis of electrical motors are another important challenge to be met for MV drive systems. In this area, a significant amount of work has been done in recent years, and a comprehensive review is presented in [65].

The interesting possibility of sensorless fault detection is to use observer-based diagnostic methods for mechanical faults detection [66]. More advanced solutions are possible to detect failures also when *LC* inverter output filter is installed on the drive [67]. Any partial discharge in the stator of high voltage rating machines can be monitored by using high-frequency current transformer sensor located at hazardous locations such as oil and gas industries [68].

10.3.3.8 Research and Industrial Trends

10.3.3.8.1 Components

The present research and industrial trends for MV drives are multidimensional encompassing the front-end converter, inverters, motor, and control. A new trend is using advanced materials for power semiconductor switches such as silicon carbide (SiC)-based switches. Simulation of SiC thyristors has given excellent conduction and switching performance. So, SiC thyristor promises device for high-power applications. It has the drawback that it needs *di/dt* snubber and also *dv/dt* snubber for turn-off. This drawback is overcome by the development of SiC emitter turn-off thyristor (ETOT), hence improving the speed and dynamic performance of the device [69]. In [70], full bridge converter was realized using SiC MOSFETs. In semiconductor research, silicon carbide power switches promise to increase the efficiency by 10% due to reduced switching losses. However, up to now no MV drive manufacturer declared to use the SiC devices. It may be due to high cost of the device and high-current switching devices are yet to come in the market.

An intelligent power module has been developed to protect IGBTs from short circuit and to provide a negative voltage to avoid false switching signal, which reduces the switching losses [71]. Furthermore, to obtain closed-loop gate control, a new technique has been proposed in [72]. The voltage across parasitic inductance of IGBT is monitored to measure the switching speed and *di/dt*, so that the gate drive voltages are adjusted to control the switching speed, and the full capability of power devices is utilized.

10.3.3.8.2 Topologies

For more than 6 kV, modular multilevel converter (MMC) is the best solution to get rid of transformers and isolated power sources; however, they have the problem of voltage fluctuation and voltage balance of capacitors. The research trends in modular multilevel cascaded inverter (MMCI) have been discussed in [73]. This chapter proposes methods to mitigate the fluctuation in AC voltage to stabilize the operation at low speed and low frequency, and those methods were compared with other two methods.

10.3.3.8.3 Control and Modulation Techniques

To control MIs at the low switching frequency, synchronous optimal PWM technique was proposed in [74]. This modulation technique brings considerable benefits such as THD level of the machine current doesn't effect from low switching frequency. The technique also keeps optimal CM voltage and neutral point potential balance.

To improve the dynamics of the control system, the hybrid direct torque control (HDTC) has been proposed in [75]. In this control, it is shown that torque is proportional to quadrature axis component of current. This can improve the error (between quadrature reference current and actual quadrature current) calculations.

Initially selected harmonic elimination has been proposed in [76] and then in [77]. After that new techniques to eliminate the selective harmonic elimination PWM schemes have been suggested such as programmed PWM [78] and multilevel selective harmonic elimination (SHE)-PWM for series-connected inverters [79, 80]. The major challenge with SHE is to solve online the algebraic equations obtained for specific harmonic cancelation [81].

10.4 Summary

The chapter discusses the problems associated with controlled MV AC drives and the trend of lowering switching frequency and improving the efficiency of the MV drive system.

While analyzing MV drives, the quality of current and voltage wave shapes at both the input and output terminals is an important factor. The application, the topology used, the size of filter components, the switching frequency, and control strategy handle the current waveforms. There will be minimum effects on the motor side if a better power factor, power quality with higher efficiency are obtained.

In this chapter, most of the existing shortcomings such as overvoltage due to *dv/dt* or wave reflection, bearing currents due to CMV, regeneration capabilities, torsional vibration, isolating transformer or transformerless solution, fault detection, and condition monitoring were discussed, and solutions were commented, whenever possible.

References

1. Wu, B. (2006) *High-Power Converters and AC Drives*. IEEE Press, Wiley Inter science, Wiley.
2. Chausovsky, A. Industrial Motors&Drives Global Market Update [online]. Available: http://www.e-driveonline.com/Conf-12/images/Presentations/IMS%20Research.pdf.
3. Fracking energies MV drives sales in the Americas [online]. Available: http://www.drivesncontrols.com/news/fullstory.php/aid/4227/Fracking_energises_MV_drives_sales_in_the_Americas.html.
4. Abu-Rub, H., Malinowski, M., and Al-Haddad, K. (2014) *Power Electronics for Renewable Energy Systems, Transportation and Industrial Applications*. IEEE Press and Wiley. ISBN: 9781118634035.
5. Abu-Rub, H., Holtz, J., Rodriguez, J., and Baoming, G. (2010) Medium-voltage multilevel converters—state of the art, challenges, and requirements in industrial applications. *IEEE Trans. Ind. Electr.*, 57(8), 2581–2596.
6. Abu-Rub, H., Lewicki, A., Iqbal, A., and Guzinski, J. (2010) Medium voltage drives - challenges and requirements. *IEEE Int. Sympo. Ind. Electr. (ISIE)*, 4–7 July 2010. Bari, Italy, pp. 1372–1377.
7. ABB Medium Voltage Drives [online]. Available: http://new.abb.com/drives/medium-voltage-ac-drives.
8. Siemens Medium Voltage Converters [online]. Available: https://new.siemens.com/global/en/products/drives/sinamics/medium-voltage-converters.html

9. TMEIC Medium Voltage AC Drives [online]. Available: https://www.tmeic.com/North%20America/259-Products%20MediumVoltageACDrives-116.
10. AMTECH Medium Voltage Drives [online]. Available: http://www.amtechelectronics.com/product.aspx?sid=3.
11. ALSTOM Medium Voltage Drives [online]. Available: http://www.mvdrives.com/alstom-mv-drives-specifications
12. Scneider Electric Medium Voltage variable Speed Drive [online]. Available: http://www.schneider-electric.com/en/product-range/61394-altivar-1200
13. Fuji Electric Medium Voltage variable Speed Drive [online]. Available: http://www.fujielectric-europe.com/components/drives-inverters/medium-voltage-frenic4600.
14. Delta Group Medium Voltage Drive [online]. Available: http://www.deltaww.com/Products/CategoryListT1.aspx?CID=060103&PID=ALL&hl=en-US.
15. Eaton Medium Voltage Adjustable Frequency Drives [online]. Available: http://www.eaton.com/Eaton/ProductsServices/Electrical/ProductsandServices/ElectricalDistribution/MediumVoltageDrives/SC9000/index.htm
16. Toshiba Medium Voltage Drives [online]. Available: https://www.toshiba.com/tic/industrial-systems/adjustable-speed-drives/medium-voltage-drives
17. WEG Equipamentos Elétricos Medium Voltage variable Speed Drive [online]. Available: http://ecatalog.weg.net/files/wegnet/WEG-mvw01-medium-voltage-drive-usamvw0109-brochure-english.pdf.
18. Hitachi Medium Voltage Drives [online]. Available: http://www.hitachi-hirel.com/products/drives-and-automation/medium-voltage-drives.
19. INGETEAM Medium Voltage Drives [online]. Available: http://www.ingeteam.com/Portals/0/Catalogo/Sector/Documento/SSE_258_Archivo_sbp15-catalogo-ingedrive.pdf.
20. Rockwell Automation Medium Voltage AC Drives [online]. Available: http://ab.rockwellautomation.com/Drives/Medium-Voltage.
21. Yaskawa Medium Voltage Drives [online]. Available: https://www.yaskawa.com/pycprd/products/mv-drives.
22. Benshaw Medium Voltage Converters [online]. Available: https://www.benshaw.com/sites/default/files/downloads/brochures/benshaw-m2l-3000-brochure.pdf
23. www.converteam.com
24. LS Ind System. https://www.lsis.com/products/category/Smart_Automation_Solution/Inverter_-∗VFD∗-/Medium_Voltage_VFD
25. Rongxin Power. www.rxhk.co.uk/solutions/voltage-source-converters-vsc/mv-variable-frequency-drives
26. https://www.gepowerconversion.com/sites/default/files/GEA30738C%20MV6%20Medium%20Voltage%20Drive.pdf
27. Severn Drives & Energy. www.severnde.co.uk/wp-content/uploads/2016/11/SDE-MV-AC-Drive-EMAIL-V3.pdf
28. Kouro, S., Malinowski, M., Gopakumar, K., Pou, J., Franquelo, L. G., Wu, B., Rodríguez, J., Pérez, M., and León, J. I. (2010) Recent advances and industrial applications of multilevel converters. *IEEE Trans. Ind. Electr.*, 57(8), 2553–2580.
29. Kouro, S., Rodriguez, J., Wu, B., Bernet, S., and Perez, M. (2012) Powering the future of industry: high-power adjustable speed drive topologies. *IEEE Ind. Appl. Mag.*, 18(4), 26–39.
30. Du, S., Dekka, A., Wu, B., and Zargari, N. (2018) *Modular Multilevel Converters: Analysis, Control, and Applications*, 1st edn. The Institute of Electrical and Electronics Engineers, Inc. Wiley.

31. Li, Y. and Quan, Z. (2017) Derivation of multilevel voltage source converter topologies for medium voltage drives. *Chin. J. Electr. Eng.*, 3(2), 24–31. doi:https://doi.org/10.23919/CJEE.2017.8048409
32. Li, J., Jiang, J., and Qiao, S. (2017) A space vector pulse width modulation for five-level nested neutral point piloted converter. *IEEE Trans. Power Electron.*, 32(8), 5991–6004. doi: https://doi.org/10.1109/TPEL.2016.2618931.
33. Narimani, M., Wu, B., Cheng, G., and Zargari, N. (2014) A new nested neutral point clamped (NNPC) converter for medium-voltage (MV) power conversion. *IEEE Trans. Power Electron.*, 29(12), 6375–5382.
34. Narimani, M., Wu, B., and Zargari, N. R. (2015) A novel seven-level voltage source converter for medium-voltage(MV) applications. *IEEE Energy Conv. Congress Exposition (ECCE).* Montreal, QC, pp. 4277–4282.
35. Chaudhuri, T., Barbosa, P., Steimer, P., and Rufer, A. (2007) Cross-connected intermediate level (CCIL) voltage source inverter. *2007 IEEE Power Electr. Spec. Conf.* doi:https://doi.org/10.1109/PESC.2007.4342036
36. Rajeevan, P. P., Sivakumar, K., Gopakumar, K., Patel, C., and Abu-Rub, H. (2013) A nine-level inverter topology for medium-voltage induction motor drive with open-end stator winding. *IEEE Trans. Ind. Electr.*, 60(9), 3627–3636.
37. TMEIC Medium Voltage Drive Evolution [online]. Available: http://www.tmeic.com/Repository/Brochures/MV%20Drive%20Evolution%20Brochure%202011%20hi-res.pdf.
38. Filsecker, F., Alvarez, R., and Bernet, S. (2013) Comparison of 4.5-kV press-pack IGBTs and IGCTs for medium-voltage converters. *IEEE Trans. Ind. Electr.*, 60(2), 440–449.
39. Tekwani, P. N., Kanchan, R. S., and Gopakumar, K. (2007) A dual five-level inverter-fed induction motor drive with common-mode voltage elimination and DC-link capacitor voltage balancing using only the switching-state redundancy—part I. *IEEE Trans. Ind. Electr.*, 54(5), 2600–2608.
40. Rodriguez, J., Bernet, S., Steimer, P. K., and Lizama, I. E. (2010) A survey on neutral-point-clamped inverters. *IEEE Trans. Ind. Electr.*, 57(7), 2219–2230.
41. Fazel, S. S., Bernet, S., Krug, D., and Jalili, K. (2007) Design and comparison of 4-kV neutral-point-clamped, flying-capacitor, and series-connected H-bridge multilevel converters. *IEEE Trans. Ind. Appl.*, 43(4), 1032–1040.
42. Fazel, S. S. (2007) Investigation and comparison of multi-level converters for medium voltage applications. Ph.D. dissertation, TechnischeUniversität Berlin, Berlin, Germany.
43. Akagi, H. and Tamura, S. (2006) A passive EMI filter for eliminating both bearing current and ground leakage current from an inverter-driven motor. *IEEE Trans. Power Electron.*, 21(5), 1459–1469.
44. Jouanne, A. (1996) Filtering techniques to minimize the effect of long motor leads on PWM inverter-fed AC motor drive systems. *IEEE Trans. Ind. Appl.*, 32(4), 919–926.
45. Akagi, H. and Isozaki, K. (2012) A hybrid active filter for a three-phase 12-pulse diode rectifier used as the front end of a medium-voltage motor drive. *IEEE Trans. Power Electron.*, 27(1), 69–77.
46. Guzinski, J. and Abu-Rub, H. (2013) Sensorless induction motor drive with voltage inverter and sine-wave filter. *IEEE International Symposium on Sensorless Control for Electrical Drives SLED 2013 (4th Symposium) and Predictive Control of Electrical Drives & Power Electronics PRECEDE 2013 (2nd Symposium)*, 17–19 October 2013. Munich, Germany.
47. Baoming, G., Peng, F.Z., de Almeida, A. T., and Abu-Rub, H. (2010) An effective control technique for medium-voltage high-power induction motor fed by cascaded neutral-point-clamped inverter. *IEEE Trans. Ind. Electr.*, 57(8), 2659–2668.

48. Franquelo, L. G., Rodriguez, J. L., Kouro, S., Portillo, R., and Prats, M. A. M. (2008) The age of multilevel converters arrives. *IEEE Ind. Electr. Mag.*, 2(2), 28–39.
49. Ferreira, F. J. T. E., Trovão, J. P., and de Almeida, A. T. (2008) Motor bearings and insulation system condition diagnosis by means of common-mode currents and shaft-ground voltage correlation. *Proc. 2008 Int. Conf. Electr. Mach.* 6–9 September 2008. Vilamoura, Portugal.
50. Tallam, R. M., Leggate, D., Kirschnik, D. W., and Lukaszewski, R. A. (2011) PWM scheme to reduce the common-mode current generated by an AC drive at low modulation index. *IEEE Energy Conversion Congress and Exposition (ECCE)*, 17–22 September 2011, pp. 3299–3305.
51. Akagi, H. and Kondo, R. (2010) A transformerless hybrid active filter using a three-level pulse width modulation (PWM) converter for a medium-voltage motor drive. *IEEE Trans. Power Electron.*, 25(6), 1365–1374.
52. Akagi, H. and Hatada, T. (2009) Voltage balancing control for a three-level diode-clamped converter in a medium-voltage transformerless hybrid active filter. *IEEE Trans. Power Electron.*, 24(3), 571–579.
53. Abu-Rub, H., Iqbal, A., and Guzinski, J. (2012) *High Performance Control of AC Drives with Matlab/Simulink Models*. Wiley.
54. Guzinski, J. and Abu-Rub, H. (2009) Nonlinear control of an asynchronous motor with inverter output LC filter. *2nd Mediterranean Conference on Intelligent Systems and Automation (CISA'09)*. Zarzis, Tunis, 23–25 March 2009.
55. Hatua, K., Jain, A. K., Banerjee, D., and Ranganathan, V. T. (2012) Active damping of output LC filter resonance for vector-controlled VSI-fed ac motor drives, *IEEE Trans. Ind. Electr.*, 59(1), 334–342.
56. Liu, L., Li, H., Hwang, S.-H., and Kim, J.-M. (2013) An energy-efficient motor drive with autonomous power regenerative control system based on cascaded multilevel inverters and segmented energy storage. *IEEE Trans. Ind. Electr.*, 49(1), 178–188.
57. Song-Manguelle, J., Nyobe-Yome, J. M., and Ekemb, G. (2010) Pulsating torques in PWM multi-megawatt drives for torsional analysis of large shafts. *IEEE Trans. Ind. Electr.*, 46(1), 130–138.
58. Corbo, M. A. and Malanoski, S. B. (1996) Practical design against torsional vibration. *Proc. Twenty-Fifth Turbomach. Symp.*, Turbomachinery Laboratory, Texas A&M University, College Station, Texas, pp. 189–222.
59. Automation, R. Maximize efficiency, space&weight savings with medium-voltage transformerless drive configuration [online]. Available: http://literature.rockwellautomation.com/idc/groups/literature/documents/ap/oag-ap002_-en-p.pdf.
60. Yang, B., Li, W., Gu, Y., Cui, W., and He, X. (2012) Improved transformerless inverter with common-mode leakage current elimination for a photovoltaic grid-connected power system. *IEEE Trans. Power Electron.*, 27(2), 752–762.
61. Hagiwara, M., Nishimura, K., and Akagi, H. (2010) A medium-voltage motor drive with a modular multilevel PWM inverter. *IEEE Trans. Power Electron.*, 25(7), 1786–1799.
62. Wu, B., Rizzo, S., Zargari, N., and Xiao, Y. (2001) An integrated DC link choke for elimination of motor common-mode voltage in medium voltage drives. *IEEE Conference Record of the Thirty-Sixth IAS Annual Meeting, Industry Applications Conference*, 3, pp. 2022–2027.
63. Hatti, N., Hasegawa, K., and Akagi, H. (2009) A 6.6-kV transformerless motor drive using a five-level diode-clamped PWM inverter for energy savings of pumps and blowers. *IEEE Trans. Power Electron.*, 24(3), 796–803.
64. Dieckerhoff, S., Bernet, S., and Krug, D. (2005) Power loss-oriented evaluation of high voltage IGBTs and multilevel converters in transformerless traction applications. *IEEE Trans. Power Electron.*, 20(6), 1328–1336.

65. Refaat, S. S., Abu-Rub, H., Saad, M. S., Aboul-Zahab, E. M., and Iqbal, A. (2013) ANN-based for detection, diagnosis the bearing fault for three phase induction motors using current signal. *IEEE International Conference on Industrial Technology (ICIT)*, pp. 253–258.
66. Guzinski, J., Abu-Rub, H., Diguet, M., Krzeminski, Z., and Lewicki, A. (2010) Speed and load torque observer application in high–speed train electric drive. *IEEE Trans. Ind. Electr.*, 57(2), 565–574.
67. Guziński, J., Abu-Rub, H., and Toliyat, H. A. (2010) Speed sensorless ac drive with inverter output filter and fault detection using load torque signal. ISIE 2010, 4–7 July 2010. Bari, Italy.
68. Renforth, L., Armstrong, R., Clark, D., Goodfellow, S., and Hamer, P. (2014) High-voltage rotating machines: a new technique for remote partial discharge monitoring of the stator insulation condition", *IEEE Ind. Appl. Mag.*, 20(6), 79–89.
69. Wang, J., Wang, G., Li, J., and Huang, A. Q. (2009) Silicon carbide emitter turn-off thyristor, a promising technology for high voltage and high frequency applications. *IEEE Twenty-Fourth Annual Applied Power Electronics Conference and Exposition (APEC)*, pp. 658–664.
70. Vaculík, P. (2014) Application note for the construction of SiC medium power converter with high switching frequency. *Proceedings of the 15th International Scientific Conference on Electric Power Engineering (EPE)*, pp. 563–568.
71. Vogler, B., Herzer, R., Buetow, S., Mayya, I., and Becker, S. (2014) Advanced SOI gate driver IC with integrated VCE-monitoring and negative turn-off gate voltage for medium power IGBT modules. *IEEE 26th International Symposium on Power Semiconductor Devices & IC's (ISPSD)*, pp. 317–320.
72. Chen, L. and Peng, F. Z. (2009) Closed-loop gate drive for high power IGBTs. *Proc. 24th Annu. IEEE Appl. Power Electron. Conf. Expo*, pp. 1331–1337.
73. Okazaki, Y., Matsui, H., Hagiwara, M., and Akagi, H. (2014) Research trends of modular multilevel cascade inverter (MMCI-DSCC)-based medium-voltage motor drives in a low-speed range. *International Power Electronics Conference (IPEC-Hiroshima 2014 - ECCE-ASIA)*, pp. 1586–1593.
74. Rathore, R., Holtz, H., and Boller, T. (2013) Generalized optimal pulsewidth modulation of multilevel inverters for low-switching-frequency control of medium-voltage high-power industrial AC drives. *IEEE Trans. Ind. Electr.*, 60(10), 4215–4224.
75. Patil, U. V., Suryawanshi, H. M., and Renge, M. M. (2014) Closed-loop hybrid direct torque control for medium voltage induction motor drive for performance improvement. *IEEE Trans. Power Electron.*, 7(1), 31–40.
76. Turnbull, F. G. (1964) Selected harmonic reduction in static DC-AC inverters. *IEEE Trans. Commun. Electr.*, 83(73), 374–378.
77. Patel, H. S. and Hoft, R. G. (1973) Generalized techniques of harmonic elimination and voltage control in thyristor inverters: part I--harmonic elimination. *IEEE Trans. Ind. Appl.*, IA-9(3), 310–317.
78. Enjeti, P. N., Ziogas, P. D., and Lindsay, J. F. (1990) Programmed PWM techniques to eliminate harmonics: a critical evaluation. *IEEE Trans. Ind. Appl.*, 26(2), 302–316.
79. Li, L., Czarkowski, D., Liu, Y., and Pillay, P. (2000) Multilevel selective harmonic elimination PWM technique in series-connected voltage inverters. *IEEE Trans. Ind. Appl.*, 36(1), 160–170.
80. Sirisukprasert, S., Lai, J.-S., and Liu, T.-H. (2002) Optimum harmonic reduction with a wide range of modulation indexes for multilevel converters. *IEEE Trans. Ind. Electr.*, 49(4), 875–881.
81. Holtz, J. and Qi, X. (2013) Optimal control of medium-voltage drives—an overview. *IEEE Trans. Ind. Electr.*, 60(12), 5472–5481.

11

Current Source Inverter Fed Drive

Marcin Morawiec and Arkadiusz Lewicki

11.1 Introduction

The power converter topologies of variable speed electric drives can be divided into two groups. The first large group is based on the voltage source inverters (VSI), while the second one consists of current source inverters (CSI) [1]. The VSIs are more often used in industrial applications than CSIs because they offer better properties of the drive system. In the 1980s, the power electronic devices were mainly based on low switching frequency thyristors. Therefore, the output current of CSIs was not sinusoidal (due to the content of higher harmonics) – it had a stepped shape. Because of the low switching frequency the inductor in the DC-link had to be large and bulky. The large content of higher harmonics in machine currents causes oscillations of the electromagnetic torque. These problems led to abandoning the development of these drive systems. Over time, the further development of power electronics introduced the IGBT transistors, connected in series to a diode in the single case. The switching frequency of IGBT power electronic modules could be many times higher than that of thyristors, for example 10, 20, and more times higher. Increasing the switching frequency has a positive influence on the CSIs dimensions. The parameters of passive elements, for example for a frequency of 10 kHz of the CSI, can value about 5–10 μF for output capacitors and about 5 mH for a DC-link inductor. This could make the CSIs a still attractive option.

High Performance Control of AC Drives With MATLAB®/Simulink, Second Edition.
Haitham Abu-Rub, Atif Iqbal, and Jaroslaw Guzinski.

Companion website: www.wiley.com/go/aburubcontrol2e

11.2 Current Source Inverter Structure

The general circuit scheme of a current source converter-based electric drive is presented in Figure 11.1. In such a configuration the bidirectional flow of energy from the AC-side to the machine and back to AC-side is satisfied. The current source converter consists of two CSIs, where in each of them the transistors are connected in a different way. In literature [1, 2], the AC-side CSI is called a current source rectifier (CSR). The CSR provides regulated voltage to the input of the DC-link which can be controlled from zero to maximum rectified voltage (it depends on the amplitude of the AC-side voltage). The CSR with an inductor in the DC-link is the controlled current source. Appropriate control of the CSR allows to achieve a unity power factor on the AC-grid as well as reactive power compensation.

The structure of a single CSI is shown in Figure 11.2a. In such a configuration only one direction of energy flow is possible. The controlled input voltage is marked by e_d and in the

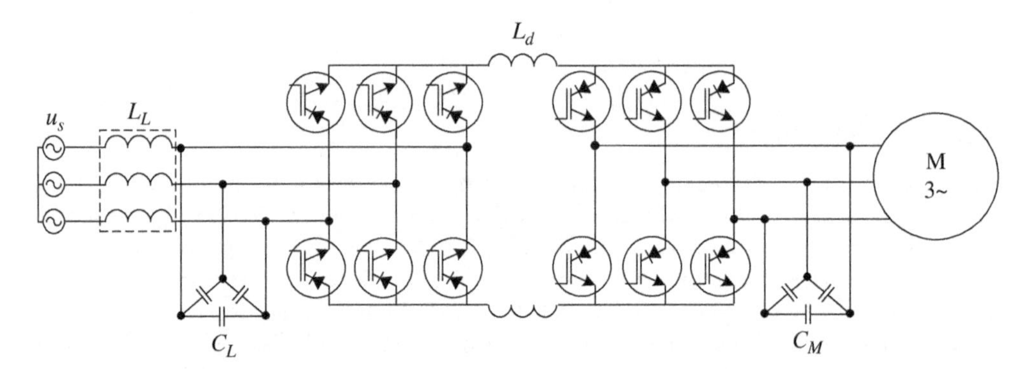

Figure 11.1 Current source converter structure with RBIGBT transistors

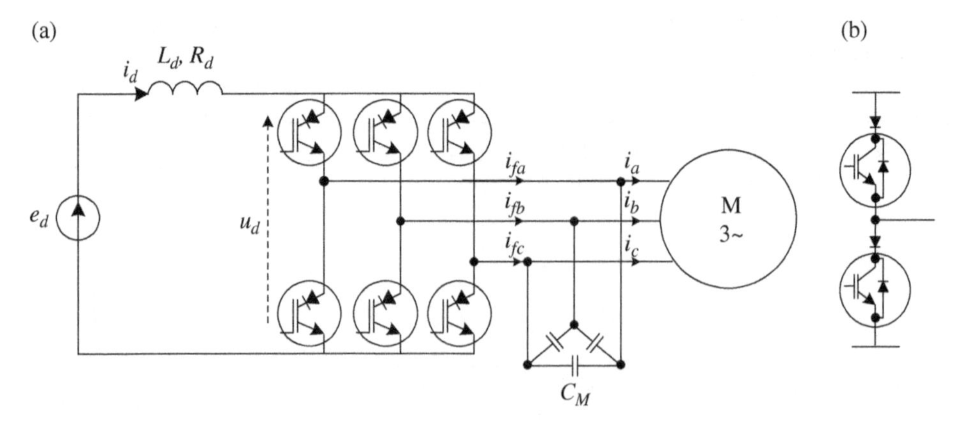

Figure 11.2 Current source inverter structure (a), single leg of current source inverter with IGBT series connected to diode (b)

practical applications it can be a chopper (based only on one transistor or on the half-bridge transistors' structure). The transistors' structure shown in Figures 11.1 and 11.2a is known in literature [2, 3] as reverse-blocking IGBT transistors (RBIGBT). The RBIGBT is a structure in which the transistor is series connected with the diode and can be seen in Figure 11.2b. The output capacitors are marked as C_M and the inductor as L_d.

The parameters of a CSI system are the inductance of the L_d choke and the capacitance of the C_M motor capacitors. These parameters can be determined with the use of a computer simulation or by calculations. The inductance of the choke can be approximated from the following formula:

$$L_d = \frac{1}{\Delta i_d} \int_{t_1}^{t_2} (e_{d\max} - u_{d\max}) dt, \tag{11.1}$$

where:

e_{dmax} is the maximum value of DC-link input voltage e_d,
u_{dmax} is the maximum value of DC-link output voltage u_d,
t_1–t_2 is the time interval, and
Δi_d is defined as in Figure 11.3.

The system parameters can be selected using an empirical method of checking the correct operation of the inverter and shaping the currents and voltages in numerical simulation. This selection method is unreliable because it can lead to an incorrect estimation of parameters, which will cause high value of oscillations in motor voltages. A better solution is to define

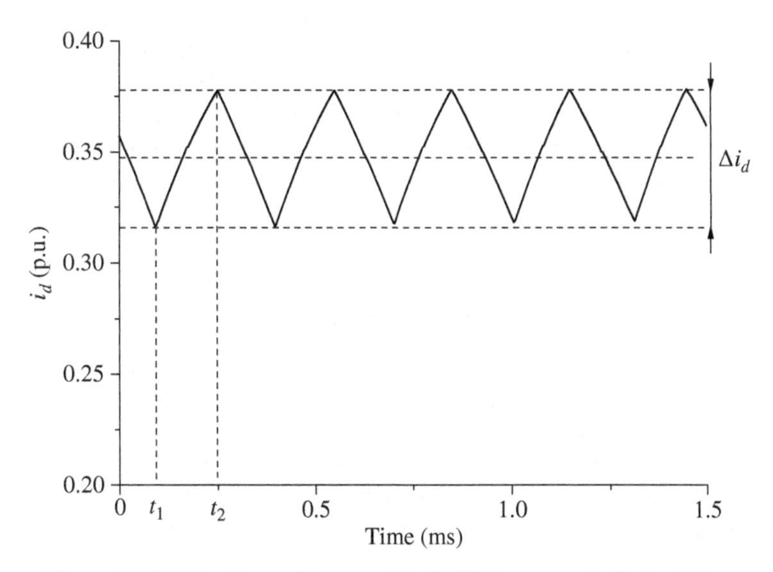

Figure 11.3 Waveform of i_d current in DC-link – numerical simulation

appropriate criteria for which the system parameters will be selected. Two criteria have been formulated:

- the criterion of minimization of ripple ratio in current i_d,
- the criterion of minimizing the weight and dimensions of the inductor.

The current ripple ratio is defined as follows:

$$w_i = \frac{\Delta i_{dmax}}{i_d}, \tag{11.2}$$

w_i = 10–20%,where w_i is the ripple ratio of the DC-link current i_d, Δi_{dmax} is the maximum change of i_d current defined in t_1–t_2.

The criterion for minimizing the current ripple ratio is to ensure the correct commutation processes in the inverter. A too high ripple ratio w_i causes a distortion of the stator current (output current), as well as affects the distortion of the stator voltage. The value of the current ripple i_d is assumed in a range of w_i = 10–15%. In literature, the value of w_i coefficient is assumed to be 15% [4–6]. The current ripple ratio $w_i < 10\%$ ensures less ripples of the electromagnetic torque of the motor. However, it may not meet the second criterion of inductor selection. Of course w_i is strictly connected to the switching frequency of the inverter transistors.

The capacitance of the output filter capacitors is the second parameter that affects the operation of the CSI. This capacitance is closely related to the commutation processes that occur during the switching of the inverter transistors. The capacitance of the output filter can be chosen by using the simulation approach.

In Figure 11.4, the waveforms of DC-link current i_d, voltages: e_d and $u_{s\alpha}$, and stator current component $i_{s\alpha}$ for different cases of chosen CSI parameters: capacitance C_M and inductance L_d are shown. The transistors' switching time was 3.3 kHz. The oscillations of the stator voltage occurred in cases (a) and (d) (small capacitance C_M was applied). Satisfactory waveforms are achieved for the cases of (c) and (d) (ripples of i_d current are smaller than 10%) but in (c) L_d = 50 mH, therefore the inductor will be large in practical application. The problem of the selection of CSI parameters will be more complicated if the total harmonic distortion (THD) of the stator current and voltage is taken into account [5].

11.3 Pulse Width Modulation of Current Source Inverter

The pulse width modulation method (PWM) is based on the rotation of the reference space vector in $\alpha\beta$ orthogonal system [2–4, 6–8]. The average value of output current vector components in the sampling time are received by turning on switching states (transistors) in cycles. In accordance with the space vector theory, the projection of three-phase currents by space vectors in the $\alpha\beta$ stationary coordinate system is possible. The three-phase currents (i_{sa}, i_{sb}, i_{sc}) can be specified by reference to the space vector module, marked I_0 and rotated with angular speed

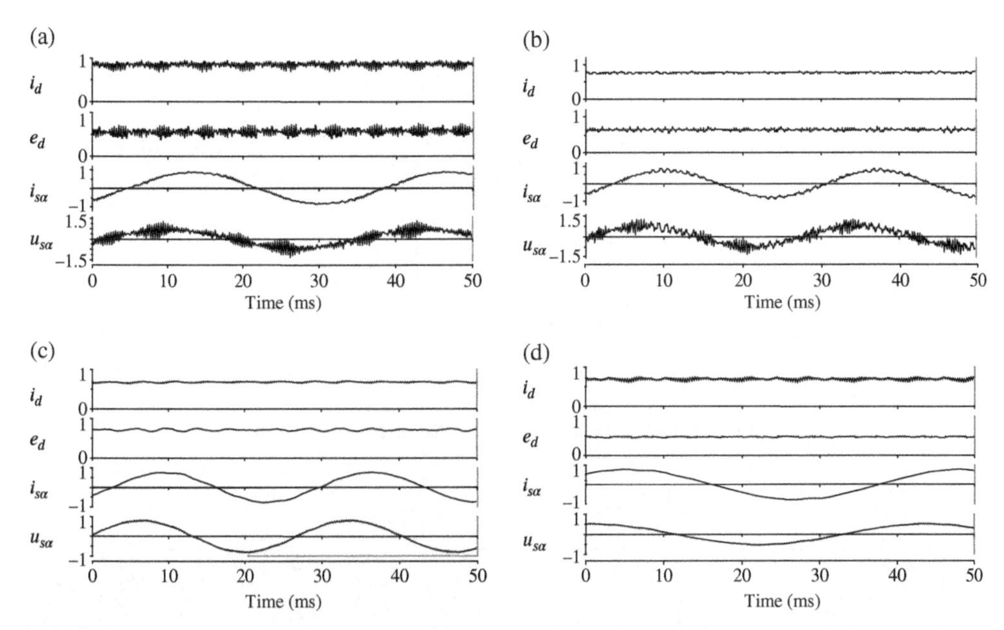

Figure 11.4 (a) Waveforms for C_M = 2 μF, L_d = 10 mH, (b) Waveforms for C_M = 2 μF, L_d = 50 mH, (c) Waveforms for C_M = 22 μF, L_d = 50 mH, (d) Waveforms for C_M = 22 μF, L_d = 5 mH. Switching frequency was 3.3 kHz and induction machine parameters presented in Table 11.1

Table 11.1 Parameters of the induction machine

Nominal power	$P_n = 5.5$ kW
Nominal voltage	$U_n = 400$ V
Nominal phase current	$I_n = 10.8$ A
Nominal rotor speed	$n_n = 1490$ rpm
Nominal frequency	$f_n = 50$ H
Number of poles	$P = 2$
Power factor	$\cos \varphi_n = 0.85$

ω_0 in the $\alpha\beta$ orthogonal system (Figure 11.5). The PWM CSI uses six unidirectional switches in two bridge legs: the first in the upper part (switches 1, 3, 5) and the second in the lower part of the bridge (switches 4, 6, 2). The switches from Figure 11.6 should be chosen in such a way that continuous current is assured in the DC-link. This condition will be fulfilled if two switches are turned on. Active vectors are denoted as I_1 to I_6. Three passive vectors (zero vectors) are denoted as I_7, I_8, I_9 in Figure 11.5. The inverter output current vectors from I_1 to I_9 are presented in Table 11.2.

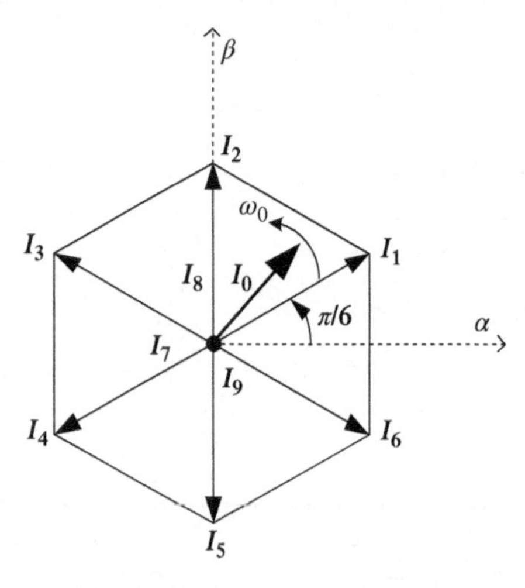

Figure 11.5 The current space vector representation

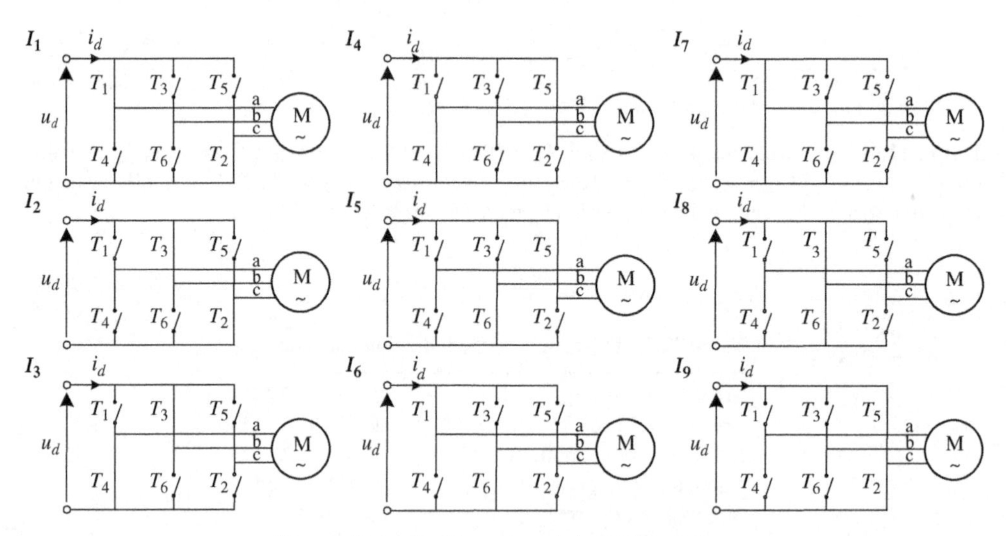

Figure 11.6 Configuration of the CSI switches

Table 11.2 The space vector states for the current vectors I_1–I_9 (transistors with the states I_1–I_9 are presented in Figure 11.6)

	Active states						Passive states		
Transistor number T_1–T_6	I_1	I_2	I_3	I_4	I_5	I_6	I_7	I_8	I_9
1	1	0	0	0	0	1	1	0	0
2	1	1	0	0	0	0	0	0	1
3	0	1	1	0	0	0	0	1	0
4	0	0	1	1	0	0	1	0	0
5	0	0	0	1	1	0	0	0	1
6	0	0	0	0	1	1	0	1	0

The reference vector $\boldsymbol{I}_0$, composed with adjacent active space vector components, is shown in Figure 11.7. In the orthogonal $\alpha\beta$ stationary system, the value of module I_0 follows the transformation of three-phase values to a two-phase system, where the invariant power will be taken into account:

$$|I_0| = \sqrt{2} \cdot i_D \tag{11.3}$$

where: $|I_0|$ is the module of output current vector, i_D, the DC-link current.

The values of the output current vectors, when the passive vectors are turned on, are equal to zero. Therefore:

$$I_{n\alpha} = I_{n\alpha}(i) + I_{n\alpha}(i+1), \tag{11.4}$$

$$I_{n\beta} = I_{n\beta}(i) + I_{n\beta}(i+1), \tag{11.5}$$

$$I_{n\alpha} = \frac{t_1(i)}{T_{imp}} I_{n\alpha}(i) + \frac{t_2(i+1)}{T_{imp}} I_{n\alpha}(i+1), \tag{11.6}$$

$$I_{n\beta} = \frac{t_1(i)}{T_{imp}} I_{n\beta}(i) + \frac{t_2(i+1)}{T_{imp}} I_{n\beta}(i+1), \tag{11.7}$$

where:

$t_{1,2}$ – the active vector times,
t_0 – the passive vector time,
T_{imp} – the cycle time, and
$I_{\alpha,\beta}$ – the space vector coefficient in ($\alpha\beta$) (index – i is the number of the active vector, n – the sector number).

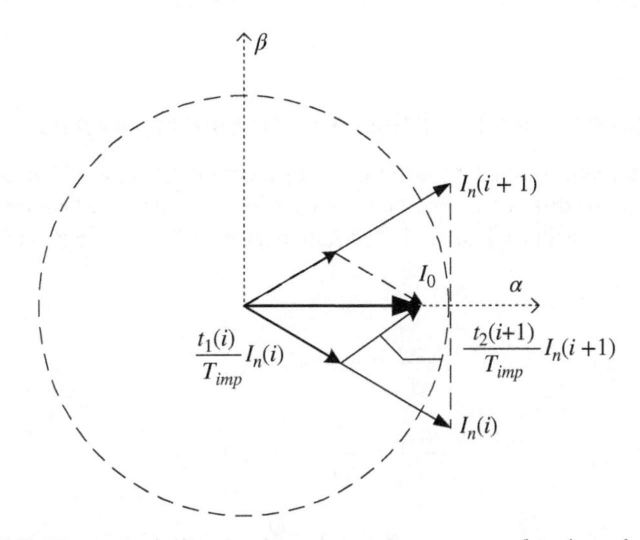

Figure 11.7 The method of computing average space vector values in each cycle time

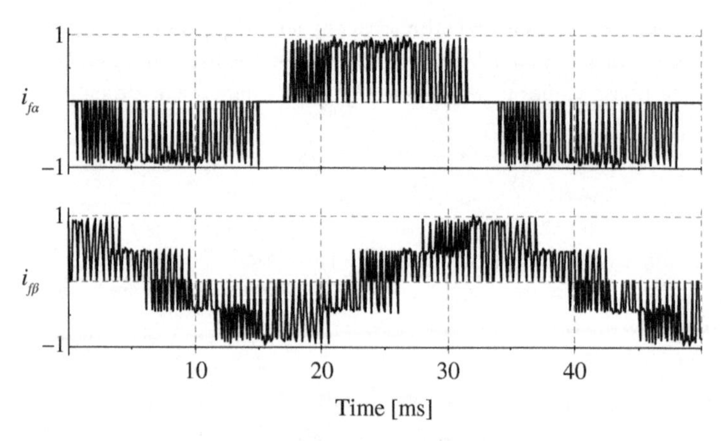

Figure 11.8 The output current vector components $i_{f\alpha}$, $i_{f\beta}$

Using (11.6) and (11.7) the active and passive vector times can be obtained from

$$t_1 = \frac{I_{n\alpha} \cdot I_{n\beta}(i+1) - I_{n\beta} \cdot I_{n\alpha}(i+1)}{I_{n\alpha}(i) \cdot I_{n\beta}(i+1) + I_{n\beta}(i) \cdot I_{n\alpha}(i+1)}, \tag{11.8}$$

$$t_2 = \frac{-I_{n\alpha} \cdot I_{n\beta}(i) + I_{n\beta} \cdot I_{n\alpha}(i)}{I_{n\alpha}(i) \cdot I_{n\beta}(i+1) + I_{n\beta}(i) \cdot I_{n\alpha}(i+1)}, \tag{11.9}$$

$$t_0 = T_{imp} - (t_1 + t_2). \tag{11.10}$$

The inverter output current components are shown in Figure 11.8.

The PWM modulation proposed in this section will be applied in the simulation example.

11.4 Mathematical Model of the Current Source Inverter Fed Drive

The mathematical model of the CSI, determined by a commutation function, was presented in [9]. In this section, the differential equations of capacitors C_M and the DC-link model are presented below. The capacitor model in the $\alpha\beta$ stationary coordinate system and DC-link model are as follows:

$$\frac{du_{c\alpha}}{d\tau} = \frac{1}{C_M}\left(i_{f\alpha} - i_{s\alpha}\right), \tag{11.11}$$

$$\frac{du_{c\beta}}{d\tau} = \frac{1}{C_M}\left(i_{f\beta} - i_{s\beta}\right). \tag{11.12}$$

$$e_d = R_d i_d + L_d \frac{di_d}{d\tau} + u_d, \tag{11.13}$$

where:

$\alpha\beta$ – stationary coordinate system,
$i_{f\alpha}$, $i_{f\beta}$ – output six transistors bridge current,
$i_{s\alpha}$, $i_{s\beta}$ – stator current vector coefficients,
$u_{s\alpha}$, $u_{s\beta}$ – stator voltage vector coefficients,
L_d – DC-link inductance,
R_d – interior inductor resistance,
C_M – output capacity of the capacitor,
i_d – the dc-link current, and
e_d, u_d – input to DC-link voltage (marked in Figure 11.2).

The vector components of stator voltage can be determined as follows:

$$u_{s\alpha} = R_c\left(i_{f\alpha} - i_{s\alpha}\right) + u_{c\alpha}, \tag{11.14}$$

$$u_{s\beta} = R_c\left(i_{f\beta} - i_{s\beta}\right) + u_{c\beta}, \tag{11.15}$$

where:

$\alpha\beta$ – stationary coordinate system,
$i_{f\alpha}$, $i_{f\beta}$ – output six transistors bridge current,
$i_{s\alpha}$, $i_{s\beta}$ – stator current vector components,
$u_{s\alpha}$, $u_{s\beta}$ – stator voltage vector components,
L_d – DC-link inductance,
R_d – interior inductor resistance,
R_c – resistance series connected to the capacitor,
C_M – output capacity of the capacitor,
C_L – input capacity of the capacitor, and
i_d – the dc-link current.

The voltage u_d in the DC-link is the input voltage to the machine-side CSI. This voltage is determined by the equation (under assumption that $u_d \cdot i_d = u_{s\alpha} i_{f\alpha} + u_{s\beta} i_{f\beta}$):

$$u_d = \frac{u_{s\alpha} i_{f\alpha} + u_{s\beta} i_{f\beta}}{i_d}. \tag{11.16}$$

The mathematical model of the CSI (11.11) to (11.16) will be used in the simulation model of the drive system.

11.5 Control System of an Induction Machine Supplied by a Current Source Inverter

11.5.1 Open-Loop Control

The induction machine supplied by a CSI can be controlled using an open-loop control system [4–6]. This method of control was named: the stator current to IM slip (*I*/*s*). This is a simple way

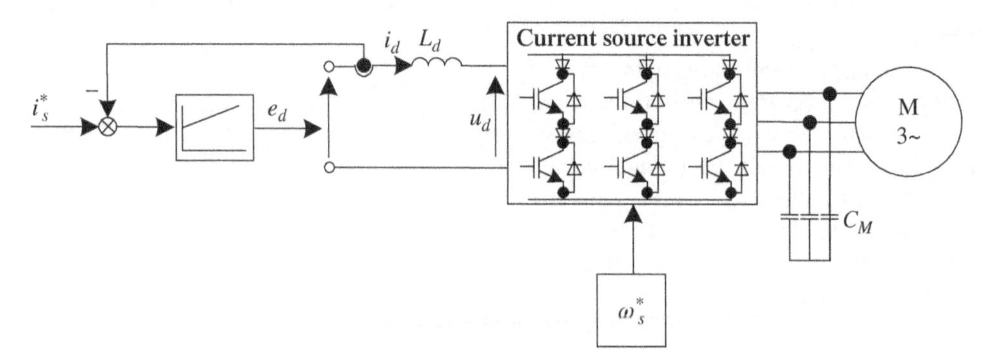

Figure 11.9 Open-loop control of IM supplied by the CSI

to control an IM, but the properties of the drive system are limited. It is impossible to independently control the rotor speed and rotor flux module. The open-loop control system structure is presented in Figure 11.9. There is only one PI (proportional–integral) controller of current i_d and the reference angular slip block.

11.5.2 Direct Field Control of Induction Machine

The mathematical model of the induction machine can be expressed in orthogonal coordinates (*dq*), which means that the rotor flux vector is oriented with the *d* axis and the *q* component of the rotor flux vector is close to zero. Adjusting the coordinate system in this way means that the Park transformation must be applied to other state variables. This is a transformation to a coordinate system rotating by an appropriate angle. In the case of an induction machine, it is a transformation that linearizes the non-linear differential equations of the machine model.

The mathematical model of a squirrel-cage induction machine in (*dq*) coordinate system has the following form [2]:

$$\frac{di_{sd}}{d\tau} = -\frac{R_sL_r^2 + R_rL_m^2}{L_rw_\sigma}i_{sd} + \frac{R_rL_m}{L_rw_\sigma}\psi_{rd} + \omega_{\psi r}i_{sq} + \frac{L_r}{w_\sigma}u_{sd}, \tag{11.17}$$

$$\frac{di_{sq}}{d\tau} = -\frac{R_sL_r^2 + R_rL_m^2}{L_rw_\sigma}i_{sq} - \omega_r\frac{L_m}{w_\sigma}\psi_{rd} - \omega_{\psi r}i_{sd} + \frac{L_r}{w_\sigma}u_{sq}, \tag{11.18}$$

$$\frac{d\psi_{rd}}{d\tau} = -\frac{R_r}{L_r}\psi_{rd} + \frac{R_rL_m}{L_r}i_{sd}, \tag{11.19}$$

$$\frac{d\omega_r}{d\tau} = \frac{1}{J}m_e - \frac{1}{J}T_L, \tag{11.20}$$

where

$$\omega_{\psi r} = \omega_r + \frac{R_r L_m}{L_r} \frac{i_{sq}}{\psi_{rd}}, \tag{11.21}$$

$$w_\sigma = L_s L_r - L_m^2, \tag{11.22}$$

i_{sd}, i_{sq} – stator current vector components,
u_{sd}, u_{sq} – stator voltage vector components,
ψ_{rd} – rotor flux vector component d,
ω_r – rotor angular speed,
R_s, R_r – stator and rotor resistances,
L_m – magnetizing inductance,
L_s, L_r – stator and rotor inductances,
J – torque of inertia, and
T_L – load torque.

The electromagnetic torque is determined by

$$m_e = \frac{L_m}{L_r} \psi_{rd} i_{sq}. \tag{11.23}$$

The idea of field-oriented control (FOC) is based on forcing the stator current vector components that affect the rotor flux module and the value of the electromagnetic torque. There are two subsystems in the control structure. In the first, the angular speed of the rotor is controlled. In the second, the rotor flux vector module is stabilized.

A block diagram of the control structure is shown in Figure 11.10. Park transformations are defined in (Chapter 2). Proportional–integral linear controllers were used. Optimization of the speed controller, according to the symmetry criterion allows for the compensation of the mechanical time constant. Optimization of the rotor flux controller, according to the module criterion, allows to compensate the time constant of the rotor. The current inverter is treated as a regulated current source and works with a variable modulation factor. This means that the DC-link current should be forced at least equal to the rated stator current of machine. Due to dynamic states, the value of the current in the DC-link circuit stabilizes at about 1.2–1.5 of the rated current of the machine. This kind of control strategy significantly increases the power losses emitted in transistors during commutation (it is named: control strategy with variable modulation index – the modulation index is the ratio of stator current module to the DC-link i_d average current value). Optimization of the DC-link current controller according to the module criterion provides the compensation of the large time constant of the DC-link circuit. The reference components of the stator current must be subjected to Park's inverse transformation. On their basis the angle φ_{is} is determined. The $i_{s\alpha}^*$, $i_{s\beta}^*$ components go to the *Modulator* block, in which the PWM is carried out, as described in Section 11.3.

In accordance with the general idea of FOC of a machine supplied by the CSI, simplifying assumption is adopted in which the capacitor model has a negligible capacitance value $C_M \approx 0$.

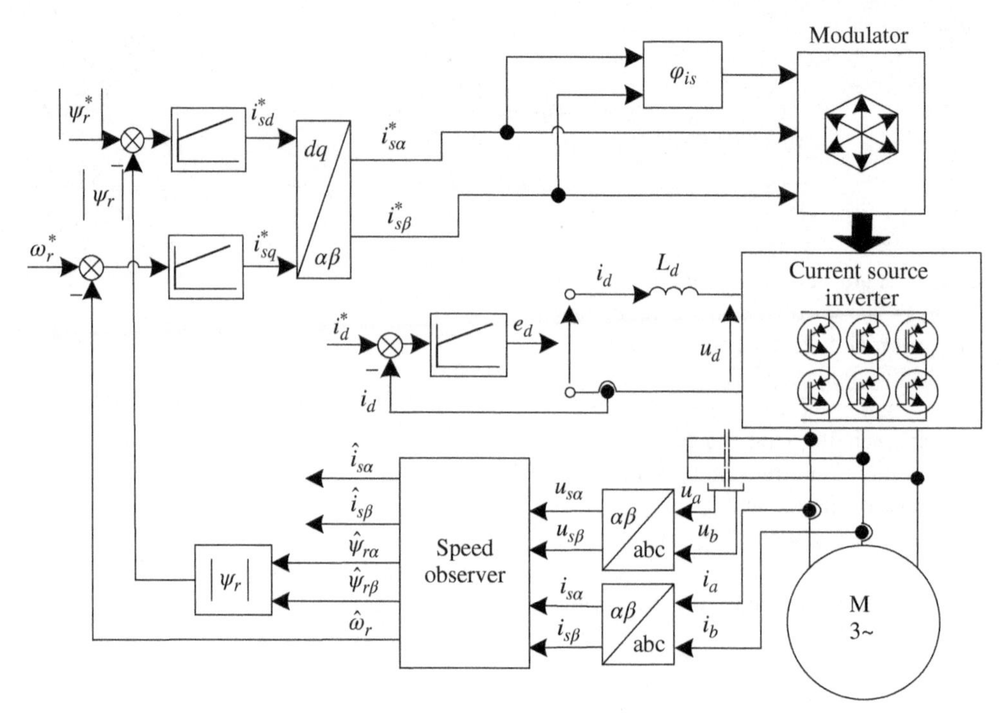

Figure 11.10 Block diagram of the control system with FOC. A control strategy with a variable modulation index was used ("^" denotes the variables estimated in the Speed Observer block)

The angle φ_{is} from Figure 11.10 can be determined as follows:

$$\varphi_{is} = arctg\left(i_{s\beta}^{*}/i_{s\alpha}^{*}\right). \tag{11.24}$$

The rotor flux vector module can be defined as:

$$\left|\psi_r^*\right| = \sqrt{\hat{\psi}_{r\alpha}^2 + \hat{\psi}_{r\beta}^2}. \tag{11.25}$$

The second control strategy of FOC is the operation of the CSI with the constant modulation index. The adjustment structure is shown in Figure 11.11. The reference current to the DC-link current controller can be calculated as follows:

$$\left|i_{ref}^*\right| = \sqrt{i_{s\alpha}^{*2} + i_{s\beta}^{*2}}. \tag{11.26}$$

The value of the modulation index will be stabilized to a constant value, while the DC-link current i_d doesn't have a constant value, but its value is dependent on the machine working point.

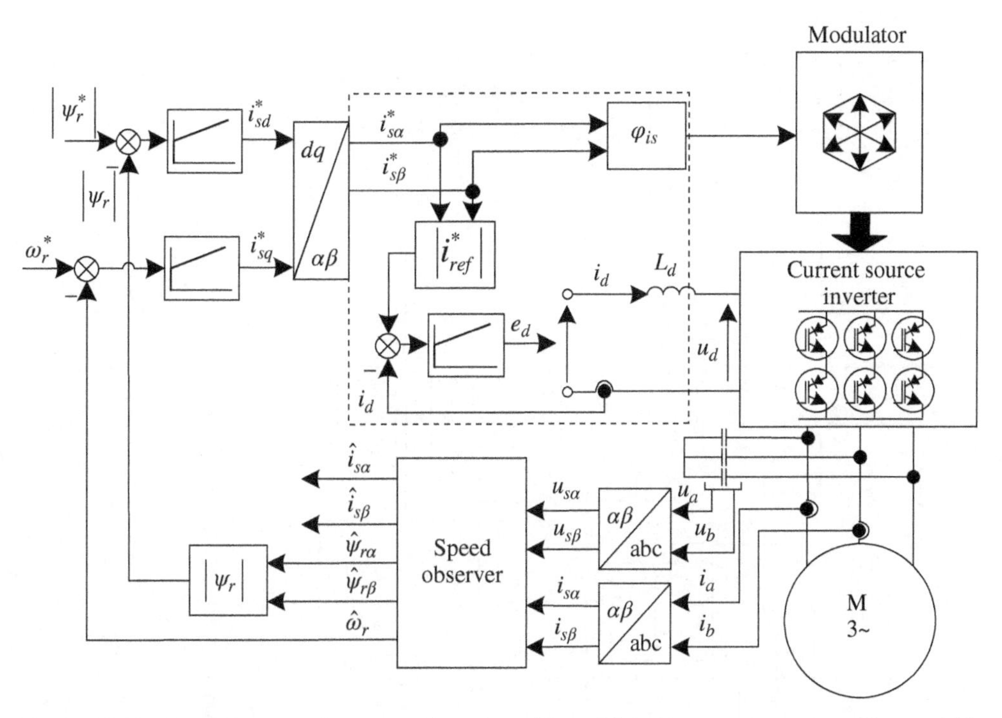

Figure 11.11 Block diagram of the control system with an FOC. (A control strategy with a variable modulation index was used)

11.6 Control System Model in Matlab/Simulink

The control system prepared in Simulink is presented in Figure 11.12. There is only one input, which was named *ramp_inc*. This signal is the influence limiting the increase of the reference speed. All functions are included in the C/C++ Code Block.

The function *motor()* contains eight right sides of differential equations of the electric drive, where DV [7], DV [8], DV [4] are for the CSI model.

```
void motor(double DV[8], double V[8])
{
isx=V[0];  isy=V[1];
frx=V[2];  fry=V[3];
omegaR=V[4];
id=V[5];
ucx=V[6];  ucy=V[7];

usx=ucx+Rc*(Ifx-isx);
usy=ucy+Rc*(Ify-isy);

if (id<0.001) id=0.001;
ud=(usx*Ifx+usy*Ify)/id;
```

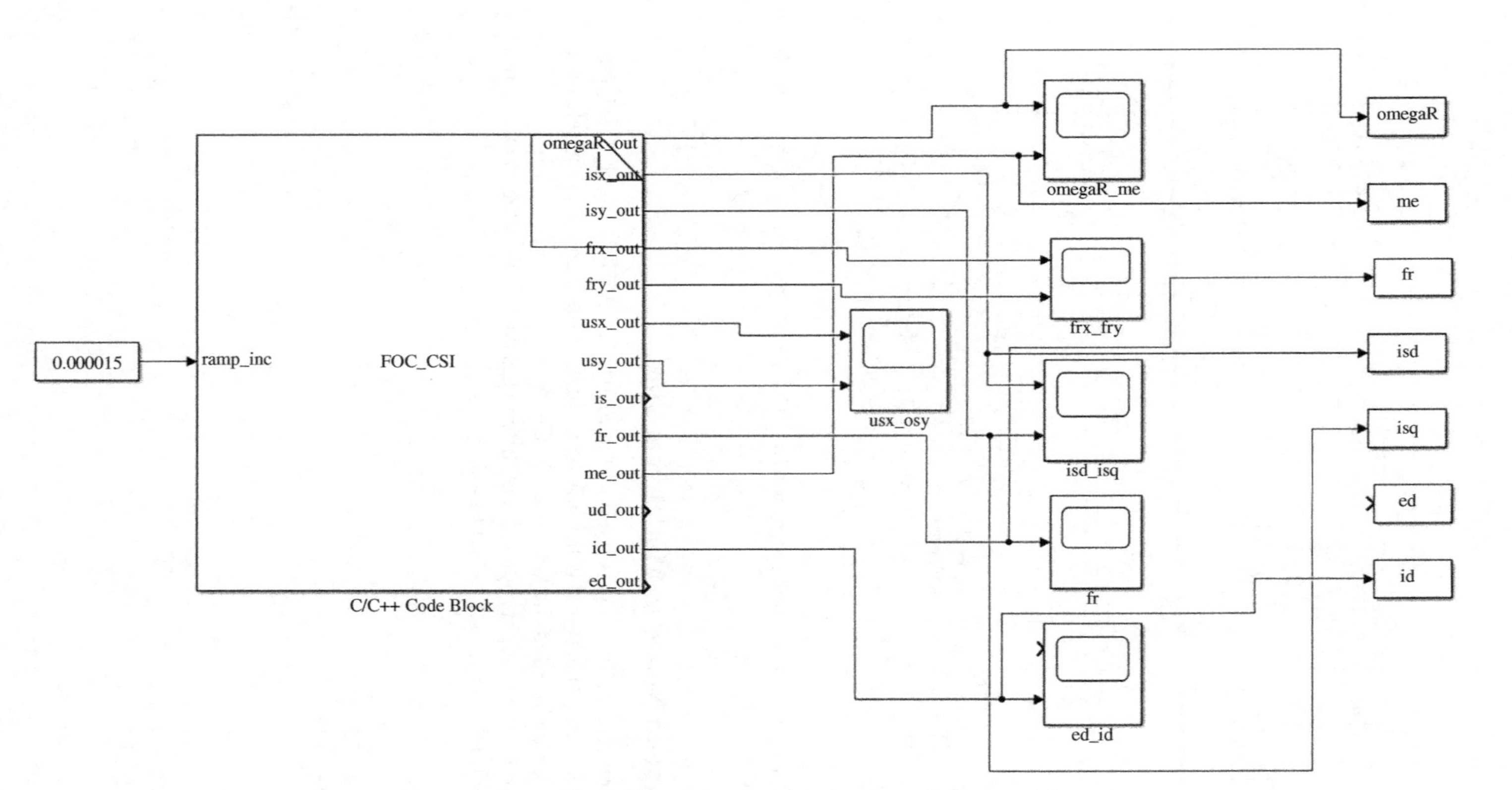

Figure 11.12 Block diagram of the FOC control system in Matlab/Simulink

```
DV[0]= -a1*isx+a2*frx+omegaR*a3*fry+a4*usx;
DV[1]= -a1*isy+a2*fry-omegaR*a3*frx+a4*usy;
DV[2]= -a5*frx+a6*isx-omegaR*fry;
DV[3]= -a5*fry+a6*isy+omegaR*frx;
DV[4]= (((frx*isy-fry*isx)*Lm/Lr)-m0)/JJ;
DV[5]= 1/Ld*(ed-Rd*id-ud);
DV[6]= 1/Cm*(Ifx-isx);
DV[7]= 1/Cm*(Ify-isy);
}
```

The function of the PWM modulation is presented below:

```
int PWM_CSI()
{double wsp;
wsp=p2;

Ix[1]=wsp*p4;    Ix[2]=0;   Ix[3]=-wsp*p4; Ix[4]=-wsp*p4; Ix[5]
=0;  Ix[6]=wsp*p4;
Iy[1]=wsp*0.5; Iy[2]=wsp*1;  Iy[3]=wsp*0.5;    Iy[4]=-wsp*0.5;
Iy[5]=-wsp*1; Iy[6]=-wsp*0.5;

NroIs=floor(roIs/(M_PI/6));

switch(NroIs)
{
case 0:  id1=6; id2=1; break;
case 1:  id1=1; id2=2; break;
case 2:  id1=1; id2=2; break;
case 3:  id1=2; id2=3; break;
case 4:  id1=2; id2=3; break;
case 5:  id1=3; id2=4; break;
case 6:  id1=3; id2=4; break;
case 7:  id1=4; id2=5; break;
case 8:  id1=4; id2=5; break;
case 9:  id1=5; id2=6; break;
case 10: id1=5; id2=6; break;
case 11: id1=6; id2=1; break;
};

wt=Ix[id1]*Iy[id2]-Iy[id1]*Ix[id2];
t1=-Timp*(ISY*Ix[id2]-ISX*Iy[id2])/(wt);
t2=Timp*(ISY*Ix[id1]-ISX*Iy[id1])/(wt);
t0=Timp-t1-t2;
if (t0<0) t0=0;
}
```

The following variables are defined:

rols – the angle defined in (11.24),
Nrols – the number of the reference output current position (it is divided into six sectors as shown in Figure 11.5), and
ISX, ISY – the reference values of the output current vector (stator current) marked in Figures 11.10 and 11.11 as $i^*_{s\alpha}, i^*_{s\beta}$.

In Figure 11.13 the IM is starting-up to 1.0 p.u. Waveforms of the rotor angular speed ω_r, rotor flux module ψ_r, DC-link current i_d and electromagnetic torque m_e are shown.

In Figure 11.14 the IM is reversed from 1.0 to –1.0 p.u. Waveforms of rotor angular speed ω_r, rotor flux module ψ_r, DC-link current i_d and electromagnetic torque m_e are shown.

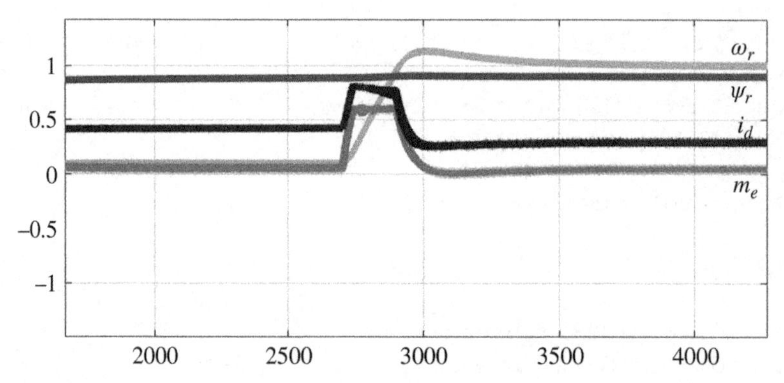

Figure 11.13 Machine start-up to 1.0 p.u. Waveforms of rotor angular speed ω_r, rotor flux module ψ_r, DC-link current i_d and electromagnetic torque m_e

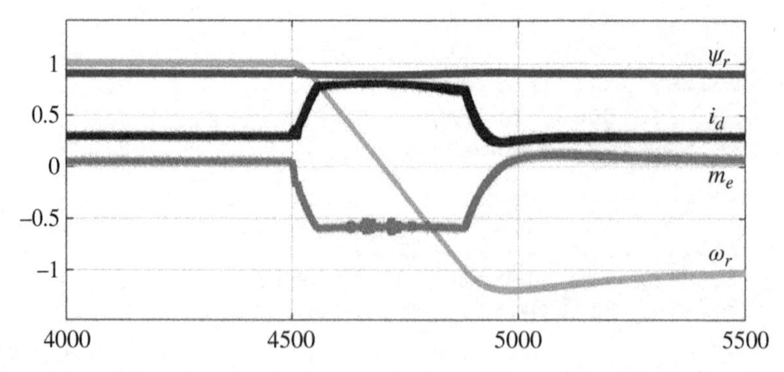

Figure 11.14 Machine reverses from 1.0 to –1.0 p.u. Waveforms of rotor angular speed ω_r, rotor flux module ψ_r, DC-link current i_d and electromagnetic torque m_e are presented

References

1. Bose, B. K. (2002) *Modern Power Electronics and AC Drives*. Prentice-Hall, Inc.
2. Abu-Rub, H., Malinowski, M., and Al-Haddad, K. (2014) *Power Electronics for Renewable Energy Systems, Transportation and Industrial Applications*. Wiley-IEEE Press, Chapter 20: M. Morawiec, Z. Krzeminski: "The Electrical Drive Systems with the Current Source Converter", pp. 630–663, DOI: 10.1002/9781118755525.ch20
3. Wu, B. (2006) *High-Power Converters and AC Drives*. Wiley-IEEE Press.
4. Morawiec, M., Krzeminski, Z., and Włas, M. (2005) The PWM current source inverter with IGBT transistors and multiscalar model control system. *11th European Conference on Power Electronics and Applications*.
5. Fuschs, F. and Kloenne, A. (2004) DC-link and dynamic performance features of PWM IGBT current source converter induction machine drives with respect to industrial requirements. *IPEMC 2004*, **3**, 14–16.
6. Li, Y. W., Pande, M., Zargari, N. R., and Bin, W. (2009) DC-link current minimization for high-power current source motor drives. *IEEE Trans. Power Electron.*, 24, 232–240.
7. Houldsworth, J. A. and Rosink, W. B. (1980) Introduction to PWM speed control for three phase AC motors. *Proc. Electr. Compon. Appl.,* **2**, 66–79.
8. Ledwich, G. (1991) Current source inverter modulation. *IEEE Trans. Ind. Electron.*, 6(4), 618–623.
9. Grbovic, P. J., Gruson, F., Idir, N., and Le Moigne, P. (2010) Turn-on performance of reverse blocking IGBT (RB IGBT) and optimization using advanced gate driver. *IEEE Trans. Power Electron.*, 25(4), 970–980.

Index

AC machines models, 19
Active power, 9, 121, 159, 194, 197, 198, 208, 446, 565
Adjacent line-to-line voltage, 11, 317, 354, 358
Arbitrary common reference frame. *see* Machine model
Artificial neural network, 5, 48, 100, 171

Back EMF. *see* Electromotive force (EMF)
Base values, 37, 38, 42, 475
Bearing
  capacitance, 477
  circulating current, 479
  current, 4, 12, 15, 62, 353, 430, 473, 478
    calculation, 197
    classification, 410–20
    reduction, 480, 481, 486
    types (*see* Bearing current classification)
  discharging current, 479
Bearing current classification, 410–20
BLDC, 14, 21, 198, 276, 278, 466
Butterworth filter, 247, 248, 265

Cable parameters, 478
Choke
  E shape, 404
  3 phase, 404
  toroidal, 404, 412–13, 422, 433
Clarke transformation. *see* Transformation Clarke
Common mode, 104, 105, 345, 379
  active zero voltage vector AZVC, AZVC-1, AZVC-2, 486–8
    three active vectors 3AVM, 486
  choke, 480, 492
    design, 433
  circuit, 477
  component, 473
  current, 477, 478, 544
    path, 478
  filter, 489, 544
    inductance, 495
    magnetic field density, 501–2
  flux, 180
  motor parameters, 5, 42, 179, 185, 190, 201, 303, 319, 329, 442, 444, 445, 454, 460, 475, 478, 506, 516
  reduction, 477–8, 480
  transformer, 414–15
    equivalent circuit, 24, 27, 33, 34, 154, 155, 162, 221, 222, 241, 249, 344, 399, 407, 427, 477, 478, 482, 483, 493–5, 499, 504
    reduction, 415
    voltage, 11, 60, 307, 309, 405–406, 408, 415–20
    waveform, 406
Comparator
  flux comparator, 278, 284
  torque comparator, 278, 280, 284, 285, 411
Compensation, 401, 435, 480, 483, 493, 514, 522, 525, 529, 576, 585
Computational, 275, 436, 443
Control problems, 420, 516
Co-ordinate transformation. *see* Motor model
Cost function. *see* Model predictive control
Current limit, 192, 544

Damper winding, 277
DC link, 10, 17, 49, 54, 66, 76, 87, 116, 118, 122, 125, 128, 132–55, 159, 163, 191, 200, 314, 351, 354, 358, 361, 366, 384, 395, 399, 473, 475, 491, 495, 525, 531, 558, 564, 575–90
DC motor, 1, 5, 14, 17, 23, 27, 182, 185, 198, 299, 301, 303, 338

High Performance Control of AC Drives With MATLAB®/Simulink, Second Edition.
Haitham Abu-Rub, Atif Iqbal, and Jaroslaw Guzinski.

Companion website: www.wiley.com/go/aburubcontrol2e

DC motor (*cont'd*)
separately excited, 17, 23, 24, 182, 183, 185, 198, 299, 301, 303, 338
analogy, 143–4
series excited, 22–4
Dead time, 48, 163, 164, 167, 168, 352–4, 403, 475, 493
dead band, 169
effect, 164, 167, 168
Differential mode. *see* Sinusoidal filter
Direct torque control, 4–6, 23, 178, 211–13, 222, 254, 275, 276, 278, 338, 404, 407, 552, 569
Double fed induction generator
autonomous generation system, 3, 193
base values, 37, 38, 42, 475
control system, 2, 8, 9, 13, 18, 179–81, 185–9, 194, 198, 201–3, 205, 213, 235, 260, 275, 277, 290, 300, 307, 309–11, 314, 316, 318, 322, 329, 332–4, 442, 450, 506, 509, 511, 517, 520, 525–40, 564, 569, 581, 583, 584, 586
grid connected system, 153–4
model, 32–5
vector control, 153–9
DSC direct self control, 275
DTC direct torque control, 4–6, 23, 178, 211–13, 222, 254, 275–6, 278, 338, 404, 407, 552, 569
DTC of five phase IM, 338
Dumping resistance, 482
dv/dt, 4, 10–18, 116, 469, 471, 479, 503, 506, 545, 550, 551, 555, 556, 561–2, 565, 567–9
effects, 545
filters, 469, 471, 480
Dwell time. *see* Space vector PWM
Dynamic model, 16, 17, 182, 220, 222, 240–3, 247, 276

Electric drive system, 2, 4, 5, 10, 11
Electromotive force (EMF), 3, 24, 27, 41, 199, 276, 422, 448–53, 512
estimation, 467
Estimation, 212, 213, 223, 234, 247, 254, 275, 280, 282, 324, 434, 436, 437, 440–8, 453, 458, 471, 509–11, 514, 518, 522, 531, 535, 536

Field oriented method, 526
direct, 188
with filter, 471, 495, 509, 511, 516
indirect, 188, 398, 399
stator oriented, 314, 316
Field weakening control, 191, 209
Five-phase, 293, 342
five-phase drive system, 10, 338, 339, 422, 426
five-phase induction motor model, 339, 399
five-phase inverter model, 395
five-phase supply, 346, 376, 378
five-phase VSI, 351, 354–64, 377–91, 422, 423, 426
Five-phase induction motor, 10, 17, 18, 337, 339, 345, 346, 351, 367, 372, 375, 396, 398–404, 412, 426–8, 454–60
Five-phase system, 337, 338, 344, 357, 368, 370, 389, 426
Flux
adaptive observer, 188, 434
air gap flux, 2, 6–7, 243, 276
estimation, 17, 458
rotor flux, 8, 9, 17, 35, 183–90, 208, 213–16, 226, 275–6, 280, 296, 300, 307–8, 310–12, 314–15, 345, 350, 397–412, 434–60, 512, 515, 525, 526, 529, 532, 534, 584–6, 590
stator flux, 188, 190, 191, 196, 212–54, 262, 264, 275, 278, 280, 282–4, 288, 292, 294, 315–18, 322, 324, 402, 409–12, 434–6, 444, 514, 515
Flux vector acceleration, 297
Fourier series, 65, 100, 107–9, 168, 358, 360, 361
Frequency modulation ratio, 52, 54, 67, 71

Gate drive signal, 55, 66, 354–7, 361, 362, 366

High performance drive, 1, 5–7, 16–18, 211
Hysteresis band, 212–14, 217–20, 223, 274, 284–6, 288, 399

Impedance, 12, 13, 16, 38, 42, 48, 150, 153–6, 159, 169, 471, 475, 483, 498, 542–4, 561, 563, 567
base, 475
characteristic, 542, 543
wave, 13, 471, 483, 563
Impedance Source or Z-Source inverter, 16, 48, 150, 153, 154, 159, 169
Induction motor, 1–3, 9–10, 17–18, 28, 33–6, 178–88, 192–3, 200, 212, 214, 216, 220, 223–8, 232, 234–6, 240–7, 253, 254, 258–62, 269, 275, 302, 306, 308, 311, 314,

318–19, 322, 334, 335, 337–9, 345–6, 351, 367, 372, 376, 396–8, 401–4, 412, 426–31, 434–5, 441, 442, 444, 445, 454, 457–60, 469, 470, 474–7, 481, 483, 495, 499, 504–6, 508, 514, 518, 522, 526–7, 531–4, 540–1, 543, 563
common mode model, 404–408
dq model, 429
five phase, 388–96
control, 388–96
parameters, 396
machine control, 139–53
per unit model, 25, 33, 37
scalar control, 6, 17, 177, 178, 207–8, 318, 319
sensorless control, 433, 434, 448, 452, 454, 457, 464–7, 539, 540
squirrel cage, 3, 16, 17, 23, 28, 36, 177, 178, 182, 197, 242, 457, 495, 499, 505, 584
stator resistance, 8, 37–8, 43, 45, 179, 189–90, 208, 212–13, 215, 224, 338, 372, 374, 412, 516
vector control, 3–9, 17, 29, 177–8, 182–5, 190–4, 200–3, 208, 209, 211–13, 241, 275, 299–300, 324, 329, 338, 397–404, 426, 434, 436, 443–5, 452, 454, 460, 486, 530
Inverter output filter, 13, 23, 469–73, 489–91, 494, 503, 513, 564, 568
control, 447–61
design, 425–33
estimation, 440–47
structures, 420–25
Iron losses, 225–6, 243–5, 247, 250–70, 546

LC filter. *see* Sinusoidal filter
Leg/pole voltage
five-phase, 307, 309–10
three-phase, 64
Load angle control, 530, 531, 533, 539
Long cable connection, 545
Look-up table, 219, 227, 230, 284, 416

Machine model, 343
Magnetizing inductance, 13, 221, 260, 262, 267, 343, 407, 563, 585
Matlab, 16–18, 21, 38, 39, 43, 45, 46, 54, 60, 63, 66, 69, 72, 74, 75, 84–90, 99, 103, 114, 118, 120, 125, 128, 137, 150, 159, 164, 169, 177, 184, 194, 196, 198, 201, 211, 222, 225, 232, 240, 243, 251, 260, 278, 280, 284, 286, 299, 303, 316, 339, 345, 352, 361, 380, 382, 386, 392, 394, 398, 422, 423, 426, 474–5, 506–8, 587
Maximum torque production, 183, 191, 192
Medium voltage drive, 10, 18, 549–51, 570
Model predictive control, 17, 420, 431
Model reference adaptive system MRAS. *see* Observer model reference adaptive system
Model transformation. *see* Five-phase
Modulation index definition, 49, 51
Motor model, 25, 28, 36, 40, 42, 179, 183, 185, 303, 308, 316–19, 327, 339, 399, 441–2, 453, 458, 474, 506, 522, 525, 529, 532, 537
Motor torque, 462
Multilevel inverters, 116, 563–4
cascaded H-bridge, 552, 556
diode clamped, 16, 106, 116, 118–19, 556
flying capacitor, 16, 48, 120–5, 555–61
Multi-loop control, 518
Multi-phase, 428–31
Multiscalar control with filter, 457–61

Non-adjacent line-to-line voltage, 354, 358, 359, 361
Nonlinear field oriented method NFOC with filter, 6, 453–6
Normal mode filter. *see* Sinusoidal filter

Observer, 397, 434–52, 506, 508, 511–26, 530, 534–40, 586
adaptive back stepping, 383–5
closed-loop, 366–75, 445–7
structure 1, 367–9
structure 2, 370–72
structure 3, 372–5
disturbance (*see* Observer speed)
flux (*see* Observer close loop)
Luenberger, 9, 17, 189–90, 440, 448
model reference adaptive system MRAS, 9, 433, 452
five-phase induction motor, 10, 17, 18, 337–9, 345, 346, 353, 372, 375, 396, 398, 3999, 401–5, 412, 426, 454, 457–8, 460
observer 1 structure, 376–8
PMSM, 448–58
speed estimator, 18, 213, 224, 225, 443, 452, 458
simple (*see* Observer close loop)
speed, 366, 442–5

Observer close loop, 366–75, 445–7
Observer model reference adaptive system, 9, 450, 452
Observer speed, 366, 442–5
Over-modulation, 51, 171, 495, 497
over-modulation I, 91
over-modulation II, 94

Parasitic
capacitances, 471, 477
current, 477
Park transformation, 29, 32–3, 184, 525, 530
Passive filter, 13, 18, 469, 479, 480, 563, 564
Permanent magnet synchronous motor, 2, 39, 46, 177, 276, 448
back EMF observer, 448, 450
control, 329
scalar, 6, 17, 177, 178, 207, 318, 526
sensorless, 380–88
model, 42, 44, 201, 203, 329, 330, 354
in αβ coordinates, 525, 526, 536
base values, 37, 38, 42, 475
in dq coordinates, 522, 525, 526
in per unit, 38–40
properties, 161
vector control, 17, 29, 177, 178, 182–5, 190–4, 200–3, 208, 209, 211–13, 241, 275, 299–300, 324, 329, 338, 397–404, 426, 434, 436, 443–5, 452, 454, 530, 460486
Per unit system, 16, 42
Phase shifting network, 565–6
Phase variable model. *see* Five-phase induction motor
Phase voltage or phase-to-neutral voltage, 62, 63, 65, 67, 69, 76, 353, 354, 356, 360, 364, 366, 367, 378, 399
PI controller. *see* Proportional-integral PI controller
PMSM, 448–58
Power calculation, 197, 446, 467
Power measurement. *see* Power calculation
Predictive control, 5–6, 17, 275, 420, 534
Predictive current control with filter, 461–71
Proportional-integral PI controller, 181, 184, 198, 199, 202, 203, 205, 206, 208, 212, 225, 301, 304, 309, 313, 319, 333, 345, 397–9, 403, 443, 453, 457, 458, 506, 525, 526, 530
cascaded, 126
Pulse width modulation PWM, 141, 184, 275, 314, 377, 384, 387, 388, 473, 566, 578
ANN based PWM, 96
carrier-based PWM, 104, 118, 126, 382, 383, 385–6
discontinuous PWM, 48–9
fifth harmonic injection, 384, 386–8
modifications, 483–4
offset addition, three-phase, 69, 71
five-phase, 336
sinusoidal PWM control, 337
space vector PWM three-phase, 72
five-phase, 294, 338, 342
synchronous and asynchronous, 51
third harmonic injection, 67, 69, 72–4, 158, 169, 384
unipolar, bipolar, 57–60

Quality factor, 497–9, 502, 542, 544
Quasi impedance source or qZSI inverter, 159, 162–4

Reactive power, 3, 121, 159, 192, 194, 197–9, 208, 446, 453, 565, 576
Ripple
flux ripple, 232, 275
Ripple inductor current, 496
Rotor flux estimation, 17, 458

Saturation, 7, 182, 220, 225, 258–62, 266, 269–75, 539–41
Sector, 8, 10, 18, 76–81, 86–107, 134, 141, 146, 213–19, 224, 278, 280, 379, 389, 391–2, 394, 405, 409–16, 486, 590
Selective Harmonics Elimination, 109
Sensorless speed control, 433
Simulation, 42, 48, 55, 60, 67, 84, 87, 104, 118, 150, 153, 164, 169, 177, 182, 186, 198, 202–6, 222, 224–6, 232, 234–5, 240–7, 251, 256, 258, 260, 262, 269, 272, 275, 278–80, 284, 306–7, 322–3, 333–4, 338–9, 345, 353, 361, 383–6, 388, 395–6, 399, 405, 414, 422, 426, 445–7, 454
Simulink, 16–18, 21, 38–9, 43, 45–6, 54, 60, 63, 66, 69, 72, 74–5, 84–90, 99, 103, 114, 118, 120, 125, 128, 137, 150, 159, 164, 169, 177, 184, 194, 196, 198, 201, 211, 222, 225, 232, 240, 243, 251, 260, 278, 280, 284, 286, 299, 303, 316, 339, 345, 352,

361, 380, 382, 386, 392, 394, 398, 422, 423, 426, 474–5, 506–8, 587
Sinusoidal filter, 470, 489, 493–501, 543
characteristics, 430
circuit, 430
design, 429–30
elements (*see* Sinusoidal filter design)
model, 437, 439
Sinusoidal filter design, 429–30
Sinusoidal wave shape, 276
Six-step mode, 63–6, 74, 94
Space vector
five-phase, 405
representations of AC machines, 182
three-phase, 72
Space Vector Modulation, 128, 145, 149, 275, 390, 486, 559
Space vector PWM, 77, 79, 392, 393, 406, 410, 564
Speed control, 10, 17, 180, 198, 201, 208, 212, 213, 223–5, 232–43, 251, 253, 254, 258, 260, 269, 272, 284, 292, 302, 304, 314, 397, 398, 403, 434, 436, 460, 540, 560
SPMSM, 276–99
Stator current relationship, 196, 435, 514
Steady state equivalent circuit. *see* Five-phase system
SVPWM, 338, 388–96, 426, 486, 488, 559
Switching combination, 120, 126, 421, 473
Switching Frequency
changing switching frequency, 275
constant switching frequency, 275, 296
Switching table, 154, 216, 219, 224, 226, 227, 244, 278, 284, 286, 395, 415

Ten step operation. *see* Five-phase system
Three-phase, 63, 64, 66–9, 87, 125, 169, 351, 383, 388, 405, 474, 506
Torque control, 4–6, 23, 178, 211–13, 222, 254, 275, 276, 278, 338, 404, 407, 552, 569
Torque ripple, 232, 234, 241, 254, 275, 339
Transformation
Clarke, 29, 31, 131, 143, 184
matrix constant
magnitude, 403
power, 404
Park, 29, 32, 35, 134, 184, 525, 530, 584, 585
variables, 9, 185, 194, 204, 313
Transformation Clarke, 29, 31, 184
Trapezoidal wave shape, 234

Uniform cylindrical surface, 276

Variable speed drives, 6, 8, 10, 277, 296
Voltage source inverter. *see* Three-phase
five-phase, 105, 125, 337, 351, 358, 361, 378, 390, 395, 404, 421
Voltage space vector, 76, 77, 93, 95, 97, 170, 213–17, 223, 228, 231, 275, 379, 381, 422, 426

Wind generation systems, 3, 193

Zero sequence component. *see* Common mode